Biological Foundations of Biomedical Engineering

Edited by
Jacob Kline, Ph.D.

Professor and Director, Biomedical Engineering Program,
School of Engineering and Environmental Design
and School of Medicine, University of Miami, Coral Gables,
Florida

Little, Brown and Company Boston

To Barbara, my wife, and to David, Jonathan, and Pamela, our children

Contents

VI. The Endocrine System

VII. The Gastrointestinal System

Preface

More and more biomedical engineering education programs at both the undergraduate and graduate levels are being introduced into universities. One problem
in this emerging field involves coupling an engineering student or physical scientist
with the area of the life sciences. The average engineering student has had minimum
contact (usually in general education) with the life sciences. He needs a background
in the biomedical sciences so that he can (1) understand the language and thus communicate effectively with his biomedical colleagues in the discussion, formulation,
and solution of interdisciplinary problems, (2) understand the biomedical literature,
(3) initiate experiments and proceedings in the areas of biology, medicine, and
health care systems in which engineering, physics, and mathematics can be applied
to the solution of interrelated problems, and (4) understand data and results of
biomedical research.

Among the existing academic programs in biomedical engineering, students commonly obtain their life science background by taking a varied assortment of disconnected courses such as biochemistry, biology, and physiology that are normally
designed and scheduled for life science majors and premedical students. The courses
may be excellent for the pure life scientist or the medical student, but they often
do not meet the specific needs of the biomedical engineer. At best, they offer
separate segments of information that he must piece together on his own. At worst,
they force him to flounder in his efforts to wrestle with technical language and ideas,
often foreign to his orientation, for which he has not been prepared. As a result, he
may not realize the greatest return for his investment of time and energy.

The University of Miami has developed a two-course, 12-credit sequence (BME
501 and 502 Unified Medical Sciences) that is uniquely designed for the study of
biomedical concepts in a manner that is perhaps more meaningful and efficient for
students with only physical science backgrounds. *Biological Foundations of Biomedical Engineering* is an outgrowth of the syllabus "Unified Systems Approach to
the Medical Sciences for Biomedical Engineers," which was written for and tested in
these courses.

The text is divided into eight sections. It begins, as do the courses, with the basic
sciences, treating the biochemical, thermodynamic, metabolic, and bioelectric topics
needed for the understanding of many of the processes covered later. The seven
sections that follow present the major mammalian organ systems from a functional
point of view, with an emphasis, when possible, on a systems approach. The
cardiovascular, respiratory, body fluids, nervous, endocrine, gastrointestinal, and
reproductive systems are considered individually. Throughout, an attempt is made to
consider relationships and interactions among the systems. The basic areas of biology,
biochemistry, physiology, etc., are intentionally not treated separately. Instead these

areas and the dynamic components of a primary life system are studied as they interact. One system is covered at a time, with emphasis on structure, function, and dynamics; quantitative analyses are stressed; minor technical details are *not* stressed; excessive repetition is avoided, and disconnected concepts are kept to a minimum.

The text assumes little in the way of background biology. Each section is developed by first presenting the basic concepts and organization of the organ system in such a manner that the reader may understand its function without any prior exposure to biology or physiology. Later chapters in each section become more sophisticated with the use of mathematics and engineering concepts, where they may apply, to quantify and make more understandable the biological functions.

Where it clarifies physiological phenomena, the use of mathematics or analog models is included. The mathematics is not beyond calculus on an undergraduate level. The block diagram and "black box" approach effective in the description and analysis of physical systems is used throughout and, where deemed appropriate, describes organ systems or their parts. Several chapters include instrumentation dealing with the measurement of physiological and diagnostic parameters, as well as therapeutic techniques.

Although this book was primarily designed for prospective biomedical engineers, its orientation should make it useful for other advanced students. It differs from most standard texts in physiology and the other life sciences, which are geared toward a typical fast-paced medical school program. The more quantitative treatment of biological phenomena presented here should be particularly useful for students in physiology and biophysics. The quantitative approach should also interest those medical school students and medical researchers who have a more mathematical orientation than the average medical student.

For students entering graduate programs in biology, physiological psychology, or other life sciences without a comprehensive premedical background, this book offers two advantages: first, it assumes that the reader does not have an extensive background in all the life sciences; second, it provides a firm foundation in those sciences upon which a more specialized education can be constructed.

Because of their quantitative training, coupled with a biological background, it is likely that biomedical engineers and medical personnel with engineering and physical science backgrounds will be responsible for many of the important developments in medicine. It is hoped that the material presented here will spark the interest of those students who wish to make a contribution to this important and growing area of biological and medical endeavor.

The goal, then, of *Biological Foundations of Biomedical Engineering* is to provide a broad-based, systematic, quantitatively oriented panorama of the life sciences, which will be useful to biomedical engineering students and other prospective specialists. It is therefore hoped that the broad background and expertise of the many contributors has resulted in a whole that is more than the sum of its parts.

We are grateful to and acknowledge the Exxon Education Foundation for the support of the syllabus which led to the writing of this book. We acknowledge the contribution made to the organization of the syllabus by Dr. Sidney Besvinick, Professor of Education and Acting Vice President for Academic Affairs at the University of Miami. Thanks are due to Drs. H. Bolooki, L. F. Dell'Osso, and Peter Tarjan for supplying material respectively on cardiac balloon pumping, arrhythmias, and pacemakers. We also thank the many graduate students, secretaries, and typists who worked with dedication to make this book possible. A final word of gratitude goes to the many contributors to this volume.

J. K.

Contributing Authors

William W. Allen, B.A.
Research Associate, Obstetrical/Gynecological Research, Loma Linda University School of Medicine, Loma Linda, California

Chapter 46

E. O. Attinger, M.D., Ph.D.
Chairman of Biomedical Engineering and Professor of Physiology, University of Virginia School of Medicine, Charlottesville, Virginia

Chapters 6, 9–13

Leon A. Cuervo, Ph.D.
Associate Professor, Department of Biological Sciences, Florida International University, Miami, Florida

Chapters 5, 26

Michael Gimpl, Ph.D.
Maytag Research Fellow, Department of Psychology, University of Miami, Coral Gables, Florida

Chapter 29

Esther P. Hill, Ph.D.
Assistant Research Physiologist, University of California, San Diego, School of Medicine, La Jolla, California

Chapter 45

Ronald S. Hosek, Ph.D.
Research Assistant Professor, Department of Neurological Surgery, University of Miami School of Medicine, Miami, Florida

Chapters 25, 29

Donald R. Humphrey, Ph.D.
Associate Professor, Department of Physiology, Emory University School of Medicine, Atlanta, Georgia

Chapters 31, 32

Jacob Kline, Ph.D.
Professor and Director, Biomedical Engineering Program, School of Engineering and Environmental Design and School of Medicine, University of Miami, Coral Gables, Florida

Editor; Chapters 6–8

Harry Lipner, Ph.D.
Professor, Department of Physiology, Florida State University Program in Medical Sciences, Tallahassee, Florida

Chapters 33–39

Lawrence D. Longo, M.D.

Professor of Physiology and Obstetrics/Gynecology, Loma Linda University School of Medicine; Perinatal Research, Loma Linda Medical Center, Loma Linda, California

Chapters 45, 46

Joel B. Mann, M.D.

Associate Clinical Professor of Medicine, University of Miami School of Medicine; Attending Physician, Cedars of Lebanon Hospital and Mercy Hospital, Miami, Florida

Chapters 20–23

Thomas J. Mende, Ph.D.

Professor, Department of Biochemistry, and Associate Professor, Department of Radiology, University of Miami School of Medicine, Miami, Florida

Chapters 1–5

D. D. Michie, Ph.D.

Professor and Chairman, Departments of Physiology and Bioengineering, Eastern Virginia Medical School, Norfolk, Virginia; Consultant, Department of Medicine, Veterans Administration Center, Hampton, Virginia

Chapters 6, 7

Richard E. Poppele, Ph.D.

Associate Professor of Physiology, University of Minnesota Medical School, Minneapolis, Minnesota

Chapter 27

Gordon G. Power, M.D.

Associate Professor of Physiology and Obstetrics/Gynecology, Loma Linda University School of Medicine, Loma Linda, California

Chapters 45, 46

Wayne S. Rogers, M.D.

Clinical Associate Professor, Obstetrics/Gynecology, University of Miami School of Medicine, Miami, Florida

Chapters 44, 45, 47–49

Marvin A. Sackner, M.D.

Professor of Medicine, University of Miami School of Medicine, Miami, Florida; Chief, Department of Medicine, and Chief, Division of Pulmonary Disease, Mount Sinai Medical Center, Miami Beach, Florida

Chapters 14–19

Neil Schneiderman, Ph.D.

Professor of Psychology, University of Miami, Coral Gables, Florida

Chapters 25, 29, 30

Daniel Weiner, M.D.

Associate Professor of Medicine, George Washington University School of Medicine; Veterans Administration Hospital, Washington, D.C.

Chapters 40–43

Ray W. Winters, Ph.D.
Associate Professor of Psychology, University of Miami; Coral Gables, Florida
Chapter 28

Matthew B. Wolf, Ph.D.
Associate Professor, Engineering and Biomathematics, and Assistant Professor, Physiology and
Biophysics, School of Engineering, University of Alabama in Birmingham, Birmingham, Alabama
Chapter 24

BASIC SCIENCES

Energetics

1

Thomas J. Mende

INTRODUCTION

The unit on basic sciences consists of a selection of physical and biochemical topics that will serve as background material to the units on the various physiologic systems. It begins with a review of traditional topics of physical chemistry, presenting a discussion of energy metabolism as approached through the methods of equilibrium thermodynamics. A chapter outlining the biologically most significant properties of aqueous solutions follows. The emphasis here is on major concepts. The view toward this treatment is that it is to be augmented in subsequent presentations on acid-base, water, and electrolyte balance.

A brief account of biologic structures and function, as a "material science" introduction to the organism, is offered in Chapters 3 and 4 of this unit. The emphasis is on structural and catalytic macromolecules. Metabolic reactions that are not directly involved in the major catabolic pathways are omitted by intent, but these are introduced in the units on the gastrointestinal, reproductive, and endocrine systems as deemed necessary. Chapter 5 offers a traditional introduction to bioelectrical and biomechanical phenomena, with an emphasis on the cellular and molecular approach. It attempts to convey the basic concepts for the comprehension of the functions of the cardiovascular and nervous systems.

THERMODYNAMIC CONCEPTS

The most significant source of energy for the mammalian organism is chemical energy supplied by food. This energy is transformed by biochemical reactions into the work that is required for normal body functions, including the maintenance of structural integrity.

Food energy can be regarded as a station in the carbon cycle, a process maintained by photosynthesis. During this cycle, the energy of light quanta is utilized to perform a series of reactions, in part photochemical, which can be exemplified by Equation 1-1, where h is Planck's constant and ν is the frequency of light:

$$6CO_2 + 6H_2O + h\nu \rightarrow C_6H_{12}O_6 + 6O_2 \tag{1-1}$$

One may regard the reaction product, glucose ($C_6H_{12}O_6$), as quantitatively the most important form of chemical energy supplied in food and as the biologic precursor of the second most important energy source, fat. Due to its central role, glucose will be used frequently as a model of energy utilization in man. The utilizable energy content from a mole of glucose is the difference between the energy content of the mole of glucose and the total "energy" of its metabolic end products, six moles of CO_2 and six moles of H_2O.

In order to understand the utilization of the chemical energy by the body, one may compare the oxidation of one mole of glucose inside and outside the body. If the heat output of this oxidation is measured in vitro or in vivo (i.e., outside or inside the body, respectively), the measurements agree within experimental error. This fact assures us that classic thermodynamic considerations are applicable to living systems. It is obvious that the heat energy liberated during a combustion process can be harnessed to produce useful work, as is evidenced by the steam or internal combustion engine. However, as the calculations of efficiency derived from the Carnot cycle indicate, this utilizable work will depend on the temperature differential, $T_2 - T_1$ (ΔT), in the system:

$$W = Q \frac{T_2 - T_1}{T_2} \qquad (1\text{-}2)$$

where Q is the amount of heat lost when the absolute temperature, T, is lowered from T_2 to T_1 and W is the maximum available useful work. As can be seen from this equation, a heat source and a heat sink are necessary requirements for any work to be produced by a heat engine.

This condition has no parallel in the living body. Although the body produces heat, it is more or less evenly distributed. Thus, the body can be regarded as an *isothermal engine,* and, in such a system, the conversion of energy to work cannot take place in a manner similar to that in a heat engine. In order to understand the energy conversion mechanisms of the body, some pertinent thermodynamic concepts will be reviewed.

The classification of thermodynamic systems depends on the nature of heat exchange and the action between a system and its environment. The human body in this regard is a "system" surrounded by its "environment," with which it can exchange both matter and energy. The same consideration applies to isolated parts of the body, such as organs or cells. A system that exchanges neither matter nor energy with its environment is called *isolated;* one that exchanges energy but not matter is *closed,* and the one previously shown to be like the body and its parts is termed an *open* system [1].

"Open" living systems exist in what is called a state of *dynamic equilibrium.* This means that the concentrations of most body constituents over a period of time show a remarkable constancy, as one would expect for a system in equilibrium. This constancy of composition, however, can be maintained only at the expense of continuously introducing into the body metabolizable compounds and withdrawing the waste products. In other words, a continuous exchange of matter with the body's environment is a mandatory requirement for the maintenance of its steady state. An interruption in the flow of the material exchange quickly alters the system's established composition. Hence, the label *dynamic* is used to describe this pseudo-equilibrium. It may be mentioned that the establishment of a true equilibrium within the body would be a sufficient criterion of death.

Certain contrasting characteristics between dynamic equilibrium and true equilibrium are worth pointing out:

True Equilibrium	Dynamic Equilibrium
1. Composition is time-invariant with no exchange of matter and energy.	Composition is time-invariant only at the expense of continuous exchange of matter with environment.
2. Addition of any one component will result in a permanent change in the concentration of all the components.	Addition of any component to the system will only cause a *temporary* change in the concentrations of the components.
3. Laws describing the behavior of a true equilibrium system contain no terms related to time (rates).	Laws describing dynamic equilibrium contain rate terms.
4. No work can be produced by a system in thermodynamic equilibrium.	A system in a "steady state" can produce useful work.

Life processes are best characterized by the "dynamic equilibrium" concept. A branch of chemical physics known as nonequilibrium thermodynamics or thermodynamics of irreversible processes is actively being developed to deal with such systems [2]. The complexity of life processes and the historical development of bioenergetics, however, have amassed a considerable body of knowledge in terms of classic or equilibrium thermodynamics. This approach helped to focus attention on individual chemical reactions that can be isolated for the purposes of observation and measurement. Despite the limited nature of this method, far-reaching conclusions have been drawn concerning energy conversions, which have resulted in the establishment of the present foundations of bioenergetics.

Formulation of quantitative bioenergetic relationships requires the selection of appropriate parameters to be utilized in the characterization of the system. These are expressed in various *equations of state.*

The equations of state can be used to calculate the net changes in the energy of the system when it changes from one state to another. The formulation of the equations of state is accomplished by the selection of appropriate energy terms that depend *only* on the initial and final equilibrium states and not on the pathway by which the final state was reached. This kind of energy function is called a *state function* or *thermodynamic property.* There are different ways the energy of the system can be expressed by state functions, depending on the macroscopic properties of the system, such as temperature, volume, pressure, and composition.

In bioenergetics it is most convenient to select an energy function that can be used for the calculation of energy changes under the conditions of *constant temperature and pressure*. This is the most convenient selection, because, as already mentioned, the human body is an isothermal system and all bodily processes take place under constant atmospheric pressure. The function that is most suitable for bioenergetic calculations is the Gibbs free energy, G, or simply *free energy*.

The meaning of free energy may be summarized as follows: the free energy is that part of the total energy which is available for doing work. The difference in the free energy, ΔG, between two states of a system is the maximum amount of useful work that theoretically can be extracted from the system when it is allowed to change in a "reversible" fashion from the initial state, with a free energy level of G_1, to the final state, G_2. The meaning of "reversibility" here indicates the (hypothetical) situation that the change proceeds through an infinite number of equilibrium steps. Clearly this condition is incompatible with finite reaction times, so the free energy change that is obtained by calculation is a theoretical maximum; actual utilizable energy is always less. When the latter is expressed as a percentage of the calculated free energy of the process, the *efficiency* is defined.

In addition to being a measure of the useful work obtainable from the system, the free energy function is capable of predicting the direction of spontaneity of a given change. It is also the criterion for the state of thermodynamic equilibrium.

For a closed system, Equation 1-3 applies:

$$\Delta G = \Delta H - T\Delta S \tag{1-3}$$

where ΔH is the enthalpy change, T is the absolute temperature, and ΔS is the entropy change.

Enthalpy change, ΔH, is the difference between the state functions H_1 and H_2. Its value may be obtained by simple calorimetric measurements, since it is defined as the *heat of reaction under constant pressure*. Entropy change, ΔS, is the difference between the entropies S_1 and S_2 of states 1 and 2. Entropy, just as G and H, is a state function. Its meaning is harder to grasp than either of the other two. It is frequently described as an index of randomness of the system; more positive values of entropy correspond to increasing disorder in the system. Entropy is related to specific heat, as shown by its definition in Equation 1-4:

$$S = \frac{Q_{rev}}{T}\bigg|_T \tag{1-4}$$

where Q_{rev} is the heat absorbed by the system at a constant absolute temperature, T, if the reaction is allowed to proceed under the condition of reversibility as defined previously. In Equation 1-3, the free energy change is shown to be the difference between an energy term, ΔH, and an entropy term, $T\Delta S$. The latter term embodies the property of the free energy as a predictive function concerning the direction of spontaneous change, because, for spontaneous reactions, the entropy has to increase. Free energy then holds the balance between energy and entropy. The latter indicates the amount of energy that is not available for useful work. Since free energy and enthalpy are energy terms, the dimensions of entropy have to be in terms of energy per unit temperature, e.g., ergs per degree.

The intuitive notion that an equilibrium system is incapable of performing work will make it clear that at equilibrium, the free energy change ΔG must be zero. A negative free energy change indicates that the reaction will be spontaneous in the direction written, whereas positive values indicate that the reaction is spontaneous in the opposite direction. The magnitude of the free energy change is a measure of the degree by which the system is removed from its equilibrium state. An important relationship exists between the equilibrium constant of a reaction and the free energy change when measured under specified conditions.

Free energy as defined by G is the available energy for performing work, excluding volume-pressure work, at constant temperature and pressure. Bioenergetics deals mainly with the computation of the free energy for biochemical transformations. Just as other forms of energy, free energy at constant temperature and pressure can also be regarded as the product of an intensity factor, μ, or *chemical potential*, and a capacity factor that is proportional to the mass of the different components, which is customarily expressed in moles, m.

Some analogies with other types of energy are (1) electrical energy = potential $\times$ charge, (2) surface energy = surface tension $\times$ surface area, and (3) heat energy = temperature $\times$ heat capacity. In all these examples, the first factor is the intensity factor and the second is the capacity factor. Considering ΔG, it should be clear that it must be a function of the number of moles and the chemical potential of the different components in the system. A statement representing these considerations, which is applicable to ideal solutions, is expressed in Equation 1-5:

$$\mu_i = \mu_i^{0} + RT \ln X_i \qquad (1\text{-}5)$$

where R = gas constant = 1.987 cal$\cdot$deg$^{-1}\cdot$mole^{-1}
$\qquad T$ = absolute temperature
$\qquad \mu_i$ = chemical potential of component i
$\qquad X_i$ = mole fraction of solute i in solution

The quantity μ_i^{0} is the chemical potential at unit mole fraction (or other concentration unit); it is a constant, a function of temperature and pressure, and called the *standard chemical potential.* The mole fraction of a substance is equal to the number of moles of the substance in question divided by the sum of the number of moles of all components present in the system. For nonideal solutions, *activities* take the place of mole fractions and we may substitute a_i for X_i. G is defined by the following relationship:

$$G = \Sigma n_i \mu_i \qquad (1\text{-}6)$$

where n_i = stoichiometric coefficient.

These relationships allow us to derive the relationship between the free energy change of a reaction and the concentrations (which are proportional to the activities) of the reaction components. Let us take the following chemical reaction:

$$aA + bB + \ldots \rightleftharpoons mM + nN + \ldots \tag{1-7}$$

where capital letters indicate the chemical species and lower case letters the appropriate stoichiometric coefficients. The chemical potentials of the reactants and products then can be substituted from Equation 1-5. The general condition of equilibrium will be:

$$a\mu_A + b\mu_B + \ldots \rightleftharpoons m\mu_M + n\mu_N + \ldots \tag{1-8}$$

Combining Equation 1-5 with Equation 1-6 and replacing X_i with a_i yields the following important relationship:.

$$\Delta G = \Delta G^0 + RT \ln \frac{(a_M)^m \, (a_N)^n \ldots}{(a_A)^a \, (a_B)^b \ldots} \tag{1-9}$$

where a_M is the effective concentration of solute M (and so on) and

$$\Delta G^0 = m\mu^0_M + n\mu^0_N + \ldots - a\mu^0_A - b\mu^0_B \ldots \tag{1-10}$$

In the most frequently used free energy value in chemical literature, ΔG^0 has the following meaning: at *equilibrium*, the free energy change equals 0, i.e.,

$$\Delta G^0 = -RT \ln K \tag{1-11}$$

where K, the *equilibrium constant,* is the value of the fraction on the right side of Equation 1-9 that is obtained by substituting the activities existing at equilibrium.

The so-called *standard free energy change,* ΔG^0, is equal to the free energy change of the reaction *if* the reactants, all present in unit activity, are converted to reaction products with unit activity, while maintaining the reactants and products at unit activity during the process. This definition indicates that ΔG^0 is a *constant,* although it is a function of T and is characteristic of the reaction. The equation shows that for the calculation of free energy changes under any desired concentration conditions, the knowledge of ΔG^0 is essential.

As seen in Equation 1-11, the measurement of the equilibrium constant allows the calculation of the standard free energy change, ΔG^0. Other ways to determine ΔG^0 are:

1. Determination of the potential of oxidation-reduction couples in electrochemical cells.
2. Experimental determination of enthalpy and entropy changes in reactions.

3. Published data on the free energies of formation of the reactants and products in question. Published values of standard enthalpies and entropies can serve the same purpose.

CALCULATION OF FREE ENERGY CHANGE FROM OXIDATION-REDUCTION POTENTIAL DATA

Oxidation-reduction reactions can be regarded as the sum of two half reactions. Half reactions, according to convention, may be written as a reduction (an electron-gaining) step, as expressed in Equation 1-12:

$$1/2 \ O_2 + 2H^+ + 2e^- \rightarrow H_2O \tag{1-12}$$

Although the potential of half reactions cannot be measured directly, two half reactions can be combined to form an electrochemical cell. An accepted reference electrode is the *standard hydrogen electrode,* in which the following equilibrium is established on the surface of a platinum electrode:

$$H^+ + e^- \rightarrow 1/2 \ H_2 \tag{1-13}$$

where H^+ is at unit activity and H_2 is at 1 atmosphere pressure. The potential of the standard hydrogen electrode is, by convention, set at zero. In this way, the potential difference measured in a cell composed of a standard hydrogen electrode half-cell is totally ascribed in sign and magnitude to the other half-cell. If the half-cell in question has the oxidized and reduced species present at unit activity or at equal concentrations (the solution pH is usually adjusted to 7), the measured potential difference will be the *standard reduction potential,* E^0 (volts). For the O_2-H_2O couple, for example, this value is +0.816 volt. Standard reduction potential values range from positive to negative in reference to the hydrogen electrode. The above statement can be verified from the Nernst equation:

$$E = E^{0\prime} + 2.303 \frac{RT}{nF} \log \frac{[\text{oxidized form}]}{[\text{reduced form}]} \tag{1-14}$$

where $E^{0\prime}$ is the standard reduction potential at pH 7, F is the Faraday constant, and n is the number of electrons transferred per mole in the oxidation-reduction reaction.

From any two selected couples, the more positively charged member can oxidize the less positively charged one *under standard conditions.* In other words, in a mixture of two oxidation-reduction systems, the more positive one will actually proceed as written, whereas the more negative one will chemically change in the opposite direction. The standard free energy change is related to the standard oxidation-reduction potential by Equation 1-15:

$$\Delta G^{0\prime} = -nF\Delta E_0{}^{\prime} \tag{1-15}$$

where $\Delta G^{0'}$ is the standard free energy change at pH 7 and n is the number of electrons transferred per molecule in the reduction reaction in question.

The reaction between nicotinamide adenine dinucleotide (NAD) and ethyl alcohol will be used as an illustration [6]. The two half reactions have the following standard reduction potentials $(E^{0'})$:

$$NAD^+ + 2H^+ + 2e^- \rightarrow NADH + H^+ \qquad E^{0'} = -0.32 \text{ volt} \qquad\qquad (1\text{-}16)$$

$$acetaldehyde + 2H^+ + 2e^- \rightarrow ethanol \qquad E^{0'} = -0.163 \text{ volt} \qquad\qquad (1\text{-}17)$$

Comparing the two $E^{0'}$ values, note that the acetaldehyde-ethanol couple is more positive (i.e., less negative) than the NAD-NADH couple. According to previous statements, if both systems are present simultaneously in their standard states and if the appropriate catalyst (enzyme) is available, a reduction of acetaldehyde to ethanol will take place at the expense of NADH oxidation. Combining the equations for these reactions yields Equation 1-18C:

$$acetaldehyde + 2H^+ + 2e^- \rightarrow ethanol - 0.163 \text{ volt} \qquad\qquad (1\text{-}18A)$$

$$NADH + H^+ \rightarrow 2H^+ + 2e^- + NAD^+ + 0.320 \text{ volt} \qquad\qquad (1\text{-}18B)$$

$$acetaldehyde + NADH + H^+ \rightarrow ethanol + NAD^+ + 0.157 \text{ volt} \qquad\qquad (1\text{-}18C)$$

It should be noted that the combined reaction should have a positive potential to give the standard free energy change in the direction of spontaneity. The standard free energy change is:

$$\Delta G^{0'} = -nF\Delta E^{0'}$$

$$= -2 \times 23{,}086 \times 0.157 \text{ cal/mole*} \qquad\qquad (1\text{-}19)$$

$$= -7249 \text{ cal/mole}$$

This result is applicable only for standard conditions (i.e., unit activity of reaction components). The free energy change for other concentrations may be calculated from the Nernst equation (Equation 1-14). At $30°C$ $(303°K)$, the value of $2.303RT/F$ = 0.06 volt; thus:

$$E = E^{0'} + \frac{0.06}{n} \log \frac{[\text{oxidized}]}{[\text{reduced}]} \qquad\qquad (1\text{-}20)$$

*The factor 23,086 was obtained by dividing 96,500 coulombs (Faraday constant) by 4.18 to convert from coulomb·volts to calories.

The potential E is related to the free energy change by the equation

$$\Delta G = -nF\Delta E \tag{1-21}$$

where $\Delta E = E_2 - E_1$

Assuming hypothetically that the NAD/NADH ratio is 100:1 and the ethanol/acetaldehyde ratio is 100:1, the ΔE and ΔG of this reaction would be calculated as follows:

$$E_1 = -0.163 - 0.03 \log 100 = -0.163 - 0.060 \tag{1-22A}$$

$$= -0.223 \text{ volt } (note: \log \frac{1}{100} = -\log 100)$$

$$E_2 = -0.32 + 0.03 \log 100 = -0.320 + 0.060 \tag{1-22B}$$

$$= -0.260 \text{ volt}$$

In order to write the reaction in the direction of spontaneity, one must first establish that E_1 is more positive than E_2. Therefore, even under these nonstandard conditions, $E_1 - E_2$ will result in a positive ΔE, which corresponds to a negative ΔG. Accordingly, the value for ΔE is +0.037 volt.

Since the acetaldehyde-ethanol reaction is more positive in terms of potential than the NAD-NADH reaction, the system will spontaneously go in the direction of reduction: acetaldehyde + NADH → ethanol + NAD⁺. Hence,

$$\Delta G = -2 \times 23,086 \times 0.037 \text{ cal/mole} \tag{1-23}$$

This value of ΔG yields the maximum useful work if one mole of the reactants is converted to one mole of the products under the above steady-state concentration levels. In body fluids, a steady-state level of solutes is generally maintained. Therefore, by knowing or determining these concentrations, calculations can be made as to the actual free energy changes at physiologic concentrations.

Standard free energy changes frequently give valuable clues to certain energetic calculations. The previous examples were drawn from oxidation-reduction reactions, but, as can be seen from Equation 1-9, free energy calculations may be carried out on other types of chemical reactions and phase transformations of the most diverse kinds.

In view of the importance of free energy calculations, the most important methods, other than the ones based on the equilibrium constant and electrical potential measurements, will be briefly discussed.

CALCULATION OF FREE ENERGY FROM EXPERIMENTAL DETERMINATION OF ENTHALPY AND ENTROPY

One method is based on the equation of state,

$$\Delta G^0 = \Delta H^0 - T\Delta S^0 \tag{1-24}$$

The value of ΔG^0 can be determined from Equation 1-24 once ΔH^0 and ΔS^0 are known. As we have seen, the heat of reaction at constant pressure, ΔH^0, may be obtained (among other ways) from calorimetric measurements. The enthalpy change, ΔH^0, is the amount of heat liberated when reactants in their standard states are converted to products in their standard states. Alternatively, the standard enthalpy change can be calculated as the difference between the standard heats of formation of the products and the reactants, as expressed in Equation 1-25. The values for the standard heats of formation of many compounds are available in several handbooks [2].

$$\Delta H^0 = \Sigma \Delta H_f{}^0 \text{ (products)} - \Sigma \Delta H_f{}^0 \text{ (reactants)} \tag{1-25}$$

The standard entropy change may be evaluated by specific heat measurements over a wide temperature range. This method is based on the third law of thermodynamics, which states that the entropy of a perfect crystal of each element and compound is zero at $0°K$. In consequence, with the help of our previous definition of the entropy function, it can be seen that the absolute entropy of a substance at temperature T can be obtained from the following relationship:

$$S - S_0 = \int_0^T \left(\frac{Q_{rev}}{T}\right) dT \tag{1-26}$$

where $S_0 = 0$.

This approach will ultimately give the absolute entropies of the reactants and products and from their difference, ΔS^0 can be obtained. Absolute entropy data are available in handbooks as well.

It is probably correct to say that most free energy calculations utilize Equation 1-9, which gives the free energy change for arbitrary concentrations of reactants and products, provided the value for the standard free energy change is known.

Since G, H, and S are state functions, algebraic manipulations with them are permissible and frequently very useful. If the heat of a certain reaction cannot be determined experimentally, it is possible to calculate it from thermochemical data pertaining to the reactions that may be combined (at least on paper) to yield the desired reaction. For example, let us take the calculation of heat of formation of hydrocarbons [4]. Carbon and hydrogen do not combine under experimentally observable conditions, but the experimentally available heats of combustion can be used to calculate the heats of formation. In order to establish H_f (heat of formation) of CH_4 (methane) from its elements (as expressed in Equation 1-27), it becomes necessary to employ the known heats of combustion of the reactants and products, as illustrated in Equations 1-28A–C:

$$C + 2H_2 \rightarrow CH_4 \qquad\qquad \Delta H_f = ? \tag{1-27}$$

$$C + O_2 \rightarrow CO_2 \qquad\qquad \Delta H_1 \text{ (measurable)} \tag{1-28A}$$

$$H_2 + 1/2\,O_2 \rightarrow H_2O \qquad\qquad \Delta H_2 \text{ (measurable)} \tag{1-28B}$$

$$CH_4 + 2O_2 \rightarrow CO_2 + 2H_2O \qquad\qquad \Delta H_3 \text{ (measurable)} \tag{1-28C}$$

Combining these equations, an expression for the enthalpy changes is obtained:

$$\Delta H_f = \Delta H_1 + 2\Delta H_2 - \Delta H_3 \tag{1-29}$$

This example illustrates a common method for obtaining ΔH values indirectly. Noting from the previous discussion that the overall energy-yielding process in the human body is oxidation, the importance of bioenergetic considerations based on oxidation-reduction reactions is evident.

Before embarking on a survey of the main energy-yielding reactions in the cell, one further point has to be made concerning the mechanistic aspects of metabolism. Again taking glucose as an example, its oxidation could be written as a one-step reaction:

$$C_6H_{12}O_6 + 6O_2 \rightarrow 6CO_2 + 6H_2O \tag{1-30}$$

If the reaction were to take this path in the body, an instant burst of 67 Kcal per mole of glucose would become available as free energy. For a number of reasons, this is not desirable. First, the energy liberated would likely exceed by far that required by the energy-consuming process. In consequence, much of the energy would be wasted as heat, or, stating it differently, the efficiency would be very low. Second, through the evolutionary process, the body has developed an elaborate enzymatic apparatus that is capable of catalyzing the oxidation of smaller moieties, mainly two-carbon-containing metabolic intermediates. This allows a wide selection of oxidizable fuels (foods) that can all be channeled by chemical interconversions to these common smaller fragments. The result is an economy in the number of necessary enzymes.

The same principle operates when the body temporarily converts food into a storable form of energy. A single energy storage form is a great simplification, and cells accomplish this by converting fuel energy to a common intermediate energy source, adenosine triphosphate (ATP). In this case, chemical energy is converted from one form to another that is more immediately utilizable than the first, although this occurs at the expense of some energy loss in the conversion process. Since ATP has such a central role in metabolism, we will examine it and the reactions in which it can participate in more detail.

The general reaction that ATP undergoes in metabolic processes is the transfer of phosphate, as shown in Equation 1-31:

$$ATP + X \rightarrow ADP + PX \tag{1-31}$$

where ADP is adenosine diphosphate and X functions as a phosphate acceptor, provided an appropriate enzyme exists for the catalysis of the reaction.

A number of other phosphate compounds besides ATP may also act as phosphate donors. They can be arranged in an order of reactivity on the basis of their *phosphate transfer potentials* (PTP) [3]. This is simply the negative value of the standard free energy of phosphate transfer (in kilocalories) with water serving as the acceptor, i.e., the free energy of hydrolysis. ATP has an intermediate value, approximately 7, between compounds of high phosphate transfer potential, such as phosphoenolpyruvate (PEP) with a value of 12 to 13, and those with low phosphate transfer potential, such as glycerophosphate with a value of 2 to 3. Compounds with higher values can transfer phosphate to acceptors, which, in their phosphorylated form, have lower values. Accordingly, ATP can transfer a phosphate group to form glycerophosphate and ADP. ADP can, in turn, accept a phosphate group from PEP. If the phosphate transfer from donor to acceptor releases free energy in the kilocalorie range, the equilibrium constants are very large and product formation is virtually quantitative. ATP, with its intermediate phosphate transfer potential, can be formed in high yield by transphosphorylation from compounds with higher PTP and can pass its terminal phosphate to low PTP compounds in a similar, irreversible reaction.

Cells depend on ATP as the key energy storage form for providing the immediate source of chemical energy to perform chemical and physical work. In order to understand the bioenergetic mechanisms that allow the utilization of ATP for driving energy-requiring reactions, recall that *if the proper catalysts are available, a coupling between an endergonic (free energy-requiring; +ΔG) and exergonic (free energy-releasing; −ΔG) reaction is possible.*

The biochemical machinery of the body utilizes the strongly exergonic ATP-mediated phosphate transfer for driving endergonic reactions, which, as we have seen, are never spontaneous. In these cases, a composite reaction equation may be written that contains all reactants. *The overall reaction has to be exergonic.*

The principle is illustrated by Equations 1-32A and 1-32B, the product of the first being the reactant of the second:

$$A \rightleftharpoons B \tag{1-32A}$$

$$B \rightleftharpoons C \tag{1-32B}$$

By coupling these reactions into a composite one, Equation 1-33 is obtained:

$$A \rightleftharpoons B \rightleftharpoons C \tag{1-33}$$

At equilibrium, the *ratios* of the participants will be the same as they would be in the uncoupled reaction, since the equilibrium constants remain unaltered:

$$K_1 = \frac{[B]}{[A]} \text{ and } K_2 = \frac{[C]}{[B]} \tag{1-34}$$

Also,

$$\frac{[C]}{[A]} = K_1 \times K_2 \tag{1-35}$$

If in the reaction of Equation 1-32A the equilibrium favors A and in the reaction of Equation 1-32B it favors C, a conversion of A to B can be forced by the conversion of the reaction product B to C. Consider the following coupled reaction [7]:

$$\text{ATP} + \text{H}_2\text{O} \rightarrow \text{ADP} + \text{P}_i + 7300 \text{ cal} \qquad \Delta G^{0\prime} = -7300 \text{ cal} \quad (1\text{-}36)$$

$$\text{glycerol} + \text{P}_i + 2200 \text{ cal} \rightarrow \text{glycerophosphate} \qquad \Delta G^{0\prime} = +2200 \text{ cal} \quad (1\text{-}37)$$

where P_i is inorganic phosphate

Combining the reactions of Equations 1-36 and 1-37 results in

$$\text{ATP} + \text{glycerol} \rightarrow \text{ADP} + \text{glycerophosphate} + 5100 \text{ cal} \quad \Delta G^{0\prime} = -5100 \text{ cal} \quad (1\text{-}38)$$

It can be seen that the endergonic reaction 1-37 can be driven at the expense of the very exergonic reaction 1-36 to virtual completion. Without this coupling, it would not proceed at all. This example illustrates the general idea of coupled reactions and the pivotal utilization of ATP in driving metabolic reactions.

REFERENCES

1. Bertalanffy, L. The theory of open systems in physics and biology. *Science* 111:23, 1950.
2. Katchalsky, A., and Curran, P. F. *Non-Equilibrium Thermodynamics in Biophysics.* Cambridge, Mass.: Harvard University Press, 1967.
3. Lehninger, A. L. *Bioenergetics.* New York: Benjamin, 1965.
4. Mahan, B. H. *Elementary Chemical Thermodynamics.* New York: Benjamin, 1964.
5. Manufacturing Chemists Association, Carnegie Institute of Technology, Pittsburgh, Pa., 1955.
6. Montgomery, R., and Swenson, C. A. *Quantitative Problems in the Biochemical Sciences.* San Francisco: Freeman, 1969.
7. Segel, I. H. *Biochemical Calculations.* New York: Wiley, 1968.
8. *Selected Values of Chemical Thermodynamic Properties.* Washington, D.C.: National Bureau of Standards (U.S.), Circular 500, 1952.

SELECTED READING

Bauman, R. P. *An Introduction to Equilibrium Thermodynamics.* Englewood Cliffs, N.J.: Prentice Hall, 1966.
Christensen, H. N., and Cellarius, R. A. *Introduction to Bioenergetics: Thermodynamics for the Biologist.* Philadelphia: Saunders, 1972.
Denbigh, K. *The Principles of Chemical Equilibrium.* New York: Cambridge University Press, 1966.
Klotz, I. M. *Energy Changes in Biochemical Reactions.* New York: Academic, 1967.
van Holde, K. E. *Physical Biochemistry.* Englewood Cliffs, N.J.: Prentice-Hall, 1971.

Physicochemical Properties of   **2**
Gases and Solutions
Thomas J. Mende

FUNDAMENTAL RELATIONSHIPS

The chemical processes that participate in intermediary metabolism are, to a large extent, solution reactions. The metabolic conversion of one chemical species into another is usually limited to nonvolatile and water-soluble components, which may or may not be electrolytes. The most fundamental of metabolic transformations in aerobic organisms is, however, oxidation-reduction reactions in which the solution phase has to interact with the atmosphere in order to maintain the oxygen supply and provide a medium for the release of the ultimate gaseous metabolic waste product, carbon dioxide.

In complex organisms, solution reactions take place in all cells of the body. These reactions then require access to the atmospheric gas supply and, at the same time, must have an effective waste disposal system. The only access route for a gas, or for that matter any other soluble reactant, to its reaction site in solution is *diffusion*. Therefore, it is necessary to study the conditions that regulate or limit this process.

Diffusion in solution in one dimension can be described by Fick's law [5]:

$$\dot{M} = \frac{dm}{dt} = -AD \frac{dc}{dx} \tag{2-1}$$

This law states that the mass, m, of solute (in grams) crossing a reference plane perpendicular to the direction of transport in unit time is proportional to the product of the area, A, and the concentration gradient, dc/dx, i.e., the rate of change of concentration with distance. The *diffusion constant* or coefficient of the solute is represented by D, which has the dimensions $cm^2 \cdot sec^{-1}$. The negative sign is a consequence of the convention that flow is the result of force, which, according to mechanics, is the negative of the gradient, and flow is from higher to lower concentrations.

Equation 2-2 expresses the mass transport per unit area and provides a definition for "flow" or *flux, J:*

$$J = -D \frac{dc}{dx} \tag{2-2}$$

This law is derived for isothermal situations. The parameter that is therefore characteristic for a diffusing solute is D, which has values that can be determined by measurements of concentration changes with time. It is normally calculated from Fick's second law [5]:

$$\frac{\delta c}{\delta t} = D \frac{\delta^2 c}{\delta x^2} \tag{2-3}$$

17

which must be integrated to convert it into a suitable expression for experimental use. Measurements of this type reveal that for solutes with molecular weights up to a few hundred, i.e., most substances that are classified as crystalloid, the diffusion constant is proportional to the reciprocal of the square root of the molecular weight [9]:

$$D(MW)^{1/2} = \text{constant} \tag{2-4}$$

This relationship holds when the solute particle radius is commensurate with the lattice dimensions, i.e., the distances between the solvent molecules. In cases where the solute radius is large in comparison with the lattice dimensions, the relationship $D(MW)^{1/3}$ = constant applies. This can also be seen from the Stokes-Einstein equation, which has been derived for large, spherical solute particles:

$$D = \frac{kT}{6\pi\eta r} \tag{2-5}$$

where k is Boltzmann's constant, T is temperature, η is the coefficient of viscosity of the solution, and r is the radius of the particle. Since D is proportional to the reciprocal of the radius, it is inversely proportional to the cube root of the molecular weight.

In those areas of the body where transport by mechanical means – e.g., blood flow, air flow in certain parts of the lungs, or peristalsis – does not take place, the exchange of solutes proceeds by diffusion provided that there are no barriers to impede it. Membranous structures are such barriers around cells, across which diffusion may or may not be the only method of exchange, depending on the nature of the solute. A variety of specialized carrier systems are known to function in these surfaces in order to assist material exchange.

In the case where diffusion is responsible for membrane penetration, a modified form of Fick's law for free solutions is applicable. In the expression of the concentration gradient, the differential dx indicates distance over which the concentration changes by dc. In most instances, membranes and membrane systems have a constant thickness, Δx, that can be incorporated into the diffusion constant, D, to yield a new constant, P. This is the *permeability constant*, which has the dimensions of $cm \cdot sec^{-1}$, and the result of substitution in Equation 2-1 is:

$$\dot{M} = PA \, \Delta c \tag{2-6}$$

where Δc is the concentration difference between the two sides of the membrane. Here, as before, transport in unit time per unit membrane area is called *flux* and it may be derived from Equation 2-6 ($J = \dot{M}/A$; $gm \cdot sec^{-1} \cdot cm^{-2}$).

The exact value of the membrane thickness is not usually known. Estimations from electron micrographs may introduce errors due to artifactual shrinkage. Therefore, the determinations of permeability coefficients are usually comparative, employing one type of cell and different penetrating solutes. For a particular type of cell, the relative rates of entry, which are functions of the permeability coefficients of the solutes, depend on a number of factors. High relative solubility in oil versus water enhances solute penetration, which is in keeping with the substantial lipid content of all cell membranes. Small nonelectrolytes usually enter faster than electrolytes, and their low oil solubility demands that their entry take place through special hydrophilic pores.

Returning to the process of diffusion in the body, a discussion of the atmospheric gases entering and leaving the respiratory system is in order. To reach the tissues, oxygen, for example, has to go into solution in the body fluids, such as the pulmonary venous blood. To follow this process, the theories regarding the solubility of gases in liquids should be reviewed.

Gas solubilities in liquids are governed by Henry's law, which states that the concentration of gas contained in physical solution in a liquid is proportional to the partial pressure of the gas to which the liquid is exposed:

$$P_p = kX \tag{2-7}$$

where P_p is the partial pressure of the gas, X is the mole fraction of the gas in the liquid, and k is a proportionality constant, or *Henry's law constant,* which is determined from empirical measurements. It is a function of the gas in question, the chemical composition of the solvent system, and the temperature.

For the sake of simplification, a slightly different expression is employed for calculations with respiratory gases. The absorption coefficient generally used in respiratory gas calculations is the *Bunsen coefficient,* α. It is defined as the number of milliliters of gas dissolved in one milliliter of the liquid in question (blood in this case) at $38°C$ if the partial pressure of the gas is one atmosphere and the volume of the gas is corrected to conditions of standard temperature and pressure (STP).

One works with partial pressures whenever mixtures of gases exist in the system, as is the case with atmospheric air. The *partial pressure* is the pressure that the gas in question would exert if it were present by itself in the volume occupied by the gas mixture. From this definition, the total (e.g., atmospheric) pressure is the sum of the partial pressures:

$$P_T = P_1 + P_2 + \ldots \tag{2-8}$$

where P_1 is the partial pressure of gas 1, and so on. Since the ideal gas law applies to gas mixtures as well as pure gases:

$$PV = nRT \tag{2-9}$$

where n = number of moles
R = gas constant
T = temperature (absolute)

Therefore

$$P_1 = \frac{n_1 RT}{V}$$

$$P_T = \frac{n_1 RT}{V} + \frac{n_2 RT}{V} + \ldots \qquad (2\text{-}10)$$

$$P_T = \frac{RT}{V}(n_1 + n_2 + \ldots)$$

and

$$\frac{P_1}{P_T} = \frac{n_1}{n_1 + n_2 + \ldots} = \text{mole fraction of gas 1 [2]}$$

Avogadro's law states that equal volumes of gas contain equal numbers of moles; therefore, the fraction of gas by volume must equal the mole fraction of the gas. In consequence, the knowledge of the volume fractions of the components of the gas mixture will yield their partial pressures, i.e.,

$$\text{Mole fraction of gas} = \frac{\text{volume \% of gas}}{100}$$

For air, $O_2 = 20.94\%$, $CO_2 = 0.03\%$, and $N_2 = 79.04\%$. Since

$$P_i = P_T \left(\frac{\text{volume \% of gas } i}{100} \right) \qquad (2\text{-}11)$$

the partial pressure of any gas can be calculated, provided the volume composition and total gas pressure are known. For example, to calculate the volume of physically dissolved oxygen in 100 ml of blood at an atmospheric pressure of 740 mm Hg, we first have to know *not* the dry inspired air composition but the dry alveolar air composition, which has the following average (resting) values: O_2, 14.0%; CO_2, 5.6%; and N_2, 80.4%. It can readily be seen that for both O_2 and CO_2, the differences are considerable. Alveolar gas composition will vary according to whether the person is active or at rest, or it may vary as a result of pathologic processes.

Since the above resting values of alveolar gas composition are given for dry air, whereas pulmonary air is saturated with water vapor, the pressure due to the listed components is equal to the atmospheric pressure minus the saturated water vapor pressure. The latter is indeed at saturation level in the lung, and, at body temperature, it has a value of 47 mm Hg. Thus, the residual pressure is 740 mm Hg − 47 mm Hg = 693 mm Hg. The partial pressure of O_2 will then be

$$P_{O_2} = 693(14.0/100) = 97.02 \text{ mm Hg}$$

The Bunsen coefficient of O_2 = 0.023; therefore, the number of milliliters of O_2 dissolved in 100 ml blood = 0.023 × (97.02/760) × 100 = 0.29 ml. The Bunsen coefficient is defined in terms of one atmosphere of pressure (760 mm Hg), so division by 760 is required. Multiplication by 100 gives the result in volume-percent.

It should be mentioned at this point that blood carries a far larger volume of oxygen than what is calculated by this method, which accounts only for the physically dissolved gas. In fact, oxygen transport must be mainly accomplished by the oxygen being in a chemically bound form. However, calculation shows that nitrogen is carried in simple physical solution, and CO_2 is transported mainly by physical solution with partial assistance from chemical binding. Without further elaboration, it can be seen that the body fluids take up gases as long as the partial pressure in the gas phase exceeds that of the blood, and they can lose it by respiratory exchange in the opposite situation. Once the gases are in solution, they travel by diffusion like any other dissolved, nonvolatile component, disregarding, of course, the effect of hydrodynamic transport in the bloodstream.

Two additional classes of solution properties will now be discussed. The first is based on the number of independent solute particles present, and the properties included in this class are collectively known as *colligative properties.* The second encompasses the specific effects connected with the properties of the hydrogen ion, H^+, or, more correctly, the hydronium ion, H_3O^+.

COLLIGATIVE PROPERTIES OF SOLUTIONS

Solutions possess certain colligative properties that are independent of the chemical composition of the system and, in dilute solution, depend only on the relative number of solute to solvent molecules present. Quantitatively, the solution concentrations are usually expressed in mole fractions or in molalities when dealing with colligative properties. The *mole fraction* of component 1 is defined as the number of moles of component 1 in solution, n_1, divided by the total number of moles in the system:

$$X_1 = \frac{n_1}{n_1 + n_2 + \ldots} \tag{2-12}$$

The mole fraction of component 2 is

$$X_2 = \frac{n_2}{n_1 + n_2 + \ldots}$$

and so on. It can be seen that for very dilute solutions,

$$X_{solute} \approx \frac{n_{solute}}{n_{solvent}} \tag{2-13}$$

The sum of the mole fractions of all components equals unity.

The concept of molality is closely related to the mole fraction. The *molality* of a solution is the number of moles of solute per kilogram of solvent. For example, a 1 molal solution of glucose contains 1 mole (180 gm) glucose per 1000 gm of water. The mole fraction of glucose in this system is

$$X_2 = \frac{n_2}{n_1 + n_2} = \frac{1}{1 + (1000/18)} = \frac{1}{56.6} = 0.0177$$

where 18 is the molecular weight of water. The mole fraction of water, on the other hand, is

$$X_1 = \frac{55.6}{56.6} = 0.982$$

For dilute solutions, the molality and mole fractions are proportional to one another; also, in dilute solutions, the molality and molarity (moles of solute per liter of solution) are approximately equal.

The colligative properties are:

1. Lowering of the vapor pressure of the solvent
2. Depression of the freezing point
3. Elevation of the boiling point
4. Osmotic pressure

The name *colligative* (meaning tied together) indicates that given any one of the four properties, the other three can be derived by calculations. The following discussion is limited to binary solutions — i.e., a solvent with a single solute — with the further restrictions that the solute is nonvolatile and that it will not associate, dissociate, or otherwise interact with the solvent. This type of solute is referred to as an *ideal non-electrolyte*.

LOWERING OF THE VAPOR PRESSURE

Of the four properties, the first is the most fundamental, though the most cumbersome to measure in practice. The degree of lowering of vapor pressure by a solute can be calculated from Raoult's law [1], which states that the degree of lowering of vapor pressure is equal to the vapor pressure of the solvent multiplied by the mole fraction of the solute. As seen in Figure 2-1, the observed linear relation between the partial pressure, P_p, and the mole fraction of the solvent, X_1, leads to Equation 2-14:

$$P_p = (\text{constant}) \, X_1 \tag{2-14}$$

The value of the constant is P_0, or the vapor pressure of the pure solvent, because $P_p = \text{constant} = P_0$ if $X_1 = 1$.

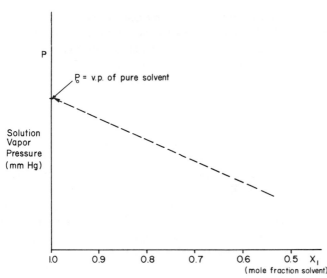

Figure 2-1
Lowering of vapor pressure according to Raoult's law.

A more useful form of this equation can be obtained if the mole fraction of the solute, X_2, is introduced:

$$P_p = P_0X_1 = P_0(1 - X_2) = P_0 - P_0X_2 \tag{2-15}$$

For dilute solutions, since n_2 is approximately equal to the molality of the solute,

$$X_2 \approx \frac{n_2}{n_1} = \frac{m}{1000/MW_1} \tag{2-16}$$

where m = molality and MW_1 = molecular weight of the solvent.

Therefore:

$$P_0 - P_p = \Delta P = \frac{P_0 m}{1000/MW_1} \tag{2-17}$$

For a 1 molal aqueous solution:

$$X_2 = 0.018$$

The values of P_0 for water at $0°$, $25°$, $37°$, and $100°C$ are 4.58, 23.75, 47.06, and 760 mm Hg, respectively, and the corresponding ΔP values at these temperatures for 1 molal solutions are 0.081, 0.43, 0.833, and 13.45 mm Hg. These small values indicate that this colligative property would be hard to measure accurately unless the concentrations were relatively high. For concentrated solutions, however, deviations from Raoult's law are severe. There are, nevertheless, clinical instruments for the measurement of the osmolal concentration of body fluids that utilize the principle of the lowering of the vapor pressure, but these employ kinetic methods.

OSMOLARITY AND OSMOSITY

For an explanation of the meaning of *osmolality* [10], or, for dilute solutions, *osmolarity,* recall the ideal nonelectrolyte as defined previously. A 1 osmolal solution shows the same depression of vapor pressure, freezing point depression, boiling point elevation, and osmotic pressure as a 1 molal solution of an ideal nonelectrolyte.

A unit sometimes employed for simple calculations is the *osmosity,* which is defined simply in terms of the molarity of an aqueous solution of NaCl with the same freezing point depression as the solution under consideration. Since the freezing point depression of 1 molar NaCl is approximately $3.46°C$, the osmosity of a solution freezing at $-x°C$ (i.e., at x degrees below zero Celsius) is given approximately by Equation 2-18:

$$\text{osmosity} = \frac{x°C}{3.46} \tag{2-18}$$

BOILING POINT ELEVATION AND FREEZING POINT DEPRESSION

The connection of these two colligative properties with the lowering of vapor pressure can be seen by inspection of the temperature-vapor pressure curves of aqueous solutions shown in Figure 2-2. The value of the boiling point elevation constant, K_{BP}, depends only on the solvent; it is, for example, $0.51°C$ for water. The boiling point elevation is used extensively for molecular weight determinations in organic chemistry, but, due to the thermal lability of many biochemicals, it is not used for biologic fluid analyses. This complicating factor is much less serious for freezing point depression, and this method is of great importance in the biologic field. As seen from the graph in Figure 2-2,

$$\Delta T_{FP} \propto \Delta P \propto m$$

and

$$\Delta T_{FP} = K_{FP}m \tag{2-19}$$

where K_{FP} (the molal freezing point depression constant or cryoscopic constant) is defined as the freezing point depression of a 1 osmolal solution. The value of K_{FP} depends only on the nature of the solvent, and, for water, it is $1.86°C$. The osmolal concentration of solutions — including those of body fluids, plasma, and urine — can be determined using Equation 2-19.

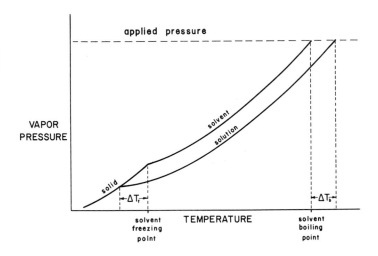

Figure 2-2
Boiling point elevation and freezing point depression.

Any colligative property is suitable for the determination of molecular weights, for example:

$$m = \frac{gm}{MW}$$

$$MW = \frac{(K_{FP})}{(\Delta T_{FP})} \times gm = \frac{1.86 \times gm}{\Delta T_{FP}} \tag{2-20}$$

where gm = weight in grams of solute per 1000 gm H_2O
  MW = molecular weight of solute

OSMOTIC PRESSURE

The last and, from a physiologic point of view, the most important of the colligative properties is osmotic pressure. This property is illustrated by the experimental setup designed for its measurement (Fig. 2-3). It shows that osmotic pressure is also governed by Raoult's law and that the solution will rise in the column to a height where the vapor pressure of the solution equals the vapor pressure of the solvent, which is the state of equilibrium. The vapor pressure diminishes with height according to the barometric formula:

$$P_p = P_0 \exp\left[-\frac{g(MW)}{RT}\right]$$

$\Delta P \propto \pi \propto m$
π = osmotic pressure
$\Delta P = P_0 - P$
m = molality

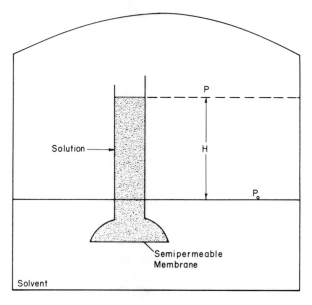

Figure 2-3
Osmotic pressure measurement.

where P_p = partial pressure of gas (water vapor) at height h
P_0 = partial pressure of gas at the surface (height = 0)
g = acceleration of gravity
MW = molecular weight
R = gas constant
T = absolute temperature

At the point of equilibrium, the hydrostatic pressure of the column is equal to the osmotic pressure. More perspicuously, osmotic pressure may be described by the force one has to apply in order to prevent the solvent from entering and diluting the solution that is separated from it by the semipermeable membrane. Osmotic pressure in very dilute solutions follows van't Hoff's law, which exhibits a formal resemblance between the pressure of an ideal gas and the osmotic pressure existing in solution. This law states that the osmotic pressure, π, is proportional to the concentration of solute and the absolute temperature. The proportionality constant is R, the gas constant:

$$\pi = cRT \qquad (2\text{-}21)$$

where the concentration, c, stands for

$$c = \frac{m}{V} \tag{2-22}$$

where m = number of moles of solute
$\quad\quad V$ = volume of solution

The concentration here is expressed in terms of molarity (M), since, under the conditions to which this law applies, the difference between molality and molarity may be neglected.

An example will serve to show the difference in magnitude between osmotic pressure and other colligative properties. To illustrate the difference between the osmotic pressure and the freezing point depression of a 1% sucrose solution (10 gm/liter; MW of sucrose = 342), let us first calculate the osmotic pressure at 25°C. The concentration is 10 gm/342 = 0.0292 moles per liter (molar). Sucrose at low concentration behaves like an ideal nonelectrolyte; thus,

$$\pi = cRT$$

$$= 0.0292 \times 0.082 \times 298 = 0.714 \text{ atm } (R = 0.082 \text{ liter} \cdot \text{atm} \cdot \text{mole}^{-1} \cdot \text{deg}^{-1})$$

$$= 54.26 \text{ cm Hg} = 738 \text{ cm H}_2\text{O} \ (1 \text{ atm} = 76 \text{ cm Hg} = 76 \times 13.6 \text{ cm H}_2\text{O})$$

This means that the solution column would rise to approximately 7.38 meters. The freezing point depression of the same solution would be:

$$\Delta T_{FP} = K_{FP}m \approx K_{FP}c$$

$$\Delta T_{FP} = 1.86 \times 0.0292 = 0.054°C$$

It can be seen that for the measurement of the osmolar concentration of a high-molecular-weight substance, osmotic pressure measurements are feasible. On the other hand, they cannot be adapted for low-molecular-weight substances, because existing membranes are not impermeable to them. Hence, the other three methods are more useful for low-molecular-weight substances.

As previously mentioned, van't Hoff's law and Raoult's law are valid only for very dilute solutions. However, the osmotic pressure of dilute solutions can be determined by making a series of measurements at higher concentrations and extrapolating the data to zero concentration. This procedure is illustrated in Figure 2-4 [3], where "reduced osmotic pressure," π/c, is plotted versus concentration, c, at several experimentally determined points, and the line of best fit is extrapolated to zero concentration. In this example, the molecular weight can be determined from

$$MW = \frac{RT}{\pi/c} \tag{2-23}$$

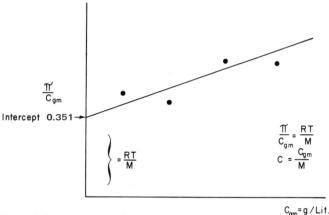

Figure 2-4
Reduction of osmotic pressure versus concentration of solute.

Determining $\pi/c = 0.351$ from the relation of Figure 2-4 and letting $T = 3°C$, the molecular weight of the test substance (in this case, hemoglobin) can be calculated:

$$MW = \frac{(84.7)(276)}{0.351} = 66,600 \qquad\qquad (2\text{-}24)$$

where R is expressed as 84.7 liter·cm H_2O/mole/deg.

As colligative properties are based on solute particle concentrations (regardless of their chemical nature), all individual ions have to be considered when calculating the osmolarity. In consequence, salts that dissociate into two ions (e.g., Na^+, Cl^-, both monovalent, or Mg^{+2}, SO_4^{-2}, both divalent) should show twice as large a freezing point depression or other colligative property as an equimolar solution of a nonelectrolyte. Such salts as $CaCl_2$ and $(NH_4)_2SO_4$, by the same reasoning, should have a three-fold effect.

Although this prediction conforms to the actual experimental results in solutions of high dilution, at moderate concentrations the deviations are considerable; values less than the expected ones are obtained. Electrostatic interactions in the more concentrated solutions diminish the ionic activities, which results in lower values. Also, in the case of weak electrolytes, incomplete dissociation explains the lower measured values.

To take these effects into account, one has to multiply the theoretical values by an empirical constant, the *cryoscopic coefficient*, G (not to be confused with the molar freezing point depression constant), expressed in Equation 2-25 [5]:

$$G_i = \frac{\text{freezing point depression of a given molal concentration of the electrolyte } i}{\text{freezing point depression of the same molal concentration of an ideal nonelectrolyte}} \qquad (2\text{-}25)$$

Some representative values of G are:

Electrolyte	Cryoscopic Coefficient
$MgCl_2$	2.7 at 0.05 M
KCl	1.89 at 0.05 M
$MgSO_4$	1.30 at 0.05 M
NaCl	1.87 at 0.05 M

The Nature of Colloid Osmotic Pressure

Colloid osmotic pressure or oncotic pressure refers to a mechanism that is mainly responsible for the maintenance of plasma volume. The endothelial walls of the capillaries are permeable to water and to the low-molecular-weight solutes of plasma. They are not, however, appreciably permeable to plasma proteins, at least not in most areas of the body. Since the only nonpermeable solutes are the protein molecules and since they have high molecular weights that are characteristic of colloids, they give rise to the colloid osmotic pressure, which tends to draw water into the capillaries from the tissues. This effect functions to counterbalance hydrostatic pressure, which drives the water out of the capillaries by filtration.

Although the hydrostatic pressure exceeds the colloid osmotic pressure at the arterial end of the capillaries, on the venous end the hydrostatic pressure is less than the oncotic pressure, which results in reabsorption of the plasma ultrafiltrate.

The Gibbs-Donnan Relation and Its Effect on Osmotic Equilibrium

Special situations arise in cases where a membrane separates a compartment containing a polyvalent, nondiffusible ion (which, for the sake of example, may be assumed to be negatively charged) from a second, external compartment that contains a monovalent salt yielding the same cation that is neutralizing the polyvalent anion inside the membrane [4]. The initial conditions may be set as follows:

Initial Concentrations (Inside)	Initial Concentrations (Outside)
$[P^{-n}]_{in}$ (nonpermeable polyvalent anion) $= w$	$[C^+]_{out} = y$
$[C^+]_{in}$ (cation $= n[P^-]$) $= z$	$[A^-]_{out} = y$

The polyvalent anion, P^{-n}, is effectively trapped inside the membrane so that only the monovalent ions and the solvent can pass through. Let us impose the further restriction that volume changes will not take place during the equilibration process. Since electrical neutrality is maintained, C^+ and A^- will diffuse from the outside together. As this buildup progresses, exit of the diffusible ions will commence.

Equilibrium for the diffusible species will be reached when the product of the concentrations of the diffusible ions inside equals their product outside the membrane; this is termed *Donnan equilibrium:*

$$[C^+]_{in} \ [A^-]_{in} = [C^+]_{out} \ [A^-]_{out} \tag{2-26}$$

The final concentration for all species can be derived by assuming that at equilibrium, x equivalents each of C^+ and A^- will have entered the inner compartment, which will result in the following concentrations:

Final Concentrations (Inside)

$[P^{-n}]_{in} = w$

$[C^+]_{in} = z + x$

$[A^-]_{in} = x$

Final Concentrations (Outside)

$[C^+]_{out} = y - x$

$[A^-]_{out} = y - x$

Expressing these results in the terms of Equation 2-26, we obtain

$$x(z + x) = (y - x)^2 \tag{2-27}$$

Solving for x, we can calculate the number of equivalents of either ion transferred from outside to inside the membrane:

$$x = \frac{y^2}{z + 2y} \tag{2-28}$$

We can see from Equation 2-27 that the product of two unequal numbers equals a square. The sum of these numbers always exceeds the sum of the square roots of the square, i.e.,

$$x + (z + x) > 2(y - x) \tag{2-29}$$

The consequence of this is that the osmolar concentrations of ions inside the membrane — disregarding even the contribution from the polyvalent, nondiffusible ion — would, at equilibrium, exceed that on the outside. This would continuously draw water into the inner compartment. Thus, osmotic equilibrium cannot be established unless work is performed on the system to oppose the flux of water.

In our example, we have postulated the existence of rigid walls, but cells have to maintain their steady-state volume by actively extruding ions and water that seep in due to the Donnan effect. Cell contents are high in polyvalent, nondiffusible ions, such as proteins and nucleic acids, whereas the surrounding interstitial fluid contains far fewer such ions. This situation is similar to the system described in our example. An additional characteristic feature of the Donnan equilibrium can also be seen by rearranging Equation 2-26 to show that the ratio of $[C^+]_{in}/[C^+]_{out}$ equals the inverse ratio for the diffusible anion. If such ratios are observed in experimental situations, it is highly suggestive of the existence of a Gibbs-Donnan equilibrium.

Osmotic Behavior of Living Cells and Tissues

Plasma membranes of cells are semipermeable to a degree, which allows the cells to act as osmometers. The osmotic behavior of cells will depend on the differences in osmolarity between the intracellular and extracellular fluids. A solution such as plasma or 0.15 M NaCl has the same osmolar concentration as mammalian cellular fluid; in other words, it is *isotonic* with it. Solutions more dilute than these are hypotonic, and the more concentrated are hypertonic. Cells swell in hypotonic and shrink in hypertonic solutions. This cellular swelling and shrinking in response to the osmotic concentration of the milieu approximately follows the analog of Boyle's law for solutions:

$$\pi V = \text{constant} \tag{2-30}$$

where π is osmotic pressure of the solution and V is the volume of the cells. The behavior of the cells conforms to this relationship very closely if the "nonsolvent volume" is taken into account. This volume is the fraction of the cell volume that is occupied by its solid content and cannot participate in volume changes. If the volume of a cell population in two solutions of different osmolarity is measured (e.g., by determining the hematocrit of red blood cells), the following relationship will hold:

$$(V - V_a)\pi = (V_n - V_a)\pi_n = \text{constant} \tag{2-31}$$

where V_n is the volume of the cells under normal conditions surrounded by an isotonic solution of osmotic pressure π_n, V is the cell volume with an osmotic pressure π, and V_a is the nonsolvent volume, which is constant for a particular cell type. V_a may be expressed by Equation 2-32:

$$V_a = \frac{\pi_n V_n - \pi V}{\pi_n - \pi} \tag{2-32}$$

Some cellular nonsolvent volumes, expressed in terms of percentages of total cellular volume [7], are:

Sea urchin eggs	12%
Erythrocytes	45%
Myelinated frog nerve	40%

The osmotic volume change of cells in different osmotic milieus permits the calculation of the intracellular osmolar concentration, which is what was meant by the above reference to cells as "osmometers." The intracellular osmolar concentration equals the osmolar concentration of the solution with which it is in osmotic equilibrium.

Osmotic Behavior in Penetrating Solutes

Plasma membranes are not completely impermeable to all low-molecular-weight solutes. Some substances, mostly nonelectrolytes and weak electrolytes in their nondissociated forms, can rapidly enter cells, although their rate of penetration will be slower than that of water. If the volume of red blood cells is observed in a medium that contains a penetrating solute such as urea, their behavior will be found to be different from the one described above.

This can be explained as follows. The red blood cell interior contains essentially impermeable ionic species (salts) at a concentration of approximately 0.3 osmolar. If we place the cells into a large volume of isosmotic* urea solution, this nonelectrolyte will enter the cell fairly rapidly. As the concentration of urea builds up inside the cell, an osmotic concentration difference builds up between the cell and its exterior due to the impermeable ions inside. The intracellular urea concentration would approach the same level as outside, since the cellular volume is assumed to be much smaller than the exterior solution volume. If we were to imagine that the cells have a rigid membrane, the end result would be that the red blood cell content would reach 0.3 Osm salt/liter + 0.3 Osm urea/liter = 0.6 Osm/liter, while the osmolarity of the surrounding solution would be 0.3 Osm urea/liter. This would result in a difference of 0.3 Osm/liter, i.e., an effective excess osmolarity would exist inside the red blood cell. Such a developing osmotic imbalance would draw water into the cell causing it to swell and burst ("osmotic hemolysis"), since the cell membrane is, in fact, a nonrigid structure.

In the case of hyperosmolar (e.g., two times isosmolar) urea, one should keep in mind that *water is the most rapidly penetrating of all substances.* In consequence, an initial imbalance of osmolarity ensues because the external solution is twice as concentrated as the cell content. Thus, the cellular water will rapidly leave and cause the cells to shrink. A cell shrinkage to approximately half its original volume — disregarding the nonsolvent volume for simplicity — would bring the intracellular concentration to twice its original value, i.e., into balance with the surrounding solution. This situation, however, would only be temporary in the presence of a penetrating solute, so that urea would build up gradually inside. The end result in this case (again assuming there are rigid membranes) would also be an effective excess of 0.3 Osm/liter due to the effect of intracellular salts. The urea concentration would be 0.6 Osm/liter on both sides, nullifying any of its own osmotic effects, which again would lead to cell lysis.

Therefore, the initial shrinkage (due to rapid water loss) and the subsequent hemolysis (due to the penetration of urea) are explained. Cells other than erythrocytes that behave like osmometers are skeletal muscle fibers, plant cells (plasmolysis), and cartilage.

Erythrocytes normally have a biconcave disk shape that allows them to swell in hypotonic solutions until they round out to spheres. This normally takes place at solution concentrations of around 0.4% to 0.5% NaCl (an isotonic solution of NaCl is 0.85% to 0.9%). In solutions more dilute than this, the 50% hemolysis point of erythrocytes can readily be determined. This is employed in the osmotic fragility test, a diagnostic procedure [8].

---

*This term should be contrasted with "isotonic," which is reserved for solutions composed of nonpermeating solutes.

ACIDS AND BASES

The Brønsted-Lowry definition for acids and bases will be employed. According to this concept, a substance is termed an *acid* if it can release hydrogen ions (protons); whereas a *base* is a substance that can accept hydrogen ions. In releasing a hydrogen ion, an acid is converted to its *conjugate base:*

$$A \rightleftharpoons H^+ + B^- \tag{2-33}$$

where B^- is the conjugate base. This concept has meaning only for weak electrolytes, i.e., those for which permeable dissociation constants exist. Table 2-1 lists some common acids and corresponding conjugate bases.

Table 2-1. Common Acids and Corresponding Conjugate Bases

Acids	Conjugate Bases
CH_3COOH	CH_3COO^-
NH_4^+	NH_3
H_3PO_4	$H_2PO_4^-$
$H_2PO_4^-$	HPO_4^{-2}
HPO_4^{-2}	PO_4^{-3}
H_2CO_3	HCO_3^-
HCO_3^-	CO_3^{-2}
$CH_2NH_3^+$ $\overset{|}{COO^-}$	CH_2NH_2 $\overset{|}{COO^-}$

It may be seen from Equation 2-33 that A and B can be molecules or ions of any sign or charge, provided A is one unit more positive than B. In aqueous solutions, the dissociation of an acid leads to the following equilibrium:

$$HA + H_2O \rightleftharpoons H_3O^+ + A^- \tag{2-34}$$

Thus, the hydrogen ion as a chemical species does not exist in solution, but rather hydronium ions, H_3O^+, are formed. In calculations involving hydrogen ion concentrations, however, the H^+ notation is retained for the sake of simplicity. The thermodynamic dissociation constants are expressed in terms of the activities of the reactants and products, but in the approximate treatment used here, the equating of activities with concentrations is a simplification that is convenient for dealing with dilute solutions. The dissociation constant, K_a, for a weak acid is defined as

$$K_a = \frac{[H^+] \cdot [A^-]}{[HA]} \tag{2-35}$$

The dissociation equation for a weak acid can also be written as

$$[H^+] = K_a \frac{[HA]}{[A^-]} \tag{2-36}$$

Taking the logarithm of both sides,

$$\log [H^+] = \log K_a + \log \frac{[HA]}{[A^-]} \tag{2-37}$$

Multiplying both sides of Equation 2-37 by -1 and noting that pH is defined as $-\log [H^+]$ and pK_a as $-\log K_a$,

$$pH = pK_a + \log \frac{[A^-]}{[HA]} \tag{2-38}$$

This is known as the Henderson-Hasselbalch equation.

THE SIGNIFICANCE OF pK_a

As can be seen from Equation 2-38, if the concentration of the conjugate base is equal to that of the acid, the pH equals the pK_a (log 1 = 0). This equation can be used to calculate pH if the pK_a and the conjugate base-to-acid ratio are given. Also, the ratio, but not the absolute concentrations, of the conjugate base to the acid can be calculated if the pH and the pK_a are given. To determine the absolute concentration of either the acid or the conjugate base, the concentration of the other one has to be stated.

An equation that is involved in the solution of some of the problems arising from proton equilibria is the expression for the ion product of water. Water, a very weak electrolyte, dissociates:

$$2HOH \rightleftharpoons H_3O^+ + OH^- \tag{2-39}$$

The concentration of water (HOH) can be taken as a constant, since it is barely affected by the minimal dissociation that occurs. It can therefore be incorporated into the dissociation constant in the following equation:

$$K_w = [H_3O^+] \cdot [OH^-] = 1 \times 10^{-14} \text{ at } 25°C \tag{2-40}$$

Equations 2-38 and 2-40 can be used for the calculation of the majority of acid-base relationships.

BUFFERS

A *buffer* is a solution that is capable of withstanding the introduction of relatively large amounts of H^+ and OH^- ions without excessive shifts in pH. It also stabilizes the pH against the effects of dilution. Buffers are composed of a weak acid (or base) and its salt. Their action is illustrated by the following experiment.

If 1 ml of 10 M HCl is added to 1 liter of distilled water (pH 7), the pH of the solution would be approximately 2, since the hydrogen ion concentration, $[H^+]$, will be 1×10^{-2}. If we were to add the same volume of 10 M NaOH to 1 liter of distilled water, the pH would be 12. The $[OH^-]$ would be 1×10^{-2}, i.e., a pOH of 2, and, since

$$pK_w = pH + pOH$$

$$14 \ \ = pH + 2$$

then

$$pH \ \ = 12$$

Let us now consider a 1 liter solution containing 0.1 mole of sodium acetate and 0.1 mole of acetic acid. The starting pH of this buffer solution can be calculated from the Henderson-Hasselbalch equation and the fact that the pK_a for acetic acid is 4.75:

$$pH = pK_a + \log \frac{0.1}{0.1}$$

$$pH = 4.75 + 0 = 4.75$$

On adding 1 ml of 10 M HCl, the ratio of conjugate base to acid changes to 0.09/0.11. The resulting calculated pH would be 4.66, a change very much smaller than that noted above. Similar considerations for the addition of 1 ml of 10 M NaOH would yield a ratio of 0.11/0.09 and a resulting pH of 4.84.

It can be seen that the acid component of the buffer disposes of the OH^- ions that are introduced by converting them to water through neutralization:

$$HAc + OH^- \rightleftharpoons Ac^- + H_2O \tag{2-41}$$

On the other hand, H^+ ion neutralization takes place according to the reverse of the dissociation equation:

$$Ac^- + H^+ \rightleftharpoons HAc \tag{2-42}$$

The essence of this step is the conversion of a strong acid (complete dissociation) to a weak one (slight dissociation). One should not forget that an undissociated acid, such as the chemical species CH_3COOH (nonionized acetic acid), does not affect the pH; only the H^+ ions (the dissociation product) are able to do so.

The *buffer capacity* or buffer value is the quantitative measure of a solution's buffering power. It may be denoted by B and calculated from the following relationship:

$$B = \frac{dB}{d\text{pH}} \approx \frac{\Delta B}{\Delta \text{pH}} \tag{2-43}$$

where ΔB = the increment of strong acid or base added

ΔpH = the corresponding increment in pH

A solution with $B = 1$ will take up 1 equivalent (i.e., mole/valence) of acid or base per liter per unit change in pH.

The buffer capacity depends on (1) the buffer concentration and (2) the relation between the pH required and the pK_a of the buffer acid. The closer the desired pH is to the pK_a of the buffer acid, the higher the buffer capacity. Both of these conditions are evident from the Henderson-Hasselbalch equation. In general the useful range of a buffer solution extends ±1 pH unit from the pK_a value of its weak acid component. A range spanning two pH units accounts for the neutralization of 99% of the acid, which follows from the definition of pH in terms of logarithms.

NEUTRALIZATION (TITRATION) CURVES AND INDICATORS

A neutralization curve is obtained by taking a certain volume of acid (or base) and titrating it with a base (or acid) and determining the pH at intervals during the addition of the neutralizing reagent. Four common types of titration curves are shown in Figure 2-5. Each of these curves can be calculated a priori by recognizing that strong acids (or bases) are completely dissociated and that the dissociations of weak acids (or bases) follow the Henderson-Hasselbalch equation. From these titration curves, it may be seen that when a weak acid is completely neutralized with a strong base, the pH of the resulting solution will be on the alkaline side of neutrality. The reverse applies to the neutralization of a weak base with a strong acid. This fact gives a clue for the selection of the neutralization indicators that are appropriate for particular acid-base titrations.

Neutralization indicators are themselves weak acids or bases. They are special only in the sense that their undissociated and dissociated forms differ in color. In this way, a specific pH can be matched to a certain color. The indicator, of course, must be highly diluted so that its own contribution to the hydrogen ion equilibrium will be negligible and its own dissociation will be governed by the pH of the solution. The transition color of the indicator is best observed at the pH where the undissociated and dissociated forms are present in equal concentrations, i.e., at the pK_a of the indicator. Ideally, the pK_a of the indicator should be at the pH where the titration is just completed, i.e., at the endpoint of the titration.

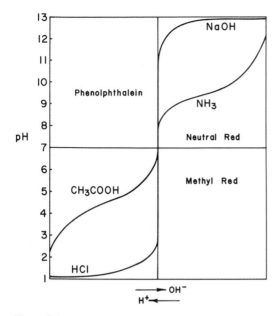

Figure 2-5
Typical titration curves.

The use of indicators and neutralization curves can be demonstrated with the titration of gastric juice [11]. This is composed of hydrochloric acid and a mixture of weak acids. The relative concentration of these two groups has diagnostic significance. The strong acid in gastric juice can be selectively titrated with the help of an indicator with an endpoint at pH 3.3. The use of a second indicator permits the completion of the titration for the weak acids, for which the endpoint occurs at a higher pH.

One of the most important buffer systems in the body is the H_2CO_3-HCO_3^- system. In contrast to what was said previously, this system operates at an unfavorable conjugate base/acid ratio of 20 to 1 at the normal pH of blood (pH 7.4). That this system is still very effective in maintaining a constant blood pH is due to the fact that the acid component is a gas, CO_2, which can be quickly excreted or retained by the lungs upon sudden changes in the HCO_3^- levels. The acid is present as dissolved CO_2, whose concentration was shown previously to be proportional to the P_{CO_2}. Therefore, the Henderson-Hasselbalch equation for this case can be written as

$$pH = pK_a + \frac{[HCO_3^-]}{\alpha P_{CO_2}}$$

(2-44)

The concentration of the bicarbonate ion, HCO_3^-, cannot be measured directly, but it can be calculated as the difference between the total CO_2 and the αP_{CO_2}:

$$[HCO_3^-] = (CO_2)_T - \alpha P_{CO_2}$$

(2-45)

Total CO_2 may, for example, be determined by the gasometric determination of CO_2 following acidification of the plasma, which converts all forms of carbonic acid into gaseous CO_2. The form of the Henderson-Hasselbalch equation that is most appropriate for summarizing all the factors is:

$$pH = 6.10 + \log \frac{(CO_2)_T - 0.0301 P_{CO_2}}{0.0301 P_{CO_2}} \tag{2-46}$$

where 6.10 is the pK_a of the $H_2CO_3 \rightarrow HCO_3^- + H^+$ dissociation reaction and 0.0301 is the value of α when $(CO_2)_T$ is given in millimoles/liter and P_{CO_2} in mm Hg.

PARTITION OF WEAK ELECTROLYTES

The pH of different body compartments has a significant effect on the partition of weak acids and bases. Under ordinary circumstances, the extracellular fluid pH is 7.4 and the cellular pH is slightly under 7. An unequal concentration of these weak electrolytes frequently may come about without any supply of energy (i.e., without active transport). This is due to the fact that nondissociated molecular species penetrate the cellular membranes with ease, whereas the dissociated forms (ions) have low permeabilities. The difference in pH between two compartments will impose on the dissolved weak electrolytes a different ratio of conjugate base-to-acid in each compartment, as predicted by the Henderson-Hasselbalch equation. The nonionized species (i.e., the undissociated acid or unprotonated base) will be equal in concentration on both sides of the cell membrane at equilibrium, whereas the charged species will differ according to the pH of the milieu, as already mentioned. It will be shown below how the accumulation of weak electrolytes inside cells is affected by the pK_a of the electrolyte, as well as the intracellular and extracellular pH. The importance of this phenomenon is clearly evident in the effective intracellular or extracellular concentration of drugs, the majority of which are weak electrolytes [6].

In order to examine the phenomenon of the partition of weak electrolytes quantitatively, let us first define the total concentration of the dissociated and undissociated forms of a weak acid in compartment 1 as c_1 and the concentration of the undissociated (permeable species) as c_n. Written in the antilogarithmic form of the Henderson-Hasselbalch equation, the following relationships apply:

In compartment 1,

$$\frac{c_1 - c_n}{c_n} = 10^{(pH_1 - pK_a)}$$

In compartment 2,

$$\frac{c_2 - c_n}{c_n} = 10^{(pH_2 - pK_a)}$$

Hence,

$$c_1 = c_n 10^{(\text{pH}_1 - \text{p}K_a)} + c_n$$

$$c_2 = c_n 10^{(\text{pH}_2 - \text{p}K_a)} + c_n$$

$$\frac{\text{Total conc. in comp. 1}}{\text{Total conc. in comp. 2}} = \frac{c_1}{c_2} = \frac{10^{(\text{pH}_1 - \text{p}K_a)} + 1}{10^{(\text{pH}_2 - \text{p}K_a)} + 1}$$

For a weak base:

$$\frac{c_n}{c_1 - c_n} = 10^{(\text{pH}_1 - \text{p}K_a)}$$

$$\frac{c_n}{c_2 - c_n} = 10^{(\text{pH}_2 - \text{p}K_a)}$$

Therefore,

$$\frac{c_1}{c_2} = \frac{\dfrac{c_n}{10^{(\text{pH}_1 - \text{p}K_a)}} + c_n}{\dfrac{c_n}{10^{(\text{pH}_2 - \text{p}K_a)}} + c_n} = \frac{\dfrac{1}{10^{(\text{pH}_1 - \text{p}K_a)}} + 1}{\dfrac{1}{10^{(\text{pH}_2 - \text{p}Ka)}} + 1} = \frac{10^{(\text{p}K_a - \text{pH}_1)} + 1}{10^{(\text{p}K_a - \text{pH}_2)} + 1} \tag{2-47}$$

The relationship just derived can be utilized in the assessment of the average intracellular pH of tissue samples or even of the whole body. A suitable weak acid that will partition between the intracellular and extracellular space is 5,5 dimethyl-2,4-oxazolidinedione (DMO). The dissociation reaction is:

$$\text{HDMO} \rightleftharpoons \text{H}^+ + \text{DMO}^-$$

where the pK_a = 6.13 at 38°C. Recognizing that HDMO_{int} = HDMO_{ext} (*int* = internal; *ext* = external),

$$\text{pH}_{int} = 6.13 + \log \frac{(\text{DMO}_T)_{int} - \text{HDMO}_{int}}{\text{HDMO}_{int}} \tag{2-48}$$

$$= 6.13 + \log \frac{(\text{DMO}_T)_{int} - \text{HDMO}_{ext}}{\text{HDMO}_{ext}}$$

where $(\text{DMO}_T)_{int}$ is the sum of the dissociated and undissociated DMO inside the cell. The total intracellular DMO can be determined directly by sampling the cellular material, and $HDMO_{ext}$ could be derived from the total plasma DMO $(DMO_1)_{ext}$ – and the plasma pH. In measurements of this type, a separate determination of the extracellular space is necessary to make appropriate corrections for the admixture in the tissue sample of extracellular DMO.

As an extension of this method, an aggregate pH of the intracellular spaces of animals or men can be estimated [12]. First, the total body water and the extracellular fluid space have to be measured, which will yield the intracellular space by difference. One can then calculate the amount of $(DMO_T)_{int}$ from the total amount administered and the amount found in the extracellular space after equilibration, corrected for the excreted fraction. The rest of the calculation is analogous to the preceding one, and it yields the average pH for all the body cells.

REFERENCES

1. Ball, H. B. *An Introduction to Physical Biochemistry* (2nd ed.). Philadelphia: Davis, 1971.
2. Davenport, H. W. *The ABC of Acid-Base Chemistry* (5th ed.). Chicago: University of Chicago Press, 1969.
3. Dawes, E. A. *Quantitative Problems in Biochemistry* (3rd ed.). Baltimore: Williams & Wilkins, 1965.
4. Dick, D. A. T. *Cell Water*. Toronto: Butterworth, 1966.
5. Giese, A. *Cell Physiology* (3rd ed.). Philadelphia: Saunders, 1968.
6. Goldstein, A., Aronow, L., and Kalman, S. M. *Principles of Drug Action*. New York: Harper and Row, 1969.
7. Netter, H. *Theoretische Biochemie*. New York: Springer-Verlag, 1959.
8. Rapaport, S. R. *Introduction to Hematology*. New York: Harper and Row, 1971.
9. Stein, W. D. *The Movement of Molecules Across Cell Membranes*. New York: Academic, 1967.
10. Wolf, A. V., and Crowder, N. A. *An Introduction to Body Fluid Metabolism*. Baltimore: Williams & Wilkins, 1964.
11. Wolf, P. L., et al. *Methods and Techniques in Clinical Chemistry*. New York: Wiley-Interscience, 1972.
12. Yamamoto, W. S., and Brobeck, J. R. *Physiological Controls and Regulations*. Philadelphia: Saunders, 1965.

SELECTED READING

Christensen, H. N. *Neutrality Control in the Living Organism*. Philadelphia: Saunders, 1971.
Edsall, J. T., and Wyman, J. *Biophysical Chemistry*, Vol. 1. New York: Academic, 1958.
Martin, R. B. *Introduction to Biophysical Chemistry*. New York: McGraw-Hill, 1969.
Williams, V. R., and Williams, H. B. *Basic Physical Chemistry for the Life Sciences*. Philadelphia: Freeman, 1967.

The Chemistry of Living Cells – Macromolecular Constituents   3

Thomas J. Mende

CHEMICAL COMPOSITION OF BIOSYSTEMS

The basic building blocks of living tissues show remarkable similarities throughout the entire biosphere. The major component of living tissue is water, and most — though not all — cellular processes can be described in terms of solution chemistry. The aqueous phase contains low-molecular-weight solutes, both organic (such as sugars and amino acids) and inorganic (ions), as well as soluble macromolecules of diverse types. Subcellular assemblies and microscopically identifiable structures are formed from organized macromolecular elements. Since organization is one of the main features of life, the characterization of these components is necessary for the understanding of complex cellular functions. Three main groups of biologic macromolecules are to be considered: (1) proteins, (2) nucleic acids, and (3) polysaccharides.

Lipids, which are not, strictly speaking, macromolecules (i.e., substances formed by stringing a large number of low-molecular-weight subunits together by covalent bonds), should be added to this list for reasons of expediency. Although they do not conform to this definition, they tend to aggregate to large units by weaker forces than covalent bonding. They also are essential components of cellular membranous structures, and they participate in organelle formation. The three groups listed above are all *condensation polymers,* i.e., they can be regarded as having been formed from their respective building blocks by the elimination of water. They differ, however, in the chemical nature of their subunits.

PROTEINS

Proteins and amino acids are formed from linear sequences of twenty different kinds of L-α-amino acids. The peptide bond (CO—NH) is characteristic for protein molecules (Fig. 3-1). This bond has a certain degree of double-bond character, which limits the free rotation around the bond axis. This fact has important consequences for protein structure.

Proteins differ in their amino acid composition. The twenty common amino acids differ from one another by their sidechains, which are indicated by "R" in Figure 3-1. The number, proportions, and the sequential arrangement of these amino acids in a linear chain allow vast variation in the kinds of proteins that could conceivably exist. Under normal conditions, the cellular apparatus for protein synthesis nevertheless forms strictly identical protein molecules within each type, which, even in a single cell, constitute a very large number of varieties. This implies that the uniformly identical sequences of amino acids are coded by the genetic apparatus of the cell.

41

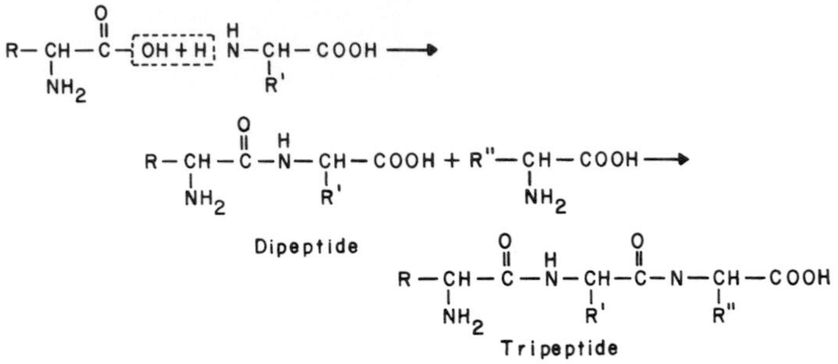

Figure 3-1
The peptide bond is a special case of an amide bond formed between an amino group of one amino acid and the carboxyl group of a second amino acid.

From a functional point of view, proteins exhibit great diversity. They form the main framework of the cellular architecture, and they are, in part, responsible for the mechanical properties of tissues. As mechanochemical devices, they act as the functional units of the molecular mechanism of muscle action. They also provide the catalytic machinery of life processes inasmuch as all enzymes, the biocatalysts, are proteins. Other areas in which proteins play determining roles include immune reactions, endocrine functions, and blood coagulation.

This great diversity in function is a reflection of an equal variety of permutations offered by the three-dimensional structure of the protein molecule. This three-dimensional structure is, somewhat surprisingly, uniquely determined by the linear sequence of the amino acid chain. A simple mapping of the amino acid sequence alone imparts no understanding of the three-dimensional structure of a protein molecule, although current efforts are being focused on this problem. It is customary to speak about the four different levels of protein molecular organization as the primary, secondary, tertiary, and, on occasion, quaternary structures.

The *primary structure* of the protein molecule is simply the linear sequence of the amino acids. All the other levels of structure, as already mentioned, ultimately depend on it.

Secondary structure applies to those structural arrangements that are formed from linear peptide chains through maximal stabilization by hydrogen bonding. This may be brought about by intrachain or interchain bonding, and it can give rise to different secondary structures.

One of the most frequently observed of such structures is the α helix. In these, the chain coils into a right-handed helix that contains 3.6 amino acid residues per turn. The CO group of each peptide bond is hydrogen-bonded to the NH group located in the third peptide bond "behind" it, which imparts a translational advance of 0.54 nanometers per turn of the coil. The sidechains project outward from the helix surface (Fig. 3-2). This structure has been shown to form portions of many globular proteins.

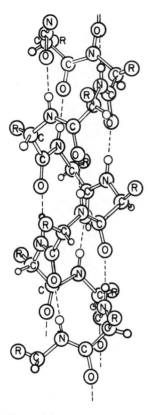

Figure 3-2
The α-helical structure for polypeptide chains, as determined by L. Pauling and co-workers. (Courtesy of Professors L. Pauling and R. B. Corey.)

The stretching of α-helical structures — for example, as occurs in the keratin molecule — may result in so-called β structures, where the helixes are pulled into "zig-zag" chains. In this case, adjacent chains form *interchain* hydrogen bonds, and a "pleated sheet" structure results. These β structures are also found as natural components in a variety of proteins. If the β form were enlarged, it would appear as a corrugated sheet with the various sidechains projecting above and below the plane (Fig. 3-3). If the chains run all in one direction, the structure is a parallel pleated sheet; if they run alternately up and down, an antiparallel structure is formed [1]. These β structures are also found in globular proteins.

In considering secondary structures, mention should be made of collagen, the most abundant protein in animals. Collagen is formed from three strands of polypeptides, each of which forms a very tight left-handed coil, and the three coiling fibers form a right-handed coiled-coil fiber bundle (Fig. 3-4). This unit, known as tropocollagen, is 300 nm long and 1.5 nm in diameter. Tropocollagen units assemble linearly, but the periods in the parallel bundles are shifted out of register, giving the

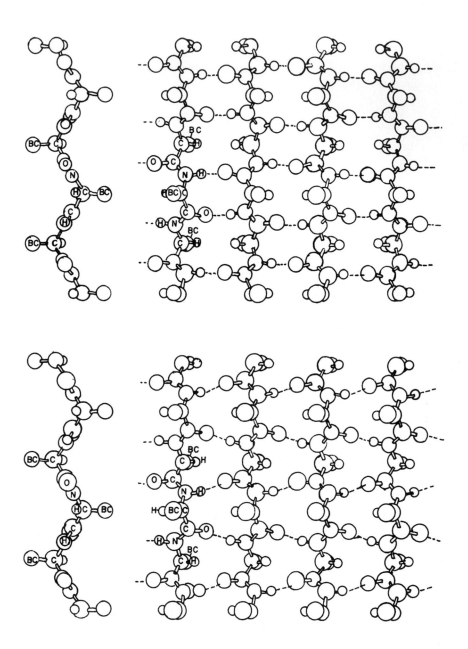

Figure 3-3
Two possible structures involving lateral intermolecular hydrogen bonds. These have been called *pleated sheet* structures. (From R. E. Marsh, R. B. Corey, and L. Pauling. *Biochim. Biophys. Acta* 16:13, 1955. Reproduced with permission.)

|←————— 86 Aº —————→|

Figure 3-4
Segment of collagen structure showing the end-to-end linking of tropocollagen molecules. (From A. White, P. Handler, and E. L. Smith. *Principles of Biochemistry* (4th ed.). New York: McGraw-Hill, 1968. P. 874. Reproduced with permission.)

fiber a banded appearance in electron micrographs (Fig. 3-5). These native collagen fibers have a characteristic periodicity of 64 to 70 nm [2].

The *tertiary structure* of protein molecules is also determined by the primary amino acid sequence, and it refers to the three-dimensional folded, final structure of the protein molecule. This structure is formed by bending and folding of the polypeptide chains, and it is stabilized by different types of forces, the most important of which appears to be hydrophobic interactions. These tend to bury the hydrophobic portions, e.g., the uncharged sidechains, internally in the molecule. Most of the hydrophilic groups are exposed on the molecular surface, which indirectly helps in the solubilization of the protein by means of hydration. Hydrophobic interactions have as their basis the minimization of the system's entropy, which occurs only if the hydrophobic groups are not permitted to protrude into the polar, aqueous environment.

Tertiary structures are immensely varied, and they are studied primarily by x-ray diffraction techniques. The tertiary structure of a protein determines its overall shape, its exposed and unexposed groups, the proximity of certain of its groups, and, since the last determines in turn the prominent features of protein behavior, its biologic

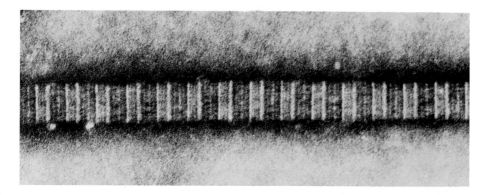

Figure 3-5
Electron micrograph of a single collagen fiber from rat tail; the periodicity results from the staggered alignment of chains of tropocollagen subunits. (Reproduced with permission of Dr. J. R. M. Moore.)

function. Tertiary and secondary structures may be destroyed, i.e., the protein may be *denatured* by a number of influences and agents, such as elevated temperatures, extremes in pH, surface energy factors, and certain chemical agents. The most probable tertiary structure is the native one, and, under careful experimental conditions, many proteins can, when denatured, be "renatured."

In addition to the effect of hydrophobic bonding alluded to previously, the stability of the three-dimensional tertiary structure also depends on specific sidechain interactions, which are electrostatic in nature. These may be ionic (opposite charge), dipole-dipole, or charge-dipole interactions. Specific amino acids that can be involved are those that carry permanently affixed positive charges at neutral pH (lysine, arginine, and, to some degree, histidine). Another group carries negative charges (aspartic and glutamic acids). The rest carry no charge. In addition, proline and the uncommon, but important, amino acid hydroxyproline should be singled out as *imino acids.* As members of a peptide chain, they cannot form hydrogen bonds due to the lack of a free NH group. Collagen contains a very high proportion of proline and hydroxyproline, and their associated incapacity for hydrogen-bond formation accounts for its unique secondary structure.

Some of the types of interactions that are implicated in the stabilization of tertiary protein structure are shown in Figure 3-6.

Quaternary structures are organized assemblies of individual protein molecules. Structures having molecular weights significantly greater than 50,000 may be suspected of being formed from an organized assembly of subunits or *protomers.* These are held together by specific forces, but they may be dissociated under certain experimental or physiologic conditions. The assembly frequently shows different functional properties from the component protomers, even if the protomers are identical. Component protomers need not, however, be identical. Hemoglobin, for example, is tetrameric, with two pairs of subunits (two α and two β chains).

The proteins discussed so far function mainly as architectural and supportive or protective components (e.g., collagen and keratin). The most important function of

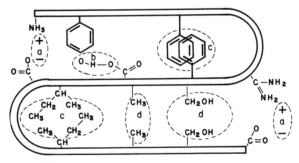

Figure 3-6
Some types of noncovalent bonds that stabilize the folded structure of proteins: *a.* ionic bond; *b.* hydrogen bonding between phenolic (tyrosine) and carboxyl groups on sidechains; *c.* and *d.* Several types of van der Waals interactions between amino acid sidechains. (From C. Anfinsen. *The Molecular Basis of Evolution.* New York: Wiley, 1959. P. 102. Reproduced with permission.)

proteins is, however, their action as catalysts of biochemical reactions, which will be discussed later. Proteins (and most other body constituents) are constantly synthesized and degraded. Some of them turn over with rapidity. Others, mainly the structural proteins, degrade very slowly.

NUCLEIC ACIDS

Cellular protein formation is ultimately controlled by and linked to nucleic acids. These comprise two chemically and functionally distinct groups, the deoxyribonucleic acids (DNA) and the ribonucleic acids (RNA). Their chemical similarities are more striking than their differences; they are similar with respect to their component monomeric units called *nucleotides* (Fig. 3-7). Nucleotides contain a

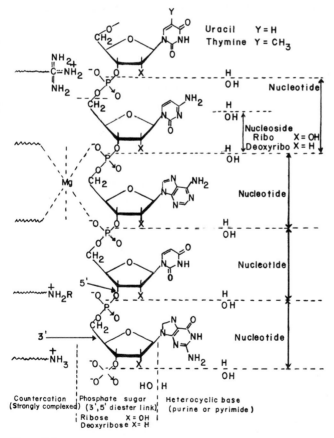

Figure 3-7
A random segment of a nucleoprotein chain and its constituent parts.

nitrogenous heterocyclic base that is attached to a five-carbon sugar unit. The sugar unit is ribose in RNA and 2'-deoxyribose in DNA. The bond is an N-glycosidic one, which connects one of the ring nitrogens of the base to carbon number 1 (the aldehydic carbon) of the sugar unit. Phosphoric acid is joined to the sugar by an ester linkage. The bases guanine, cytosine, and adenine are common to both DNA and RNA. The base thymine is characteristic of DNA; its counterpart in RNA is uracil. The deoxyribonucleotides and ribonucleotides form linear (unbranched) condensation polymers. The coupling is furnished by the phosphoric acid moiety, which, in both DNA and RNA, is present as a $3'$, $5'$ diester of the adjacent sugar units.

There are four participating bases (with minor exceptions) in both types of nucleic acids, and their sequences and proportions vary widely, as do the molecular weights of the complete molecule. The primary structure of all protein molecules is uniquely determined by the sequence of bases composing DNA.

The structure of the heterocyclic bases allows them to form hydrogen bonds with one another. This happens optimally when specific base pairs are in apposition, in particular, adenine with thymine in DNA, adenine with uracil in RNA, and guanine with cytosine in either. In the case of DNA, these specific base pairings give rise to the Watson-Crick double helix structure, where two strands of DNA fibers (which are strictly complementary) run coiled in an *antiparallel* fashion (i.e., the $3'$, $5'$ phosphodiester bonds run down on one and up on the other strand). This structure is stabilized by a layering or "stacking" of the bases that project into the center channel of this column. (This helix is different from the protein α helix, where the sidechains are exposed to the outside.) The structure of double-stranded DNA is shown in Figure 3-8. In contrast to DNA, the RNA molecule, which is capable of forming double strands in vitro, is usually found in vivo to be singly stranded.

The ultimate genetic substance is the DNA molecule, the double-stranded helicity of which can be looked upon as its secondary structure. The genetic information content is coded in toto in one of the two strands, since the complementary strand is uniquely determined by the other. In fact, the utilization of this information for genetic copying via an RNA molecule, a process called *transcription*, takes place from only one of the fibers. The main features of information storage and transfer in living cells involve two problems: (1) How is the duplication of its own structure, which is required for the generation of daughter cells, accomplished by DNA? (2) How does DNA structure determine protein structure?

It has been found that duplication of DNA takes place through a copying process involving both strands of DNA by what is called *semiconservative replication.* This replication requires the separation of the complementary strands and the formation of two new ones, each complementary to one of the parental ones. The mechanism of this copying process, which was studied in a cell-free system, appears to require, in addition to the DNA template to be copied, the presence of *all four* deoxynucleotide triphosphates and the enzyme DNA polymerase. This is a simplified model of the process of replication. It is even possible that the two parental strands are copied by different mechanisms. The copying results in two identical double-stranded DNA molecules in which one strand is from the old molecule and one strand is new.

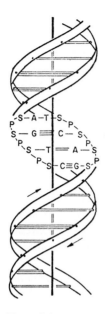

Figure 3-8
The Watson-Crick model. P represents phosphate; S, sugar; A, adenine; T, thymine; G, guanine; C, cytosine. Horizontal parallel lines symbolize hydrogen bonding between complementary bases. (From G. W. Beadle, *Missouri Agric. Exp. Station Res. Bull.*, p. 588, 1955. Reproduced with permission.)

This duplication takes place with the strictest base pairing; e.g., adenine is matched with thymine, guanine with cytosine, and vice versa. The preservation of a unique sequence is thereby accomplished, since with the next round of duplication, other identical DNA double-strands will be formed. In the enzymatic polymerization process, as the 3', 5' phosphate-diester bonds are formed, pyrophosphate is liberated.

The second phase of the transmission of the genetic message is *transcription*. This means the transfer of the DNA code to an RNA molecule. It takes place by a process similar to that of DNA duplication. Underlying this reaction is again a specific base-pairing process that is enforced by hydrogen bonding. In this case, however, the adenine of DNA pairs with uracil instead of thymine to generate an RNA polymer counterpart, while the pairing of guanine and cytosine takes place as in DNA duplication. The point should be stressed that since only one strand of the DNA double helix is required for the coding of the complete message, *only one* of the strands is transcribed to RNA.

Transcription again requires an enzyme, DNA-dependent RNA polymerase, which in turn requires the simultaneous presence of all four ribonucleotide triphosphates plus the DNA template for its action. The reaction products are the complementary RNA strand and pyrophosphate.

A thorough study of RNA polymerase has been possible only in bacteria (*E. coli*). Here, the enzyme possesses multiple subunits and has a molecular weight of approximately 5×10^5. One of the subunits is required for proper strand selection and recognition of the initiation site (σ factor). A special protein – the ρ factor, which is not part of the enzyme – recognizes a termination marker on DNA, and it thereby permits correct copying of the gene onto RNA.

RNA polymerase has not been well characterized in cells of higher organisms, due mostly to difficulties in solubilization. There is evidence that at least two types of the enzyme may exist, one that functions in the nucleolar region, and the other at alternate sites on the chromatin of the cell nucleus. Whether subunits with similar functions to those of the bacterial enzyme also exist in higher organisms has yet to be demonstrated.

The last step in the transmission of the genetic message is *translation,* i.e., the transformation of the code represented by nucleotide sequences into another type of macromolecule (a protein) by the unique designation of the location of the amino acids in the primary structure of the protein. Translation takes place by an extremely complicated mechanism, which has not yet been clarified in all of its details, although its outlines are known.

In order to explain translation, the different types of RNAs found in the mammalian cell must be described. The RNA that is a copy of one or more cistrons and the instrument of coding for proteins is known as *messenger RNA* (mRNA). It is formed in the nucleus, but it subsequently appears in the cytoplasm, where it combines with cytoplasmic organelles called *ribosomes.* One strand of mRNA is often found attached to several ribosomes, and such a structure is known as a *polysome.* The number of ribosomes increases with the size of the messenger.

Ribosomes are composed of an RNA and a variety of proteins. *Ribosomal RNA* (rRNA) is the most abundant RNA component in cells and (as is *all* RNA) the product of a transcription process, i.e., it is of nuclear origin. Ribosomes have two functional parts and can be separated into a small and large fragment, which both contain RNA and different but characteristic proteins.

In order to understand the mechanism of translation, it is essential to introduce another important type of RNA molecule, *transfer RNA* (tRNA). Its function is to bring the appropriate amino acids and the rest of the protein-synthesizing system together. There is, in all cells, one or more specific and distinct tRNA for *each* protein amino acid. These smallest of RNA molecules have a common $3'$-terminal sequence of three nucleotides with the base sequence cytosine–cytosine–adenine, to which the corresponding amino acid becomes attached. The attachment is enzymatically effected; each tRNA and its associated amino acid are brought together by an enzyme that is specific for both, called an amino acid-activating enzyme. The bond formed between the adenine riboside terminal of the tRNA and the amino acid is an ester-type bond, and ATP is required by the enzyme to achieve the coupling. AMP and pyrophosphate are the reaction products.

Assuming that all the necessary amino acyl-tRNAs are available and ready to incorporate their amino acid charges into the final product, protein, there is still the problem of accounting for the extraordinary fidelity inherent in translation. To explain this, we have to recall that a DNA chain has a unique base sequence that carries the genetic message. This sequence in DNA determines the location of a specific amino acid in the protein. Each amino acid is coded by a sequential base triplet. Three bases in sequence provide a necessary and sufficient condition for determining the location of a specific amino acid, as is shown by the following consideration. There are four common bases in DNA: adenine, thymine, guanine, and cytosine. Obviously, one base could not specifically determine each amino acid, since there would be only four bases for twenty amino acids. Four bases in combination as sets of three can code for $4^3 = 64$ amino acids, whereas combinations of two bases would allow unique sequences for only 16 amino acids. This shows that a nonoverlapping triplet code is the most likely explanation for reasons of economy. It allows more than one triplet code for a single amino acid (i.e., redundancy) and some triplet sequences that do not code for any amino acid. Some of these triplets, however, may have importance for an unambiguous method of protein synthesis, e.g., as signals for chain termination.

The final link in this complicated process is the recognition of a specific nucleotide triplet, called the *codon,* on the mRNA by the respective amino acid-carrying tRNA. This is achieved through the presence of a unique triplet sequence that is complementary to the codon, called the *anticodon,* which, through its hydrogen-bonding capability, allows the attachment of the amino acid-tRNA to the mRNA.

In considering an overview of translation, the mRNA attaches itself at its 5′-terminal (the 5′-hydroxyl group of free ribose) to the smaller subunit of the ribosome. To this is added a methionine-carrying tRNA at the initiation codon of adenine—uracil—guanine or guanine—uracil—guanine, which thus specifically selects the initiation site. (Another methionine-carrying tRNA responds to the same codon internally.) The process is completed in the presence of supernatant protein factors and guanosine triphosphate, GTP, by the addition of the larger microsomal component. Adjacent to the initiation site, the next codon attracts the appropriate amino acid-loaded tRNA. Additional sites are not occupied, because each ribosome has two tRNA binding sites. The first site may be occupied by the chain-initiating, methionine-carrying tRNA or a peptidyl tRNA in the process of chain elongation. In either case, a transfer of the appropriate residue (methionine or peptidyl) to the sidechain of the second site will occur, changing the bond from an ester to an amide linkage in the process.

The enzyme that catalyzes the transfer, peptidyl transferase, is part of the larger microsomal fragment. Following its action, the first site is occupied by an unsubstituted tRNA, and the second by a peptidyl tRNA. For further elongation, three elongation factors and GTP are necessary. For the complete system, a translocation of the ribosome with respect to the messenger will take place in such a way that the second site, with its occupying group, will move to the first site and expose a new codon in the second site, while the unsubstituted RNA floats off. Repetitions of this cycle will elongate the peptide chain until the messenger exposes a chain-terminating codon (uracil—adenine—adenine, uracil—adenine—guanine, or uracil—guanine—adenine). These are read by protein factors instead of tRNAs. The protein chain will then cleave from the anchoring tRNA. The chain-initiating methionine is, in most instances, removed by a specific peptidase. It is to be noted that protein synthesis proceeds from the amino to the carboxyl end.

The preceding discussion emphasizes the fact that the informational macromolecules — DNA, RNA, and proteins — must have strictly determined primary structures, i.e., definitive monomer sequences and molecular weights.

POLYSACCHARIDES

These biologic polymers are usually found to have a statistical molecular-size distribution, and they may or may not have fixed sequences. For the purposes of classification, it is conventient to distinguish between homo- and heteropolysaccharides. The former contain only one type of monomeric sugar unit, whereas the latter contain more than one. There is a large variety of polysaccharides, but this introductory discussion of human metabolic and structural components will be limited to only a few.

The monomer units are usually aldose sugars, which exist mainly in an internally cyclized, hemiacetal form, as exemplified by glucose (Fig. 3-9). The structural formula of glucose shows that the reducing aldehyde group is masked in the cyclic form, but it possesses a hydroxyl group that is accessible for reaction. This group can form a condensation product with any of the free hydroxyl groups of another sugar unit by the elimination of water. The bond thus formed — e.g., a 1,4 bond (Fig. 3-10) — is called the *glycosidic bond* and is fundamental in the formation of polysaccharides, since all their monomeric units are coupled in this fashion.

Of the homopolysaccharides, the only one of significance in the human body is glycogen. It is a high-molecular-weight compound with a branching structure that is

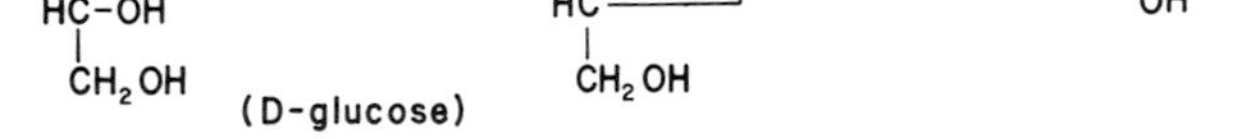

Figure 3-9
Glucose, an aldose sugar, in open chain and cyclized hemiacetal forms.

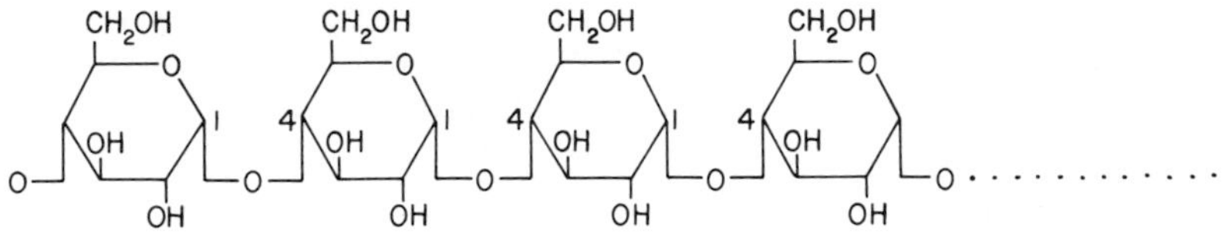

Figure 3-10
The glycosidic bond (1,4 bond) is fundamental in the formation of polysaccharides.

composed entirely of glucose units. The molecules have predominantly 1,4 glycosidic bonds (see Fig. 3-10) except at the branch points, which have 1,6 bonds. Glycogen serves to store energy in animals and is found principally in liver and muscle tissue. Storage of glucose in the form of glycogen helps to decrease intracellular osmotic pressure.

The other polysaccharides found in the body are heteropolysaccharides, which have no energy-providing function, but rather serve as structural components. These polysaccharides in part constitute connective tissues, which in turn are responsible for shape and mechanical stability of the body. They are distributed extracellularly, where they form part of a gel-like matrix, the ground substance in which the cells of the connective tissues (e.g., fibroblasts) are embedded. Due to the presence in them of carboxyl or sulfuric acid groups, they are negatively charged. All occur as unbranched chains, and all contain a nitrogen-containing sugar derivative, either glucosamine or galactosamine (Fig. 3-11). The amino sugar component is responsible for their classification as *mucopolysaccharides.* Most, but not all, of the heteropolysaccharides contain hexose units with a carboxylic acid on C-6.

Some important mucopolysaccharides are depicted in Figures 3-12 and 3-13. With the exception of hyaluronic acid, they are covalently linked to protein chains. The

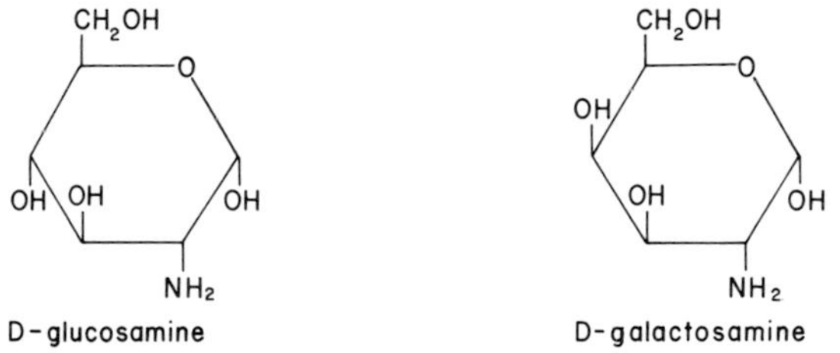

CH_2OH CH_2OH

OH   OH   OH   OH   OH   OH

NH_2 NH_2

D-glucosamine   D-galactosamine

Figure 3-11
D-Glucosamine and D-galactosamine are components of heteropolysaccharides. The latter have no energy-providing function but serve in the formation of structural components.

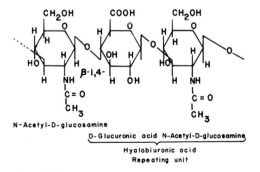

N-Acetyl-D-glucosamine

D-Glucuronic acid N-Acetyl-D-glucosamine

Hyalobiuronic acid
Repeating unit

Figure 3-12
Example of normal glucosidic bond linkage in hyalobiuronic acid.

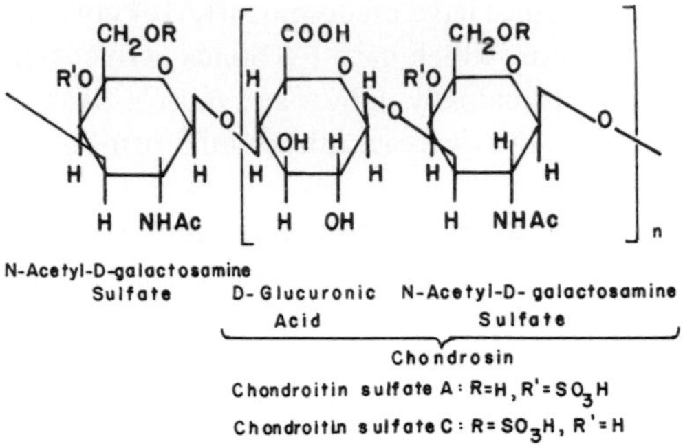

Figure 3-13
Chondroitin sulfate contains a disaccharide, which is composed of a uronic acid and a D-galactos-amine derivative, as the repeating unit.

ground substance has a high water content and is crossed by collagen bundles and elastic fibers. The mucopolysaccharides of the ground substance are polyvalent anions; the positive ions associated with them are the usual extracellular cations, e.g., Na^+ and Ca^{+2}. Electrical neutrality requirements permit this tissue component to function as a cation exchanger. In addition, it is subject to Donnan equilibrium conditions.

LIPIDS

The lipids serve in a dual capacity: they act as a fuel source for the body and as components of all the membranous structures of cells. Chemically, they are very heterogeneous; their only common characteristic is their preferential solubility in organic solvents. The lipids that are of structural importance either contain long-chain aliphatic carboxylic acids (fatty acids) as nonpolar components of more complex molecules or belong to a group of organic solvent-soluble polycyclic alcohols, the sterols (e.g., cholesterol). Free fatty acids, in contrast, serve as metabolic fuel in the body; they are liberated according to need from the adipose tissues, where they are present as esters of the trihydric alcohol glycerol, whence the name *triglyceride*.

Among structural lipids, the most widely distributed are the phospholipids, glycolipids, and sphingolipids (Fig. 3-14). They are distinguished by the simultaneous presence of polar and nonpolar groups in a single molecule. Figure 3-15 shows molecules of this type (amphophils), which tend to form oriented monolayers in interfaces and micelles in solution, where the hydrophobic groups tend to be isolated from the aqueous environment. Structural lipids are usually present as lipoproteins. Membranes contain, on the average, 40% lipids and 60% specific proteins. The cohesion between these chemically different structural components appears to depend mainly on hydrophobic interactions.

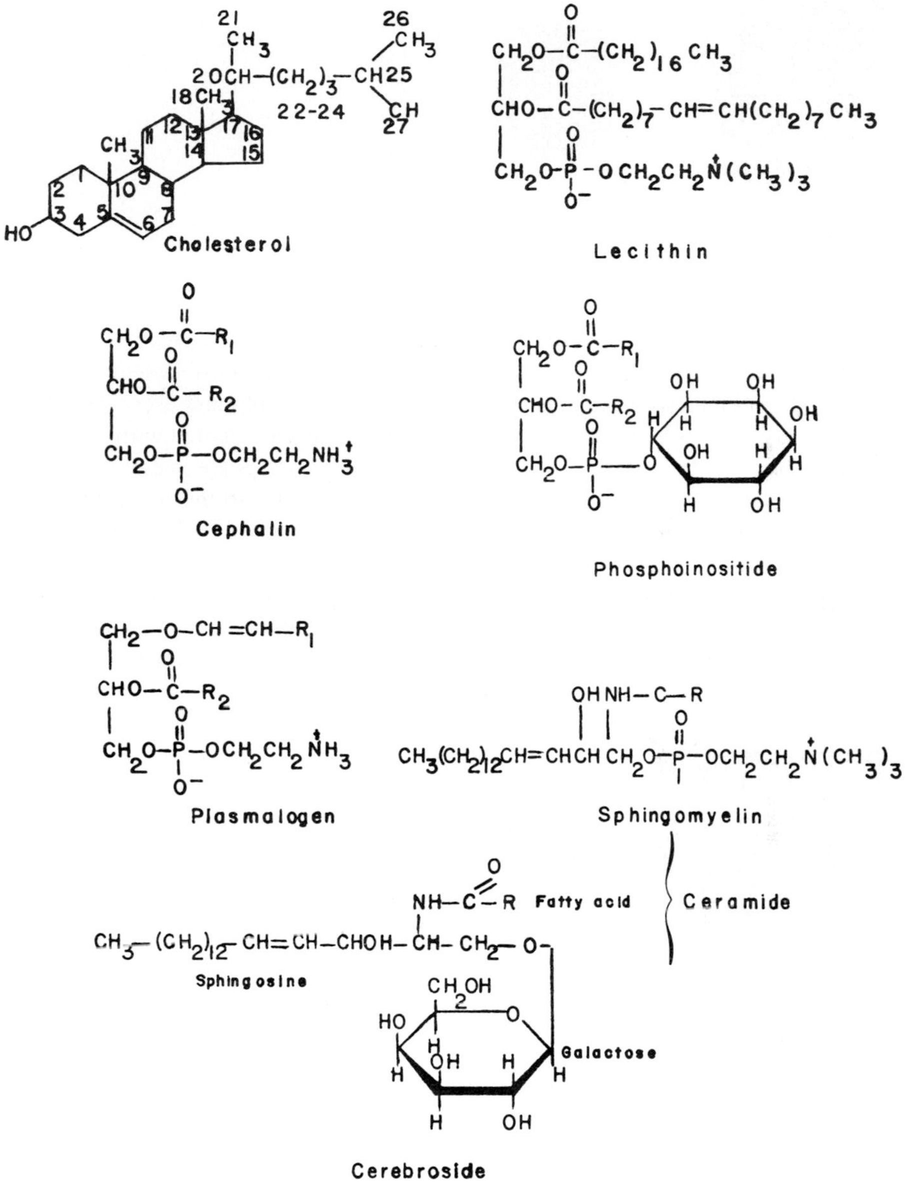
Cholesterol
Lecithin
Cephalin
Phosphoinositide
Plasmalogen
Sphingomyelin
Fatty acid
Ceramide
Sphingosine
Galactose
Cerebroside

Figure 3-14
Structures of various lipids and phospholipids.

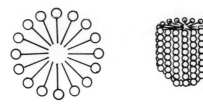

Figure 3-15
Arrangement of amphophil molecules in micelles.

The *unit membrane hypothesis,* which is partially supported by electron microscopic evidence, may explain the characteristic appearance of most biologic membranes. This hypothesis postulates that the structure of membranes is composed of
two monolayers of polar lipids (e.g., phospholipids) that are oriented with their
polar groups toward the exterior (Fig. 3-16). The polar groups are in close association with the membrane protein on both polar sides. Modifications of this model
are necessary, however, if membrane behavior is to be described with greater accuracy.
Some of these modified models allow for more hydrophobic interaction between the
hydrocarbon chains of the lipid and the hydrophobic regions of the protein. These
models may or may not depict continuous structures. Some membranes (e.g., that
of the red blood cell) contain cholesterol admixed with other polar lipids.

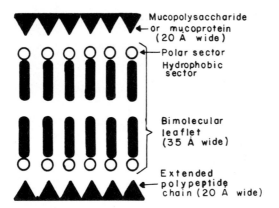

Figure 3-16
The unit membrane structure as proposed by Robertson. (From J. D. Robertson. In M. Locke (Ed.),
Cellular Membranes in Development. New York: Academic, 1964. P. 13. Copyright of Academic
Press; reproduced with permission.)

REFERENCES

1. Marsh, R. E., Corey, R. B., and Pauling, L. An investigation of the structure of silk fibrin. *Biochim. Biophys. Acta* 16:1, 1965.
2. White, A., Handler, P., and Smith, E. L. *Principles of Biochemistry* (4th ed.). New York: McGraw-Hill, 1968. P. 873.

SELECTED READING

Barry, J. M., and Barry, E. M. *An Introduction to the Structure of Biological Molecules.* Englewood Cliffs, N.J.: Prentice-Hall, 1969.
Bittar, E. E. *Cell Biology in Medicine.* New York: Wiley, 1973.
Haschemeyer, R. H., and Haschemeyer, A. E. V. *Proteins, A Guide to Study by Physical and Chemical Methods.* New York: Wiley, 1973.
Watson, J. D. *Molecular Biology of the Gene* (2nd ed.). New York: Benjamin, 1970.

Enzymes and Pathways for ATP Formation

4

Thomas J. Mende

BIOCHEMICAL REACTIONS

A substantial portion of the cellular macromolecular content consists of enzymes. These proteins perform their biochemical catalytic functions under narrowly defined physiologic conditions, e.g., in aqueous solution, at body temperature, at atmospheric pressure, and, for the most part, at neutral pH. Proteins, with their variable three-dimensional structures, ionic charges, and hydrophobic regions, are singularly suited for acting as catalysts.

The reactions that they catalyze may arbitrarily be grouped into six categories:

1. Oxidoreductases: catalyze electron transfers, hydrogen transfers, or both
2. Transferases: catalyze chemical group transfers between donor and acceptor molecules
3. Hydrolases: catalyze the addition of H_2O, associated with single-bond breaking
4. Lyases: catalyze the addition of certain groups to double bonds
5. Isomerases: catalyze the interconversion of isomers, i.e., chemical species with identical overall chemical composition but different forms
6. Ligases: catalyze the formation of covalent bonds at the expense of ATP cleavage

All enzymes are proteins, but many contain additional nonprotein components that are essential for their function, such as coenzymes and metallic or nonmetallic cofactors. Enzymes, like all catalysts, increase the *rates* of reactions, but they do not affect the final equilibrium of the system. They function by lowering the activation energy barrier between the initial state and the transition state on a pathway from the reactants to the products (Fig. 4-1). This may be accomplished either by placing the transition state into an environment that lowers its free energy or by raising the energy of the ground state. Alternatively, the reaction mechanism could be changed via a different transition state. It has been suggested that on any pathway, more than one activated complex can and practically always does occur.

The rate increase caused by enzyme catalysis can be discussed in terms of a number of factors that are not necessarily independent. The key underlying supposition is that an enzyme-substrate complex must be formed; the existence of such complexes has been proved in many cases. This complex will favorably orient the substrate to the catalytic site of the enzyme. This, in some ways, resembles intramolecular reactions, which usually have higher rates than the corresponding intermolecular ones. For example, if a catalytic group (e.g., carboxylic acid) and a substrate group (e.g., an ester or amide) are brought together in one molecule and in close proximity, the resultant hydrolysis reaction will be faster than if they are located on separate molecules.

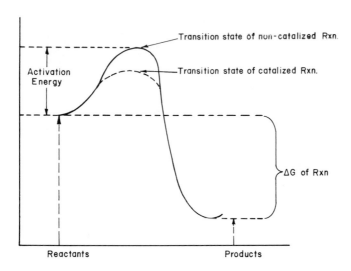

Figure 4-1
Catalytic effect of an enzyme on the rate of a thermodynamically feasible reaction (ΔG = change in free energy).

However, the rate enhancement that is calculated on the basis of such intramolecular steric effects frequently is much smaller than the enzymatic reaction rate, and, accordingly, additional factors have been taken into consideration. One such factor, for example, is the additional rate acceleration that is due to bond distortion of the substrate, resulting in a configuration that more closely resembles the transition state.

Generally, the activation energies of biochemical reactions are so high that they would not be observed to occur in the absence of enzymes. This is a very important factor in the orderly process of intermediary metabolism, where a large number of reactions that could possibly occur simultaneously are channeled by their appropriate catalysts, whereas others simply do not take place because of the lack of the necessary enzymatic machinery.

As pointed out above, enzyme action involves the initial binding of the reactants to specific binding sites to form an enzyme-substrate complex *(ES)*. The transformation of the reactants to products takes place at the enzyme's active site, which participates in the reaction either by forming transient covalent intermediates with certain groups from the substrate or by noncovalent, generally acid-base, catalysis.

Following the release of the products from the enzyme, the cycle starts again. During the binding process, the enzyme may even change its tertiary structure, enhancing the chances for a favorable geometry. This is known as the *induced fit* hypothesis. As a result of the participation of the enzyme in the formation of an enzyme-substrate complex, its concentration will affect the rate of the reaction. The total enzyme concentration is not changed during the reaction, as is true of any catalyst.

The simplest treatment of the kinetics of enzyme reactions was provided by Michaelis and Menten [2]. Their approach describes enzyme reactions in the following equation:

$$E + S \underset{k_{-1}}{\overset{k_1}{\rightleftharpoons}} ES \overset{k_2}{\longrightarrow} E + P \tag{4-1}$$

where $E = [E_T] - [ES]$ = free enzyme
E_T = total enzyme
S = substrate
ES = enzyme-substrate complex
P = product
k = reaction rate constant

The rate of formation of ES is represented by:

$$\frac{d[ES]}{dt} = k_1([E_T] - [ES])[S] \tag{4-2}$$

The rate of decomposition of ES is described by:

$$\frac{d[ES]}{dt} = k_{-1}[ES] + k_2[ES] \tag{4-3}$$

When the reacting system is in a steady state,

$$\frac{d[ES]}{dt} = 0$$

and

$$k_1([E_T] - [ES])[S] = k_{-1}[ES] + k_2[ES]$$

$$\frac{[S]([E_T] - [ES])}{[ES]} = \frac{[S][E]}{[ES]} = \frac{k_{-1} + k_2}{k_1} = K_M \tag{4-4}$$

where K_M is the Michaelis constant, which, in the particular but not infrequent case when $k_2 \ll k_{-1}$, represents the dissociation constant of the enzyme-substrate complex. Under these conditions, K_M expresses the affinity of the enzyme for its substrate.
Further rearrangement of Equation 4-4 yields:

$$[ES] = \frac{[E_T][S]}{K_M + [S]} \tag{4-5}$$

The velocity, v, of the reaction is a function of $[ES]$:

$$v = k_2[ES] \tag{4-6}$$

It is evident that the maximum concentration of ES is reached when $[ES] = [E_T]$; this occurs when all the enzyme molecules are saturated with substrate. Under this condition, the velocity is maximum:

$$v_{max} = k_2 [E_T] \tag{4-7}$$

Therefore,

$$v = \frac{k_2 [E_T] [S]}{K_M + [S]} \tag{4-8}$$

and

$$\frac{v}{v_{max}} = \frac{\left(\dfrac{k_2 [E_T] [S]}{K_M + [S]}\right)}{k_2 [E_T]} \tag{4-9}$$

$$v = \frac{v_{max} [S]}{K_M + [S]} \tag{4-10}$$

Thus, if $[S] = K_M$,

$$v = \frac{v_{max}}{2} \tag{4-11}$$

or K_M equals the substrate concentration at half maximum velocity.

The Michaelis constant and the v_{max} are applicable in the analysis of the mechanism of enzyme inhibition. Some substances that are structurally related to the substrate can compete with the latter by forming dissociable complexes themselves with the enzyme. Inhibitions of this type are reversed by an excess of normal substrate, and the v_{max} is unchanged under these conditions. On the other hand, since, when an inhibitor is present, more substrate is required to reach $v_{max}/2$ than in the uninhibited situation, the Michaelis constant increases in the presence of the inhibitor.

Another important type of inhibition is noncompetitive, and it may be due to a variety of causes. This is not reversible by excess substrate. Such inhibition decreases the v_{max}, but it does not affect K_M. Numerous drugs are inhibitors of specific enzymes, thereby exerting an effect on the function of the whole organism.

A number of enzymes do not exhibit the characteristic hyperbolic velocity-versus-substrate concentration curves that are anticipated from the Michaelis-Menten formulations; instead they often yield sigmoid curves. Most of these enzymes, which are termed *allosteric,* are regulatory in function. They contain two or more subunits. The reaction speeds up considerably when a critical concentration of the substrate is reached. This type of behavior is shown by homotropic enzymes. These have at

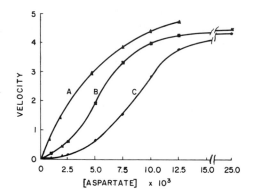

Figure 4-2
Reaction velocity of aspartate transcarbamylase as a function of substrate concentration: *A*, in the presence of an activator ATP; *B*, substrate alone; *C*, in the presence of an inhibitor CTP.

least a second site for binding substrate. This binding affects the other site and increases or decreases it.

In some regulatory enzymes, i.e., heterotropic enzymes, the *allosteric site* can accept a molecule that is structurally unrelated to the substrate. This bound molecule may activate or inhibit the action at the catalytic site and both K_M and v_{max} may be affected. This type of activity is depicted in Figure 4-2. Heterotropic regulation is found in feedback inhibition, where the final product, A, of a long sequence of chemical conversions inhibits the first enzyme in the sequence of catalytic steps that functions to produce A.

Allosteric mechanisms are usually encountered in the regulation of opposing pathways that operate in the synthesis or breakdown of key metabolites in the body. These reactions may go on simultaneously in the same location. The only way this can be accomplished is by having two different pathways that include essentially irreversible steps, since thermodynamic considerations would otherwise weigh against the existence of such a phenomenon. Some of the steps are shared by both pathways, but the mandatory endergonicity of the synthetic (anabolic) pathways demands that some steps should be different and that these be coupled to exergonic reactions. Enzymes that are not used in common by the two pathways are usually the ones that are subject to allosteric regulation by the appropriate metabolites.

In the following, we will outline the most important features of the central catabolic pathways. Anabolic pathways will not be discussed.

CENTRAL PATHWAYS FOR ENERGY PRODUCTION

Cells achieve the total oxidation of their primary fuel sources by channeling them through a metabolic mill, the *Krebs cycle* or tricarboxylic acid cycle. The importance of this cyclic chemical process resides in the fact that it permits the total oxidation, through a common mechanism, of such chemically diverse substances as glucose, fatty acids, and amino acids. The first two are the dominant fuels in the energy metabolism of the body. The Krebs cycle can function in this way because, during the metabolic breakdowns that take place in all three groups, a common key intermediate is generated. This is a two-carbon fragment, a derivative of acetate, which appears in its metabolically active form as the S-acetyl derivative of coenzyme A (acetyl-CoA) (Fig. 4-3).

The pathways leading to production of acetyl-CoA and to the formation of other derivatives that can directly enter the Krebs cycle will be described first. As the starting point, we will choose the initial steps in the metabolism of glucose.

Glucose may be obtained from glycogen through the action of the enzyme phosphorylase, which yields glucose 1-phosphate as its product:

$$(glucose)_n + PO_4^{-3} \underset{\text{phosphorylase}}{\rightleftharpoons} (glucose)_{n-1} + \text{glucose 1-phosphate} \qquad (4\text{-}12\text{A})$$

where glycogen = $(glucose)_n$. This is followed by the reaction:

$$\text{glucose 1-phosphate} \underset{\text{phosphoglucomutase}}{\rightleftharpoons} \text{glucose 6-phosphate} \qquad (4\text{-}12\text{B})$$

Free glucose may also be phosphorylated by hexokinase:

$$\text{glucose} + ATP \underset{\text{hexokinase}}{\rightleftharpoons} \text{glucose 6-phosphate} + ADP \qquad (4\text{-}13)$$

Glucose 6-phosphate is the common product in both reactions.

Glucose 6-phosphate is converted into lactic acid by enzymes that are found in solution in the cytoplasm of cells. This transformation can take place in the absence of oxygen, and it is known, therefore, as *anaerobic glycolysis* (Fig. 4-4).

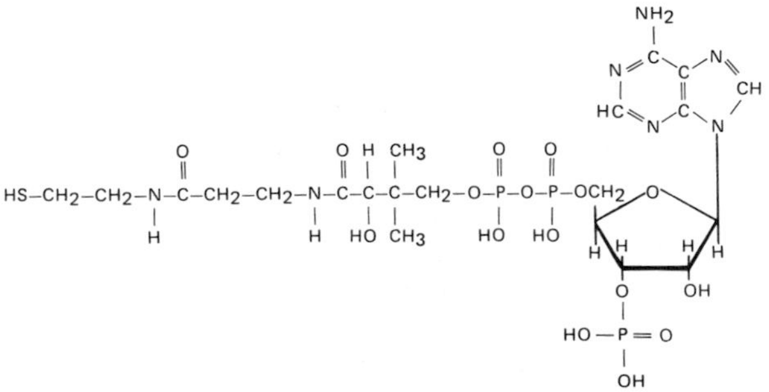

Figure 4-3
Coenzyme A is the most prominent acyl-group transfer coenzyme in living systems. This molecule is quite complex structurally, and it contains a multiplicity of possible functional groups.

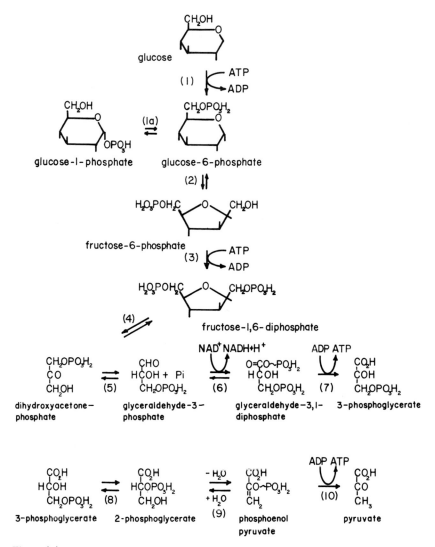

Figure 4-4
Embden-Meyerhof-Parnas sequence of reactions during the degradation of glucose. Two molecules of ATP are consumed in stage I and four are produced in stage II. The enzymes involved in the reactions are (1) hexokinase, (1a) phosphoglucomutase, (2) phosphoglucoisomerase, (3) phosphofructokinase, (4) aldolase, (5) triose phosphate isomerase, (6) phosphotriose dehydrogenase, (7) phosphoglycerokinase, (8) phosphoglyceromutase, (9) enolase, and (10) pyruvate kinase [1].

Under anaerobic conditions, there is no net change in the oxidation-reduction state of the coenzyme nicotinamide adenine dinucleotide (NAD^+) (Fig. 4-5), which acts as a hydrogen donor or acceptor in its reduced and oxidized forms, respectively. This reaction sequence — which starts with one mole of glucose and ends in two moles of

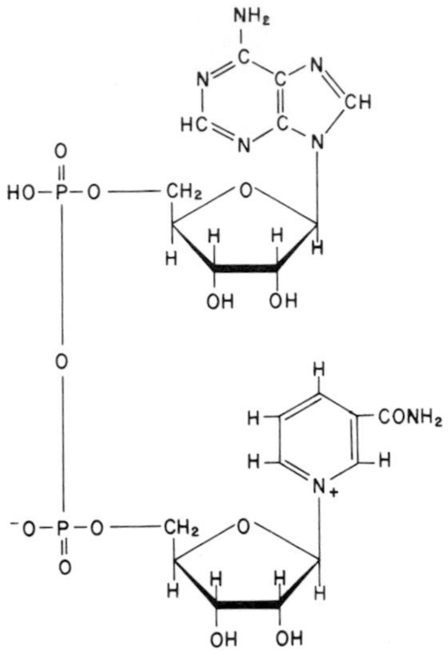

Figure 4-5
Coenzyme NAD^+.

lactic aid — can be simply regarded as a disproportionation reaction of the glucose molecule:

$$C_6H_{12}O_6 \rightarrow 2C_3H_6O_3 \qquad (4\text{-}14)$$
(glucose)   (lactic acid)

In the reaction sequence of Figure 4-4, it can be seen that for every two ATP molecules invested in this pathway, four molecules of ATP are formed. The standard free energy change of the reaction of one glucose molecule to form two lactate molecules is 48 Kcal/mole. There is also a net synthesis of two ATP molecules (a recovery of approximately 14 Kcal).

This process, though reasonable in terms of efficiency (close to 30%), would be wasteful because of the large demand on glucose and the accumulation, without further possibility of utilization, of an energetically rich metabolite, lactic acid. Glycolysis, however, provides all the energy in red blood cells (they lack mitochondria, the organelles that contain the Krebs cycle enzymes) and in skeletal muscle when oxygen demand exceeds the supply, i.e., during strenuous exercise.

Glycolysis is also a requisite for the generation of acetyl-CoA from glucose. This takes place via the oxidative decarboxylation of pyruvate, a glycolytic intermediate. This reaction is the normal follow-up step in the presence of oxygen instead of the reduction of pyruvate to lactate. Under these conditions, NADH is not reoxidized, as can be seen from Figure 4-4. The decarboxylation of pyruvate is a complex reaction that is given only schematically in Figure 4-6 in order to show the stoichiometry of the reaction. The intermediates as shown do not actually appear in the pathway.

$$\begin{array}{c}
\text{CH}_3 \\
| \\
\text{C}=\text{O} \\
| \\
\text{CO}_2\text{H}
\end{array}
\longrightarrow
\begin{array}{c}
\text{CH}_3 \\
| \\
\text{C}=\text{O} \\
| \\
\text{H} \\
+\text{CO}_2
\end{array}
\xrightarrow{\text{HOH}}
\begin{array}{c}
\text{CH}_3 \\
| \\
\text{C} \\
\diagup \quad \diagdown \\
\text{H} \quad \text{OH}
\end{array}
\xrightarrow{1/2\,\text{O}_2}
\begin{array}{c}
\text{CH}_3 \\
| \\
\text{C}=\text{O} \;+\; \text{HOH} \\
| \\
\text{OH}
\end{array}$$

Figure 4-6
Decarboxylation of pyruvate. The intermediates as shown do not appear in the actual pathway.

Following the metabolic decarboxylation step, the acetate is recovered as the CoA ester, $\text{CH}_3\text{CO}\sim\text{S}-\text{CoA}$ (acetyl-CoA). During this step, NAD^+ is reduced to NADH.

This reaction, as well as the following steps in the metabolism of acetate via the Krebs cycle (Fig. 4-7), take place in cellular organelles called *mitochondria*. The initial reaction is the condensation of the acetate fragment with oxalacetic acid to yield citric acid. In the subsequent steps, the completion of a cycle leads to the recovery of oxalacetic acid. In a single cycle, the acetate fragment will be oxidized to CO_2, three moles of NAD^+ will be reduced to NADH, and one mole of another oxidation-reduction coenzyme, flavin adenine dinucleotide (FAD), will be reduced to FADH_2.

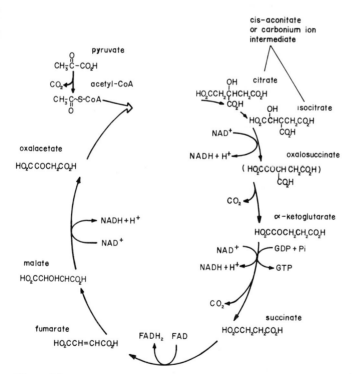

Figure 4-7
Citrate (Krebs) cycle. The reactions accompanied by the reduction of coenzymes are shown. Other details of the cycle are discussed in the text.

In addition, one high-energy phosphate group (potentially ATP) is produced in the Krebs cycle. Thus, by combining the processes of glycolysis with the Krebs cycle, the total oxidation of glucose to CO_2 is achieved. The reoxidation of the reduced coenzymes NADH and $FADH_2$ to NAD^+ and FAD by the electron-transport chain (see below) will produce water, which is the other end product of glucose oxidation.

As in the case of the Krebs cycle, the terminal sequence of events in biologic oxidation also takes place in the mitochondrion. These events involve the reoxidation of the reduced coenzymes by the electron-transport chain. This entails a highly complex chain of events in which iron-containing electron carriers of increasingly positive oxidation-reduction potentials pass electrons to one another. These carriers are the *cytochromes*; the ultimate one, cytochrome oxidase, combines with and reduces atmospheric oxygen to O^{-2}, which in turn forms water with hydrogen ions. These hydrogen ions may be considered to have originated from the reduced coenzymes by having been stripped of their electrons by the electron-transport chain (e.g., $FADH_2 \rightarrow FAD + 2e^- + 2H^+$).

The reoxidation of reduced coenzymes by the electron-transport chain is coupled to *oxidative phosphorylation*, another mitochondrial function. The details of this mechanism are still obscure, but it accomplishes the rephosphorylation of ADP to ATP at specific oxidation-reduction steps where the energy yield is favorable; i.e., the energy yield has to exceed the energy requirement of the endergonic reaction of ADP to ATP, which is approximately 7 Kcal/mole. Three such steps exist in the reoxidation of NADH, resulting in the formation of three moles of ATP, while the reoxidation of $FADH_2$ provides two such steps. A balance sheet can now be set up for the energy yield of the total oxidation of one mole of glucose to CO_2 and water in terms of ATP produced:

	Moles Reduced Coenzyme	Moles ATP
ATP formed in glycolysis		2
ATP formed in Krebs cycle (two acetate units)		2
NADH formed in glycolysis	2 giving	6
NADH formed in two pyruvate decarboxylations	2 giving	6
NADH formed in two acetate oxidations in Krebs cycle	6 giving	18
$FADH_2$ formed in two acetate oxidations in Krebs cycle	2 giving	4
Total		38

For the oxidation of a mole of glucose, the total yield expected is 38 moles of ATP,* which corresponds to an efficiency of 266/680 = 39% (since $\Delta G°$ for the reaction ATP $\rightarrow$ ADP + P_i is 7 Kcal/mole, and for glucose + $6O_2 \rightarrow 6CO_2 + 6H_2O$, it is 680 Kcal/mole).

Fatty acids contribute a major proportion of fuel energy. Their catabolism occurs by fragmentation to two-carbon acetyl-CoA units through a special oxidation pathway (Fig. 4-8). The energy yield of this pathway, which also is carried out in the mitochondria, can be calculated on the basis of the previous considerations.

*In fact, it is found to be slightly less than this for reasons that are beyond the scope of this presentation.

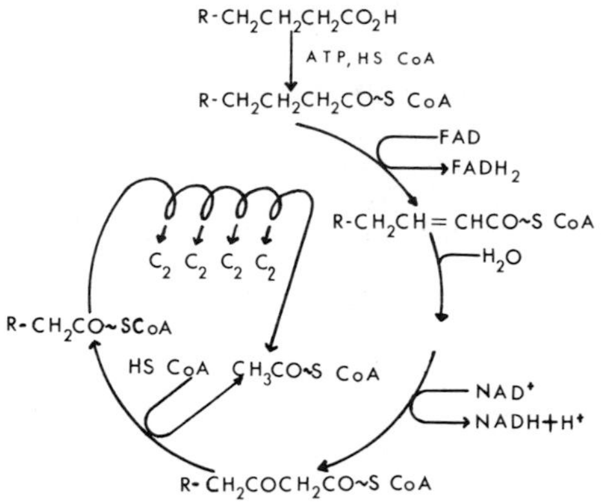

Figure 4-8
Mitochondrial oxidation of fatty acids. The process is repetitive and is schematically viewed as a spiral.

Finally, it should be mentioned that complex pathways exist for the catabolism of all 20 amino acids, which ultimately yield acetyl-CoA, a Krebs cycle intermediate, or both. Since any Krebs cycle intermediate can yield oxalacetate, and since the latter can be converted to pyruvate through enzymatic decarboxylation,

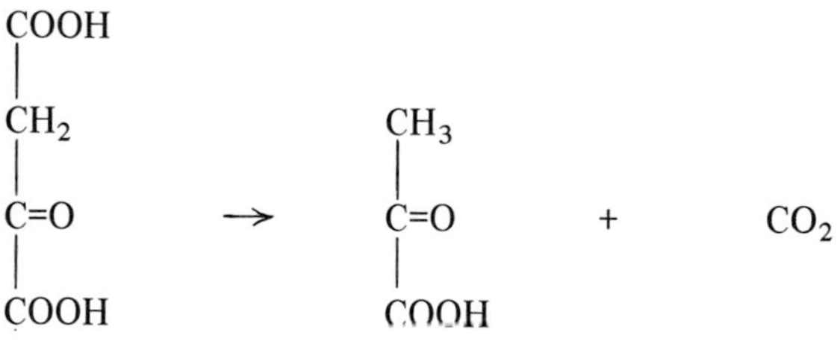

$$
\begin{array}{ccccc}
\text{COOH} & & \text{CH}_3 & & \\
| & & | & & \\
\text{CH}_2 & & \text{C=O} & & \\
| & \rightarrow & | & + & \text{CO}_2 \\
\text{C=O} & & \text{COOH} & & \\
| & & & & \\
\text{COOH} & & & & \\
\end{array}
$$

(oxalacetate)   (pyruvate)

it follows that the terminal steps of glucose, fatty acid, and amino acid metabolism all converge into the Krebs cycle and lead to CO_2 and water through oxidations.

REFERENCES

1. Neuhaus, O. W., and Orten, J. M. *Biochemistry* (8th ed.). St. Louis: Mosby, 1970.
2. White, A., Handler, P., and Smith, E. *Principles of Biochemistry* (4th ed.). New York: McGraw-Hill, 1968. Pp. 224–232.

SELECTED READING

Bartley, W., Birt, L. M., and Banks, P. *The Biochemistry of Tissues.* New York: Wiley, 1968.

Bernhard, S. *The Structure and Function of Enzymes.* New York: Benjamin, 1968.

Lehninger, A. L. *Biochemistry.* New York: Worth, 1970.

Mahler, H. R., and Cordes, E. H. *Biological Chemistry* (2nd ed.). New York: Harper and Row, 1971.

Rafelson, M. E., Jr., and Binkley, S. B. *Basic Biochemistry* (2nd ed.). New York: Macmillan, 1968.

Westley, J. *Enzymic Catalysis.* New York: Harper and Row, 1969.

White, A., Handler, P., and Smith, E. L. *Principles of Biochemistry* (5th ed.). New York: McGraw-Hill, 1973.

Properties of Excitable and Contractile Tissue

5

T. J. Mende and L. Cuervo

THE ORIGINS OF THE RESTING POTENTIAL

A common characteristic of nerve and muscle cells is their ability to propagate an electrical impulse along their membranes. The main function of the impulse in nerve cells is communication, whereas in muscle cells, it functions as the initiating stimulus that leads to the activation of the contractile mechanism.

In order to understand the mechanisms of propagated impulses (which are described in more detail in Chapter 26), we must first look at the *resting transmembrane potential,* which arises as a consequence of the ionic charge distributions on both sides of the membrane. The extracellular and intracellular ionic compositions are drastically different, as is shown in Table 5-1. Sodium is the chief extracellular cation, and K^+ is the most abundant internal cation. Chloride is the main extracellular anion, whereas organic acids of different types make up the internal anionic components. The separating structure between the intracellular and extracellular compartments is the *plasma membrane.*

The plasma membrane, whose detailed structure is still under investigation, is frequently depicted as a double layer of lipid molecules and associated proteins or "unit membrane" (see Chap. 3, p. 56). These molecules are oriented with their hydrophilic polar ends toward the intracellular and extracellular aqueous phases, and they are covered by a protein layer on both sides. The hydrophobic hydrocarbon parts of the lipid molecules are stacked in parallel arrays in a head-to-head orientation in this "sandwich." This simple description, however, fails to explain many characteristics of cell membranes that seem to depend on specialized areas of this general membrane. The unit membrane hypothesis is based mainly on electron microscopic evidence, which shows an electron-dense double layer, presumably the hydrophilic part of the lipids, that is separated by a transparent area that would then correspond to the hydrophobic double leaflet. A number of alternative membrane models have been proposed in place of the basic unit membrane; the differences among these models range from slight to extensive. It is certain that all membranes do not have identical structures, since their chemical compositions vary widely.

Cell membranes do not behave as isotropic structures. For example, they exhibit transport characteristics that are directionally dependent, provided an energy supply is available. The unequal distribution of different ions is due in part to this directional dependence.

If an electrode consisting of a nonpolarizable substance such as silver-silver chloride is used to penetrate the cell membrane without causing damage to it, a potential difference can be shown to exist with reference to an externally placed electrode. This potential difference varies greatly between cell types, but it is always negative inside the cell relative to the outside and may be assumed to be approximately 90 mV in the cell types of interest here.

Table 5-1. Approximate Steady-State Ionic Concentrations and Potentials in Mammalian
Muscle Cells and Interstitial Fluid

Interstitial Fluid (micromoles/ml)		Intracellular Fluid (micromoles/ml)		$E_{ion} = \dfrac{[ion]_0}{[ion]_1} \dfrac{61}{z} \log \dfrac{[ion]_0}{[ion]_1}$ (mV)	
Cations		Cations			
Na^+	145	Na^+	12	12.1	66
K^+	4	K^+	155	1/39	−97
H^+	3.8×10^{-5}	H^+	13×10^{-5}	1/3.4	−32
pH	7.43	pH	6.9		
others	5				
Anions		Anions			
Cl^-	120	Cl^-	4	30	−90
HCO_3^-	27	HCO_3^-	8	3.4	−32
others	7	A^-	155		
Potential	0		−90 mV		−90 mV

From Ruch, T. C., and Patton, H. D. *Physiology and Biophysics.* Philadelphia: Saunders, 1965. P. 4.

The value of this "resting" or "steady" potential can be calculated from the ionic distribution on both sides of the cell membrane. Accordingly, the behavior of the electrolyte solutions forms the basis of the theory dealing with the origin of bioelectrical potentials.

There is no thermodynamic equilibrium across cell membranes. An unequal distribution of ions alone is no proof for this statement, because if a membrane restricts the free diffusion of some ions (e.g., due to size), an unequal distribution for the diffusible ones will result (due to Gibbs-Donnan equilibrium). It will be seen, however, that the steady-state distribution of ions across cell membranes does not correspond strictly to Donnan equilibrium conditions.

In order for an ion to be subject to the Gibbs-Donnan equilibrium, a thermodynamic parameter − the *electrochemical potential* − must be the same on both sides of the membrane. The electrochemical potential is determined by the chemical activity of the ion and the effect of electric field present:

$$\mu_j = \mu_j{}^0 + RT \ln (\gamma c_j) + z_j FV \qquad (5\text{-}1)$$

where

μ_j = electrochemical potential of ion j
$\mu_j{}^0$ = partial molal free energy of ion j (a constant if the solvent is the same on both sides of the membrane)
z_j = number of charges carried by ion j
V = electrical potential

c_j = concentration of ion j
γ = activity coefficient of ion j
R = gas constant
T = temperature (absolute)
F = Faraday constant

Expressing the condition of equilibrium as $(\mu_j)_{in}$ (inside of membrane) = $(\mu_j)_{out}$ (outside of membrane) and assuming the activity coefficient, γ, is the same inside and outside, we have

$$RT \ln(c_j)_{out} + z_j F(V_j)_{out} = RT \ln(c_j)_{in} + z_j F(V_j)_{in} \tag{5-2}$$

The potential difference between the inside and outside of the membrane, V_m, can be derived from Equation 5-2:

$$V_m = (V_j)_{in} - (V_j)_{out}$$

$$V_m = \frac{RT}{z_j F} \ln \frac{(c_j)_{out}}{(c_j)_{in}} \tag{5-3}$$

Equation 5-3 is a form of the *Nernst equation*. It may be used to establish whether an ion is distributed according to electrochemical equilibrium. If the Nernst potential is determined for the ions K^+, Na^+, and Cl^- as indicated in Table 5-1, it may be seen that the calculated potential is close to equilibrium for K^+. The chloride ion is distributed exactly according to the electrochemical equilibrium condition, whereas the concentration noted for Na^+ is far from the equilibrium value. This finding is but one piece of the evidence in favor of the hypothesis that there is an active process, known as the *sodium pump,* that extrudes this ion from the interior of the cell against both concentration and electrical gradients.

Returning to the question of the origin of the resting potential, it may be stated that it can be identified almost exclusively with the passive fluxes of the diffusible ions inside and outside the excitable cells. It is a well-known physical principle that ions in solution tend to equalize their distribution by diffusion, so that there will be a movement of particles from a region of higher concentration to a lower one as well as movement of water molecules in the opposite direction. When the ionic mobilities of the anions and cations present are different, which is generally the case, a charge separation takes place and an electrical potential gradient develops that exerts an influence on the velocities of the particles according to their charge. This influence is, of course, superimposed on that of the concentration gradient.

The movement of ions that are subjected simultaneously to concentration and electrical gradients has been investigated according to either a kinetic or thermodynamic point of view. Both approaches arrive at almost equivalent differential equations.

ADDITION OF IONIC FLUXES

Consider a solution of electrolytes in which the forces acting on the system will produce a net movement of ions in a given direction. If it is assumed that an imaginary cylindrical segment of cross-sectional area A exists within this solution and that the average velocity of ions is such that they will traverse the segment in unit time, Δt, then all of the ions contained within the cylindrical segment at a particular time t will have moved out of the segment at time $t + \Delta t$.

The flux, J_i, is the number of moles of substance i, m_i, that cross a unit area per unit time:

$$J_i = \left(\frac{1}{A}\right)\frac{dm_i}{dt} = \bar{v}_i c_i \quad \frac{\text{moles}}{\text{sec}\cdot\text{cm}^2} \tag{5-4}$$

where A = cross-sectional area
$\quad c_i$ = concentration of the solution (moles/cm^3)
$\quad \bar{v}_i$ = average velocity of the ions (cm/sec)

The average velocity $\bar{v}_i$ is constant because, when particles move in a viscous medium, a frictional force that opposes this movement is generated such that

$$F_{fric} = b\bar{v} \tag{5-5}$$

where b is a frictional coefficient. When a movement-generating force is applied, the velocity increases until the frictional force becomes equal to the driving force.

The *electrical mobility* of an ionic species, u_i, is defined as the average velocity per unit electrical field:

$$u_i = \frac{\bar{v}_i}{\Psi} \text{ (cm·coulomb·sec}^{-1}\text{·newton}^{-1}) \tag{5-6}$$

where Ψ = electrical field strength (volt/meter).

The *absolute mobility*, $\bar{u}_i$, is defined as the average velocity per unit "generalized field":

$$\bar{u}_i = \frac{\bar{v}_i}{F_i/m} \text{ (cm·mole·sec}^{-1}\text{·newton}^{-1}) \tag{5-7}$$

where F_i/m = generalized field (generalized driving force per mole). Then Equation 5-4 becomes

$$J_i = \bar{u}_i(F_i/m)c_i \tag{5-8}$$

When a concentration gradient exists within the solution, the resulting diffusion of solute generates an electrical field because of the different mobilities of the cations and anions present. Both of these driving forces, the diffusional and the electrical, would determine the fluxes $(J_i)_a$ and $(J_i)_e$ if acting independently. The total flux J_i is given by:

$$J_i = (J_i)_a + (J_i)_e \tag{5-9}$$

Also, the corresponding fields are equal to the respective negative potential gradients.* Choosing the x-axis in the direction of movement,

$$\Psi = -\frac{dV}{dx}$$

$$(F_i)_a/m = -\frac{d\mu_i}{dx} \tag{5-10}$$

where V = electrical potential (volts)
μ_i = chemical potential of species i (joules$\cdot$mole^{-1})

The chemical potential is a thermodynamic property given by

$$\mu_i = \mu_i{}^0 + RT \ln a_i \tag{5-11}$$

where R = gas constant (8.314 joules$\cdot$deg$^{-1}\cdot$mole^{-1})
T = absolute temperature
$\mu_i{}^0$ = chemical potential when $a = 1$
a_i = activity of i

The activity, a_i, is equal to the product of the concentration, c_i, and an activity coefficient, γ_i:

$$a_i = c_i \gamma_i \tag{5-12}$$

Therefore,

$$(F_i)_a/m = -\frac{d\mu_i}{dx} = -RT\frac{d\ln a_i}{dx} = -RT\left(\frac{1}{c_i}\frac{dc_i}{dx_i} + \frac{d\ln\gamma_i}{dx}\right) \tag{5-13}$$

*The derivation followed here is based on the additivity of the fluxes (Eq. 5-9). An identical result is obtained by considering instead the additivity of the fields using the definition of electrochemical potential and its gradient.

and the flux due to the diffusional force is:

$$(J_i)_d = RT\bar{u}_i c_i \left(\frac{1}{c_i} \frac{dc_i}{dx} + \frac{d\ln\gamma_i}{dx} \right)$$

(5-14)

When $\gamma_i = 1$ (ideal solutions*) this becomes:

$$(J_i)_d = -RT\bar{u}_i \frac{dc_i}{dx}$$

(5-15)

Equation 5-15 is analogous to Fick's first law of diffusion:

$$(J_i)_d = -D_i \frac{dc_i}{dx}$$

By substitution:

$$D_i = RT\bar{u}_i$$

(5-16)

where Equation 5-16 is the Nernst-Einstein equation that relates the diffusion coefficient to the mobilities.

Next, the electrical force per mole of ions is given by

$$(F_i)_e/m = z_i F\Psi = -z_i F \frac{dV}{dx}$$

(5-17)

where z_i = valence of ion i
F = Faraday constant (96,489 coulombs·mole^{-1})
V = electrical potential (volts)

The corresponding flux is:

$$(J_i)_e = -z_i \bar{u}_i c_i F \frac{dV}{dx}$$

(5-18)

From Equations 5-9, 5-15, and 5-18, the total flux is obtained:

$$J_i = -RT\bar{u}_i \frac{dc_i}{dx} - z_i \bar{u}_i Fc_i \frac{dV}{dx}$$

(5-19)

which is the Nernst-Planck flux equation.

Let us now consider systems that consist of two different solutions of electrolytes separated by a membrane. The movement of ions from one solution to the other through the membrane obeys the Nernst-Planck equation. The set of flux equations

*Also, if γ_i is constant throughout the gradient, the same result is obtained. The assumption of ideality, although frequently made to simplify equations, is usually unwarranted. In actual cases of biologic membranes where several ions with concentration gradients in opposite directions are involved, the probability of a nearly constant activity coefficient is good.

(one for each ionic species present) that describes a given system is subject to the condition of *macroscopic electrical neutrality*, which, in bulk solution or in membranes lacking fixed charges, may be written as:

$$\sum_i z_i c_i = 0 \tag{5-20}$$

This means that in the solution or within the membrane, it will not be possible to find macroscopic regions that are electrically charged. All charge separations that give rise to the observed potential difference must occur at the molecular level.

Consider a simple system consisting of a membrane that separates two solutions of different but constant concentrations of NaCl. If the concentration gradient of NaCl within the membrane is linear, then the ionic fluxes through the membrane may be easily determined. If the x direction is taken as perpendicular to the membrane and if the ionic gradient is assumed to be positive in the positive x direction, then the ions will move in a negative direction:

$$J_{Na} = -RT\bar{u}_{Na}\frac{dc_{Na}}{dx} - z_{Na}\bar{u}_{Na}Fc_{Na}\frac{dV}{dx} \tag{5-21}$$

$$J_{Cl} = -RT\bar{u}_{Cl}\frac{dc_{Cl}}{dx} - z_{Cl}\bar{u}_{Cl}Fc_{Cl}\frac{dV}{dx} \tag{5-22}$$

The valences $z_{Na} = +1$ and $z_{Cl} = -1$, and, because of the condition of electrical neutrality, $c_{Na} = c_{Cl} = c$, where c is the concentration of NaCl. Also, the condition of electrical neutrality demands that both fluxes be equal at all points, otherwise a macroscopical charge separation would develop. Then,

$$-RT\bar{u}_{Na}\frac{dc}{dx} - \bar{u}_{Na}Fc\frac{dV}{dx} = -RT\bar{u}_{Cl}\frac{dc}{dx} + \bar{u}_{Cl}Fc\frac{dV}{dx} \tag{5-23}$$

$$RT(\bar{u}_{Cl} - \bar{u}_{Na})\frac{dc}{dx} = Fc(\bar{u}_{Cl} + \bar{u}_{Na})\frac{dV}{dx} \tag{5-24}$$

or

$$\frac{dc}{c} = \frac{F}{RT}\frac{\bar{u}_{Cl} + \bar{u}_{Na}}{\bar{u}_{Cl} - \bar{u}_{Na}}dV \tag{5-25}$$

If we assume (1) that the mobilities are constant throughout, (2) that the concentrations are continuous, and (3) that there exists no potential at the membrane-solution interfaces, we may integrate Equation 5-25 from the conditions of one solution (I) to those of the other (II):

$$\int_I^{II}\frac{dc}{c} = \frac{F}{RT}\frac{\bar{u}_{Cl} + \bar{u}_{Na}}{\bar{u}_{Cl} - \bar{u}_{Na}}\int_I^{II}dV \tag{5-26}$$

Then we have:

$$\Delta V = V_{\mathrm{II}} - V_{\mathrm{I}} = \frac{RT}{F} \frac{\bar{u}_{\mathrm{Cl}} - \bar{u}_{\mathrm{Na}}}{\bar{u}_{\mathrm{Cl}} + \bar{u}_{\mathrm{Na}}} \ln \frac{c_{\mathrm{II}}}{c_{\mathrm{I}}} \tag{5-27}$$

This equation defines a diffusion potential whose sign and magnitude are dependent upon the mobilities of the two ions involved. This simple case is never found in actual biologic situations and is presented only to illustrate the importance of ionic mobilities. Notice that if one of the mobilities is zero, Equation 5-27 reduces to the Nernst equation.

Actual biologic systems are not in equilibrium, since there is a continuous and spontaneous net transfer of matter from one side of the membrane to the other. This movement of ions across cell membranes, as well as the fluxes of each ion in both directions, has been amply demonstrated and measured in experiments using radioactive ions. It has been found that for many different cells under normal conditions, the measured influx of K^+ and efflux of Na^+ are much greater than those predicted theoretically, except when the energy sources of the cells have been curtailed by metabolic inhibitors. When the latter is done, the agreement between experiment and theory is very good. This fact supports the contention that an energy-dependent, active transport process serves to move Na^+ and K^+ across the membrane to maintain the ionic gradients.

This *sodium-potassium pump* is indirectly responsible for the resting potential, since it maintains the concentration gradients that, through diffusion, give rise to the microscopic charge separations. The exact nature of this "pump" is still unknown. Some membrane-bound enzymes, however, have been isolated and have been proved to be important parts of it. In some cells, there seem to be "pumps" for ions like Cl^- as well. In other cells, these ions are passively distributed, i.e., their concentrations adjust to the potential difference set by K^+ and Na^+. This means that the ratio of the concentrations of these ions on both sides of the membrane is found to be equal to that predicted by the Nernst equation if the system were at equilibrium. The electrochemical potential differences for the ions in such cases are zero, whereas the electrochemical potential differences for ions that are pumped have a finite value. In other words, when an ion is passively distributed, the force acting on it as a result of the electrical field is completely balanced by the force due to the concentration gradient, and no net transfer of that ion occurs.

THE GOLDMAN EQUATION

For more complicated systems involving several ions, different approaches to the problem of membrane potentials have been followed by several investigators. We will limit this discussion to the case of a membrane that lacks fixed electrical charges, as treated by David E. Goldman [2]. The condition of electrical neutrality still holds in the same manner as before.

Let us assume that the membrane contains many dipoles that are able to balance the microscopic charge separations between ions and fixed charges, the net effect being that the space charge density, ρ, is zero throughout the membrane.

Since

$$\frac{d^2 V}{dx^2} = \frac{-\rho}{\epsilon} = 0 \tag{5-28}$$

where ρ is expressed in coulomb·meter^{-3} and ϵ is the permittivity (coulomb·volt^{-1}· meter^{-1}) of the medium, it follows that

$$\frac{dV}{dx} = \Psi = \text{constant} \tag{5-29}$$

(It is from this that the expression we will derive obtained the name *constant field equation*.) When the system reaches a steady state, $\delta c_i/\delta t = 0$. This implies:

$$\frac{\delta J_i}{\delta x} = 0 \tag{5-30}$$

The boundary conditions are:

$c_i = (c_i)_1$, $V = V_1$ at $x = x_1$

$c_i = (c_i)_2$, $V = V_2$ at $x = x_2$

The coordinates x_1 and x_2 are chosen so that they fall within the membrane phase precisely at its boundaries with the two solutions. By so choosing, we are able to refer to transmembrane fluxes and potentials while avoiding the necessity of defining the relationship between the concentrations in the membrane and those in the adjacent solutions. Thus, the thickness of the membrane is $h = x_2 - x_1$. Let $\Delta V = V_1 - V_2$. Then,

$$J_i = -RT\bar{u}_i \frac{dc_i}{dx} + z_i \bar{u}_i Fc_i \frac{\Delta V}{h} \tag{5-31}$$

which follows from Equations 5-19 and 5-29.

Notice that the sign of the second term on the right side of the equation is now positive. Since the positive direction is now from x_1 to x_2, we have:

$$\frac{dV}{dx} = \frac{V_2 - V_1}{h} \tag{5-32}$$

whereas ΔV was defined in the opposite direction. The definition of ΔV is arbitrary and reflects one way of making the measurement. If the change were defined to go in the other direction, the final equation would have a form slightly different from the one originally derived by Goldman; both equations, however, are equivalent.

By separation of variables, Equation 5-31 becomes:

$$RT\bar{u}_i \; \frac{dc_i}{z_i\bar{u}_iF \dfrac{\Delta V}{h} c_i - J_i} = dx \tag{5-33}$$

or

$$\frac{RT}{z_iF \dfrac{\Delta V}{h}} \; \frac{d\left(z_i\bar{u}_iF \dfrac{\Delta V}{h} c_i - J_i\right)}{\left(z_i\bar{u}_iF \dfrac{\Delta V}{h} c_i - J_i\right)} = dx \tag{5-34}$$

Then, by integrating from x_1 to x_2 and assuming that the mobilities are constant,

$$\ln \frac{z_i\bar{u}_iF \dfrac{\Delta V}{h} (c_i)_2 - J_i}{z_i\bar{u}_iF \dfrac{\Delta V}{h} (c_i)_1 - J_i} = \left(\frac{z_iF \dfrac{\Delta V}{h}}{RT}\right) h \tag{5-35}$$

from which we get:

$$J_i = z_i\bar{u}_iF \frac{\Delta V}{h} \; \frac{(c_i)_2 \exp\left(\dfrac{-z_iF\Delta V}{RT}\right) - (c_i)_1}{\exp\left(\dfrac{-z_iF\Delta V}{RT}\right) - 1} \tag{5-36}$$

This is the *Goldman equation*. It predicts a nonlinear current-voltage relationship, i.e., a rectifying capability that is qualitatively similar to that observed in some axonal membranes.

GOLDMAN-HODGKIN-KATZ EQUATION

As already stated, the concentrations in Equation 5-36 are those within the membrane phase. At present, it seems impossible to perform measurements of these quantities on biologic membranes, but attempts have been made to relate their values to those in the solutions in contact with the membrane. Hodgkin and Katz assumed that they are directly proportional, i.e., that

$$(c_i)_m = (c_i)_s \beta_i \tag{5-37}$$

This assumption is valid for neutral membranes, but for fixed-charge membranes, the relationship will be more complex. The quantity β_i may be termed the *partition coefficient* of i between the membrane and the aqueous solution.

The current for a particular ion is given by:

$$I_i = J_i z_i F \tag{5-38}$$

For K^+, this is:

$$I_K = \bar{u}_K F^2 \frac{\Delta V}{h} \beta_K \frac{(c_K)_{out} \exp\left(-\dfrac{F\Delta V}{RT}\right) - (c_K)_{in}}{\exp\left(-\dfrac{F\Delta V}{RT}\right) - 1} \tag{5-39}$$

where $(c_K)_{in}$ = concentration inside cell
$(c_K)_{out}$ = concentration outside cell

The relevant ions in determining the electrical behavior of most cell-surface membranes are K^+, Na^+, and Cl^-. Hodgkin and Katz applied equations similar to 5-39 to calculate the current carried by each of these ions.

Since the positive direction, originally from 1 to 2, is from inside to outside in this equation, ΔV becomes automatically the potential difference that is measured between the inside and the outside of the cell ($\Delta V = V_{in} - V_{out}$).

The *permeability coefficient* of an ion through a membrane is defined as:

$$P_i = \frac{D_i}{h}$$

where D_i = diffusion coefficient for ion i ($cm^2 \cdot sec^{-1}$)
h = thickness of the membrane (cm)

The partition coefficient β may be included in the definition of the permeability coefficient, so that $P_i = D_i \beta_i / h$. Making use of the permeability coefficient and the Nernst-Einstein equation,

$$I_K = P_K \frac{F^2 \Delta V}{RT} \frac{(c_K)_{out} \exp\left(-\dfrac{F\Delta V}{RT}\right) - (c_K)_{in}}{\exp\left(-\dfrac{F\Delta V}{RT}\right) - 1} \tag{5-40}$$

For Na^+ and Cl^-, we have:

$$I_{Na} = P_{Na} \frac{F^2 \Delta V}{RT} \frac{(c_{Na})_{out} \exp\left(-\dfrac{F\Delta V}{RT}\right) - (c_{Na})_{in}}{\exp\left(-\dfrac{F\Delta V}{RT}\right) - 1} \tag{5-41}$$

$$I_{Cl} = P_{Cl} \frac{F^2 \Delta V}{RT} \frac{(c_{Cl})_{in} \exp\left(-\dfrac{F\Delta V}{RT}\right) - (c_{Cl})_{out}}{\exp\left(-\dfrac{F\Delta V}{RT}\right) - 1} \tag{5-42}$$

The total current $I = I_K + I_{Na} + I_{Cl}$, or

$$I = P_K \frac{F^2 \Delta V}{RT} \left\{ \frac{\left[(c_K)_{out} + \dfrac{P_{Na}}{P_K}(c_{Na})_{out} + \dfrac{P_{Cl}}{P_K}(c_{Cl})_{in}\right]\left[\exp\left(-\dfrac{F\Delta V}{RT}\right)\right]}{\exp\left(-\dfrac{F\Delta V}{RT}\right) - 1} - \frac{(c_K)_{in} + \dfrac{P_{Na}}{P_K}(c_{Na})_{in} + \dfrac{P_{Cl}}{P_K}(c_{Cl})_{out}}{\exp\left(-\dfrac{F\Delta V}{RT}\right) - 1} \right\} \tag{5-43}$$

It is conceivable that when no external current is applied to the cell, the different ionic currents predicted by these equations (i.e., passive currents) balance each other. The active transport systems of the cell would then maintain the constancy of the intracellular and extracellular concentrations by returning the ions that carry the passive currents back to their original compartments.

If an ion does not carry any current (i.e., if $J_i = 0$), either its permeability is zero or its distribution follows the equilibrium equation

$$\Delta V = \frac{RT}{z_i F} \ln \frac{(c_i)_{out}}{(c_i)_{in}} \tag{5-44}$$

If the total passive ionic current is not zero, the maintenance of the steady state of the cell demands that the active transport systems of the cell should not be electrically neutral.

For the case when $I = 0$,

$$[P_K(c_K)_{out} + P_{Na}(c_{Na})_{out} + P_{Cl}(c_{Cl})_{in}] \exp\left(-\frac{F\Delta V}{RT}\right) =$$

$$P_K(c_K)_{in} + P_{Na}(c_{Na})_{in} + P_{Cl}(c_{Cl})_{out} \tag{5-45}$$

$$-\frac{F\Delta V}{RT} = \ln\frac{P_K(c_K)_{in} + P_{Na}(c_{Na})_{in} + P_{Cl}(c_{Cl})_{out}}{P_K(c_K)_{out} + P_{Na}(c_{Na})_{out} + P_{Cl}(c_{Cl})_{in}}$$

$$\Delta V = \frac{RT}{F}\ln\frac{P_K(c_K)_{out} + P_{Na}(c_{Na})_{out} + P_{Cl}(c_{Cl})_{in}}{P_K(c_K)_{in} + P_{Na}(c_{Na})_{in} + P_{Cl}(c_{Cl})_{out}} \tag{5-46}$$

which is the *Goldman-Hodgkin-Katz* (GHK) equation.

In spite of the several unwarranted assumptions made in the derivation of this equation, it is extensively used by electrophysiologists since it has shown remarkable agreement with experimental data. For example, the value of the resting potential of frog muscle fibers calculated from known permeabilities and concentrations of K^+, Na^+, and Cl^- is −89 mV inside the cell, and the value measured directly by intracellular recording is −90 mV.

Independent measurements of fluxes have yielded values for the permeabilities of the major ions and have shown that under resting conditions, the permeability of potassium is much greater than that of other ions in many cells. Thus, the potassium concentration gradient exerts the main influence in determining the value of the resting potential. Equation 5-46 predicts a negative state inside the cell because of this higher potassium permeability, which is in accordance with experimental results (Table 5-2). Also, in tissues for which enough information is available, the magnitudes of the calculated resting potentials usually agree within 1 or 2 mV with the measured values.

Table 5-2. Resting Membrane Potentials in Various Tissues

Tissue	Resting Potential (mV)
Frog myelinated axon	−70
Cat spinal motoneuron	−70
Squid giant axon	−60
Rat skeletal muscle fiber	−100
Dog atrium	−85
Frog skeletal muscle fiber	−90

From an analysis of Equation 5-46, it becomes evident that the state of polarization of the cell could be changed either by altering the concentrations of the major ions or by varying their permeabilities. The ionic composition of a solution that bathes isolated tissues can be modified at will, and, in some giant cells like the squid axon, even the internal composition can be precisely controlled.

Figure 5-1 shows the results of an experiment in which the external sea water bathing a giant axon of the squid was altered by substituting sodium for potassium in varying amounts. The curve was calculated from Equation 5-46 by taking the following permeability ratios: $P_K/P_{Na}/P_{Cl} = 1/0.025/0.3$. The excellent fit with the experimental points over a wide range of external potassium concentrations further indicates the usefulness of the equation.

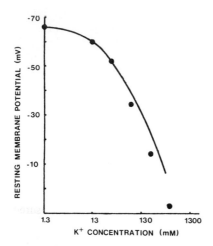

Figure 5-1
Resting potential of the squid giant axon versus the external potassium concentration. The potential was measured inside the cell relative to the outside medium. The normal potassium concentration of sea water is 13 mM.

It has also been observed for many neurons that the net Cl^- current through the membrane is zero in the steady state, that is, the Cl^- concentration ratio is adjusted to balance the membrane potential difference, as indicated by Equation 5-44. Therefore, in those cases, I_{Cl} is zero (Eq. 5-42), and Equation 5-46 can be rewritten as:

$$\Delta V = \frac{RT}{F} \ln \frac{P_K(c_K)_{out} + P_{Na}(c_{Na})_{out}}{P_K(c_K)_{in} + P_{Na}(c_{Na})_{in}} \tag{5-47}$$

$$= \frac{RT}{F} \ln \frac{(c_K)_{out} + P_{Na}/P_K(c_{Na})_{out}}{(c_K)_{in} + P_{Na}/P_K(c_{Na})_{in}} \tag{5-48}$$

It can be seen that if the Na^+ and K^+ concentrations are maintained constant, ΔV depends only on P_{Na}/P_K. When this ratio varies from zero to infinity, ΔV varies from

$$\Delta V = \frac{RT}{F} \ln \frac{(c_K)_{out}}{(c_K)_{in}} \tag{5-49}$$

to

$$\Delta V = \frac{RT}{F} \ln \frac{(c_{Na})_{out}}{(c_{Na})_{in}} \tag{5-50}$$

Processes whereby the membrane potential is altered by changes in the sodium-to-potassium permeability ratio are discussed in detail in Chapter 26.

INTRODUCTION TO PROPAGATING ACTIVITY

Changes in the resting potential will take place if the permeability coefficients or the ion concentrations on both sides of the membrane assume new values. The ionic permeability characteristics and the changes that take place during the activity of nerves and muscles may be shown in an equivalent circuit diagram employed by Hodgkin [3] (Fig. 5-2B). In this circuit, the reciprocal resistances (i.e., conductances) are related to the permeabilities of each particular ionic species. The lumped parameter model (Fig. 5-2C) represents passive membrane behavior.

Figure 5-2B illustrates the equilibrium potential for the particular ion of each ionic channel. This would be the transmembrane potential if the permeability of the other ions were negligible. The analogy with the Goldman equation is evident. It may be noted that the measured membrane potential must have a value between the Na^+ and the K^+ equilibrium potentials.

The ionic channels (Fig. 5-2A) in this system may become more permeable (i.e., increase in conductance) independently of one another. The nature of the "gates" through which ions move is not known, but their existence is accepted. The opening of a gate will increase the conductance of the appropriate ion. An increase of Na^+ conductance, for example, would allow Na^+ ions to move down the electrochemical gradient, carrying their positive charges into the cell and thereby lowering the membrane potential.

The lowering of the membrane potential by either chemical or physical means is called *depolarization.* It is reversible up to a certain minimum voltage, the *threshold voltage,* without initiating a self-propagating depolarization, the *action potential.*

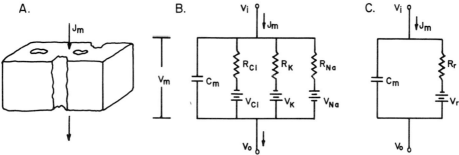

Figure 5-2
Equivalent circuits representing the electrical properties of a unit area of resting neuronal membrane. *A.* The schematic membrane patch is shown with a number of pores, which some authors believe to correspond to ionic conductance channels. *B.* Representation in terms of equivalent ionic resistances and equilibrium potentials. *C.* Lumped-parameter representation of the membrane's electrical properties. C_m = the membrane capacitance per unit area ($\mu F/cm^2$); R_r = the resting, transverse resistance of a unit area of membrane (ohm·cm^2); V_r = the resting potential; J_m = the transverse membrane current density; V_i = the potential on the inner surface of the membrane, defined with respect to a remote, extracellular reference point; and V_o = the potential on the external surface of the membrane. The resting potential values for the ions shown are V_{Na} = +80 mV, V_K = −90 mV, and V_{Cl} = −70 mV.

Once the threshold is surpassed, however, the membrane potential will continue to decrease without the further application of a stimulus until the polarity reverses and the membrane becomes positive inside and negative outside in the area of stimulation.

As a consequence of this depolarization and reversal of polarity, the membrane depolarization will spread into the neighboring, unstimulated areas. The potential difference generated between the stimulated and unstimulated areas of the membrane will give rise to a current that draws off charges from the unstimulated membrane area. This process lowers the membrane potential, and, when the threshold voltage is reached, the events described for the stimulated area will repeat themselves. Meanwhile, the area originally stimulated will reverse its sign again, and it will return to the base line following a mild overshoot. This series of events — depolarization, charge reversal, and repolarization — is called a *spike* or *action potential* that is propagated along the length of the nerve fiber. Since the action potential either is of full height or is not generated at all (i.e., when the stimulus is subthreshold), the response is termed an "all or none" response.

The initiating event for the generation of the action potential is a sudden, transient increase in Na^+ conductance. This phase of the process ends within one millisecond. It partially overlaps with and is followed by a delayed, transient rise in K^+ conductance, which persists even after the resting Na^+ conductance level is reestablished and lasts three or four times longer. During this period, K^+ ions move down the electrochemical potential gradient; i.e., they leave the cell, carrying positive charges out, and restore the resting potential in the process.

A composite curve of the time sequence for the two conductance (G) changes is shown in Figure 5-3, which accurately reproduces the actual record of an action potential. The chemical changes following the occurrence of the action potential in the cell include a small loss of K^+ and an approximately equal gain of Na^+. A very large number of action potentials may be generated before these changes become

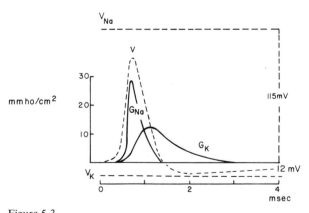

Figure 5-3
Theoretical reconstruction of a propagated action potential (V) and sodium and potassium conductances (G_{Na} and G_K). (From J. Hodgkin and H. E. Huxley. *J. Physiol.* (Lond.) 117:500, 1962. Reproduced with permission.)

chemically observable and affect the resting potential. In the normal cell that performs metabolic work, the Na^+ pump will continuously restore the steady-state ion concentrations. It will, however, in most cases have no direct effect on the membrane potential, which, as we have seen, is the result of passive ionic diffusion.

Under experimental conditions, stimulation may be achieved electrically. Under natural conditions, the nerves, muscles, or glands are stimulated by specific chemicals that are liberated from nerve endings. These agents, called *transmitter substances,* act by combining with specialized membrane areas, e.g., the postsynaptic membrane in nerve-nerve stimulation or the motor end-plate in nerve-muscle stimulation. Such chemical agents induce a change in ion permeability, which in turn results in depolarization and action potential initiation in the manner described previously. A much more detailed view of the action potential and the associated nerve phenomena is presented in Chapter 26.

Accompanying the action potentials are *action currents* that flow not only within the cell and across the membrane, but also in the medium surrounding the cell and, in some cases, extend throughout the body. The active membrane then may be termed a *bioelectrical source* that generates action currents. These currents flow within the body, which acts as a *volume conductor.* Although the magnitude of the volume conduction currents generated by an individual membrane is very small, the currents generated by populations of membranes that are active simultaneously are additive, and their effects may be measured by appropriate recording apparatus, e.g., in electrocardiography (ECG), electroencephalography (EEG), and electromyography (EMG). In addition, the body behaves as a volume conductor when electrical fields are applied via external electrodes. This is done in the techniques of electrical stimulation, electrosleep, electroanesthesia, and clinical impedance techniques. A quantitative description of volume-conduction phenomena requires the employment of electrical field theory, which is beyond the scope of this presentation.

CHEMICAL AND PHYSICAL PROCESSES IN MUSCLE CONTRACTION

There are three different types of muscles: (1) *skeletal* muscle with locomotor function, which is under voluntary control; (2) *cardiac* or heart muscle; and (3) *visceral* or smooth muscle, which is found in the walls of the alimentary canal and accessory organs, the vascular system, the reproductive and excretory systems, the eyes, and the skin, and which is under involuntary control. The fundamentals of skeletal muscle action are relevant for all muscle groups, and, because most of the experimental data were derived from skeletal muscle, it will be considered here.

Skeletal muscle has a striated appearance under the microscope. In human muscles, the cells are cylinders of variable lengths that range up to 4 cm, with a diameter of 10 to 100 microns. Muscle cells are also called *muscle fibers,* and their specialized sheath, which contains the plasma membrane, is called the *sarcolemma.* Skeletal muscle fibers are multinucleated. The contractile element of the muscle cell is localized in specialized threads within the cell, called *myofibrils.* Myofibrils in turn are composed of linear chains of a fundamental repeating unit, the *sarcomere*. A parallel alignment of the sarcomeres in adjacent myofibrils gives rise to the striations observed under the microscope. Sarcomeres contain threads of two types, called *myofilaments*, which are the basic structural and functional units of the contractile system. Figure 5-4 shows the arrangement of myofilaments and myofibrils in a skeletal muscle fiber. The myofilaments of the sarcomere may be either thick or thin (Fig. 5-5).

It has been established that the thick filaments are composed of *myosin* and the thin filaments are made up mainly of *actin*, both of which are proteins. The area occupied by myosin in the sarcomere is the *A band* (anisotropic). Its center area, the *H zone*, is free of the cross-bridges that are characteristic of those regions where the thick filaments overlap with the thin filaments. The H zone is bisected by the *M line*. The borders of a sarcomere can be identified by the *Z lines*, which appear to cut

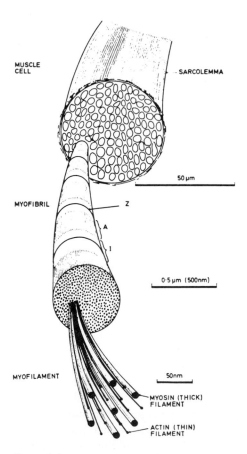

Figure 5-4
Arrangement of myofilaments and myofibrils in a skeletal muscle fiber. (From R. Passmore and J. S. Robson. *A Companion to Medical Studies,* Vol. 1. Oxford: Blackwell, 1968. P. 285. Reproduced with permission.)

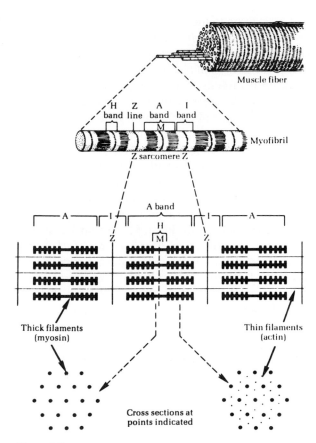

Figure 5-5
Structure of myofibrils. (From A. L. Lehninger. *Biochemistry*. New York: Worth, 1970. P. 585. Reproduced with permission.)

across the *I bands* (isotropic). The I regions contain only thin filaments. The regularity of the cross-sectional arrangement of thick and thin filaments is remarkable, exhibiting an almost crystalline hexagonal pattern, with six thin filaments surrounding each thick one.

The basic mechanism of muscle contraction depends on an antiparallel sliding of thick and thin filaments. During this process, cross-bridges are made and broken between thick and thin filaments, and tension develops. It should be noted that although the sarcomere shortens, the myofibrils do not.

In order to elucidate the molecular events underlying this mechanism, the main protein components of the sarcomere have to be described.

The myosin molecule has a total length of approximately 160 nanometers; it has a bulbous, double head with a combined diameter of 35 to 60 nm and a tail with a diameter of 2.0 nm. Its molecular weight is about 460,000 to 500,000. It has two subunits of similar size, and the tail appears to be made up of their intertwined α-helixes. There is evidence that in addition to the two main components of the myosin molecules, there are several lower molecular weight components (less than 30,000 MW) apparently present in the head region of the molecule. The stoichiometry and function of these smaller components are still obscure.

Muscle contraction depends on the hydrolysis of ATP as a direct energy source. ATP hydrolysis in turn is associated with the presence of the enzyme ATPase. It has been shown that the ATPase activity of muscle fibers is located in the head region of the myosin molecule.

Myosin molecules are packaged into bundles that form the heavy filaments found in the A band of the contractile unit. When examined with the electron microscope, the heavy filaments show regularly repeating bilateral projections along the fiber axis. X-ray diffraction data indicate that the projections form a spiral with adjacent pairs displaced at a 120° angle. This would result in identical positioning at every third projection, which would have an interval of 43.0 nm (Fig. 5-6). The thick filament is a bidirectionally polarized structure. The projections have been identified as the bulbous ends of individual myosin molecules, while the tails of the molecules are packaged into the thick-filament axis. The center area of the thick filament has no projections. Since the head parts are bound reversibly to the thin filaments during contraction and appear as the cross-bridges mentioned previously, only the projection-bearing regions contribute to the tension of the contracting muscles.

The thin filaments are mainly composed of actin. Two other components, tropomyosin and troponin, are also present and participate in an important way in the

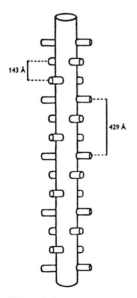

Figure 5-6
Cross-linkage arrangement of thick, myosin-containing filaments. (From H. E. Huxley. *Science* 164:1359, 1969. Reproduced with permission.)

function of the muscle. In addition, two other proteins, α- and β-actinin, are found in association with the thin filaments; α-actinin is found in the Z band, and β-actinin in the A-I junction region of the myofibril.

Actin can be obtained by suitable isolation procedures as *G-actin* (globular actin) which is the monomeric unit found in the thin filaments. In the latter, G-actin units appear to be polymerized into strings, which are then called *F-actin* (fibrous actin). The main component of the thin filaments is F-actin.

Polymerization of G-actin to F-actin requires ATP:

$$n(\text{G-actin}-\text{ATP}) \longrightarrow (\text{G-actin}-\text{ADP})_n + n\text{P}_i \qquad (5\text{-}51)$$

$$\Updownarrow$$

F-actin

F-actin contains one mole of bound ADP per G-actin unit. In the thin filaments, F-actin fibers form a double helix.

Contraction is the result of alternate formation and breaking of bonds between actin and myosin, i.e., between the thin and thick filaments. It has been shown that one molecule of H-meromyosin (a fragment obtained from myosin by proteolytic enzymatic digestion, which includes the head region of the myosin molecule) bonds to one actin monomer. In the contractile system, then, heavy filaments, through their protruding myosin "heads," interact with actin monomers, which are found in a double helical chain in the thin filaments. The myosin heads contain the enzyme ATPase, which is activated by Ca^{+2}, a very important feature of muscle contraction.

The troponin-tropomyosin-B complex functions as the relaxing protein of the contractile system. These proteins are found in the I band, i.e., the thin filament region of the sarcomere. Their essential function lies in the regulation of the interaction of the contractile system with Ca^{+2} ions, which in turn determine the ATPase activity of myosin. The Ca^{+2}-binding component is troponin. Tropomyosin-B interacts with F-actin to form a complex. It appears that grooves of the actin helix are occupied by tropomyosin. Troponin is localized on the thin filament at 40 nm intervals (Fig. 5-7).

How this system functions can be gleaned from experiments where the purified components have been brought together in vitro. Actin reacts with myosin in solution to form a viscous complex, *actomyosin*. Actomyosin is itself an ATPase, which normally requires small amounts of Ca^{+2} for its activity, although purified actomyosin will act as an ATPase independently of Ca^{+2}. However, if "relaxing protein" is added to the system, the ATPase activity of actomyosin in the absence of Ca^{+2} will be inhibited, but, in the presence of this ion, ATP splitting will occur.

The ATPase activity of pure myosin in contrast is highest in the absence of divalent cations. It is slightly inhibited by Ca^{+2} and strongly by Mg^{+2}.

Dissociation of the actomyosin complex can be brought about by Mg-ATP, and it is accompanied by a drastic decrease of the viscosity of its solutions. If sufficiently high Mg-ATP concentrations are provided, the usually simultaneous ATPase activity can be inhibited, but the dissociation will still take place, i.e., Mg-ATP can dissociate the bonds between actin and myosin.

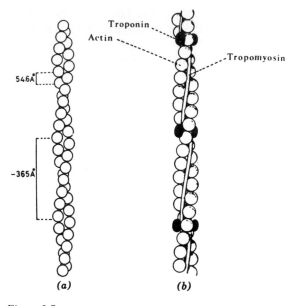

Figure 5-7
a. Structure of the thin actin filament. Two strands of F-actin are wound around each other in a helical configuration. The pitch of the filament (approximately 36.5 nm) provides for a non-integral number of G-actin subunits per turn. (From H. E. Huxley and W. Brown. *J. Mol. Biol.* 30:383, 1967. Reproduced with permission.) *b.* Structure of thin filament showing a possible arrangement of the molecules of tropomyosin-B and troponin. (From S. Ebashi, M. Endo, and Y. Ohtsula. *Q. Rev. Biophys.* 2:351, 1969. Reproduced with permission.)

With this background, the sequence of events in muscle contraction can now be considered. The initiation of contraction requires the occurrence of a nerve impulse at the neuromuscular junction. The details of the complex events that take place in the region where the nerve joins the muscle (the motor end-plate) will be omitted for the present. The end result, however, will be a propagated action potential in the muscle cell. This in turn triggers a series of reactions that culminate in the contraction and eventual relaxation of the muscle. These intervening reactions are called *excitation-contraction coupling.*

Electron microscopic studies of the muscle cell have done much to elucidate the nature of these reactions. In addition to the myofibrils, an elaborate network of tubular structures, which have pores opening to the exterior and are continuous with the plasma membrane, is present in muscle cells. The channels with outside openings are called the *T system* because they run transversely, i.e., perpendicularly, to the muscle cell axes. The T system tubules are found periodically throughout the muscle fiber, but their exact location in relation to the sarcomere varies according to the animal species. In the frog muscle, for example, they run along the Z lines of the sarcomeres. The T tubes are bracketed on both sides by vesicular structures, called *terminal cisternae,* and the entire unit is thus termed a *triad.* Appendages of the terminal cisternae, the *intermediate cisternae,* converge to form a structure halfway between the T tubes, called the *fenestrated collar* (Fig. 5-8). The terminal cisternae, their connecting tubules to the fenestrated collar, and the fenestrated collar itself form the *L system* (longitudinal). The T system and terminal cisternae are in close apposition through their respective membranes, but there are no direct interconnections between them. It is highly probable that the T system functions as an extension of the plasma membrane of the muscle cell and conducts its action potential inward in order to elicit contraction without delay in the deeper lying myofibrils.

The T and the L systems, which are also known as the *sarcoplasmic reticulum,* can be isolated by centrifugal sedimentation of muscle tissue homogenates. When added to carefully washed muscle fibers, this sedimented component prevents both the splitting of ATP and the concomitant contraction of the fibers. Both of these activities occur if this component, called the *relaxing factor* (not to be confused with the "relaxing protein"), is not present. This relaxing system can be shown to accumulate Ca^{+2} actively if Mg^{+2} and ATP (as an energy source) are present. The function of the sarcoplasmic reticulum is thus seen to be the regulation of the intracellular Ca^{+2} level. Intracellular concentrations of Ca^{+2} in the inactive muscle are very low.

Dissociation of the actin-myosin links is accomplished by ATP, as is shown by the viscosity decrease in actomyosin solutions in vitro. ATP accordingly acts as a "plasticizer."

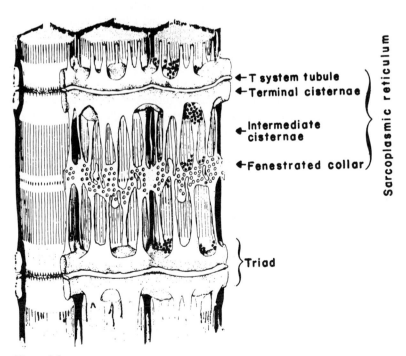

Figure 5-8
Internal membrane system of a frog sartorius muscle fiber. (From D. J. Aidley. *Physiology of Excitable Cells.* New York: Cambridge University Press, 1971. P. 193. Reproduced with permission.)

In light of the above observations, let us now examine the events occurring in a single muscle fiber following activation of the muscle cell membrane by an arriving nerve impulse that sets off a propagated action potential. This potential spreads along the fiber in both directions from the motor end-plate and travels inward through the T system. Through its surface contact with the vesicles of the sarcoplasmic reticulum, the T system elicits a change in the terminal cisternae that results in the release of Ca^{+2} ions into the cytoplasm. When the Ca^{+2} concentration reaches a threshold value that is sufficient to activate myosin ATPase, the process of contraction will begin with myosin and actin forming cross-bridges. This will cause a sliding of the thick and thin filaments in relation to each other, and the cross-bridges will again be broken by Mg-ATP. The motion is unidirectional. Contraction will proceed as long as the cytoplasmic Ca^{+2} concentration is maintained in excess of the critical level. As a result of the pumping action of the L system of the sarcoplasmic reticulum, the intracellular Ca^{+2} concentration will eventually fall below the threshold level. This will then allow the dissociation of the cross-links, and lengthening of the sarcomere, with relaxation of the muscle fiber, will occur.

This model correlates well with a key requirement imposed upon it by the shape of the length-tension diagram of muscle. This diagram shows that muscle, when fixed at different lengths, will exert different degrees of tension under maximum stimulation. The maximum point on this curve is the *rest length* of the muscle (Fig. 5-9). Since the sliding filament model bases its explanation of tension development on the number of cross-links produced, it predicts that maximum tension development will coincide with the maximum number of cross-bridges, which would occur at rest length. Electron microscopic studies give strong support to this claim.

The biophysical behavior of muscle can be studied either in whole muscle or in isolated muscle cells under different experimental conditions. One method is to apply a single stimulus and observe a single, brief contraction, called a *twitch*. In a second approach, the contraction is initiated with high-frequency, multiple stimuli, which give rise to *tetanus*, i.e., the state of a muscle in steady contraction. Tension measurements have been carried out under both conditions. Tetanic tension exceeds that produced by a single twitch (Fig. 5-10).

Another variable is introduced by permitting or preventing contraction during tension development. If the muscle is allowed to shorten with a load, we speak of *isotonic contraction,* whereas if the muscle is held fixed during stimulation, it is said to exert *isometric tension.*

As with all chemical changes, muscle activity involves heat changes. The heat changes associated with muscular function can be grouped into three major categories:

1. Resting muscle produces *resting heat,* which is a reflection of its basic metabolism.
2. The *initial heat,* which is the heat produced during contraction and relaxation, reflects multiple muscle activities.
3. A *heat of recovery,* which is usually somewhat delayed in comparison to the initial heat and may take 20 to 30 minutes for the full course of its liberation, is found following muscle activity.

Skeletal Muscle

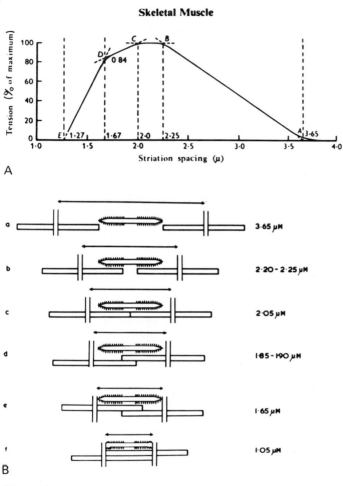

A

B

Figure 5-9
A. Variation in tension with sarcomere length in a single frog muscle fiber as a result of stimulation. (From A. M. Gordon, A. F. Huxley, and F. J. Julia. *J. Physiol.* (Lond.) 184:170, 1966. Reproduced with permission.) *B.* Arrangement of the overlap between actin and myosin filaments at various sarcomere lengths. (Modified from Gordon et al. 1966.)

The heat of recovery is liberated during the restoration of the muscle to its pre-activation state and is approximately equal to the total initial energy expenditure. This in turn equals the mechanical work done by the muscle plus the initial heat. The heat of recovery is clearly connected with aerobic metabolism, since it can be affected by decreasing the oxygen supply.

The initial heat may be regarded as the resultant of two thermal components: the *heat of activation* and the *heat of relaxation.* These events can be observed under isometric conditions. When muscle is allowed to undergo isotonic contraction, an amount of heat, the *heat of shortening,* is liberated in excess of that measured under

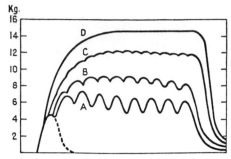

Figure 5-10
Genesis of tetanus. Response of mammalian nerve-muscle preparation. Isometric contraction records. Tension (ordinate) in kg. Lowest curve – response of muscle to single maximal stimulus to motor nerve (simple twitch). *A, B, C, D* – responses to rapidly interrupted maximal repetitive stimuli: *A* at 19, *B* at 24, *C* at 35, and *D* at 115 stimuli per second. Curves *A, B, C* show partial tetanus; curve *D* shows full tetanus. Note as the frequency of stimulation rises the tension developed becomes greater and is sustained more steadily. (After Cooper and Eccles (1930). Reprinted with permission of *J. Physiol.* (Lond.).

isometric conditions. The heat of shortening is directly proportional to the distance that the muscle is shortened.

In the case of the initial heat, the heat of activation must be related to the chemical events responsible for tension development. This heat appears before any tension is measurable, and it apparently correlates with the establishment of the condition required for shortening. This condition is called the *active state*. Its strict definition is the tension that can be elicited by the contractile element if it is neither shortening nor lengthening. The heat of activation rises rapidly to its maximum value, but it decays with the disappearance of the active state. If the heats of activation are summed up in tetanus, the result is called *maintenance heat.*

The origin of the heat of relaxation is revealed by the fact that it can be practically eliminated if the muscle is allowed to contract under a load while elongation of the relaxing muscle is prevented by blocking the load during relaxation. This finding is a clear indication that the heat of relaxation is, in great part, due to the dissipation in the form of heat of work done *on* the relaxing muscle by the stretching force.

Observations of contracting muscle under both isometric and isotonic conditions indicate that as the contractile elements undergo contraction, they do so at the expense of an elastic component, which develops its own tension under extension. This component of muscle is called the *series elastic element*. In part, it is extraneous to muscle: approximately half of it can be associated with the tendons that attach the muscle to bone, and the other half is apparently part of the muscle tissue itself. Besides the series elastic component, there is also a *parallel elastic component*, whose presence is revealed when resting muscle is stretched. It is probably at least partly identifiable with the connective tissue sheath that surrounds the muscle. An appropriate mechanical model of these factors in muscle contraction is shown in Figure 5-11.

Muscles at rest and muscles during shortening exhibit different elastic behavior. When resting muscle is stretched, it behaves as a simple elastic body with a small modulus of elasticity, i.e., its tension will be proportional to the length. On activation,

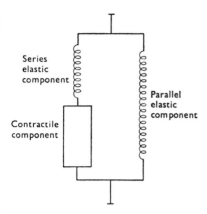

Figure 5-11
Modification of the two-component model to include a third component, the parallel elastic component, which is responsible for the resting tension of a stretched muscle. (From D. J. Aidley. *Physiology of Excitable Cells.* New York: Cambridge University Press, 1971. P. 233. Reproduced with permission.)

however, it changes its elastic properties. At a characteristic force, F_o, that opposes the tension developed by the muscle, equilibrium will be reestablished and neither shortening nor extension will take place, i.e., a state of isometric tension will obtain. If less than F_o is applied, the muscle will contract, whereas a force in excess of F_o will extend it. This contraction (or extension) will not come to a stop, but will continue. Different opposing forces (weights) will only affect the *velocity* of shortening, not its extent (at least on first approximation).

This change in the elasticity of muscle that occurs on activation is also reflected in the heat production of muscle. Active muscle shows the thermal characteristics of the majority of elastic bodies (although, as we have just shown, it does not behave like one); that is, it cools down on expansion and heats up on release. In contrast, resting muscle — which, like rubber, is an exceptional elastic body — behaves in the opposite way.

When muscle shortens under isotonic conditions, it liberates extra energy, the magnitude of which is the sum of the mechanical work done lifting a load F a distance x plus the heat of shortening. The heat of shortening equals $a \cdot x$, where a is a proportionality constant defined as the heat of shortening per unit distance. In symbolic terms,

$$\text{Extra energy} = (F + a)x \tag{5-52}$$

The time derivative of this energy — the liberation rate of extra energy — has been shown experimentally by Hill to be a linear function of the tension developed:

$$(F + a)\frac{dx}{dt} = (F + a)v = (F_o - F)b \tag{5-53}$$

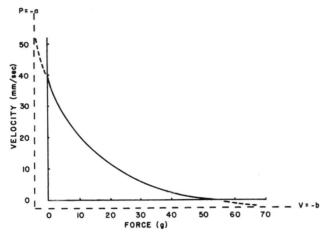

Figure 5-12
Isotonic load plotted against the initial velocity of shortening. (Adapted from J. A. V. Butler and J. T. Randall. *Progress in Biophysics,* Vol. 4. New York: Academic, 1954. P. 306. Reproduced with permission.)

where F_o is the isometric tension, v is the velocity of shortening, and b is a constant. Restated, the velocity of shortening is a function of the force of contraction:

$$(F + a)(v + b) = \text{constant} = (F_o + a)b \qquad (5\text{-}54)$$

This form of Hill's equation yields a rectangular hyperbola when force is plotted against velocity of shortening (Fig. 5-12). The asymptotes of this hyperbola are $F = -a$ and $v = -b$.

The constant a, as was shown, is a thermal constant. Recently, it was found that the heat of shortening is not only a function of distance but also one of the load. This finding makes the determination of a from heat measurements problematical. The use of the Hill equation for the formal description of the force-velocity relationship, however, is still correct.

The theoretical considerations underlying the force-velocity relationship imply the existence of the series elastic component.

A consequence of Hill's analysis is that the more rapidly a muscle shortens, the less force it can generate. The sliding filament model explains this finding by the contention that the faster the fibers slide by one another, the less time, and therefore less opportunity, will be available for cross-bridge formation.

A number of theoretical and molecular models attempting to summarize the known facts about muscle have been proposed. Their detailed discussion is beyond the scope of this book. A prototype is A. F. Huxley's model (*Progress in Biophysics,* 17, 1957) and some of its more recent modifications. These and other models are continuously under review as new findings come to light.

REFERENCES

1. Braunwald, E., Ross, J., and Sonnenblick, E. H. *Mechanism of Contraction of the Normal and Failing Heart.* Boston: Little, Brown, 1968. Pp. 31--48.
2. Goldman, D. E. Potential, impedance, and rectification in membranes. *J. Gen. Physiol.* 27:37, 1943.
3. Hodgkin, A. L. The Croonian lecture: Ionic movements and electrical activity in giant nerve fibers. *Proc. R. Soc.* [Series B] 148, 1958.

SELECTED READINGS

Aidley, D. J. *Physiology of Excitable Cells.* New York: Cambridge University Press, 1971

Ashley, C. E. Mechanisms of Muscle Contraction. In E. E. Bittar (Ed.), *Cell Biology in Medicine.* New York: Wiley, 1973.

Bourne, G. H. *The Structure and Function of Muscle* (2nd ed.). Vol. 3, Physiology and Biochemistry. New York: Academic, 1973.

Dowben, R. M. The Mechanical Latent Period of Frog's Striated Muscle at $0°C$. In *General Physiology (A Molecular Approach).* New York: Harper and Row, 1969.

Hosek, R. S. An experimental and theoretical analysis of effects of volume conduction in a non-homogeneous medium on scalp and cortical potentials generated in the brain. Ph.D. Dissertation, Marquette University, Milwaukee, Wis., 1970.

Huxley, H. E. The mechanism of muscular contraction. *Science* 164, 1969.

Lehninger. A. L. *Biochemistry.* New York: Worth, 1970.

Lowey, S. Protein Interactions in the Myofibril. In *Polymerizations in Biological Systems.* Amsterdam: Associate Scientific Publishing, 1972.

Passmore, R., and Robson, J. S. *A Companion to Medical Studies.* Oxford: Blackwell, 1968.

Plonsey, R. *Bioelectric Phenomena.* New York: McGraw-Hill, 1969. Chapter 5.

Stevens, C. F. *Neurophysiology: A Primer.* New York: Wiley, 1966.

THE CARDIOVASCULAR SYSTEM   II

11 THE CARDIOVASCULAR SYSTEM

D. D. Michie, J. Kline, and E. O. Attinger

OVERALL FUNCTION AND ORGANIZATION

In terms of mass and heat transport, the cardiovascular system represents the major transport system of the body. It also serves as the primary channel for the information flow that is carried by humoral substances. Its involvement in metabolic processes includes the gas and fluid exchange between the external and internal environment, the uptake of food from the gastrointestinal tract, the elimination of nongaseous metabolites through the kidney, and the dissipation of heat from various surfaces. Adequate blood flow to the various organs is, therefore, of primary importance for the maintenance of normal biologic functions and homeostasis.

The uniquely integrated components of the system essentially consist of the heart and a vast network of blood vessels. The heart functions as a pump to supply the propelling energy for the flow of blood. The blood vessels, on the other hand, function as conduits to deliver the blood to various portions of the body and to return it to the heart.

The heart, from a functional standpoint, may be divided into two parts: a volume pump and a pressure pump. Each of these pumps is composed of two chambers: the *atrium* and the *ventricle*. The unidirectional flow of blood into and out of the pumping chambers is rigidly maintained by a series of one-way valves. The volume pump, or right ventricle, pumps unoxygenated blood throughout the pulmonary bed, which is a low-resistance, high-compliance system. On the other hand, the pressure pump, or left ventricle, is responsible for pumping oxygenated blood throughout the systemic circulation. Figure 6-1 shows the series arrangements of these components.

During each heart cycle, the chambers of the heart are periodically filled and emptied. During *diastole,* the filling period, the cardiac muscles are relaxing; during *systole,* the emptying period, the muscles are contracting.

Figure 6-2 shows the direction of blood flow through the heart and the major components of the pump itself. Unoxygenated blood enters the right atrium and passes through the tricuspid valve into the right ventricle. The tricuspid valve, a unidirectional valve, prevents the reflux of blood from the right ventricle back into the right atrium during the time the pumping chamber contracts. When the right ventricle contracts, blood passes through the pulmonary, or pulmonic, valve into the pulmonary arteries and subsequently into the pulmonary vascular bed. The blood is oxygenated during its passage through the lungs and returns via the pulmonary veins to the left atrium (Fig. 6-3). The oxygenated blood passes from the left atrium through the mitral, or bicuspid, valve into the left ventricle.

The bicuspid and tricuspid valves are also referred to as the *atrioventricular (AV) valves.* They are operated primarily by pressure differentials between an atrium and

103

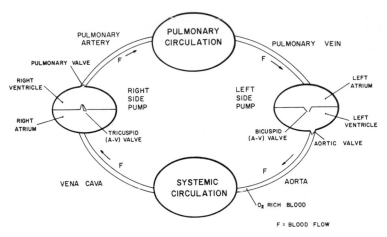

Figure 6-1
Series circuit of cardiovascular system.

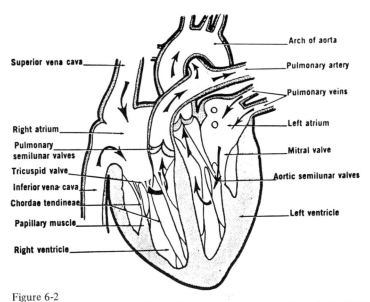

Figure 6-2
Chambers, valves, and blood flow directions in the heart. (From R. M. DeCoursey. *The Human Organism* [4th ed.]. New York: McGraw-Hill, 1974. Reproduced with permission of McGraw-Hill Book Company.)

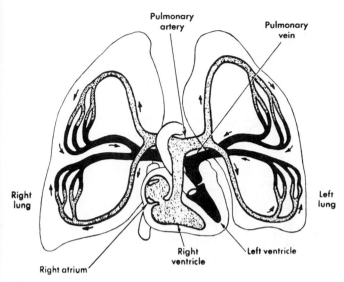

Figure 6-3
Pulmonary circulation. (From L. L. Langley and E. Cheraskin, *Physiology of Man* [3rd ed.].
Copyright © 1965 by Litton Educational Publishing Inc. Published by Van Nostrand Reinhold
Company.)

the ventricle. Reversal of the AV valves is prevented by the chordae tendineae, of
which one end is attached to the valves and the other to the papillary muscles (see
Fig. 6-2).

During contraction of the left ventricle, blood is forced through the aortic valve
into the aorta and is distributed throughout the systemic, or greater, circulation via
a network of vessels referred to as the *arterial tree* (Fig. 6-4). The forces created by
the heart develop pressures throughout the cardiovascular system that differ with
location. These pressures at a given time are determined by the contractile properties
of the heart muscles and the state of the vessels. They are influenced by neural and
humoral mechanisms, dynamic feedback from chemical (chemo-) and pressure (baro-)
receptors in the cardiovascular system, and feedback from the kidney-body fluids
system.

As blood leaves the heart, its flow is pulsatile, moving through the major arteries
with a pulse wave velocity [4]. When the smaller arterioles and the capillary bed are
reached, these vessels essentially act as a low-pass frequency filter, and the flow becomes
steady.

The pressures within the right side of the heart and the pulmonary bed are considerably lower than those in the left ventricle and in the systemic circulation. Thus,
the pulmonary circulation has been called the "lesser" circulation and the systemic
circulation the "greater" circulation. These designations refer to the pressures that
are characteristic of these systems and not to the functional importance of the systems.
A discussion of the pressures within the greater and lesser circulations, a dynamic
analysis of cardiac functions, and their physiologic significance is found in Chapter 7.

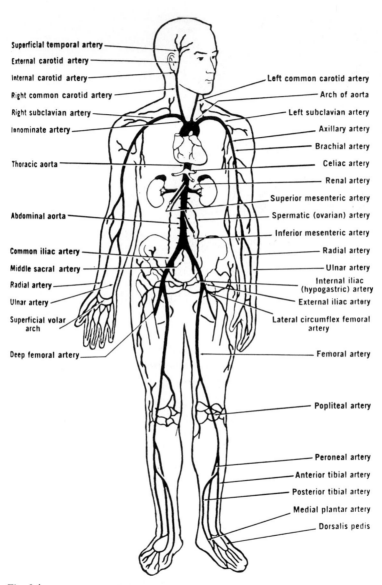

Superficial temporal artery

External carotid artery

Internal carotid artery

Right common carotid artery

Right subclavian artery

Innominate artery

Thoracic aorta

Abdominal aorta

Common iliac artery

Middle sacral artery

Radial artery

Ulnar artery

Superficial volar
arch

Deep femoral artery

Left common carotid artery

Arch of aorta

Left subclavian artery

Axillary artery

Brachial artery

Celiac artery

Renal artery

Superior mesenteric artery

Spermatic (ovarian) artery

Inferior mesenteric artery

Radial artery

Ulnar artery

Internal iliac
(hypogastric) artery

External iliac artery

Lateral circumflex femoral
artery

Femoral artery

Popliteal artery

Peroneal artery

Anterior tibial artery

Posterior tibial artery

Medial plantar artery

Dorsalis pedis

Fig. 6-4
Arterial tree. (From R. M. DeCoursey. *The Human Organism* [4th ed.]. New York: McGraw-Hill, 1974. Reproduced with permission of McGraw-Hill Book Company.)

The conduits or vessels through which the blood is pumped are of several types. The designation given to these structures is based upon their structural and functional importance. Specifically, when blood leaves the heart (either the right ventricle or the left ventricle), it travels through a series of large arteries, small arteries, arterioles, capillaries, venules, and veins. It is then returned to the heart. Figure 6-4 identifies and indicates the location of the major arteries [1], and Figure 11-1 (Chap. 11) shows the organization of the microcirculation [5]. The walls of the major blood vessels are composed of an inner layer, the endothelium, which is surrounded, depending on the type of vessel, to varying degrees by elastic tissue, smooth muscle, and fibrous tissue (collagen).

DISTRIBUTION OF BLOOD FLOW AND OXYGEN CONSUMPTION

The blood in the closed circulatory system does not directly bathe the tissue cells. All exchange between the various cells of the body and the blood occurs across the walls of the very small vessels, i.e., at the level of the microcirculation. Blood does not flow continuously through all parts of any given section of the capillary bed, but it is redistributed sequentially through various parts of the capillary network. This redistribution is regulated by specialized structures, the precapillary sphincters, which consist of smooth muscle tissue.

The cardiac output (CO), the mean flow of blood from the heart in one minute, can be derived from the product of the stroke volume (SV), the blood ejected from one ventricle in one contraction measured in milliliters per beat, and the heart rate (HR) measured in beats per minute:

$$CO = SV \times HR \tag{6-1}$$

Table 6-1 shows the distribution of blood flow, blood volume, and oxygen consumption to the various major organs in comparison with their relative weights for a normal 70-kg man at rest. The adjustment of blood flow according to the varying needs of individual organs is regulated by a hierarchically arranged control system that involves both neural and humoral mediators. The feedback patterns through which this control system operates are only partly known. The mechanisms, on the

Table 6-1. Distribution of Blood to Various Organs for a Normal 70-kg Man at Rest

Percentage	Residual[a]	Muscle	Skin	GI Tract	Brain	Kidney	Heart
Weight	46.0	41.0	5.0	4.0	2.5	1.0	0.5
Blood volume	62.5	10.0	1.5	23.0	0.5	2.0	0.5
Blood flow	5.0	17.0	7.0	27.0	13.0	26.0	5.0
O_2 consumption	22.0	21.0	6.0	22.0	8.5	8.0	12.5

[a]Bone, fat, connective tissue, pulmonary circulation, the heart chambers, and the larger peripheral arteries and veins.

other hand, by which the necessary blood flow is maintained are better understood and, for purposes of analysis, can be divided into two groups:

1. A change in the overall delivery per unit time from the pump, i.e., a change in cardiac output. This may be achieved by altering either the stroke volume, the frequency, or both.
2. A redistribution of blood flow within and between organs. The available flow is directed preferentially into channels that supply those parts of organs most active at a given moment and is partially diverted from regions with lesser metabolic requirements. Such functional shunts can be obtained by selective changes in the cross section of different parallel beds by means of vasomotor regulation.

The relationships between these two groups are extremely intricate and vary under different conditions. Both mechanisms are usually associated with shifts in blood volume from one part of the system to another.

Only a fraction, approximately 25%, of the oxygen that is carried by the blood to the various organs is extracted. Blood returning to the heart from the systemic circuit therefore contains a major portion of the arterial oxygen. A significant reserve is thus established in case oxygen requirements suddenly increase.

Since most metabolic processes ultimately require a supply of oxygen, the oxygen consumption per gram of tissue can serve as a rough indicator of the requirements for blood flow in different tissues. In vitro experiments indicate that the kidney, liver, and brain have a high metabolic activity (0.5 ml O_2/gm wet weight/hour); spleen and muscle at rest (including heart muscle) are moderately active (0.4 to 0.8 ml O_2/gm wet weight/hour); while fat, skin, and bone are relatively inactive (0.3 to 0.2 ml O_2/gm wet weight/hour) [3]. By multiplying these values by the relative weight of the tissues in the whole body, some estimate of the oxygen consumption of individual organs can be obtained, which can be compared to the corresponding distribution of the blood flow and blood volume given in Table 6-1. The percentages in Table 6-1 have been averaged from many sources, and, at best, they only indicate the orders of magnitude. Furthermore, drastic changes in these distributions occur during selective activities of the organs. Nevertheless, it is apparent that the kidney and the skin are over-perfused, whereas the residual regions are considerably under-perfused with respect to their oxygen requirements. Hence, in addition to cellular metabolic requirements, other factors must affect the blood flow distribution, such as the maintenance of constant temperature, the clearance of waste products, and so on.

REFERENCES

1. Goss, C. M. (Ed.). *Gray's Anatomy of the Human Body* (29th ed.). Philadelphia: Lea & Febiger, 1973.
2. Hurst, J. W., and Logue, R. B. *The Heart* (2nd ed.). New York: McGraw-Hill, 1970.
3. Martin, A. W., and Fuhrman, F. A. The relationship between summated tissue respiration and the metabolic rate in the mouse and the dog. *Physiol. Zool.* 28:18, 1955.

4. Noordergraaf, A. *Biological Engineering.* New York: McGraw-Hill, 1969. Chap. 5.
5. Zweifach, B. W. *Functional Behavior of the Microcirculation.* Springfield, Ill.: Thomas, 1961.

SELECTED READING

Burton, A. C. *Physiology and Biophysics of the Circulation.* Chicago: Year Book, 1968.
Conn, H. L., and Horwitz, O. *Cardiac and Vascular Diseases,* Vol. 2. Philadelphia: Lea & Febiger, 1972.
DeCoursey, R. M. *The Human Organism* (3rd ed.). New York: McGraw-Hill, 1974.
Langley, L. L., and Cheraskin, E. *Physiology of Man* (3rd ed.). New York: Reinhold, 1965.
Montagna, W., and Ellis, R. A. *Advances in Biology of Skin.* (Blood Vessels and Circulation, Vol. 2.). New York: Pergamon, 1961.
Reeve, E. B., and Guyton, A. C. *Physical Basis of Circulatory Transport: Regulation and Exchange.* Philadelphia: Saunders, 1967.
Resnekov, L. (Ed.). *The Medical Clinics of North America Symposium on Coronary Heart Disease.* Philadelphia: Saunders, 1973.
Selkurt, E. E. (Ed.). *Physiology* (3rd ed.). Boston: Little, Brown, 1971.

The Heart as a Muscle and a Pump　　　7

D. D. Michie and J. Kline

THE MYOCARDIAL CONTRACTILE SYSTEM

In the cardiovascular system, the heart serves as the central pump and provides the driving force for supplying the power to move the blood through the systemic and pulmonary circulation. It consists of excitable tissue and responds to a variety of electrical, mechanical, thermal, and chemical stimuli. Feedback from neural, chemical, thermal, and pressure sensors provides regulation for the heart rate, cardiac output, and myocardial contractile force. Because of the inherent rhythmicity of the cardiac muscle, the outflow from the pump is pulsatile.

The reserve capacity of the heart insures the capability of providing for a wide range of outputs from about 5 liters per minute at rest to as much as 35 liters per minute during heavy exercise for well-trained athletes. The reliability of the pump is amazing if one considers the staggering amount of work it performs over the lifetime of an individual — some 25 billion heartbeats, each amounting to about 13 million ergs of external work.

Cardiac muscle exhibits primarily the cross-striated structure and the characteristics of skeletal muscle that were discussed in Chapter 5; however, its involuntary functional control properties resemble those of smooth muscle. As will be shown in the next chapter, each cardiac muscle group is related, and the contractions of these muscles are controlled in a sequential manner by an electrical signal as it conducts through the heart tissues.

In similarity to the skeletal muscle cell, the myocardial muscle cell is made up of sarcomeres (from Z line to Z line) that contain interdigitating thick myosin threads (A bands) and thin actin threads (I bands). In contrast to the skeletal muscle cell, however, the cardiac muscle resembles a syncytium with branching interconnecting fibers. As a consequence, a wave of depolarization is followed by contraction of the entire myocardium if a suprathreshold stimulus is applied to the atrium. The myocardial muscle fiber is also characterized by an abundance of mitochondria, which is related to the fact that the cardiac muscle is, for all practical purposes, incapable of operating anaerobically, in contrast to skeletal muscle. Provision of adequate oxygen and substrates for the metabolic machinery is further assured by a rich capillary supply (one capillary per muscle fiber), resulting in very short diffusion distances.

The heart comprises four muscles: the superficial sinospiral, superficial bulbospiral, deep sinospiral, and deep bulbospiral muscles (Fig. 7-1). With the exception of the lateral aspect of the wall of the apex of the ventricle, there is no apparent fiber-orientation rearrangement with an increase in ventricular wall thickness due to compression of the chamber [13]. Thus, an earlier concept that tension or force is stored in the form of intervascular tension between myocardial layers during systole

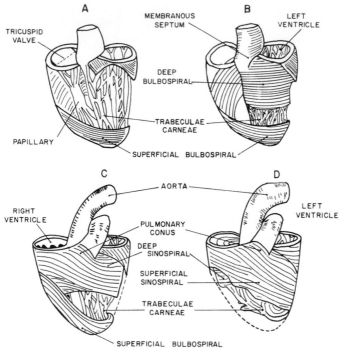

Figure 7-1
Anatomy of the ventricular walls. *A*. The superficial bulbospiral muscle bundles arise principally from the mitral ring and from the external investment for portions of the left and right ventricles as they spiral toward the apex. Emerging from the vortex on the inside of the chambers, these muscle bundles spiral back toward the valve rings either as trabeculae carneae or as papillary muscles, which are joined to the valves through chordae tendineae. *B*. The deep bulbospiral muscle fibers encircle the basilar portions of the left ventricle. *C*. The deep sinospiral muscle encircles both the right and the left ventricular chambers. *D*. The superficial sinospiral muscle is a counterpart of the superficial sinospiral and bulbospiral muscles; it is functionally unimportant. (Modified from R. F. Rushmer, *Cardiovascular Dynamics* [3rd ed.]. Philadelphia: Saunders, 1970.)

and released during diastole may be questioned [8, 9]. Figure 7-2 shows the changes in length of ventricular muscle fibers during systole and during diastole. Figure 7-3 shows a comparison of right ventricular and left ventricular pressures and ejection velocities.

Figure 5-11, which shows a mechanical analog of striated muscle, may be applied to cardiac muscle. Hill's [5] concepts have been applied to the isolated papillary muscle from the right ventricle of the cat. This preparation was selected because it provides a small segment of ventricular muscle with fibers arranged in a linear fashion so that they may be studied under well-controlled conditions. Figure 7-4 shows the system used in these studies. (In this figure, *P* represents the afterload and *dl/dt* the initial velocity of shortening of the muscle.) Preload is somewhat analogous to end-diastolic pressure in the intact ventricle, and it determines the initial muscle length. The afterload is the equivalent to aortic pressure in the intact heart.

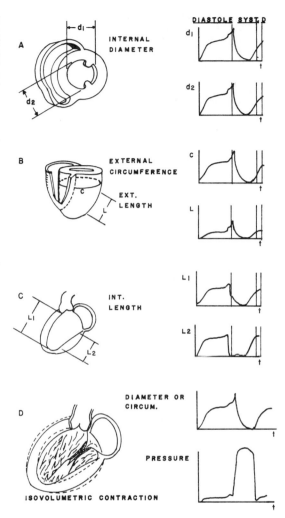

Figure 7-2

Cyclic changes in left ventricular dimensions. During ventricular diastole, all dimensions rapidly increase at first, but the increase is more gradual during the latter part of the filling interval. Atrial contraction adds a slight additional increment of blood. At the onset of ventricular systole, the internal diameters, d_1 and d_2, in part A and the external circumference, c, and length L, in part B abruptly increase because the internal lengths, L_1 and L_2, in C shorten during this interval, which is called *isovolumetric contraction* (D). During the interval when the internal length abruptly diminishes, the external length in part C increases because of the outward bending of the thick-walled ventricle. (From R. F. Rushmer, *Cardiovascular Dynamics* [3rd ed.]. Philadelphia: Saunders, 1970. Reproduced with permission.)

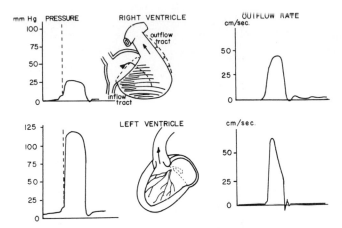

Figure 7-3
Right and left ventricular ejection. In contrast with the left ventricle, right ventricular ejection produces a much more gradual pressure rise to a lower peak ventricular pressure and a more gradual decrease in pressure at the end of systole. Similarly, right ventricular outflow begins earlier, accelerates more gradually to a lower peak velocity, which occurs near midsystole, and decelerates more gradually than the left ventricular outflow. (From R. F. Rushmer, *Cardiovascular Dynamics* [3rd ed.]. Philadelphia: Saunders, 1970. P. 64. Reproduced with permission.)

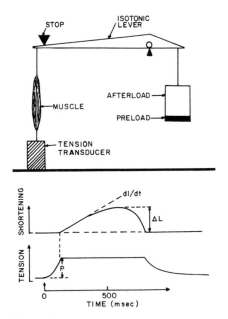

Figure 7-4
Isotonic lever system used for the study of isolated muscle (*top*). Representative recordings of shortening and tension of afterload isotonic contraction (*bottom*), where P is afterload and dl/dt denotes the initial velocity of shortening of muscle. (From E. H. Sonnenblick et al., *Ventricular Function: Evaluation of Myocardial Contractility in Health and Disease*. New York: Grune & Stratton, 1970. P. 451. Reproduced with permission.)

As the muscle contracts, an amount of force is developed to match the afterload. The process in which the muscle develops force without altering its length is called *isometric contraction.* Only if the force of the load is overcome by the contractile force of the muscle can the latter contract; this is termed *isotonic contraction.* If the afterload is increased in increments, the initial velocity of shortening decreases progressively. This relationship is described by the isotonic force-velocity curve for muscle. If the force-velocity curve is extrapolated to zero load, the theoretical maximum velocity of shortening, v_{max}, can be calculated (Fig. 7-5).

There are several factors that alter the force-velocity relationship. Figure 7-6 shows the effect of changes in length (preload) as well as of changes in load (afterload) upon the velocity of shortening. It may be seen from these diagrams that the family of curves is shifted to the right by an increase in preload. Simultaneously, the amount of actively developed force increases, while v_{max} remains essentially unchanged. Positive inotropic agents, such as norepinephrine, calcium, glucagon, and cardiac glycosides, increase both v_{max} and the force developed (Fig. 7-7).

Thus, v_{max} can be used as an indicator of the contractile state, since it is not affected by changes in initial muscle length but is augmented by an increase in the contractile state. The hyperbolic relationship between the velocity of muscle shortening and the load describes the contractile properties of the contractile element when it is activated. During isometric contraction when the contractile element is activated and shortens, the series elastic element is stretched, and the rate of force development, translated to pressure (dp/dt), depends upon (1) the shortening velocity of the contractile element (dl/dt) and (2) the instantaneous stress-strain relationships of the series elastic element (dp/dl). Therefore,

$$\frac{dp}{dt} = \left(\frac{dl}{dt}\right)_{ce} \times \left(\frac{dp}{dl}\right)_{se} \qquad (7\text{-}1)$$

The slope of the curve $(dp/dl)_{se}$ is a function of load p:

$$\frac{dp}{dl} = kp + c \qquad (7\text{-}2)$$

where k and c are constants.

During isometric contraction, v_{ce} can therefore be computed if the following are known: (1) the rate of development of force (dp/dt) and (2) the instantaneous stress-strain relationships of the series elastic element (dp/dl):

$$v_{ce} = \frac{dl}{dt} = \frac{dp/dt}{dP/dl} = \frac{dp/dt}{kP + c} \qquad (7\text{-}3)$$

Since c is very small, v_{ce} for the contractile element is more simply obtained as:

$$v_{ce} = \frac{dp/dt}{kP} \qquad (7\text{-}4)$$

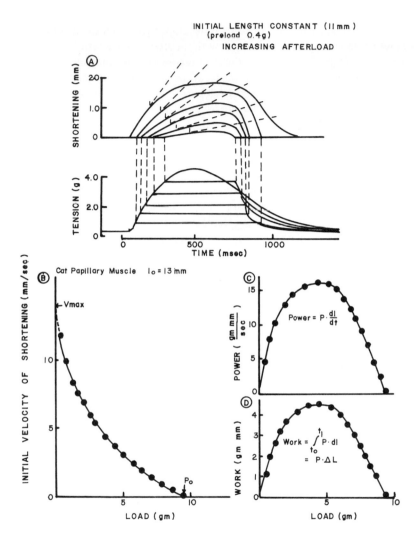

Figure 7-5

A. Superimposed recordings of shortening and tension development from a series of afterloaded isotonic contractions. As contraction starts, the tension rises (*lower panel*). When the tension reaches the level of the load, the muscle shortens (*upper panel*). Dashed lines on the upper tracings indicate the initial velocity of shortening. As the afterload is increased, the velocity and extent of shortening decrease. B. Inverse relation between the initial velocity of isotonic shortening and load (force-velocity relation) of cat papillary muscle. C. Power, which is product of load (P) and the velocity of shortening (dl/dt), as function of load. D. Work, which is product of load (P) and the extent of shortening (l), as function of load. The curves in parts C and D are both calculated from the experimental data shown in part B. (From E. H. Sonnenblick et al., *Ventricular Function: Evaluation of Myocardial Contractility in Health and Disease.* New York: Grune & Stratton, 1970. P. 452. Reproduced with permission.)

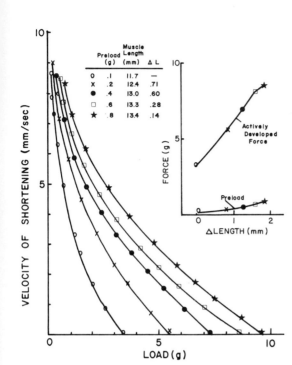

Figure 7-6
Effects of increasing the initial length (by increasing the preload) on the force-velocity relation of cat papillary muscle. Increasing the preload leads to an increase in the actively developed force, but it does not alter the maximum velocity of shortening at zero load (v_{max}). Insert shows the length-tension curves appropriate to these preloads. (From E. H. Sonnenblick et al., *Ventricular Function: Evaluation of Myocardial Contractility in Health and Disease*. New York: Grune & Stratton, 1970. P. 454. Reproduced with permission.)

As seen from Equation 7-4, an increase in dp/dt at any load will occur when there is an increase in (1) the initial muscle length and (2) the contractility, since both of these factors affect dl/dt of the contractile element.

Although the concepts and the results of the experiments described in the previous paragraphs are extremely helpful for the understanding of cardiac dynamics, their application to cardiac contraction is complicated because of the geometrical configuration of the heart. The integrated tension developed by all the myocardial fibers as ventricular pressure as well as the relation between tension and pressure in a hollow spheroid depend on the radii of curvature and the wall thickness. These relationships are not well worked out for the heart, and the mechanical properties of this organ are therefore usually described in terms of a volume-pressure diagram (Fig. 7-8).

The force of ventricular contraction is a function of the initial length of the myocardial fibers. As the ventricles fill with blood, the myocardial fibers are stretched. With an increase in stretch, the heart, within its normal operating range, develops more tension. This relationship between length and tension is called *Frank-Starling's law of the heart* [12] and is depicted in Figures 7-8 and 7-9. It may be seen from

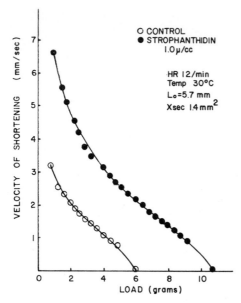

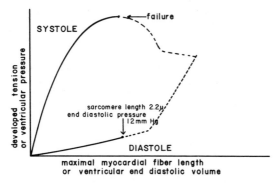

Figure 7-7
Effects of strophanthidin on the force-velocity relation. A symmetrical shift in the curve is induced; both v_{max} and the maximum force of contraction increase. (From E. H. Sonnenblick et al., *Ventricular Function: Evaluation of Myocardial Contractility in Health and Disease.* New York: Grune & Stratton, 1970. P. 455. Reproduced with permission.)

Figure 7-8
Volume-pressure diagram of heart muscle.

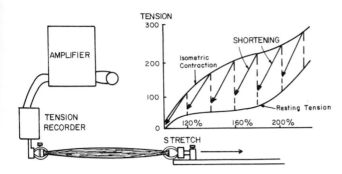

Figure 7-9
Effects of shortening on myocardial contractile tension. The tension above the resting tension point that is developed by myocardial strips contracting under isometric conditions increases progressively from resting length to about 160% to 180% of resting length, then diminishes. If the myocardial strips shorten by 20% during contraction, the contractile tension falls off sharply. (From R. F. Rushmer, *Cardiovascular Dynamics* [3rd ed.]. Philadelphia: Saunders, 1970. P. 87. Reproduced with permission.)

Figure 7-8 that as the fiber length increases, there is an increase in tension developed by the myocardium. The abscissa of Figure 7-8 is also labeled "ventricular end-diastolic volume" because an increase in the length of the myocardial fiber results from an increase in the filling volume. This relationship is valid only up to a critical level. Beyond this point (the dashed line in Figure 7-8), a further increase in fiber length does not result in an increase in tension. Rather, the tension that is developed gradually decreases as the fiber length increases beyond this critical point. When the heart enters this phase, it is considered to be in failure.

In summary, the ascending limb of the curve demonstrates that the heart is able to compensate for an increase in fiber length by developing more tension, whereas the descending limb of the curve demonstrates that the heart is uncompensated and is therefore said to be in failure.

Proper ionic balance is essential for the proper functioning of the myocardium. Disturbances in electrolyte balance are primarily encountered in pathologic states; therefore, only a few examples as to the effects of an imbalance in the electrolytic state will be mentioned. *Hyponatremia* (a decrease in sodium ion concentration) causes a decrease in myocardial contractility due to the fact that some of the myocardial fibers cannot become excited. On the other hand, hypernatremia (an increase in sodium ion concentration) has little or no effect on myocardial contractility. *Hyperkalemia* (an increase in potassium ion concentration) inhibits repolarization and therefore brings about cardiac arrest during diastole. A moderate increase in calcium ion concentration causes a bradycardia of vagal origin; a large increase in calcium ion causes cardiac arrest during systole. The fact that a certain concentration of the calcium ion is essential for proper myocardial contraction has already been observed. The calcium ion, in proper concentration, can also have a positive inotropic effect (Fig. 7-10).

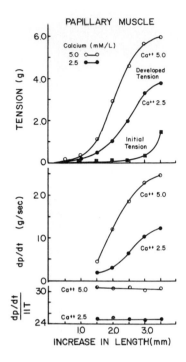

Figure 7-10
Effect of calcium on myocardial time-tension relations (isometric papillary muscle, initial length 10 mm, cross section 0.9 mm, 23°C). (Modified from J. H. Siegel and E. H. Sonnenblick. *Circ. Res.* 12:606, 1963. Reproduced with permission of the American Heart Association, Inc.)

Another attempt at analyzing ventricular function was proposed by Sarnoff [11] in the early 1960s. This analysis of ventricular function is used to correlate the left ventricular end-diastolic pressure with the left ventricular stroke work (Fig. 7-11). An increase in the functioning ability of the heart is reflected by a shift in the curves to the left, whereas myocardial failure is reflected by a shift downward and to the right. In other words, an improvement in cardiac function is demonstrated by an increase in stroke work for a given level of left ventricular end-diastolic pressure.

There are obviously several primary factors that influence the performance of the heart. These can be summarized as follows: (1) *preload* — this term describes the initial fiber length, which is determined by the intraventricular pressure and volume; (2) *afterload* — this is determined (a) by the intraventricular pressure and wall thickness and (b) by the physical properties of the peripheral circulation (the height to which intraventricular pressure must rise during systole will depend on the aortic pressure); and (3) the *contractile state* of the myocardium as reflected by the force-velocity relationship.

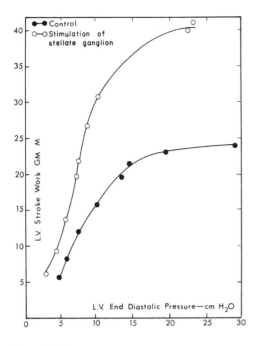

Figure 7-11
Improvement in cardiac function as demonstrated by an increase in stroke work for a given level of left ventricular end-diastolic pressure. (From S. J. Sarnoff and J. Mitchell. In W. F. Hamilton and P. Dow [Eds], *Handbook of Physiology, Section 2, Circulation,* Vol. 1. Washington, D.C.: The American Physiological Society, 1962. P. 509. Reproduced with permission.)

CARDIODYNAMICS

Assessment of myocardial contractility and the performance of the intact left ventricle depends upon the evaluation of preejection and ejection phenomena. The preejection phenomena involve calculations from pressure measurements and are relatively independent of afterload. On the other hand, ejection phenomena relate to fiber shortening and reflect the Frank-Starling mechanism as well as the contractile state of the heart. It may be seen from Figure 7-12A that an increase in the preload (left ventricular end-diastolic pressure) produces an increase in the maximal tension without a change in the v_{ce}. An increase in the contractility of the heart is thus reflected by an increase in the v_{ce}, as shown in Figure 7-12B.

End-diastolic pressure is critically dependent on the amount of blood returning to the heart from the periphery (venous return). Except for transient adjustments, venous return and cardiac output must be equal, as must be the outputs of the left and right ventricles. Cardiac performance thus depends not only on the state of the cardiac muscle but on that of the peripheral circulation as well. Figure 7-13 summarizes the relationships between the right atrial pressure and the cardiac outputs of hypoeffective and hypereffective hearts. As in the case of the ventricular function

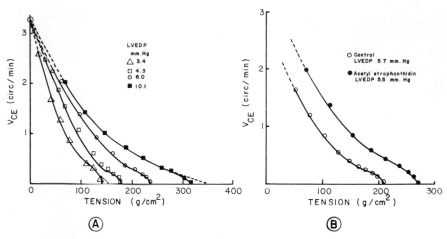

Figure 7-12
Relation between v_{ce} (contractile element velocity) and ventricular wall tension (or stress) in the ventricle of the dog. Each curve is obtained from single isovolumetric contractions. *A.* Effects of increasing the preload (*LVEDP*) on the force-velocity relation. The tension is increased without any change in v_{max} (i.e., the velocity of shortening at zero load). *B.* Effects of acetyl strophanthidin on the force-velocity relations of the normal dog ventricle. Both the v_{max} and the isovolumetric tension are increased. (From E. H. Sonnenblick et al., *Ventricular Function: Evaluation of Myocardial Contractility in Health and Disease.* New York: Grune & Stratton, 1970. P. 458. Reproduced with permission.)

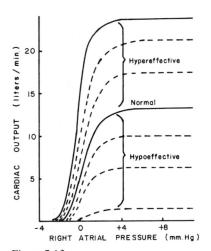

Figure 7-13
Family of cardiac output curves for hypoeffective and hypereffective hearts. (From A. C. Guyton. In *Circulatory Physiology.* Philadelphia: Saunders, 1963. P. 226. Reproduced with permission.)

curves proposed by Sarnoff, the curves that reflect hypereffectiveness of the heart are shifted to the left and above the values for the normal heart. On the other hand, curves that are representative of the hypoeffective heart are shifted to the right and downward from the control values.

The instantaneous pressures in the left ventricle during the cardiac cycle, the corresponding electrical activity of the heart as characterized by the electrocardiogram (ECG), and the associated heart sounds are shown in Figure 7-14. The electrical activity of the heart, which is discussed comprehensively in Chapter 8, precedes the mechanical events. The heart sounds, which may be heard on the body's surface with a stethoscope and recorded by a phonocardiograph, are caused by the movement, oscillations, and turbulence of the blood, as well as by the action of the heart valves and walls.

The pressure in the left ventricle (curve *A*, Fig. 7-14) at the start of systole rises from practically zero to a value slightly greater than the atrial pressure at point *E* when the AV valves close. Then the pressure continues to rise during a phase of iso-volumetric or isometric contraction of the ventricle to *D*. At this point, the aortic valve opens and the first heart sound occurs. This sound, which travels in all directions

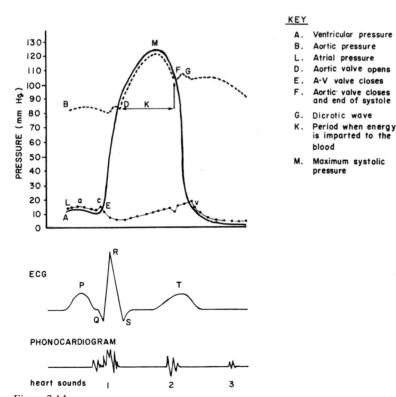

KEY

A. Ventricular pressure
B. Aortic pressure
L. Atrial pressure
D. Aortic valve opens
E. A-V valve closes
F. Aortic valve closes and end of systole
G. Dicrotic wave
K. Period when energy is imparted to the blood
M. Maximum systolic pressure

Figure 7-14
Instantaneous pressure curves during the cardiac cycle, with corresponding electrocardiogram (ECG) waves and heart sounds.

through the chest, is caused primarily by the abrupt closure of the atrioventricular valves and is produced by vibrations of the valve leaflets and of the surrounding fluid. These vibrations are low in pitch; they begin with the onset of isometric contraction and last into the early portion of the period of maximum ejection. The left ventricular pressure rises to a maximum of 120 mm Hg, which occurs at point M (maximum systolic pressure), and then begins to drop.

At point F, when the ventricular pressure falls just below the aortic pressure, the aortic valves close, and a second heart sound, a rapid "snap," is produced that indicates the closure of the aortic and pulmonic valves and the slight reverse of blood flow in the aorta. This point corresponds to the onset of the diastolic period. The pressure continues to drop toward zero as the ventricles undergo a phase of isovolumetric relaxation. The AV valves then open and the ventricles begin to fill. There is also an atrial sound that is associated with the flow of blood into the ventricle. A third heart sound is associated with the flowing of blood into an almost-filled ventricle. These sounds can usually be recorded with either an intracardiac phonocatheter or precordial microphones, but they can be heard by direct auscultation using a stethoscope only with great difficulty. For most mammals, the waveforms and the magnitude variations in both blood pressure and the ECG over the cardiac cycle are similar.

As one looks at the changes in intraatrial pressure during the cardiac cycle (curve L, Fig. 7-14), three rather prominent waves become apparent. These are referred to as the A, C, and V waves. The A wave is caused by atrial contraction and slightly augments the intraventricular end-diastolic volume. Once again it will be noted that the mechanical event of atrial systole follows the electrical event — depolarization of the atria — as represented by the P wave on the ECG. The C wave occurs as the ventricles begin to contract. This rise in intraatrial pressure is produced in part by the bulging of the AV valves into the atria because of an increasing pressure in the ventricles. The V wave occurs near the end of ventricular systole. This slight elevation in pressure results from the slow buildup of blood in the atria and a subsequent pressure rise due to the fact that the AV valves are themselves closed at this time.

For the dog, the normal resting heartbeat ranges from 70 to 120 beats per minute, whereas in man, it ranges from 60 to 80 beats per minute. The instantaneous pressure curve for the right ventricle has the same form as the one for the left ventricle shown in Figure 7-14. The pressure levels in the right ventricle, however, are considerably lower (see Fig. 7-3).

The external work performed by the heart per minute is expressed as the integral of the product of instantaneous ejection pressure and expelled volume. The mechanical efficiency of the heart is equal to the external work done divided by the energy expenditure. The external, or useful work, is the energy imparted to the blood by the ventricles and is composed of two parts: pressure energy and velocity energy. These are expressed in the following equations:

$$\text{Pressure energy} = \int_{V_1}^{V_2} (P)\, dV \tag{7-5}$$

$$\text{Velocity energy} = \int_{V_1}^{V_2} \frac{\rho v^2}{2} \, dV \tag{7-6}$$

where P = ventricular pressure during ejection
   V_1 = ventricular end-diastolic volume
   V_2 = ventricular end-systolic volume
   ρ = density of blood
   v = velocity of blood at the outflow valve

In 1958 Sarnoff [10] demonstrated the fallacy of taking external work as an index of myocardial energy expenditure when both the stroke volume and the peripheral resistance change in different proportions and in opposite directions. The transformation of chemical to mechanical energy by the contractile machinery of the heart is reflected only to a small extent by the external work; most of it is dissipated as heat. As discussed previously, elastic structures arranged in series and parallel with the contractile elements attach the muscle to the external load. Activation of cardiac muscle causes the contractile elements to shorten and the series elastic elements to stretch, thereby developing tension on the muscle and performing internal work. The muscle can perform external work only if the afterload is such that the force developed by the contractile elements allows shortening of the muscle as a whole. Thus, the total mechanical work of the contractile elements equals the internal work (tension development) plus the external work (that done in moving blood).

The magnitude of internal and external work varies with the load. For example, if there is no external load, there is no external work performed, and the internal work will be minimal, with little shortening of the elastic elements. If the afterload is very high, the external work is almost nil, but the internal work becomes significant because of significant tension development. This occurs during isovolumetric and isometric contraction. Under conditions of moderate afterload, the contractile elements perform both internal and external work.

Sarnoff and his co-workers reported that there was a good correlation between left ventricular oxygen consumption (V_{O_2}) and the product of mean ventricular ejection pressure and the duration of ejection. This is referred to as the *tension-time index* (TTI). The factors that contribute to the overall oxygen consumption of the pumping heart are (1) the energy utilized to maintain the sodium/potassium ratio, (2) the excitation-contraction coupling, i.e., the energy of activation, and (3) the mechanical or contractile activity. This last category includes the performance of internal work when the contractile elements shorten, the energy expended to eject blood (external work), and the energy associated with relaxation (this is very small, approximately 9%). The factors that regulate the oxygen consumption per beat by the heart have been described by Badeer [2] and are shown in the following equation:

$$\frac{V_{O_2}}{\text{beat}} = \text{resting } V_{O_2} + \text{activation } V_{O_2} + \text{contraction } V_{O_2} + \text{relaxation } V_{O_2}$$

The component "contraction V_{O_2}" is made up of the V_{O_2} required to develop tension (internal work) plus the V_{O_2} required to eject blood (external work).

The work done per beat by the left ventricle in ejecting blood may be expressed as follows:

$$W = P \int dV + \tfrac{1}{2}mv^2 + E_m \qquad (7\text{-}7)$$

where $P \int dV$ = the work done to overcome the arterial pressure
(P is the maximum pressure during one stroke and the pressure curve is considered to be rectangular; $\int dV$ is the stroke volume)
$\tfrac{1}{2}mv^2$ = the kinetic energy required to accelerate the blood
E_m = the metabolic and mechanical energy requirements of the musculature during the contraction process

The external work done and the power output by the entire heart under resting conditions for man can be calculated by considering the systolic pressure, P_{sys}, to be 120 mm Hg, or 160×10^3 dynes per cm^2, and the stroke volume to be 70 ml at a heart rate of 72 beats per minute. For the left ventricle, the calculation of work is:

$$\begin{aligned} P_{sys} \int dV &= (1.60 \times 10^5)(70) \\ &= 1.12 \times 10^7 \text{ ergs} \\ &= 1.12 \text{ joules} \end{aligned}$$

Since the resting heart rate is assumed to be 72 beats per minute, the corresponding power delivered by the left ventricle to overcome arterial pressure is:

$$\text{Power} = \frac{1.12}{(60/72)} = 1.34 \text{ watts} \qquad (7\text{-}8)$$

The pressure in the right ventricle is approximately one-sixth of that of the left. Since the average volume ejected by the right ventricle must be the same as that of the left, the power required to overcome the pulmonary artery pressure is $(1/6)(1.34) = 0.223$ watt. The total PV power becomes:

$$\begin{aligned} \text{Power } (PV) &= \text{power}_{sys} + \text{power}_{pul} & (7\text{-}9) \\ &= 1.34 + 0.223 & (7\text{-}10) \\ &= 1.563 \text{ watts} & (7\text{-}11) \end{aligned}$$

For a 10-kg dog under basal conditions, an average maximum systolic pressure would be 120 mm Hg and an average stroke volume could be 14 ml. Using these figures, the external work done by the left ventricle in overcoming arterial pressure during systole is 0.224 joule. Based upon a heart rate of 100 beats per minute, the corresponding power is 0.373 watt. The power for the right ventricle is $(1/6)(0.373) = 0.062$ watt. The total power, then, is 0.435 watt.

Under resting conditions, the theoretical work done by the heart to overcome arterial pressure, when calculated on the basis of blood pressure and stroke volume, accounts for only approximately 95% of the external work done by the heart. Additional work must be done to accelerate the blood. For one ventricle, the kinetic energy, *KE*, required to accelerate the blood is:

$$KE = \frac{m\bar{v}^2}{2} \qquad (7\text{-}10)$$

where *m* = the mass (in grams) of blood ejected during systole
$\bar{v}$ = the average velocity (cm/sec) of the blood ejected

For both ventricles, the energy is approximately twice this amount. During heavy exercise, this kinetic energy factor may add approximately 20% to the total theoretical value for the static work calculated above.

Under severe conditions of heavy exercise, the maximum systolic pressure in the human may exceed 200 mm Hg and the heart rate may increase to 150 beats per minute. If we consider the stroke volume to be nearly the same as at resting conditions, then to overcome the arterial pressure (afterload), the required power to deliver the external pressure energy becomes about 6.0 watts. Adding the power provided by the right ventricle results in a total of 7.0 watts. Taking into account the kinetic-energy power factor of 20% results in a total of about 8.4 watts.

For the dog under similar conditions of exercise stress and a heart rate of 180 beats per minute, the power required to overcome the afterload for the left ventricle is 1.68 watts and for the right ventricle, 0.28 watt. The total power, including that required for blood acceleration, is about 3.00 watts.

The mechanical efficiency of the heart, i.e., the mechanical work done divided by the total energy transformed, is quite low (3% to 15%). It rises significantly when the external work is increased, but it never exceeds 10% to 15%. Therefore, under exercise conditions, the heart must be capable of supplying a total power of at least 9.00 watts for man and 3.00 watts for the dog to perform its total mechanical work. The power levels required for the external work are certainly within an order of magnitude that makes it feasible to consider the use of mechanical assist devices for diseased or damaged hearts.

CARDIOVASCULAR ASSIST DEVICES

In diseases hearts with weakened contractile forces and in cases of reversible or irreversible cardiac failure, systems are available to assist the mechanical action of the organ or to replace the natural heart. When reversible cardiac arrest occurs and other therapy (such as drugs or oxygen) fails or is unavailable, it is possible to revive the heart by mechanical massage. With the chest closed, pressure may be applied to the area over the heart, either manually or with a device that exerts forces to alternately compress and relax the thoracic cavity and thus, indirectly, to activate the ventricles.

If external massage is ineffective, mechanical massage may be applied directly to the heart muscles either by hand or by using a myocardial prosthetic system (MPS) [1]. To effect direct massage, first the chest is opened by performing a thoracotomy with the incision being made either between the ribs or through the sternum. The MPS consists of a prosthetic device (Fig. 7-15A) that may be cupped around the heart. The device has two sections: an inner liner, which is relatively flexible, bonded to an outer shell, which is also flexible but more rigid. Gas (CO_2 or air) pressure is programed between the two sections, which causes the inner liner to compress the heart, forcing blood from the ventricles. Alternately, the gas pressure is removed, which relaxes the inner liner, allowing the heart to fill. The pneumatic system is electronically controlled

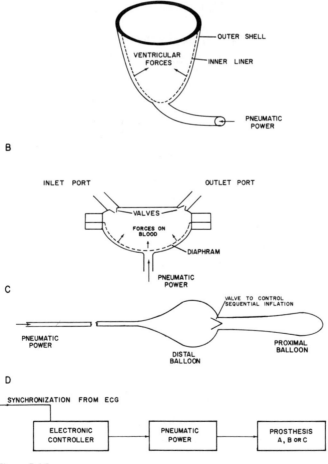

Figure 7-15
Cardiac assist devices.

and synchronized with the electrocardiogram (ECG). (Chapter 8 offers a comprehensive account of the ECG.) The heart rate and periods that correspond to diastole and systole may be varied. This device is available for clinical use.

In cases of weakened hearts or diseased hearts with acute myocardial insufficiency, it is clinically possible to apply, on a temporary basis, a single-ventricle bypass device. This device (Fig. 7-15B), which is composed of a chamber, a diaphragm, and inlet and outlet ports with valves, is also powered pneumatically. The chamber may be connected into the cardiovascular system in several ways. In a typical application, the outlet is anastomosed (connected) to the aorta at the location of the arch, and the inlet is anastomosed to the left atrium. After the chamber is filled with blood, gas pressure moves the diaphragm, forcing the blood to be ejected from the outlet port at a time when the inlet port valve is closed. When the gas pressure is removed, the diaphragm relaxes, the valve on the outlet port is closed, and the chamber is allowed to fill. The cycling and the period of the pneumatic pressure are controlled electronically and are synchronized with the P wave of the ECG. When the natural heart is in diastole, blood from the chamber is ejected into the aorta. During systole, blood is drained from the left atrium. The pumping action parallel to the left heart relieves stresses and strains on the left ventricle during systole, thus making it easier for the diseased or damaged myocardium to recover.

An intraaortic balloon-type diastolic cardiac augmentator is available for clinical use that does not require a thoracotomy or other major surgery for its application [3]. The balloon (Fig. 7-15C) is constructed over a 3-mm diameter catheter and is threaded into the central aorta from the femoral artery. The balloon is positioned in the descending thoracic aorta distal to the subclavian artery. The device consists of two chambers: (1) a spherical, distal balloon that inflates during the onset of diastole and occludes the aorta and (2) a cylindrical, proximal balloon that inflates immediately afterward and forces the flow of blood in the aorta at a time when the aortic valves are closed. Both chambers are completely deflated during systole. This action causes an increase in the diastolic pressure, thus increasing the coronary blood flow and decreasing the afterload, which in turn decreases the work of the left ventricle. The deflation is synchronized with the QRS complex of the ECG during left ventricular contraction. The gas is therefore pushed in and out of the balloon in synchrony with the heartbeat.

In cases of completely irreversible heart failure, it may soon be possible to employ a fully implantable artificial heart system. At the present time, several research groups are working on the development of a system of this kind [1, 6]. Basically, such a device would consist of a prosthesis that has two chambers simulating the left and right sides of the heart, a device to provide the pneumatic or hydraulic power, an electronic controller, and an energy source, which may be either electrical or nuclear.

The artificial heart devices that are available today consist of an implantable prosthesis with the major part of the support system remaining external to the body. Research is being conducted on the development of a fully implantable artificial heart system using a nuclear-powered energy source [4] for both the prosthesis and the electronic controller. It is predicted that within five to ten years, an artificial heart system for implantation will be available for clinical use.

REFERENCES

1. Aldrich, J. G. A Totally Implantable Electrically Actuated Left Ventricular Assist Device (LVAD): In Vivo Testing. In *Proceedings of the 26th Annual Conference on Engineering in Medicine and Biology,* Vol. 15. Arlington, Va.: Alliance for Engineering in Medicine and Biology, 1973.
2. Badeer, H. S. Work and energy expenditure of the heart. *Acta Cardiol.* (Brux.) 24:227, 1969.
3. Gregman, D., Kripke, D. C., and Goetz, R. H. The Effect of Synchronous Unidirectional Intra-aortic Balloon Pumping on Hemodynamics and Coronary Blood Flow in Cardiogenic Shock. In *Proceedings of the American Society for Artificial Internal Organs,* Vol. 26. Washington, D.C., 1970. P. 439.
4. Cole, D. W., Holman, W. S., and Mott, W. E. Status of the USAEC's Nuclear-Powered Artificial Heart. In *Proceedings of the American Society for Artificial Internal Organs,* Vol. 19. Washington, D.C., 1973. P. 537.
5. Hill, A. V. Heat of shortening and dynamic constants of muscle. *Proc. R. Soc. Lond.* (Biol.) 126:136, 1938.
6. Kito, Y., et al. Recent Results in Total Artificial Heart. In *Proceedings of the American Society for Artificial Internal Organs,* Vol. 19. Washington, D.C., 1973. P. 573.
7. Kline, J. Biomedically Engineered Myocardial Prosthetic Systems. In *International Conference Proceedings of IEEE,* Vol. II, Part 9. New York, N.Y., 1963. Pp. 141–147.
8. Rushmer, R. F. The mechanics of ventricular contraction – A cinefluorographic study. *Circ. Res.* 4:219, 1951.
9. Rushmer, R. F. Factors influencing stroke volume: A cinefluorographic study of angiography. *Am. J. Physiol.* 168:509, 1952.
10. Sarnoff, S. J., Braunwald, E., and Stainsby, W. N. Determination of duration and mean rate of ventricular ejection. *Circ. Res.* 6:319, 1958.
11. Sarnoff, S. J., and Mitchell, J. The Control of the Function of the Heart. In W. F. Hamilton and P. Dow (Eds.), *Handbook of Physiology, Section 2, Circulation,* Vol. 1. Washington, D.C.: The American Physiological Society, 1962.
12. Siegel, J. H., and Sonnenblick, E. H. Myocardial contractility. *Circ. Res.* 12:597, 1963.
13. Streeter, D. D., Jr., et al. Fiber orientation in the canine left ventricle during diastole and systole. *Circ. Res.* 24:339, 1969.

SELECTED READING

Brehnan, K., and Kline, J. Prevention of deep-vein thrombosis due to stasis. *IEEE Trans. Biomed. Eng.* 21:232, 1974.

Guyton, A. C. *Circulatory Physiology: Cardiac Output and Its Regulation.* Philadelphia: Saunders, 1963.

Hamilton, W. F., and Dow, P. (Eds.). *Handbook of Physiology, Section 2, Circulation,* Vol. 1. Washington, D.C.: The American Physiological Society, 1962. P. 509.

Kline, J. Regulation and Control of Cardiac Assist Devices and Artificial Hearts. In *Proceedings of the 8th Annual IEEE Region 3 Conference,* Huntsville, Ala., Nov., 1969. No. 69, c46. Pp. 41–42.

Electrical Activity of the Heart 8

J. Kline and D. D. Michie

CARDIAC CONDUCTION

The action of the heart is excited by an electrical signal, a repetitive pulse that is comparable to the output of a free-running multivibrator. This signal originates in the sinoatrial (SA) node, which is also known as the *pacemaker*. The SA node is located in the upper portion of the right atrium. The heart has a well-defined conduction path through which the signal generated by the SA node propagates. First, the electrical impulse spreads from the SA node throughout the nodes of the atria at a rate of approximately one meter per second. It then converges at the atrioventricular (AV) node. The electrical impulse does not travel immediately to the conduction system of the ventricles; instead, it pauses briefly at the AV node. After a time delay, the impulse spreads to the atrioventricular bundle or bundle of His, then to the bundle branches, the Purkinje network. Finally, the impulse traverses the myocardium from the endocardial surface to the epicardial surface of the ventricles. The specialized network of the ventricles is characterized by a rate of impulse propagation (4 to 5 meters per second) that is much faster than elsewhere in the heart. Consequently, the intraventricular septum is excited first. The impulse then appears at the apex of the heart and spreads within the ventricular walls toward the base of the ventricles. This accounts for the "wringing motion" of the heart contraction, and it is probably important insofar as ejection of blood from the ventricles is concerned.

Figure 8-1 shows the sequential development with time of the action potentials in the various sections of the heart as an electrical impulse travels through it. The rate of impulse conduction through the atrium, approximately one meter per second, is slower than that of the ventricles, which is due in part to the lack of any discrete pathway. There are other areas of the heart that will also depolarize spontaneously, but these will do so at a much slower rate than the sinoatrial node.

Although the heart will respond to an externally applied stimulus by contracting, the nervous and muscle tissues possess an inherent, spontaneous rhythmicity that allows the heart to continue beating without the influence of externally applied stimuli. Following the application of a stimulus of at least threshold intensity to the myocardium, the muscle goes through a sequence of events that is similar to that observed in skeletal muscle following similar stimulation. In other words, following depolarization, an absolute refractory period occurs and is followed by a relative refractory period. The absolute refractory period is of particular importance. It lasts through the period of contraction; therefore, it is impossible for a second stimulus to become effective during the period of contraction. For this reason, summation of contraction, a characteristic of skeletal muscle, is not possible in cardiac muscle. The myocardium will not, therefore, undergo a *tetanic* or sustained contraction.

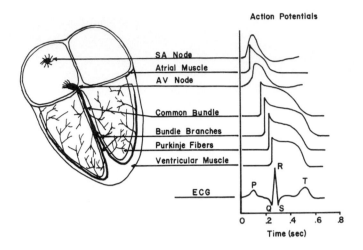

Figure 8-1
Action potentials from distinct neuromuscular regions of the heart. (Modified from the Ciba Collection of Medical Illustrations.)

THE ELECTROCARDIOGRAM (ECG)

The summation of the individual action potentials generated by the various segments of the heart's conduction circuit during a cyclic interval results in a signal that is displayed by the electrocardiogram (ECG). The waveform of this periodic ECG signal is shown in Figure 8-1. At the location of the ventricles, its peak value is in the order of 25 mV. The source of the electrocardiographic signal is generated and contained within a conducting medium, the heart, as well as the entire body. A system of this kind is termed a *volume conductor* [8]. Through volume conduction, the ECG potentials appear at the surface of the heart and also on the surface of the body at the skin. At the skin surface, the peak ECG potentials are in the order of 1.0 mV in a normal adult.

Electrodes may be applied to the skin's surface to detect these potentials, and the signals may then be amplified and recorded. Figure 8-2 shows the time intervals in a typical clinical ECG recording for a normal adult. The space-volume conduction relationship between the potential source generated by the integrated components of the heart's action potentials (see Fig. 8-1) and a measuring point p on the skin surface having the potential $V(p)$ is:

$$V(p) = \rho_t G \Omega \phi \tag{8-1}$$

where $V(p)$ = potential at any point p
ρ_t = a constant representing tissue resistivity
G = a geometrical constant

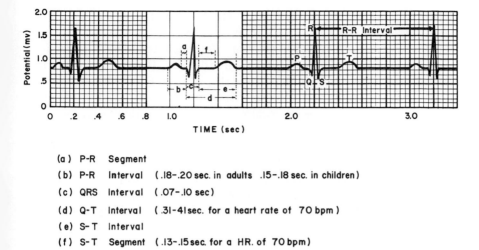

(a) P-R   Segment
(b) P-R   Interval   (.18-.20 sec. in adults .15-.18 sec. in children)
(c) QRS   Interval   (.07-.10 sec)
(d) Q-T   Interval   (.31-41sec. for a heart rate of 70 bpm)
(e) S-T   Interval
(f) S-T   Segment   (.13-.15 sec. for a HR. of 70 bpm)

Figure 8-2
Typical electrocardiogram (ECG) for a normal adult.

The quantity Ω is the solid angle subtended between the measuring point, p, and the area covered by the depolarizing wave front. The value of ϕ represents a voltage that is proportional to the charge density per unit solid angle.

Equation 8-1 states that the potential at the measurement point increases as the solid angle, Ω, increases. The solid angle may be thought of as being the apparent size of an object when viewed from the measurement point, and thus moving closer to the object increases the solid angle. This simply means that the potential increases as the measurement point gets closer to the source.

It is important to remember that a surface potential is generated only when cells are active and actually undergoing the dynamic process of depolarization. As pointed out in Chapter 5, a nerve axon that is negative inside relative to the external interstitium does not generate an external potential as long as it remains in the static state. Thus, the ECG potentials measured on the surface of the skin represent dynamic changes (depolarizations) that are occurring at various locations and magnitude levels within the heart.

The principal components of the ECG are the P wave, the QRS complex, and the T wave. These represent the events of atrial depolarization, ventricular depolarization, and ventricular repolarization, respectively. The U wave, a small positive deflection that is occasionally observed following the T wave, is caused by an afterpotential, possibly from a papillary muscle.

The magnitude of the waves, the complexes, and the intervals between the waves are all important for the interpretation of the electrical activity of the heart. The P-R interval represents the time between atrial depolarization and ventricular depolarization. The S-T segment indicates the interval between the end of ventricular depolarization and the beginning of repolarization. The junction between the QRS complex and the S-T segment is referred to as the *J point*. The duration of the S-T segment is influenced by cardiac rate.

In calculating the electrical axis of the heart, the body is considered to be contained within an equilateral triangle (Einthoven's triangle). The right arm, left arm, and left leg serve as the corners of the triangle (Fig. 8-3). The conventional electrical connections used for recording the ECG are also noted in Figure 8-3. The different types of leads used to record the electrical axis of the heart, its rhythm, and so on, are classified into several general groups: limb leads (Fig. 8-3A), augmented limb leads (Fig. 8-3B), and precordial leads (Fig. 8-3C). The three limb leads designated as I, II, and III are bipolar in that they detect differences in electrical potential between two points on the frontal plane of the body. The algebraic sum of the potential difference at each of these equidistant points at the angles of this triangle is zero. On the other hand, the augmented limb leads may be considered unipolar, since they detect electrical variations at one point with respect to another point that does not vary significantly in electrical potential during contraction of the heart.

The precordial or V leads (see Fig. 8-3) are used in a unipolar recording situation with stationary leads attached to the left arm and leg. The movable, or exploring, lead is applied sequentially to a number of specifically designated intracostal spaces. The

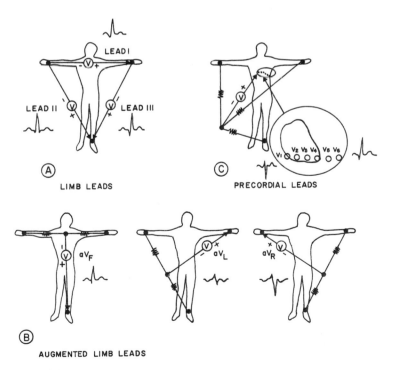

Figure 8-3
Lead placements for electrocardiography. (Modified from the Ciba Collection of Medical Illustrations.)

precordial lead technique creates axes that approximate a horizontal plane through the dipole center. Precordial leads lack the predetermined mathematical relationships to one another that the limb leads have, since the detected potentials depend more strongly on variations in body structure.

Precordial leads are routinely placed at six different chest positions, which are deisgnated as V_1 through V_6, respectively. This method registers the electrical potential directly under the electrode against a "V" or central terminal connection that is made by interconnecting the wires from the right arm, left arm, and left leg. This can be done because the electrical potential of this central terminal does not vary significantly throughout the cardiac cycle.

THE VECTOR CARDIOGRAM

The vector cardiogram is another way to depict graphically the net effect of the different directions that the depolarizations may be taking at a given instant. This instrument is capable of plotting forces in two directions simultaneously. From the three projections (frontal, horizontal, and sagittal), a spatial loop of three dimensions may be visualized. With the aid of computer technology, it is possible to record the magnitude of the spatial vector as well as the projections in the different planes. The development of a vector loop may be seen in Figure 8-4. This figure shows the sequential atrial and ventricular depolarizations and the resultant development of electrical vectors within the heart. The depolarization begins at the sinoatrial node and travels through the atrial myocardium toward the atrioventricular node. The propagation of this impulse results in an electrical wave front that may be treated as a vector. The "dipole hypothesis" treats the resulting electrical potential as a vector. Several other assumptions are incorporated into this hypothesis, and a more detailed account of this technique is given in the literature [6].

The initial atrial depolarization vector indicates the magnitude of the voltage generated by the propagating wave front. Voltage resulting from late atrial depolarization is represented by a second vector, whose length is a measure of the potential generated at a single point in time. By connecting the heads of these instantaneous vectors with their points of origin, the P loop is obtained. A resultant vector can be constructed (see Fig. 8-4) from any two instantaneous vectors with a common point of origin by constructing a parallelogram with the component vectors as two of its sides. The resultant vector is the diagonal formed across the parallelogram from the point of common origin to the opposite corner. The atrial and ventricular depolarization vectors can be analyzed against the Einthoven-triangle reference frame to show the relationship of the resulting vector loop to the standard ECG potentials.

CLINICAL CONSIDERATIONS

The electrocardiogram is one of the important sources of information for the diagnosis and treatment of myocardial infarction, which is due to the lack of blood supply to

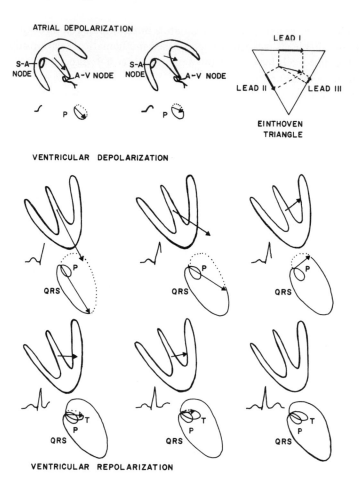

ATRIAL DEPOLARIZATION

S-A NODE — A-V NODE

S-A NODE — A-V NODE

P

P

LEAD I

LEAD II   LEAD III

EINTHOVEN TRIANGLE

VENTRICULAR DEPOLARIZATION

QRS   QRS   QRS

QRS   QRS   QRS

VENTRICULAR REPOLARIZATION

Figure 8-4
Development of the vector cardiogram.

the myocardium. It is by means of continuous ECG monitoring that the arrhythmias that usually complicate infarctions are detected and treated. The waveforms of the ECG that represent several different types of abnormal cardiac activity are shown in Figure 8-5. These arrhythmias may be divided into three general, medically significant groups: minor, major, and lethal. Minor arrhythmias are represented by premature atrial beats (less than six per minute), occasional premature ventricular beats, and wandering pacemakers (where the origin of depolarization in the atrium changes from beat to beat). Some major arrhythmias are atrial tachycardia (rapid heart rate; 140 to 250 beats per minute), atrial flutter with block (due to an open electrical conductive path), slow or rapid atrial fibrillation, premature ventricular beats occurring in conjunction with acute infarction (four to six beats per minute or greater), a vulnerable premature beat, ventricular tachycardia, nodal ectopic beats (abnormal in time and

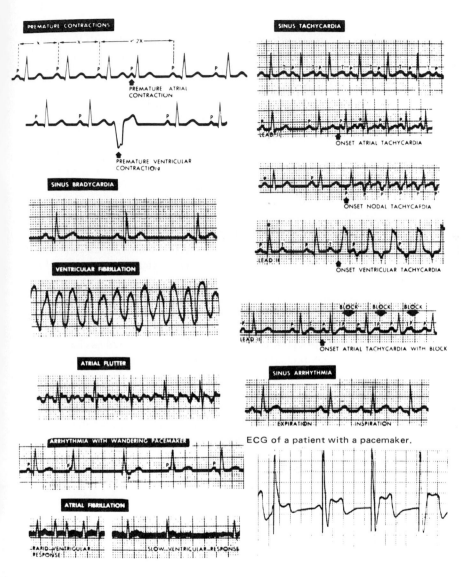

Figure 8-5
Examples of abnormal ECG patterns.

origin; more than six beats per minute), nodal tachycardia (where the AV node becomes the primary pacemaker), sinus arrest (where the primary pacemaker is inhibited or stops), prolonged AV conduction first-degree block, complete AV dissociation (ventricular rate of usually less than 40 beats per minute), bigeminy (double beats in one period), trigeminy (triple beats in one period), quadrigeminy, and so on, as well as multiformed or multifocal ectopic beats. The more prominent lethal arrhythmias are ventricular fibrillation (disorganized electrical and, consequently, mechanical activity of heart with zero cardiac output, which is usually reversible with the application of a high-voltage, short-duration pulse of energy in the order of 75 to 300 watt-seconds to the thoracic area or directly to the heart), cardiac asystole (no contraction), or ventricular standstill (cardiac arrest) [5].

The criteria used for detecting such abnormalities are based on the relative timing and shape of the ECG waveforms. The absolute criterion for a given abnormality in terms of either timing or wave shape is still a matter of study and debate. In general, the morphologic criteria that are relied upon are the QS width, the area of the QRS complex, and the polarity of a given beat relative to the patient's normal polarity. The amplitude of the waveforms is used to detect the presence of atrial or ventricular hypertrophy (overgrowth or enlargement), as well as hypokalemia (subnormal potassium levels in the blood).

Although the exact physiologic implications of a given detected arrhythmia or anomalous beat depend on the patient involved, his specific condition, and the therapy being applied, the following general observations are diagnostically important: (1) Transient changes in rhythm, i.e., premature beats, may be indicative of an increase in the excitability of the tissue such that new foci are generating the depolarization waveforms. This occurs in both the atria and the ventricles. (2) Changes in the QRS complex may also reflect changes in the locus of origin of the depolarization. Changes in the conduction pathways through the heart muscle will be detected as changes in the QRS complex, even though the locus of the origin remains the same. The position and polarity of the T wave in relation to the QRS complex is also a diagnostic indicator, as is the position of the S-T segment.

Extensive utilization of computers for the analysis of abnormal electrical activity is currently being undertaken by several cardiology groups [1, 2] for the diagnosis of cardiac rhythm disturbances and other pathologic conditions.

ELECTRONIC CARDIAC PACEMAKERS

Atrioventricular block is a conduction defect that may be due to various causes. It is manifested by dissociation of the P waves and the QRS complexes in the ECG, and it usually results in a low ventricular contraction rate, low cardiac output, and dizziness. The AV block may be either temporary (lasting up to a few weeks) or permanent. Since the early 1960s, electrical stimulation of the heart with battery-powered electronic pacemakers has been widely used for the treatment of low ventricular rates. The ECG of a patient using a pacemaker (see Fig. 8-5) is modified from the normal standard pattern.

External pacemaker units can be used for temporary control. For long-term treatment, totally implantable battery-powered (and recently radioisotope-powered [7]) pacemakers are used. The stimulating electrodes may be sewn to the epicardium (the outside layer of cardiac muscle), but more frequently these are introduced into the right side of the heart via the venous system. The simplest pacemaker delivers the stimuli in the form of an uninterrupted train of pulses that occur at a fixed rate in the range of 60 to 100 pulses per minute, with an amplitude of 2 to 10 milliamperes (or 2 to 8 volts) and a duration ranging from 0.3 to 3 msec.

When spontaneous ventricular activity is present, "competition" may develop between the natural and artificial electrical foci [4] . If it persists, this competition is not only uncomfortable but is possibly dangerous. For this reason, "noncompetitive" demand or standby electronic pacemakers are available that sense the electrical activity of the ventricles and adjust their own output pulses to avoid the concurrence of competitive stimuli. A third type of pacemaker senses the spontaneous activity of the atrium, which in turn is controlled by the oxygen demand of the body, and the pacemaker's stimuli synchronize the contractions of the ventricles with the atria. This type of pacemaker acts as an artificial AV node and achieves optimization of the cardiac output [4] . Some models are available now in which either the pulse repetition rate, the pulse strength, or both can be varied noninvasively in order to match the pacemaker output characteristics to the needs of the patient and thus conserve battery power.

A fourth type of pacemaker sequentially stimulates the atria and the ventricles. This type is particularly suited for treating patients with slow atrial rates and AV block [3] . The most recent developments in cardiac pacing are related to the control of arrhythmias due to factors other than AV block [10] . The longevity of pacemakers is primarily determined by the availability of power sources. These are usually primary chemical cells. Occasionally, secondary, rechargeable cells or radioisotopic power sources are used.

Recently, experimental automatic implantable defibrillators have begun to receive serious consideration for the prevention of "sudden cardiac death" attributable to ventricular fibrillation [9] .

REFERENCES

1. Beaumont, J. O. On-Line Patient Monitoring System. In *Datamation 50*. Pasadena, Calif.: Technical Publishing, 1969.
2. Budkin, A., and Warner, H. R. Computer Assisted Teaching of Cardiac Arrhythmias. In H. R. Warner (Ed.), *Computers and Biomedical Research*, Vol. 2. New York: Academic, 1968. P. 145.
3. Castillo, C. A., et al. Bifocal demand pacing. *Chest* 59:360, 1971.
4. Furman, S., and Escher, D. J. W. Choice of cardiac pacemaker. *Ann. N.Y. Acad. Sci.* 167:557, 1969.
5. Goldman, M. J. *Principles of Clinical Electrocardiography*. Los Altos, Calif.: Lange Medical Publishers, 1970.
6. Massie, E., and Walsh, J. J. *Clinical Vectorcardiography and Electrocardiography*. Chicago: Year Book, 1960. Pp. 3–16.
7. Parsonnet, V. Editorial: Nuclear powered pacemakers. *Med. Instrum. J.* 7:170, 1973.
8. Plonsey, R. *Bioelectronic Phenomena*. New York: McGraw-Hill, 1969. Pp. 202–275.

9. Schruder, J. C., et al. Ventricular definitions in the dog with a bioelectrode intravascular catheter. *Arch. Intern. Med.* 132:286, 1973.
10. Wellins, H. J. H., and Hein, J. J. Cardiac Pacing in the Study and Treatment of Arrhythmias and Tachycardias. In H. J. T. Thalen (Ed.), *Proceedings of the Fourth International Symposium in Cardiac Pacing.* The Netherlands: Van Gorcum, 1973.

The Structure and Rheology of Blood 9

E. O. Attinger and D. D. Michie

INTRODUCTION

In contrast to simple solids and fluids, the components of the cardiovascular system
exhibit complex physical properties that determine their mechanical behavior both
at rest and under stress. Stimulated in part by those complexities dealing with the
physical laws involving the deformation and flow of matter, rheologists have developed
concepts during the past few decades that are readily applicable to the analysis of the
physical properties of biologic matter. Most of the dynamic aspects of blood flow,
called *hemodynamics,* in the cardiovascular system are governed by the laws of fluid
flow. These laws are augmented by theories regarding physiologic, and sometimes
pathologic, cardiovascular mechanisms, as well as by special laws governing the pressures and volume distribution in the system. In this chapter, the physical makeup of
blood is described. The basic principles of rheology are introduced and are then applied
to the analysis of the properties of the vasculature that are related to hemodynamics.

The theoretical concepts discussed in this chapter are generally limited to linear
fluid behavior and small deformations. Nevertheless, they have proved to be extremely
useful in giving a better understanding of the mechanical properties of biologic
materials and systems (which are somewhat nonlinear) as well as models thereof,
provided the simplifying assumptions underlying the application of these concepts
are clearly recognized.

STRUCTURE OF BLOOD

Blood is a suspension of cellular elements in an aqueous solution of electrolytes and
nonelectrolytes called *plasma.* Its major components are listed in Table 9-1. The
cellular elements consist of the red blood cells (RBC) or *erythrocytes,* which play a
major role in gas transport, the white blood cells (WBC) or *leukocytes,* which provide
resistance against infection, and the platelets or *thrombocytes,* which are major contributors to the clotting reactions of the blood. Because the RBC represent the over-
whelming proportion of formed elements of the blood, the effects of the white blood
cells and the platelets can be neglected as far as their effects on the hemodynamic
aspects of the circulation are concerned, except in certain disease processes.

CELLULAR ELEMENTS

Red Blood Cells

If blood is left to stand in a test tube or is centrifuged, the red blood cells settle at
the bottom because of their greater specific gravity (1.08 versus 1.03 for plasma).

141

Table 9-1. Composition of Blood

	Concentration	Shape	Size or Mol. Weight
Cellular Elements			
Erythrocytes	5×10^6 cells/mm³	discoid	$8~\mu$ diam.
Leukocytes	5 to 8×10^3 cells/mm³		
Monocytes		polymorph.	$16-22~\mu$ diam.
Granulocytes		polymorph.	$10-12~\mu$ diam.
Lymphocytes		spherical	$7-12~\mu$ diam.
Platelets	2.5 to 5×10^5 cells/mm³	–	$2.5~\mu$ diam.
Plasma Molecular Constituents			
Albumin	3.5–5.3 gm/100 ml	prism	69×10^3 gm/mole
Globulin	2.1–3.3 gm/100 ml	ellipsoid	$(41-1000) \times 10^3$ gm/mole
Lipoprotein	–	spherical	$(200-13000) \times 10^3$ gm/mole
Fibrinogen	0.2–0.4 gm/100 ml	dumbbell	400×10^3 gm/mole
Glucose	70–120 mg/100 ml	–	180 gm/mole
Ionic Content of Plasma			
Na^+	145 mEq/liter		
K^+	4.2 mEq/liter		
Ca^{+2}	4.8 mEq/liter		
Cl^-	103 mEq/liter		
HCO_3^-	29 mEq/liter		
HPO_4^{-2}	2 mEq/liter		
Other cations	6 mEq/liter		
Other anions	21 mEq/liter		

The rate at which they settle is called the *sedimentation rate* and amounts normally to only a few millimeters per hour. In certain diseases, the sedimentation rate is much higher, primarily because of a change in the size of the sedimenting cells or because of aggregates formed by clumping or rouleau formation.

Normally, the RBC occupy, on the average, about 40% to 45% of the whole blood by volume. This percentage is called the *hematocrit* and may be expressed as:

$$\text{hematocrit} = \frac{\text{RBC volume}}{\text{total blood volume}} \times 100\% \qquad (9\text{-}1)$$

The hematocrit is overestimated by 1% to 2% in conventional centrifugal determination because some plasma is trapped between the cells.

The hematocrit may vary from one vascular bed to another. This is particularly true for microvessels, which, in general, have a lower hematocrit than their supply vessels. This is associated with the fact that the transit time of RBC through a vascular bed is shorter than that of plasma. The dependence of RBC concentration on flow appears to be related to the fractionation of blood at branch points of the vasculature. If the flow is unequally distributed between branches, the branch receiving the smaller volume is supplied primarily from the peripheral layers of the feeding vessels, which contain proportionately more plasma than the central layers.

ROULEAU

Figure 9-1
Red blood cell and rouleau structure.

Human RBC have the shape of a biconcave disc. Their dimensions are approximately 8 μ in diameter with a thickness of 2.4 μ at the edges and 1 μ in the center (Fig. 9-1). The cells may orient themselves side by side to form a rouleau structure [1], as shown in Figure 9-1. The bidiscoidal shape permits the membrane to deform while maintaining a constant surface area without stretching. The ease with which RBC can be deformed (the flexibility of the RBC) is an extremely important property, since many capillaries are smaller than the cell diameter. Although initially the RBC contain a nucleus, the latter is shed as the cells mature. Maturation is associated with shrinkage and increased fragility, and, after a life span of about 80 days, the RBC are trapped in the lung and the spleen and are broken down by the reticuloendothelial system. The RBC are in osmotic equilibrium with the plasma at an osmotic pressure corresponding to that of a 0.85% to 0.9% (0.15 M) sodium chloride solution. If placed in a hypotonic solution, they swell to spheres and rupture, releasing hemoglobin through their membrane. If such cells are later transferred into an isotonic solution, they resume their original shape, although they have lost most of their hemoglobin content; they are then termed *ghost cells.*

Gas transport is the primary function of the RBC. Its main constituent, hemoglobin* (a protoporphyrin-iron-globin complex), associates easily with oxygen and CO_2. Because of this chemical reaction, which accounts for the sigmoidal shape of the oxygen dissociation curve, nearly 100 times more O_2 can be transported than if O_2, according to Henry's law, were only carried in physical solution.

Anemias are diseases that are associated with an inadequate O_2-carrying capacity of the blood due to a lack of adequately functioning RBC. Because of the complexities in the structure of the RBC, there are many factors that may lead to anemia and must be considered in choosing an appropriate treatment. The most frequent causes of anemia are blood loss (hemorrhagic anemia); excessive destruction of RBC (hemolytic

*The hemoglobin molecules are extremely tightly packed within the RBC; they constitute about 25% of the available volume, the remainder being taken up by water (70%) and other constituents (5%). One gram of hemoglobin can combine with 1.34 ml of O_2. Hence, at a hemoglobin concentration of 15 gm/100 ml blood, the O_2-carrying capacity of blood is about 20 volume-percent.

anemia); loss of hemoglobin from the RBC because of osmotic imbalance or membrane damage; inadequate supply of iron, proteins, or vitamins for the formation of hemoglobin; and inadequate blood formation due to toxic states or to the occupation of bone marrow by other than hematopoietic tissue. An increase in RBC (polycythemia) occurs in certain states associated with chronic hypoxia, such as chronic pulmonary disease or during chronic exposure to high altitudes.

White Blood Cells

The leukocytes (white blood cells; WBC) are present in blood in a concentration of approximately 5000 to 10,000 cells/mm^3. They are referred to as either granular or agranular, depending upon the presence or absence of granules in their cytoplasm. The three types of granular leukocytes are the *neutrophil,* which is important in phagocytic activity; the *eosinophil,* which is important in allergic reactions; and the *basophil,* which is also thought to be important in allergic reactions. There are two types of agranular leukocytes: *lymphocytes* and *monocytes,* which play an important role in the immune tissue reaction and are composed of DNA. Thus, they may supply essential "building blocks" for cell manufacture.

Platelets

The platelets are the smallest cellular elements in the blood and are essential for the hemostatic or clotting process. They function in several ways: (1) They form platelets plugs. These small aggregations of platelets actually stop hemorrhage from the very small vessels, in particular the capillaries. (2) The platelets contain clotting factors that are released during the process of clotting. The platelets also undergo viscous metamorphosis at the time of clotting. These clotting factors help to accelerate the formation of fibrin (clots). (3) The platelets are essential for retraction of the clot. In some unknown manner, the platelets actually apply tension to the fibrin strands and decrease the size of the clot. (4) Platelets carry several substances either within their protoplasm or adsorbed on their surfaces. It has been reported that serotonin and epinephrine are released from the platelets during clotting, and that these are important in causing vasoconstriction of small vessels. (5) Substances such as thrombin, adenosine diphosphate, thrombocyte-agglutinating factor, and others appear to bring about platelet stickiness and viscous metamorphosis.

PLASMA

The plasma, the medium in which the cells are suspended, contains about 7% plasma proteins, which are usually divided into three main groups: (1) Fibrinogen represents one of the largest plasma protein molecules and is the most asymmetric. Plasma from which fibrinogen has been removed through clotting is called *serum.* In the clotting process, fibrinogen polymerizes to fibrin, which is responsible for the contraction of the blood clot. (2) The globulins are classified into α, β, and γ groups, and their molecular weights range over a broad spectrum. They are carriers of lipids and other water-insoluble components. The γ-globulins in particular contain the antibodies that provide immunity from infections due to bacteria and viruses. (3) Albumin, the most abundant plasma protein and the one with the lowest molecular weight, is responsible for about 80% of the total osmotic pressure of the plasma proteins. Hence, it plays an important role in maintaining the balance of body water. Because of their size, proteins pass through capillary walls far less readily than glucose and electrolytes. However, under pathologic conditions associated with increased permeability of the capillary walls, proteins may traverse them more easily.

BASIC RHEOLOGIC RELATIONSHIPS

The fundamentals of rheology provide the theoretical background that underlies the principles of blood flow. The concepts discussed are basic in nature, and their specific application to blood flow is presented in the next section.

The classic theory of viscous fluids relates stress to the rate of strain. If a material does not change its form when a stress is applied, it is termed a *rigid solid*. Normally, however, some deformation will occur. If the original form is resumed when the stress is removed, the material is called *elastic*. The ratio of the stress applied to the displacement or strain defines the *modulus of elasticity*; if the displacement is completely recovered when the stress is removed and the modulus of elasticity has a constant value, the material is termed *completely elastic* or Hookean in behavior. A large stress may, however, cause the solid to flow or to fracture, irreversibly disturbing the relative position of the solid elements (e.g., bone fractures or aneurysms).

Consider a small volume element ΔV of material in the shape of a parallelepiped (Fig. 9-2). The surface force acting on any plane of this element is

$$\mathbf{F} = \mathbf{S}\,dA \tag{9-2}$$

where dA is the area of the plane and $\mathbf{S}$ is a function of position, time, and orientation of the element; it is called a *stress vector* (force per unit area), or simply *stress*. The vector can be resolved into three components: one perpendicular to dA (the normal stress, tension, or pressure; $\mathbf{S}_{2,2}$ in Figure 9-2)* and two parallel to dA (the shearing stresses, $\mathbf{S}_{2,1}$ and $\mathbf{S}_{2,3}$ in Figure 9-2). By proper orientation of the volume element within the framework of the coordinate system, the shearing stresses can be made to disappear so that only the normal stress remains (Fig. 9-2B).

Figure 9-3 depicts the strain (deformation) experienced by a volume element when it is subjected to various types of stresses. Figure 9-3A shows a longitudinal, tensile, or compressive stress ($\mathbf{S}_{1,1}$). The ratio of this stress to the resulting strain is called the *elastic modulus* (E_{mod}) or Young's modulus. The length change Δh transverse to the direction of force is not directly proportional to the longitudinal length change ΔL,

*In this notation, the first subscript indicates the axis perpendicular to the surface subjected to the stress; the second subscript denotes the direction of the stress. Vector quantities are indicated by boldface type.

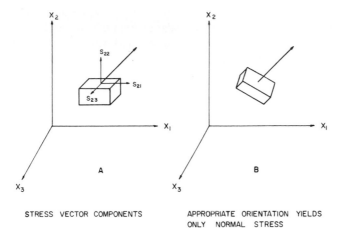

STRESS VECTOR COMPONENTS

APPROPRIATE ORIENTATION YIELDS
ONLY NORMAL STRESS

Figure 9-2
Stress vector components.

and therefore the volume does not remain constant. The ratio of the compressive strain $\Delta h/h$ associated with the tensional strain $(\Delta L/L)$ is known as *Poisson's ratio, σ.* Under compression, the volume decreases, whereas a tensile stress will produce an expanded volume.

A shearing stress, $S_{2,1}$, is applied to the volume element in Figure 9-3B. The volume remains the same for small deformations, but the originally vertical sides are displaced by an angle, θ. The shear modulus, S_{mod}, that relates the stress to the angular displacement is defined in the figure.

The bulk modulus, B_{mod}, relates the volume change to the applied normal stress, S_n, at all faces of the volume, as illustrated in Figure 9-3C. The bulk modulus is the ratio of the applied pressure to the change in volume of the element.

TYPES OF FLOW

If a material is unable to sustain shearing forces (i.e., no restraining forces develop inside the material when the force producing a deformation ceases), it is called a *fluid*; it may be either a liquid or a gas.

Since fluids are unable to sustain shear forces, stresses are related to the rate of change of strain. The tangential stress required to maintain a velocity difference of one unit, v, between two parallel planes at unit distance apart, x, is equal to the *coefficient of viscosity, η*:

$$\eta = \frac{S_{tan}}{\dot{\epsilon}} = \frac{S_{tan}}{v/x} \tag{9-3}$$

where $\dot{\epsilon}$, the *rate of shear strain,* represents the difference in displacement per unit time between planes situated unit distance apart (Fig. 9-4). The shear rate, $\dot{\epsilon}$, is equal to

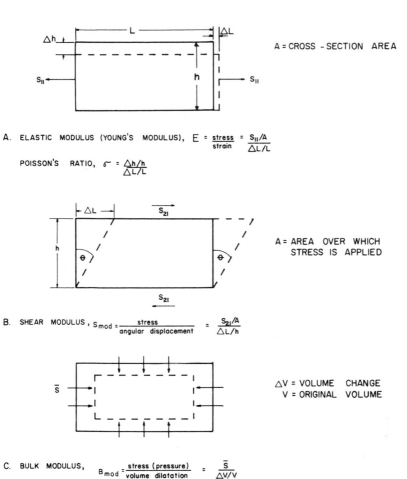

A. ELASTIC MODULUS (YOUNG'S MODULUS), $E = \dfrac{stress}{strain} = \dfrac{S_{\parallel}/A}{\Delta L/L}$

   POISSON'S RATIO, $\sigma = \dfrac{\Delta h/h}{\Delta L/L}$

A = CROSS - SECTION AREA

A = AREA OVER WHICH STRESS IS APPLIED

B. SHEAR MODULUS, $S_{mod} = \dfrac{stress}{angular\ displacement} = \dfrac{S_{21}/A}{\Delta L/h}$

ΔV = VOLUME CHANGE
V = ORIGINAL VOLUME

C. BULK MODULUS, $B_{mod} = \dfrac{stress\ (pressure)}{volume\ dilatation} = \dfrac{\bar{S}}{\Delta V/V}$

Figure 9-3
Deformations of a Hookean volume element.

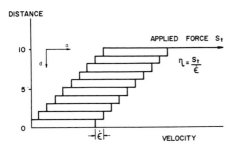

Figure 9-4
Laminar flow between two plates. The rate of shear strain, $\dot{\epsilon}$, is given by the ratio of the velocity (*a*) of the uppermost layer to its distance (*d*) from the lower plate, which is stationary.

the ratio of the velocity of the uppermost fluid layer and the separation distance to the lower plate, as shown in Figure 9-4. However, only when the lines of shear are straight and parallel are the rate of shear strain, $\dot{\epsilon}$, and the velocity gradient, v/x, synonymous. For a Newtonian fluid (see Fig. 9-5B), the relation between stress and the rate of shear strain is linear, and the coefficient of viscosity of such a fluid is therefore a constant.

This type of flow is called *laminar* or *viscous,* since the viscous drag, through which the energy supplied to the fluid is dissipated, is the result of the slippage of parallel fluid layers over each other. If individual fluid particles are tagged with a dye, they will move along so-called streamlines, whose tangents are, at every point, in the direction of flow. As the velocity of the fluid increases, the fluid elements are no longer able to follow continuous streamlines. They move in and out of layers, producing mixing of the fluid. During the mixing, some elements are accelerated, others are decelerated, and some of the energy in the fluid is directly transformed into heat. This direct translation of velocity energy to heat energy is due to inertial forces, and it must be distinguished from the viscous forces that cause the translation of velocity energy to heat energy due to the viscous drag *between* adjacent fluid elements.

If the flow disturbances generated by the inertial forces grow as the fluid moves downstream, *turbulent flow* occurs. The development of turbulent flow depends on the ratio of inertial to frictional forces, as Reynolds [10] first recognized. He expressed the limit below which turbulent flow cannot occur by the *Reynolds number, Re*:

$$Re = \frac{\text{inertial force}}{\text{viscous force}} = \frac{\rho}{\eta}vd = \frac{vd}{\nu} \qquad (9\text{-}4)$$

where ρ = fluid density

η = coefficient of viscosity of fluid

ν = kinematic viscosity (η/ρ)

v = velocity

d = "hydraulic depth" (in the case of a circular tube, d corresponds to the tube diameter)

Generally, turbulent flow does not occur when the value of the Reynolds number computed for a given situation is less than 2000.

Turbulent flow is characterized by completely random motion of the individual fluid elements. Modern views on turbulence relate the various observed phenomena to the original observation of Reynolds by means of the *boundary-layer theory.* Although a well-defined mean velocity may be present, random fluctuations of the velocity occur about the mean at each point. There is no sudden transition from smooth, laminar flow to fully developed turbulent flow. At first, small zones of instability appear that alternately grow and shrink as the flow velocity increases within a given geometry. As the velocity is increased, these zones become larger and more persistent until a sufficiently random motion is superimposed on the forward flow to classify it as true turbulence. In this transition zone, there is a sudden change in the pressure-flow relationship that appears between two limiting values of the Reynolds number: the lower value characterizes the first appearance of turbulent flashes and the higher value, the presence of fully established turbulence.

For many fluids, the rate of shear strain is not proportional to the applied stress, and therefore these fluids cannot be characterized by a single, constant coefficient of viscosity; these are termed *non-Newtonian fluids.* Values calculated from the ratio of the applied tangential stress to the shear rate are called *apparent viscosities,* and these determine the flow characteristics of the material. For a Newtonian fluid, the flow characteristic is represented by a straight line passing through the origin (Fig. 9-5).

Vascular walls are viscoelastic, and the stress-strain relationships depend on the speed with which the stress is applied. In cardiovascular physiology, the viscoelastic properties are frequently expressed in terms of stress relaxation and creep [11]. *Stress relaxation* may be defined as a decrease in stress when a constant strain is suddenly

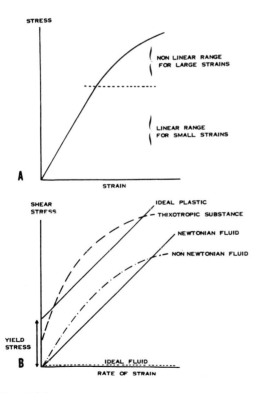

Figure 9-5
A. Relationship between stress and strain for an elastic material. *B.* Relationship between shear stress, **S,** and the rate of shear strain, $\dot{\varepsilon}$, for several classes of fluids. (From E. O. Attinger [Ed.]. *Pulsatile Blood Flow.* New York: McGraw-Hill, 1964. Reproduced with permission of McGraw-Hill Book Company.)

imposed, and *creep* may be defined as a change in strain associated with the application of a constant stress. If either stress or strain is applied with a continuously varying, bidirectional amplitude (such as a sine wave), then a hysteresis loop results. The width of the hysteresis loop is an indication of the stress relaxation or the creep that would have occurred if a step function of either the stress or the strain of corresponding magnitude had been used as the force.

VELOCITY PROFILES

Consider a rigid, straight, cylindrical tube of radius R in which a Newtonian fluid is flowing with constant velocity. Since the tube walls are rigid, radial flow is not possible and the driving pressure must therefore be uniform throughout any cross section of the tube. If the pressure difference over a length l of the tube is ΔP, the pressure gradient maintaining the steady flow is equal to $\Delta P/l$. Imagine a cylindrical fluid element of unit length and radius r situated on the tube axis. For steady flow, the viscous drag on the surface of the cylinder must exactly balance the driving force:

$$\frac{\Delta P}{l} \pi r^2 = S_{tan} 2\pi r \qquad (9\text{-}5A)$$

or

$$S_{tan} = \frac{\Delta P r}{2l} \qquad (9\text{-}5B)$$

The left side of Equation 9-5A expresses the driving force on the fluid cylinder and the right side, the viscous drag force (S_{tan} is the tangential stress). At the tube wall, $r = R$ and the shear stress is given by

$$S_{wall} = \frac{\Delta P R}{2l} \qquad (9\text{-}6)$$

Hence, the rate of shear strain changes continuously as a function of the tube radius; it is zero in the center of the tube and reaches a maximum at the wall. Since the streamlines are not curved, the rate of shear strain is equal to the velocity gradient dv/dr. The velocity, v, decreases as r increases; therefore,

$$S_{tan} = -\eta \frac{dv}{dr} \qquad (9\text{-}7)$$

Substituting Equation 9-7 in Equation 9-5B and simplifying,

$$dv = -\frac{\Delta P}{2l\eta} r\,dr \qquad (9\text{-}8)$$

Using the boundary condition that $v = 0$ at $r = R$ (i.e., assuming there is no slip at the wall), Equation 9-8 can be integrated to yield:

$$v = \frac{\Delta P}{4 l \eta} (R^2 - r^2) \qquad (9\text{-}9)$$

The parabolic velocity profile predicted by Equation 9-9 for Newtonian fluid flow inside a rigid cylinder is shown in Figure 9-6.

Integrating again over the cross section of the tube from $r = 0$ to $R = r$, we obtain the volume flow, $\dot{Q}$:

$$\dot{Q} = \frac{\Delta P \pi R^4}{8 l \eta} \qquad (9\text{-}10)$$

From Equation 9-10, it follows that the coefficient of viscosity of the fluid is:

$$\eta = \frac{\Delta P \pi R^4}{8 l \dot{Q}} \qquad (9\text{-}11)$$

Equation 9-11 is known as the *Poiseuille equation* after the French physician who was the first to establish these pressure-flow relations experimentally [8].

In turbulent flow, the velocity profile is no longer parabolic. As the average velocity increases, the velocity profile becomes increasingly blunt. The erratic motion of the fluid particles in turbulent flow transfers momentum through the fluid and thereby creates apparent shear stresses that may be much greater than those due to viscosity

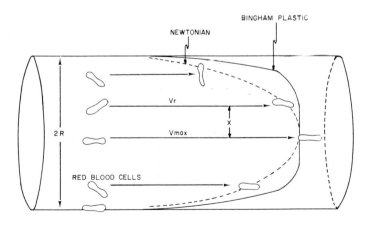

Figure 9-6
Velocity profiles for laminar flow of Newtonian and non-Newtonian fluids (i.e., those with Bingham plastic characteristics, such as blood).

alone. The linear relation between pressure drop and flow velocity in laminar flow becomes exponential in turbulent flow:

$$\Delta P = av^n \tag{9-12}$$

The value of the exponent n varies between 1.7 and 2, depending upon the material of the tube or the conditions of the wall. (Note that the pressure-flow relations in vascular beds are of a similar form but for entirely different reasons, as discussed in Chapter 10.)

The behavior of non-Newtonian fluids has been characterized by a variety of flow equations, most of which are empirically derived.

In some cases, the flow properties of a material depend on its history. This time-dependence is called *thixotropy*.

In pulsatile flow, there is a continuous interaction of viscous and inertial forces. As a result, the velocity profile changes throughout the pulsatile cycle. In the central core, inertial forces dominate and the profile is relatively blunt. In the peripheral layers surrounding this core, viscous forces are much more dominant and the velocity profile is much steeper.

Poiseuille's equation is valid only when the parabolic velocity profile has been developed. In order to establish this profile, additional energy is necessary. For example, consider the pressure drop at the origin of a tube connected to a reservoir (Fig. 9-7). The reservoir is assumed to be so large that the velocities inside can be

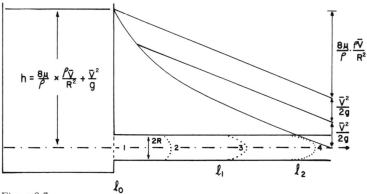

Figure 9-7
Pressure drop at steady flow within the inlet length of a circular rigid tube. At point 1, the velocity profile is uniform throughout the cross section. The development of the parabolic velocity profile (point 2) requires a pressure drop in addition to the gradient accounted for by the Poiseuille relation. The latter is valid for fully developed flow, i.e.,

$$P_0 - P_1 = 8\eta \left[(l_1 - l_0)\bar{v}/R^2 \right] + \rho \bar{v}^2$$

and

$$P_1 - P_2 = 8\eta \left[(l_2 - l_1)\bar{v}/R^2 \right]$$

(From E. O. Attinger [Ed.]. *Pulsatile Blood Flow*. New York: McGraw-Hill, 1964. Reproduced with permission of McGraw-Hill Book Company.)

neglected. The pressure head in the reservoir at the elevation of the center line of the tube is $P_0 = \rho h$. At the entrance l_0, there is a loss of pressure energy that is equivalent to the gain in the kinetic energy of the flow at l_0, which amounts to $\frac{1}{2}\rho v^2$. The flat velocity profile at l_0 is gradually transformed to a parabola due to an increase in the kinetic energy of the flow of $\frac{1}{2}\rho v^2$. (For a parabolic velocity distribution, the flow of kinetic energy through a circular cross section is larger by a factor of 8/3 than that of a constant velocity distribution if the mean velocities are equal.) A total pressure drop of ρv^2 is therefore necessary to generate the parabolic velocity profile, and within the entrance length, the total pressure drop is:

$$\Delta P = 8\eta \frac{l\bar{v}}{r^2} + \rho\bar{v}^2 \qquad (9\text{-}13)$$

Once the parabolic velocity profile has been established as at position l_2 in Figure 9-7, the pressure-flow relations obey Poiseuille's relationship and the term $\rho\bar{v}^2$ drops out.

At large values of the Reynolds number, the flow field can be split into two regions. The first is represented by a thin layer where the velocity gradient perpendicular to the wall is very large and where the transition from zero velocity at the wall to the free stream velocity takes place. In the second region outside this boundary layer (the central core), the velocity gradient is so small that the effect of viscosity may often be neglected. In general, it can be stated that the larger the Reynolds number, the thinner the boundary layer. As flow disturbances, such as vortices, diffuse from the boundary into the free stream, the boundary layer grows in the direction of flow and may separate itself from the wall. Boundary-layer separation is always associated with the formation of vortices and large energy losses in the region of the strongly decelerated flow behind the separation zone, the so-called wake.

SOLUTIONS AND SUSPENSIONS

The flow properties of liquids change significantly when particles or large molecules are dispersed within them. Since such particles displace some of the liquid in the original volume of the suspending medium, the mean rate at which the remainder of the liquid is sheared by a given velocity gradient is greater than it would be if no particles were present, and the flow resistance is therefore increased. The exact flow properties depend on the concentration, the shape, and the rigidity of the suspended particles. For rigid spheres at low concentrations, for example, the coefficient of solution viscosity, η_s, is given by:

$$\eta_s = \eta_0 \, (1 + 2.5c) \qquad (9\text{-}14)$$

where c is the volume concentration of the particles and η_0 is the coefficient of viscosity of the solvent. At normal values of the hematocrit, the viscosity of blood is therefore much more dependent (3 to 5 times) on the shear rate than is that of plasma.

PRESSURE-FLOW RELATIONSHIPS

To characterize fully the general pressure-flow relationships in a flow channel, five different relationships are needed:

1. The equations of motion for the fluid
2. The equations of motion for the walls of the tube
3. The equations of continuity (conservation of mass)
4. An equation of state relating pressure, density, and temperature (for compressible fluids only)
5. An equation expressing the balance between heat and mechanical energy, as well as an equation relating viscosity to temperature for nonisothermal processes.

For cardiovascular dynamics under normal environmental conditions, only the first three equations are important. The fourth must be considered in the analysis of gas flow as it relates to respiratory mechanics, and the fifth may become important under certain conditions, such as hypothermia.

For an incompressible fluid, the forces associated with the pressure gradient (the driving pressure) are balanced by the sum of the inertial forces, the frictional forces, and the body forces. For our purposes, the last can be neglected. As illustrated in Figure 9-8 for a cylindrical tube (inside radius R_i, outside radius R_o), the total pressure gradient acting on a volume element of fluid is equal to the total acceleration of the fluid element minus the force necessary to overcome viscous resistance. Note that if the vessel wall is able to move radially, both the pressure gradient and velocity, v, have two components: one parallel to the longitudinal vessel axis $(\delta P/\delta z, v_z)$ and one in the radial direction $(\delta P/\delta r, v_r)$. The inertial forces depend upon the mass that is being accelerated. In steady flow, the acceleration term disappears and the only force necessary to maintain flow is that required to balance the viscous resistance. Under

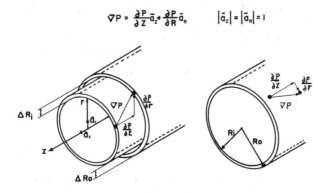

Figure 9-8
Equations of motion for a fluid and a wall element in an elastic tube. (From E. O. Attinger. In Ven Te Chow [Ed.], *Advances in Hydroscience*, Vol. 3. New York: Academic, 1966.)

these conditions, there is no motion of the elastic vessel wall and the pressure-flow relationship reduces to Poiseuille's equation (Eq. 9-11) if the flow is laminar and the fluid is Newtonian.

In pulsatile flow in the bloodstream, the vessel wall moves during each cardiac cycle; the radius increases during systole and decreases during diastole. The forces involved in the motion of the vessel wall are illustrated by the diagram on the left side of Figure 9-8. The pressure gradient associated with the transmural pressure (the inside pressure minus outside pressure) is balanced by the inertial forces minus the elastic and viscous losses. Except for the addition of a term to characterize the elastic forces of the vessel wall, the form of the equations for wall motion is identical to those for the motion of liquid.

For a closed, untethered cylindrical tube with a transmural pressure, P, the stresses in the three orthogonal directions are:

Radial stress $S_r = -P$

Tangential stress (hoop stress) $S_{tan} = Pr/h$ $\qquad\qquad$ (9-15)

Longitudinal stress $S_{lo} = Pr/2h$

where r and h are the mean radius and wall thickness, respectively, of the vessel at pressure P. It turns out that the physical properties of the vessel walls are frequency-dependent and the static elastic modulus is somewhat lower than the modulus at the frequency of the heart. At the boundary between the blood and the vessel wall, the equations for the wall motion and the fluid motion must hold simultaneously. By making a number of simplifying assumptions,* they can be solved in terms of volume flow, $\dot{Q}$ [6]. For a sinusoidal pressure gradient $M \cos(\omega t - \phi)$, the solution is

$$\dot{Q} = \frac{\pi r^4}{\eta \alpha^2} \frac{M'}{M} \sin(\omega t - \phi + \psi) \qquad\qquad (9\text{-}16)$$

*These assumptions are:

1. Blood is incompressible and Newtonian
2. Blood flow is laminar
3. The vessel is cylindrical, uniform, and infinitely long
4. The nonlinear terms in the Navier-Stokes equation (equation of motion) are small compared with the linear terms
5. The physical properties of the vessel wall are linear
6. The vessel wall is thin ($h/r \ll 1$)
7. The wall material is isotropic and homogeneous
8. Tangential velocities can be neglected
9. The displacements of the vessel wall and their derivatives are small
10. The imaginary components of $E_{mod}*$ and $\sigma*$ are much smaller than the real components (i.e., the viscous properties of the vessel wall can be neglected with respect to the elastic properties)
11. All quantities of the order $2\pi r/\lambda = r/v$ are negligible (λ is wavelength and v is the wave velocity)
12. The densities of the blood and the vessel wall are equal

By dimensional analysis, however, it can be shown, using experimental data on vascular tethering, that assumption 11 leads to Equation 9-16 without requiring assumptions 6, 7, 8, 10, and 12.

where $\alpha^2 = r^2 (\omega/\nu)$, ω is the angular frequency, ν is the kinematic viscosity, and M' and ψ are basically the ratio of and the phase-angle difference between two Bessel functions of zero and first order with complex argument $\alpha j^{3/2}$. These last functions depend on wall thickness, the Poisson ratio, and the longitudinal tethering of the vessel.

HEMORHEOLOGIC CHARACTERISTICS

Due to its complicated structure, blood does not behave in a Newtonian fashion under all circumstances. Since it is a variable quantity that depends upon numerous parameters, the term "anomalous" has been used to describe the viscosity of whole blood. Factors that influence the viscosity of blood in addition to those of Newtonian fluids are:

1. The viscosity of plasma
2. The hematocrit
3. Suspension stability
4. Emulsion stability
5. The flow velocity

Other factors that influence hemorheologic parameters are:

1. Coherence of the suspended particles with one another
2. The "wall" effect in the formation of a slippage layer
3. Orientation of highly anisodiametric particles with one another
4. Shear-rate dependency of blood viscosity
5. Lindqvist-Fahraeus effect
6. The microscopic picture of blood in capillaries
7. Anticoagulants
8. Scott-Blair effect
9. In vitro versus in vivo measurements
10. Temperature
11. Arterial blood pressure

Several of these factors will be discussed here. A more detailed and comprehensive discussion can be found in the literature [7].

FLOW PROPERTIES OF BLOOD

Plasma behaves practically as a Newtonian fluid with a viscosity of 0.011 to 0.016 poise. In contrast, the flow properties of whole blood are significantly non-Newtonian. Rheologically, blood is best described as a Bingham plastic [2], and most of the

experimental data are best fitted (but not explained) by the velocity-profile characteristic predicted by the Casson equation [3], as shown in Figure 9-6. At shear rates
greater than about 100 sec^{-1}, blood behaves as a Newtonian fluid at normal values
of the hematocrit, with a viscosity of 4 to 5 centipoise. This value decreases by 2%
to 3% per degree Celsius rise in temperature. The dependence of viscosity on the
hematocrit value and temperature is shown in Figure 9-9. A suspension in plasma of
hard spheres having the same size and distribution as red blood cells exhibits a viscosity
that is nearly twice that of blood. This difference in behavior is attributable primarily
to the ease with which RBC deform.

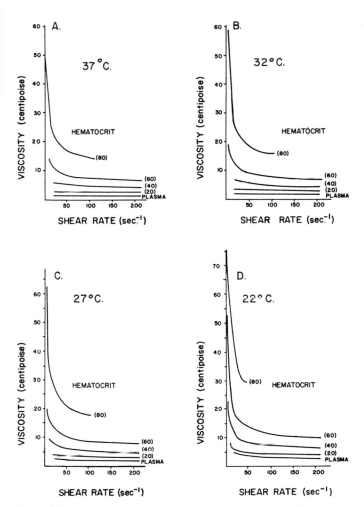

Figure 9-9
Viscosity–shear rate relationships of reconstituted blood from 37.0 to 22.0°C. (From P. W. Rand
et al. *J. Appl. Physiol.* 19:117, 1964. Reproduced with permission.)

In large vessels, the viscosity is mainly a function of the shear stress. However, since the shear rates at which the transition occurs from the flat area to the steep slope of the shear rate–dependency curves of Figure 9-9 are rarely reached in those vessels, the use of asymptotic viscosity values, corrected for the existing hematocrit value, will introduce errors of less than 5% for pressure-flow calculations.

The relation of flow resistance to the hematocrit value is shown in Figures 9-10 and 9-11. For each unit increase in the hematocrit values, the flow resistance increases by about 3%.

Figure 9-12 depicts the effect of the size of the tube on the anomalous viscosity of blood, i.e., the *Lindqvist-Fahraeus effect*. It can be seen from this curve that as the vessel or tube diameter increases, there is an increase in the anomalous viscosity of blood. This effect is particularly noticeable when the vessel radius is approximately 2.5×10^{-2} cm or less. From a physiologic standpoint, this is important because it favors the movement of fluid through the microcirculation. As the anomalous viscosity of blood decreases, less driving force (i.e., hydrostatic pressure) is required to move a given volume of fluid per unit time in a given vessel. This is indeed fortunate, since at the level of the microcirculation (the capillary bed), the pressures are extremely low. Figure 9-12 also shows that as the vessel radius exceeds approximately 5×10^{-2} cm, any further change in the radius has very little effect upon the anomalous viscosity of blood. Again, from a physiologic standpoint, this is of significance since the pressure drop from the aorta through the small arteries is relatively small.

There have been two hypotheses set forth to explain the Lindqvist-Fahraeus effect: (1) the axial streaming of cellular components and (2) the sigma effect of Dix and Scott-Blair. Each of these will be discussed individually.

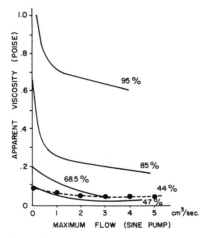

Figure 9-10
Values for apparent viscosity versus peak flow rate for sinusoidal blood flow (based on Womersley's theory for various hematocrits. Dashed line indicates steady flow. (From A. L. Kunz and N. R. Coulter. *Biophys. J.* 7:25, 1967. Reproduced with permission.)

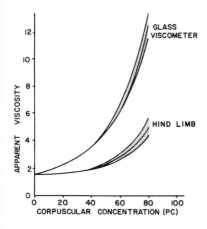

Figure 9-11
Increase in relative viscosity (viscosity of blood/viscosity of H_2O at the same temperature) with increasing hematocrit. The values were determined using a viscosimeter with a tube diameter greater than 1 mm (*above*) and using the vascular bed of the hind limb of a dog (*below*). (From S. R. F. Whittaker and F. R. Winton. *J. Physiol.* 78:339, 1933. Reproduced with permission.)

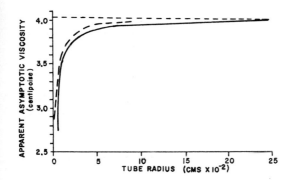

Figure 9-12
Lindqvist-Fahraeus effect of tube size on viscosity measurement. The values were calculated according to the "sigma" theory for red blood cell diameters of 6.0 μ (*solid line*) and 6.6 μ (*dotted line*). (Data from Kumin. Calculations from Haynes, *The Rheology of Blood.* Ph.D. Thesis, University of Western Ontario, 1957.)

As has been mentioned previously, the erythrocytes in flowing blood tend to accumulate in the axis of the blood vessel and leave a very thin zone near the wall of the blood vessel that is essentially cell-free. There is debate over the cause of this axial streaming. Bayliss [1] stated that there is no existing physical force to move the cells toward the axis. Starkey's work [12], which appeared later, elucidated this matter. A particle, even if it is spherical, that is in a stream where there is a velocity gradient across the tube will rotate in a complicated manner. As a result, the path of the particle will be modified by a Bernouilli force that will cause the particle to swerve. Thus, the hematocrit values near the wall will be reduced, and the hematocrit values for the core will be increased.

The sigma effect of Dix and Scott-Blair [5] can be explained simply as follows. If it is assumed that a blood vessel is so narrow that there is only room enough for five erythrocytes abreast, then it would not be possible to integrate for a series of infinite annuli, as is done in deriving Poiseuille's equation. Thus, there would be five terms to be summated. On the basis of this, the effect of viscosity in a tube of radius R is equal to that in a tube of infinite radius.

$$\eta_R = \frac{\eta_0}{(1 + d/R^2)} \tag{9-17}$$

$$v_r = v_{max}(1 - r^2/R^2)$$

where d = diameter of particle
   R = tube radius
   v_r = velocity at any radius r from the axis
   r = radial distance from axis to any point
   η_R = coefficient of viscosity for a tube of radius R

The broken line in Figure 9-12 shows how well the experimental curve for blood can be fitted to this formula if the value chosen for d is 6 microns, the diameter of the ox erythrocytes used in the experiment.

In capillaries, where the diameter is frequently less than that of a red blood cell, considerations derived from continuum mechanics no longer apply. It is meaningless, for example, to speak about pressure-flow relationships of the type discussed in the previous section except in terms of averages over large ensembles of microvessels. Direct observation of a capillary bed reveals that at any instant there are large variations in hematocrit values and flow velocities among the different capillaries that constitute such an ensemble of vascular pathways. Erythrocytes tend to move through individual capillaries orientated edgeways. Because of geometrical restraints, the cells are deformed, and the deformed cell shape becomes more axisymmetric as the cell speed increases. The surprisingly small pressure drop observed in such networks appears to be related to the great flexibility of the RBC membrane. The plasma that is trapped between successive RBC proceeding in tandem through a capillary moves as "bolus flow." Along the vessel axis, the plasma travels at about twice the speed of the RBC and turns toward the wall as it approaches the preceding RBC. Although such a mechanism facilitates (at least theoretically) mixing within this segment of plasma, its effect on transport and the exchange of materials by the blood is negligible.

Since anticoagulants of various types are used so frequently in clinical medicine, brief mention will be made of the effect of the various anticoagulants on the anomalous viscosity of blood. Rand et al. [9] claimed that heparin did not alter the anomalous viscosity of blood and that it was the agent of choice because (1) it is effective in minute quantities, (2) sample dilution is negligible, (3) the shape of cellular components is not altered, and (4) samples may be transported in the original syringe without further exposure to air or foreign surfaces. However, to the contrary, Wells and Merrill [13] have suggested that not only is the anomalous viscosity of blood altered by the addition of heparin, but the plasma viscosity is altered also. Copley and Blair [4] have studied the effects of various types of anticoagulants upon the anomalous viscosity of blood. They have arranged the anticoagulants in the following order with respect to their effect in decreasing the apparent viscosity of blood: versene gives the highest apparent viscosity, followed by ammonium oxalate, sodium citrate, and, lastly, heparin.

Measurement of the anomalous viscosity of blood is further complicated by the fact that the values obtained are dependent upon the type of viscometer that is used. It is beyond the scope of this chapter to delve into a discussion of various types of viscometers, and the reader is referred to review articles on this subject for more details.

REFERENCES

1. Bayliss, L. E. In A. Frey-Wyssling (Ed.), *Deformation and Flow in Biological Systems.* Amsterdam: North Holland, 1952.
2. Bingham, E. G., and Roephe, R. R. The rheology of blood. III. *J. Gen. Physiol.* 28:79, 1944.
3. Casson, N. A Flow Equation for Pigment-Oil Suspensions of the Printing Ink Type. In C. Mill (Ed.), *Rheology of Disperse Systems.* London: Pergamon, 1959.
4. Copley, A. L., and Blair, S. Haemorheological Method for the Study of Blood Systems and of Processes in Blood Circulation. In A. L. Copley and G. Stainsby (Eds.), *Flow Properties of Blood.* London: Pergamon, 1960. Pp. 412–417.
5. Dix, F. J., and Scott-Blair, G. W. On the flow of suspensions through narrow tubes. *J. Appl. Phys.* 11:574, 1940.
6. Hardung, V. Die Bedeutung der Anisotrope und Inhomogenität bei der Bestimmung der Elastizität der Blutgefässe. *Angiologica* 1:185, 1961.
7. Mason, S. G., and Bartok, W. *Rheology of Disperse Systems.* London: Pergamon, 1959.
8. Poiseuille, J. L. M. Recherches expérimentales sur le mouvement des liquides dans les tubes de très petits diamètres. *C. R. Acad. Sci.* (Paris) 11:961, 1840.
9. Rand, P. W., Lacombe, E., Hunt, H. E., and Austin, W. H. Viscosity of normal human blood under normothermic and hypothermic conditions. *J. Appl. Physiol.* 19:117, 1964.
10. Reynolds, O. Experimental investigation of the circumstances which determine whether the motion of water shall be direct or sinuous and of the law of resistance in parallel channels. *Phil. Trans. R. Soc.* (Lond.) 174:935, 1883.
11. Schlichting, H. *Boundary Layer Theory.* New York: McGraw-Hill, 1968.
12. Starkey, T. V. Viscosity and action. *J. Appl. Phys.* 7:448, 1956.
13. Wells, R. E., and Merrill, E. W. Shear rate dependence of the viscosity of whole blood and plasma. *Science* 133:763, 1961.

The Vascular System

10

E. O. Attinger

ORGANIZATION AND STRUCTURE

The arterial tree starts at the outflow channels of the left and right ventricles of the heart, branches into a number of arteries, and then diverges into approximately four billion capillaries. Each capillary has an extremely small cross-sectional area, which is in the order of 5×10^{-7} cm^2. Because of the vast difference in the total cross-sectional areas of the arteries and capillaries, the outflow tract to the microcirculation is increased approximately 500 times. The series-circuit characteristics of the system requires the average blood flow to be constant. Thus, the linear velocity of blood flow in the capillary bed must decrease correspondingly because volume flow is equal to the product of linear velocity and cross-sectional area. The mean velocity of blood flow in the capillaries is in the order of 0.05 to 0.15 cm/sec, which corresponds to a volume flow of 5×10^{-8} cm^3/sec. It thus takes approximately six hours for one cubic millimeter of blood to pass through one capillary. Figure 10-1 shows how the average linear velocity, cross-sectional area, and percentage of blood volume change in the systemic circulation.

Normally, only a fraction of the capillary bed is perfused. Additional capillaries open up when local blood flow rates increase. During exercise, for example, the number of perfused capillaries in muscle may increase tenfold or more. Hence, all the estimates of total capillary cross section that are based on calculations from figures for cardiac output and linear velocities must underestimate the total number of capillaries by a sizeable factor. On the return path from the microcirculation, the cross section decreases to a value that is about 50% larger than that of the outflow tract. (There are two venae cavae feeding the blood back into the right atrium and five to seven pulmonary veins for the left atrium.) The average velocity increases correspondingly, and it is only slightly lower in the large veins than in the arteries. The largest fraction of the blood volume is found on the venous side of the vascular bed (see Fig. 10-1). Recent data indicate that the major part of this fraction is contained in the smaller veins, contrary to the statements often found in physiology textbooks.

The *arterial bed* is a distribution system that transports oxygenated blood to the organs of the body and regulates their blood supply according to individual demands. The distribution of blood flow is shown in Table 6-1, as are the relative blood volumes and oxygen requirements of key organs. The distensible vessel walls are composed of an inner layer (the endothelium), elastic tissue, smooth muscle, and fibrous tissue (collagen) (Fig. 10-2). The relative proportions of these components vary with regard to both location and age. Toward the periphery, the vessel walls become thicker compared to the vessel radius, and the relative amount of muscle and fibrous tissue

163

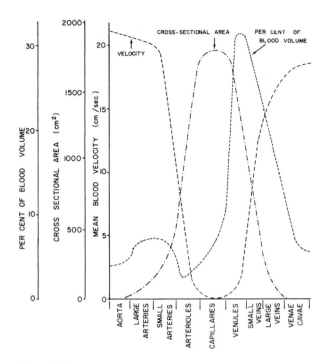

Figure 10-1
Changes in mean blood velocity (cm/sec), vascular cross-sectional area (cm²), and percentage of blood volume (%) in different segments of the systemic circulation. Note the large decrease of velocity from the large vessels to the microcirculation, which results from the corresponding increase in cross-sectional area ($v = \dot{Q}/A$, where v = velocity, $\dot{Q}$ = volume flow, and A = cross-sectional area). The major fraction of the blood volume is contained within the venous system, primarily in the smaller veins. (Adapted from R. M. Berne and M. N. Levy. *Cardiovascular Physiology* [2nd ed]. St. Louis: Mosby, 1972. Reproduced with permission.)

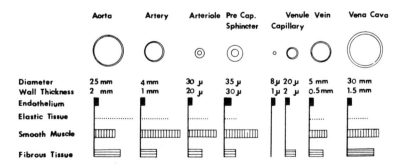

Figure 10-2
Size, wall thickness, and composition of the vascular walls of various types of blood vessels. (Adapted from A. C. Burton. *Physiol. Rev.* 34:619, 1954. Reproduced with permission.)

increases while that of the elastic tissue decreases. Correspondingly, the vessel walls become stiffer and the ability of the vessel to change its cross section by means of changes in the contractile state of the vascular smooth muscle becomes greater. The aging process is also associated with progressive changes in the structure and composition of the arterial wall. The water content of the tissues decreases, increased amounts of calcium and cholesterol are deposited, and the elastic and collagenous tissues degenerate, resulting in a stiffer and more brittle vessel, particularly in cases of arteriosclerotic disease.

The *capillary bed* represents the diffusion system of the circulation. An average capillary has a length of a few millimeters at most and a diameter of about 8 μ, so that the red blood cells (RBC) are frequently deformed during their passage through the capillary. The walls consist only of a single layer of endothelial cells; this permits easy diffusion of gases and metabolites to near equilibrium conditions in the surrounding tissues in a fraction of the transit time of the blood.

The *venous bed* is a collection system that drains all the capillaries and carries the blood back to the heart. The veins are provided with valves that prevent backflow. Their walls are much thinner than those of the vessels on the arterial side, although they consist of the same structural components. Since the venous walls are so thin, they offer little resistance to collapse if either the intravascular pressure decreases (e.g., during periods of transient flow acceleration) or the extravascular pressure increases (e.g., during muscular contraction). The major structural characteristics of the walls of different types of vessels are summarized in Figure 10-2.

PHYSICAL CHARACTERISTICS

The cross section, A, of a vessel is a function of the transmural pressure. The physical properties of the vascular walls and the wall thickness are also controlled by the transmural pressure, which is the difference between the pressure acting on the inside and that acting on the outside of the wall. The former, or lateral pressure, represents the total intravascular pressure minus the kinetic energy of the fluid. This follows from the Bernouilli principle, which states that the sum of the pressure and the kinetic energy per unit volume at a given point in a fluid is constant:

$$P + 1/2 \, \rho v^2 = k \tag{10-1}$$

For a blood velocity of 100 cm/sec in steady flow, the kinetic energy term amounts to a pressure of approximately 4 mm Hg ($0.5 \times 1.05 \times 100^2$ gm·cm^{-1}·sec^{-2} = 5250 dynes/cm^2 or about 4 mm Hg, since 1 mm Hg = 1333 dynes/cm^2). The outside pressure is usually assumed to be constant for the peripheral arterial bed, although it may change somewhat if pressure is exerted by outside structures in close proximity to the vessel (e.g., by muscular contraction or by an increase in the size of the bladder or the uterus). Since the mean transmural pressure of the systemic arterial bed is generally in the order of 100 mm Hg, small changes in the outside pressure can be neglected.

The situation is quite different for the venous flow channels and the pulmonary artery, since the transmural pressures in these systems are low.

At low transmural pressure, the cross section of these latter vessels becomes elliptical and finally collapses (Fig. 10-3). The extravascular pressure changes abruptly at two points in the low-pressure systems: at the entrance of the large veins into the thoracic cavity (here the vessels are exposed to subatmospheric intrapleural pressure — see Chap. 21) and at the level of the small pulmonary vessels, which are exposed to alveolar pressure. At these points, vascular collapse may occur, depending upon the difference in transmural pressure at the discontinuity and upon the flow rate.

The pressure-flow relationships are thus governed primarily by the local pressure situation (the "waterfall" phenomenon) and *not* by the difference between arterial and venous pressure. The local pressures determine the cross section, and hence the resistance, through the pulmonary microcirculation. The cyclic variation in the subatmospheric pressure and the changes in lung volume that are associated with respiratory

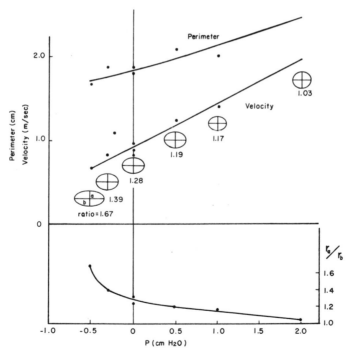

Figure 10-3
Change in cross section, perimeter, and wave velocity of a jugular vein segment at low distending pressures. The shape of the cross section varies from an ellipse, with a ratio of major to minor semiaxis of 1.67 at a transmural pressure of −0.5 cm H_2O, to a near circle at 2 cm H_2O. Over this pressure range, the change in perimeter is small, but the wave velocity increases by a factor of about 3. The ratio of the major to minor semiaxes, r_a/r_b, decreases asymptotically toward unity with increasing transmural pressure. (From E. O. Attinger. *IEEE Trans. Biomed. Eng.* BME-16:253, 1969. Reproduced with permission.)

motion thus alter both the geometry and the blood flow through the pulmonary vasculature. The increased venous return during inspiration results in the elevation of right ventricular stroke volume, and it is transmitted to the left ventricle within a few heartbeats. The time lag between the responses of the two ventricles to a change in venous return is due to the capacity and the flow resistance of the pulmonary vasculature. With increasing heart rate, the time lag increases and the stroke volume modulation decreases. Although the right side of the heart acts to moderate pulmonary flow by temporarily slowing part of the large inflow during inspiration, a change in either the blood volume or the capacity of the vascular bed leads to a dissociation between vascular and intrapleural pressure changes.

The transmural pulmonary artery pressure increases during inspiration, partly because of an increase in the right ventricular stroke volume and partly because of an increase in vascular resistance. The latter can be inferred from the fact that the pulmonary capillaries flatten out at larger lung volumes. If the respiratory volume changes become large or are associated with excessive transpulmonary pressure changes (as in bronchial asthma or in chronic pulmonary emphysema), the normal balance between cardiac output and pulmonary vascular resistance is upset.

The pressure-flow relationships are very sensitive functions of the instantaneous radius of the vessel wall. The radius is determined by the stresses applied to the wall and by the stiffness of the wall itself. Applying the principles discussed in Chapter 9 and by considering the applied stresses, it is apparent that the intravascular and extravascular pressures act in the radial direction, while the primary effective stress that tends to stretch or compress the wall is tangential. For a cylindrical vessel under equilibrium conditions, this relationship is:

$$S_{tan} = \frac{R_{in}^2 P_{in} - R_{out}^2 P_{out}}{R_{out}^2 - R_{in}^2} + \frac{(P_{in} - P_{out})R_{in}^2 R_{out}^2}{r^2 (R_{out}^2 - R_{in}^2)} \qquad (10\text{-}2)$$

where S_{tan} = tangential stress (force per unit area)
$\quad\quad P$ = pressure
$\quad\quad R$ = radius
$\quad\quad r$ = radial coordinate
$\quad\quad h$ = wall thickness $(R_{out} - R_{in})$

When P_{out} equals zero, Equation 10-2 becomes:

$$S_{tan} = \frac{R_{in}^2 P_{in}}{R_{out}^2 - R_{in}^2} \left(1 + \frac{R_{out}^2}{r^2}\right) \qquad (10\text{-}3)$$

Equation 10-3 shows that the tension is largest at the innermost layers of the vascular wall and decreases in a curvilinear fashion toward the outside layers. As the wall thickness h becomes small with respect to r, Equation 10-3 can be further simplified to:

$$S_{tan} = \frac{\bar{r}P_{in}}{h}$$
(10-4)

It must be emphasized that Equation 10-4 applies only to thin-walled vessels (wall thickness/radius < 0.1). If this ratio becomes larger, the errors associated with assuming uniform tangential stress can no longer be neglected. The use of the so-called Laplace equation ($S_{tan} = P_{in}r$ is inappropriate if it is applied to blood vessels since it neglects wall thickness, and the concept of critical closing pressure that is derived from it does not apply to real vessels.

The elastic modulus, E_{mod} (see Fig. 9-3A), for a cylindrical, isotropic vessel takes the form:

$$E_{mod} = \frac{\Delta S_{tan}}{\Delta \bar{R}/R_0} = (1 - \sigma^2)\left(\frac{\Delta P}{\Delta \bar{R}}\right)\frac{\bar{R}^2}{h}$$
(10-5)

where $\bar{R}$ = mean radius
$\quad R_0$ = initial radius
$\quad \sigma$ = Poisson ratio

Because the stress-strain relationships of blood vessel walls are significantly nonlinear, the elastic modulus is a function of the distending pressure and is frequently expressed as an *incremental modulus* (Fig. 10-4), where strain is defined as the ratio of each increment in length change to the mean length during that change. Figure 10-4 indicates that the pressure-dependence of the elastic modulus can be approximated by a straight line on a log-log plot. It will be seen that the elastic modulus varies by almost a decade in the range of the normal operating pressure of a blood vessel. At pressures above the normal range, the veins and the pulmonary artery appear to be stiffer than the systemic arteries.

These differences in the stress-strain behavior of vascular walls are due primarily to differences in structural arrangements rather than to those in composition. In thin-walled vessels, the passive wall elements (elastin and collagen) exert tension at much lower distending pressures than do arteries. Values for the elastic modulus of passive wall elements range from 100×10^6 to 1000×10^6 dyne·cm^{-2}, that of elastic tissue averages 6×10^6 dyne·cm^{-2}, and in smooth muscle, values of the elastic modulus are found to range from 0.1×10^6 to 2.5×10^6 dyne·cm^{-2}. The elastic modulus of smooth muscle, however, depends upon the degree to which the muscle is contracted.

When strain is expressed as change in volume rather than as change in radius, as is frequently done in the clinical literature, the pressure-volume curves obtained from dilated venous segments are curvilinear and display convexity toward the volume axis. In constricted veins, on the other hand, the curves are characterized by a sigmoid

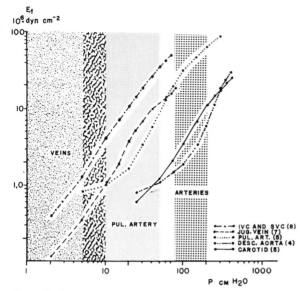

Figure 10-4
Pressure-dependence of the elastic modulus for arteries and veins in anesthetized dogs. The curves represent the average from all the experiments (the number of experiments is in parentheses in the key of the figure), and the standard errors of the mean are less than 5% of the mean at each distending pressure. The shaded areas represent the normal range of operating pressures for various vessels. (From E. O. Attinger. In S. N. Levine (Ed.), *Advances in Biomedical Engineering and Medical Physics*. New York: Wiley, 1968. Reproduced with permission.)

shape. Because of these differences in the shape of the curves of the stress-strain relationships, which are due primarily to changes in the activity of the vascular smooth muscle (usually, and unfortunately, called "vascular tone" in the medical literature), the measurement of venous pressure alone reveals very little about the venous system as long as it cannot be related to its state of filling. The same objection can, of course, be made with respect to the measurement of arterial pressure. In this case, however, the degree of uncertainty is considerably less because of the significant differences in the degree of distensibility between the two systems. With increasing age, progressively larger pressure increments are required to produce the same volume change, which indicates that the stiffening of the vascular wall is associated with age (Fig. 10-5).

The low-pressure venous system, in contrast to the arterial system, may undergo large changes in capacity without major pressure changes. As fluid is added to a collapsed vein, the vessel first exhibits an elliptical and then a circular cross section without increasing its circumference (see Fig. 10-3). This filling phase is then followed by distension of the segment by means of an increase in both circumference and length. In this latter phase, most of the volume increase within the physiologic pressure range results from radial distension.

It was pointed out in Chapter 9 that the classic theory of elasticity cannot be strictly applied to the analysis of the physical properties of the vessel walls, because

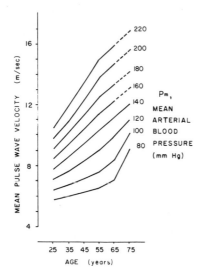

Figure 10-5
Mean aortic pulse wave velocity as a function of age and mean arterial blood pressure. Since the elastic modulus changes in proportion to the square root of pulse wave velocity, the curves clearly demonstrate the progressive stiffening of arteries with age or with increasing blood pressure, e.g., in hypertension. (Modified from T. Kenner. *Foundations and Objectives in Biomechanics*. Reproduced with permission.)

they are viscoelastic, nonlinear, and anisotropic [1]. As a result of viscoelasticity, the dynamic modulus of elasticity is considerably larger than the static modulus [8]; the dynamic modulus is obtained, for example, by applying a sinusoidal stress, in contrast to the static modulus, where a constant stress is applied. This difference increases toward the periphery of the arterial tree and is larger in young people than in older persons [5]. As a result, the arteries are stiffer with respect to pulsatile pressure than they are with respect to changes in mean pressure. The age-dependence of the stress-strain relationship also leads to stress relaxation, creep, and hysteresis phenomena (see Chap. 9).

Blood vessels are sometimes classified as either *resistance* or *capacitance* vessels. The latter are able to accommodate changes in blood volume easily; the former control the pressure-flow relationships by altering the resistance to blood flow. The significance of these properties for the motion of blood in the cardiovascular system will be discussed in the next section.

VASCULAR IMPEDANCE

The physical characteristics of the vasculature can be defined using measurable parameters — peripheral resistance (R), vessel compliance (C), and blood inertance (L) — which are related to the hemodynamic variables of pressure and flow:

$$R = \frac{\Delta P}{\dot{Q}} \tag{10-6}$$

$$C = \frac{\Delta V}{\Delta P} \tag{10-7}$$

$$L = \frac{\Delta P}{\Delta \dot{Q}/\Delta t} \tag{10-8}$$

where, for an arterial segment, ΔP is the mean pressure drop, ΔV is the volume change, $\dot{Q}$ is the mean flow rate, and $\Delta \dot{Q}/\Delta t$ is the incremental time rate of change in volume flow.

Since both pressure and flow in the cardiovascular systems are pulsatile, the input impedance $Z(\omega)$, expressed by Equation 10-9, must be frequency dependent, in contrast to peripheral resistance, which is not.

$$Z(\omega)_{in} = \frac{P(t)}{\dot{Q}(t)} \tag{10-9}$$

where $P(t)$ is the pulsatile driving pressure at time t and $\dot{Q}(t)$ is the volume flow at time t.

Correspondingly, the expression for pressure-flow relations contains a term related to the viscous resistance of the fluid (as does the Poiseuille equation), a term related to the viscoelastic properties of the vascular walls, and a term related to the inertial forces associated with the mass of blood being moved. The values of all three terms vary as a function of the heart rate, which is reflected in the parameter α of Equation 9-16.

The viscous resistance and the inertial terms constitute the *longitudinal impedance*, which can be expressed as:

$$Z_{lo} = \frac{8\eta l}{\pi r^4} + j\omega \frac{\rho l}{\pi r^2} \tag{10-10}$$

The viscoelastic properties of the wall determine the *transverse impedance*, which is defined as:

$$Z_{tr} = \frac{1}{j\omega} \frac{(E_{mod})_{tan}(2a + 1)}{3\pi r^2 (a + 1)^2} \tag{10-11}$$

where $\quad \omega$ = angular frequency
$\quad\quad\quad \rho$ = density of fluid
$\quad\quad\quad \eta$ = coefficient of viscosity of fluid
$\quad\quad\quad r$ = vessel radius
$\quad\quad\quad l$ = length of vessel segment
$\quad (E_{mod})_{tan}$ = tangential modulus of elasticity
$\quad\quad\quad a$ = vessel radius/wall thickness ratio
$\quad\quad\quad j = \sqrt{-1}$

Note that the contribution of viscous resistance to the impedance varies as a function of the fourth power of the radius and the inertial contribution varies as a function of the second power. This emphasizes the importance of vascular geometry, i.e., the cross-sectional area, on the pressure-flow relationships.

The longitudinal impedance, Z_{lo}, is frequency-dependent not only because of its inertial component but also because the viscous and inertial forces interact. The transverse impedance, Z_{tr}, is frequency-dependent because of the viscoelastic properties of the vascular wall (the complex modulus of elasticity shows an increase, in terms of both amplitude and phase, from its static value at frequencies between 0 and 3 Hz, but it remains relatively constant at higher frequencies).

In the cardiovascular system, the major pressure drop occurs in the arterioles, which have a radius of 10 to 100 μ. In the large arteries such as the aorta, the inertial effects predominate; on the other hand, the importance of viscous resistance increases as the cross section of the individual vessels decreases. The flow resistance through the arterioles greatly exceeds that of large arteries, although the total cross section of the former is much larger. This may, at first glance, appear to be paradoxical, since in an arrangement of parallel tubes, the total resistance is related to the individual resistances by:

$$\frac{1}{R_T} = \frac{1}{R_1} + \frac{1}{R_2} + \frac{1}{R_3} + \ldots \tag{10-12}$$

where R_T is the total resistance and R_1, R_2, and so on, are the resistances of the individual parallel pathways. A simple numerical example may explain this apparent discrepancy. Assume that five narrow tubes, each of cross-sectional area A_i and length l, are connected in parallel. Their total cross-sectional area thus equals $A_T = 5A_i$. The viscous resistance of a simple tube of cross-sectional area $A_w = A_T$ is $R_w = k/A_w^2$, and the resistance of a single narrow tube equals $R_i = k/A_i^2$ (since the viscous resistance is inversely proportional to the fourth power of the radius, it will be inversely proportional to the square of the area). The total resistance of the five narrow tubes thus amounts to $R_T = k/5A_i^2$, whereas the resistance of the single tube is only $R_w = k/A_w^2 = k/(5A_i)^2 = k/25A_i^2$, i.e., the resistance of the five tubes in parallel is five times as great as that of a single tube of equal total cross-sectional area.

The relationships discussed thus far are valid only if the vessel diameter is much larger than the dimension of the particles (red blood cells) that are suspended in the liquid. When the dimensions of the red blood cells are similar to those of the flow channel, conventional hydrodynamic concepts, such as pressure-flow relationships, become meaningless. In such a case, the individual cells are deformed and travel at velocities that are different from those of the normal cells in the suspending medium. These phenomena will be discussed in more detail in Chapter 11.

WAVE PROPAGATION

One of the characteristics of the behavior of the vascular system is the dispersion of the pulse wave during its travel. The simple harmonic components of the pulse wave are propagated at a finite velocity. Only if the shortest wavelength of the disturbance is very large compared with the maximal dimensions of the system can the wave motion be considered to have the same values at all points in the system. In this case, the system can be represented by lumped parameters, such as in the Windkessel model [2]. If the wavelength is short with respect to the dimensions of the system, the model has to be described by partial differential equations, and one thinks of the various parameters as being distributed throughout the system. Analogs of the circulation that are based on transmission line theory provide an example of such a concept.

In order to get a clearer picture of wave mechanics, consider a water wave of simple harmonic shape traveling at a constant velocity. A cork floating on the wave and restrained from moving horizontally will be observed to bob up and down in simple harmonic motion. It is apparent that the velocity of propagation of the harmonic wave, the wavelength (i.e., the distance from crest to crest), and the frequency with which the cork bobs up and down are not independent. The frequency with which the cork bobs is equal to the ratio of wave velocity over wavelength. Hence, the wave, which is propagated at a definite velocity, not only is a simple harmonic function of distance, but, at a fixed point, also represents a simple harmonic function of time.

The assumption that the pressure P in a cylindrical vessel is a function only of the axial coordinate z and the time t is implicit in all present methods for measuring blood pressure. This assumption is not strictly true, but it provides a good approximation since the radial dimension of the vessel is less than 2.5% of the wavelength at the highest frequencies of interest. Under these conditions, the propagation of the wave can be described by:

$$P(z,t) = f_1(z - vt) + f_2(z + vt) \tag{10-13}$$

where $P(z,t)$ is the amplitude of the pressure wave, v is its speed of travel, and f_1 and f_2 are arbitrary functions.

A function such as $f_1(z - vt)$ represents a disturbance traveling along the z-axis with a velocity v, whereas $f_2(z + vt)$ corresponds to a wave that is propagating in the negative z direction. The mathematical form of the functions f_1 and f_2 depends upon the type of disturbance creating the wave. For sine waves, they will be sine or cosine functions of $z + vt$.

In the presence of frictional losses, a damping term appears in the wave equation:

$$\frac{\delta^2 P}{\delta z^2} = \frac{1}{v^2}\left(\frac{\delta^2 P}{\delta t^2} + K\frac{\delta P}{\delta t}\right) \tag{10-14}$$

For a Newtonian fluid and laminar flow in a linear system, the damping term is directly proportional to the viscosity of the conducting medium. The frictional losses result in an exponential decay of the wave as it propagates away from the source. The solution for a sinusoidal disturbance, $P_0 \sin \omega t$, in the direction of the positive z-axis then becomes:

$$P(z,t) = P_0 e^{-\alpha z} \sin(\omega t - \beta z) \tag{10-15}$$

This is more conveniently written in exponential form:

$$P(z,t) = P_0 e^{-\gamma z} e^{j\omega t} \qquad (10\text{-}16)$$

where $\gamma = \alpha + j\beta$ = the propagation constant
α = damping constant
β = phase constant

For the case of the superposition of two waves where one is traveling downstream and the other upstream, the pressure at any point is given by adding the solutions for each wave. This results in an expression of the form:

$$P(z,t) = P_1 e^{(j\omega t - \gamma z)} + P_2 e^{(j\omega t + \gamma z)} \qquad (10\text{-}17)$$

where $P_1 e^{j\omega t}$ and $P_2 e^{j\omega t}$ represent the disturbances creating the waves traveling downstream and upstream, respectively.

The propagation constant is, of course, a function of the physical properties of the conducting medium and is always a complex number. Its real part, α, the damping constant, describes the amplitude variation with distance, and its imaginary part, β, the phase constant, indicates the way in which the phase of the wave changes, i.e., the velocity at which the transient disturbance is propagated.

The velocity at which a simple harmonic disturbance travels is the *phase velocity*, which is mathematically described by:

$$v_p = \frac{\Delta l}{\Delta \theta} \omega = \frac{\Delta l}{\Delta \theta} (2\pi f) = \frac{\omega}{\beta} = \lambda f \qquad (10\text{-}18)$$

where v_p = phase velocity
Δl = distance over which the phase velocity is measured
$\Delta \theta$ = difference in phase angle between the two measurements
λ = wavelength
f = frequency

The phase constant β is inversely proportional to the wavelength and is experimentally evaluated by measuring the phase difference of the wave over a given distance. The damping constant α is roughly proportional to frequency.

In a simple elastic system, the phase velocity is related to the physical properties of the system by the Moens-Korteweg equation [7]:

$$v_p{}^2 = \frac{E_{mod} h}{2\rho r} \qquad (10\text{-}19)$$

where E_{mod} = modulus of elasticity of vessel wall
ρ = density of blood
r = vessel radius
h = thickness of vessel wall

Theoretically, the determination of phase velocity offers an easy experimental approach to the evaluation of the elastic properties of a blood vessel in vivo. In contrast to the phase velocity, the *group velocity* expresses the speed at which the compound wave, and hence the energy, travels. Skilling [9] gives the following analogy for the two velocities: the ripples along the back of a caterpillar travel at the phase velocity, whereas the bulk of the beast moves at the group velocity. In general, the group velocity may be expected to be close to the familiar foot-to-foot velocity of the pressure pulse. However, because the pressure pulse changes its shape considerably along the arterial bed, points of equal pressure on the rising and falling parts of the pressure curve will have different transmission times, and therefore the foot-to-foot velocity is only an approximation of the group velocity.

PRESSURE-FLOW CHARACTERISTICS

Although the classic Poiseuille equation is not strictly applicable to the circulatory system, it emphasizes the importance that must be placed on the size of the vascular cross section in determining the resistance to blood flow.

The outstanding feature of arterial blood flow is its pulsatile character. Figure 10-6 illustrates the relative pressures found in the various parts of the circulatory system; all pressures are referred to the same level, the right atrium. The height of the bars indicates the range in magnitude of the pressure pulse in a given section, and the dashed line indicates the mean pressure. The latter changes little in the systemic arteries between the aorta and the periphery, but it falls rapidly (by about two-thirds) across

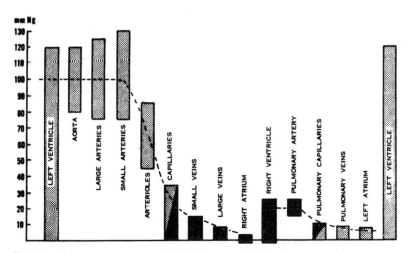

Figure 10-6
Pressure relations in various segments of the vascular system. The dashed line indicates the mean pressure. The dark bars represent the part of the system containing deoxygenated blood. (From E. O. Attinger [Ed.]. *Pulsatile Blood Flow.* New York: McGraw-Hill, 1964. Reproduced with permission.)

the arterioles, which again emphasizes the predominant role of vascular geometry with respect to pressure-flow relationships. This striking fall in pressure occurs in a very short distance and thus points to the arterioles as the natural site where blood flow distribution can be controlled most effectively. From the smaller veins to the right atrium, the change in mean pressure is quite small. Although the absolute pressure level in the pulmonary circulation is considerably less (approximately one-fifth of that of the systemic circulation), the spatial and temporal relations of the hydraulic gradient to the different vascular segments of the pulmonary circulation are similar to those in the peripheral vasculature.

During ventricular ejection, the proximal part of the arterial system accommodates the propelled stroke volume. The arterial pressure rises from its diastolic value of approximately 80 mm Hg to a systolic value of 120 mm Hg, and the pulmonary arterial pressure rises from about 10 to 25 mm Hg. The difference between the systolic and the diastolic pressure is called the *pulse pressure*, and, for a given stroke volume, it depends primarily on the distensibility of the proximal arterial tree (the Windkessel model [2]). A number of indirect estimates of stroke volume are based on the incorrect assumption that this distensibility and its relation to the other components of vascular impedance remain constant and that the stroke volume can therefore be calculated from the pressure pulse [4].

As the valves in the outflow tract close at the end of ventricular ejection, there is a short period of backflow that occurs primarily into the coronary arteries and secondarily into the space made available by the closing valves. The drainage from the arteries into the arterioles continues during this period, transforming a highly oscillatory flow into a less pulsatile flow in the smallest vessels (Windkessel effect). However, even in the capillaries, the blood flow is still pulsatile. The pressure wave that is generated by the ventricular contraction is propagated through the arterial bed at a speed that is 10 to 20 times greater than the blood velocity. The transmission of the pressure pulse is a function of the physical properties of the vessel wall (Eq. 10-19), and the dispersion of the pulse wave during its travel is one of the characteristics of the vascular system (Fig. 10-7).

On the venous side, the pressure and flow pulses are much smaller than on the arterial side. They are barely discernible in the periphery, but grow somewhat larger toward the atria. This indicates that they do not represent a transmission of the arterial pressure pulse, but rather they are generated by atrial contraction and travel against the direction of flow. Whereas the characteristics of aortic impedance are determined almost entirely by the properties of the arterial system, the venous impedance depends largely on the behavior of the generator on its outflow side, namely the right atrium. The impedance to forward flow is only about half as large as the impedance to backward flow. This may be related to the presence of valves in the peripheral veins, and it illustrates another feature of optimal design in a system where large amounts of blood are moved rapidly by small driving forces.

Figure 10-7 illustrates how the pressure and flow pulses change from the aorta to the peripheral part of the arterial system. The amplitude of the pressure pulse increases, the shape becomes narrower, the sharp notch in the downstroke disappears,

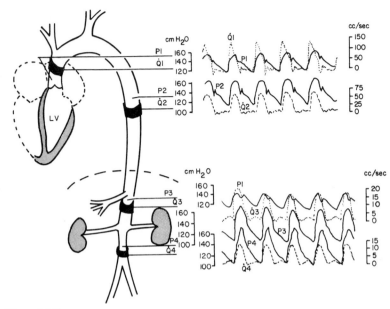

Figure 10-7
Pressure and flow tracings obtained simultaneously in the aorta of an anesthetized dog at the points indicated on the diagram at left. (From E. O. Attinger et al. *Circ. Res.* 19:230, 1966. Reproduced with permission of the American Heart Association, Inc.)

and a dicrotic wave appears. Although the pressure pulse increases in size, the amplitude of the flow pulse decreases significantly, its shape becomes broader, and the backflow component begins to disappear. This dispersion of the pulse waves is related to the progressive stiffening of the arterial tree toward the periphery, the viscoelastic behavior of the vascular walls, the complex branching pattern of the vasculature, and the wave reflections arising from mismatches in impedances.

The progressive change of the amplitude of pressure and flow pulses in the opposite direction as they travel along the arterial tree implies, of course, that the impedance in the peripheral part of the arterial bed must be considerably larger than in the proximal part. The magnitude of the pressure pulse usually amounts to less than one-half the mean pressure, while the magnitude of the flow pulse may be five to eight times as large as the mean flow rate (Fig. 10-8A). Because of this difference in the relative magnitude of the pulsatile components of pressure and flow, the frequency-dependent part of the vascular impedance (Fig. 10-8B) represents only a fraction of the so-called DC impedance (the latter is usually referred to as *peripheral vascular resistance* and is not frequency-dependent). This fraction is in the order of 5% to 10% in the ascending aorta and femoral artery, but it is considerably higher in the abdominal aorta (20%) and the mesenteric artery (45%), indicating that the physical properties of the vascular beds supplied from these sites are considerably different.

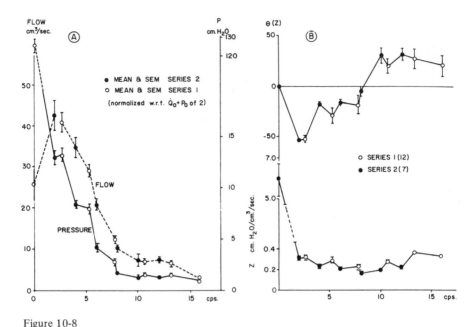

Figure 10-8
A. Frequency spectra of pressure and flow in the ascending aorta (mean and SE) in two groups
of anesthetized dogs. B. The pressure and flow data were subjected to Fourier analysis, and the
magnitude of the harmonic components is plotted as a function of frequency. The quotient of
these harmonic components determines the input impedance at the frequency of these harmonics,
which, being a vector, is characterized by a magnitude and a phase angle. (A negative phase angle
implies that pressure is leading flow.) Note that the DC value (total peripheral resistance) is in
the order of 6 cm H_2O/cm^3/sec, whereas the frequency-dependent impedance is much less
(approximately 5% of the DC values) and changes very little with frequency, which indicates
that the *external* cardiac work associated with pulsatile flow is minimal at any heart rate. (From
E. O. Attinger et al. Flowpatterns in the Peripheral Circulation of the Anesthetized Dog. *Angiologica* 4:1–27. Karger, Basel, 1967. Reproduced with permission.)

It is somewhat puzzling that while the significance of the pulsatile components of
pressure was recognized long ago as an important factor in the control of blood pressure there is still no agreement about the role that the much larger pulsatile flow components play in the overall function of the cardiovascular system. Although some
investigators [1] feel that the function of an organ suffers if it is perfused with a nonpulsatile flow, the evidence is by no means unequivocal. This question is, of course, of
extreme importance for the design of an artificial heart. As indicated in Figure 10-8B,
the magnitude of the vascular impedance in the ascending aorta is relatively independent
of the heart rate. As a consequence, the costs of providing pulsatile flow in terms of
external cardiac work are minimal at any heart rate. The differences between the
central and the peripheral arterial impedance constitute an essential feature of the
optimal design of the cardiovascular system.

In the previous section, the distensibility of blood vessels was emphasized. As the
transmural pressure rises, the vessel caliber increases, and the flow resistance decreases
as an exponential function of the radius. As discussed in connection with pressure-volum

curves, equal pressure increments have progressively less influence on vessel caliber and therefore on flow resistance. At high pressures, the system behaves more and more as if it were rigid. Changes of this type are characteristic of the aging process (see Fig. 10-5). Because of the complexities of the geometry of vascular beds and the physical properties of their components, Equations 9-10 and 9-16 for single, uniform tubes cannot be directly applied to the pressure-flow relationships in entire vascular beds. For this reason, an empirical expression has been proposed [3] :

$$\text{Flow} = \dot{Q} = cP^n \tag{10-20}$$

where c and n are experimentally determined parameters. In perfusion experiments of dog hind limbs, flow rates varying from 0.000062 (virtual occlusion) to 8.7 cm^3/min can be obtained with a perfusion pressure of 10 mm Hg and from 1.29 to 1.25 cm^3/min with a perfusion pressure of 200 mm Hg. The curvilinear relationships of Equation 10-20 are illustrated in Figure 10-9 for three different levels of the contractile state of the vascular walls. As the latter increases, the curves approach the pressure axis, so that for any given level of pressure, the flow is less than it was at lower levels of the contractile state of the vasculature. The exponent n can be interpreted as an index of vascular reactivity to changes in perfusion pressure, and it varies from about 1 to 3. Its value is strongly affected by the tissue metabolism—blood flow ratio. The lowest

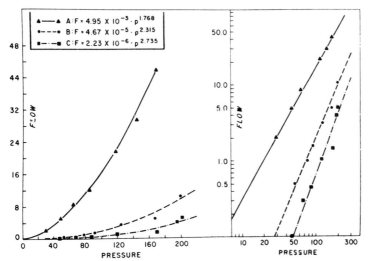

Figure 10-9

Relationship between driving pressure and flow in a cutaneous vascular bed at three levels of spontaneous, contractile states of the vasculature (the triangles represent the lowest, and the squares the highest level). The insert lists the values for the parameters c and n of Equation 10-20 as they pertain to the straight lines in the log-log plot (*right*) and to the curvilinear lines in the linear plot (*left*). (From H. D. Green, C. E. Rapela, and M. C. Conrad. In W. F. Hamilton [Ed.], *Handbook of Physiology, Section 2, Circulation*, Vol. 2. Washington, D.C.: The American Physiological Society, 1963. P. 935. Reproduced with permission.)

values for n and the highest values for c are found in the relaxed vascular bed. When maximal dilation is induced by a 10-minute period of ischemia and subsequent perfusion with hypoxic blood, the value of n is close to 1.0 for cutaneous beds. Many vascular beds tend to keep blood flow reasonably constant even in the presence of abrupt pressure changes (autoregulation).

The effects of changes in transmural pressure and its dependence on the physical properties of vascular walls are clearly apparent from the changes in flow patterns associated with respiration. In the dog with its chest opened and with artificial ventilation being supplied, there are no respiratory changes in transmural pressure and no effects on blood flow can be seen. However, during positive-pressure respiration with the chest closed, the cardiac output decreases (from 27 to 18 cm^3/sec) as does the arterial pressure (from 100 to 75 cm H_2O). Respiration has a pronounced influence on venous flow; in the superior vena cava, the mean flow is 5.1 cm^3/sec, compared to 7.5 cm^3/sec in the open chest. The flow in the superior vena cava varies from zero at positive intrapleural pressures to 10 cm^3/sec during expiration, and a significant amount of backflow appears during the peak of inspiration. The inferior vena cava mean flow is also reduced from 19.5 to 12.6 cm^3/sec in this experiment. This flow under closed-chest conditions varies between 10 and 15 cm^3/sec depending upon the phase of the respiratory cycle; the pulsatile components become larger during expiration. During spontaneous respiration as compared to positive-pressure respiration, the ascending aorta flow rate rises again to 23 cm^3/sec, the superior vena cava flow rate to 8 cm^3/sec, and the inferior vena cava flow rate to 15 cm^3/sec. The venous return in both venae cavae becomes considerably larger and the pulsatile components become smaller during inspiration as compared to expiration. The changes in arterial pressure and the pulsatile components of arterial flow during respiration are quite small compared to the changes in the venous bed.

BLOOD VOLUME AND ITS DISTRIBUTION

The maintenance of an adequate blood volume is of prime importance for normal cardiovascular function. It has been emphasized in the previous sections that estimates of blood volume based on intravascular pressure measurements alone are extremely unreliable because of the distensibility of the vascular beds (particularly on the venous side) and because of changes in the vascular contractile state. The latter may occur under different forms of stress and are not easily assessed even by direct methods. Normally, the blood volume as determined by dilution techniques is in order of 8% of the body mass, as compared to about 70% for total body water.[*] Of the total blood volume, about 22% is found in the pulmonary circulation, 14% in the heart, and 64% in the peripheral circulation. The distribution of blood volume with respect to different organ systems is shown in Table 6-1. The largest fraction is contained in the category of residual areas, which include the pulmonary circulation,

*This value depends strongly on the fat content of the body and is therefore better related to lean body mass.

the heart chambers, and the larger peripheral arteries and veins. The circulation of the gastrointestinal tract and of the skin serves as a major buffer for rapid shifts in blood volume caused by rapid changes in environmental temperature or by acute blood loss.

Estimates of the blood volume distribution within the individual segments of the systemic vasculature have, in the past, been primarily based on extrapolation of the anatomic data obtained by Mall [6] regarding the dog's intestinal vascular bed. This led to the assumption that three-fourths of the peripheral blood volume is contained in the venous system, most of it in vessels with a diameter greater than 1 mm. More recent data, however, appear to indicate that the major fraction of the peripheral blood volume is located in vessels with a diameter less then $200\,\mu$ [10]. The implications of these findings will become obvious when the mechanism of fluid balance is discussed in Chapter 11.

Circulatory shock of various types (e.g., hemorrhagic, toxic, or infectious) is characterized by the fact that the available blood volume becomes inadequate for the size of the vascular bed, as determined by its contractile state. Such discrepancies occur if there is massive blood loss from the circulation through either hemorrhage or sequestration. They also occur under conditions of general vasodilation, where the capacity of the vascular bed has become too large with respect to the remaining blood volume for the preservation of blood pressures that are compatible with normal cardiovascular function.

CORRELATIONS OF BIOMEDICAL AND PHYSICAL PROPERTIES

The blood flow to any given organ is a function of the difference between the pressures at the inflow and the outflow sites, the cross-sectional area, the length of the vascular channels, and the blood viscosity. Since the driving pressure is the same for all the vascular beds, the cross-sectional area emerges as the most important of these factors, being determined by the transmural pressure and the distensibility of the vessel walls. Because of their stiffness and the richness of smooth muscles in their walls, the arterioles are the main determinants of vascular resistance, whereas the distensible veins are able to accommodate large changes in blood volume with little change in pressure. This latter property provides the reason why the measurement of venous pressure per se is of little physiologic significance unless it is correlated with some index of venous volume. Blood flow is pulsatile in practically all segments of the vasculature, although the degree of pulsatility varies depending upon the generating source and the physical properties of the vascular walls. Because of the parallel arrangement of the vascular beds to individual organs, the measurement of arterial pressure is no indication of adequate distribution of blood flow or of adequate cardiac output (blood pressure is tightly controlled by the baroreceptors). Although estimates of peripheral resistance yield information about the condition of the peripheral vasculature, only flow measurements provide reliable data regarding the adequacy of the blood supply.

The flow behavior of blood in a uniform segment of the vascular tree is determined by two parameters: the characteristic impedance Z_0 and the propagation constant γ. The characteristic impedance relates pressure to flow at the entrance of an infinitely long, uniform vessel. In contrast to the vascular resistance — which, in the hemodynamic literature, is defined as the ratio of mean pressure to mean flow and is represented by a real number — impedance includes the effects of the viscoelastic and inertial properties of the vascular system and its contents. Like the propagation constant, it is characterized by both a magnitude and an angle.

The relation between pressure and flow at the end of a uniform segment of the vasculature is called the *load impedance*. Like the characteristic impedance, it is a function of viscoelastic, resistive, and inertial components, and it therefore has a complex value. As the vasculature is traced farther toward the periphery of the arterial bed, the influence of the imaginary components — viscoelasticity and inertia — becomes less and the more closely the load impedance approaches a pure resistance. Since the circulatory system is a distributed parameter network, each segment to be considered is very small, and its load impedance represents the input impedance of the following segment. If two consecutive segments have different mechanical properties, there is a mismatch between the characteristic and load impedances, which gives rise to reflected waves. The wave observed in the steady state will be the result of superposition of the incicent wave traveling downstream and the reflected wave traveling upstream.

Figure 10-10 shows a diagram of a blood vessel ending in two branches. A steady-state sinusoidal pressure generator is assumed to be at the vessel entrance. The pressure wave that is drawn with solid dots at the bottom of the diagram travels for two wavelengths along the vessel to the right, and it is partly transmitted and partly reflected at the branch point. The dashed line represents the transmitted wave; the starred line,

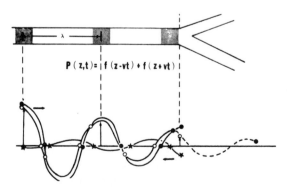

SUPERPOSITION OF INCIDENT AND REFLECTED WAVES (TIME FIXED)

Figure 10-10
Wave reflection at a branch point resulting from the application of a sinusoidal pressure at the vessel origin. The sum of the amplitudes of the incident wave (*solid dots*) and of the reflected wave (*stars*) at the branch point is equal to the amplitude of the transmitted wave (*dashed line*). Upstream, the incident and reflected waves appear superimposed as a single wave (*open circles*). (Adapted from E. O. Attinger. *Circ. Res.* 12:623, 1963. Reproduced with permission.)

the reflected wave. The line indicated by open circles is the actual pressure wave upstream of the branch point, i.e., the measured pressure, which is the vector sum of the incident and reflected waves. The amount and the phase of reflection are determined by the relation between the characteristic impedance and the load impedance. Since frictional losses occur, the amplitude of the incident wave decreases along the vessel toward the branch point, and the amplitude of the reflected wave decreases toward the origin of the vessel. Because of the frictional losses, no true standing waves can be produced.

REFERENCES

1. Attinger, F. M. Two dimensional in vitro studies of femoral arterial walls of the dog. *Circ. Res.* 22:829, 1968.
2. Bergel, D. H. *Cardiovascular Fluid Dynamics.* New York: Academic, 1972.
3. Green, H. Circulatory System: Physical Principles. In O. Glasser (Ed.), *Medical Physics,* Vol. 1. Chicago: Year Book, 1944.
4. Ketterer, E., and Kenner, T. *Dynamik des Arterien Pulses.* Berlin: Springer, 1968.
5. Learoyd, B. M., and Taylor, M. G. Alterations with age in the visco-elastic properties of human arterial wall. *Circ. Res.* 18:278, 1966.
6. Mall, F. Blut und Lymphanege in Oilnndarm des Hundes. *König Sachs Ges. Wiss., Abt. Math. Phys. Klasse.* Vol. 14, 1888.
7. Moens, A. I. *Die Pulskurve.* Leiden: E. J. Brill, 1878.
8. Patel, D. J., Greenfield, J. C., and Fry, D. L. In vivo pressure-length-radius relationship of certain blood vessels in man and dog. In E. O. Attinger (Ed.), *Pulsatile Blood Flow.* New York: McGraw-Hill, 1964.
9. Skilling, H. H. *Fundamentals of Electric Waves.* New York: Wiley, 1948.
10. Wiedman, M. P. Dimensions of blood vessels from distributing artery to collecting vein. *Circ. Res.* 12:375, 1963.

Material Exchange and Fluid Balance 11

E. O. Attinger

INTRODUCTION

The transport function of the cardiovascular system is an essential stage in the adequate exchange of nutrients and waste products between the body cells and the outside environment. This exchange takes place through the microcirculation, which is ideally suited for such a task: (1) its large total cross section provides a huge surface area for exchange and allows a low flow velocity (see Fig. 10-1), (2) the capillary walls are thin and easily permeated (Fig. 10-2), and (3) the large number of capillaries minimizes the distance between the blood and the individual cells over which the exchange takes place.

STRUCTURE OF THE MICROCIRCULATION AND ITS ENVIRONMENT

The exchange system consists of four components: the microcirculation, the interstitial spaces, the lymphatic system, and the tissue cells. The last represents the site of metabolic processes.

The microcirculation is arranged in a complex network of vascular channels that differ greatly in various regions of the circulation (Fig. 11-1). Although the blood flow is still somewhat pulsatile in the microcirculation, the path of an individual red blood cell through this maze seems to follow a stochastic pattern. Within the capillaries, there are periods of forward movement, backward flow, and stagnation that follow each other in a random fashion. These occur as a result of local vasomotion involving the smallest arteries and veins as well as the capillary sphincters. The sphincters are located at the entrance to most capillaries.

In contrast to the arteriolar and venular walls, which contain a rich muscle coat, the capillary walls consist of a single layer of endothelial cells (Fig. 11-2). Usually, the cells themselves are tightly joined by a cement-like substance, but intracellular pores also exist (see Fig. 11-2). One of the most prominent features of the capillary endothelium is the presence of a large number of circular and oval inclusions of cell membrane, the so-called *micropinocytotic vesicles.* The structure of capillaries differs considerably among different organs and even among different areas of the same organ. These differences have led to a broad classification of capillaries:

1. Those with a continuous endothelium and a continuous basement membrane (heart, muscle, lung, skin, and nervous system)
2. Those with a fenestrated endothelium and a continuous basement membrane (glomerulus, endocrine organs, and gastrointestinal tract)
3. Those with an open endothelium and a discontinuous or absent basement membrane (the sinusoids of the liver, spleen, and bone marrow)

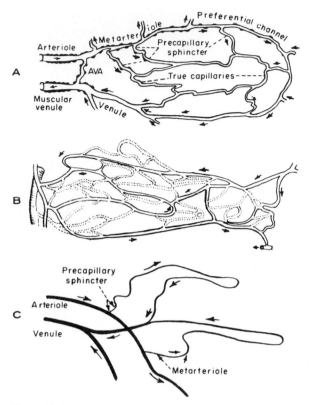

Figure 11-1
Three types of microcirculation patterns. *A*. Muscle with preferential channels. The location
of smooth muscle cells in the vascular walls is indicated by the black dots (*A VA* = arteriovenous
anastomosis). *B*. Mesenteric microcirculation. The true capillaries are shown by the dotted lines.
C. Hairpin capillary loops in human nailbeds. (From B. Zweifach. In B. W. Zweifach and E. Schorr
[Eds.], *Transactions of the 3rd Conference on Factors Regulating Blood Pressure.* New York:
Josiah Macy, Jr. Foundation, 1949. Reproduced with permission.)

The capillaries themselves are imbedded in the interstitial space, which can be
considered as a two-phase system consisting of a fluid phase and the meshwork of
the ground substance. The ground substance comprises a heterogeneous group of
materials, some of which arise by local synthesis in fibroblasts and other cells, whereas
others are derived from the blood. Among the former, the mucopolysaccharides
appear to be particularly important with respect to fluid exchange.

Figure 11-2 shows the interstitial space between a venous capillary (bottom) and a
lymphatic capillary (top). The capillary shown has a 25-nm wide pore, through which
plasma that contains water and protein is filtered. On the left side of the illustration,
there is a tight, 4-nm wide junction in the capillary, through which reabsorption occurs.

The system of lymphatic vessels exists only in the vertebrates, and, in the higher
orders, it consists of a network of closed lymphatic capillaries and efferent lymphatic

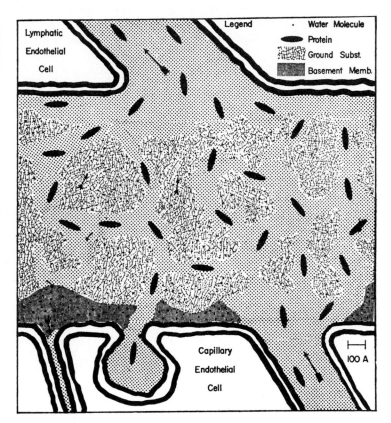

Figure 11-2
Interrelationships of ground substance, free fluid, and fluid and protein exchange in the interstitial space. The large pore and the tight junction between the capillary endothelial cells as well
as the size of protein molecules are drawn approximately to scale. (From C. A. Wiederhielm.
J. Gen. Physiol. 52:29, 1968. Reproduced with permission.)

vessels [7]. The lymph is collected from the various organs by the large lymphatic
trunks, which, like the veins, are equipped with valves. Some of these trunks empty
independently into the large veins, whereas others join the most important lymph
channel, the thoracic duct, which enters the vena cava. Total lymph flow in man is
in the order of 1 ml/min.

MECHANISMS OF EXCHANGE

The exchange of materials (solute and solvents) between blood and tissue cells takes
place through a combination of free and restrictive diffusion as well as through filtration and, in certain beds, active transport. After filtration or diffusion out of the

capillary, the water and solutes diffuse along the fibers of the connective tissues (the structural elements of the ground substance of the interstitial space) and are mixed with the extracellular fluid. They are then reabsorbed by cells, capillaries, or the lymphatic system.

FACTORS DETERMINING THE PASSAGE OF MATERIAL ACROSS THE CAPILLARY WALL

Characteristics of Exchanged Molecules

The size and shape of the solute and solvent molecules are important determinants of the diffusion coefficient because of their effects on the possible choices of diffusion paths (e.g., selective passage through pores or passage through the cells and intracellular cement substance) and on the ease with which this passage can be achieved. The range of pressures involved is shown in Table 11-1.

Table 11-1. Hydrostatic and Osmotic Pressures in the Microcirculation of Man

		Capillary			
		Arteriolar End (mm Hg)		Venous End (mm Hg)	
Direction	Force	Old	New	Old	New
Out of capillary	Blood pressure	32	32	15	25
	Tissue osmotic pressure	5	5 to 10[a]	5	5 to 10[a]
	Total	37	37 to 42	20	30 to 35
Into capillary	Tissue pressure	0 to 10	−5 to 0[b]	0 to 10	−5 to 0[b]
	Plasma osmotic pressure	25	25	25	25
	Total	25 to 35	20 to 25	25 to 35	20 to 25
Balance favoring filtration, arterial end		2 to 12	12 to 22	−	−
Balance favoring reabsorption, venous end		−	−	5 to 15	−5 to −15

[a]If there is no interaction between proteins and hyaluronic acid, the colloid osmotic pressure is 5 mm Hg. If the acid displaces the protein, the colloid osmotic pressure is 10 mm Hg.
[b]Guyton's capsule value is −5 mm Hg [3] and Wiederhielm's microneedle value is 0 mm Hg [9].

Additional factors in the exchange of molecules include the diffusion coefficient of the solute in serum, the solubility coefficient of the solute (this is not a limiting factor for the exchange of gases), and the electrochemical properties of the traveling molecules and their surrounding media. Furthermore, the chemical affinity between the materials to be exchanged and the capillary wall tissues plays an important role. Substances such as gases, which are lipid-soluble, have access to the entire surface of the endothelial cells, whereas materials that are only water-soluble have to pass through the intracellular junction or through pores.

Finally, the rate of utilization (or production) of metabolites is a prime determinant for one of the major driving forces, the concentration gradient.

Characteristics of the Capillary Wall

It has been pointed out that the microvascular structure exhibits considerable local differences. The size and shape of pores, their total area, and the pinocytotic mechanisms have to be considered in the evaluation of the local exchange of metabolites. Other factors include the chemical characteristics of the endothelial cells with respect to the exchanged molecules, the metabolism and the electrical activity of the wall, and changes in the physical, electrical, and chemical characteristics of the wall that occur due to neural and humoral changes.

Molecules up to a molecular weight of about 5000 appear to pass readily through the capillary walls. For example, every minute a volume of water equivalent to two-thirds of the blood volume is exchanged between blood and tissue. The exchange of larger molecules, such as proteins, is more limited. Nevertheless, in 24 hours a total amount of protein that is equal to the mass of protein in the circulating blood passes from the blood into the interstitial space, from whence it is either absorbed by individual cells, recirculated, or removed via the lymphatic system.

DRIVING FORCES RESPONSIBLE FOR MATERIAL EXCHANGE

External Forces

The hydrostatic pressure in capillaries represents one of the major driving forces (Fig. 11-3). It should be apparent from the discussion in Chapter 10 that this pressure is quite sensitive to small changes in arterial or venous pressure. It thus follows that it is also most readily changed by short-term control mechanisms.

Local Forces

Local forces, which are determined by the comparative characteristics of the exchanged material and the capillary wall, significantly affect the colloidal osmotic pressure of both plasma and interstitial fluid, as well as influence the Donnan potential for charged particles.

Concentration Gradient

The concentration gradient of the various molecules in the solute depends on the utilization and production rates of these molecules as well as on external and local forces. The exchange of small suspended particles through pores can be increased or decreased with respect to the exchange of the solvent depending on their respective

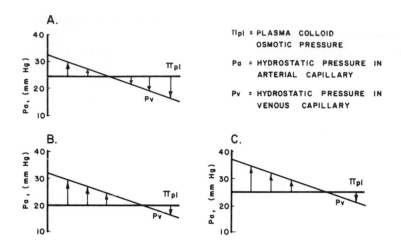

Figure 11-3
Hydrostatic and colloidal osmotic pressures along a capillary. *A*. Normal values. *B*. Hypopro-
teinemia. *C*. Elevated venous and arterial capillary pressure. (From C. A. Wiederhielm. *J. Gen.
Physiol.* 52:29, 1968. Reproduced with permission.)

concentration gradients. The transmural exchange rate for the entire capillary wall
is determined primarily by the electrochemical affinity between the exchanged molecules and the wall material, and the concentration gradient for each species of molecule provides the major driving force. For large molecules, additional mechanisms
are involved, such as pinocytosis and the stochastic sieving mechanism that is driven
by solvent passage.

The rate and the amount of exchange for any substance are functions of the combined effects of all these factors, which are, as already pointed out, weighted differently in various beds with respect to particular metabolites. Fluid exchange will be
discussed in more detail as a specific example of these general principles.

FLUID EXCHANGE IN THE MICROCIRCULATION

The classic formulation for the exchange of fluid across the capillary walls was derived
by Starling in the 1890s [8]. The transcapillary flow, $\dot{Q}$, of fluid is governed by the
difference in the hydrostatic and colloidal osmotic pressures between plasma and
tissues (see Fig. 11-3):

$$\dot{Q} = KA \left(Pc - Pt - \pi_{pl} + \pi_t \right) \tag{11-1}$$

where $\dot{Q}$ = flow (ml/min)

K = capillary permeability; K ranges from 0.01 to 0.06 μ/sec/mm Hg. (The permeability coefficient is larger on the venous side than on the arterial side of
the capillary.)

A = surface area of the capillary
Pc = hydrostatic pressure in the capillary
Pt = hydrostatic pressure in tissue
π_{pl} = colloidal osmotic pressure of plasma
π_t = colloidal osmotic pressure of tissue

The generally accepted values, as well as those more in agreement with recent experimental findings, for the four pressure parameters in Equation 11-1 are given in Table 11-1. Although the classic theory postulates the existence of a balance between filtration and reabsorption, the newer findings seem to indicate that fluid is lost throughout the capillary bed. In order to maintain homeostasis, fluid has to be reabsorbed by some channel other than the lymphatic system, whose flow rate is inadequate for this purpose. The most likely paths for this reabsorption are the venules, the larger veins, or both.

Ultrafiltration is the process that is chiefly responsible for maintaining fluid balance between the blood and the interstitial compartments. It contributes very little to the exchange of nutrients and metabolic waste; this process is governed primarily by the mechanism of molecular diffusion. In a network of capillaries, the contribution of each capillary to the total value of KA (Eq. 11-1) is largely independent of its share of total blood flow, because the flow is ordinarily much greater than the filtration rate [7].

The rate of transcapillary diffusion of any solute is proportional to the difference in concentration (or activity) of that substance on either side of the capillary wall:

$$\frac{dm}{dt} = \dot{M}_{1 \to 2} = PA_S\,(c_1 - c_2)$$

(11-2)

The proportionality constant, PA_S ("capillary diffusion capacity" or permeability coefficient–surface area product), depends on the permeability coefficient for the particular solute (P) and the effective capillary surface area (A_S). Since the mean concentration gradient across the wall of an individual capillary depends on the rate of blood flow relative to its permeability–surface area product, the contribution of each capillary of a network to the total PA_S value is highly dependent on its share of the total blood flow:

$$PA_S = -\dot{Q}\ln(1 - E)$$

(11-3)

where $\dot{Q}$ is the blood flow and $E = 1 - \exp(-PA_S/\dot{Q})$ is the fractional extraction rate. It follows that the fractional extraction rate decreases with increasing blood flow at a constant value of PA_S. Capillary clearance increases with increasing flow, but not proportionately. An increase in the fractional extraction rate may be brought about by the opening of capillaries that are ordinarily closed, the closure of functional shunts (see Fig. 11-1), the improvement of the uniformity of capillary circulation by readjustment of capillary sphincters (see Fig. 11-5), as well as by increases in capillary

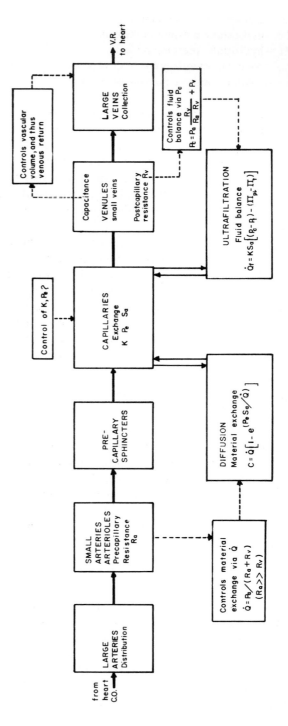

Figure 11-4
Summary of the exchange function of the microcirculation. (Modified from illustration by Dr. Eugene M. Renkin.)

permeability. These essential functions of the microcirculation are summarized in Figure 11-4.

The explanation of the equilibrium of fluid exchange according to the traditional model (Eq. 11-1) involves three generally unstated assumptions: (1) the permeability to water and solutes is uniform throughout the capillary, (2) the surface areas of the arterial and venous ends of the capillary network are identical, and (3) the hydrostatic and colloidal osmotic pressures of the tissue fluid are negligible compared to the corresponding blood values. Recent experiments, however, indicate that the venous capillaries have a larger surface area and a higher permeability to water than arterial capillaries and that tissue fluid pressures (as measured by means of implanted capsules) may be subatmospheric. This subatmospheric pressure could result from a redistribution of proteins in an essentially two-phase system (see Fig. 11-2). Protein molecules are restricted to the free fluid phase in the interstitial space, whereas water freely permeates the meshwork of the ground substance. The arrows in Figure 11-2 indicate the net drift of the solvent through the free fluid spaces as well as through the basement membrane and the ground substance. Since the proteins are restricted to the free fluid phase, they cause a higher osmotic pressure than if the bound fluid were available for their dilution. The protein concentration in the free fluid phase of the interstitial space may thus be considerably higher than the concentration in either the lymph or the capillary filtrate, and this higher protein concentration may therefore bring about a correspondingly higher osmotic pressure.

Mixtures of plasma protein and connective-tissue mucopolysaccharides, particularly hyaluronate, exhibit higher osmotic pressures than the sum of the osmotic pressures due to the two components individually. This synergistic effect appears to be related to steric exclusion of solvent by the long, rod-shaped mucopolysaccharide molecules. The extravascular fluid protein pool of a 70-kg man is in the order of 250 gm; assuming an interstitial volume of approximately 15 liters, the average plasma protein concentration in the interstitium amounts to about 1.7%. In a mixture of 1% hyaluronate (a reasonable value for its concentration in the interstitial space) and 1.7% plasma proteins, the excluded volume amounts to 52% of the total fluid volume. The resulting reduction of solvent for the proteins leads to a protein concentration of about 3.6%, which is in close agreement with the protein concentration of lymph from most organs [9]. A mixture of hyaluronate and plasma proteins in these proportions would exert an osmotic pressure of about 10 mm Hg, which is considerably higher than the values commonly used to describe the Starling-Landis [4, 8] hypothesis of capillary fluid balance.

Taking into account the facts that the surface area of the venous capillaries is considerably larger than that on the arterial side and that the surface of the venous capillaries is more permeable to water, the two-phase hypothesis of Wiederhielm [9] seems to be more realistic than the one of Starling. It also provides a convenient and practical explanation of the mechanism for fluid control. The development of edema has usually been explained in terms of swelling of the tissue spaces because of a rise in interstitial pressure associated with an imbalance of the Starling equilibrium. In view of the new findings, it seems more likely that the pressure that causes the swelling of tissues is osmotic in nature.

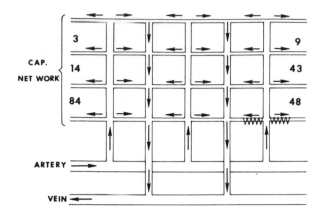

Figure 11-5
Flow distribution in a capillary network assuming a total inflow of 100 cm/min. Because of the unequal length of the individual channels, the perfusion is very inhomogeneous (*left side*). If the resistance of the shorter flow channels is increased (*right side*), the perfusion becomes more homogeneous (the numbers in the individual cells of the network indicate the magnitude of the local blood flow). This example clearly illustrates the practical use of simple theoretical concepts. (From E. O. Attinger. In Y. C. Fung [Ed.], *Biomechanics: Its Foundation and Objectives.* Englewood Cliffs, N.J.: Prentice-Hall, 1972. Reproduced with permission.)

IMPLICATIONS OF MICROCIRCULATORY FUNCTION FOR CARDIOVASCULAR CONTROL

Because of the parallel arrangement of the different vascular beds, an increase of blood flow to one bed can be achieved only if the supply to another bed (or beds) is correspondingly decreased or if the cardiac output increases. The concept of the fractional extraction rate (Eq. 11-3) implies that the transport of metabolites to and from tissues can be enhanced if the blood flow distribution becomes more uniform, for example, by means of appropriate contraction of the capillary sphincters.

Because of the complex geometry of the microcirculatory bed (see Fig. 11-1), the transit times for individual fractions of a unit volume of blood entering a vascular bed must be quite different (Fig. 11-5). If all the capillaries have the same cross section, the amount of blood flowing through any particular flow channel is inversely proportional to its length (see Eq. 9-10). Not only is the transit time of blood flowing through the uppermost channels in Figure 11-5 the longest, but they also receive the least blood supply. If the blood supply becomes too small, the oxygen consumption within a cell on such a channel must decrease because the tissue oxygen falls below its critical value. The blood flow distribution is improved if the resistance in the lowermost channels is increased, as indicated on the right side of Figure 11-5. As a consequence, the trans-capillary exchange of material (such as O_2) due to molecular diffusion is improved, whereas there is relatively little change in water transport, which is determined primarily by osmotic processes.

This suggests that there are at least three hierarchical control levels for the regulation of an adequate metabolite supply for individual tissues. The first, and lowest of these, is concerned with changes resulting in a more adequate intraorgan flow distribution, the second with changes in interorgan blood flow distribution, and the third with the adjustment of cardiac output to the metabolic requirements of the body as a whole.

Intraorgan blood flow distribution is clearly a matter of autoregulation. *Autoregulation* refers to the intrinsic tendency of an organ to maintain a constant blood flow despite changes in perfusion pressure. Three main theories have been advanced to explain the mechanisms involved:

1. The *tissue pressure hypothesis* assumes that autoregulation occurs in the low-pressure vessels because of the tissue pressure imposed on them. Since more recent findings indicate that the tissue pressure is subatmospheric, this hypothesis does not appear to be particularly promising.

2. The second hypothesis supposes that there is an intrinsic mechanism in the smooth muscle cells of the arteries or arterioles that responds to an increase in internal pressure or tension by contraction. Given the fact that the driving pressure for the various vascular beds of the systemic circulation is virtually identical, it is difficult to see how such a mechanism could result in a flow distrubution that is clearly determined according to a priority system, which changes continuously as a function of the physical and mental stresses to which the body is subjected.

3. The *metabolic hypothesis* is based on the supposition that a decrease of the blood flow/metabolism ratio causes vascular relaxation by the accumulation of vasodilatory metabolites in the tissue or by decreased nutrient supply. Experimental evidence, including quantitative data, clearly favors this hypothesis [1].

Many of the arterial and venous capillaries are arranged in loops so as to permit counter-current exchange. This mechanism has a well-established role in improving the efficiency of renal function as well as in temperature homeostasis in animals living in cold environments, but it seems to be even more ubiquitous than originally thought. For example, recent data indicate that a significant fraction of the total oxygen supply to the tissues leaves the vascular channels by diffusion from precapillary vessels [2]. Such an observation would not only be difficult to explain in the absence of a counter-current mechanism, but it also provides the basis for an extremely effective mechanism for the control of intraorgan blood flow distribution.

Interorgan blood flow distribution is partially controlled by means of autoregulation and partially by means of the nervous system (for example, through the baroreceptor and chemoreceptor reflexes). The local controls for intraorgan and interorgan blood flow distribution are, of course, in competition with higher level control loops that maintain the integrity of the organism in the presence of excessive local demands. Although there are also autonomous components in the control of cardiac output (as exemplified by Starling's law of the heart), changes in cardiac output, particularly under conditions of stress, are mediated primarily at various levels of the nervous system, either directly or through humoral agents.

CORONARY BLOOD FLOW AND OXYGEN CONSUMPTION

The coronary circulation is not only the shortest vascular bed, but it is also characterized by the highest oxygen extraction ratio. Hemodynamically, coronary blood flow is particularly complex because it is embedded in the cardiac muscle, which, during cardiac systole, exerts compression on the coronary vessels equivalent to (or higher than) the arterial pressure. As a consequence, the phasic flow in the coronary arteries is minimal during systole and generally reaches its peak just after the closure of the aortic valves.

One would therefore expect that coronary flow increases during bradycardia, since the diastolic period lengthens proportionately more than the systolic period. Conversely, during tachycardia, the maintenance of adequate flow requires vasodilation of the coronary bed. Normal blood flow is in the order of 85 $cm^3/100$ gm myocardium/minute, and oxygen consumption is about 10 $cm^3/100$ gm tissue/minute. These values may increase to nearly 600 for blood flow and to about 30 for oxygen consumption during exercise. Heart muscle has little capacity for storing oxygen or contracting an oxygen debt, and increased oxygen requirements must therefore be met by an increase in coronary flow. The control of the coronary vessels is obscured by the fact that the energy required to drive the blood through the coronary bed is provided by the myocardium, the function of which is in turn influenced by the adequacy of its perfusion. Metabolic factors, however, appear to play a major role in this control. If these special aspects are taken into account, the regulation of coronary blood flow has many features in common with that of skeletal muscle [5].

REFERENCES

1. Attinger, E. O. Biomechanics, Patient Care and Rehabilitation. In Y. C. Fung (Ed.), *Foundations and Objectives of Biomechanics.* New York: Academic, 1971.
2. Duling, B. R., and Berne, R. M. Longitudinal gradients in periarteriolar oxygen tension, a possible mechanism for the participation of O_2 in local regulation of blood flow. *Circ. Res.* 27:669, 1970.
3. Guyton, A. C. A concept of negative interstitial pressure based on implanted perforated capsules. *Circ. Res.* 12:399, 1963.
4. Landis, E. M., and Pappenheimer, J. R. Exchange of Substances Through the Capillary Walls. In W. F. Hamilton (Ed.), *Handbook of Physiology, Section 2, Circulation,* Vol. 2. Washington, D.C.: The American Physiological Society, 1963.
5. Mellander, S. Interaction of local and nervous factors in vascular control. *Angiologica* 8:187, 1971.
6. Renkin, E. M. Transcapillary exchange in relation to capillary circulation. *J. Gen. Physiol.* 52:96, 1968.
7. Rusznyak, I., Foldi, M., and Szabo, G. *Lymphatics and Lymph Circulation.* London: Pergamon, 1960.
8. Starling, E. H. On the absorption of fluids from the connective tissue spaces. *J. Physiol.* (Lond.) 19:312, 1896.
9. Wiederhielm, C. A. Dynamics of transcapillary fluid exchange. *J. Gen. Physiol.* 52:29, 1968.

Cardiovascular Control 12

E. O. Attinger

GENERAL CONSIDERATIONS

Cardiovascular control systems basically affect three variables: cardiac output, peripheral resistance (general or local), and circulating blood volume. Normally, blood pressure is maintained relatively constant either by an appropriate adjustment of all these variables or by overcompensation on the part of the peripheral resistance component. Changes in these variables are achieved either *passively* by means of mechanical feedback, such as the distension of a blood vessel with an increase in the transmural pressure, or *actively* by means of local control mechanisms (autoregulation) or neural or humoral control loops. Thus far, most of these control systems have been studied as isolated control loops, although it is clear that they interact extensively during any adjustment of cardiovascular performance.

Changes in cardiac output require changes in stroke volume, heart rate, or both. Changes in peripheral resistance are associated with either vasodilation, vasoconstriction, or both. Since blood flow varies in proportion to the fourth power of the vessel radius, small changes in dimensions may affect the flow resistance significantly. The circulating blood volume can also be altered by selective changes in the relative dimensions of blood vessels in series, as well as by changes in the osmotic pressure within the blood or the tissues (see Eq. 11-1).

It is thus clear that the effectors of the different cardiovascular control systems are either the heart itself or the smooth muscle in the walls of the vascular tree. Changes in the contractile state of the resistive vessels determine regional blood flow, and changes in the activity of the precapillary sphincters determine the degree of homogeneity of flow through the microcirculation (see Fig. 11-5). The state of the vascular smooth muscle thus affects most of the factors governing nutritional exchange, such as the magnitude and distribution of capillary flow, the flow velocity, the size of the capillary exchange surface, and the diffusion distances to tissue cells. Vasomotor adjustments of the ratio of precapillary to postcapillary resistance influence the hydrostatic capillary pressure and hence the net transcapillary fluid movements and plasma volume (see Chap. 11). Changes in the dimensions of the capacitive vessels are associated with redistributions of regional blood volumes, thus affecting venous return and cardiac filling.

Figure 12-1 illustrates an example of the full range of effects elicited by strong vasodilation (white columns) and vasoconstriction (shaded columns) in cat muscle. Using the corresponding data calculated for human skeletal muscle, it may be seen that the blood flow may vary by a factor of 100, the capillary surface area by a factor of nearly 7, and the rate of volume changes may reach values of 170 ml/min in the capillaries and 600 ml/min in the capacitive vessels. The magnitude of these

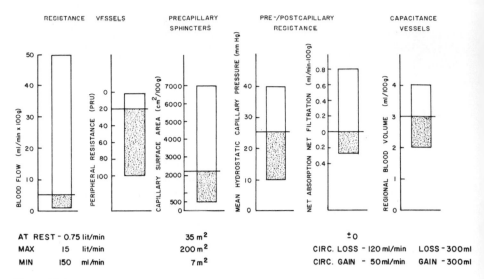

Figure 12-1
Maximum effects in the different consecutive vascular sections elicited by active changes of the smooth muscle tone in 100 gm of cat skeletal muscle. (Shaded columns: constriction; white columns: dilatation.) Corresponding data calculated for 30 kg of human skeletal muscle are shown at bottom. (From S. Mellander. *Angiologica* 8:187, 1971. Reproduced with permission.)

values indicates the importance of the vascular musculature for overall cardiovascular performance.

It is important to realize that the response of local vascular beds to any given stimulus may be quite different [9]. For example, the pulmonary vascular bed constricts when exposed to low alveolar PO_2, whereas the peripheral vasculature dilates. Another example is the effect of the baroreceptor reflex on local vascular resistance, which is quite different for individual vascular beds.

In this chapter, the control aspects of the cardiovascular system are arbitrarily divided into four sections: local regulation, nervous control, humoral control, and integrative control.

LOCAL CONTROL

LOCAL CONTROL OF THE HEART

The heart has the inherent ability to initiate its own beat in the absence of any nervous and humoral control (Chap. 8). It also has the capacity to adapt to changing hemodynamic conditions by virtue of mechanisms that are intrinsic to cardiac muscle itself. Adaptive changes that involve changes in myocardial fiber length (in accordance with Starling's law of the heart) have been called *heterometric autoregulation,* whereas those that are independent of changes in fiber length are termed *homeometric regulation.*

It is important to remember that the adaptations and control mechanisms discussed in this and the following two sections have usually been investigated under highly artificial conditions. Hence, they may indicate only potential control mechanisms, since in the intact animal under stress, they are often either overridden or modified by other control loops.

Heterometric Autoregulation

From isolated heart-lung preparations, Frank and later Starling and his co-workers [4] concluded that the heart compensates for increases in venous return or in peripheral resistance by a change in diastolic muscle fiber length. An increase in fiber length facilitates ventricular contraction, until an optimal fiber length is reached beyond which contractions become impaired. The dilated heart, however, requires considerably more oxygen for a given amount of external work than the normal heart.

Since subsequent investigators were not always able to demonstrate the Frank-Starling mechanism, some doubt arose about its significance in the intact animal. Some of this controversy has been clarified by Sarnoff and his co-workers [10], who suggested that the Frank-Starling mechanism should be represented by a family of ventricular function curves (see Fig. 7-11) in which ventricular stroke work is plotted as a function of end-diastolic ventricular or mean atrial pressure. A shift in the ventricular function curve to the left (as, for example, during the stimulation of the stellate ganglion) indicates an improvement in ventricular performance, whereas a shift to the right signifies impairment and a tendency to cardiac failure. Because the two ventricles are arranged in series and their outputs must, on the average, be exactly equal, a reliable mechanism that is capable of matching cardiac output with changes in venous return is essential for normal cardiovascular performance.

Homeometric Autoregulation

It has now been established that the mammalian ventricle is capable of adapting to changes in venous return or peripheral resistance without a progressive increase in myocardial fiber length. Although transient heterometric regulation frequently precedes homeometric regulation, the former is by no means a prerequisite for the latter. Furthermore, homeometric regulation avoids the mechanical disadvantage of ventricular dilation upon myocardial fibers.

Rate-Induced Regulation

Bowditch [3] first demonstrated the importance of the time interval between heart beats for myocardial performance. Since a longer time interval permits more adequate ventricular filling, part of this regulatory mechanism can be explained by heterometric autoregulation. This explanation is, however, invalid for the increased contractility observed after premature beats in isovolumetrically contracting hearts, in which neither filling nor ejection takes place. Recently, this mechanism has been exploited for the improvement of ventricular performance in cardiac patients by the induction of repetitive postextrasystolic potentiation through stimulating electrodes. Various hypotheses have been advanced to explain ventricular autoregulation, but the basic mechanisms are still poorly understood.

LOCAL CONTROL OF PERIPHERAL BLOOD FLOW

The fact that imposed changes in perfusion pressure at constant levels of tissue metabolism are met with vascular resistance changes that tend to maintain a constant blood flow is commonly referred to as *autoregulation* (Fig. 12-2). An abrupt, step-like change in perfusion pressure is followed by a change in blood flow (solid circles in Fig. 12-2). The blood flow, however, returns to control levels within 30 to 60 seconds, so that over a range of 20 to 120 mm Hg, steady-state flow is nearly constant (open circles). Autoregulatory phenomena of this type have been described for the vascular beds of the kidney, brain, heart, muscle, and intestine, and three hypotheses have been advanced for their explanation [7]:

1. The *tissue pressure hypothesis* links an increase in perfusion pressure to an increase in blood volume and in net transfer of fluid from intravascular to extravascular compartments. The resultant increase in tissue pressure then compresses the thin-walled vessels and thereby reduces the local blood flow. A reduction of perfusion pressure would have the opposite effect. Physically, however, this hypothesis requires relatively rigid encapsulation of the tissue, for which there is little evidence.

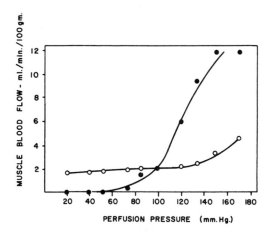

Figure 12-2
Pressure-flow relationship in the skeletal muscle vascular bed of the dog. The solid circles represent the flows obtained immediately after abrupt changes in perfusion pressure from the control level (the point where the lines cross). The open circles represent the steady-state flows obtained at the new perfusion pressure. (From R. M. Berne and M. N. Levy. *Cardiovascular Physiology* [2nd ed.]. St. Louis: Mosby, 1972. Redrawn from R. D. Jones and R. M. Berne. *Circ. Res.* 14:126, 1964.)

2. The *myogenic hypothesis* states that vascular smooth muscle contracts in response to stretch and relaxes as tension is reduced. The initial flow increase produced by an abrupt rise in pressure (which passively distends the blood vessels) would thus be followed by a return of flow to control levels, since the contraction of the smooth muscle would reestablish the geometry of the flow demands required for this.

3. The third hypothesis links autoregulation to the *metabolic activity* of the tissue. Vasodilatory metabolites are thought to be released from tissue as a function of its metabolic rate, and these metabolites would dilate the vessels locally. Blood flow would thus be automatically adjusted to meet local metabolic requirements, since a decrease in metabolic activity would result in a decrease in the local concentration of the metabolite (because of decreased production and washout or inactivation) and a subsequent increase in resistance due to the intrinsic contractile state of vascular smooth muscle. This tonic activity of the vascular smooth muscle is independent of the nervous system, but the factors that may be responsible for the basic contractile state of the muscle are not known. Although a large number of substances (such as oxygen, lactic acid, potassium, and adenine nucleotides) have been shown to possess vasodilatory effects, the agents mediating metabolic vasodilation have not yet been identified. There may be several factors involved, and they may carry different weights for different beds under different circumstances.

Reactive hyperemia is closely linked to autoregulation of blood flow. If the arterial inflow to a bed is occluded for a short period, there will be a sizeable overshoot of flow after the release of the occlusion before the flow returns to its control value. The amount of overshoot is a function of the duration of the occlusion as well as of the metabolic state of the tissue (Fig. 12-3).

The low-pressure system of the pulmonary circulation responds to low alveolar oxygen tension and high CO_2 tension by vasoconstriction, thus shunting blood away from poorly ventilated areas. The site of these resistance changes is still controversial.

NERVOUS CONTROL OF THE CARDIOVASCULAR SYSTEM

EFFECTORS

Neural control of the cardiovascular system is mediated through the autonomous nervous system, i.e., the sympathetic and parasympathetic pathways.

Heart Rate

Normally, the sinoatrial node is under the tonic influence of both divisions of the autonomic nervous system. The sympathetic system exerts a facilitative influence and the parasympathetic system an inhibitory influence upon the rhythmicity of the pacemaker. Changes in the heart rate, however, usually involve reciprocal action by the two systems, i.e., cardiac acceleration is produced by a diminution of parasympathetic activity, with a concomitant increase in sympathetic activity. Ordinarily,

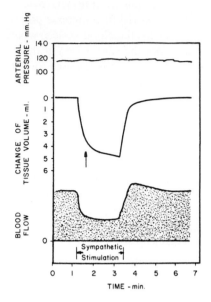

Figure 12-3

Effect of sympathetic nerve stimulation (2 impulses/sec) on blood flow and tissue volume in the hindquarters of the cat. The arrow denotes the change in slope of the tissue volume curve where the volume decrease due to the emptying of capacitance vessels ceases and the loss of extravascular fluid becomes evident. (From R. M. Berne and M. N. Levy. *Cardiovascular Physiology* [2nd ed.]. St. Louis: Mosby, 1972. Redrawn from S. Mellander. *Acta Physiol. Scand.* 50 [Suppl. 176]:1, 1960.)

parasympathetic influences are predominant, particularly in well-trained athletes, whose heart rate under resting conditions (50 to 60 beats per minute) is considerably lower than that of untrained persons (70 beats per minute or more).

Myocardial Contractility

Sympathetic nervous activity facilitates myocardial performance primarily by enhancing the contractility of the individual cardiac muscle cells. In addition, the spread of excitation through the musculature is accelerated, thus making the contraction of the individual fibers more synchronous. As a result, greater pressures along with a greater maximum rate of change of pressure are generated, and the abridgment of systolic duration as well as more rapid ventricular relaxation enhance ventricular filling. The positive inotropic influence of sympathetic activity upon atrial contraction also facilitates filling. The importance of these mechanisms increases as the heart rate increases. The multiplicity of facilitative sympathetic influences is easily appreciated in terms of ventricular function curves (see Fig. 7-11): for any given end-diastolic pressure, the ventricle is capable of performing more work as the level of sympathetic nervous activity is progressively raised.

The parasympathetic system exerts profound depressant effects upon the cardiac pacemaker, the atrial myocardium, and the atrioventricular conduction tissue. The vagus nerves also impose a negative inotropic effect upon the ventricular myocardium. The latter seems in part to be achieved by antagonism of the facilitative influences engendered by the prevailing level of sympathetic nervous activity. Thus, at low levels of sympathetic activity, the negative inotropic effects produced by vagal stimulation are small, whereas at high levels of sympathetic activity, the negative inotropic effects produced by vagal stimulation become considerable.

The Vascular Tree

The sympathetic system mediates both vasoconstriction and vasodilation throughout the entire peripheral circulation, whereas the effects of the parasympathetic system are limited to the blood vessels of the head, the viscera, and the genitalia.

The sympathetic vasoconstrictor fibers originate from the vasomotor center in the medulla oblongata, descend the spinal cord, and synapse at different levels of the thoracolumbar region. They then emerge with the ventral roots and join the paravertebral chain via the white rami communicantes. The preganglionic fibers pass up or down the sympathetic chain and synapse in the sympathetic ganglia or pass via the splanchnic nerves to the sympathetic ganglia in the abdomen. The postganglionic fibers from the chain join the corresponding spinal nerves and innervate the peripheral arteries and veins. The postganglionic fibers from the splanchnic and cervical ganglia join the large arteries and follow them as an investing network to the resistive and capacitive vessels.

The vasomotor center itself is tonically active, and this activity may be enhanced by a number of stimuli. An increase in activity results in an increase in the frequency of impulses reaching the peripheral vessels, where a neurohumor (norepinephrine) is released and elicits vasoconstriction. Inhibition of the tonic activity reduces the impulse traffic through the constrictor system, which results in vasodilation.

The major effects of the constrictor system are observed in the vasculature of the microcirculation. The relative contributions of the resistive and capacitive vessels are illustrated in Figure 12-4. At constant arterial pressure, sympathetic nerve stimulation results in a reduction in blood flow due to constriction of the resistive vessels. The abrupt decrease in tissue volume is due to movement of blood out of the capacitive vessels, while the late and slow volume decrease is associated with transcapillary fluid exchange (see Chap. 11). Cessation of stimulation results in the relaxation of the smooth muscles of the capacitive vessels and a restoration of their previous blood volume. The stimulus frequency at which the capacitive vessels reach maximal constriction is considerably lower (1 to 10 Hz) than that of the resistive vessels (15 to 25 Hz). The mobilization of blood from the capacitive vessels is important for circulatory adjustments (for example, during exercise), since it enhances venous return and thereby cardiac output.

The fibers of the cholinergic sympathetic dilatory system arise in the cerebral cortex and pass through the hypothalamus before joining the sympathetic outflow from the

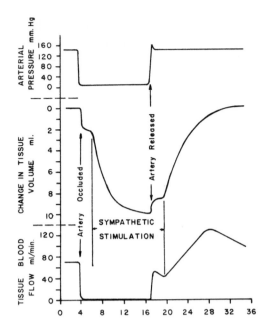

Figure 12-4

Effect of arterial occlusion and sympathetic nerve stimulation (6 impulses/sec) on the tissue volume in the hindquarters of the cat. Note the small changes in tissue volume with arterial occlusion and release in comparison to the large volume change obtained with sympathetic nerve stimulation. (From R. M. Berne and M. N. Levy. *Cardiovascular Physiology* [2nd ed.]. St. Louis: Mosby, 1972. Redrawn from S. Mellander. *Acta Physiol. Scand.* 50 [Suppl. 176] :1, 1960.)

spinal cord. In contrast to the constrictor system, the activation of the dilation system is transitory and has been associated primarily with the anticipatory muscular vasodilation that occurs prior to exercise (fight or flight).

The effects of the parasympathetic system on peripheral resistance are small because of its limited distribution to the peripheral vessels. Stimulation of these cholinergic fibers may induce, for example, vasodilation to the salivary glands.

PERIPHERAL INPUTS

The activity of autonomic effector neurons depends on the nature of the afferent information that reaches them from the peripheral inputs. Frequently, this information is modified during transit, either in the central nervous system itself or by means of the effects of competing inputs. In particular, generalized stresses, such as exercise, hypoxia, or hemorrhage, tend to alter the inputs from several cardiorespiratory receptor groups so that the autonomic reflex effects become a function of the entire input profile.

Mechanoreceptors

These include the stretch receptors in the carotid sinus and aortic arch, the pulmonary arterial baroreceptors, and the cardiac mechanoreceptors. Impulses arising in the carotid sinus receptors travel up the sinus nerve (nerve of Hering) and pass via the glossopharyngeal nerve to the medulla, whereas impulses arising in the aortic arch receptors reach the medulla via the afferent fibers in the vagus nerve. These impulses respond to changes in mean pressure as well as to the rate at which arterial pressure changes. In the case of the carotid sinus receptors, for example, the relationship between the firing rate $F(t)$ and the arterial pressure $Pa(t)$ has been expressed as:

$$F(t) = AdP^+/dt + BdP^-/dt + C[Pa(t) - P_{th}] \qquad (12\text{-}1)$$

where P_{th} is the pressure threshold for the receptor unit, dP^+/dt and dP^-/dt are the time derivatives of pressure during the rising and falling phases of the pressure pulse, and A, B, and C are sensitivity coefficients [8]. The magnitudes of the threshold pressure and of all the coefficients vary with the level of mean pressure. The relationship between the mean sinus pressure and the average carotid sinus baroreceptor nerve firing rate is sigmoid, and, at any mean pressure, the firing rate is greater when the pressure is pulsatile than when it is constant (Fig. 12-5).

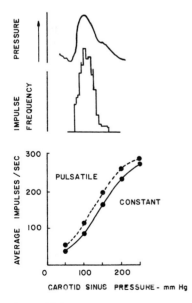

Figure 12-5
Impulse firing frequency in a single carotid baroreceptor unit and the pressure pulse at normal arterial pressure (*top*). Average impulse frequency recorded from a multifiber preparation of the carotid sinus nerve in relation to the mean carotid sinus pressure while the pressure in the sinus is either constant or pulsatile (*bottom*). (From P. I. Korner. *Physiol. Rev.* 51:312, 1971. Reproduced with permission.)

The pulmonary arterial baroreceptors are located in both main branches of the pulmonary artery. They respond to changes in mean pressure, but they appear to be particularly sensitive to changes in the rate of pressure with each pulse.

The sensors in the left and right atrium provide rate-sensitive information at the time of the A wave of atrial pressure (see Fig. 7-14) that is related to the booster action of the atrium on ventricular filling. They also provide information at the time of the V wave of atrial pressure about the degree of filling of the pulmonary and systemic venous systems. For this reason, they have been called *volume receptors* by Gauer [5]. Receptors with somewhat similar properties are present in the coronary sinus. Ventricular receptors signal changes in systolic tension; their firing rate can be increased by alkaloids extracted from *Veratrum* species.

Chemoreceptors

The arterial chemoreceptors in the aortic arch and the carotid bodies are sensitive to changes in arterial PO_2, PCO_2, and pH, to certain abnormal constituents, such as carbon monoxide, and to changes in their blood supply. The small chemoreceptor discharges that are found at a PO_2 of 100 mm Hg increase exponentially with a reduction in PO_2 or a rise in PCO_2 (Fig. 12-6). Their discharge rate fluctuates with respiratory frequencies up to about 1 Hz.

Lung Inflation Receptors

The pulmonary stretch receptors represent the largest group of vagal afferent sources. They are sensitive primarily to changes in transpulmonary pressure (and might thus be stretch receptors), but they are little influenced by pulmonary congestion.

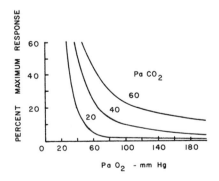

Figure 12-6
Average chemoreceptor impulse frequency recorded from a multifiber preparation in relation to changes in arterial PO_2 and PCO_2. (From P. I. Korner. *Physiol. Rev.* 51:312, 1971. Reproduced with permission.)

Somatic, Visceral, and Special Sensory Inputs

Somatic inputs through the trigeminal nerve are important in relation to diving and nasal circulatory reflexes. Spinal somatic and visceral afferent impulses, which travel in small to medium-sized fibers, originate in the skin, muscle, gastrointestinal tract, kidney, urinary tract, and the ventricular wall.

CENTRAL PATHWAYS OF THE NEURAL CARDIOVASCULAR CONTROL SYSTEM

The integration of the afferent input profile into a coherent, autonomous nervous system output takes place in the central nervous system. The vasomotor center is located bilaterally in the reticular substance of the lower pons and upper medulla (Fig. 12-7). Its lateral portions are tonically active and maintain a partial state of contraction of the blood vessels (the so-called *vasomotor tone*). The medial part of the center does not participate in the excitation of the vasomotor fibers, but it transmits inhibitory impulses into the lateral part, thus producing vasodilation. The two parts of the center thus inhibit each other reciprocally. The lateral part of the vasomotor center also transmits impulses through the sympathetic fibers to the heart that result in cardiac acceleration and increased contractility, whereas the inhibitory center affects heart rate and myocardial function via the parasympathetic fibers of the vagus nerve. The vasomotor center may show rhythmic changes in tonic activity; these changes may be manifested as oscillations of arterial pressure that occur at the frequency of respiration (Traube-Hering waves) or that are independent of respiration and occur at lower frequencies (Mayer waves).

Although the vasomotor center still operates effectively in the absence of higher centers (e.g., after transection of the pons), it is normally under the influence of

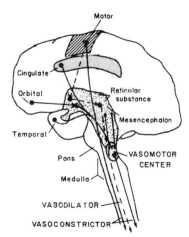

Figure 12-7
Areas of the brain that play important roles in the neural regulation of the circulation. (From A. C. Guyton. *Textbook of Medical Physiology* [4th ed.]. Philadelphia: Saunders, 1971. P. 286. Reproduced with permission.)

neural impulses arising in the baroreceptors and chemoreceptors, the hypothalamus, the cerebral cortex, the viscera, and the skin. The operation of the vasomotor center can also be altered by changes in the blood concentration of O_2 and CO_2. Many areas throughout the reticular substance of the pons, mesencephalon, and diencephalon can either excite or inhibit the vasomotor center (see Fig. 12-7). For example, stimulation of the anterior hypothalamus results in bradycardia and a fall in blood pressure, wherea stimulation of the posterolateral regions has the opposite effect. Through the temperature-regulating center in the hypothalamus, skin vessels are selectively affected in order to either preserve heat by vasoconstriction or enhance heat losses by vasodilation.

Some of the information transfer between afferent and efferent sources is also made at the level of the spinal cord, where somatic and visceral dorsal-root afferents make polysynaptic contact with the preganglionic autonomic motoneurons. However, only a fraction of the spinal autonomic motoneuron pool participates in the transfer along these pathways, and the remainder becomes activated only if the supraspinal centers are intact. The supraspinal control of these motoneurons tends to leave influences descending from the higher centers undisturbed while attenuating the reflex effects produced by segmental spinal afferent inputs. As a consequence, relatively large spinal afferent stimuli are required to produce significant autonomic reflex effects.

It is thus apparent that every type of autonomous effect can be reflectively elicited provided the cardiovascular centers in the medulla and pons are intact. This activity can be influenced from many suprabulbar regions, particularly those to which the principal cardiorespiratory afferents project. The divergence of these projections provides the anatomic basis for the differential activation of the suprabulbar centers with varying input profiles.

CARDIOVASCULAR REFLEXES

Systemic Arterial Baroreceptor Reflex

Although the reflex effects that are elicited by the activity of the four receptor zones are qualitatively similar, the magnitude of the overall effect does not correspond to the algebraic sum of the individual effects. Furthermore, the open-loop gain, G_o, which is defined as the change in systemic arterial pressure divided by the pressure change in the receptor area, is different for each receptor site. For example, the G_o of the aortic arch is about one-fourth to one-half of that of the carotid sinus receptors. The integration of these four inputs into a single control mechanism represents an excellent example of the redundancy of design in biologic control systems.

The mean pressure in the isolated carotid sinus is inversely related to the mean systemic arterial pressure, and the latter is lower at any given sinus pressure if that pressure is pulsatile (Fig. 12-8). Because of the sigmoid nature of the relationship between the two pressure variables, G_o is a continuously varying function and must therefore be defined with respect to a specific operating point on the sigmoid curve.

The effects of the baroreceptor reflex on the various local beds are quite different. It produces significant changes in the vascular resistance of the splanchnic and muscular beds but practically none in the skin vessels.

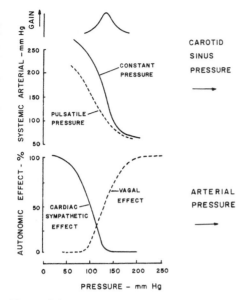

Figure 12-8
Relationship between mean carotid sinus pressure and mean systemic arterial pressure when the pressure in the isolated sinus is either constant or pulsatile (*top*). (The relationship of steady sinus pressure to the gain of the reflex is shown at top of graph.) Relationship between cardiac sympathetic and vagal autonomic effects at different arterial pressure levels (*bottom*). (From P. I. Korner. *Physiol. Rev.* 51:312, 1971. Reproduced with permission.)

A rise in systemic pressure produces a decrease in the heart rate because of increased vagal efferent activity, whereas the tachycardia that occurs after reduction in sinus pressure is predominately the result of increased cardiac sympathetic activity. At normal resting arterial pressures, there is some overlap between the two efferent systems (see Fig. 12-8). The response of the vagal effectors is considerably faster (the time-constants are in the order of 1 second) than that of the sympathetic effectors (10 to 20 seconds).

The operation of the baroreflex appears to be relatively independent of influences from higher centers. Despite the many anatomic connections (see Fig. 12-7), practically every autonomic effector may be involved during changes in this input. The degree of involvement is highly nonuniform, and it depends on differences among central autonomic mechanisms and on the resting neurogenic vascular tone.

Pulmonary Artery Baroreceptor Reflex

The effects of this reflex are more pronounced in the respiratory system than in the circulatory system, since it contributes to the adjustment of the respiratory pump to the venous return during changes in right ventricular output. The relationship between a rise in pulmonary pressure and ventilation is an inverse one, with an approximately

20% transient change in ventilation per 1 cm H_2O pressure change. Cardiovascular effects become significant only if arterial pressure changes exceed 60 mm Hg.

Cardiac Mechanoreceptor Reflexes

The main group of cardiac receptors that are active with each heart beat appears to be the atrial receptors. The effects of their reflexes are qualitatively similar to those of the carotid sinus reflex, although they are considerably weaker except for the effects on heart rate and the renal vasculature.

Normally, little information reaches the central nervous system from epicardial and ventricular receptors. If they are strongly stimulated, they produce reflex brady-cardia, hypotension, and, in the case of the epicardial mechanoreceptors, a reduction in respiration.

Arterial Chemoreceptor Reflexes

The most important reflex effects elicited by these receptors pertain to the stimula-tion of ventilation, the carotid body receptors being 3 to 6 times as effective as the aortic body receptors.

The cardiovascular reflex effects of chemoreceptor stimulation are dependent on the level of ventilation (Fig. 12-9). During controlled ventilation, they may involve every autonomic effector. Hypoxia plus hypercapnia result in (1) bradycardia (due to increased vagal and reduced cardiac sympathetic activity), (2) an increase in arterial blood pressure and in total peripheral resistance (sympathetic constrictor nerve activity) (3) increased constrictor effects in all major beds, including muscle, skin, gastrointes-tinal, and renal beds, and (4) increased secretion of catecholamines. The cardiac

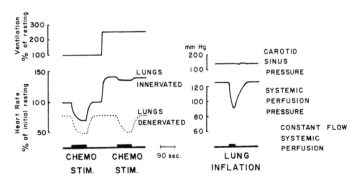

Figure 12-9
Stimulation of isolated carotid chemoreceptors with hypoxic blood at two levels of ventilation in a dog with normally innervated lungs (solid lines) and after section of the pulmonary vagi (dotted lines) (*left*). Effect of pulmonary inflation of the lung (at arrow) on the systemic perfusion pres-sure in a dog in which the systemic circulation is perfused at constant volume flow while the carotid sinus pressure is maintained constant (*right*). (From P. I. Korner. *Physiol. Rev.* 51:312, 1971. Reproduced with permission.)

autonomic effects are reinforced by the action of the baroreceptor reflexes associated with the rise in arterial pressure.

Although the respiratory effects of the arterial chemoreceptor reflex are not significantly altered after decerebration, the cardiovascular responses require an intact diencephalon.

Lung Inflation Reflex

The normal respiratory cycle produces phasic alterations in the input from the pulmonary stretch receptors, which are abolished after section of the vagal afferents.

Increased pulmonary stretch tends to suppress cardioinhibitory reflex effects from the stimulated chemoreceptors, as well as the peripheral sympathetic constrictor tone in all major peripheral beds.

Respiration also has cyclic effects on the susceptibility of the central autonomic mechanisms to the input from arterial baroreceptors, and it renders the sympathetic constrictor mechanisms and the autonomic mechanisms that control the heart rate less susceptible to incoming baroreceptor inputs during inspiration than during expiration. The tonic activity of the respiratory centers can influence the activity of the vasomotor center directly, even in the absence of respiratory movements, and may produce typical Traube-Hering waves [1].

Spinal Reflexes

Strong stimulation of many somatic and visceral receptor groups evokes a reflex-induced increase in heart rate, blood pressure, regional constrictor effects, and secretion of adrenal catecholamines.

Central Ischemic Reflex

A reduction of the blood supply to the brain leads to a rise in arterial pressure and a significant increase in sympathoadrenal activity. Vagal activity, however, decreases unless the cerebral ischemia is sudden and extreme. The basis of this reflex may be associated with direct asphyxial excitation of central sympathetic mechanisms.

INTEGRATION OF NEURAL VASCULAR CONTROL

Changes in the input from all cardiorespiratory receptors can alter the activity of every circulatory autonomic effector, although there tends to be preferential engagement in the case of a particular input. Most autonomic effectors have a significant degree of resting autonomic activity, so that a change in a particular direction in the input will produce either a reflex increase or decrease in effector activity, whereas a change in the opposite direction will evoke the opposite autonomic effect. The setpoint about which these directional changes for a particular autonomic effect occur, as well as the gain of the control loop (i.e., the change in autonomic output per unit change in input), provide a reference for quantitative analysis, as Korner has suggested [8]. Since every major cardiorespiratory input converges on the autonomic motoneuron pool to every effector, the effects of multiple inputs (and of their interactions) can be quantified by comparing the combined effects with that of the primary input (e.g., from the baroreceptors) in terms of changes in the set-point and the gain of the control loop under consideration. Figure 12-10 shows an example of the changes in set-point without a change in gain for the baroreceptor reflex for heart rate during arterial hypoxia; these changes are due to the added effects of increased activity of the arterial chemoreceptors. Thus far, this type of analysis has been limited to the static characteristics of the baroreceptor reflex.

Alterations in the excitability of most of the neurons in a particular effector pool and changes in the pool size can account for changes in the set-point, whereas changes in gain occur as a result of facilitative and occlusive convergence onto a certain fraction of motoneurons that are common to two or more inputs, as a result of changes in effective pool size, or as a result of very large changes in the excitability of most neurons in the pool.

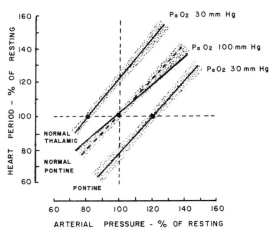

Figure 12-10
Regression lines expressing the mean relationship between the mean arterial pressure and the heart period (reciprocal of heart rate) in normal, thalamic, and pontine unanesthetized rabbits while breathing air and during arterial hypoxia. Data at 100 mm Hg Pa_{O_2} from ten normal rabbits (dashed line; 56 observations) and five pontine animals (solid line; 20 observations) were obtained by raising the arterial pressure with phenylephrine infusion or lowering it with trinitroglycerine, neither of which has direct local effects on heart rate. Data at 30 mm Hg Pa_{O_2} were calculated from changes occurring between the second and thirtieth minute of hypoxia in normal, thalamic, and pontine animals with all effectors intact and in normal animals pretreated with phenoxybenzamine to reduce the arterial pressure during hypoxia (14 normal rabbits, 70 observations; 10 thalamic animals, 40 observations; and 10 pontine animals, 50 observations). Resting arterial pressures of different preparations were closely similar, and there was no significant difference between the slope of any of the lines. Shading represents ±1 SE about the lines; at 100 mm Hg Pa_{O_2}, the SE refers to normal data. (From P. I. Korner. *Physiol. Rev.* 51:312, 1971. Reproduced with permission.)

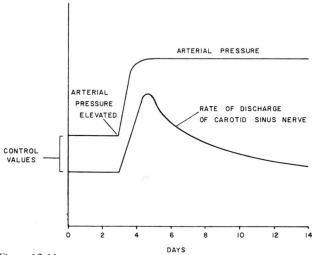

Figure 12-11
Adaptation of the carotid sinus baroreceptors following initial strong stimulation by increased arterial pressure. (From A. C. Guyton. *Textbook of Medical Physiology* [4th ed.]. Philadelphia: Saunders, 1971. P. 301. Reproduced with permission.)

The autonomic nervous system represents a powerful control system for cardiovascular performance under conditions of acute stress, such as exercise or hemorrhage. Because of the adaptation of the receptors as well as of the cardiovascular system itself, the effectiveness of these control systems decreases progressively if the disturbance persists for long periods of time. In arterial hypertension, for example, the baroreceptor activity returns to normal values a few days after the hypertension has been induced (Fig. 12-11). It therefore appears that long-term regulation is mediated by the kidneys and humoral mechanisms, rather than by the nervous system [6].

HUMORAL CONTROL

Apart from the humoral factors through which both local and neural controls are mediated (i.e., metabolites and the neural transmitter substances acetylcholine, epinephrine, and norepinephrine), cardiovascular performance is also dependent on substances that are transported to the effectors by the bloodstream. Although this aspect of cardiovascular control is less well-understood than local or neural control, its importance for long-term control has been firmly established.

NOREPINEPHRINE AND EPINEPHRINE

Both of these hormones are secreted by the adrenal medulla. When injected in small quantities, they produce vasoconstriction throughout the peripheral circulation. In large doses, epinephrine causes vasodilation in skeletal and cardiac muscle, which is related in part to the fact that it increases muscular metabolism.

THE KININS

The kinins are small polypeptides that are split off from an α-globulin in plasma or tissue fluids by proteolytic enzymes. The kinins are powerful vasodilators. Bradykinin is believed to be present in exocrine glands, but its physiologic significance is still a matter of controversy.

ANGIOTENSIN

The hormone angiotensin is formed in the blood in response to renin release from the juxtaglomerular cells of the kidney. This release is stimulated by low arterial pressure or low sodium concentrations in the blood. Angiotensin exerts a powerful vasoconstrictor effect on the arterioles. Because these vessels can adapt to angiotensin rather rapidly, it is unlikely that angiotensin is an important factor in long-term vascular control.

VASOPRESSIN

The vascular action of the hormone vasopressin is similar to that of angiotensin. Since it is secreted in minute quantities, its significance for direct vascular control is doubtful, but it is of primary importance for the control of water reabsorption in the renal tubules.

SEROTONIN

Serotonin is present in large concentrations in the chromaffin tissues of abdominal structures and in the platelets. Depending upon the state of the circulation, it may produce either vasodilation or vasoconstriction. Its role in vascular control is not understood at present.

HISTAMINE

Histamine is released by damaged tissue. It constricts the arterioles and dilates the veins. In many pathologic conditions, this leads to fluid losses into the tissues and to edema because of elevated capillary pressures.

INTEGRATIVE CARDIOVASCULAR CONTROL

The dual control of the cardiovascular system by extrinsic and intrinsic mechanisms provides a wide range of cardiovascular adjustments with a maximum of efficiency.

In addition to changes in cardiac output, the blood flow can be redistributed to organs that by virtue of their activity have higher metabolic requirements and shunted away from tissues with low activity. Such a priority system requires a hierarchically structured control system that is flexible enough to assess priorities under a wide variety of changes in the internal and external environments [2].

The brain and the heart are both vital structures with a very limited tolerance for reduced blood supply. In these organs, intrinsic flow-regulating mechanisms are usually dominant.

In contrast, the control of skin blood flow is dominated primarily by extrinsic mechanisms, since the cutaneous vessels participate strongly in general vasoconstrictor responses and also respond selectively to information from the temperature-control system in the hypothalamus.

The vasculature of skeletal muscle is a prime example of the interplay and changing balance between extrinsic and intrinsic control mechanisms. In resting muscle, neural control is dominant; section of the sympathetic nerves to the tissue is followed immediately by a significant increase in blood flow, which, however, persists for only a few hours (adaptation). In anticipation of and at the start of exercise, the blood flow increases, which is probably mediated by activation of the cholinergic sympathetic dilator system. Following the onset of exercise, the intrinsic flow-regulating mechanisms assume control, and the vasodilation resulting from local metabolic effects completely overrides the general sympathetic discharge that manifests itself by general vasoconstriction in inactive tissues.

Although some of the features of the interaction between the various control systems under different types of stress are now perceived in a qualitative way, their quantitative analysis is still in its infancy.

REFERENCES

1. Attinger, E. O. Atmung und Atmungsarbeit. *Dtsch. Med. Wochenschr.* 86:288, 1961.
2. Attinger, E. O., and Attinger, F. M. Hierarchy levels on the control of blood flow to the hind limb of a dog. *Kybernetik* 13:195, 1973.
3. Bowditch, H. I. *The Young Stethoscopist.* New York: Hafner, 1964.
4. Braunwald, E., Ross, J., and Sonnenblick, E. H. *Mechanisms of Contraction of the Normal and Failing Heart.* Boston: Little, Brown, 1968.
5. Gauer, O. H., Henry, J. P., and Siker, H. P. Cardiac Receptors and Fluid Volume Control. In A. M. Master and E. Doneso (Eds.), *Progress on Cardiovascular Disease.* New York: Grune & Stratton, 1961.
6. Guyton, A. C., Coleman, T. G., and Granger, H. I. Circulation: Overall regulation. *Annu. Rev. Physiol.* 34:13, 1973.
7. Johnson, P. C. (Ed.). Autoregulation of blood flow. *Circ. Res.* 15(Suppl. 1):1, 1964.
8. Korner, P. I. Integrative neural cardiovascular control. *Physiol. Rev.* 41:312, 1971.
9. Levy, M. N. (Ed.). Neural regulation of the cardiovascular system: Symposium. *Fed. Proc.* 31:1226, 1972.
10. Sarnoff, S. J., and Mitchell, J. The Control of the Function of the Heart. In W. F. Hamilton and P. Dow (Eds.), *Handbook of Physiology, Section 2, Circulation,* Vol. 1. Washington, D.C.: The American Physiological Society, 1962.

SELECTED READING

Berne, R. M., and Levy, M. N. *Cardiovascular Physiology*. St. Louis: Mosby, 1967.
Guyton, A. C. *Textbook of Medical Physiology*. Philadelphia: Saunders, 1968.

Models of the Cardiovascular System   13

E. O. Attinger

GENERAL ASPECTS OF MODELING

Models are a prerequisite for human communication. They range from concepts and hypotheses expressed in pictorial or descriptive form to complex mathematical formulations, sophisticated physical constructions, and extensive simulations by computer techniques. The primary goals for constructive models are (1) the achievement of a better understanding of the nature of the thing modeled, (2) the quantitative formulation of its phenomena, and (3) the prediction of the behavior of the system on the basis of a few select parameters. A physician, for example, diagnoses and treats a *model* of a patient when he interprets the patient's condition. Such interpretations are often biased because of the subjective components in the process of interpretation, i.e., the process of establishing a correspondence between the model and that which is modeled. Because of the unknown and variable characteristics of many of the links by which information is gathered and interpreted, individual judgments about the performance of physical, biologic, and socioeconomic systems can differ widely.

In this chapter, we are concerned with models of the physical characteristics of the cardiovascular system. Models for other physiologic systems are presented elsewhere in this book. These models may be mathematical expressions of dynamic physiologic behavior, or they may be in the form of electrical or mechanical analogs.

Such models, if based on adequate experimental data, lead not only to better understanding of cardiovascular function, but, even more importantly, to more reliable indicators of cardiovascular performance. Throughout the previous discussions, the convenience of using simplifying assumptions for the analysis of biologic systems has been emphasized. A judicious selection of these assumptions represents one of the most critical points of departure for any analysis, and it requires considerable insight into biologic phenomena if the simplification brought about by such assumptions is not to distort the results of the analysis.

With these points in mind, the following factors have to be considered:

1. The model must be physically realizable, and it must be established in terms of parameters that are significant and measurable in the living system.
2. The model has to include all the available pertinent information regarding the system under study. Assumptions that conflict with existing data must be evaluated and justified with particular care.
3. The simpler a model, the easier it is to evaluate its behavior. It is also easier to assess the contribution of its individual components to the differences between the performance of the biologic system and its analog.

4. It is preferable, although not always possible, to construct the model in a form that permits alterations in assumptions and system parameters without excessive effort. The model must, of course, be simpler to manipulate than the system under study.

5. A good model should serve as a guide for the experimental investigator, suggest certain experiments, and rule out the necessity of others. It should also form a basis for extrapolation, both beyond the range of observed data and to more general properties of biologic systems, i.e., it should have predictive power.

Some of the limits of existing models are best illustrated by examples from clinical medicine, where the inadequacy of presently available noninvasive measurements for the management of critically ill patients has been amply demonstrated.

LUMPED PARAMETER MODELS

Lumped parameter models are readily suitable to simulate the entire cardiovascular system or any section of it. In these models, the space-dependence of the variables is neglected, although the nonlinear pressure-volume relations of the various vascular chambers are frequently included. The cardiovascular system can be divided into a few segments, and the physical parameters of each segment may be concentrated in one point. The diagram of such a model consists, then, of a number of interconnected blocks (Fig. 13-1), each of which is described by a set of equations. Since this type of model must involve infinite wave velocities (i.e., the model dimensions are too small with respect to the real wavelengths), it cannot be used to simulate the changes in pressure and flow pulses along the various segments of the vascular tree. On the other hand, it is quite adequate for a study of the relations between cardiac output and peripheral load or for the evaluation of the factors that determine the overall

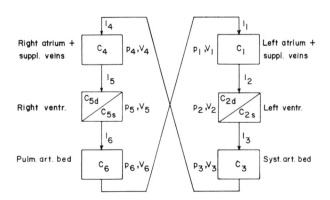

Figure 13-1
Warner's model of the closed circulatory system. The 18 simultaneous equations were solved on an analog computer. (From H. R. Warner. *Fed. Proc.* 21:87, 1962. Reproduced with permission.)

behavior of the cardiovascular system and its control. The differences among the various models of this type that have been proposed consist of the number of blocks contained within and the number or type of equations expressing the physical behavior of each block.

Because of the difficulties associated with building hydraulic models, electrical analog models are being increasingly used. Electrical analogs also can easily be programed for analog or digital computers [3, 9]. These techniques are much more suitable for the evaluation of the performance of a system by varying individual parameters over a wide range than are the more clumsy hydraulic models. Hydraulic models are easily translated into equivalent electrical networks using the analogies listed in Table 13-1.

Table 13-1. Analogies Between Hydraulic Parameters and Electrical Parameters

Cardiovascular Parameters	Electrical Circuit Analogs
P = pressure (mm Hg)	V = voltage (volts)
V = volume (ml)	q = charge (coulombs)
$\dot{Q} = \delta V/\delta t$ = flow (ml/sec)	I = current (amperes)
$C = \Delta V/\Delta P$ = compliance (ml/mm Hg)	C = capacitance (farads)
$R = \dfrac{P}{\dot{Q}}$ = vascular resistance (mm Hg·sec/ml)	$R = V/I$ = electrical resistance (ohms)
$L = \dfrac{P}{\delta \dot{Q}/\delta t}$ = inertance (mm Hg·sec^2/ml)	$L = \dfrac{V}{\delta I/\delta t}$ = inductance (henries)

Figure 13-1 illustrates the analog computer model of Warner [7], which he developed to study cardiovascular control. Warner subdivided the system into six sections, in which flow, $\dot{Q}$, is defined as I, pressure as P, and volume as V, as indicated in the diagram.

In order to solve for the 18 unknown values of I_i, P_i, and V_i, eighteen equations are required:

1. Six equations of motion of the form:

$$P_{i-1}(t) - P_i(t) = L_{i-1}\frac{dI_i(t)}{dt} + R_{i-1}I_i(t) \tag{13-1}$$

where I_3 and I_6 are zero during diastole; I_2 and I_5 are zero during systole

2. Six continuity equations of the form:

$$V_i(t) = V_i(t = 0) + \int [I_i(t) - I_{i+1}(t)]dt \tag{13-2}$$

where $i = 1, 2, \ldots, 6$

3. Six equations of state of the form:

$$P_i = \frac{V_i^m}{C_i} \tag{13-3A}$$

where $m > 1$ for each segment representing an atrium plus the supplying veins ($i = 1, 4$) and $m = 1$ for the arterial trees ($i = 3, 6$)

$$P_i = \frac{V_i^n}{(C_i)_{dia} \text{ or } (C_i)_{sys}} \tag{13-3B}$$

where $n > 1$ for each ventricle; $(C_i)_{dia}$ and $(C_i)_{sys}$ hold during diastole and systole, respectively ($i = 2, 5$).

Pumping is thus defined in terms of the instantaneous changes in C_2 and C_5, the resistances R_2 and R_5 incorporate the aspects of the force-velocity relationship of the myocardium, and inertial terms are introduced to account for the effect of the blood mass.

The equations used by various investigators in their lumped parameter models are similar to Equations 13-1 and 13-2, except for the representation of cardiac function and myocardial mechanics. During recent years, considerable attention has been devoted to the latter areas. As an example, Robinson's model [5] will be discussed briefly (Fig. 13-2).

In muscle mechanics, a muscle of length l may be conceptually divided into two lengths: l_{se}, the length of the series elastic component, and l_{ce}, the length of the contractile elements (see Chaps. 5 and 7). With the onset of isometric contraction, the contractile element shortens at a rate that is limited by the force-velocity relationship, building up tension and stretching the elastic component. Conceptually, the "active state" pressure $P_{sys}(V)$ in the contractile portion V_{ce} forces blood into the elastic portion V_{se}, which builds up pressure there. Blood is transferred from V_{ce} to

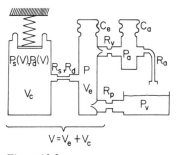

$$V = V_e + V_c$$

Figure 13-2
A hydraulic, lumped-parameter model of the cardiac pump. Subscripts d and s refer to diastole and systole, respectively; V_c refers to the volume of the contractile element, and V_e to that of the series elastic element. (Modified from D. A. Robinson. *Circ. Res.* 17:207, 1965. Reprinted by permission of the American Heart Association.)

V_{se} at a rate that is limited by the force-velocity relationships of the fibers, whose role is reproduced by a constrictive resistance between the two hypothetical chambers (see Fig. 13-2). This concept is formalized by the following equations:

Ventricular systole:

$$P_{iv} = P_{sys}(V) + R_{sys}\frac{dV}{dt} - R_{sys}C_{se}\frac{dP_{iv}}{dt} \tag{13-4}$$

Ventricular diastole:

$$P_{iv} = P_{dia}(V) + [(R_{sys} - R_{dia})e^{-t/\tau} + R_{dia}]\, C_{se}\frac{dP_{iv}}{dt} \tag{13-5}$$

where P_{iv} = intraventricular pressure
     $P_{sys}(V)$ = isometric pressure-volume function during systole
     V = total ventricular volume = $V_{se} + V_{ce}$
     $R_{sys} = (\eta/dV)_{sys}$ = coefficient of myocardial viscosity per unit volume during systole
     dV/dt = cardiac output during systole; venous return during diastole
     C_{se} = compliance of V_{se}
     $P_{dia}(V)$ = isometric pressure-volume function during diastole
     $R_{dia} = (\eta/dV)_{dia}$ = coefficient of myocardial viscosity per unit volume during diastole
     τ = relaxation time-constant of myocardial viscosity

The system is completed by providing an arterial load and a venous source, the equations for which are similar to Equations 13-1 and 13-2.

Robinson estimated the values of the critical parameters for a 10-kg dog to be:

$$R_{sys} = 3.2 \times 10^3 \text{ dyne} \cdot \text{cm}^{-5} \cdot \text{scc}$$
$$R_{dia} = 0.26 \times 10^3 \text{ dyne} \cdot \text{cm}^{-5} \cdot \text{sec}$$
$$C_{se} = 18.8 \times 10^{-6} \text{ cm}^5 \cdot \text{dyne}^{-1}$$

As a result of this analysis, he concluded that the force-velocity relationship of muscle is manifested in ventricular dynamics as a hydraulic resistance R_{sys} across which a pressure drop occurs from the developed isometric pressure (about 3.9×10^5 dyne·cm^{-2}) to the pressure actually seen in the ventricle (about 1.9×10^5 dyne·cm^{-2}). Its magnitude is comparable to that of the "direct current" (DC) impedance of the vasculature (i.e., the total peripheral resistance), and it is considerably higher than that of the dynamic impedance of the arterial load. This implies that the left ventricle acts as a flow pump rather than a pressure pump and that cardiac output is relatively independent of the peripheral load.

DISTRIBUTED PARAMETER MODELS

A distributed parameter model was first proposed by Witzig [8], and it has been extensively developed during the past decade [4]. The parameters are time- and space-dependent, which permits the simulation of finite wave velocities as well as changes in the shapes of pressure and flow pulses that occur between any two points in the system. These models are thus ideally suited for studying the details of the pressure-flow relationships in vascular channels, but, because of their complexity, they are inferior to good lumped parameter models for the investigation of the over-all performance of the cardiovascular system and its control. They are usually based on the linearized Navier-Stokes equation and the continuity equation (Fig. 13-3A):

$$-\frac{\delta P}{\delta x} = R'\dot{Q} + L'\left(\frac{\delta\dot{Q}}{\delta t}\right) \tag{13-6}$$

$$-\frac{\delta\dot{Q}}{\delta x} = G'P + \frac{dA}{dP}\frac{\delta P}{\delta t} \tag{13-7}$$

where R' = viscous resistance of the fluid per unit length
        L' = inertance per unit length
   dA/dP = distensibility per unit length
        G' = parallel conductance (reciprocal resistance; due to leakage current) per unit length
        A = vascular cross section

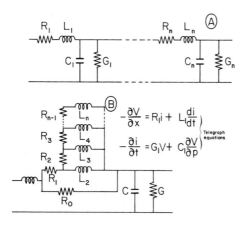

Figure 13-3
A. Transmission line analog of the arterial tree. B. Refinement of the transmission line analog taking into consideration the Bessel-type velocity profile and the sleeve effect. The equations are based on the analogies between hydraulic and electrical models that are expressed in Table 13-1. (From E. O. Attinger and A. Anné. *Ann. N.Y. Acad. Sci.* 128:810, 1966. Reproduced with permission.)

These equations are analogous to the "telegraph equations" given in Figure 13-3B, taking into account the analogies defined in Table 13-1. In order to account for the Bessel-type velocity profile, Noordergraaf [4] introduced the parallel R-L ladder shown in Figure 13-3B. The resistance R_0 has been introduced to take into consideration the *sleeve effect*, which is the concept that close to the vascular wall, there is a relatively cell-free layer whose viscosity is significantly less than that of the main stream in which the bulk of the blood cells is suspended.

Noordergraaf's model [3] consists of some 130 similar segments for the arterial tree, whose parameters include elastic and geometrical tapering, and it has been programed for an analog computer.

Another example of a distributed parameter model, which comprises the entire systemic circulation and which has been programed for a digital computer, is illustrated in Figure 13-4. The model is based on Womersley's theory [9], using a modified formulation of his equations to express the pressure-flow relationships. The vascular impedances were translated into their electrical analogs.

In its present form (see Fig. 13-4), the model is capable of quantitatively simulating the mechanical behavior of the peripheral circulation of the anesthetized dog. The cardiac output is distributed between eight parallel vascular beds (two for the head, one for the gut, two for the kidney, two for the legs, and one for the pelvis). Each individual bed consists of four parts: the large arteries, the small arteries, the capillaries, and the veins. The aorta is divided into eight segments of 5-cm length each, and the major artery for each vascular bed originates at the appropriate aortic segment. Similarly, the venous outflows from the individual beds merge with the vena cava flow at the appropriate sites. The inferior vena cava consists of six segments

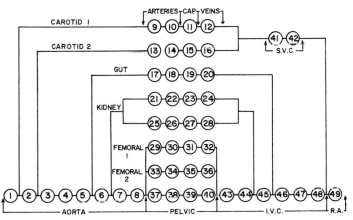

Figure 13-4
Model of the peripheral circulation consisting of eight parallel vascular beds. Each numbered block represents one vascular segment and contains a number of four-terminal networks. Each vascular bed is subdivided into four functional units: large arteries, small arteries, capillaries, and veins. (From E. O. Attinger et al. In A. Copley [Ed.], *Hemorrheology*. New York: Pergamon, 1967. Reproduced with permission.)

and the superior vena cava of two segments of 5-cm length each. The two venae cavae join at the right atrium, where the model terminates.

It has already been shown (Eq. 9-16) that for pulsatile flow, the pressure-flow relations depend greatly on the nondimensional parameter, α. Because of the interaction of inertial and viscous forces, the resistance and inductance, R_0 and L_2, of the equivalent network in Figure 13-3B become frequency-dependent: $R(\omega)$ and $L(\omega)$. The longitudinal impedance (see Eq. 10-10) thus changes to:

$$Z_{lo} = \left(\frac{8\eta l}{4\pi r^4} + j\omega \frac{\rho l}{\pi r^2} \right) (1 - F_{1,0})^{-1} = R(\omega) + j\omega L(\omega) \tag{13-8}$$

where $(1 - F_{1,0}')$, $R(\omega)$, and $L(\omega)$ are as shown in Figure 13-5. Since the vascular wall is viscoelastic, its compliance C and the equivalent capacitance in the network must be complex, i.e., C^* or $C(\omega)$. The complex modulus of elasticity E_{mod}^* shows

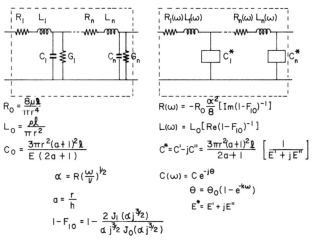

Figure 13-5
Two four-terminal networks used as building blocks for the model of the peripheral vascular bed. The values of the network parameters are calculated according to the expressions below the diagrams. On the left, these parameters are frequency-independent, on the right, frequency-dependent. Two possible choices for a frequency-dependent compliance are indicated by C^*. The parameters in the parallel branch are given by:

$$C(\omega) = C' = K \frac{E'}{(E')^2 + (E'')^2}$$

$$C(\omega) = \omega C'' = \omega K \frac{E''}{(E')^2 + (E'')^2}$$

The second expression for $C(\omega)$ is that proposed by Taylor [6]. The terms J_0 and J_1 are Bessel functions of the zero and first order, respectively. (From E. O. Attinger et al. In A. Copley [Ed.], Hemorrheology. New York: Pergamon, 1967. Reprinted with permission.)

an increase from its static value of between 0 and 3 Hz and remains relatively constant. Similarly, the phase angle increases significantly in the low-frequency range. On the right side of Figure 13-5, two possible expressions for the complex compliance are shown.

From model studies of this type, the following conclusions can be drawn:

1. On the basis of the changes in magnitude of the harmonics of the pressure pulse and of the phase velocity with distance, McDonald proposed that wave reflections are the primary mechanism involved in the deformation of the pressure pulse [2]. According to this hypothesis, one would expect pronounced oscillations of the input impedance, with maxima and minima occurring alternately at quarter-length intervals; such behavior is characteristic of uniform transmission lines and rubber-tube models. Experimental evidence, however, indicates that this is clearly not the case. Hence, additional factors must be invoked to explain the frequency behavior of the impedance. These factors are:

   A. The geometrical taper, the elastic taper, the viscous wall properties, and the viscous properties of the blood are each capable of damping the oscillatory behavior of the impedance. The relative importance of these four factors has not yet been established, but it seems reasonable to assume that their effects are cumulative.

   B. The distributed and scattered nature of the terminations of an assembly of randomly branching elastic tubes greatly reduces the influence of reflections upon the behavior of the input impedance [6].

2. The model of the peripheral circulation [1] illustrates that pressure pulse transmission from the systemic arteries to the systemic veins is effectively prevented.

3. The behavior of the input impedance is determined primarily by the parameters chosen to describe the aorta and its major branches. The frequency dependence of $R, L,$ and C has little effect on the behavior of the aortic input impedance, but it is important for the analysis of the pressure-flow relations in smaller arteries. The effects of convective acceleration in the thoracic aorta and deceleration in the abdominal aorta, which are related to the particular geometry of that vessel, can be adequately simulated by dividing the aorta into eight uniform segments.

4. The wave velocity in the venous system is a fraction of that in the arterial system. Measurable pulsatile pressures are present only in the thoracic venae cavae. These originate in the right atrium and are rapidly dampened out, but pulsatile flow persists throughout the venae cavae.

5. By introducing the nonlinear pressure-dependence of the various parameters into such models, it has been possible to simulate the behavior of the peripheral circulation at any distending pressure. In contrast to the venous system, the mechanical behavior of the arterial system is relatively insensitive to changes in arterial and venous distending pressures as far as the frequency behavior of the impedance is concerned. On the other hand, the transmission of the pressure pulse is significantly influenced by the distending pressures in terms of both wave velocity and damping.

6. In the pulmonary circulation, the arterial pressure pulse has been found to be transmitted into the pulmonary veins by the use of model studies.

MODEL PERFORMANCE

The comparison of the behavior of models of the circulatory system with experimental results has, in general, been surprisingly good. Such agreement, however, may sometimes be fortuitous. For example, if one introduces geometrical and elastic tapering, one can always obtain a "physiologic" behavior of the input impedance, provided a sufficient number of segments and branches are used. There are two reasons for this: (1) by increasing the number of networks (or equations) representing the total transfer function, the overall effects of mismatched individual impedances are minimized, and (2) errors in the assumption of network and transmission line parameters tend to cancel out easily. Furthermore, the greater the number of networks, the more difficult it becomes to associate a given network parameter with its specific physiologic counterpart.

With the experimental data available at this time, only a few vascular segments can be evaluated objectively, and most of the selected values for the model parameters are based, at best, on educated guesses. In most models, the equations that describe the behavior of the circulatory system may be of quite different form from those used in the model, and the evaluation of the validity of such models must be based on more than one type of comparison with experimental data. Even for the characterization of a simple transmission line, two parameters — the input impedance and the propagation constant — are required. It is possible, for example, to obtain an excellent fit between the experimental data for the frequency spectrum of vascular impedance and those obtained from a transmission-line model, but the transmission characteristics of the pressure pulse may be quite different in the two sets of data. Under these circumstances, it is apparent that further refinements of the theoretical aspects of pulsatile blood flow are of no practical value until better measurement techniques for the collection of experimental data become available.

Although the overall physical aspects of flow behavior and the wall properties in the larger arteries and veins are now reasonably well-understood, there is, as yet, no available comprehensive analysis of flow behavior in smaller vessels. These vessels, however, represent the site at which the exchange of blood gases and metabolites between the body tissues and the external environment takes place, and the time-dependent geometry of these small vessels determines the distribution of blood flow in the body. Very few of the assumptions that have been used in the analysis of pulsatile flow in large vessels can be applied to the microcirculation. In addition, the problems associated with measurement techniques for the microcirculation are greater than those of the large vessels by at least an order of magnitude.

Similarly, a comprehensive analytical treatment of the cardiac pump is still lacking. Although in this case an abundance of experimental data is available, the problems that are related to the nonlinear properties of the flow generator and its complex geometry have defied any but the crudest attempts at a mathematical treatment.

REFERENCES

1. Attinger, E. O. Wall properties of veins. *IEEE Trans. Biomed. Eng.* BME-16:253, 1969.
2. Dintenfass, L. Rheologic approach to thrombosis and atherosclerosis. *Angiology* 15:333, 1964.
3. Noordergraaf, A. *Biological Engineering.* New York: McGraw-Hill, 1969.
4. Noordergraaf, A., Verdow, P. D., Van Brummelen, A. G. W., and Wiegel, F. W. Analog of the Arterial Bed. In E. O. Attinger (Ed.), *Pulsatile Blood Flow.* New York: McGraw-Hill, 1964.
5. Robinson, D. A. Quantitative analysis of cardiac output in the isolated left ventricle. *Circ. Res.* 17:207, 1965.
6. Taylor, M. D. The input impedance of an assembly of randomly branching elastic tubes. *Biophys. J.* 6:29, 1966.
7. Warner, H. R. Use of analogue computers in the study of control mechanisms in the circulation. *Fed. Proc.* 21:87, 1962.
8. Witzig, K. Über erzwungene Wellenbewegungen zäher, inkompressibler Flüssigkeiten in elastischen Rohren. Doctoral Thesis, University of Bern, 1914.
9. Womersley, J. R. *Mathematical Analysis of the Arterial Circulation in a State of Oscillatory Motion.* Dayton, Ohio: Wright Air Development Center, Tech. Rep. WADC-TR-56-614, 1958.

THE RESPIRATORY SYSTEM III

Pulmonary Structure and Pathology   14

Marvin A. Sackner

INTRODUCTION

The respiratory system provides the mechanisms for the exchange of gases between the organism and the outside environment. The lung, which is the major component of this system, serves as the intermediary for the transport of oxygen into the blood. In addition to this activity, the lung is an active metabolic organ. Because of its interposition between the venous and arterial circulations, metabolic substances that are released from the organs into the venous circulation must first pass through the lung before exerting their effects on distant sites in the body. During this passage, the lung tissue often modifies and inactivates many of these substances. In future years, the assessment of the metabolic activity of the lung will assume increasing importance. In this unit, however, attention will be mainly directed toward describing the lung in terms of its function as a gas exchanger with the outside environment.

The study of respiratory function is probably based upon more engineering concepts than any other branch of physiology. The mechanical properties of the lung, for example, are thought of by the physiologist in terms of the following electrical analogies: (1) the stiffness or compliance of the lungs corresponds to capacitance, (2) the resistance of air flowing through the trachea and other conducting pathways is analogous to resistance, and (3) the inertia of air in the tracheobronchial tree is viewed as an inductance. Systems-control theory plays an important role in the understanding of the control of ventilation; e.g., the maintenance of a normal level of carbon dioxide tension in the blood is possible because of a negative feedback loop to the respiratory center that sends impulses to the respiratory muscles to increase or decrease pulmonary ventilation. Furthermore, the pulmonary physiologist often borrows and modifies devices that were designed for industrial use – e.g., servo systems, analog and digital computers, pressure switches, solenoid valves, logic devices, and so on – to measure the functions of the respiratory system, e.g., pressure, flow, and volume waves. In this unit, an effort will be made to include electrical and mechanical analogs, which those versed in the physical sciences will soon appreciate. Such backgrounds are ideal for a clearer understanding of this material, but the life scientist should also appreciate this approach.

Recent advancements in the methods for studying pulmonary structure have made us realize the necessity of examining the structural components of the functional "black box" we call the lung. Many physiologic concepts and theories have been corroborated or disconfirmed by direct visualization of the lung structures.

231

EMBRYOLOGY

The lung develops in the twenty-fourth day of fetal life as an outpouching of the embryonic gut. Figure 14-1 shows the chronologic sequence of its development. A few days later, this outpouching divides into two branches. The branching then continues in an irregular, dichotomous, branching pattern for about two months, when most of the airways become lined with ciliated columnar epithelium, i.e., a layer of rectangular cells whose short edge faces into the airway. These cells bear fine projections, called *cilia,* that beat with a wave-like motion in order to move a mucous blanket from the deep recesses of the lungs to the trachea. This mucous blanket travels in the trachea of normal adults at the rate of approximately 20 mm/min, and it serves to eliminate inhaled foreign particulate matter [8]. Mucous glands develop beneath the ciliated columnar epithelium, and ducts or channels from them penetrate the columnar epithelium so that the mucus can be discharged into the airways.

In the fifth fetal month, small airways develop deep within the lung that are lined by nonciliated cuboidal epithelium; these airways become invested by capillaries that arise from vascular structures in the mesenchyme. A month later, the *alveoli,* the ultimate gas-exchanging units of the lungs, appear from these airways. The cuboidal epithelium becomes attenuated and finally cannot be recognized in the terminal alveoli. Capillaries proliferate around these terminal air spaces. At birth, there are

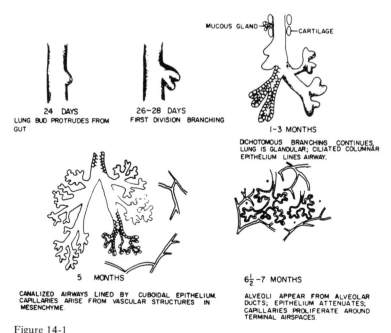

Figure 14-1
Embryology of the lung. (From M. D. Avery. *The Lung and Its Disorders in the Newborn Infant.* Philadelphia: Saunders, 1968. Reproduced with permission.)

about 24 million alveoli in the lungs; the number of alveoli increases to about 300 million at 8 years of age, at which time the growth of the lung ceases [1].

BIOLOGIC STRUCTURE

LARGE AIRWAYS

The airways serve as the conduits for air passage from the external environment to the gas-exchanging units of the lungs. These passages include the oral and nasal cavities and the pharynx. The lungs proper begin with the larynx and the trachea. The larynx contains fibromuscular structures, the vocal cords, which upon vibration produce sound. The trachea is a fibromuscular tube about 10 to 12 cm in length that lies half in the neck and half in the thorax (Fig. 14-2). Its width is approximately 1.4 to 2.0 cm. The anterior and lateral surfaces are composed of approximately 20 U-shaped cartilages; the posterior surface is a fibrous membrane that has the property of invaginating during a violent cough or exhalation. The average cross-sectional area of the trachea is 1.5 cm^2 during quiet breathing; with a violent cough, the area may be reduced to 0.25 cm^2. The termination of the trachea within the thorax is called the *carina*; at this point there is a division into the right and left main bronchi. The right bronchus deviates less from the axis of the trachea than does the left (this is the explanation given for the higher incidence of inhaled foreign bodies lodging in the right bronchus or one of its branches than in the left).

The right lung accounts for 55% and the left lung for 45% of the total gas volume. Each lung has incomplete separations that produce divisions called lobes; each lobe is supplied by a division of the main bronchi called the *lobar bronchi*. The right lung has three lobes (upper, middle, and lower), the left lung two (upper and lower).

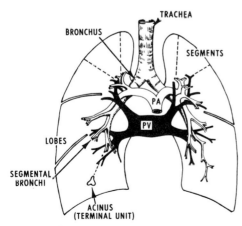

Figure 14-2
Pulmonary anatomy.

Tertiary branching of the bronchi allows for further classification into bronchopul
monary segments, of which there are ten on the right and nine on the left [17].
Since certain pulmonary diseases are often limited to a particular lobe, segment, or
segments, knowledge of the gross anatomy of the lung is often an important consideration in diagnosis and management.

SMALLER AIRWAYS

The bronchial tree branches in an irregular, dichotomous pattern through 16 generations until the terminal bronchioles are reached. These airways have ciliated columnar
epithelium and function as passageways to conduct air in and out of the lungs: they
have been termed the *conductive zone* (Fig. 14-3). Branching then proceeds quite
irregularly, either dichotomously or trichotomously, through a transitional zone of
three generations of respiratory bronchioles. In the transitory zone, two functions
are carried out: conduction as well as a limited amount of gas exchange. The latter
takes place because the respiratory bronchioles have a few outpouchings of alveoli in
their walls.

The terminal unit of the lung, the *acinus,* functions as the respiratory zone. It
includes the continuum of passage through three generations of respiratory bronchioles
and the alveolar ducts to the alveolar sacs (Fig. 14-4). The acinus is therefore composed

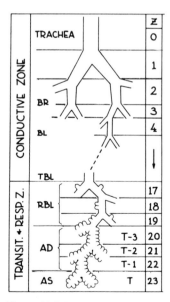

Figure 14-3
Branching pattern of the tracheobronchial tree (*Z* denotes the generation of branching, *BR* =
bronchi, *BL* = bronchiole, *TBL* = terminal bronchiole, *RBL* = respiratory bronchiole, *AD* = alveolar
duct, and *AS* = alveolar sac). (From E. R. Weibel. *Morphometry of the Human Lung.* New York:
Academic, 1963. Reproduced with permission.)

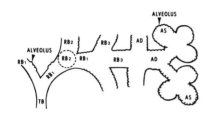

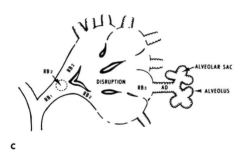

Figure 14-4
Normal and pathologic states of the terminal unit, the acinus. *A*. Normal acinus. *B*. Panacinar emphysema; dilatation and destruction of the entire acinus. *C*. Centrilobular (centriacinar) emphysema; destruction and dilatation of the respiratory bronchioles with sparing of the alveolar ducts and sacs. (*TB* denotes the terminal bronchiole, the passage that leads into the acinus; RB_1, RB_2, and RB_3 denote the branches of the respiratory bronchioles; *AD* indicates the alveolar duct.) (From G. Cumming. *Form and Function in the Human Lung.* Edinburgh: Livingstone, 1967. Reproduced with permission.)

of the respiratory bronchioles, which are the 17th to 19th generation airway branches, the alveolar ducts, which are the 20th to the 22nd generations, and the alveolar sac, which is the 23rd and final generation.

An alveolar duct may be likened to a long corridor with a succession of rooms opening into it on either side as well as from the ceiling and floor. Just as each alveolus shares the entrance frame of the next, so each shares a wall with its neighbor, and the two rooms are separated by a single wall. Smooth muscle is distributed throughout the respiratory bronchioles and alveolar ducts. There is no smooth muscle in the last alveolar duct, which terminates in one to several rotunda-like structures called *alveolar sacs*. These structures bear the terminal alveoli [12].

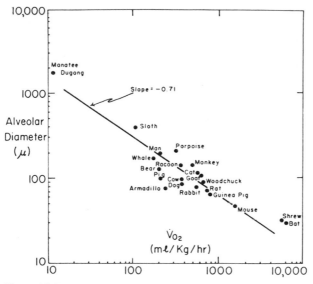

Figure 14-5
Alveolar diameter as a function of oxygen consumption for different animal species. (From S. M. Tenney and J. E. Remmers. *Nature* [Lond.] 197:54, 1963. Reproduced with permission.)

ALVEOLI

In humans, the alveolar size at resting lung volume ranges between 200 and 300 microns in diameter. At small lung volumes, such as after a maximum exhalation, the alveolar walls fold up to below some critical resting length, whereas in well-inflated lungs, the alveolar septa become fairly uniform in thickness and their diameter is much larger. The wall thickness, exclusive of capillaries, ranges from about 2 to 8 μ [12].

From a teleologic standpoint, the alveolar size in different animals appears to be a function of physical activity. The smaller the alveolar size, the more efficient the transfer of oxygen from the alveolus into the capillary blood. The bat, with a high oxygen consumption (i.e., the total oxygen requirement of all tissues of the body) of 10,000 ml/kg/hour, has an alveolar diameter of 20 μ, whereas the manatee, with a low oxygen consumption of 12 ml/kg/hour, has an alveolar diameter of 1800 μ [11]. Figure 14-5 shows the spectrum of oxygen requirements as a function of various alveolar sizes in various animals.

PULMONARY VASCULAR CIRCULATION

The lungs are perfused by a dual vascular circulation, the pulmonary and bronchial systems. The pulmonary artery arises from the right ventricle of the heart. It branches with the airways, and it continues to divide into five additional irregular, dichotomous generations (28 total generations) to terminate in the pulmonary capillary network [12]. This circulation serves to convey deoxygenated blood from the tissues to the alveoli for gas exchange; i.e., oxygen diffuses into the capillary blood from the alveoli and carbon dioxide passes out through the alveoli. The pulmonary circulation also supplies the nutrition to the acini or terminal units.

The structure of the pulmonary vessels differs considerably from the systemic counterparts. The main pulmonary artery has a much higher fraction of smooth muscle than the aorta. Pulmonary vessels less than 100 μ in diameter (e.g., arterioles) are practically devoid of smooth muscle, whereas systemic arteries of comparable size have thick smooth muscle coats [5].

The bronchial arterial system consists of several small vessels that arise from the aorta or intercostal arteries; their total flow is normally less than 1% of the total cardiac output [5]. This system supplies the nutritional needs of the large conducting airways and pulmonary vessels; these vessels are not distributed to the acinus.

RELATION OF STRUCTURE TO FUNCTION

In artificially ventilated, anesthetized, open-chest animals, the pouring of liquid propane cooled by liquid nitrogen to $-180°C$ over the lungs almost instantaneously freezes pulmonary tissue [10]. With special processing techniques, the form and structure of the tissue can be preserved for observation almost as it had been during life. This technique provides an opportunity to observe directly the effects of pharmacologic and nervous stimulation on the airways and blood vessels, which can only be surmised from indirect physiologic observations. For example, the breathing of low inspired oxygen mixtures produces a rise in pulmonary arterial pressure, a response that implies vasoconstriction (the narrowing of vascular caliber); such a phenomenon can be visualized directly using the rapid-freezing technique [9].

CAPILLARIES

In thick, histologic sections, the pulmonary capillaries appear as a sheet-like network that covers about 65% to 75% of the septum. There are over 300 billion capillary segments in the lung. However, not all capillaries are filled with red blood cells, and diffusion of oxygen into the blood only becomes meaningful when red blood cells are present. Capillary filling and the number of capillaries per unit area are functions of gravity and the mechanical forces in the lungs [4]. In the erect position, the hydrostatic pressure is higher at the bottom than at the top of the lungs, and this facilitates better filling of the lower-zone capillaries with red blood cells. As the lung becomes more inflated, the capillaries at any given hydrostatic level become stretched and thinned out, and the red blood cell density becomes significantly reduced.

AIRWAYS

The conducting-zone airways are innervated by the vagus nerve and perfused by the bronchial arteries (Fig. 14-6). Stimulation of the vagus nerve promotes the release of

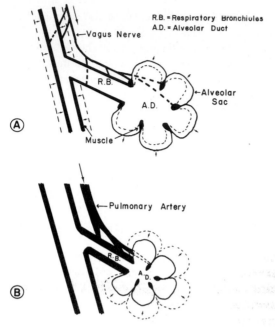

Figure 14-6
Innervation and perfusion of airways. *A.* Stimulation of the vagus nerve. *B.* Injection of histamine or serotonin into the pulmonary artery. (Modified from J. A. Nadel et al. *J. Appl. Physiol.* 19:387, 1964.)

acetylcholine, an agent that produces constriction of the smooth muscle coat of the large airways (bronchoconstriction). Injection of acetylcholine or agents with a similar pharmacologic effect (such as histamine) into the bronchial artery also results in bronchoconstriction [3]. If these alterations in structure were measured by functional tests, one would predict an increase in the resistance to air flow and a compensatory overinflation of the terminal units that were not directly affected by the stimulus. Indeed, such functional disturbances have been detected.

In contrast to the conducting airways, the airways of the transitory zone of the terminal unit (which are also called the small or peripheral airways) are not perfused by the bronchial circulation but rather by the pulmonary arterial system. Injection of acetylcholine, histamine, or serotonin into the latter produces entirely different structural and physiologic effects from those that occur when such agents are injected into the bronchial circulation, since in the terminal unit, only the smooth muscle contracts. Figure 14-7 shows a microscopic section that indicates the constriction of these ducts as a result of histamine injection. The lung volume becomes smaller, and there is an apparent increase in lung stiffness. It should be noted that although the vagus nerve innervates the small airways, the absolute magnitude of bronchoconstriction due to vagal stimulation is greater in the large airways.

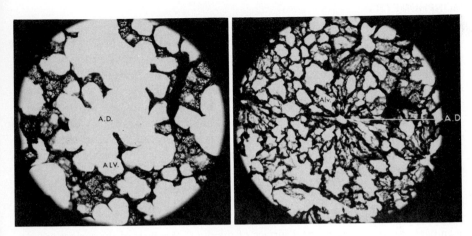

Figure 14-7
Thick microscopic section from a rapidly frozen cat lung after injection of histamine into the pulmonary artery. The left section is from a control lung, whereas the right section is from a histamine-injected lung. Both have a similar magnification. The pronounced alveolar duct (*A.D.*) constriction is readily apparent. (From H. J. H. Colebatch et al. *J. Appl. Physiol.* 21:217, 1966. Reproduced with permission.)

MUSCLES OF RESPIRATION

The diaphragm, which is the muscle dividing the thoracic and abdominal cavities, is the principal muscle involved in respiration during quiet breathing [2]. Contraction of this muscle causes descent of the dome and expansion of the base of the thorax. Both these actions become less efficient if the diaphragm is depressed (as in emphysema), and the second action may actually be reversed. The dome of the diaphragm, as visualized by roentgenographic examination, moves 1.5 cm during quiet breathing and 6 to 10 cm during maximal breathing. During quiet breathing, the diaphragm contracts only during inspiration; in normal subjects, expiration is a passive event.

The intercostal muscles, which are situated between the ribs, contract during inspiration in quiet and increased breathing, and they help to enlarge the thoracic cavity by their action on the ribs. The abdominal muscles are almost always inactive during quiet breathing. They contract vigorously in all voluntary expiratory maneuvers. such as coughing, straining, and vomiting. During exercise, they come into play when ventilation is in the order of 40 liters per minute (about five times the resting ventilation).

The anterior neck muscles, the scalene and sternocleidomastoids, are, in inspiration, the most important accessory muscles to the diaphragm and intercostal musculature. They contract only during deep breathing in normal subjects. When there is impairment of diaphragmatic and intercostal contraction (as in patients with emphysema), the neck muscles become an important muscle group in maintaining pulmonary ventilation [2].

MORPHOMETRY

Morphometrics of the lungs deals with the design, geometry, and relative and absolute dimensions of the spaces through which air and blood are transported and between which gases are exchanged [12]. Measurement of the dimensions of the airways has facilitated the analysis of certain disease processes. For example, air is moved to the periphery mainly by bulk flow until it reaches the terminal unit. At the latter site, diffusion of gases in the airways assumes increasing importance (Fig. 14-8). If a lesion is produced at this level, gas exchange in the lung might be limited by a longer diffusion pathway in the airways [7]. Indeed, such a lesion exists in centrilobular emphysema (see Fig. 14-4).

The geometrical shape of the alveoli is determined by their arrangement around the alveolar ducts and by their close, foam-like packing against each other. This results in a model in which the alveoli resemble the cells of a honeycomb that has been bent around the cylindrical surface of the duct. Analysis of the surface area-to-volume ratio of the alveolus yields a coefficient that corresponds closely to that of a sphere. This means that the alveolar diameter will increase by a factor of 1.3 when the inflation of the lungs doubles. Measurement of a physiologic parameter that reflects changes in surface area of the lung, the *diffusing capacity* of the alveolocapillary membrane, follows these anatomic predictions [6].

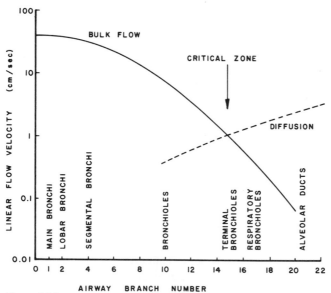

Figure 14-8
Linear velocity of flow in airways plotted against the airway branch number. Bulk flow is more important than diffusion in gas transport until the 15th generation is reached. At that point, diffusion in the airways becomes important in gas transfer to and from the alveoli. (From D. C. F. Muir. *Br. J. Dis. Chest* 60:169, 1966. Reproduced with permission.)

According to morphometric data, the human pulmonary capillary blood volume ranges from 150 to 200 ml. These values are about twice the resting capillary volumes as determined by physiologic techniques. They agree well with physiologic values obtained during maximal exercise, however, which indicates that the anatomic volume sets the limit on the expansion of the pulmonary capillary bed [6].

REFERENCES

1. Avery, M. E. *The Lung and Its Disorders in the Newborn Infant.* Philadelphia: Saunders, 1968.
2. Campbell, E. J. M. *The Respiratory Muscles and the Mechanics of Breathing.* Chicago: Year Book, 1958.
3. Colebatch, H. J. H., Olsen, C. R., and Nadel, J. A. Effect of histamine, serotonin and acetylcholine on peripheral airways. *J. Appl. Physiol.* 21:217, 1966.
4. Glazier, J. B., Hughes, J. M. B., Maloney, J. E., and West, J. B. Measurements of capillary dimensions and blood volume in rapidly frozen lungs. *J. Appl. Physiol.* 26:65, 1969.
5. Harris, P., and Heath, D. *The Human Pulmonary Circulation.* Baltimore: Williams & Wilkins, 1962.
6. Miller, J. M., and Johnson, R. L., Jr. Effect of lung inflation on pulmonary diffusing capacity at rest and exercise. *J. Clin. Invest.* 45:493, 1966.
7. Muir, D. C. F. Bulk flow and diffusion in the airways of the lung. *Br. J. Dis. Chest* 60:169, 1966.
8. Santa Cruz, R., Landa, J., Hirsch, J., and Sackner, M. A. Tracheal mucous velocity in normal man and patients with obstructive lung disease; effects of terbutaline. *Am. Rev. Resp. Dis.* 109:458, 1974.
9. Staub, N. C. Microcirculation of the lung utilizing very rapid freezing. *Anesthesiology* 12:469, 1961.
10. Staub, N. C., and Storey, N. J. Relation between morphological and physiological events in lung studied by rapid freezing. *J. Appl. Physiol.* 17:381, 1962.
11. Tenney, S. M., and Remmers, J. E. Comparative quantitative morphology of the lungs: Diffusing areas. *Nature* (Lond.) 197:54, 1963.
12. Weibel, E. R. *Morphometry of the Human Lung.* New York: Academic, 1963.

Survey of Pulmonary Functions   15

Marvin A. Sackner

INTRODUCTION

The major functions of the respiratory system are depicted in Figure 15-1. The process of breathing is also called *ventilation*. In order for the lungs to function efficiently as an organ of gas exchange, the inspired air must be distributed to alveoli that are perfused with blood. The resting breathing pattern is usually composed of an inspiratory and expiratory phase in the ratio of 2:3. The volume of air displaced, the *tidal volume*, averages about 500 ml in the adult. The resting frequency of respiration ranges from 10 to 16 breaths per minute. The distribution of gas is determined by both active and passive mechanisms; these mechanisms operate to insure optimal matching of ventilation with perfusion. After fresh air has been distributed to the alveoli, oxygen diffuses across the alveolocapillary membrane, through the plasma, and across the red blood cell membrane, and it finally reacts with reduced hemoglobin within the red blood cell. Similarly, carbon dioxide diffuses from the blood into the alveolar air to be eliminated during expiration. Alveoli are supplied by capillaries of the *pulmonary circulation* so that gas exchange between the air and blood can take place. The interrelationship of pressure, volume, and flow that is developed in the lungs and pleural space by the movements of the chest muscles and diaphragm constitutes the *mechanics of breathing* [4].

Figure 15-2 illustrates the simplest model of the lung: a cylinder that divides into branches, which give rise to spheres. The tubes represent the tracheobronchial tree or the conductive system, and the spheres, the alveoli or gas-exchanging units. Each sphere receives venous blood through a channel representing the pulmonary arterial system; an effluent tube represents the pulmonary venous system, which normally contains oxygenated blood. Frequently, the respiratory model contains electrical analogs, particularly when it deals with the mechanics of breathing. The conductive zone is depicted as an ohmic resistance, and the gas-exchanging units are represented as a capacitance.

LUNG VOLUMES

The terminology for the various lung volumes is given in Figure 15-3. The *functional residual capacity* (FRC) is the volume of gas in the lungs at the end of a normal expiration. It is sometimes called the "resting lung volume" because it is the level where the elastic recoil of the lungs and that of the thoracic cage are equal and opposite in sign. Its value ranges from about 2 to 5 liters, and it is a direct function of age and height. The *total lung capacity* (TLC) is the volume of air contained in the lungs after

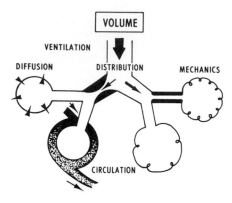

Figure 15-1
Major functions of the respiratory system.

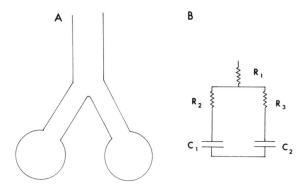

Figure 15-2
Respiratory models. *A.* The lung is represented as divided in two elements; the tubes are analogous to the tracheobronchial tree, the spheres analogous to the alveoli. *B.* The corresponding electrical analog utilizes resistances for the conductive zones and capacitances for the gas-exchanging zones.

maximum inspiration. It ranges from about 3.5 to 9 liters, and it is a direct function of height and an inverse function of age. The volume of air obtained after a maximum inspiration from the FRC point is termed the *inspiratory capacity* (IC), and that released after a maximum expiration from the FRC point, the *expiratory reserve volume* (ERV). In the upright subject, these volumes are usually present in the ratio of 2:1, respectively. The sum of the IC and ERV is called the *vital capacity* (VC). This volume usually ranges from 2 to 6 liters; it is directly dependent on height and inversely dependent upon age. The *residual volume* (RV) is defined as the volume of air remaining in the lungs after a maximal forced expiratory effort. In normal adults it ranges from 1.0 to 3.5 liters, and it, like the FRC, is directly proportional to age and height. In general, lung volumes are smaller in females than in males of comparable age and height [4].

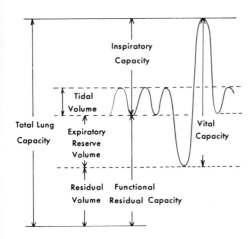

Figure 15-3
Nomenclature and relations of the various lung volumes.

DEAD SPACE AND VENTILATION

The *anatomic dead space* is the volume of air contained within the conducting pathways up to the site of the terminal bronchioles. Figure 15-4 includes the types of
dead space. The dead space includes the nasopharynx and the larynx, as well as the
tracheobronchial tree. In normal subjects, it has a value of about 1 ml per pound of
body weight. It is termed "dead space" because no appreciable gas exchange takes
place between it and the alveoli [2]. The *physiologic dead space* includes the anatomic
dead space and, in addition, those volumes of the lung that should anatomically participate in gas exchange but functionally do not. The latter consist of alveoli with no
blood flow and alveoli with ventilation in excess of blood flow. This additional space —
i.e., the areas of the lung that do not significantly participate in gas exchange — is also
called the *parallel dead space*. Under normal circumstances, the parallel dead space
in the erect position amounts to only 2 to 10 ml [1]. The sum of the physiologic
dead space and the alveolar volume constitutes the *tidal volume*.

The physiologic dead space can be determined from Bohr's equation by using the
principle $V_1 c_1 = V_2 c_2$ (where V is volume and c is concentration) and by measuring
the mixed expired carbon dioxide tension and the arterial or mean alveolar carbon
dioxide tension. Bohr's equation is expressed as:

$$V_T \times F_{E_{CO_2}} = (V_A \times F_{A_{CO_2}}) + (V_D \times F_{I_{CO_2}}) \qquad (15\text{-}1A)$$

where V is volume, F is fractional concentration of gas, V_T is tidal volume, $F_{E_{CO_2}}$
is the mixed expired CO_2 concentration, V_A is alveolar volume, $F_{A_{CO_2}}$ is the mean
alveolar CO_2 concentration, V_D is physiologic dead space volume, and $F_{I_{CO_2}}$ is the
inspired CO_2 concentration. Making substitutions as is shown below in Equations 15-1B

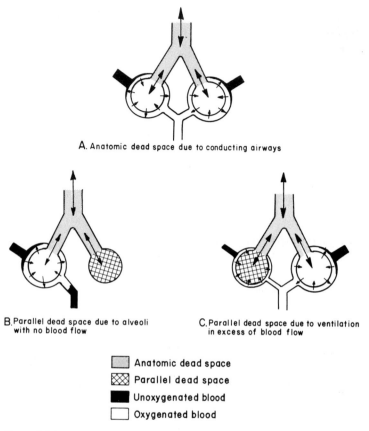

A. Anatomic dead space due to conducting airways

B. Parallel dead space due to alveoli
with no blood flow

C. Parallel dead space due to ventilation
in excess of blood flow

☐ Anatomic dead space
☒ Parallel dead space
■ Unoxygenated blood
☐ Oxygenated blood

Figure 15-4
Types of respiratory dead spaces.

through 15-1D, the product of the volume and the fractional concentration of gas as expressed by Equation 15-1A may now be expressed as is shown in Equation 15-1E:

$$V_A = V_T - V_D, F_{I_{CO_2}} = 0 \tag{15-1B}$$

$$V_T \times F_{E_{CO_2}} = (V_T - V_D) F_{A_{CO_2}} + (V_D \times 0) \tag{15-1C}$$

$$P = F \times (P_B - 47) \tag{15-1D}$$

$$V_T \times P_{E_{CO_2}} = (V_T - V_D) \times P_{A_{CO_2}} \tag{15-1E}$$

where P is the partial pressure of gas, P_B is the barometric pressure in mm Hg, 47 is the vapor pressure of water in mm Hg at 37°C, and $P_{A_{CO_2}}$ is the arterial CO_2 tension.

Equation 15-1F expresses the ratio of physiologic dead space to tidal volume:

$$\frac{V_D}{V_T} = \frac{Pa_{CO_2} - Pe_{CO_2}}{Pa_{CO_2}} \tag{15-1F}$$

The ratio V_D/V_T is normally less than 0.3 and decreases during exercise.

Hyperpnea is an increase of the tidal volume, *tachypnea* an increase in respiratory frequency, and *dyspnea* the subjective sensation of breathlessness. In this respect, a hyperpneic, tachypneic subject may or may not be dyspneic. *Hyperventilation* is an increase in ventilation of the alveolar volume above the normal, whereas *hypoventilation* is the converse. The partial pressure of carbon dioxide tension in the arterial blood (Pa_{CO_2}) ranges from 38 to 42 mm Hg, and it corresponds to the true mean alveolar carbon dioxide tension. A decrease in Pa_{CO_2} is a sign of alveolar hyperventilation, an increase, a sign of alveolar hypoventilation. Because the dead space in a normal individual increases only slightly with increases in tidal volume, it is this parameter, rather than the respiratory frequency, that is the more important determinant of *alveolar ventilation* ($\dot{V}_A$). Thus, if the minute-ventilation (i.e., the sum of dead space and alveolar ventilation per minute) is measured and is found to be 6400 ml and if the dead space is 150 ml, then $\dot{V}_A$ will be 1600 ml/min at a respiratory frequency of 32 and 5200 ml/min at a frequency of 8 breaths per minute (Fig. 15-5).

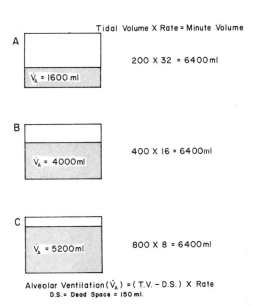

Tidal Volume X Rate = Minute Volume

A $\dot{V}_A$ = 1600 ml   200 X 32 = 6400 ml

B $\dot{V}_A$ = 4000 ml   400 X 16 = 6400 ml

C $\dot{V}_A$ = 5200 ml   800 X 8 = 6400 ml

Alveolar Ventilation ($\dot{V}_A$) = (T.V. – D.S.) X Rate
D.S.= Dead Space = 150 ml.

Figure 15-5
Relationship of respiratory frequency and tidal volume to alveolar ventilation.

DISTRIBUTION OF INSPIRED GAS

During inspiration, the tidal volume is added to the functional residual capacity; for the whole lung, the ratio of the final to initial volume, the *expansion ratio,* may range from just above 1 up to 6. However, within the lung, expansion ratios have an even wider range, since the expansion of lung tissue does not take place uniformly. Variation occurs among lobes, lobules, and even alveoli due to differences in the intrapulmonary forces that regulate lung expansion. *Spatial inequality* is the unevenness that occurs when some alveoli have larger expansion ratios than others. *Temporal inequality* occurs when some alveoli empty and fill earlier than others [2]. Some causes of uneven distribution of ventilation (Fig. 15-6) include regional variations in elasticity due to gravity or disease, regional obstruction, and regional check valves. Severe unevenness of distribution of ventilation is the hallmark of pulmonary emphysema.

DIFFUSING CAPACITY OF THE LUNG

The term *diffusing capacity of the lung* was originally introduced to describe the efficiency of the transfer of oxygen from the alveoli air through the alveolocapillary membrane into the red blood cell [5]. This diffusion process is shown in Figure 15-7. Because of technical problems, it is easier to measure the transfer of trace amounts of carbon monoxide, since red blood cells handle CO in a way similar to O_2. Values for the diffusing capacity of the lung that are obtained using CO directly reflect the diffusing capacity for O_2. Investigators in the early twentieth century believed that the rate

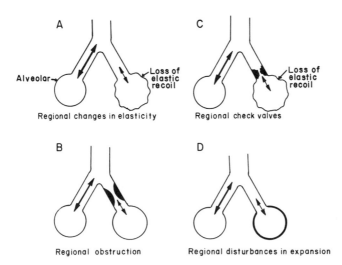

Figure 15-6
Causes of uneven ventilation. The size of the arrows indicates the volume of gas ventilating each region. The thick line in part *D* represents alveoli that expand less than normally.

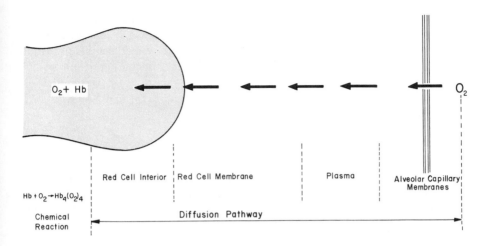

Figure 15-7
Diffusion pathway of oxygen from the alveoli into the red blood cell.

of combination of O_2 or CO with hemoglobin was practically instantaneous, so that the diffusing capacity measured the integrity of the alveolocapillary membrane. However, the reaction time of oxygen with hemoglobin is relatively slow and must be considered in the interpretation of diffusing capacity measurements. Under resting circumstances, the average transit time of the red blood cell through the pulmonary capillaries is 0.8 to 1.0 second, but this is sufficient time to achieve equilibration between the O_2 in the alveolus and that within the red blood cell. In diseased lungs, particularly when there is an increase of pulmonary blood flow as in exercise, there may be insufficient time for O_2 equilibration to be achieved. The relation among the diffusing capacity of the lung, the diffusing capacity of the pulmonary membrane, and the capillary blood volume has been established [8] :

$$\frac{1}{D_l} = \frac{1}{D_m} + \frac{1}{\theta Vc} \tag{15-2}$$

where D_l is the diffusing capacity of the lungs, D_m is the diffusing capacity of the pulmonary membrane, and θVc is the diffusing capacity of the red blood cells in the pulmonary capillary bed at any moment. All these factors are expressed in ml CO/min $\times$ mm Hg partial pressure gradient of CO. In turn, θ is the rate of gas uptake by 1 ml of whole blood per minute for a partial pressure of 1 mm Hg, and Vc is the average volume of the capillary bed in milliliters [8].

The membrane diffusing capacity D_m depends upon (1) the total surface area of the capillary bed, (2) the average thickness of the pulmonary membrane, and (3) the diffusion coefficient of the pulmonary membrane, which is proportional to the solubility of a given gas in the membrane and inversely proportional to the square root of the molecular weight of the gas. The capillary bed volume Vc depends upon the average diameter and total length of the capillaries. The factor θ depends upon (1) the red blood cell concentration in the blood and (2) the rate of gas uptake per red blood cell per 1 mm Hg gas tension. In addition to the preceding, the experimental lung diffusing capacity D_l is influenced by (1) nonuniformity of gas distribution within the lung and (2) possible technical errors in the methods used to measure D_l. During the breathing of air at sea level, D_l depends almost equally upon D_m and Vc [8].

PULMONARY CIRCULATION

The pulmonary arterial system presents venous blood to the capillaries for oxygen uptake and carbon dioxide elimination. The supine basal pulmonary blood flow ranges from about 3.0 to 3.3 liters/min/meter^2 body surface area. In an average size adult, this amounts to about 6 liters/min. In the erect position, values may be 40% to 60% less than in the supine position [9]. During strenuous exercise, pulmonary blood flow may rise to as much as 25 liters/min [6]. The resting systolic (peak) pressure in the pulmonary artery has an upper limit of about 30 mm Hg, and the diastolic (trough) pressure is about 10 mm Hg. This contrasts to the average systemic arterial pressure of 120/80 mm Hg.

The pulmonary capillaries drain oxygenated blood into the pulmonary veins, which in turn empty into the left atrium. This oxygenated blood is then pumped by the left ventricle into the systemic circulation for delivery to the tissues. The partial pressure of O_2 in the mixed venous blood (pulmonary arterial blood) is about 40 mm Hg during passage through the capillaries, it rises to about 95 mm Hg. The partial pressure of CO_2 in mixed venous blood is about 46 mm Hg; during passage through the capillaries, it falls to 40 mm Hg.

MECHANICS OF BREATHING

Movement in the chest-lung system can be described in terms of Newton's law of motion. Equation 15-3 expresses the pressure in the chest-lung system [7]:

$$P_{appl} = P_2 - P_1 = \frac{1}{C} V + R\dot{V} + L\ddot{V} \tag{15-3}$$

where $P_2 - P_1$ is the change in pressure (which is analogous to voltage), C is the compliance (capacitance), V is the volume (electrical charge in coulombs), R is the resistance (ohmic resistance), $\dot{V}$ is the air flow (current), L is the inertance (inductance), and $\ddot{V}$ is the acceleration of the air (current per unit time).

In the chest-lung system under atmospheric conditions and with a normal breathing pattern, the pressure component due to the acceleration of air is so small that it can be neglected; however, in deep-sea diving, the inertance factor becomes appreciable and must be taken into account. About 75% of the pressure that is applied by the chest muscles during inspiration is expended in changing the lung volume. Expiration normally proceeds passively. The compliance ($\Delta V/\Delta P$) of the lung is linear to about 1.5 liters above the FRC, and it averages about 0.2 liter/cm H_2O. The total compliance of the chest-lung system is about 0.1 liter/cm H_2O. The chest wall compliance is about 0.2 liter/cm H_2O. The resistance to air flow during quiet breathing with peak flows of about 0.7 liter/sec ranges from 1 to 2 cm H_2O/liter/sec. During such breathing, the air flow in the tracheobronchial tree is probably a function of laminar flow ($P = K_1 \dot{V}$) and turbulent flow due to the larynx ($P = K_2 \dot{V}^2$). Figure 15-8 illustrates resistance to flow and types of air flow.

The *maximum breathing capacity* (MBC) is the volume of air that can be respired during a short period — e.g., 10 to 15 seconds — of voluntary hyperventilation; this value is extrapolated to one minute. It is a direct function of height and an inverse function of age. Its value ranges from 50 to 250 liters/min. Although it is common to relate this value to flow resistance in the airways, this test as well as other indirect tests of flow resistance is also dependent on the integrity of the respiratory muscles, the lung compliance, and patient cooperation. Nevertheless, the MBC is generally reduced in obstructive airway diseases and is normal or slightly reduced in restrictive lung disorders.

The $FEV_{1.0}$ is the *forced expiratory volume delivered in 1 second after a maximal inspiration*. It is popular as an epidemiologic survey test and is generally more reproducible and easier to perform than the MBC test. It has proved to be a useful prognostic index in pulmonary emphysema and chronic bronchitis, since it mirrors the severity of these disease processes. In such diseases, an $FEV_{1.0}$ of less than 0.5 liter is generally associated with a shortened life expectancy, whereas an $FEV_{1.0}$ of greater than 1.5 liters carries a good prognosis [3]. This is because of the fact that if these patients develop a respiratory infection, the individual with an $FEV_{1.0}$ of less than 0.5 liter has almost no pulmonary reserve to combat this stress.

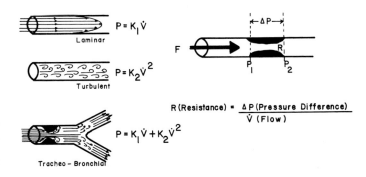

Figure 15-8
Types of air flow.

PERFORMANCE OF PULMONARY FUNCTION TESTING

No other branch of laboratory medicine demands so much cooperation from the patient as pulmonary function testing. Indeed, it is also demanding on the operator or the technician. Since many tests require much patient effort, the technician has to give a great deal of vocal encouragement to the subject. It has been stated that in order to measure the MBC correctly, the technician must do an MBC test along with the patient. The operator must develop rapport with the patient to ensure successful testing; tests may become invalid if the operator cannot achieve this rapport. Some patients require quiet, soothing encouragement, but most have to be badgered into doing their utmost. However, after poking barbs at the patient during a particular test — such as "You can do better!" or "You did better last time!" (whether he did or didn't) — the technician should always compliment the patient after the test is completed with "You're the best we've ever had!" and "You did a great job!" (even if he didn't) in order to maintain the rapport. Clear explanations of the test and the breathing maneuvers that must be performed are essential. Most pulmonary function tests require that the subject breathe through his mouth into a mouthpiece while his nose is clipped shut by a nose piece. This produces a certain degree of discomfort, and if the subject is not apprised as to how long he must breathe in such a fashion during the test, he may become frightened and irritable (for example, if the subject has to breathe five minutes, he should be told so before the test, and, while he is breathing on a mouthpiece, the operator must tell him, "You've done one minute, only four to go!" "Two minutes, only three to go!" and so on).

Screening tests for pulmonary function vary considerably throughout the country. A vital capacity test and an $FEV_{1.0}$ as aids in the detection of patients with obstructive lung diseases can be performed in a physician's office. It is generally neither technically nor economically feasible to set up more complicated tests of pulmonary function in such an office. In a hospital laboratory, the determination of lung volumes, the distribution of ventilation, the dynamic mechanics of the lung, diffusing capacities, and arterial blood gases may constitute routine studies. Other tests, such as lung compliance and capillary blood volume, are usually carried out only in pulmonary research laboratories.

REFERENCES

1. Bouhys, A. Respiratory Dead Space. In W. O. Fenn and H. Rahn (Eds.), *Handbook of Physiology, Section 3, Respiration,* Vol. 1. Washington, D.C.: American Physiological Society, 1964. P. 699.
2. Bouhys, A. Distribution of Inspired Gas in the Lung. In W. O. Fenn and H. Rahn (Eds.), *Handbook of Physiology, Section 3, Respiration,* Vol. 1. Washington, D.C.: American Physiological Society, 1964. P. 715.
3. Burrows, B., and Earle, R. H. Course and prognosis of chronic obstructive lung disease. *N. Engl. J. Med.* 280:397, 1969.
4. Comroe, J. H., Jr., et al. *The Lung: Clinical Physiology and Pulmonary Function Tests* (2nd ed.). Chicago: Year Book, 1962.

5. Forster, R. E. Exchange of gases between alveolar air and pulmonary capillary blood: Pulmonary diffusing capacity. *Physiol. Rev.* 37:391, 1957.
6. Johnson, R. L., Jr., Taylor, H. F., and Lawson, W. H., Jr. Maximal diffusing capacity for carbon monoxide. *J. Clin. Invest.* 44:349, 1965.
7. Mead, J., and Milic-Emili, J. Theory and Methodology in Respiratory Mechanics with Glossary of Symbols. In W. O. Fenn and H. Rahn (Eds.), *Handbook of Physiology, Section 3, Respiration,* Vol. 1. Washington, D.C.: American Physiological Society, 1964. P. 363.
8. Roughton, F. J. N., and Forster, R. E. The relative importance of diffusion and chemical reaction rates in determining the rate of exchange of gases in the human lung, with special references to the volume of blood in the lung capillaries and the true diffusing capacity of the pulmonary membrane. *J. Appl. Physiol.* 11:290, 1957.
9. Segel, N., Dougherty, R., and Sackner, M. A. Effects of tilting on pulmonary capillary blood flow in normal man. *J. Appl. Physiol.* 35:244, 1973.

Gas Volumes 16

Marvin A. Sackner

GAS LAWS

Volumes of gas within the lungs are conventionally reported under BTPS conditions, i.e., body temperature (37°C), barometric pressure (approximately 760 mm Hg), and saturated with water vapor [2]. In order to convert volumes measured under ambient temperature and pressure (ATP) to BTPS, one may employ the gas laws of Boyle, Charles, and Dalton, assuming that air behaves as an ideal gas:

$$\text{Boyles' law:} \quad V_1 P_1 = V_2 P_2 \tag{16-1}$$

$$\text{Charles' law:} \quad V_1 T_2 = V_2 T_1 \tag{16-2}$$

$$\text{Dalton's law:} \quad P_T = P_1 + P_2 + \ldots + P_n \tag{16-3}$$

where V is volume, P is pressure, T is temperature in degrees Kelvin, P_T is the total pressure in a system, and P_n is the partial pressure of the nth component of a mixture of gases.

When applied to the lung, Dalton's law can be rewritten as:

$$P_B = P_{O_2} + P_{CO_2} + P_{N_2} + P_{H_2O} \tag{16-4}$$

where P_B is the barometric pressure, P_{O_2} the partial pressure of oxygen, P_{CO_2} the partial pressure of carbon dioxide, P_{N_2} the partial pressure of nitrogen, and P_{H_2O} the partial pressure of water vapor. The partial pressure of water vapor varies with relative humidity and temperature. For gas in the lung that is saturated with water vapor at 37°C, the P_{H_2O} is 47 mm Hg. Thus, for alveolar gas, Dalton's law may be expressed as:

$$P_B - 47 = P_{A_{O_2}} + P_{A_{CO_2}} + P_{A_{N_2}} \tag{16-5}$$

where P_A is alveolar pressure. Under ordinary circumstances at sea level, P_B can be taken as 760 mm Hg, since only insignificant errors are introduced when P_B varies from 750 to 770 mm Hg.

To convert gas volumes from ambient temperature and pressure to BTPS conditions, Equation 16-6 is employed:

$$V_{BTPS} = V_{ATPS} \times \frac{273 + 37}{273 + T} \times \frac{P_B - P_{H_2O}}{P_B - 47} \tag{16-6}$$

255

where V_{ATPS} is the gas volume at ambient temperature and pressure, 237 + 37 is body temperature (degrees Kelvin), T is the ambient temperature, P_B is barometric pressure in mm Hg, P_{H_2O} is the water vapor pressure in the collecting device, and 47 is the water vapor pressure in mm Hg in the alveolar air.

Although the gas volumes are expressed in BTPS terms, the oxygen consumption of the body is conventionally reported under STPD conditions, i.e., standard temperature (0°C), standard pressure (760 mm Hg), and dry:

$$V_{STPD} = V_{ATPS} \times \frac{273}{273 + T} \times \frac{P_B - P_{H_2O}}{760} \qquad (16\text{-}7)$$

SPIROMETRY

The vital capacity of the lung and its subdivisions is most often measured with a device called a *spirometer* (Fig. 16-1). There are two different types of spirometers: wet and dry. The former is more commonly used, probably because of its historically

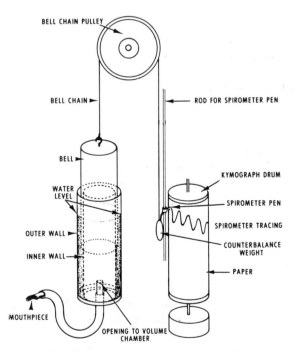

Figure 16-1
Water-sealed spirometer. A lightweight bell cylinder is placed over a volume chamber and rendered airtight by means of a water seal. The bell is free to move through a pulley system that is counterweighted. Near the counterweight is a pen that records the tracing on chart paper as the bell is displaced upward and downward due to respiration.

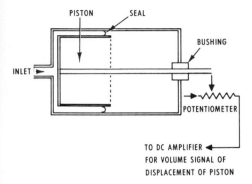

Figure 16-2
Dry spirometer. The piston is displaced by respiration; an airtight seal is obtained by means of a silicone rubber rolling seal. As the piston is displaced, it moves the arm of an infinite-resolution potentiometer. The output of the potentiometer is fed simultaneously to two chopper-stabilized, high-gain amplifiers. One produces the DC output volume signal; the other is used in a differentiating circuit to provide an output that is proportional to flow.

earlier introduction into pulmonary laboratories. Air is trapped under a water seal by a lightweight, floating drum whose linear displacement can be recorded by a pen on moving paper (kymograph) to give an analog trace of the breathing pattern. In the dry type of spirometer, a collecting piston is rendered airtight by means of a silicone rubber, rolling seal. The dry spirometers (Fig. 16-2) offer much less inertia to movement and generally have a better dynamic frequency response than the wet spirometers.

Although the spirometer is the simplest and best device for recording volumes on a short-term basis, its disadvantage is that the gases are rebreathed. This makes long-term measurements of ventilation impossible unless carbon dioxide is removed and oxygen is added to the system as a function of the subject's oxygen consumption. Carbon dioxide is removed by inserting a canister filled with soda lime or barium hydroxide into the breathing circuit; proper addition of oxygen is obtained by maintaining the breathing level constant to match the carbon dioxide absorbed. Such a technique assumes that the RQ — i.e., the *respiratory quotient* (the carbon dioxide output divided by the oxygen consumption) — is equivalent to 1.0. This is rarely valid for normal subjects, in whom the resting RQ usually equals 0.7 to 0.8. Therefore, a large error may result if ventilation is monitored in this manner over a long period. However, for resting ventilation measurements of a few minutes duration, this method is satisfactory.

If ventilation must be measured over a long interval of time without rebreathing the inspired gas mixture, the bag in-box principle may be employed (Fig. 16-3). This method requires the use of unidirectional breathing valves that direct the expired gas into a container that is different from the one used for the inspired mixture. Usually, the inspired mixture is taken from a bag located within an airtight chamber attached to a spirometer. The expired gas is then directed into another bag within the chamber or into the air in the chamber surrounding the bag for inspired gas. During inspiration from the box, the resultant negative pressure produces a decrease in the level recorded

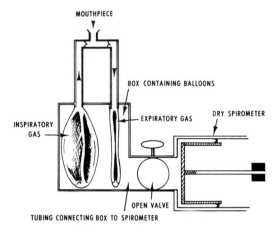

MOUTHPIECE

BOX CONTAINING BALLOONS

EXPIRATORY GAS

DRY SPIROMETER

INSPIRATORY GAS

OPEN VALVE

TUBING CONNECTING BOX TO SPIROMETER

Figure 16-3
Bag-in-box system. A rigid container is connected by large-bore tubing through a valve to a spirometer. As gas is inspired from the inspiratory line, the spirometer piston moves inward; the reverse occurs during expiration. No mixing of inspiratory and expiratory gas occurs. A record is obtained of the volumes of both inspiration and expiration.

by the spirometer, whereas the spirometer recording is deflected upward during expiration. The frequency response of this apparatus is a function of the time-constant (RC product) of the tubing that connects the chamber to the spirometer:

$$V_{sp} = \frac{V_{box}}{\sqrt{(\omega RC)^2 + 1}} \qquad (16\text{-}8)$$

where V_{sp} is the displacement of the spirometer, V_{box} is the volume of air displaced in the box, $\omega = 2\pi f$ where f is the frequency, and R and C are the resistance and compliance of the tubing. In practice, 1-inch diameter tubing gives a critically damped system at less than 1 Hz, whereas 8-inch diameter tubing gives a system whose response is flat to about 9 Hz if a resistor is put in the line for damping.

PNEUMOTACHOGRAPHY

A *pneumotachograph* is a device for measuring air flow [7]. It employs a tube that is designed to permit laminar flow through a linear resistance. This flow is a function of differential pressure across the resistance. The resistance must be of low magnitude so as not to produce nonelastic loading on the subject. The resistance that is usually employed consists either of a number 400 Monel mesh screen or of small tubes placed longitudinally within the pneumotachograph.

　　The advantages of the pneumotachograph over the spirometer are: (1) there is virtually no inertia in the pneumotachograph and hence there is an increased response time, (2) there is no loading on the respiratory system, and (3) no significant problem results from rebreathing, since the dead space of the pneumotachograph is small with respect to the tidal volume.

The disadvantages of the pneumotachograph include: (1) drift problems occur with existing analog integrating devices, which become a problem in long-term monitoring, (2) the differences in temperature and humidity between inspired and expired air result in different calibrations for inspired and expired gas, (3) changes in calibration are necessary with different gas mixtures because of viscosity and density effects, and (4) the rebreathing of a special gas mixture cannot be prevented unless two pneumotachographs are used: one on the inspiratory line and the other on the expiratory line.

The internal diameter of the pneumotachograph determines its DC linearity: the smaller the diameter, the less dead space, and the less the detection ability for high flow rates. This is because the calibration assumes that the flow is laminar, not turbulent. Hence, pneumotachographs are made in a variety of diameters, and their choice is determined by the peak flow that is expected during the test. The frequency response of the pneumotachograph using the Monel-mesh resistance (Lilly type) or the longitudinally placed tubes (Fleisch type) may be adjusted so that linearity from DC to better than 15 Hz is achieved [7, 14]. In contrast, dry spirometers have a frequency response that is flat up to 30 to 50 Hz.

Another problem with the pneumotachograph is that the phase angle between the applied signal and the sensed signal will change if there are different resistances loading the pneumotachograph, which occurs when patients with various lung diseases are tested. Except in research work, however, these small phase-angle shifts are of little concern.

In addition to the pneumotachographic methods, Pitot tubes have been employed instead of the resistances, but these systems require special electronic circuitry for linearization and they do not discriminate very low flow rates (less than 1 liter/min) as well as the conventional pneumotachographs.

MISCELLANEOUS METHODS FOR MEASURING GAS VOLUMES

Other less commonly employed methods for the estimation of gas volumes include (1) the volume-body plethysmograph, (2) thoracic and abdominal impedance plethysmographic methods, and (3) thoracic and abdominal magnetic methods.

In the volume-body plethysmograph method, the subject is seated within a small, airtight chamber of about 300 liters capacity. He breathes to the outside via a valve arrangement [9]. As air is compressed and expanded within the chamber by the breathing motions of his chest-lung system, a spirometer attached to the chamber senses these changes in a way analogous to the bag-in-box system. The advantages of this system include long-term monitoring of ventilation with good stability, but a major disadvantage is the extremely cramped space of the chamber, which results in discomfort to the patient.

Because direct methods for measuring gas volumes involve mouthpieces, face masks, or neck seals, they are cumbersome and impractical for the long-term, continuous monitoring that would be desirable in acutely ill patients. They also tend to make the subject aware that his breathing is being measured, which in itself influences the breathing pattern. Finally, they limit the subject's ranges of activity. For these reasons, indirect methods have been sought to measure ventilation. Attempts have been made to measure the changes in electrical properties of the chest wall and abdomen through the use of electrical impedance plethysmographs or magnetometers to detect changes in magnetic fields [1, 10]. Thus far, most of these methods correlate well with spirometric determinations in normal subjects, but they have not proved very useful for diseased patients or for normal subjects who change their body position frequently.

MEASUREMENT OF FUNCTIONAL RESIDUAL CAPACITY

The volume of air remaining in the lungs after a maximum expiration, the *residual volume*, cannot be estimated by spirometric methods. Gas-dilution or manometric techniques must be employed. Although it is possible to measure residual volume directly, it is more convenient to measure the functional residual capacity and to subtract the expiratory reserve volume in order to estimate the residual volume [2].

CLOSED CIRCUIT METHOD

In this method, an insoluble inert gas, usually helium, is rebreathed from a spirometer circuit of known volume and a known concentration of the inert gas. The initial concentration of the gas in the lungs is zero; then the subject rebreathes until mixing is complete, i.e., until the concentration is the same in both lungs and spirometric circuit. The initial volume of gas in the lungs can then be computed. If the test begins at the moment of maximum expiration, the residual volume is measured; if it begins at maximal inspiration, the total lung capacity is measured; and if it begins at the end of a normal expiration, the functional residual capacity is measured. The last is the preferred starting point because the resting lung volume has a more constant value than do the levels at either full inspiration or expiration. The test generally takes about five minutes to perform. In normal subjects, it can be repeated in about 10 to 15 minutes; in patients with an abnormal distribution of inspired air who eliminate helium from the lungs more slowly, a longer waiting period is necessary.

  The volume of the spirometer circuit is determined prior to testing the patient. The volume of the spirometer is calculated using Equations 16-9 and 16-10 and is expressed in Equation 16-11:

$$V_1 c_1 = V_2 c_2 \qquad\qquad\qquad (16\text{-}9)$$

$$V_{sp} \times (c_{He})_i = (V_{sp} + V_{air}) \times (c_{He})_f \qquad\qquad\qquad (16\text{-}10)$$

$$V_{sp} = \frac{V_{air} \times (c_{He})_f}{(c_{He})_i - (c_{He})_f}$$ (16-11)

where V_{sp} is the volume of the spirometer circuit, $(c_{He})_i$ is the initial helium concentration in the spirometer circuit, V_{air} is the known volume of air that is added to the spirometer circuit, and $(c_{He})_f$ is the final helium concentration in the circuit after the air has been added.

The functional residual capacity (FRC) is calculated using Equations 16-12, 16-13, and 16-14:

$$V_1 c_1 = V_2 c_2$$ (16-12)

$$V_{sp} \times (c_{He})_i = (V_{sp} + V\text{FRC}) \times (c_{He})_f$$ (16-13)

$$V\text{FRC} = \frac{V_{sp} \times [(c_{He})_i - (c_{He})_f]}{(c_{He})_f}$$ (16-14)

where $V\text{FRC}$ is the functional residual capacity [8].

OPEN CIRCUIT METHOD

Although the volume of gas in the subject's lungs is unknown, the nitrogen concentration is 79% when the subject is breathing air. If the total amount of N_2 in the lungs can be determined, the gas volume can be easily calculated. The principle of this method is based on washing out the N_2 from the lungs by the inspiration of 100% O_2 and collecting the expired gas in a spirometer that has been previously flushed with O_2 so that it is free of N_2. In healthy young subjects, the N_2 is washed out almost completely by O_2 within 2 to 3 minutes. In patients with obstructive lung disease whose lung areas may be poorly ventilated, a longer period of O_2 breathing is required (at least 7 minutes). The open circuit method is less frequently employed in pulmonary laboratories than the closed circuit method, although the results are similar. The length of time required to reach the endpoint of the open and closed circuit methods, taking the tidal volume, the breathing rate, and the FRC into consideration, is a function of the distribution of the inspired gas [4].

SINGLE-BREATH DILUTION

In addition to the conventional closed circuit methods, a fairly good estimate of the FRC can be obtained in normal subjects by having them take a single inspiration of a helium mixture to total lung capacity and hold their breath for 10 seconds; an alveolar sample is then analyzed. This method measures the *effective alveolar volume,* i.e., the volume of lung that is in good communication with the major airways. The FRC is calculated as follows:

$$V_1 c_1 = V_2 c_2 \tag{16-15}$$

$$VI_{He} \times cI_{He} = VTLC \times cA_{He} \tag{16-16}$$

$$VTLC = VI_{He} \times \frac{cI_{He}}{cA_{He}} \tag{16-17}$$

$$VFRC = VTLC - VI_{He} \tag{16-18}$$

where VI_{He} is the volume of the helium mixture inspired (the inspiratory capacity), cI_{He} is the helium concentration of the inspired gas, $VTLC$ is the total lung capacity, and cA_{He} is the alveolar helium concentration [13].

CRITIQUE OF GAS DILUTION METHODS

If all the alveoli are in good communication with the major airways and reasonably well ventilated, then either open or closed circuit methods would provide accurate values for the FRC. In patients with certain pulmonary diseases, however, some areas of the lung are poorly ventilated and act as if they were almost completely obstructed during the testing procedure. If the N_2 in the poorly ventilated areas is not completely washed out during the conventional 7-minute period of the test or if the He is not evenly distributed, the value of the FRC may be underestimated. Prolongation of the test period and mathematical extrapolation may yield the correct value for the FRC, but, in general, such techniques are not practical for routine testing [6]. Instead, the body plethysmographic method is utilized to determine the true value of the FRC, including areas that are well ventilated, poorly ventilated, and nonventilated.

The FRC determined by body plethysmography and the value obtained by conventional gas dilution methods are equivalent in normal subjects, but in patients with lungs where pronounced uneven distribution of gas exists, the FRC measured by the body plethysmograph will be larger. The difference between the two values is termed *trapped gas*. It must be appreciated, however, that there cannot be a gas-containing space in the lung that remains completely nonventilated, because the gas would eventually be absorbed by the blood flowing around it.

BODY PLETHYSMOGRAPHIC METHOD

The principle of this method is based upon Boyle's law, which states the relationship between the pressure and volume of a gas if its temperature remains constant:

$$P_1 V_1 = P_2 V_2 \tag{16-19}$$

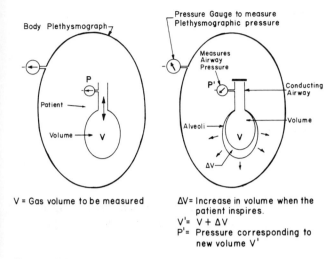

V = Gas volume to be measured

ΔV= Increase in volume when the patient inspires.

V'= V + ΔV

P'= Pressure corresponding to new volume V'

Figure 16-4
Body plethysmographic estimation of functional residual capacity. The patient is represented as his alveoli and conducting airway.

Figure 16-4 depicts a plethysmograph. The subject sits within the airtight chamber. The chamber has a volume of between 600 and 1200 liters (about the size of a telephone booth); the average volume of an adult is about 70 liters. Pressure is measured in the plethysmograph by means of a very sensitive, differential-pressure transducer, which, when calibrated for volume, is capable of detecting a 0.5 ml change in the volume of the chamber. The subject breathes the air about him through a mouthpiece. At end-expiration, the mouthpiece is occluded by an electrically controlled shutter; the patient then continues to make breathing efforts against this obstruction by panting at 1 to 2 Hz, thereby compressing and expanding the alveolar air. Because of the large volume of gas in the plethysmograph, there is insufficient time to exchange heat with the walls during the panting movements, so that the decompression and expansion is an adiabatic process. This is allowed for by calibrating the box adiabatically by repeatedly injecting and withdrawing a volume of air with a pump that cycles at the same rate that is expected to be produced by the subject's panting during the test. The change in pulmonary gas pressure is readily measured, since the mouth pressure equals the alveolar pressure in a closed system. The change in pulmonary gas volume is reflected by a reciprocal change in the plethysmographic volume [5].

From the original pulmonary gas pressure and the change in pulmonary gas pressure, the original volume of gas in the chest can be calculated using Boyle's law:

$$P_1 V_1 = (P_1 + \Delta P)(V_1 - \Delta V) \tag{16-20}$$

where P_1 is atmospheric pressure minus the vapor pressure of water or 970 cm H_2O, V_1 is the FRC (or the thoracic gas volume if the measurement is at a point other than end-expiration), $(P_1 + \Delta P)$ is the final pressure, and $(V_1 + \Delta V)$ is the final volume. Solving for V_1,

$$V_1 = \frac{-\Delta V}{\Delta P}(P_1 + \Delta P) \tag{16-21}$$

However, ΔP is small compared to P_1, e.g., $\Delta P \approx 5$ cm H_2O and $P_1 = 970$ cm H_2O. Therefore, $(P_1 + \Delta P) \approx P_1$, and, disregarding signs,

$$V_1 = V\text{FRC} = 970\frac{\Delta V}{\Delta P} \tag{16-22}$$

The FRC can then be measured from the slope of $\Delta V/\Delta P$, which may be displayed on a cathode-ray oscilloscope after appropriate calibrations have been made [5].

EFFECT OF POSTURE ON LUNG VOLUMES

Most of the effect of posture depends upon the influence of gravity [12]. The largest variation occurs in the resting lung volume (FRC). The FRC is much smaller in the supine position than in the upright position. During gradual tilting from standing to a supine position, the change in FRC is linearly related to the degree of the angle of tilt from the vertical to the horizontal. Beyond the horizontal, the FRC actually increases slightly. The changes are due to interactions between the hydrostatic and recoil pressures of the diaphragm and rib cage. Although tall individuals have a larger FRC than short individuals in the upright posture, the FRC is similar in both in recumbency, which is to be expected on hydrostatic grounds. The TLC and RV are smaller in the recumbent position than in the upright position due to an increase in the amount of blood in the thoracic cavity. These changes in lung volumes with posture are shown in Figure 16-5.

BREATH HOLDING

For a given alveolar oxygen concentration, the rate at which the lung volume will decrease with breath holding is a function of (1) the metabolic rate or oxygen consumption, (2) the pulmonary blood flow, and (3) the ambient pressure. The last factor determines the number of molecules of oxygen present in the lungs. During the first 100 seconds of breath holding in an atmosphere of 100% oxygen, the lung volume decreases nonlinearly as a function of the changing respiratory quotient, but thereafter it falls linearly. In a normal subject, it takes about 13 minutes after hyperventilation with oxygen to go from total lung capacity to residual volume. At an altitude of 18,000 feet, where the absolute quantity of oxygen is one-half of that at sea level, this time would be reduced to 6.5 minutes. For holding one's breath in an

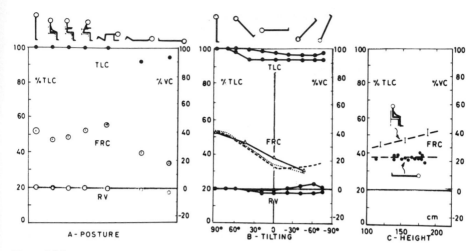

Figure 16-5
Changes in lung volumes with posture. *A.* Effects of posture. *B.* Effects of tilting. *C.* Effects of height in the seated and supine positions. (From E. Agostini and J. Mead. Status of the Respiratory System. In W. O. Fenn and H. Rahn [Eds], *Handbook of Physiology, Respiration,* Section 3, Vol. 1. Baltimore: Williams & Wilkins, 1964. P. 398. Reproduced with permission.)

atmosphere of air, the mean time in normal subjects is 80 seconds with a lung volume of 6 liters, 40 seconds with a lung volume of 3 liters, and 25 seconds with a lung volume of 1.5 liters [11].

CLINICAL INTERPRETATION OF LUNG VOLUME VALUES

VITAL CAPACITY (VC)

The vital capacity is the maximum volume of air that can be expired after a maximal inspiration. It varies directly with the height of the subject and indirectly with age; a value greater than 80% of the predicted value is considered to be normal. It may range from 2.5 to 6.5 liters. It is reduced in restrictive lung diseases (e.g., pulmonary fibrosis, congestive heart failure, and mitral stenosis) and in restrictive thoracic cage disorders (e.g., kyphoscoliosis and obesity). It is reduced in symptomatic asthmatic patients and in patients with chronic bronchitis. In pulmonary emphysema, the vital capacity may be reduced or normal.

RESIDUAL VOLUME (RV)

This is the volume of air left in the lung after a maximal expiration. It is a direct function of the subject's height and age; 70% to 140% of the predicted value is considered to be normal. Its value may range from 1 to 3.5 liters. It is reduced in restrictive lung diseases and thoracic cage diseases. It is considerably increased in patients with emphysema and certain symptomatic asthmatic patients, but it is generally only slightly increased in patients with chronic bronchitis.

TOTAL LUNG CAPACITY (TLC)

This is the sum of the vital capacity and the residual volume. It varies directly with one's height; 90% to 110% of the predicted value is considered to be normal. It is reduced in restrictive lung disorders. It is increased in emphysema and symptomatic asthmatic patients, but it is usually normal or low in those with chronic bronchitis.

REFERENCES

1. Allison, R. D., Holmes, E. L., and Nyboer, J. Volumetric dynamics of respiration as measured by electrical impedance plethysmography. *J. Appl. Physiol.* 19:166, 1964.
2. Comroe, J. H., Jr., et al. *The Lung: Clinical Physiology and Pulmonary Function Tests* (2nd ed.). Chicago: Year Book, 1962.
3. Comroe, J. H., Jr., Botelho, S. Y., and DuBois, A. B. Design of a body plethysmograph for studying cardiopulmonary physiology. *J. Appl. Physiol.* 14:439, 1959.
4. Darling, R. C., Cournand, A., and Richards, D. N., Jr. Studies on the intrapulmonary mixture of gases. III. An open circuit for measuring residual air. *J. Clin. Invest.* 19:609, 1940.
5. DuBois, A. B., et al. A rapid plethysmographic method for measuring thoracic gas volume; a comparison with a nitrogen washout method for measuring functional residual capacity in normal subject. *J. Clin. Invest.* 35:322, 1955.
6. Emmanuel, G., Briscoe, W. A., and Cournand, A. Method for the determination of the volume of air in the lungs: Measurements in chronic pulmonary emphysema. *J. Clin. Invest.* 40:329, 1961.
7. Finucane, K. E., Egan, B. A., and Dawson, S. V. Linearity and frequency response of pneumotachographs. *J. Appl. Physiol.* 32:121, 1972.
8. Gilson, J. C., and Hugh-Jones, P. The measurement of total lung volume and breathing capacity. *Clin. Sci.* 7:185, 1948.
9. Mead. J. Volume displacement body plethysmograph for respiratory measurements in human subject. *J. Appl. Physiol.* 15:736, 1960.
10. Mead, Jere, Peterson, N., Grimby, G., and Mead, J. Pulmonary ventilation measured from body surface movements. *Science* 156:1383, 1967.
11. Mithoefer, J. C. Breath Holding. In W. O. Fenn and H. Rahn (Eds.), *Handbook of Physiology, Section 3, Respiration,* Vol. 2. Washington, D.C.: American Physiological Society, 1965. P. 1011.
12. Moreno, T., and Lyons, H. A. Effect of body posture on lung volume. *J. Appl. Physiol.* 16:27, 1961.
13. Ross, J. C., Ley, G. D., Krumholz, R. A., and Rahbari, H. A technique for evaluation of gas mixing in the lung: Studies in cigarette smokers and non-smokers. *Am. Rev. Respir. Dis.* 95:447, 1967.
14. Yamashiro, S. M., Karuza, S. K., and Hackney, J. D. Phase compensation of Fleisch pneumotachographs. *J. Appl. Physiol.* 36:493, 1974.

Distribution, Diffusion, and Gas Exchange

Marvin A. Sackner

DISTRIBUTION OF INSPIRED GAS

Two physical phenomena determine the entry of inspired air into the alveoli: bulk flow and molecular diffusion. The former is the more important factor; it is responsible for transporting air from the mouth to the respiratory bronchioles, which are the 17th generation of airways. Diffusion is the main determinant of gas distribution within those units that are smaller than the respiratory bronchioles; in such small units, the gas flow rates are minimal and diffusion distances are very short.

Early investigators supposed that the more deeply situated spaces within the terminal unit, the alveolar sacs, would receive less inspired air than the superficially situated structures, the alveolar ducts. This concept has been denoted the *stratification theory,* since it suggests that there is layering of gas within the terminal units. The term *series ventilation theory* has also been used, since such a theory represents the lung as being ventilated like two or more spaces in series, each with a different ventilation rate. However, experimental data and calculations using diffusion equations indicate that the terminal unit is the basic ventilatory unit of the lung, in which gas tensions are nearly equal at all physiologic breathing rates. This conclusion holds only for normal lungs: nonhomogeneous gas composition probably occurs in terminal units affected in emphysema because the diffusion times are prolonged due to the long diffusion distances.

Modern investigators generally agree that nonuniform gas distribution is caused mainly by differences in ventilation between the regions of the lung that are larger than the terminal unit. This *parallel ventilation theory* states that separately located spaces in the lungs may be ventilated at different rates, independently of one another and in parallel. Such different ventilation rates can be produced by several mechanisms: (1) some areas may expand more than others during inspiration, i.e., their dV/V ratios may differ, where dV is the inspiratory volume increase of any lung part and V is the preinspiratory volume of that part; (2) local differences may exist in the time course of inspiratory filling and expiratory emptying; and (3) some areas may receive proportionally more dead-space gas and less inspired air than other areas. Regional volume expansion differences might be caused by local variations in pulmonary tissue distensibility or by a nonuniform distribution of inspiratory forces in relation to the hydrostatic pressure differences down the lung. Asynchronous filling and emptying may be associated with regional differences in airway resistance, lung compliance, or both. Nonuniform distribution of dead-space gas may occur because of asynchronous filling or because of regional differences in airway path lengths [2].

SINGLE-BREATH TESTS OF DISTRIBUTION

To illustrate the physiologist's concept of uneven gas distribution, consider the following situation. Assume that the functional residual capacity (FRC) is 2000 ml and that the subject is breathing air that has a N_2 concentration of approximately 80%. If he inspires 2000 ml of O_2 into his lung and if this O_2 is distributed evenly to the alveoli, each alveolus will have an O_2 concentration of 40%. However, if the O_2 is distributed unevenly to the lung (i.e., if some areas receive more oxygen and others less oxygen), the composition of the alveolar gas will be decidedly nonuniform at end-inspiration.

It is not possible to put sampling needles into thousands of alveoli to determine the alveolar gas composition. However, an indirect assessment of the distribution of ventilation can be made by measuring the expired N_2 concentration continuously at the mouth. Modern nitrogen meters have a 90%-response time of about 0.06 second and can therefore accurately record rapid changes in N_2 concentration. Such a device subjects the gas sample to electrical excitation so that the gas glows. A special filter is used so only the most intense N_2 emission bands are allowed to fall upon a photoelectric cell [3].

In practice, the subject, who has previously been breathing room air, inspires a single breath of O_2; the N_2 meter records 0% N_2 (100% O_2) during inspiration (Fig. 17-1). At end-inspiration, the dead space is filled with O_2 that has just been inspired. At the beginning of expiration, the N_2 meter continues to record 0% N_2, since the first gas to leave the lungs is the dead-space gas. The last phase is the N_2 concentration in the alveolar gas. If the inspired O_2 were distributed evenly to all the alveoli so that each contained 40% N_2 rather than 80% N_2, the first, middle, and last parts of the last phase of the N_2 record would all be 40% N_2, and it would be recorded as a perfectly horizontal record, as shown in Figure 17-1. However, if the inspired O_2 is distributed unevenly (and this occurs to a small extent even in normal subjects), the end-inspiratory N_2 concentrations would vary throughout the lung, and the concentrations of the expired alveolar N_2 would not be recorded as a horizontal line. Usually, the first part of the curve shows a lower concentration of N_2 than in the last portion of the expired gas. This phenomenon of uneven distribution also emphasizes the difficulties that result from considering a single, spot sample of alveolar gas as representative of all alveolar gas. It is obvious that the first part of the expired alveolar gas may differ significantly in composition from the last part, and the alveolar gas remaining in the unexpired residual volume may have yet a different composition. The sloping third stage (alveolar gas stage) of a single-breath record can only be explained if lung regions with different ventilatory rates do not empty in phase [7].

The lung model of Otis and his associates [16], which will be described in further detail in Chapter 19, has been offered as an explanation for uneven gas distribution. This model consists of a number of parallel units, each consisting of a compliance element (the alveoli) and a resistance element (the airways) in series, and each characterized by a unique time-constant (RC product). Units having different time-constants

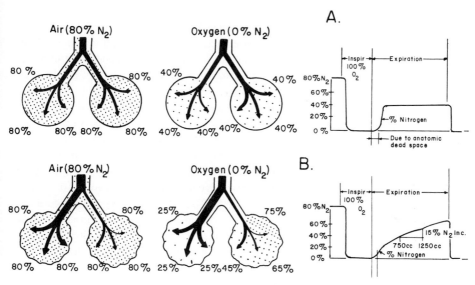

Figure 17-1
Distribution of inspired air. Nitrogen molecules are represented by black dots. *Left:* Alveolar N_2 during breathing of air. *Center:* Alveolar N_2 immediately after a single breath of O_2. *Right:* Actual records of N_2 concentration and gas flow during the next expiration. Record A is for uniform alveolar distribution and record B for nonuniform alveolar distribution. (Modified from J. H. Comroe, Jr., et al. *The Lung* [2nd ed.]. Chicago: Year Book, 1962. P. 62.)

ventilate asynchronously: early inspiring regions with small time-constants would be well ventilated and would expire early; late-inspiring regions with large time-constants would be poorly ventilated and would expire late. This is called the *first in, first out theory,* and it is in agreement with the pattern of single-breath records in patients with obstructive airway diseases.

Spirometric studies of a lobe of the lung indicate, however, that at least under some conditions, an area that fills relatively early in inspiration may empty late in expiration [12]. Such a filling and emptying pattern, as predicted by a *first in, last out theory* or *sequential ventilation theory,* has been proposed to account for the sloping third portion of the third phase of single-breath records. A region that fills early would receive a larger proportion of dead-space gas and less inspired gas than parts that fill later. After inspiration of O_2, the early filling region would have a higher N_2 concentration than it would later in expiration.

That such a phenomenon occurs has been corroborated by measurements of the nitrogen slope in association with information derived from radioisotopic scanning methods. It has been shown, for example, that inhalation of radioactive xenon (^{133}Xe) from the residual volume point to total lung capacity at slow inspiratory flow rates (less than 0.5 liter/sec) results in the following pattern of distribution: From the residual volume to the functional residual capacity level, the bulk of the inspired gas is distributed to the upper zones of the lungs in the erect subject [14, 19] (this is related mainly to gravitational influences). Later in the process of inhalation, the lower zones receive most of the inspired gas. A slow expiration produces emptying of the lower zones, and, as the residual volume level is approached, the upper zones are emptied. However, rapid inspiratory or expiratory flow rates produce an even distribution of alveolar filling or emptying [14].

Through such measurements, it has been deduced that at the residual volume level, units of the lower zones are closed off in normal subjects. This volume of air, termed the *closing volume,* can be estimated by the following modification of the single-breath nitrogen test using the principle of the "first in, last out" theory. The subject breathes O_2 until almost all the N_2 is washed out of his lungs. (In the normal subject, it takes about 3 minutes to reach an alveolar N_2 concentration of 1%.) He then expires to the residual volume level, and a 100-ml bolus of N_2 is injected into the inspiratory line at the rate of 500 ml/sec. He inspires O_2 to total lung capacity at a slow flow rate, and he then expires slowly while a continuous expired N_2 concentration and expired volume is recorded. Since the bolus of N_2 is distributed mostly to the upper zones, an expired N_2 curve, as shown in Figure 17-2, is obtained. On the expired N_2 curve, E indicates expiration; phase 2, the sloping sigmoid curve, represents a mixture of gas from the dead space and alveoli; and phase 3, a gently sloping plateau, represents gas from alveoli of mostly the lower zones. Phase 4 represents the N_2 from the upper zones. When the 100-ml bolus of N_2 is delivered at residual volume, the N_2 concentration rises in phase 4 because the upper zones have received the bulk of the bolus of N_2. When the 100-ml bolus of N_2 is delivered at functional residual capacity level, the N_2 concentration falls in phase 4 because the lower zones have received the bulk of the bolus of N_2. In Figure 17-2, *closing volume* indicates the portion of the vital capacity at the beginning of phase 4. In obese subjects or aged subjects, the closing volume may actually be above the FRC; these subjects have units of lungs that are collapsed at their resting lung volume.

At first glance, the "first in, first out" and "first in, last out" theories appear difficult to reconcile. Both mechanisms can exist, however, and whichever predominates is a function of the inspiratory and expiratory flow rates and the time-constants of lung units. The "first in, last out" theory seems to be more important in accounting for the ventilation pattern in the normal subject, in whom the distribution of airway resistance is quite constant and of small magnitude when slow ventilatory rates are employed. Rapid flow rates promote an even distribution of inspired air. The parallel model of the "first in, first out" theory, with its different time-constants, becomes important only in disease, and its ventilation pattern is accentuated by increases in flow rates. When there are extreme differences in time-constants, *pendelluft* ("pendulum breath") may occur. This is a phenomenon in which it is possible for gas to continue to flow out of one compartment of the lung into the other, and vice versa, without passing to the outside environment (Fig. 17-3). Such internal circulation of gas between regions of short and long time-constants reduces the effective tidal volume. *Pendelluft* between two lungs may occur when a major bronchus is partially obstructed by carcinoma.

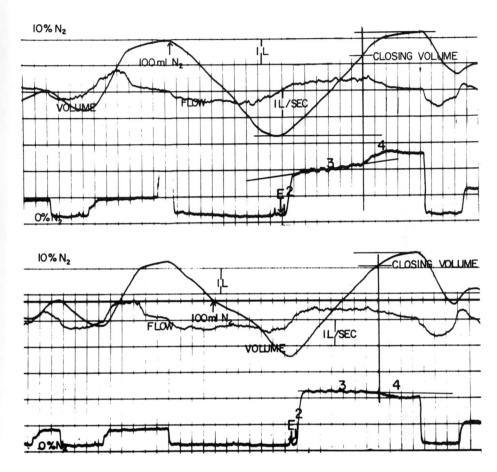

Figure 17-2.
Distribution of a bolus of nitrogen inspired at the residual volume position (*upper tracing*) and the functional residual capacity position (*lower tracing*).

CLINICAL SINGLE-BREATH NITROGEN TEST

In the pulmonary laboratory, the measurement of the nitrogen concentration difference (ΔN_2) between two volumes of the expired gas — 750 and 1200 ml — after inspiring O_2 is used as a screening test for gas distribution [3]. These volume points were chosen because they provided the best way of distinguishing between diseased and normal subjects. In young subjects, ΔN_2 does not usually exceed 2.5%; in older subjects, 3.5%. It may be elevated to as high as 15% in severe pulmonary emphysema. Indeed, it is difficult to suggest the physiologic diagnosis of emphysema without confirmation by the finding of uneven distribution. This test, however, is nonspecific, and the ΔN_2 may be elevated in such diverse conditions as asthma, bronchitis, pulmonary fibrosis, and congestive heart failure.

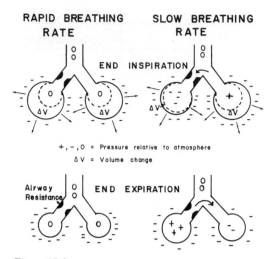

RAPID BREATHING   SLOW BREATHING
      RATE   RATE

+, −, 0 = Pressure relative to atmosphere

ΔV = Volume change

Figure 17-3
Uneven distribution of resistance and consequent differences in time-constants. Pendelluft effect is illustrated in the diagrams on the right.

NITROGEN WASHOUT CURVES

In the simplest model, the lungs may be represented by a bellows that is uniformly ventilated, i.e., the inspired gas is distributed evenly to and mixed instantly with all the gas previously present in the bellows. We will assume that the inspired gas contains no N_2 and that there is no transfer of N_2 from the blood and tissue. In this system, Equation 17-1A holds:

$$FA_n = FA_o w^n \tag{17-1A}$$

where F is the fractional gas concentration of N_2, FA_o is the alveolar N_2 concentration before O_2 inhalation (which is assumed to be 80% during the breathing of air), FA_n is the alveolar N_2 concentration after n breaths of pure O_2, and w is the alveolar dilution factor:

$$w = \frac{V_{AD}}{V_{AD} + (V_T - V_D)}$$

where V_{AD} is the volume in which N_2 is contained at functional residual capacity or the alveolar volume plus dead space volume, V_T is the tidal volume, and V_D is the respiratory dead space volume.

Taking logarithms of Equation 17-1A,

$$\log FA_n = \log FA_o + n \log w \tag{17-1B}$$

The logarithm of FA_n plotted against n yields a straight line; its intercept at $n = 0$ is log FA_o and its slope is log w.

These equations, however, which are based on the model of the ventilation of a single bellows, are inadequate for experimental purposes because the lungs, even in normal subjects, are not evenly ventilated [8]. The lungs, therefore, are best described as a *multicompartment system*, according to Equation 17-2:

$$FA_n = (FA_1)_o\, w_1^n + (FA_2)_o\, w_2^n + \ldots + (FA_i)_o\, w_i^n \tag{17-2}$$

In practice, the N_2 washout curve (Fig. 17-4) is plotted on semilog paper and a graphical method is used to solve for the number of different compartments. In normal subjects, there are one or two compartments, with the slower compartment having the larger volume. Recent studies suggest that normal, young nonsmokers generally have one compartment, whereas otherwise healthy smokers have two compartments. In cases

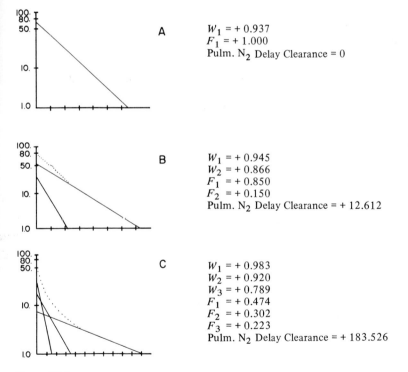

A
$W_1 = +0.937$
$F_1 = +1.000$
Pulm. N_2 Delay Clearance $= 0$

B
$W_1 = +0.945$
$W_2 = +0.866$
$F_1 = +0.850$
$F_2 = +0.150$
Pulm. N_2 Delay Clearance $= +12.612$

C
$W_1 = +0.983$
$W_2 = +0.920$
$W_3 = +0.789$
$F_1 = +0.474$
$F_2 = +0.302$
$F_3 = +0.223$
Pulm. N_2 Delay Clearance $= +183.526$

Figure 17-4
Nitrogen washout curves. These are plotted semilogarithmically; W indicates the slope of the curve, and F represents the fraction of the various compartments. The end-tidal N_2 concentration is shown on the ordinate; the number of breaths is given on the abscissa. The pulmonary nitrogen delay clearance is an overall expression of the curve (less than 50% is normal). The slowest compartment is 1, and 3 is the fastest. A normal subject washes out N_2 rapidly and does not have more than two compartments, whereas a patient with emphysema has a minimum of two compartments and usually three compartments.

of emphysema, the fast compartment usually accounts for less than 50% of the lung volume, and the remainder is divided into two or more slower compartments.

It is tempting to think of the exponential function in Equation 17-2 in terms of anatomic entities, yet it must be appreciated that these components are only the result of a method of analysis that attempts to explain the experimentally determined washout behavior of the lungs according to the properties of an oversimplified model. For example, it has been shown that a curve that yields two or three components by graphical analysis may be constructed by the addition of up to 20 components with half-times varying along an approximate, continuous distribution. Indeed, new method of mathematical analysis have shown that the N_2 washout curve can be analyzed in terms of an infinite number of compartments. Nevertheless, no further information from the viewpoint of the analysis of uneven distribution has been obtained from these more complicated mathematical techniques.

FACTORS INFLUENCING GAS DISTRIBUTION

AGE

Distribution of ventilation becomes more uneven with age. This is probably related to an increased proportion of lower-zone airways that are closed at the FRC position as a result of alteration in the elastic properties of the lung with aging in association with hydrostatic differences in pressure down the lung.

BREATHING RATE

Increases in the breathing rate do not usually affect N_2 washout in normal nonsmokers, but they tend to make the washout curves nonlinear in otherwise normal smokers [23]. Breath holding tends to make nonlinear washout curves more linear.

TIDAL VOLUME

Increases in tidal volume in normal subjects do not appear to affect gas distribution significantly. In patients with emphysema, there may be a slight improvement in distribu tion because of ventilation of the slow space when the tidal volume is increased volun- tarily or with the aid of mechanical ventilators.

CHEST-WALL COMPRESSION

The N_2 washout curve has been used to demonstrate that chest strapping or compres- sion produces airway closure. If the subject breathes at a constant tidal volume, there is an exponential decrease in N_2 concentration. When the strapping is released and the subject takes a deep breath, there is an immediate rise in N_2 concentration from those zones of the lungs whose airways had been closed by the strapping procedure.

DISEASES

Uneven gas distribution, though most pronounced in patients with emphysema, is a fairly frequent finding in patients with cardiopulmonary diseases. Although it is not pathognomonic of any particular disease, it often is a useful screening test for pulmonary function.

DIFFUSING CAPACITY

DIFFUSION THEORY

The movement of gas molecules from a region of higher chemical activity to one of lower chemical activity (which is nearly synonymous with partial pressure) takes place according to Fick's laws:

$$\frac{\delta Q}{\delta t} = -AD\frac{\delta c}{\delta x} \tag{17-3}$$

$$\frac{\delta c}{\delta t} = D\frac{\delta^2 c}{\delta x^2} \tag{17-4}$$

where $\delta Q/\delta t$ is the instantaneous flow rate of a gas (ml STPD per sec) perpendicularly across an area A (cm^2) under a concentration gradient $\delta c/\delta x$, which is the rate of change in the concentration of gas, c (ml STPD per ml total volume), in the x direction (cm). The *diffusion constant*, D (cm^2 per sec), is a property of the material making up the medium from which the diffusion takes place.

In biologic work, diffusion takes place under the influence of the pressure or fugacity gradient. Therefore, in Fick's equations, one must substitute αP for c, where P is the partial pressure of gas (mm Hg) and α is the solubility coefficient (ml gas STPD per ml fluid at 1 mm Hg partial pressure and a given temperature). Equations 17-5 and 17-6 are obtained by substituting αP for c in Equations 17-3 and 17-4:

$$\frac{\delta Q}{\delta t} = -AD\frac{\delta P}{\delta x} \tag{17-5}$$

$$\frac{\delta P}{\delta t} = D\frac{\delta^2 P}{\delta x^2} \tag{17-6}$$

The *diffusing capacity, D, of the lung* is defined as the milliliters of gas at STPD diffusing across the alveolocapillary membrane per minute per 1 mm Hg of pressure difference between the alveolar air and the pulmonary capillary blood. In electrical terms, it is analogous to conductance. Such gases as CO, CO_2, and O_2 react chemically within the red blood cell. The diffusing capacity, D_l, for these gases includes the entire diffusion path of the gas from the alveolus to the hemoglobin molecule.

It may be apportioned between the diffusing capacity of the pulmonary membrane D_m — i.e., the rate of gas diffusion (ml STPD per min) divided by the pressure difference (mm Hg) between alveolar air and the plasma outside the red blood cell — and the diffusing capacity, θVc, of the red blood cells in the pulmonary capillary bed, where θ is the rate of gas uptake (ml per min) at a 1-mm Hg pressure gradient for 1 ml of normal blood and Vc is the pulmonary capillary blood volume (ml) [17].

DIFFUSION-LIMITED VERSUS FLOW-LIMITED UPTAKE OF GASES

Since it is technically easier to deal with the mathematics of the diffusing capacity equations for CO than those for O_2, the former is used in clinical pulmonary laboratories to assess the diffusing capacity of the lung. Carbon monoxide has 210 times the affinity for hemoglobin (Hb) as does O_2. Thus, a partial pressure of CO of only 0.48 mm Hg produces the same percentage saturation of Hb as does a partial pressure of 100 mm Hg of O_2. For this reason, any CO in the vicinity of Hb becomes bound to it. The Hb capacity is quite large for CO, and even if pulmonary blood flow were to be stopped, the transfer of CO would continue (Fig. 17-5). Therefore, the transfer of CO is not limited by the pulmonary blood flow, but by diffusion across the alveolar capillary membrane as well as by diffusion and chemical reactions within the red blood cell [17].

In Figure 17-5, the black dots signify gas molecules. The bottom of each alveolus represents the alveolocapillary membrane, which has numerous pores that permit the

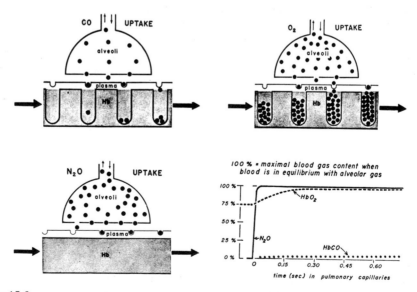

Figure 17-5
Uptake of gases showing diffusion-limited CO and flow-limited N_2O. Graph shows the blood content of gases in the pulmonary capillaries. (From J. H. Comroe, Jr., et al. *The Lung* [2nd ed.]. Chicago: Year Book, 1962. P. 118. Reproduced with permission.)

diffusion of gas molecules. The blood is shown as moving from left to right through the pulmonary capillary. The bloodstream (solely for purposes of illustration) is divided into two portions: the upper portion represents plasma, with a number of pockets to indicate its capacity for dissolving each gas; the lower portion represents hemoglobin. The large pockets in the representation of hemoglobin indicate its capacity for CO and O_2.

Gases such as N_2O and acetylene are equally soluble in the alveolocapillary membrane and the blood, since they do not combine with Hb. These gases diffuse across the alveolocapillary membrane and saturate the entire plasma compartment. Further diffusion is prevented until the blood is replaced by fresh blood, which means that the uptake of these gases is *flow-limited* rather than diffusion-limited. Therefore, the uptake of these gases can be used to estimate the pulmonary capillary blood flow.

The graph in Figure 17-5 shows the gas content in the blood plotted against the time that the blood spends in the pulmonary capillary. The horizontal line at the top of the graph indicates the maximum gas pressure maintained in the alveoli. In the case of CO, there is only a very slight increase in the Hb-CO concentration along the capillary; it never approaches the maximum value because of the low partial pressure of CO in the alveoli. It must be remembered that 100 ml of blood when saturated with CO will contain about 20 ml of CO, yet the alveolar gas at any instant contains only about 1.3 ml CO, assuming that the FRC is 2000 ml and contains 0.065% CO. In the case of N_2O, the blood acquires its maximum concentration of this gas before it has gone 1/20 of the distance along the capillary. In the case of O_2, saturation increases from 75% to 97% along the capillary [6].

METHODS FOR MEASURING DIFFUSING CAPACITY

In order to measure the transfer of CO or O_2 across the alveolocapillary membrane, one needs to know the rate of uptake of the gas and the alveolar capillary gradient. In the case of CO, the capillary CO concentration usually is so small that it can be assumed to be zero. However, since O_2 combines with Hb in a nonlinear way during its passage through the capillaries, the calculation of the mean capillary O_2 tension is fraught with difficulties. This is the reason why clinical pulmonary laboratories generally employ CO as the test gas for determination of diffusing capacity.

Steady-State Diffusing Capacity – $(D_{CO})_{ss}$

The steady-state methods are based on the assumption that no transient changes take place in the organism during the period of measurement. They depend upon measuring the difference between the product of the inspired minute-volume times its CO concentration and that of the expired minute-volume times its CO concentration (the amount taken up) and dividing the result by the alveolar partial pressure of $CO(P_{A_{CO}})$ [1]. The patient breathes a mixture of 0.1% CO in air for a few minutes, during which time the CO exchange has reached a steady state. Methods for estimating the value of $(D_{CO})_{ss}$ differ with regard to the way of obtaining the $P_{A_{CO}}$. Steady-state methods have an advantage over the other methods for the determination of diffusing capacity in that they are easy for the patient to perform, either at rest or during exercise. The test, however, cannot be repeated without taking into account the amount of CO carried as carboxyhemoglobin in the blood after performing the first test. In general, steady-state methods are highly affected by ventilation-perfusion disturbances, and they do not provide a true estimate of the diffusing capacity, particularly in diseased patients. The values obtained with this test reflect the overall diffusing capacity as well as ventilation-perfusion abnormalities.

The *Bates end-tidal sampling method* [1] is based on the assumption that an end-tidal sample is representative of the PA_{CO} of the lung as a whole. It suffers in that it considers any spot sample as representative of the alveolar gas fraction, even though uneven distribution of gas may be present. It may give a falsely low value for the diffusing capacity when the physiologic dead space is increased by disease. Because collecting an end-tidal sample depends upon clearing the subject's dead space and the apparatus dead space, a certain minimum tidal volume is necessary, usually at least 450 ml. Such a volume may not be attainable with resting patients with restrictive lung diseases. Because the ratio of dead space to tidal volume becomes small during exercise, an anatomic dead space can be assumed to exist under these conditions, and the alveolar CO concentration can be calculated with no sacrifice in accuracy.

In the *Filley method,* the physiologic dead space for CO_2 is calculated from the Bohr equation, and the dead space is used to calculate the PA_{CO} [5]. It has the disadvantage that an arterial sample is needed, and the calculated PA_{CO} can become extremely sensitive to slight changes in gas concentration, which may produce errors up to 40% in the $(D_{CO})_{ss}$ value at rest. The method is most often used with the subject exercising, in which case the ratio of dead space to tidal volume becomes small.

Single-Breath Technique – $(D_{CO})_{SB}$

In this method, the subject inspires a mixture of CO and He in air to total lung capacity, holds his breath 10 seconds, and then expires while an alveolar sample is collected [15]. Helium is used as an insoluble tracer in order to calculate the initial alveolar CO concentration, so that two points can be obtained in the equation for diffusion capacity. Historically, He was used as the insoluble tracer because of the availability of thermal conductivity analyzers for its determination. More recently, neon has been used as the tracer gas in those laboratories that have a gas chromatograph. The rate of CO disappearance from the alveolar volume will be a single exponential function for up to about 30 seconds duration. After this time, the CO will be taken up at a different rate-constant, which indicates some degree of D_l/VA unevenness in the lung. The diffusing capacity of the lung for carbon monoxide, $(D_l)_{CO}$ (ml STPD/min × mm Hg), is expressed as:

$$(D_l)_{CO} = \frac{VA \times 60}{t \times (PB - 47)} \times \ln \frac{(FA_{CO})_i}{(FA_{CO})_f} \qquad (17\text{-}7)$$

where V_A is alveolar volume (ml STPD), t is the time (sec) that the breath is held, P_B is the barometric pressure, $(FA_{CO})_i$ is the initial CO fraction at time 0 (inspired CO concentration times the ratio of He concentration expired to He concentration inspired), and $(FA_{CO})_f$ is the expired CO fraction.

This method involves the assumption that the slope of the disappearance curve for CO has an intercept at 1.0 (Fig. 17-6), and this assumtpion appears to be valid for normal subjects. A typical CO disappearance curve is shown in Figure 17-6. In diseased subjects, however, this assumption may not be true because of timing errors in the collection of a single alveolar sample, so that for research work, a CO disappearance curve with multiple points must be obtained to calculate $\ln [(FA_{CO})_i/(FA_{CO})_f]$. Even if the slope of this curve does not intercept at 1.0, the diffusing capacity calculated by the multiple-point method will still be correct, whereas the method using only the 10-second point for the sample and assuming an intercept of 1.0 might be grossly in error.

As compared to the steady-state methods, the single-breath test gives D_{CO} values that are about 30% to 40% higher at rest (15 to 24 versus 30 to 40 ml/min/mm Hg). The single-breath test is quick and requires little cooperation from most subjects, although dyspneic or exercising patients may find it difficult to hold their breath. The test can be repeated several times in rapid succession without building up a

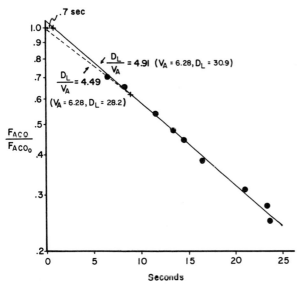

Figure 17-6
Carbon monoxide disappearance curve in a normal subject obtained by multiple breath-holding tests. The dots represent individual data points; the cross represents another data point that might be used to calculate a single-breath diffusing capacity in which the period of breath-holding varies between 8 and 12 seconds. The dashed line indicates the slope used for the calculation of the latter on the basis of the assumption that the zero intercept is unity. Because the value at unity is actually +0.7 second, the standard calculation underestimates the true diffusing capacity by 9%.

significant "back pressure" of CO in the capillary blood that must be corrected for in the calculation, unless there is a great number of tests. However, the test is somewhat sensitive to abnormal distribution of ventilation, and it therefore occasionally gives a picture of only the well-ventilated parts of the lung.

Rebreathing Technique – $(DCO)_{RB}$

This method combines certain of the advantages of the steady-state method with those of the single-breath method. A bag containing 2 to 4 liters of a mixture of CO, He or Ne, O_2, and N_2, which is a volume equal to the subject's one-second vital capacity, is rebreathed at a rate of 20 to 30 breaths per minute for 15 to 20 seconds [13, 20]. The subject endeavors to empty the bag on each inspiration. Gas concentrations within the bag are continuously recorded either with rapidly responding gas analyzers or with a mass spectrometer; sequential samples may be obtained for gas chromatography. After the first 5 seconds of rebreathing, the decrease in CO becomes exponential with respect to time. Thus, the diffusing capacity can be calculated from the CO disappearance curve in the same manner as the single-breath technique using the multiple breath-holding points. For the construction of a CO disappearance curve, however, the rebreathing tests take much less time (20 seconds) than the single-breath method with multiple points (30 minutes). Furthermore, the rebreathing method is much easier to perform than the single-breath test if the subject is exercising.

EFFECT OF EXERCISE ON DIFFUSING CAPACITY

In normal subjects, the diffusing capacity rises during exercise and parallels increases in cardiac output [10, 11]. Endurance athletes, such as swimmers, appear to have larger increases in the diffusing capacity than untrained subjects, a phenomenon that facilitates the transfer of oxygen into the blood. Patients with pulmonary fibrosis show smaller increases than normal subjects, which suggests that they cannot expand their already limited capillary bed. During maximum exercise in normal subjects, the values for both the single-breath and steady-state tests rise to similar D_{CO} values of 60 to 70 ml/min/mm Hg.

DIFFUSING CAPACITY OF THE PULMONARY MEMBRANE AND CAPILLARY BLOOD VOLUME

Equation 17-8 describes the subdivisions of diffusing capacity of the lung (D_l):

$$\frac{1}{DL} = \frac{1}{D_m} + \frac{1}{\theta Vc} \tag{17-8}$$

where D_m is the diffusing capacity of the pulmonary membrane, θ is the reaction rate of CO with hemoglobin per 1 ml of blood (which is obtained by in vitro measurements at 37°C on dilute red blood cell suspensions in a rapid reaction–velocity apparatus), and Vc is the pulmonary capillary blood volume [17].

Since θ can be decreased by increasing the alveolar P_{O_2} (apparently without changing D_m or Vc), it follows that if D_l is measured (as DCO) at different values of the alveolar P_{O_2}, Equation 17-8 can be solved. This is done by plotting $1/D_l$ against $1/\theta$. The reciprocal of the slope is Vc, and the reciprocal of the intercept is D_m. Normal subjects have D_m values by the single-breath technique of 40 to 65 ml/min/mm Hg; D_m values based on steady-state estimates are generally unreliable. The values of Vc by both methods range from 70 to 115 ml. During exercise, the Vc increases up to a maximum of about 200 ml, whereas the D_m stays relatively unchanged.

DIFFUSION LIMITATIONS

In normal subjects at rest, the hemoglobin equilibrates with the alveolar P_{O_2} within 0.2 second after the passage of blood through the pulmonary capillaries. The average transit time, $\bar{t}$, through the capillaries is:

$$\bar{t} = \frac{Vc}{\dot{Q}} = \frac{100 \text{ ml}}{100 \text{ ml/sec}} = 1 \text{ sec} \tag{17-9}$$

Diffusion, therefore, does not limit oxygen transfer at rest. However, at high altitudes where alveolar P_{O_2} might be 47 mm Hg, O_2 equilibrates with hemoglobin in about 0.5 second at rest. During heavy exercise, there is not enough time for the hemoglobin in the capillaries to achieve equilibration with O_2 at high altitudes, despite expansion of the capillary bed. This causes incomplete equilibration of the pulmonary venous blood with respect to the alveolar P_{O_2}, and an alveolar-arterial diffusion gradient results [21].

DISEASES AFFECTING THE DIFFUSING CAPACITY

The diffusing capacity is reduced in pulmonary emphysema and pulmonary fibrosis. It is generally normal in chronic bronchitis and bronchial asthma. It is reduced in any condition that produces pulmonary vascular disturbances, such as pulmonary emboli, pulmonary hypertension, and certain valvular cardiac defects. It tends to be low in anemia and elevated in polycythemia. Reduction of the diffusing capacity produces an alveolar-arterial oxygen gradient, which may become particularly pronounced during exercise, and an attendant decrease in arterial oxygen saturation.

ARTERIAL BLOOD GASES

OXYGEN TRANSPORT

Oxygen is absorbed by blood in two ways: (1) in physical solution and in the plasma and (2) in chemical combination with hemoglobin. The amount taken up in physical solution is relatively small, being in the order of 0.003 ml O_2/100 ml blood/mm Hg P_{O_2}.

This amount, however, can become appreciable if the atmospheric pressure is raised above that at sea level. For example, experiments carried out on pigs in a compression chamber involved the removal of cells from the blood and the reinfusion of plasma so that no red blood cells were present (the blood was a clear color). The animals were kept alive at 10 atmospheres pressure of O_2 for several hours. This produced a concentration of approximately 20 volume-percent O_2 in physical solution in the blood, an amount equivalent to that combined with hemoglobin under sea level conditions.

One gram of hemoglobin (Hb) is capable of combining chemically with 1.34 ml of O_2; the maximum amount of O_2 that can be combined in this way is termed the *oxygen capacity*. Thus, if 100 ml of blood contains 15 gm of Hb, it can combine with $15 \times 1.34 = 20.1$ ml O_2 (oxygen capacity). However, the amount of O_2 combined with Hb is not linearly related to the Po_2, as in the case of dissolved O_2. A plot of O_2 content against Po_2 results in a sigmoid curve that has a very steep slope between 10 and 50 mm Hg Po_2 and a flat slope between 70 and 100 mm Hg Po_2. The volume content of oxygen in the blood divided by the oxygen capacity and multiplied by 100 is called the *oxygen saturation value* [18, 24].

ARTERIAL OXYGEN SATURATION

This value normally is above 94%, but it is decreased in many cardiopulmonary disorders. It is most accurately measured by the manometric method of Van Slyke [22] by means of the estimation of arterial O_2 content and capacity. A spectrophotometric method that measures the ratio of oxyhemoglobin to hemoglobin is less accurate, but it is often employed clinically because of its simplicity [24].

ARTERIAL OXYGEN TENSION

This may be measured directly by a membrane-covered oxygen electrode that permits O_2 but not red blood cells or plasma protein to diffuse through its tip. In young subjects, values greater than 90 mm Hg are normal; in normal subjects older than 70 years, it is common to see values of 75 mm Hg. Because of the shape of the oxyhemoglobin dissociation curve, the determination of arterial Po_2 is a sensitive test of blood abnormalities when the O_2 saturation is normal. The oxyhemoglobin curve is unchanged in shape, but it is shifted to the right by acidemia and to the left by alkalemia. In alkalemia, for example, the O_2 saturation value is higher at a given arterial Po_2 than it is in blood with a normal pH [23].

CAUSES OF HYPOXEMIA

Diminution of the arterial oxygen tension can occur as a result of (1) a deficiency of O_2 in the atmosphere because of high altitude or because of smoke inhalation, (2) hyperventilation, (3) ventilation-perfusion disturbances, (4) diffusion impairment, and (5) venous-to-arterial shunts. Inadequate transport and delivery of O_2 can occur in anemia and localized or generalized circulatory deficiencies. Inadequate tissue O_2 levels may occur in tissue edema with abnormal tissue demand (e.g., thiamine deficiency) or because of poisoning of cellular enzymes (e.g., by cyanide).

CARBON DIOXIDE TRANSPORT

Carbon dioxide diffuses from the cells into the plasma. Some of it dissolves, and the rest reacts slowly with water to form carbonic acid:

$$H_2O + CO_2 \rightleftharpoons H_2CO_3 \rightleftharpoons H^+ + HCO_3^- \qquad (17\text{-}10)$$

The H^+ is accommodated by plasma buffering systems. Some of the dissolved CO_2 in plasma reacts with the amino group of plasma proteins to form carbamino compounds. Most of the plasma CO_2 passes into the red blood cell, where: (1) some remains in the red blood cell as dissolved CO_2, (2) some forms carbamino compounds, and (3) most reacts rapidly with water because of the presence of the catalyst, carbonic anhydrase, to form H_2CO_3. The bicarbonate then diffuses from the red blood cell into the plasma to reestablish the equilibrium of HCO_3^- between the cells and the plasma. Although plasma contains much more CO_2 (in all forms) than do the red blood cells and although the plasma transports 60% of the CO_2, the red blood cell, by means of carbonic anhydrase, provides practically all the additional HCO_3^- that is transported in the plasma.

CARBON DIOXIDE CONTENT AND TENSION

In contrast to the result of the oxyhemoglobin dissociation curve, in the physiologic range the relationship between CO_2 concentration and tension is almost linear. The relationship between P_{CO_2}, CO_2 concentration, and pH is expressed by the Henderson-Hasselbalch equation:

$$pH = pK + \log \frac{[HCO_3^-]}{[CO_2]} \qquad (17\text{-}11)$$

The usual laboratory determinations measure CO_2 concentration, P_{CO_2}, and pH but not the HCO_3^- concentration. This can be resolved by substituting "total $CO_2 - [CO_2]$" in the numerator of Equation 17-11:

$$pH = pK + \log \frac{\text{total } CO_2 - [CO_2]}{[CO_2]} \qquad (17\text{-}12)$$

Since $[CO_2]$ equals the partial pressure of CO_2 times the solubility coefficient α, Equation 17-12 becomes:

$$pH = pK + \log \frac{\text{total } CO_2 - \alpha P_{CO_2}}{\alpha P_{CO_2}} \qquad (17\text{-}13)$$

Since pK and α are constants, Equation 17-13 becomes:

$$pH = 6.1 + \log \frac{\text{total } CO_2 - 0.0301 \ (P_{CO_2})}{0.0301 \ (P_{CO_2})} \qquad (17\text{-}14)$$

Therefore, if one knows two of the three variables, P_{CO_2}, pH, or total CO_2 content of the blood, one can readily calculate the HCO_3^- concentration. This is important because the $[HCO_3^-]$ represents the metabolic expression of acid-base control in the blood, while the P_{CO_2} represents the respiratory expression of acid-base control.

ACID-BASE BALANCE

The lungs are the most important organ in the body for acid excretion. Under ordinary circumstances, the kidney excretes about 40 to 80 mEq of fixed acid, whereas the lungs excrete about 13,000 mEq of carbonic acid in the form of CO_2. The ratio $[HCO_3^-]/[CO_2]$ in the Henderson-Hasselbalch equation determines the pH of the blood. At a normal pH of 7.40, the ratio is approximately 20 to 1. Whenever the P_{CO_2} tends to rise, the respiratory center in the medulla is stimulated, and the alveolar ventilation rate is increased to restore the P_{CO_2} to normal, provided the respiratory system and neural mechanisms are normal. Whenever hypoventilation occurs because of respiratory disease or respiratory center depression, the P_{CO_2} rises and an acute respiratory acidosis is produced (i.e., the pH of the blood falls from its normal value of 7.40). The kidneys over a 24- to 48-hour period will retain bicarbonate, and the pH will be restored to normal, thus producing a chronic respiratory acidosis.

If bicarbonate is lost from the gastrointestinal tract, e.g., as a result of vomiting, the immediate response will be a metabolic acidosis. The respiratory center, however, will be stimulated, and alveolar ventilation will be increased to lower the P_{CO_2} below its normal value of 40 mm Hg and hence return the pH to normal. Respiratory adjustments in response to metabolic disturbances are usually quite rapid, whereas renal adjustments to respiratory disturbances are relatively slow, starting in about 6 hours and not being completed until 24 to 48 hours [18].

REFERENCES

1. Bates, D. V., Boucot, N. G., and Dormer, A. E. Pulmonary diffusing capacity in normal subjects. *J. Physiol.* 129:237, 1955.
2. Bouhys, A. Distribution of inspired Gas in the Lung. In W. O. Fenn and H. Rahn (Eds.), *Handbook of Physiology, Section 3, Respiration*, Vol. 1. Washington, D.C.: The American Physiological Society, 1964. P. 715.
3. Comroe, J. H., Jr., and Fowler, N. S. Lung function studies. VI. Detection of uneven alveolar ventilation during a single breath of oxygen; a new test of pulmonary disease. *Am. J. Med.* 10:408, 1951.
4. Dollfuss, R. E., Milic, J., and Bates, D. V. Regional ventilation of the lung, studies with boluses of [133]xenon. *Respir. Physiol.* 2:234, 1967.

5. Filley, G. F., MacIntosh, D. J., and Wright, G. W. CO uptake and pulmonary diffusing capacity in normal subjects at rest and during exercise. *J. Clin. Invest.* 33:530, 1954.
6. Forster, R. E. Diffusion of Gases. In W. O. Fenn and H. Rahn (Eds.), *Handbook of Physiology, Section 3, Respiration,* Vol. 1. Washington, D.C.: The American Physiological Society, 1964. P. 839.
7. Fowler, W. S. Lung function studies. III. Uneven pulmonary ventilation in normal subjects and in patients with pulmonary disease. *J. Appl. Physiol.* 2:283, 1949.
8. Fowler, W. S., Cornish, E. R., and Kety, S. S. Lung function studies. VIII. Analysis of alveolar ventilation by pulmonary N_2 clearance curves. *J. Clin. Invest.* 31:40, 1952.
9. Gomez, D. M., Briscoe, W. A., and Cumming, G. Continuous distribution of specific tidal volume throughout the lung. *J. Appl. Physiol.* 19:683, 1964.
10. Johnson, R. L., Jr., Spicer, W. S., Bishop, J. M., and Forster, R. E. Pulmonary capillary blood volume, flow and diffusing capacity during exercise. *J. Appl. Physiol.* 15:893, 1960.
11. Johnson, R. L., Jr., Taylor, H. F., and Lawsen, H. N., Jr. Maximum diffusing capacity of the lung for carbon monoxide. *J. Clin. Invest.* 44:349, 1965.
12. Koler, J. J., Young, A. C., and Martin, C. J. Relative volume changes between lobes of the lung. *J. Appl. Physiol.* 14:347, 1959.
13. Lewis, B. M., Lin, T.-H., Noe, F. E., and Hayford-Welsing, E. J. The measurement of pulmonary diffusing capacity for carbon monoxide by a rebreathing method. *J. Clin. Invest.* 38:2073, 1959.
14. Millette, B., Robertson, P. L., Ross, W. R. D., and Anthon, H. R. Effect of expiratory flow rate on emptying of lung regions. *J. Appl. Physiol.* 27:587, 1969.
15. Ogilvie, C. M., Forster, R. E., Blakemore, W. S., and Morton, J. W. A standardized breath holding technique for the clinical measurement of the diffusing capacity of the lung for carbon monoxide. *J. Clin. Invest.* 36:1, 1957.
16. Otis, A. B., et al. Mechanical factors in distribution of pulmonary ventilation. *J. Appl. Physiol.* 8:427, 1956.
17. Roughton, F. J. W., and Forster, R. E. Relative importance of diffusion and chemical reaction in determining rate of exchange of gases in the human lung with special reference to the true diffusing capacity of the pulmonary membrane and volume of blood in the lung capillaries. *J. Appl. Physiol.* 11:290, 1957.
18. Sackner, M. A. Arterial blood gas analyses. *Med. Times* 95:79, 1967.
19. Sackner, M. A., et al. Distribution of ventilation during diaphragmatic breathing in obstructive lung disease. *Am. Rev. Respir. Dis.* 109:331, 1974.
20. Sackner, M. A., et al. Diffusing capacity, membrane diffusing capacity, capillary blood volume, pulmonary tissue volume and cardiac output measured by a rebreathing technique. *Am. Rev. Respir. Dis.* 111:157, 1975.
21. Staub, N. C. Alveolar-arterial oxygen tension gradient due to diffusion. *J. Appl. Physiol.* 18:673, 1963.
22. Van Slyke, D. D., and Plazin, J. *Micromanometric Analyses.* Baltimore: Williams & Wilkins, 1961. Pp. 46-55.
23. Wanner, A., et al. Relationship between frequency dependence of lung compliance and distribution of ventilation. *J. Clin. Invest.* 54:1200, 1974.
24. Woolmer, R. F. (Ed.). *A Symposium on pH and Blood Gas Measurement.* Boston: Little, Brown, 1959.

Pulmonary Circulation   18

Marvin A. Sackner

INTRODUCTION

The direct measurement of the hemodynamics of the pulmonary circulation in human subjects is generally not undertaken in a pulmonary laboratory; these investigations are usually performed in a cardiac catheterization laboratory. Nevertheless, a great deal of indirect knowledge about the pulmonary circulation is obtained in the pulmonary laboratory. For example, as discussed in Chapter 17, measurements of the diffusing capacity of the lung at different alveolar oxygen tensions are used to calculate the pulmonary capillary blood volume. Estimates of the overall distribution of pulmonary perfusion are obtained through the measurement of anatomic and physiologic dead-space volumes. Methods are available for measuring the pulmonary capillary blood flow from the uptake of soluble inert gases. Estimates of regional perfusion are obtained by scanning the lung for radioactivity after venous injection of such gases as ^{133}Xe. Catheters can be placed in lobes or bronchopulmonary segments to sample oxygen and carbon dioxide, thereby enabling the calculation of ventilation-perfusion ratios on an anatomic basis.

CARDIAC CATHETERIZATION

TECHNIQUE

With the subject under local anesthesia, a vein in the antecubital fossa or, less often, in the femoral region is selected for the introduction of a cardiac catheter. The catheter is introduced into the vein under direct vision or by a percutaneous approach. The catheter is flushed with a heparinized saline solution to prevent clotting; its size usually ranges from No. 6 to 8 (French). It is guided into the right atrium and visualized roentgenographically by a fluoroscope or image intensifier. The electrocardiogram is monitored concurrently to detect cardiac arrhythmias as the catheter is passed across the tricuspid valve, into the right ventricle, across the pulmonic valve, and into the pulmonary artery. The left atrial pressure may be measured either (1) by inserting a special catheter into the right saphenous vein, passing it up the inferior vena cava, and puncturing the septum of the right atrium to enter the left atrium or (2) by inserting a cardiac catheter into a peripheral artery, passing it retrogradely into the aorta, across the aortic valve, into the left ventricle, and across the mitral valve into the left atrium.

PRESSURE RECORDING

In most laboratories, pressures are recorded from the catheter by means of a fluid-filled tubing that is connected to a strain-gauge transducer. Although the transducers are capable of accurate recording up to frequencies greater than 100 Hz in an air-filled system, when filled with water they may resonate at frequencies as low as 7 Hz. The characteristics of the catheter and the tubing connections further limit the frequency response. In addition, these systems are frequently underdamped, which produces artifactual high-frequency oscillations on the pressure-pulse recording. Recently, catheters have become available in which the manometer is placed at the tip of the catheter. Since there is no fluid-filled system to limit the frequency response in such instruments, they are critically damped to frequencies of about 40 Hz and fall off slightly thereafter so that at 200 Hz, the amplitude is down about 20%.

Another problem in recording the pulmonary arterial pressure is that of relating the zero level to an external site in order to calibrate the pressure in absolute terms. Generally, the plane of the transducer is set in relationship to both the heart and a thoracic landmark. There is no standard reference level that is agreed upon by various cardiac catherization laboratories: the most popular reference levels are the upper portion of the sternum or 5 cm below it, the middle of the chest in the lateral projection, and 10 to 20 cm above the table top. Since the distribution of pulmonary arterial pressures is gravity-dependent, some workers set the reference level at the most dependent portion of the lung. The latter practice, however, has not been generally carried out in most cardiac catheterization laboratories.

PULMONARY CIRCULATORY PRESSURES

The pulmonary arterial pressure in normal supine subjects ranges from about 20 to 30 mm Hg systolic and from 5 to 12 mm diastolic (Fig. 18-1) [3]. The mean pulmonary pressure ranges from 10 to 15 mm Hg. It characteristically shows a rapid rise to a rounded peak during systole, and a brisk small incisura and a gradual decrease in pressure occur during diastole. The *pulmonary arterial wedge pressure* is recorded by advancing a cardiac catheter until its tip occludes a terminal branch of the pulmonary arterial flow and stops in the segment beyond the branch. The mean wedge pressure closely approximates the left mean atrial pressure of 5 to 10 mm Hg. The reserve of the pulmonary circulatory bed is so large that temporary occlusion of a main pulmonary arterial branch by a balloon catheter produces an increase in pressure of only a few mm Hg in the centrolateral lung despite a doubling of blood flow. In a diseased lung, however, such a procedure is associated with a significant rise in pulmonary arterial pressure and is used as a test for prognosis in pulmonary resections.

PULMONARY BLOOD FLOW

The two standard ways of measuring pulmonary blood flow in the cardiac catheterization laboratory are based upon Fick's principle and the dye-dilution method [18].

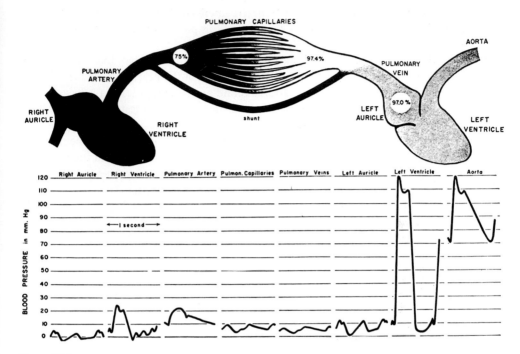

Figure 18-1
Pressures in the pulmonary and systemic circulations. Venous blood that flows past well-ventilated alveoli becomes arterialized. Some venous blood does not come in contact with ventilated alveoli, and some flows past poorly ventilated alveoli; this blood is not arterialized or is incompletely arterialized (physiologic shunt). The numbers 75%, 97.4%, and 97% refer to the percentage of saturation of hemoglobin. (Modified from J. H. Comroe, Jr., et al. *The Lung* (2nd ed.). Chicago: Year Book, 1962. P. 80.)

The former method is based upon the principle that the oxygen consumption divided by the arteriovenous oxygen content difference across the lungs equals the pulmonary blood flow:

$$\dot{Q}_p = \frac{\dot{V}_{O_2}}{ca_{O_2} - c\bar{v}_{O_2}} \tag{18-1}$$

where $\dot{Q}_p$ is the pulmonary blood flow in milliliters per minute, $\dot{V}_{O_2}$ is the oxygen consumption in deciliters per minute, ca_{O_2} is the systemic arterial oxygen content in volume-percent, and $c\bar{v}_{O_2}$ is the mean mixed venous oxygen content in volume-percent. Although the preferred mixed venous sampling site is the pulmonary artery, the right ventricle is almost as satisfactory. In this method, a steady state must be maintained by the patient for usually four minutes or longer in order to insure the validity of the measurement. From a technical standpoint, there are two major difficulties: (1) cardiac catheterization of the pulmonary artery must be performed in order to obtain a mixed venous oxygen sample and (2) measurements of the oxygen content of the expired air and blood are time-consuming, tedious procedures that limit the number of observations on any given subject.

In the indicator-dilution method, a dye or other indicator that does not diffuse out of the circulation is injected into the right side of the heart or a venous tributary, The indicator later appears in the arterial blood and its concentration is measured (Fig. 18-2). The time-course of changes in the arterial concentration of the indicator may be evaluated by the analysis of separate, timed samples or by means of continuous analysis employing optical densitometry or scintillation counting. The most commonly employed detection device is the optical densitometer, and the most commonly used indicator is indocyanine green dye, which is also known as Fox Green dye. The injected substance in normal subjects begins to appear after a delay of 6 to 15 seconds (the appearance time). Its concentration curve then builds up to a peak (the buildup time) and rounds off. The descending limb of the curve appears to be a

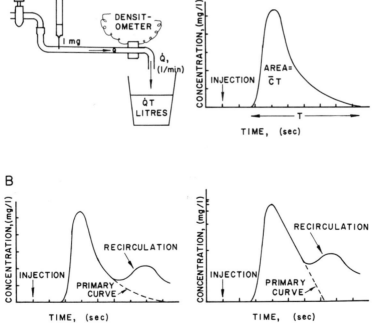

Figure 18-2
Dye dilution curves. *A.* Injection of dye into a flowing stream and the recording of concentration versus time by a densitometer. *B.* Dye dilution curve determined from arterial blood after the injection of dye into the pulmonary artery. In order to exclude the effects of recirculation, the curve is replotted on semilogarithmic graph paper and a linear extrapolation is made. (From P. Harris and D. Heath. *The Human Pulmonary Circulation.* Baltimore: Williams & Wilkins, 1962. P. 64. Reproduced with permission.)

single exponential function. After the exponential descent has progressed for a time, the concentration builds up again as a result of recirculation of the indicator that has returned to the heart. The pulmonary blood flow is calculated from the time-concentration curve according to Equation 18-2:

$$\dot{Q}_p = \frac{60 \, m_I}{\bar{c}_I t} \tag{18-2}$$

where $\dot{Q}_p$ is the pulmonary blood flow in liters per minute, m_I is the amount of indicator injected (mg), $\bar{c}_I$ is the average concentration of the indicator (mg per liter), and t is the time of passage in seconds of the indicator on its first circulation.

The problem of separating the concentration of indicator on the first circulation from the concentration due to recirculation is handled in the following way. When the indicator enters a ventricle, turbulent flow in the ventricle causes homogeneous mixing of the indicator. Indicator washout from this chamber occurs with each heartbeat at an exponential rate. Therefore, a simple graphic linear extrapolation of the descending indicator curve plotted on semilogarithmic graph paper allows the enclosure of an area that defines the time-concentration product of all the indicator on its first passage and none due to its second circulation. Advantages of the dye-dilution method over the method based on Fick's laws include ease of analysis and the ability to calculate the tracings by means of analog or digital computers. Since indocyanine green dye is inert with respect to the patient, there is virtually no limit on the number of tests that can be performed. Other indicators that have been used include saline, ^{131}I-tagged albumin, and [22]Na solutions.

The normal basal supine value for pulmonary blood flow ranges from about 3.20 to 3.45 liters per minute per square meter of body surface area. During maximal exercise, the pulmonary blood flow may increase by a factor of four to five times. In the resting, sitting position, the pulmonary blood flow is approximately 15% to 20% less than in the supine position. In diseases affecting the heart, such as arteriosclerotic and rheumatic heart disease, the pulmonary blood flow may fall to as low as 1.0 liter per minute per square meter of body surface area. In normal subjects, however, the acute response to passive tilting may also produce similar low values for pulmonary blood flow (Fig. 18-3) [16]. Under these circumstances, no ill feelings are usually experienced by the subjects. In obstructive and restrictive lung diseases, the value for pulmonary blood flow is quite variable.

PULMONARY VASCULAR RESISTANCE

Traditionally, the resistance of the pulmonary vascular bed has been thought of as analogous to that of a continuous flow through a rigid tube [18], as expressed by:

$$R = \frac{\overline{Pa}_p - \overline{P}_{la}}{\dot{Q}_p} \tag{18-3}$$

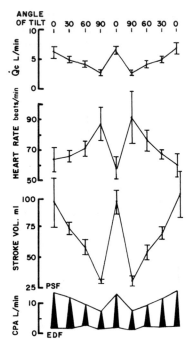

Figure 18-3
Effects of tilting from supine position (0 degrees) to the vertical position (90 degrees) on the pulmonary capillary blood flow ($\dot{Q}c$), heart rate, stroke volume, peak systolic flow (PSF), end-diastolic flow (EDF), and capillary pulse amplitude (CPA). Mean values were obtained in four normal subjects; standard deviations are shown by vertical bars. (From N. Segel et al. *J. Appl. Physiol.* 35:244, 1973. Reproduced with permission.)

where R is the resistance in arbitrary units, $\overline{Pa}_p$ is the mean pulmonary arterial pressure (mm Hg), $\overline{P}_{la}$ is the mean left atrial pressure (mm Hg), and $\dot{Q}_p$ is the pulmonary blood flow in liters per minute. Thus, under normal circumstances, a representative calculation of resistance in arbitrary units might be:

$$R = \frac{10 \text{ mm Hg} - 4 \text{ mm Hg}}{6 \text{ liters/min}} = 1.0$$

In diseases that significantly increase the resistance of the pulmonary vascular bed — such as mitral stenosis, pulmonary hypertension, and pulmonary fibrosis — the value for resistance might increase to 15 to 20 units.

The calculation of resistance according to Equation 18-3 must be considered to be overly simplified. First, the blood flow is not steady but pulsatile, and therefore the impedance, not ohmic resistance, should be measured. Second, the pressure drop across the lung from the pulmonary artery to left atrium (or pulmonary veins) is not constant for all regions of the lungs because of gravitational influences. Regional estimation of the distribution of perfusion by radioactive scanning and histologic evidence from rapidly frozen lung specimens indicate that there are at least three zones in the lung. In the most gravitationally independent region, *zone 1* (in the sitting position, at the apices), the alveolar pressure (PA) is greater than the pulmonary arterial pressure (Pa_p), which in turn is greater than the pulmonary venous pressure (Pv_p). At the next hydrostatic level, *zone 2*, $Pa_p > PA > Pv_p$. This zone has also been termed the "waterfall" or "sluice" zone because the flow rate is determined by the degree of collapse at the venous end of the capillaries from inward compression due to the surrounding alveolar pressure. Regardless of the level of pulmonary venous pressure, the flow rate is set by this restriction. The third zone, *zone 3,* comprises most of the lung volume. In this zone, $Pa_p > Pv_p > PA$. Zone 3 is the conventional model for resistance measurements [20]. In the supine, resting subject, probably greater than 70% of the lung region is zone 3 and there is minimal zone 1. Therefore, the conventional way of calculating resistance is not in great error from this standpoint, even when one considers that the entire concept of steady blood flow is an oversimplification.

PULMONARY VASCULAR IMPEDANCE

To date, measurements of pulmonary vascular impedance have not been possible in conscious human subjects. This is because it is necessary to record the pressure and flow from the pulmonary artery simultaneously. It is possible, however, to obtain these measurements in an experimental animal [1]. To calculate the input impedance to the pulmonary vascular bed it is necessary to develop a Fourier series for the flow and pressure waveforms:

$$\dot{Q}(t) = \bar{\dot{Q}} + \sum_{n=1}^{\infty} (A_n \cos n\omega t + B_n \sin n\omega t) \tag{18-4}$$

where $\bar{\dot{Q}}$ = mean flow, $\omega = 2\pi f$, t = any arbitrary point in time during the wave motion, and n = an integer corresponding to the number of the harmonic ($n = 1$ corresponds to the first harmonic, or fundamental frequency, and so on).

Each pair of terms in the summation represents a sinusoidal wave with frequency of $n\omega$ radian/sec, or $n\omega/2\pi$ cycles/sec, and can be expressed by:

$$A_n \cos n\omega t + B_n \sin n\omega t = Mod_n \sin (n\omega t + \phi n) \tag{18-5}$$

where

$$Mod_n^2 = A_n^2 + B_n^2$$

$$\tan \phi_n = \frac{A_n}{B_n}$$

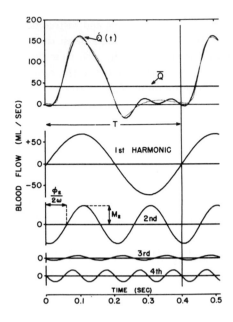

Figure 18-4
Experimentally determined pulmonary arterial blood flow as a function of time and its representation by a Fourier series. The measured blood flow (. . . . . .) and mean flow, $\overline{Q}$, appear in the top tracing, together with the sum, $Q(t)$ (———), of the first four harmonics of the Fourier series plus the mean. Discrepancies between $Q(t)$ and the observed flow diminish rapidly as additional harmonics are added, but, for the purpose of this illustration, only four harmonics are used. The frequency of the first harmonic is $\omega = 2\pi/T$ radians. In this example, T = one cardiac cycle (0.40 sec), so ω = 15.71 radians/sec or 2.5 cycles/sec, which is the heart rate. The frequency of the nth harmonic is $n\omega$. Each harmonic is represented by a sinusoidal wave of amplitude M_n. This, for example, is indicated on the tracing of the second harmonic, where M_2 = 44.2 ml/sec. The phase angle, ϕ_n, indicates the time relations between harmonics. The sine wave that represents the first harmonic, or fundamental harmonic, has arbitrarily been assigned a phase angle of zero at $t = 0$, and the second harmonic, with $\phi_2 = -1.80$ radians, lags behind the first by $\phi_2/2\omega$ = 0.057 sec. (From D. H. Bergel and W. R. Milnor. *Circ. Res.* 16:401, 1955. Reproduced with permission of the American Heart Association, Inc.)

The expression Mod_n is the modulus of the nth sinusoidal or harmonic in the series, and ϕ_n is its phase angle. Figure 18-4 illustrates pulmonary arterial blood flow and its Fourier series representation.

The modulus of impedance (Fig. 18-5) for each harmonic is expressed as:

$$|Z| = \frac{|P|}{|\dot{Q}|}$$

(18-6)

To measure the wave velocity, catheters for pressure recording are placed a known distance apart, Δx. The difference in phase (in radians) between these two points

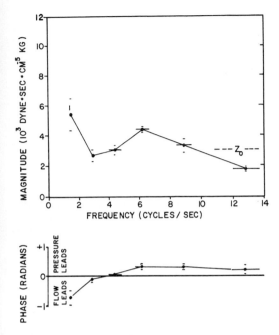

Figure 18-5
Average pulmonary vascular impedance in control experiments on 13 anesthetized, open-chest dogs (Z_0 = estimated characteristic impedance). The horizontal and vertical bars represent the standard errors of the means for frequency and amplitude or phase for each point. The input impedance at zero frequency (mean pulmonary arterial pressure divided by the mean flow rate) is 14.2×10^3 dyne·sec·cm^{-5} per kilogram. The impedance modulus falls to a minimum at 2.93 cycles/sec and rises to a peak at 6.22 cycles/sec. The impedance phase angle changes from negative (flow-leading) to positive (pressure-leading) at the same frequency as the first minimum. (From D. H. Bergel and W. R. Milnor. *Circ. Res.* 16:401, 1955. Reproduced with permission of the American Heart Association, Inc.)

represents the time for the wave to travel between the two points. The *apparent phase velocity*, v_ϕ, averaged over this interval is:

$$v_\phi = \frac{2\pi f \Delta x}{\phi} \tag{18-7}$$

REFLECTED WAVES

In a system of branching tubes like the pulmonary arterial system, the phenomenon of wave reflection must take place. In this situation, the input impedance will be determined not only by the caliber of vessels and their elastic properties, but also by the length of the system from the origin to the reflecting sites. This type of reflection is known as *closed-end reflection*; in such cases, the impedance increases across a junction (when the junction is completely closed, of course, the terminal impedance becomes infinite).

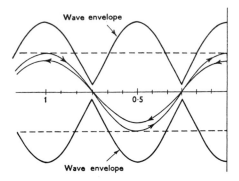

Figure 18-6
Interaction of a centrifugal and a reflected wave at a closed end. The individual components are
shown at one instant of time, with arrows to identify their direction of travel. The abscissa is
marked in fractions of a wavelength. (From D. A. McDonald. *Blood Flow in Arteries.* Baltimore:
Williams & Wilkins, 1960. P. 199. Reproduced with permission.)

The characteristics of pulmonary vascular input impedance may be interpreted in
terms of simple harmonic oscillations in a closed tube (Figs. 18-6 and 18-7). In simple
harmonic oscillation, there are two components, the incident and the reflected waves,
which travel in opposite directions. In the example of Figure 18-6, only 80% of the
wave is reflected (i.e., the reflection coefficient = 0.8). At the point of reflection and
at one-half the wavelength of the component waves, the waves are in phase, and the
oscillations of the two waves add together by superposition to form a *node.* At one-
quarter and three-quarters of the wavelength of the component waves, the waves cancel
each other and there will be no oscillation at all; this point is called an *antinode.* The
whole wave envelope that results from these superpositions is termed a *stationary* or
standing wave. When dealing with liquids, however, *damping* must be taken into
account. In such a system, true standing waves cannot occur, because there are no
points where no oscillation occurs and so there are no true antinodes. In the presence of damping, the reflected wave is smaller than the incident wave and therefore
cancellation cannot occur. However, points of maximal and minimal oscillation do
occur, which may be called *relative* antinodes and nodes and which are spaced at
approximately one-quarter wavelength intervals [1, 7].

Near to a "closed end," the incident and reflected pressure waves are nearly in
phase and add together. Here, the flow is negligible and the impedance is at a maximum. At a quarter wavelength distance, the incident and reflected waves are 180
degrees out of phase and tend to cancel each other out, so that the pressure is at a
minimum, the flow is maximal, and there results a minimal input impedance. In a
system of fixed anatomic length, the wavelength is dependent on the frequency of
oscillation, so that the first minimum to be found after performing Fourier analysis
and plotting impedance as a function of frequency can be used, if the wave velocity
is known, to define the distance at which the main reflections are occurring. In the
dog, the average impedance minimum occurs at 2.9 Hz (see Fig. 18-5), which implies

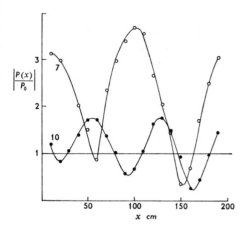

Figure 18-7
Amplitude of the pressure oscillation at various points along a tube closed at 200 cm from the
pump. The ordinate gives $P(x)/P_0$, the ratio of the pressure amplitude (modulus) to that in the
tube without reflections; the abscissa indicates the distance from the pump. The open circles
indicate the curve at a frequency of 7 Hz; the solid circles, 10 Hz. The wave velocity is approximately 1400 cm/sec. At 7 cm/sec, the length of this section of tube is equivalent to one wavelength and the tube is resonating. There are antinodes of pressure at both ends, with another at
the half wavelength position; all are of large amplitude. At 10 Hz, the tube length represents
1.43 wavelengths and there is no resonance. There are still antinodes and nodes at quarter wavelength intervals, but the variation in amplitude is much less. (From M. G. Taylor. *Phys. Med.
Biol.* 1:321, 1957. Reproduced with permission.)

that the distance from the measuring site to the reflecting site equals one-quarter of
the wavelength of the oscillations at this frequency. If the pulse wave velocity in the
dog pulmonary artery is 200 cm/sec, the major reflecting site is $(200/2.9) \times 0.25 =$
17 cm downstream from the measuring site. In the dog lung, this point corresponds
to that where the arteries are 0.5 to 1.0 mm in diameter [1].

REFLECTION COEFFICIENT

The ratio of reflected to incident pressure waves at the peripheral termination can be
calculated from the characteristic impedance (Z_0) and the terminal impedance (Z_{term}):

$$R_Z = \frac{(Z_{term} - Z_0)}{(Z_{term} + Z_0)} \tag{18-8}$$

The characteristic impedance is the impedance that would exist if no reflection had
occurred and is taken as the mean impedance of the higher harmonics (8 to 18 Hz).
The terminal impedance is the impedance of the bed beyond the reflecting site; the
best estimate of it is probably made by assuming that the impedance at very low frequencies approximates the terminal impedance. Calculations from data obtained in
dogs suggest that R_Z averages about 0.30 [1].

If a vasoconstrictive agent were infused into the pulmonary circulation, the first minimum of the impedance-frequency plot would shift to the right. In the dog, there is a shift of the impedance minimum from a control value of 3 Hz to a value between 5.5 and 7.5 Hz during hypoxia, serotonin infusion, or sympathetic nerve stimulation. This change in major wave reflection could be caused by alterations in the pulmonary vascular bed so that the main sites of wave reflection were closer to the point of measurement, or it could be due to an increase in the pulmonary arterial pulse wave velocity.

The pulmonary arterial pulse wave velocity can be calculated from the Bramwell-Hill equation [10]:

$$v_0 = \left(\frac{\Delta P \times Va_p \times g}{\Delta V \times \rho} \right)^{1/2} \tag{18-9}$$

where v_0 = pulmonary arterial pulse wave velocity (cm/sec)
ΔV = change in volume of the pulmonary arterial system (ml)
ΔP = corresponding change in distending pressure (cm H_2O)
Va_p = total pulmonary arterial volume (ml)
g = gravitational acceleration constant (cm/sec^2)
ρ = density of blood (gm/ml)

However,

$$D = \frac{\Delta V \times 100}{\Delta P \times Va_p}$$

where D = pulmonary arterial distensibility (% volume change/cm H_2O). Therefore, if the density of blood is ρ,

$$v_0 = \left(\frac{100 \times g}{D\rho} \right)^{1/2}$$

The distance from the site of measurement at the origin of the pulmonary artery to the site of wave reflection is obtained from Equation 18-10:

$$\lambda = \frac{v_0}{f} \tag{18-10}$$

where λ is the wavelength of the pulmonary arterial system (cm) at a frequency f (Hz) and v_0 is the pulmonary arterial pulse wave velocity (cm/sec). However, if the distance, d, from the site of measurement to the site of wave reflection is one-quarter wavelength at the frequency of the first impedance minimum, then:

$$d = \frac{\lambda'}{4} \tag{18-11}$$

and substitution yields:

$$d = \frac{v_0}{4f'}$$
(18-12)

where d = distance from the site of measurement to the reflection site (cm)
λ' = wavelength of the pulmonary arterial system (cm) at a frequency f' (Hz)
f' = frequency corresponding to the first impedance minimum (Hz)

During the control state, the pulmonary arterial pulse wave velocity in a group of dogs was calculated to be 189 cm/sec and the first impedance minimum was found to occur at a frequency of 3.5 Hz. Therefore, the distance from the origin of the main pulmonary artery to the reflecting site was $189/(4 \times 3.5) = 13.5$ cm. During hypoxic ventilation, the pulse wave velocity was found to be 228 cm/sec, the first impedance minimum occurred at a frequency of 5.6 Hz, and the distance from the origin of the pulmonary arteries to the site of reflection was 10.2 cm. In experiments with serotonin infusion and sympathetic nerve stimulation, the distance from the origin of the pulmonary artery to the site of reflection was determined to be 12.2 and 8.7 cm, respectively. The reflection coefficient during the control state was 0.31, during hypoxia 0.49, and during serotonin infusion 0.53. These results are consistent with the concept that the large pulmonary arterial vessels are capacle of vasoconstriction [10].

PULMONARY CAPILLARY BLOOD FLOW

The only practical, bloodless technique to measure the instantaneous pulmonary blood flow in human subjects is to measure their pulmonary capillary blood flow after inspiration of nitrous oxide (N_2O) [13]. The subject is placed within an airtight chamber (a body plethysmograph), and pressure changes within the chamber are recorded (Fig. 18-8A). The subject on signal inspires a mixture of 80% N_2O and 20% O_2 into the alveoli (Fig. 18-8B). Nitrous oxide dissolves in the pulmonary capillary blood, and its instantaneous rate of uptake is measured from the derivative of the pressure oscillations (calibrated as volume) in the plethysmograph (Fig. 18-8C). A control recording with the subject breathing air is needed to subtract the mechanical cardiogenic oscillations (see Fig. 18-9) from the pressure oscillations due to N_2O uptake. The pulmonary capillary blood flow can then be calculated using Equation 18-13:

$$\dot{Q}c = \frac{\dot{V}_{N_2O}}{F_{A_{N_2O}} \times \alpha_{N_2O}{}^{37°}}$$
(18-13)

where $\dot{Q}c$ is the instantaneous pulmonary capillary blood flow in liters per minute, $\dot{V}_{N_2O}$ is the rate of uptake of N_2O in liters per minute, $F_{A_{N_2O}}$ is the alveolar fraction of N_2O, and $\alpha_{N_2O}{}^{37°}$ is the solubility coefficient of N_2O in blood at $37°C$.

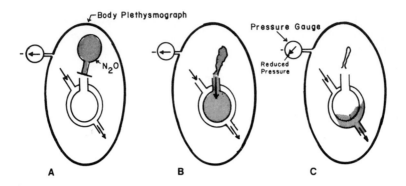

Figure 18-8
Measurement of the pulmonary capillary blood flow by the body plethysmographic technique.
The bag is filled with an 80% N_2O, 20% O_2 mixture.

Although the pulmonary capillary blood-flow method is an attractive one — it is
bloodless, is easily repeatable, and does not depend upon a steady state — technical
difficulties have prohibited its clinical use until recently. The subtraction of the air
records from the N_2O records is tedious and time-consuming, but this has been rem-
edied by a computer program. Figure 18-9 shows the display of the pulmonary
capillary blood-flow pulse output. Another difficulty with plethysmographic methods
has been that they cannot be employed with acutely ill patients. By using the chest
of the patient as a "plethysmograph," however, the rate of uptake of N_2O at the
mouth can be followed by a sensitive spirometer or pneumotachograph. This necessi-
tates that the subject hold his breath without chest movement and that the glottis be
opened. Unfortunately, such a maneuver cannot be carried out on patients except
in the operating room when the patient is anesthetized and paralyzed as part of a

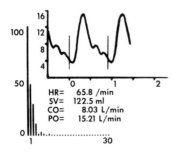

Figure 18-9
Computer display of the average pulmonary capillary blood-flow pulse from six supine normal
subjects, together with a Fourier analysis. The two vertical bars denote the R-R interval of the
electrocardiogram, the ordinate gives the blood flow in liters per minute, and the abscissa, the
time in seconds. (HR is heart rate, SV is stroke volume, CO is mean cardiac output, and PO is
peak output.) The bar graph depicts the Fourier analysis, in which the ordinate indicates the
percentage of the total harmonic content, and the abscissa, the harmonics from 1 to 30 Hz.
(From M. A. Sackner et al. Bull. Physiopathol. Respir. 9:1189, 1973. Reproduced with permission.)

surgical procedure or in the intensive care area when the patient is undergoing artificial ventilation with a respirator.

The pulmonary capillary blood flow is significantly pulsatile, and, in this respect, it differs somewhat from the relatively steady flow seen in the systemic capillaries. The pulsatility of this blood flow is diminished by an increase in heart rate and by an increase in the time-constant of the pulmonary arterial tree (the product of its compliance and resistance).

TOTAL PULMONARY BLOOD VOLUME

The total pulmonary blood volume represents the volume of blood between the pulmonic valve and the entrance of the pulmonary veins into the left atrium. Its measurement is relatively inexact in human subjects. The method is based on estimating the mean transit time (MTT) from dye-dilution curves recorded after injections of the dye into the pulmonary artery and left atrium and sampling from a systemic artery. The difference measured is the mean transit time through the lungs. The product of the MTT (Eq. 18-14) and the pulmonary blood flow is the total pulmonary blood volume [4].

$$MTT = \frac{\Sigma (ct)}{\Sigma c} \qquad (18-14)$$

where MTT is the mean transit time in seconds, c is the concentrations read at 1-second intervals from the indicator-dilution curve extrapolated through three logarithmic cycles, and t is the time at which each concentration occurred (in seconds) from the midpoint of the injection period.

The total pulmonary blood volume includes its subdivisions: the pulmonary arterial blood volume, the pulmonary capillary blood volume, and the pulmonary venous blood volume. Its value ranges from 300 to 500 ml in normal subjects. It may increase in congestive heart failure. Its clinical diagnostic significance, however, has been disappointing both because of its poor reproducibility and the apparent lack of major change in its value by physiologic, pharmacologic, and pathologic stimuli.

PULMONARY ARTERIAL BLOOD VOLUME

The pulmonary arterial blood volume has been measured in dogs and humans using a plethysmographic technique (Figs. 18-10 and 18-11) [6, 14]. If the chest is held in a fixed position or if the subject is expiring at a constant, slow flow rate, the rapid injection of a small volume of ether dissolved in alcohol just above the pulmonic valve will result in the evolution of gaseous ether into the alveoli as the ether injection reaches the capillaries. The time-course of ether evolution from the time of injection can be followed as an increase in pressure within the plethysmograph. By calculating the MTT from the tracing and multiplying by the pulmonary blood flow, an estimate of

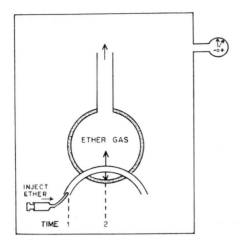

Figure 18-10
Ether plethysmographic method. Ether is injected as an alcoholic solution just above the pulmonic valve. When it reaches the alveolocapillary membrane, it diffuses into the alveoli and causes an increase in the plethysmographic pressure as it is transformed into the gaseous form.

the pulmonary arterial blood volume can be obtained. In the dog, this amounts to 25% to 40% of the total pulmonary blood volume. In humans, this volume ranges from about 175 to 225 ml.

PULMONARY ARTERIAL COMPLIANCE

Due to the presence of elastic tissue attachments within the lung, the pressure-volume characteristics of one portion of the lung are influenced by the state of inflation of adjacent portions. This phenomenon is called *interdependence* [19]. Thus, the change in the volume of the pulmonary artery (ΔVa_p) is dependent upon both the change in transmural or distending pressure of the pulmonary artery (ΔP_{tm}), which can be approximated by subtracting the pleural pressure from the intravascular pressure ($\Delta P_{tm} = Pa_p - P_{pl}$), and the change in transpulmonary pressure (ΔP_{tp}), or:

$$\Delta Va_p = C_1 \Delta P_{tm} + C_2 \Delta P_{tp} \tag{18-15}$$

If P_{tm} can be changed while keeping the lung volume constant, ΔP_{tp} becomes zero, and Equation 18-15 reduces to:

$$\Delta Va_p = C_1 \Delta P_{tm} \tag{18-16}$$

Rearranged,

$$C_1 = \frac{\Delta Va_p}{\Delta P_{tm}} \tag{18-17}$$

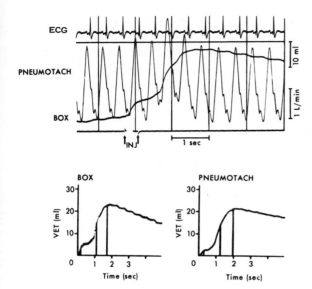

Figure 18-11
Comparison of ether plethysmographic and pneumotachographic techniques in the anesthetized, apneic dog. The upper tracing shows the raw data; the lower two, the final display on the cathode-ray oscilloscope after processing of the signal by a digital computer. The upper analog record shows the electrocardiogram (*ECG*), the pneumotachographic (*PNEUMOTACH*) tracing of the flow at the endotracheal tube along with the flow calibration, the plethysmographic (*BOX*) pressure (calibrated as a volume), and the time of injection (*INJ*) of saline to flush the ether solution into the pulmonary artery. The record shows an initial rise in the plethysmographic pressure due to the saline injection. There is then a second rise in both the volume within the plethysmograph and the flow at the airway due to the evolution of gaseous ether; this is followed by a gradual decay in both parameters as the gaseous ether redissolves into the pulmonary capillary blood. The computer displays show a record of the evolution of ether gas. The first vertical line at the left indicates the end of the injection period, the second vertical line, the midvolume point of gaseous ether evolution, and the final vertical line, the peak volume point. In this example, the following data were obtained for the plethysmographic record: median transit time, 0.96 sec; mean transit time, 1.00 sec; volume of ether, 19.3 ml; and lung density, 0.09. For the pneumotachographic record, the median transit time was 0.99 sec; mean transit time, 1.09 sec; volume of ether, 18.9 ml; and lung density, 0.10. (From M. A. Sackner et al. *Circ. Res.* 34:761, 1974. Reproduced with permission.)

which is the expression for *pulmonary arterial compliance*. Conversely, lung inflation without changing the transmural arterial pressure ($\Delta P_{tm} = 0$) allows the determination of an interdependence term between P_{tp} and Va_p:

$$C_2 = \frac{\Delta Va_p}{\Delta P_{tp}}$$
(18-18)

The magnitude of C_2 is therefore an index of the degree of interdependence between lung inflation and the expansion of the pulmonary arterial tree. It determines how much Va_p changes for a given change in lung volume.

Three different breathing maneuvers permit independent or concomitant changes of these pressure parameters. The Mueller maneuver (inspiration against a closed glottis) produces significant changes in P_{tm} without allowing P_{tp} to change. On the other hand, the quasi-Valsalva maneuver (relaxation against a closed glottis), which creates a mild positive intrathoracic pressure at higher lung volumes by passive compression of the relaxed chest wall against the closed airway, results in an essentially unchanged P_{tm}, although the P_{tp} increases with lung inflation. Finally, during held inspiration with the glottis open, both P_{tp} and P_{tm} change simultaneously.

Thus, the relationship between P_{tm} and Va_p on the one hand and P_{tp} and Va_p on the other permits the calculation of a static pulmonary arterial compliance term $(\Delta Va_p/\Delta P_{tm})$ and an interdependence term $(\Delta Va_p/\Delta P_{tp})$, from which changes in Va_p during open airway inflation can be predicted. The fact that the calculated changes in Va_p are fairly close to the actual, measured changes during open airway inflation supports the contention that the volume of the pulmonary artery is governed by both the elastic recoil of the lung itself and the elastic recoil of the pulmonary artery. Further, the relative contributions of P_{tm} and P_{tp} are considerably different. For half-inspiration, the mean value of C_1 in dogs is 3.9 ml/cm H_2O and the mean value of C_2 is 2.5 ml/cm H_2O, indicating that for a 1-cm H_2O increase in P_{tp} or P_{tm}, the Va_p should increase by 3.9 or 2.5 ml, respectively.

Another way to calculate the pulmonary arterial compliance is to record the pulmonary arterial pressure in conjunction with the instantaneous capillary blood-flow curve [5]. The latter represents the outflow of blood into the pulmonary capillaries from the pulmonary arterial tree. During diastole, the pulmonic valve is closed, and the fall in pressure in the pulmonary artery is due to elastic recoil of the vessel wall. Thus, blood passively discharges into the pulmonary capillaries from the pulmonary arterial tree. By taking two points on the pressure curve and dividing this into the integral of the capillary blood-flow pulse between these points, the compliance of the pulmonary arterial tree can be determined. In human subjects, the compliance determined by this method was found to be between 5 and 6 ml/mm Hg. It should be recognized that values calculated in such a way are lumped for the entire bed. It has been shown by radiographic methods that large pulmonary arteries are much more compliant than the smaller ones.

Knowledge of the compliance of the pulmonary arterial tree helps to localize sites of pulmonary vasomotor activity (Fig. 18-12). Hypoxia, for example, is known to produce an elevation in pulmonary arterial pressure. Since physiologic measurements indicate that most of the arterial blood volume is in the larger vessels [12] rather than in the small arteries and arterioles, changes in the arterial volume can be interpreted as changes in the caliber of the large vessels. If the pulmonary arterial pressure rises as a result of hypoxia, then if a small vessel is constricted, the pulmonary arterial volume would increase passively as a function of the compliance of the proximal bed. If the mean pulmonary arterial pressure rose 10 mm Hg above the control values and if the compliance was 2 ml/mm Hg, then the pulmonary arterial blood volume would increase by 20 ml. Thus, if the volume increase is less than 20 ml or even unchanged in response to hypoxia, one can deduce that active constriction of the large pulmonary vessels must have taken place.

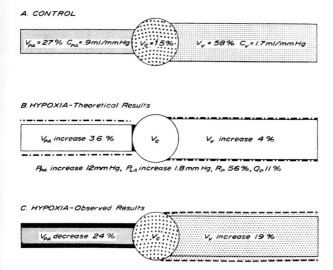

A. CONTROL

$V_{PA} = 27\%$ $C_{PA} = .9\,ml/mmHg$ $V_c = 1.5\%$ $V_v = 58\%$ $C_v = 1.7\,ml/mmHg$

B. HYPOXIA - Theoretical Results

V_{PA} increase 36 % V_c V_v increase 4 %

P_{PA} increase 12mm Hg, P_{LA} increase 1.8mm Hg, R_p 56%, Q_p 11 %

C. HYPOXIA - Observed Results

V_{PA} decrease 24 % V_c V_v increase 19 %

Figure 18-12
Distribution of the subdivisions of pulmonary blood volume in dogs during the control state and hypoxia. *A.* Normal partition of values for pulmonary arterial blood volume and capillary blood volume (calculated from the predicted formulas based on body weight and the venous blood volume, which in turn is calculated by subtraction of the pulmonary arterial and capillary blood volumes from the total pulmonary blood volume). *B.* Theoretical changes in the subdivisions of pulmonary blood volumes (assuming a compliance of 0.9 ml/mm Hg for the pulmonary arterial tree and 1.7 ml/mm Hg for the pulmonary venous tree) if the site of the vasoconstriction is located in the pulmonary arterioles. *C.* Observed changes during hypoxia assuming that the pulmonary capillary blood volume is unaltered. The significant decrease in the pulmonary arterial volume, even though a sizeable increase was predicted, indicated the presence of vaso-constriction along the major length of the pulmonary arterial tree. The greater volume on the venous side then could be accounted for by passive distension due to the 1.8 mm Hg increase in left atrial pressure, which suggests the occurrence of active venous vasodilation. (From M. A. Sackner and A. B. DuBois. *Med. Thorac.* 22:146, 1965. Reproduced with permission.)

PULMONARY TISSUE VOLUME

The volume of tissue in the lung may be correlated with the water content of the lungs. It can be estimated by the N_2O plethysmographic method [11], indicator-dilution techniques [17], the ether plethysmographic method [14], or the rate of uptake of acetylene by lung tissue by an extrapolation method [2, 15]. In the first method, the rate of uptake of N_2O after rapid inspiration of the gas is measured prior to a significant uptake of N_2O by the capillary blood flow (Fig. 18-13). In this instance, the N_2O equilibrates with the pulmonary tissue and capillary blood volumes according to Equation 18-19:

$$V(t + c) = \frac{V_{N_2O}}{FA_{N_2O} \times \alpha_{N_2O}^{37°}} \tag{18-19}$$

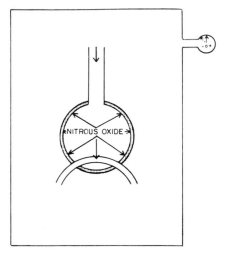

Figure 18-13
Principle used to measure pulmonary tissue volume. The inspiration of N_2O produces a fall in the plethysmographic pressure as a volume of N_2O dissolves almost instantaneously in the pulmonary tissue (stippled area) and the capillary blood. (From M. A. Sackner and A. B. DuBois. *Med. Thorac.* 22:147, 1965. Reproduced with permission.)

and

$$Vt = V(t + c) - Vc$$

where $V(t + c)$ is the combined pulmonary tissue and capillary blood volume (ml), V_{N_2O} is the volume of N_2O taken up by the lungs, FA_{N_2O} is the alveolar fraction of N_2O, and $\alpha_{N_2O}{}^{37°}$ is the solubility coefficient of N_2O in lung tissue. The mean value in normal young subjects for pulmonary tissue volume as determined by this method is 440 ml [11].

If one uses an indicator that stays in the circulation and another that diffuses into the pulmonary extravascular water space on a single passage, the difference between the quantities of these indicators as determined from indicator-dilution curves can be used to calculate the extravascular water space (Fig. 18-14). Fox Green or Evans Blue dye may be used as the intravascular indicator, and deuterated or tritiated water is the diffusible indicator. This method gives values that range from 150 to 250 ml in normal humans. This figure is much smaller than that obtained by the N_2O uptake method, perhaps because the N_2O equilibrates with a much larger water space and its uptake is not dependent upon the unevenness of vascular perfusion. In experimental animals, the tritiated-water method for the determination of extravascular water space reflects changes in tissue water, but it underestimates the gravimetrically determined water content by 40% to 60% [17].

Using the volume of gaseous ether evolved into the alveoli as detected by the body plethysmograph after injection of an ether solution into the pulmonary artery, it is

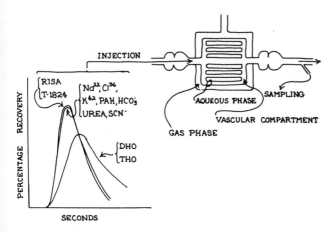

Figure 18-14
Method for the determination of the water space of the lung. Two indicator dilution curves are inscribed: the first from substances that remain in the intravascular space; the second from deuterium-labeled (*DHO*) or tritium-labeled (*THO*) water that equilibrates with the water space of the lung. The latter curve gives a larger volume (mean transit time times flow), and the difference between the volumes determined from the two curves represents the volume of the water space of the lung. (From F. P. Chinard et al. *Trans. Assoc. Am. Physicians* 72:253, 1962. Reproduced with permission.)

possible to estimate the pulmonary extravascular water space (tissue volume). Liquid ether, when it becomes gaseous, expands according to Equation 18-20:

$$\frac{V_{ether\,gas}}{V_{liquid\,ether}} = \frac{\rho_{liquid\,ether}}{MW_{ether}} \times \frac{273 + T}{273} \times \frac{760}{713} \times 22,400 \text{ ml/mole} \tag{18-20}$$

Assuming a body temperature (T) of 37°C, a barometric pressure of 760 mm Hg, a density of ether (ρ) of 0.71 gm/ml at 20°C, and a molecular weight (MW) of ether of 74 gm/mole, the volume of gas evolved per volume of liquid ether injected would be 260 ml BTPS per milliliter. However, only a fraction of the ether that is injected forms gas, and the rest remains in solution in the capillary blood and the tissues of the lungs. The volume of gas evolved equals the volume of liquid ether injected times 260 times the alveolar volume divided by the partition coefficient for ether (12) times the sum of the capillary and tissue volume plus the alveolar volume. Thus,

$$\frac{V_A}{V(c+t)} = \frac{12 \times V_{gas}}{(260 \times V_{liquid}) - V_{gas}} \tag{18-21}$$

where V_A is the alveolar volume, $V(c+t)$ is the capillary plus tissue volume, V_{gas} is the volume of gas evolved, and V_{liquid} is the volume of liquid ether injected.

If alveolar volume is measured by the plethysmographic method and capillary blood volume is measured by diffusing capacity measurements at different alveolar oxygen tensions, the tissue volume may be calculated from Equation 18-21. Recent studies in humans indicate that the tissue volume as determined by this method ranges from 250 to 400 ml. In a patient with mild pulmonary edema, the tissue volume was 700 ml [14].

The most promising practical way to determine pulmonary tissue volume in man is based on the rebreathing method described previously. The subject rebreathes into a bag containing a test mixture of CO, He, C_2H_2, O_2, and N_2 while continuous sampling of these gases is carried out by a mass spectrometer. Mixing takes place between the bag and the lung after a few breaths, and it is then assumed that rebreathing takes place at an infinite rate although the subject rebreathes at 24 to 40 breaths per minute. There is an exponential decline in the alveolar concentrations (FA) of CO and C_2H_2 with time. The initial alveolar concentrations, $(FA)_i$, of CO and C_2H_2 prior to absorption by lung tissues or blood are estimated by helium dilution in the alveolar sample. The decline in $FA/(FA)_i$ with time for these gases is exponential. Acetylene is absorbed in two phases. Initially, there is rapid uptake of this gas by the lung tissue and pulmonary capillary blood; this is followed by a slower, exponential descent in concentration due to its dissolution in the blood that perfuses the alveolar capillaries. The initial phase is represented by a depressed zero intercept of the extrapolated relationship between $\ln [FA_{C_2H_2}/(FA_{C_2H_2})_i]$ and the time of the rebreathing period. The value of this intercept is related to the partition of C_2H_2 between the alveolar air and the combined lung tissue and capillary blood as follows:

$$\frac{(FA_{C_2H_2})_{intercept}}{(FA_{C_2H_2})_i} = \frac{VA}{VA + \frac{(PA - P_{H_2O})}{760} \alpha_{t+c} V(t+c)} \tag{18-22}$$

where α_{t+c} is the Bunsen solubility coefficient (liters of gas reduced to STP dissolving in one liter of solvent at 1 atmosphere) for C_2H_2 (0.927), $V(t+c)$ is the combined pulmonary tissue and capillary blood volume, VA is the alveolar volume, and PA and P_{H_2O} are the alveolar pressure and water vapor pressure [2].

In normal young subjects, after subtraction of the pulmonary capillary blood volume, the pulmonary tissue volume (Vt) averages 311 ml with a standard deviation of 73 at rest, and it rises to 352 ml with a standard deviation of 61 at a moderate exercise load of 75 watts. Values in anesthetized dogs are quite similar to those estimated by the ether plethysmograph method (15).

VENTILATION-PERFUSION RATIO ($\dot{V}A/\dot{Q}$)

In order for gas exchange to be efficient, alveolar gas and pulmonary blood should be delivered to the pulmonary parenchyma in approximately equal proportions. This ventilation-to-perfusion proportion is known as the *ventilation-perfusion ratio, $\dot{V}A/\dot{Q}$*. In order to calculate $\dot{V}A/\dot{Q}$, two assumptions are made: (1) the inspired N_2 volume is equal to the expired N_2 volume and (2) no gas exchange occurs in the dead space. Equation 18-23 expresses the first assumption,

$$V_{N_2} = V_I F_{I_{N_2}} - V_E F_{E_{N_2}} = 0 \qquad (18\text{-}23)$$

Equation 18-24 expresses the second assumption,

$$\dot{V}_{AI} F_{I_{N_2}} - \dot{V}_A F_{A_{N_2}} = 0 \qquad (18\text{-}24)$$

where $\dot{V}_{AI}$ is the inspired alveolar ventilation, $F_{I_{N_2}}$ is the fractional concentration of inspired nitrogen, $\dot{V}_A$ is the expired alveolar ventilation, and $F_{A_{N_2}}$ is the alveolar fraction of nitrogen. Rearranging Equation 18-24 yields:

$$\dot{V}_{AI} = \dot{V}_A \frac{F_{A_{N_2}}}{F_{I_{N_2}}} \qquad (18\text{-}25)$$

Carbon dioxide ventilation may be described as:

$$\dot{V}_{CO_2} = \dot{V}_E F_{E_{CO_2}} - \dot{V}_I F_{I_{CO_2}} = \dot{V}_A F_{A_{CO_2}} - \dot{V}_{AI} F_{I_{CO_2}} \qquad (18\text{-}26)$$

Substituting from Equation 18-25 gives the CO_2 output:

$$\dot{V}_{CO_2} = \dot{V}_A \left(F_{A_{CO_2}} - F_{I_{CO_2}} \frac{F_{A_{N_2}}}{F_{I_{N_2}}} \right) \qquad (18\text{-}27)$$

The oxygen uptake may be derived similarly:

$$\dot{V}_{O_2} = \dot{V}_A \left(F_{I_{O_2}} \frac{F_{A_{N_2}}}{F_{I_{N_2}}} - F_{A_{O_2}} \right) \qquad (18\text{-}28)$$

The *respiratory exchange ratio* is defined as:

$$R = \frac{\dot{V}_{CO_2}}{\dot{V}_{O_2}} \qquad (18\text{-}29)$$

Substitution from Equations 18-27 and 18-28 into Equation 18-29 yields:

$$R = \frac{F_{A_{CO_2}} - F_{I_{CO_2}} \dfrac{F_{A_{N_2}}}{F_{I_{N_2}}}}{F_{I_{O_2}} \dfrac{F_{A_{N_2}}}{F_{I_{N_2}}} - F_{A_{O_2}}} \qquad (18\text{-}30)$$

The fractional nitrogen concentrations, $F_{A_{N_2}}$ and $F_{I_{N_2}}$, can be eliminated by substituting $F_{A_{N_2}} = 1 - F_{A_{O_2}} - F_{A_{CO_2}}$ and $F_{I_{N_2}} = 1 - F_{I_{O_2}}$ when $F_{I_{CO_2}} = 0$. Since the inspired CO_2 may be considered negligible for present purposes, the last condition is justified.

The fractional concentration of any gas, F_x, can be changed to its partial pressure, P_x, since $F_x = (P_x/P_B) - 47$. With these substitutions, one may now solve for $P_{A_{CO_2}}$, $P_{A_{O_2}}$, or R.

$$R = \frac{P_{A_{CO_2}} (1 - F_{I_{O_2}})}{P_{I_{O_2}} - P_{A_{O_2}} - (P_{A_{CO_2}} F_{I_{O_2}})} \qquad (18\text{-}31)$$

Since CO_2 is assumed to be absent in the inspired gas, Equation 18-27 yields:

$$\dot{V}_{CO_2} = \dot{V}_A F_{A_{CO_2}} \qquad (18\text{-}32)$$

or

$$\dot{V}_A = \frac{\dot{V}_{CO_2}}{F_{A_{CO_2}}} \qquad (18\text{-}33)$$

Converting Equation 18-33 to appropriate units,

$$\dot{V}_A \text{ (ml/min BTPS)} = \frac{\dot{V}_{CO_2} \text{ (ml/min STPD)} \times 0.863}{P_{A_{CO_2}}} \qquad (18\text{-}34)$$

Employing Equation 18-29 gives $\dot{V}_A$ in terms of $\dot{V}_{O_2}$:

$$\dot{V}_A = \frac{\dot{V}_{O_2} \times R \times 0.863}{P_{A_{CO_2}}} \qquad (18\text{-}35)$$

By combining Equation 18-1 for blood flow with the alveolar ventilation Equation (18-35), one may derive an expression for the ratio of the alveolar ventilation to blood flow. Rearranging Equation 18-1 yields:

$$\dot{V}_{O_2} = \dot{Q} (ca_{O_2} - c\bar{v}_{O_2}) \qquad (18\text{-}36)$$

where $\dot{V}_{O_2}$ is the oxygen consumption, $\dot{Q}$ the blood flow, and $(ca_{O_2} - c\bar{v}_{O_2})$ the arteriovenous concentration difference across the lungs for oxygen. Equation 18-35 can be rewritten as:

$$\dot{V}_{O_2} = \frac{\dot{V}_A \times P_{A_{CO_2}}}{0.863R} \qquad (18\text{-}37)$$

Equating the right-hand sides of Equations 18-36 and 18-37 and rearranging yields:

$$\frac{\dot{V}_A}{\dot{Q}} = \frac{0.863R \, (c^a_{O_2} - c^{\bar{v}}_{O_2})}{P_{A_{CO_2}}} \qquad (18\text{-}38)$$

A resting normal subject may show an alveolar ventilation of 4.5 liters per minute, giving an overall $\dot{V}_A/\dot{Q}$ of 0.9. However, there may be a wide range of $\dot{V}_A/\dot{Q}$ ratios in the lung, which are associated with a number of functional compartments. The *anatomic dead space* is ventilated but not perfused. Therefore, its $\dot{V}_A/\dot{Q}$ ratio is *infinite. Overventilated alveoli*, i.e., those with a $\dot{V}_A/\dot{Q}$ ratio of greater than 5, would constitute only a small fraction of the normal lung, the *parallel dead space*. The parallel dead space may be increased in pulmonary embolism or in grossly uneven distribution of ventilation, as in pulmonary emphysema. *Underventilated alveoli* have a $\dot{V}_A/\dot{Q}$ ratio of less than 0.4 Of course, the complete obstruction of an alveolus would result in a $\dot{V}_A/\dot{Q}$ ratio of zero. Normally, less than 3% of the cardiac output flows through such a region. This constitutes *venous admixture* or *shunt* and is the cause of systemic hypoxemia, since mixed venous blood enters the systemic circulation without becoming oxygenated. Intrapulmonary shunting is increased in conditions in which either the airway resistance is greatly increased or the lung compliance is greatly reduced. Figure 18-15 illustrates different types of abnormalities in ventilation-perfusion ratios.

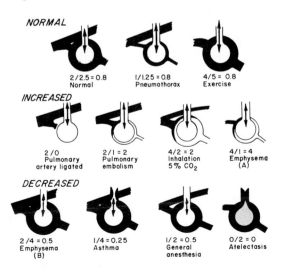

Figure 18-15
Clinical abnormalities in ventilation-perfusion ratios. Each circle represents one lung with its pulmonary capillary blood flow.

When a subject breathes room air, the inspired PO_2 is 150 mm Hg and the PCO_2 is effectively zero. The alveolar PO_2 is 100 mm Hg and the PCO_2 is 0 mm Hg when the $\dot{V}A/\dot{Q}$ ratio is normal. If the $\dot{V}A/\dot{Q}$ is zero, e.g., as in complete airway obstruction, the alveolar gas initially equilibrates with mixed venous blood so that the PO_2 is 40 mm Hg and the PCO_2 is 45 mm Hg. If the $\dot{V}A/\dot{Q}$ is infinite, the alveolar gas has the same composition as the inspired air. In the normal erect subject, there is a wide distribution of $\dot{V}A/\dot{Q}$ ratios as a result of gravitational effects. Despite these differences, the overall O_2 uptake and CO_2 output are reduced by only 2% to 3% by such ventilation-perfusion inequalities, causing alveolar-arterial differences of 4, 1, and 3 mm Hg for O_2, CO_2, and N_2, respectively.

Extensive nomograms for the solution of the $\dot{V}A/\dot{Q}$ relationship under a variety of circumstances have been published [8, 9].

REFERENCES

1. Bergel, D. H., and Milnor, W. R. Pulmonary vascular impedance in the dog. *Circ. Res.* 16:401, 1955.
2. Cander, L., and Forster, R. E. Determination of pulmonary parenchymal tissue volume and pulmonary capillary blood flow in man. *J. Appl. Physiol.* 14:541, 1959.
3. Comroe, J. H., Jr., et al. *The Lung: Clinical Physiology and Pulmonary Function Tests.* Chicago: Year Book, 1962. P. 77.
4. Dock, D. S., et al. The pulmonary blood volume in man. *J. Clin. Invest.* 40:317, 1961.
5. Engelberg, J., and DuBois, A. B. Mechanics of pulmonary circulation in isolated rabbit lungs. *Am. J. Physiol.* 196:401, 1959.
6. Feisal, K. A., Seni, J., and DuBois, A. B. Pulmonary arterial circulation time, pulmonary arterial blood volume and the ratio of gas to tissue volume in the lungs of dogs. *J. Clin. Invest.* 41:390, 1962.
7. McDonald, D. A. *Blood Flow in Arteries.* Baltimore: Williams & Wilkins, 1960.
8. Olszowka, A. J., Rahn, H., and Farki, L. E. *Blood Gases: Hemoglobin, Excess and Maldistribution.* Philadelphia: Lea & Febiger, 1973.
9. Rahn, H., and Fenn, W. O. *A Graphical Analysis of the Respiratory Gas Exchange; The O_2-CO_2 Diagram.* Washington, D.C.: The American Physiological Society, 1955.
10. Reuben, S. R., Swadling, J. P., Gersh, B. J., and G. de J. Le. Impedance and transmission properties of the pulmonary arterial system. *Cardiovasc. Res.* 5:1, 1971.
11. Sackner, M. A., Feisal, K. A., and DuBois, A. B. Determination of tissue volume and carbon dioxide dissociation slope of the lungs in man. *J. Appl. Physiol.* 19:374, 1964.
12. Sackner, M. A., Feisal, K. A., and Karsch, D. N. Size of gas exchange vessels in the lung. *J. Clin. Invest.* 43:1847, 1964.
13. Sackner, M. A., et al. Techniques of pulmonary capillary blood flow determination. *Bull. Physiopathol. Respir.* 9:1189, 1973.
14. Sackner, M. A., et al. Pulmonary arterial blood volume and tissue volume in man. *Circ. Res.* 34:761, 1974.
15. Sackner, M. A., et al. Diffusing capacity, membrane diffusing capacity, capillary blood volume, pulmonary tissue volume and cardiac output measured by a rebreathing technique. *Am. Rev. Respir. Dis.* 111:157, 1975.

16. Segel, N., Dougherty, A., and Sackner, M. A. Effects of tilting on pulmonary capillary blood flow in normal man. *J. Appl. Physiol.* 35:244, 1973.
17. Turino, G. M., et al. The volume of extravascular water of the lung in normal man and disease. *Bull. Physiopathol. Respir.* 7:1161, 1971.
18. Wade, O. L., and Bishop, J. M. *Cardiac Output and Regional Blood Flow.* Philadelphia: Davis, 1962.
19. Wanner, A., Zarzecki, S., and Sackner, M. A. Effects of lung inflation and transmural pulmonary arterial pressure on pulmonary arterial blood volume in intact dogs. *J. Appl. Physiol.* 8:675, 1975.
20. West, J. B. Regional Differences in Blood Flow and Ventilation in the Lung. In C. G. Caro (Ed.), *Advances in Respiration Physiology.* Baltimore: Williams & Wilkins, 1966. P. 198.

Mechanics of Breathing 19

Marvin A. Sackner

INTRODUCTION

The breathing mechanism is analogous to a reciprocating bellows pump. The walls
of the pump have two concentric parts: the lungs and chest wall. The "bellows
chamber" is made up mostly of the gas-exchanging units of the lungs. The remainder
of the gas in the chamber is located in a system of tubes, the *tracheobronchial tree*.
The tracheobronchial tree has an intrathoracic portion that extends well inside the
bellows (an unusual feature for a pump, which has important implications for flow
resistance) as well as an extrathoracic portion that consists of the upper trachea,
larynx, glottis, nasopharynx, mouth, and nose. Most pumps have three parts:
(1) an energy source, (2) passive elements that couple the energy source to the fluid
to be pumped, and (3) the fluid itself, along with the channels through which it
moves. The physical distinctions in the respiratory system among these elements are
vague; the chest wall, for example, intermingles motor and coupling functions, whereas
the lungs combine coupling functions and fluid conduction.

Newton's third law of motion states that a force applied to a body is met by an
equal opposing force developed by the body. Opposing forces depend upon three
aspects of motion in the lungs: instantaneous position, instantaneous velocity, and
instantaneous acceleration. The opposing forces attributable to each depend upon
the mechanical properties of elasticity, frictional resistance, and inertia, respectively.
In the case of a mass resting on a horizontal plane and attached to a fixed point by a
spring (Fig. 19-1), Newton's third law can be written as:

$$F_{appl} = F_{elas} + F_{fric} + F_{inert} \tag{19-1A}$$

or

$$F_{appl} = k_1 x + k_2 \dot{x} + m\ddot{x} \tag{19-1B}$$

where F_{appl} is the forced applied, F_{elas} is the elastic force, F_{fric} is the frictional resistance, and F_{inert} is the inertial force; k_1 is a proportionality constant, x is the displacement of the spring, k_2 is another constant of proportionality, $\dot{x} = dx/dt$ = the velocity
of the mass, m is the mass, and $\ddot{x} = d^2x/dt^2$ = the linear acceleration of the mass.

The preceding example of the mass and spring represents a single-branched rectilinear system. The respiratory system, of course, is multibranched, i.e., it constitutes
a distributed system. However, as long as the motions of the individual parts remain
identical, one can lump the distributed parameters. Although the motion of the parts
may be the same, motions within the parts may not. A portion of a spring nearest its

315

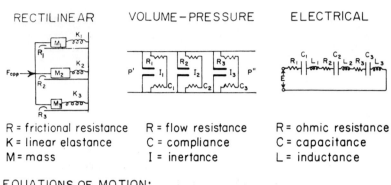

RECTILINEAR   VOLUME-PRESSURE   ELECTRICAL

R = frictional resistance   R = flow resistance   R = ohmic resistance
K = linear elastance   C = compliance   C = capacitance
M = mass   I = inertance   L = inductance

EQUATIONS OF MOTION:

$$F_{opp} = K\ell + R\dot\ell + M\ddot\ell \qquad P_{opp} = P' - P'' = \frac{1}{C}V + R\dot V + I\ddot V \qquad E = \frac{1}{C}q + R\dot q + L\ddot q$$

$$K = K_1 + K_2 + K_3 \qquad \frac{1}{C} = \frac{1}{C_1} + \frac{1}{C_2} + \frac{1}{C_3} \qquad \frac{1}{C} = \frac{1}{C_1} + \frac{1}{C_2} + \frac{1}{C_3}$$
$$R = R_1 + R_2 + R_3 \qquad R = R_1 + R_2 + R_3 \qquad R = R_1 + R_2 + R_3$$
$$M = M_1 + M_2 + M_3 \qquad I = I_1 + I_2 + I_3 \qquad L = L_1 + L_2 + L_3$$

Figure 19-1
Newton's law of motion: rectilinear, volume-pressure, and electrical analogies. (From J. Mead
and J. Milic-Emili. In W. O. Fenn and H. Rahn [Eds.], *Handbook of Physiology, Section 3,
Respiration,* Vol. 1. Washington, D.C.: The American Physiological Society, 1964. P. 366.
Reproduced with permission.)

fixed attachment, for example, clearly moves at smaller velocities and accelerations
than do portions farther away. In such a system, the relative motions of various
points within the parts, however, are related to the motions of all other points in all
the parts, as illustrated by the mechanical example in Figure 19-1. These machines
have only one way in which to move, i.e., they have only a *single degree of freedom.*
The respiratory system is a three-dimensional system with one degree of freedom.
Here, volume is analogous to motion (displacement) and pressure to force. The
pressure applied to the system is the difference in pressures at the two boundaries of
the system:

$$P_{appl} = P_2 - P_1 = \frac{1}{C}V + R\dot V + L\ddot V \tag{19-2}$$

where C is compliance, R is resistance, and L is inertance.

   Although it can be easily shown that the chest wall has a variety of ways to move,
it is a useful simplification to treat the respiratory system as one with a single degree
of freedom. Furthermore, it should be appreciated that the proportionality constants
are nonlinear, and often graphic means offer the best way to describe them.

   In volume-pressure systems like the respiratory system, the static component of
the opposing pressure, P_{stat}, is the summation of all the forces that depend upon
position: (1) elasticity, (2) gas compressibility, (3) weight, and (4) surface tension.
The dynamic components of the opposing pressure, P_{dyn}, relate to the frictional

resistance (mainly of flowing air but also, in the case of the lungs, perhaps the lung tissue itself) and to the inertia of all masses being accelerated [20, 21].

STATIC MECHANICS OF BREATHING

PRESSURE-VOLUME RELATIONSHIPS OF THE TOTAL RESPIRATORY SYSTEM

When the volume of gas in the lungs and the airway is held constant so that the flow and volume acceleration are zero, all the pressures related to flow resistance and inertance are also zero. Under these conditions, the pressures remaining are those developed by the respiratory muscles (or by some external device) and those operating due to elastic, surface-tension, and gravitational forces existing in the lungs and chest wall. This relationship is expressed by:

$$P_{ext} + P_{musc} = P_{lung} + P_{wall} \qquad (19\text{-}3)$$

where P_{ext} is the externally applied pressure, P_{musc} is the pressure applied by the respiratory muscles, P_{lung} is the pressure opposing the elastic, surface-tension, and gravitational forces in the lung, and P_{wall} is the pressure opposing similar forces in the chest wall (this includes the thorax and diaphragm).

If the muscles are completely relaxed so that $P_{musc} = 0$, then $P_{ext} = P_{lung} + P_{wall} = P_{rs}$, where P_{ext} is the externally applied pressure expressed as the difference between the pressure at the airway opening and the pressure at the body surface and $P_{rs} = f(V)$ for the total respiratory system. The slope of the volume-pressure plot, $\Delta V / \Delta P_{rs}$, is termed the *total respiratory compliance* and is usually linear up to about 10 cm H_2O applied pressure. Since compliances in the chest-lung system are analogous to capacitors in series, they are related as their reciprocals:

$$\frac{1}{C_{rs}} = \frac{1}{C_{lung}} + \frac{1}{C_{wall}} \qquad (19\text{-}4)$$

In normal subjects, C_{rs} equals about 0.1 liter per cm H_2O, C_{lung} equals 0.2 liter per cm H_2O, and C_{wall} is 0.2 liter per cm H_2O [15].

PRESSURE-VOLUME RELATIONSHIPS OF THE LUNGS AND CHEST WALL

The externally applied pressure across the lungs under static conditions is the difference between the intrapleural pressure that surrounds the lungs and the pressure at the mouth. This is called the *transpulmonary pressure*. Except in unusual circumstances, it is not technically possible to measure the intrapleural pressure directly in human subjects. It has been found, however, that the pressure in the lower third of the esophagus as recorded by an air-filled balloon system adequately reflects intrapleural pressure changes. The difference between the transpulmonary pressure and the transthoracic pressure is the pressure of the chest wall component. Figure 19-2 illustrates the pressure-volume curves for the lung and chest wall.

For the total respiratory system at the functional residual capacity (FRC) point, $P_{rs} = 0$ and the static pressures developed by the lungs and chest wall are equal in magnitude (about 5 cm H_2O) and opposite in sign. The resting volume of the lung is at less than zero percent vital capacity (VC); the resting volume of the chest wall is about 55% VC. Above this volume, both the chest and lungs recoil inward, and below this volume, the chest recoils outward. As the extremes of lung volume are approached static pressures increase rapidly with further volume change. At high lung volumes, this is due to the action of the lung, and at low volumes, to action of the chest wall.

Relaxation Pressures

This method of measuring lung and chest wall pressures has been employed in awake or anesthetized subjects. The conscious subject uses his own respiratory muscles to establish a given volume level by taking air from a spirometer and then relaxing against a closed airway. The pressure difference across the obstructed airway is a measure of the combined static force developed by the lungs and the chest wall. The pressure shows a slow small decrease that takes place over several seconds and may last up to 1 minute at the extremes of inspiration and expiration (see Fig. 19-2). This respiratory maneuver is repeated at several different lung volumes in order to construct a static volume-pressure curve. Such a procedure is quite difficult for even a trained subject to perform because of the inability to relax the respiratory muscles completely. Very few patients can be adequately studied by this method without their being extensively trained [22]. Because of these reasons, few laboratories even attempt to obtain relaxation pressures in patients.

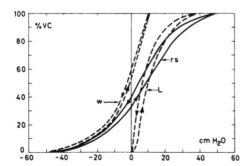

Figure 19-2
Pressure-volume curves of the respiratory system (*W* indicates the pressure-volume curve of the chest wall; *L*, that of the lung; and *rs*, that of the total respiratory system). Inflation and deflation curves do not follow identical paths due to hysteresis effects. (From E. Agostoni and J. Mead. In W. O. Fenn and H. Rahn [Eds.], *Handbook of Physiology, Section 3, Respiration*, Vol. 1. Washington, D.C.: The American Physiological Society, 1964. P. 398. Reproduced with permission.)

On the other hand, relaxation pressures are easily obtained in the operating room in anesthetized, paralyzed subjects. The volumes are easily changed by means of a calibrated giant syringe. Measurements on such subjects have shown a slightly lower value for C_{rs} than values obtained from conscious subjects. Some investigators have suggested that this implies that conscious subjects are incapable of completely relaxing their respiratory muscles; however, changes secondary to anesthesia and paralysis have not been excluded as possible causes of this phenomenon [15].

Externally Applied Pressures

Methods other than the relaxation pressure method for ascertaining the behavior of the respiratory system include (1) the application of steady pressures either at the mouth or around the body and (2) the slow cycling of pressure in a tank respirator while the subject attempts to relax and let the respirator breathe for him. The latter method is as difficult to perform as the relaxation pressure technique and therefore is rarely used. On the other hand, the steady applied-pressure methods can be used in patients for the determination of C_{rs} with some degree of reproducibility. It is assumed that the respiratory muscles are fully relaxed at end-expiration during the application of the pressure. In one method of applying the pressure, the subject rebreathes into a waterless spirometer enclosed within a steel drum that can be pressurized or evacuated of air [5]. The pressure at the mouth then constitutes the externally applied pressure (transthoracic pressure). In another method, the subject is enclosed within a chamber and breathes to the outside atmosphere. Steady pressures can be applied to the chamber to alter the transthoracic pressure [23].

In these methods, the changes in end-expiratory volume are plotted against the corresponding transthoracic pressures to obtain the pressure-volume curve for the total respiratory system. However, if simultaneous recordings are made of the transpulmonary pressure, there may not be agreement with the statically or dynamically determined lung compliance when the lung compliance is calculated from the changes obtained by applied pressures. This suggests that the respiratory muscles cannot be consciously fully relaxed even in well-trained subjects, and therefore it is again dubious if meaningful information about chest wall compliance can be obtained except in anesthetized subjects [15].

Factors Affecting Chest Wall Compliance

The average compliance of the chest wall in the adult subject is 0.2 liter per cm H_2O in the midvolume range. When it is expressed as a function of lung volume or weight, the chest wall compliance is highest in the newborn subject and shows a steady small decline with age. The chest wall compliance may be reduced in obese subjects and in normal volunteers by placing weights on their lower chest. Patients with thoracic deformities such as kyphosis may also have a reduced chest wall compliance.

Factors Affecting Lung Compliance

The normal predicted value for lung compliance is 0.05 FRC in liters; the range is from 0.038 FRC to 0.070 FRC [17]. This computed value is called *specific compliance.* Lung compliance increases with aging [26]. The maximum negative static transpulmonary pressure or elastic recoil (pressure at total lung capacity) that can be developed in young adults ranges from 23 to 40 cm H_2O; in elderly subjects, this value ranges from 15 to 25 cm H_2O. Pulmonary emphysema is characterized by an increased lung compliance and a low elastic recoil. In emphysema, values for the latter often are as low as 8 to 10 cm H_2O. This is understandable if one considers the pathologic abnormalities in pulmonary emphysema: destruction and dilatation of the terminal units. On the other hand, pulmonary fibrosis gives rise to a stiff lung, i.e., one characterized by low lung compliance and increased elastic recoil.

STATIC HYSTERESIS OF THE RESPIRATORY SYSTEM

The pressure-volume curves of the respiratory system differ somewhat depending upon the previous volume history of the lungs. Static pressures, for example, tend to be less positive after deep inspiration and more positive after deep expiration. Furthermore, the inflation and deflation curves do not follow identical pathways (see Fig. 19-2) This phenomenon, which is called *hysteresis,* is exhibited by other tissues in the body and by such materials as metal and rubber. It depends upon many mechanisms and is not well understood. In the lungs, however, it is due, in large part, to surface-active forces and, to a very small degree, to the plastic behavior of tissue elements.

When the lungs are inflated from a gas-free condition (e.g., in the first breath of a baby or under experimental conditions), another factor, *recruitment,* plays a role in hysteresis. When the alveoli are collapsed, a minimum pressure is required before any alveoli are filled at all. In newborn humans, this opening pressure is about 8 cm H_2O, and subsequent inflation is very irregular until complete expansion occurs at a pressure of 20 cm H_2O. The irregular filling is characterized by small regions of the lungs inflating while others remain gas-free. This indicates that there is a variable distribution of opening pressures. The phenomenon is also time-dependent, i.e., with each increment of pressure above the minimum opening pressure, 20 to 30 minutes may elapse before volume equilibrium is reached. Thus, at equal pressures, the number of open units is much greater during deflation than inflation.

The degree of hysteresis (i.e., the ratio of loop width to loop height) is comparatively small in the respiratory system, which is a highly desirable feature since hysteresis represents a loss of energy from the system [18].

SURFACTANT EFFECTS

When pressure-volume curves are obtained on isolated degassed lungs, the behavior observed differs considerably depending upon whether the lungs are filled with liquid or air. During liquid filling, there is an absence of hysteresis and the lungs begin to open just above zero pressure. Therefore, the presence of an air-filled interface is a very important determinant of lung elasticity. A surface-active film lining the alveoli that changes its properties during inflation and deflation is absolutely necessary for alveolar stability. If the alveoli are assumed to be spherical in configuration, they obey a solution of Laplace's equation, as expressed in Equation 19-5:

$$\text{Pressure} = \frac{2 \times \text{surface tension}}{\text{radius}} \qquad (19\text{-}5)$$

If the alveoli were lined by a plasma layer, the surface tension would be 48 dynes/cm. If an alveolus with a radius of 20 μ was contiguous to an alveolus of 160 μ, the smaller alveolus would empty into the larger one because of the pressure gradient:

$$P_1 = \frac{2 \times 48}{20} = 4.8 \text{ cm } H_2O$$

$$P_2 = \frac{2 \times 48}{160} = 0.6 \text{ cm } H_2O$$

However, because of surfactant — which is probably a lipoprotein, a phospholipo-polysaccharide, or a phospholipid-rich mixture — the value for the surface tension changes in accordance with the surface area of the alveolus. Thus, for the preceding example, the following would occur:

$$P_1 = \frac{2 \times 5}{20} = 5 \text{ cm } H_2O$$

$$P_2 = \frac{2 \times 40}{160} = 5 \text{ cm } H_2O$$

Therefore, the presence of surfactant helps to prevent alveolar collapse, which would result from smaller alveoli emptying into larger ones. Surfactant is not the sole force that promotes alveoli stability; tissue retractive forces are also important.

Surfactant can be extracted or washed from the lung. When this substance is spread on a Wilhelmy balance and the film is compressed and expanded, a definite hysteresis curve is obtained, which has a configuration that is quite similar to that of the pressure-volume curves obtained on air-filled isolated degassed lungs (Figs. 19-3 and 19-4). The minimum surface tension of normal, extracted surfactant is less than 10 dynes/cm.

Deficiency of surfactant has been demonstrated (1) in hyaline membrane disease of the newborn, (2) after prolonged cardiac bypass, and (3) when the pulmonary circulation is interrupted to a certain region of the lung, as in pulmonary embolism [1, 6].

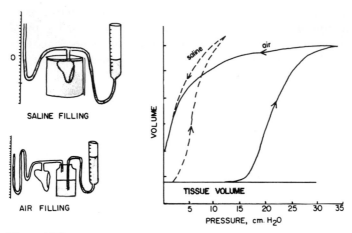

Figure 19-3
Pressure-volume curves of excised dog lungs using either saline or air for filling the lung. There is a much greater hysteresis phenomenon when air is used. (From M. A. Avery and S. Said. *Medicine* (Baltimore) 44:503, 1965. Reproduced with permission.)

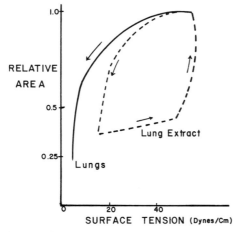

Figure 19-4
Changes in surface tension as a function of surface area. The dashed line represents the measurements of surface tension of lung extract using a modified Wilhelmy balance; the solid line represents the surface tension as calculated from pressure-volume data obtained in rat lungs.

DYNAMIC COMPLIANCE

In the clinical pulmonary function laboratory, measurements of static lung compliance are not commonly made, because of the time needed to train patients in the respiratory maneuvers. Measurements of lung compliance and elastic recoil, however, are useful clinical tests for emphysema and pulmonary fibrosis. Therefore, most clinical laboratories measure *dynamic compliance,* i.e., compliance during slow tidal breathing. The subject swallows an esophageal balloon and breathes from a spirometer or through a pneumotachograph. It is assumed that the transpulmonary pressure comes to equilibrium at that instant of tidal breathing when the flow equals zero. This may not be a valid assumption, particularly in the case of patients with obstructive lung disease who have a wide distribution of time-constants throughout the lungs. However, if one views the information as approximate, the measurement of dynamic compliance together with that of the elastic recoil (maximum pressure at total lung capacity) can provide useful clinical data.

According to Newton's third law, the transpulmonary pressure may be expressed as:

$$\Delta P_{appl} = \Delta P_{tp} = \Delta P_C + \Delta P_R + \Delta P_L \tag{19-6}$$

where ΔP_{tp} is the transpulmonary pressure, ΔP_C is the pressure due to compliance of the lung, ΔP_R is the pressure due to resistance, and ΔP_L is the pressure due to inertance [19].

During tidal breathing at atmospheric pressure, ΔP_L is so small that it can be dropped from the equation. At end-inspiration and end-expiration, the flow is zero; therefore, at these points, ΔP_R equals zero. Hence, by selecting these two points on the volume trace and intersecting two corresponding points on the transpulmonary pressure tracing, one can calculate the dynamic compliance. Figure 19-5 shows a dynamic compliance record. This measurement becomes frequency-dependent in patients with obstructive lung disease and will be further discussed in a succeeding section [27].

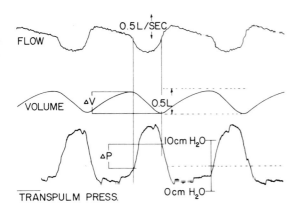

Figure 19-5
Dynamic compliance record. The flow, volume, and transpulmonary pressure are recorded during tidal breathing. The vertical lines are drawn through the points of zero flow. At these points, the transpulmonary pressure is entirely related to the elastic component of pressure and $\Delta V / \Delta P$ equals the lung compliance.

DYNAMIC MECHANICS OF BREATHING

PULMONARY RESISTANCE

Pulmonary resistance is a measure of the resistance to air flow in and out of the tracheobronchial tree along with the viscous resistance of pulmonary tissue. The latter constitutes about 20% of the total and is probably not true viscous resistance but may rather be due to hysteresis phenomena [2]. The determination of pulmonary resistance is carried out at the same time that the lung compliance is measured, using recordings of transpulmonary pressure, volume, and flow:

$$\Delta P_{tp} = \Delta P_C + \Delta P_R \tag{19-7A}$$

If the lung compliance is constant in the tidal range,

$$\Delta P_C = \left(\frac{1}{C}\right) V$$

Therefore,

$$\Delta P_R = \Delta P_{tp} - \left(\frac{1}{C}\right) V \tag{19-7B}$$

In most laboratories, P_{tp} is displayed on the x scale and $\dot{V}$ on the y scale of a cathode-ray oscilloscope. A signal proportional to $(1/C)V$ is electrically subtracted from $\dot{V}$ to close the pressure-flow loop [19]. This signal provides a proportionality constant and is set by hand using a potentiometer as the loop is observed on the oscilloscope (Fig. 19-6). However, this method for measuring the resistance to air flow is not convenient to employ routinely in patients, because of the need to insert an esophageal balloon. An alternative method is the plethysmographic method for airway resistance, which obviates the necessity for esophageal pressure measurements.

AIRWAY RESISTANCE

Airway resistance is defined as the ratio of alveolar pressure to air flow. Measurements of alveolar pressure are not possible by direct alveolar sampling. Instead, alveolar pressure is measured indirectly by means of a body plethysmograph, and it is calculated using Boyle's law [8]. If the volume of gas in the lungs is 3.5 liters, then if there were to be an alveolar pressure change of 1 cm H_2O, the amount by which the gas would be compressed is:

$$\frac{3.5 \times 10^3 \text{ ml} \times 1 \text{ cm } H_2O}{0.970 \times 10^3 \text{ cm } H_2O} = 3.6 \text{ ml}$$

where 0.970×10^3 cm H_2O is the pressure under BTPS conditions in lungs. Conversely, if this change in the volume of gas in the lungs is known, then the alveolar pressure

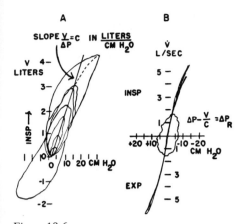

A

SLOPE $\dfrac{V}{\Delta P}$ =C IN $\dfrac{\text{LITERS}}{\text{CM H}_2\text{O}}$

V 4—
LITERS
3—
2—
INSP ↑
1
0
0 10 20 CM H$_2$O
-1—
-2—

B

V̇
L/SEC

5 —

INSP
3 —

$\Delta P - \dfrac{\dot{V}}{C} = \Delta P_R$

+20 +10   -10 -20
CM H$_2$O

— 3
EXP
— 5

Figure 19-6
A. Pressure-volume tracings on a cathode-ray oscilloscope during breaths of increased tidal volume. The slope may be obtained electronically by subtracting a factor that is proportional to the flow-resistive pressure by means of a potentiometer-amplifier network. B. Pressure-flow loops. Here, a factor that is proportional to the elastic pressure is subtracted in order to give a closed loop for pulmonary resistance when the subject is breathing at higher flow rates. (From J. Mead and J. L. Whittenberger. *J. Appl. Physiol.* 5:779, 1953. Reproduced with permission.)

change can be calculated. To determine this change in volume in practice, the subject pants in and out of a pneumotachograph while seated within a body plethysmograph, and the slope of $\dot{V}/P_{box}$ is observed on a cathode-ray oscilloscope. The air flow, $\dot{V}$, is measured at the mouth, and P_{box} is proportional to the change in gas volume (ΔV) in the plethysmograph due to the compression and expansion of alveolar gas. Figure 19-7 illustrates the body plethysmographic technique.

In this technique, the airway may be occluded by means of a solenoid shutter while the subject is panting; he then continues to attempt to pant against a closed airway. It is assumed that the mouth pressure equals the alveolar pressure (P_A) during compressional changes while there is no air flow at the mouth, because pressure changes are equal throughout a static system (Pascal's principle). The slope of the curve of mouth pressure (P_{mouth}) versus box pressure (ΔV) is noted on the cathode-ray oscilloscope. The airway resistance (R_{aw}) is then calculated from:

$$R_{aw} = \left(\frac{P_{mouth}}{\Delta V}\right)\left(\frac{\Delta V}{\dot{V}}\right) = \frac{P_{mouth}}{\dot{V}} = \frac{P_A}{\dot{V}} \tag{19-8}$$

The panting maneuver is adopted for the following reasons: (1) rebreathing the dead-space air minimizes temperature and humidity changes, (2) panting opens the upper airways, (3) the higher frequency of panting minimizes the restrictive effects on air flow resulting either from the uneven expansion of parts of the lungs and thorax or from disease, (4) shallow panting minimizes the changes in the diameter of air passages, and (5) rapid air-flow rates are obtained to facilitate the accurate measurement of alveolar pressures, which may be quite small in normal subjects [8].

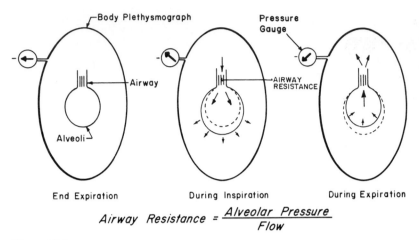

Airway Resistance = $\dfrac{\text{Alveolar Pressure}}{\text{Flow}}$

Figure 19-7
Measurement of airway resistance by a body plethysmographic technique.

Since continuous data are obtained by the plethysmographic method for measuring airway resistance, one has an opportunity to calculate the instantaneous resistance as a function of flow. However, because resistance is generally linear in the range of flow rates between 0 and 0.5 liter/sec, this is conventionally taken as the range over which airway resistance is reported. This enables the comparison of data on different subjects and from different laboratories. The airway resistance that is measured plethysmographically is not the average of unequal resistances throughout the lungs but is the average alveolar pressure change divided by the average air flow at the mouth. This corresponds to the average of airway conductances, G_{aw} and a conductance is the reciprocal of resistance $G_1 = \dfrac{1}{R}$.

$$G_{aw} = \frac{G_1 + G_2 + \ldots + G_n}{n} \qquad (19\text{-}9)$$

Conductances in parallel are additive, whereas resistances in parallel add according to their reciprocals:

$$\frac{1}{R_{aw}} = \frac{1}{R_1} + \frac{1}{R_2} + \ldots + \frac{1}{R_n} \qquad (19\text{-}10)$$

For this reason, pulmonary physiologists often report the results from the plethysmographic method as *mean airway conductance* rather than mean airway resistance.

The airway resistance in normal adults at resting lung volume ranges from about 0.8 to 1.8 cm H_2O/liter/sec at a flow rate of 0.5 liter/sec. Airway resistance is a function of lung volume: the larger the lungs, the lower the air resistance [3]. A newborn infant, for example, may have an airway resistance of 20 to 35 cm H_2O/liter/sec. Furthermore, in an adult, the airway resistance measured at total lung capacity is much smaller than that measured at residual volume. As a first approximation, the airway conductance is proportional to the volume, as expressed in Equation 19-11:

$$G_{aw} = 0.24V \qquad\qquad (19\text{-}11)$$

where V is the thoracic gas volume (TGV) in liters at the level of the measurement. As a second approximation, the conductance-volume curve is shaped like a reversed S, which sometimes intercepts the volume axis at a volume that represents the closure of some or all of the airways at maximum expiration.

The relation between G_{aw} and TGV, however, is actually dependent on the lung tissue tension, which is reflected in the transpulmonary pressure rather than the volume of gas. This may be demonstrated by the maneuver of chest strapping, which causes some terminal units to collapse and not to reopen until a deep breath is taken [4]. After such a collapse or in pulmonary fibrosis, the tissue tension is increased, and the airway conductance is increased for a given lung volume because greater tension is exerted by the tissue, which pulls outward on the airway walls. Conversely, the loss of tissue tension (as in emphysema) or the narrowing of the airway lumen (as in asthma) results in decreased conductance at each volume.

Agents such as isoproterenol and epinephrine will cause the airways to dilate, whereas histamine and fine dusts will cause airway constriction [4, 10]. The physiologic regulation of airway resistance maintains a balance that prevents the airways either from becoming too large (which would lead to an excessively large anatomic dead space and therefore increased tidal volume, minute-volume, and work of breathing) or from becoming too small (which would lead to an excessively large airway resistance and therefore an increased work of breathing).

Recent evidence indicates that airway resistance measurements primarily reflect the resistance in the large airways, i.e., those airways greater than 2 to 3 mm in diameter [16].

AIRWAY OBSTRUCTION

Primary airway obstruction is defined as a reduction in airway caliber at a given lung volume and alveolar pressure compared to normal calibers at the same lung volume and alveolar pressure. The reduction in caliber may be due to an abnormality within the lumen of the airway, within its wall, or within its supporting structures. Examples of such disorders include asthma, chronic bronchitis, emphysema, cystic fibrosis, and bronchiolitis. *Secondary airway obstruction* occurs when the airway caliber is reduced below that obtained at a normal functional residual capacity because of an abnormality outside the lung that causes a decrease in the FRC. In this category are included obesity, pleural effusion, pneumothorax, and fibrothorax.

OSCILLATORY MECHANICS OF THE CHEST-LUNG SYSTEM

The simplest electrical analog of the chest-lung system is shown in Figure 19-8. The number *1* indicates the mouth, and *2* the atmospheric pressure. In this figure, R_{aw}

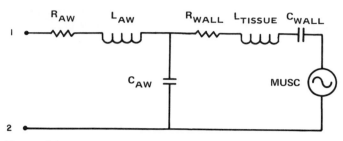

Figure 19-8
Electrical analog of the chest-lung system. (See text for key to symbols and description.)

represents the airway resistance, L_{aw} the airway inertance, C_{aw} the alveolar capacitance or lung compliance, R_{wall} the resistance of the chest wall, L_{tissue} the tissue inertance, C_{wall} the chest wall compliance, and the signal source, and $MUSC$, the respiratory muscles [9].

In this system, the signal generator (the respiratory muscles) can be removed if, in the case under consideration, the subject is relaxed. A sine-wave pump or loud-speaker can be put across terminals 1 and 2 (the mouth), and if the frequency is varied, the impedance of this system can be calculated, which corresponds to the relation between pressure and flow at the mouth. When the respiratory system is studied in this manner, it can be shown that there is a high degree of damping. The damping or *quality factor* is between 0.7 and 1.5, and there is a broad resonance band at about 6 Hz. This resonance frequency can be reduced to 4 Hz by mass-loading the lower chest, as one would predict from the analog model.

The total respiratory impedance decreases with increasing frequencies. This effect is more pronounced in patients with obstructive lung diseases and is in part related to the wide distribution of time-constants throughout the lung; the chest wall imped-ance is not frequency-dependent.

The method of using a pump to oscillate an air flow at the mouth may be employed clinically as an indirect way to estimate the airway resistance. The chest wall resistance remains relatively constant at 1.5 to 2.5 cm H_2O/liter/sec, and the total respiratory resistance normally ranges from 2.5 to 4.0 cm H_2O/liter/sec. If one oscillates the pump at resonance, then by definition the capacitive reactance and the inductive reactance are equal and opposite in sign. Therefore, one can divide the peak-to-peak mouth pressure by the peak-to-peak flow to obtain the flow resistance of the total respiratory system (Fig. 19-9). At frequencies below or above resonance, the mouth pressure at intervals where the acceleration is zero (peaks of flow) can also be used to calculate the total respiratory resistance [11]. As is always the case with determina-tions of airway resistance, the values obtained must be related to either the lung volume or the transpulmonary pressure.

PARTITIONING OF AIRWAY RESISTANCE

Airway resistance as measured by the plethysmographic method reflects the resistances of both the large and small airways. Recently a "retrograde catheter technique" has

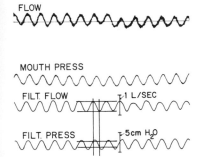

Figure 19-9
Total respiratory resistance. In the accompanying tracing, flow and mouth pressure signals are used as the input to a band-pass filter that permits frequencies of 3 to 15 Hz to pass; this eliminates low-frequency shifts due to respiration and high-frequency noise due to mechanical vibrations. These data are then analyzed as described in the text.

been employed in open-chest animals and postmortem human specimens that permits the separation of airway resistance into central and peripheral divisions. Using this method, it has been found that for normal adult lungs, the peripheral resistance (R_{periph}), which is seen in airways less than 2 mm in diameter, is negligible when measured from 60% to 100% of the vital capacity and amounts to only about 15% of the total resistance even at residual volume (Fig. 19-10). The total resistance increases at residual volume primarily because of an increase in central airway resistance (R_{cen}), which includes the airways from 2 mm diameter to the trachea. Airways that are 3 to 8 mm in diameter produce this rise in airway resistance, primarily because of increased bronchomotor tone. The low value of R_{periph} favors equality of gas distribution, but, as will be discussed later, the constriction of these small airways might affect distribution and gas exchange, with a minimal change in total resistance.

In centrilobular emphysema, practically all the resistance is located in the peripheral airways [12]. In lungs from autopsied patients less than 6 years of age, R_{periph} ranges from 30% to 75% of the total. This increase in peripheral resistance in the younger age group is probably related to the smaller number of respiratory bronchioles at this age; further development of the lung takes place up to 12 years of age. Bronchiolitis before the age of 5 years is frequently a life-threatening condition in which symptoms of severe airway obstruction are present. Perhaps this is due to the fact that the peripheral resistance is such a large percentage of the total resistance. In older children and adults, acute bronchiolitis is rarely diagnosed by means of peripheral resistance measurements, because considerable obstruction would have to be present in the small airways before it would influence the total pulmonary resistance.

FREQUENCY-DEPENDENT COMPLIANCE

The values for dynamic compliance at any breathing frequency are similar to those obtained from static compliance measurements in normal subjects breathing up to

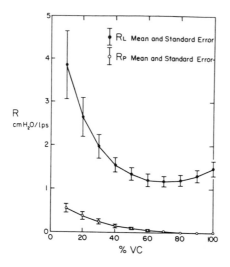

Figure 19-10

Values for the pulmonary resistance (R_l) and the small airway resistance (R_p) in dogs plotted against the lung volume. Note that the small airways contribute only a small fraction to the total pulmonary resistance. Further, this contribution is seen only at low lung volumes. (From P. J. Macklem and J. Mead. *J. Appl. Physiol.* 22:295, 1967. Reproduced with permission.)

about 1 liter above FRC. However, the dynamic compliance decreases with frequency at lung volumes below FRC and in patients with obstructive airway diseases, even when the airway resistance is not significantly elevated.

In resent years, the meaning and the application of the concept of frequrncy-dependent compliance have recieved wide attention. It may be illustrated, strictly on a descriptive basis, in terms of the model shown in Figure 19-11 for the following conditions.

1. If $R_2C_1 = R_3C_2$ and the tidal volume is 0.5 liter, the transpulmonary pressure change might be 2.5 cm H_2O, so that the static and dynamic compliance would be identical (0.1 liter/cm H_2O) and the total would be 0.2 liter/cm H_2O.

2. If R_2 was infinite (blocked airway), the tidal volume of 0.5 liter would produce a transpulmonary pressure change of 5 cm H_2O. This increase in P_{tp} occurs because only half of the units are available to receive the tidal volume. The total static and dynamic compliance would equal 0.1 liter/cm H_2O.

3. However, if R_2 is greater than R_3 (airway narrowed), then under dynamic conditions, air will pass preferentially to the unit with the more open airway, an effect that increases as the respiratory rate increases. Thus, 0.45 liter might go to R_3C_2, while only 0.05 liter would go to R_2C_1. The dynamic compliance would approach values of about 0.1 liter/cm H_2O. Under static conditions, however, the more distended units of C_2 will have a greater elastic recoil, which will cause the air to be redistributed to the partially obstructed unit until both are uniformly distended and the pressure within them is equalized. The P_{tp} will fall, and the static compliance will rise to 0.2 liter/cm H_2O.

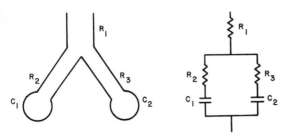

Figure 19-11
Models illustrating the dynamic compliance of the lung.

If two pulmonary pathways connected in parallel are subjected to a sinusoidal driving pressure and if their separate impedances are identical, the magnitude of the total admittance (i.e., the reciprocal of impedance) will be the arithmetic sum of the separate admittances [24]:

$$\frac{1}{Z} = \frac{1}{Z_1} + \frac{1}{Z_2} \tag{19-12}$$

If the time-constants of the parallel pathways are different, the phase angles between flow and pressure will be different and the flows will be out of phase with each other. The instants at which zero flow occurs in the separate pathways will not correspond, and gas will flow between the parallel pathways at instants of zero flow in the common path. In this case, the total admittance must be expressed in vector form:

$$\frac{1}{\mathbf{Z}} = \frac{1}{\mathbf{Z}_1} + \frac{1}{\mathbf{Z}_2} \tag{19-13}$$

As demonstrated by Otis and his associates [24], one can derive expressions for the *effective compliance* and the *effective resistance* of this two-pathway system. In the development of the expressions for effective compliance and effective resistance for two-pathway systems, each pathway is assumed to consist of a single flow resistance and volume elastic elements in series. The admittance of a pathway, Y, is the reciprocal of the pathway impedance, Z:

$$Y = \frac{1}{Z} \tag{19-14}$$

As mentioned previously, the admittance of parallel systems is equal to the sum of the individual pathway admittances. Since we must deal with vector quantities, let $\mathbf{Y}_T$ equal the total admittance and $\mathbf{Y}_1$ and $\mathbf{Y}_2$ equal the separate pathway admittances. Then,

$$\mathbf{Y}_T = \mathbf{Y}_1 + \mathbf{Y}_2 \tag{19-15}$$

To solve for Y_T in terms of the angular frequency (ω), it is convenient to use complex functions, which permit an algebraic solution. In complex form, $Z = R - j(1/\omega C)$ where j is equal to $\sqrt{-1}$ and $\omega = 2\pi f$; all terms with the coefficient $-j$ lag behind the remaining terms by 90 degrees. Expressed in complex form, the pathway admittances are:

$$Y_1 = \cfrac{1}{R_1 - j\cfrac{1}{\omega C_1}} = \frac{\omega C_1}{\omega R_1 C_1 - j} \qquad (19\text{-}16)$$

$$Y_2 = \cfrac{1}{R_2 - j\cfrac{1}{\omega C_2}} = \frac{\omega C_2}{\omega R_2 C_2 - j} \qquad (19\text{-}17)$$

The total admittance then is:

$$Y_T = Y_1 + Y_2 = \frac{\omega C_1}{\omega R_1 C_1 - j} + \frac{\omega C_2}{\omega R_2 C_2 - j}$$

$$= \frac{\omega^2 C_1 C_2 (R_1 + R_2) - j\omega(C_1 + C_2)}{(\omega^2 R_1 C_1 R_2 C_2 - 1) - j\omega(R_1 C_1 + R_2 C_2)} \qquad (19\text{-}18)$$

In terms of impedance:

$$Z_T = \frac{1}{Y_T} = \frac{(\omega^2 R_1 C_1 R_2 C_2 - 1) - j\omega(R_1 C_1 + R_2 C_2)}{\omega^2 C_2 C_1 (R_1 + R_2) - j\omega(C_1 + C_2)} \qquad (19\text{-}19)$$

Let $A = \omega^2 R_1 C_1 R_2 C_2 - 1$
  $B = \omega(R_1 C_1 + R_2 C_2)$
  $K = \omega^2 C_2 C_1 (R_1 + R_2)$
  $M = \omega(C_1 + C_2)$

Then,

$$Z_T = \frac{A - jB}{K - jM} \qquad (19\text{-}20)$$

Multiplying the numerator and denominator of Equation 19-20 by $(K + jM)$ yields:

$$Z_T = \frac{AK + BM}{K^2 + M^2} - j\frac{BK - AM}{K^2 + M^2} \qquad (19\text{-}21)$$

From the definition of impedance,

$$Z_T = R_{eff} - j \frac{1}{\omega C_{eff}} \qquad (19\text{-}22)$$

where R_{eff} is the *effective resistance* and C_{eff} is the *effective compliance*. Thus,

$$R_{eff} = \frac{AK + BM}{K^2 + M^2} \quad \text{and} \quad C_{eff} = \frac{K^2 + M^2}{\omega(BK - AM)}$$

Equations 19-23 and 19-24 are obtained by substituting the values for A, B, K, and M in these expressions and letting $R_1 C_1 = \tau_1$ and $R_2 C_2 = \tau_2$:

$$C_{eff} = \frac{\omega^2(\tau_2 C_1 + \tau_1 C_2)^2 + (C_1 + C_2)^2}{\omega^2(\tau_1^2 C_2 + \tau_2^2 C_2) + (C_1 + C_2)} \qquad (19\text{-}23)$$

$$R_{eff} = \frac{\omega^2 \tau_1 \tau_2 (\tau_2 C_1 + \tau_1 C_2) + (\tau_1 C_1 + \tau_2 C_2)}{\omega^2(\tau_2 C_1 + \tau_1 C_2)^2 + (C_1 + C_2)^2} \qquad (19\text{-}24)$$

When the time-constants, τ, of the pathways are equal, i.e., when $\tau_1 = \tau_2$, Equations 19-23 and 19-24 are simplified, and C_{eff} and R_{eff} become independent of frequency:

$$C_{eff} = C_1 + C_2 \qquad (19\text{-}25)$$

$$R_{eff} = \frac{\tau}{C_1 + C_2} = \frac{1}{\dfrac{1}{R_1} + \dfrac{1}{R_2}} = \frac{R_1 R_2}{R_1 + R_2} \qquad (19\text{-}26)$$

The R_{eff} and C_{eff} of a two-pathway system are shown as functions of frequency in Figure 19-12.

Further, it can be shown from mechanical analogs that the equations derived from electrical theory hold for parallel pathways, i.e., that frequency-dependent compliance occurs. In addition, patients with obstructive lung diseases and normal volunteers given the bronchoconstrictor agent, histamine, also show frequency-dependent compliance because of changes in the resistance through the lung.

The measurement of compliance depends upon the absolute lung volume, the lung volume history, and the frequency of respiration; it is a difficult measurement to make accurately, particularly at rapid frequencies. Frequency-dependence may be defined arbitrarily as a fall in compliance at a frequency of no more than 20% of the static value. Since inertial effects become significant at higher frequencies, it is not practical to measure dynamic compliance above 100 breaths per minute. Most normal subjects do not exhibit frequency-dependent compliance below 70 breaths per minute.

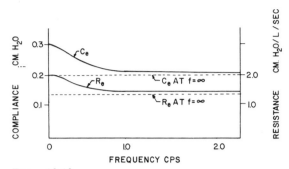

Figure 19-12

Effective compliance, C_e, and effective resistance, R_e, of a two-pathway system as functions of frequency ($R_1 = 2, C_1 = 0.1, \tau_1 = 0.2, R_2 = 4, C_2 = 0.2, \tau_2 = 0.8$).

PENDELLUFT

The proportion of gas transferred between pathways is dependent on the relative impedances of the overall system. An expression for calculating this transfer, termed *pendelluft,* as a proportion of the overall tidal volume is:

$$\frac{V_{pendelluft}}{VT} = \frac{Z(Z_1 + Z_2)}{Z_1 Z_2} \tag{19-27}$$

Since the values for the impedances vary with frequency, the proportion of pendelluft will also vary accordingly, and for any system with different time-constants, there will be a frequency at which the pendelluft will be maximal. The greatest possible pendelluft (41%) occurs in systems consisting of a pure resistance R_1 ($\tau = \infty$) and a pure capacitance C_2 ($\tau = 0$) connected in parallel and driven so that $R_1 = 1/(2\pi fC)$ [24].

SMALL AIRWAY DISEASE

The time-constant of lung units that are distal to airways of less than 2 mm in diameter is about 0.01 sec. If there is a fourfold difference between such units, the dynamic compliance (C_{dyn}) will fall as a function of increasing frequency. In analyzing the theoretical bases for frequency-dependent compliance, it is found that regional differences in the elastic properties of the lungs that are sufficient to cause a detectable fall in compliance at rapid breathing rates should also produce a definitely abnormal static pressure-volume curve. Similarly, time-constant discrepancies due to obstructions in the central airways should lead to a significant increase in the airway resistance.

Thus, if a patient with normal values for airway resistance and elastic recoil exhibits frequency-dependent compliance, then it is reasonable to conclude that there is peripheral airway obstruction [27]. Such findings have been observed in patients who smoked cigarettes or who had mild symptoms of chronic bronchitis with otherwise normal pulmonary function [14]. Patients with bronchial asthma who are symptom-free may also show frequency-dependent compliance. Furthermore, the theory that peripheral airway resistance can be tested by dynamic compliance measurements is supported by studies carried out after the insufflation of 2-mm diameter beads into the lungs of experimental animals. These animals have normal airway resistance but show frequency-dependent compliance.

COLLATERAL VENTILATION

In postmortem lung studies, it has been shown by Hogg and his co-workers [12, 13] that airways less than 2 mm in diameter are responsible for the bulk of the increased airway resistance seen in emphysema. The fact that many of the small airways are occluded without evidence of atelectasis (pulmonary collapse) led to a study of collateral ventilation in the lung. In the model shown in Figure 19-13, R_4 is a collateral pathway that permits air to flow between the alveoli.

Although collateral pathways exist in normal human lungs, their resistance is so high that no significant gas flow takes place under physiologic conditions. The major collateral pathways in normal lungs are probably interalveolar pores, the pores of Kohn, which are holes 3 μ to 13 μ in diameter situated in intercapillary regions in the alveolar walls. Such pores are numerous in the dog, less numerous in humans, and even less common in the pig and sheep [28]. Accordingly, collateral ventilation is much greater in the dog lung than in the pig lung.

In emphysema, the destruction of the lung tissue results in fenestrations between the alveoli that are much larger than the pores of Kohn; the resistance of these collateral channels is then less than the resistance of the airways. The electrical analog of this condition in emphysema is shown in Figure 19-14. In this model, R_1 and R_3 are common to both capacitors and will not influence any phase shifts between C_1 and C_2. The time-constant that determines the phase and the charge on C_1 will be R_4C_1. Because the branch containing C_2 has no resistor, its time-constant will be close to zero as an approximation. If values are assigned for the effective compliance equations, it is found that when the time-constant R_4C_1 is 0.1 sec, C_{dyn} falls 8% at 1 Hz and 22% at 2 Hz; when this time-constant is 5.0 sec, C_{dyn} is down 50% at 0.2 Hz. Therefore, ventilation of lung units via collateral channels could be one of the causes for frequency-dependent compliance in emphysema (Fig. 19-15).

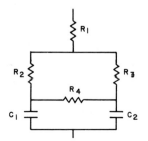

Figure 19-13
Electrical analog illustrating collateral ventilation in the lung. (Modified from J. C. Hogg et al. *J. Clin. Invest.* 48:421, 1969.)

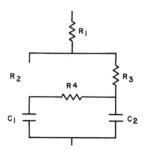

Figure 19-14
Electrical analog of collateral pathways in emphysema.

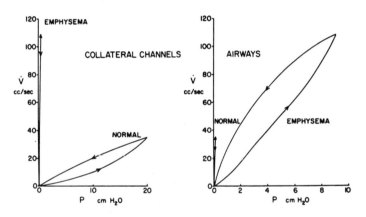

Figure 19-15
The flow through collateral channels is plotted on the ordinate of both graphs. The pressure observed through collateral channels is depicted on the left, and that through the airways on the right. The data were collected during a quasi-static inflation and deflation on one normal lung and one emphysematous lung. Note that there is less resistance to flow through the collateral channels than through the airways in the emphysematous lung. (From J. C. Hogg et al. *J. Clin. Invest.* 48:421, 1969. Reproduced with permission.)

Collateral ventilation will certainly lead to an abnormal distribution of ventilation and, presumably, to impaired gas exchange. In the model of Figures 19-13 and 19-14, C_2 functions as a nondistensible, series dead-space and is analogous to the centrilobular space in emphysema. Air entering the obstructed areas via collateral channels will have already exchanged O_2 and CO_2, and therefore such units will have low $\dot{V}_A/\dot{Q}$ ratios. The unobstructed units will have high $\dot{V}_A/\dot{Q}$ ratios, since additional air must pass through them to reach obstructed units. Furthermore, significant stratification of gases is probably present because of the diffusion distances in the collateral channels [13].

WORK OF BREATHING

The respiratory muscles do work against two main types of forces: (1) elastic forces in the lungs and chest wall when volume changes occur and (2) resistive forces due to gas flowing in the airways and the nonelastic deformation of tissues. For normal lungs, less than 15% of the total work is done against nonelastic forces (Fig. 19-16). The work of breathing is expressed as:

$$W = \int P dV \qquad\qquad (19\text{-}28)$$

where W is work, P is pressure, and dV is the derivative of volume [25].

Another estimate of work can be obtained from methods that measure the oxygen cost of breathing. This gives an estimate of the efficiency of the respiratory muscles. The values range from 0.5 to 1.0 ml O_2 per liter of ventilation. These values may be significantly increased in patients with obstructive lung diseases.

The work of breathing, when alveolar ventilation is kept constant, is mostly due to work against elastic forces at low breathing frequencies and to work against nonelastic forces at high frequencies. In terms of work, the optimal breathing rate at rest is about

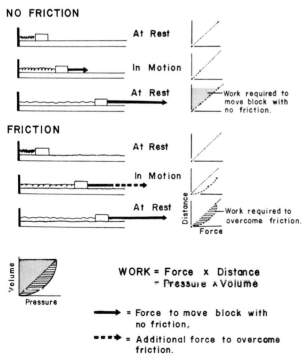

Figure 19-16
Work of breathing during inspiration.

16 breaths per minute. However, in patients with restrictive lung disease, the work of breathing can be minimized by breathing at shallow, rapid rates. Here, the abnormality is due to the stiff lungs; the airway resistance is normal or decreased.

Alveolar gas is compressed and expanded during every breathing cycle. The volume displacement measured at the mouth (the tidal volume) may therefore be smaller than the tidal volume measured on the basis of the change in thoracic gas volume. In the lung, work related to the compressibility of gases is usually insignificant. It only becomes appreciable in emphysematous patients during periods of hyperpnea.

REFERENCES

1. Avery, M., and Said, S. Surface phenomena in lungs in health and disease. *Medicine* (Baltimore) 44:503, 1965.
2. Bachofen, H. Lung tissue resistance and pulmonary hysteresis. *J. Appl. Physiol.* 24:296, 1968.
3. Briscoe, W. A., and DuBois, A. B. The relationship between airway resistance, airway conductance and lung volume in subjects of different age and body size. *J. Clin. Invest.* 37:1279, 1958.
4. Butler, J., Caro, C. G., Alcala, R., and DuBois, A. B. Physiological factors affecting airway resistance in normal subjects and in patients with obstructive respiratory disease. *J. Clin. Invest.* 39:584, 1960.
5. Cherniacky, R. M., and Brown, E. A simple method for measuring total respiratory compliance: Normal values for males. *J. Appl. Physiol.* 20:87, 1965.
6. Clements, J. A. Pulmonary surfactant. *Am. Rev. Respir. Dis.* 101:984, 1970.
7. Crosfill, M. L., and Widdicombe, J. G. Physical characteristics of the chest and lungs and the work of breathing in different mammalian species. *J. Physiol.* (Lond.) 158:1, 1961.
8. DuBois, A. B., Botelho, S. Y., and Comroe, J. H., Jr. A new method for measuring airway resistance in man using a body plethysmograph: Values in normal subjects and in patients with respiratory disease. *J. Clin. Invest.* 35:327, 1956.
9. DuBois, A. B., Brody, A. W., Lewis, D. H., and Burgess, B. F. Oscillation mechanics of lung and chest in man. *J. Appl. Physiol.* 8:587, 1956.
10. DuBois, A. B. Resistance to Breathing. In W. O. Fenn and H. Rahn (Eds.), *Handbook of Physiology, Section 3, Respiration,* Vol. 1. Washington, D.C.: The American Physiological Society, 1964. P. 451.
11. Goldman, M., et al. A simplified measurement of respiratory resistance by forced oscillation. *J. Appl. Physiol.* 23:113, 1970.
12. Hogg, J. C., Macklem, P. T., and Thurlbeck, W. M. The site and nature of airway obstruction in chronic obstructive lung disease. *N. Engl. J. Med.* 278:1355, 1968.
13. Hogg, J. C., Macklem, P. T., and Thurlbeck, W. M. The resistance of collateral channels in excised human lungs. *J. Clin. Invest.* 48:421, 1969.
14. Ingram, R. H., Jr., and O'Cain, C. F. Frequency dependence of compliance in apparently healthy smokers versus nonsmokers. *Bull. Physiopathol. Respir.* 7:195, 1971.
15. Lith, P. V., Johnson, F. N., and Sharp, J. T. Respiratory elastances in relaxed and paralyzed states in normal and abnormal men. *J. Appl. Physiol.* 23:475, 1967.
16. Macklem, P. T., and Mead, J. Resistance of central and peripheral airways measured by a retrograde catheter. *J. Appl. Physiol.* 22:395, 1967.

17. Marshall, R. The physical properties of the lungs in relation to the subdivisions of lung volume. *Clin. Sci.* 16:507, 1957.
18. Marshall, R., and Widdicomb, J. G. Stress relaxation of the human lung. *Clin. Sci.* 20:19, 1960.
19. Mead, J., and Whittenberger, J. L. Physical properties of human lung measured during spontaneous respiration. *J. Appl. Physiol.* 5:779, 1953.
20. Mead, J. Mechanical properties of lungs. *Physiol. Rev.* 41:281, 1961.
21. Mead, J., and Milic-Emili, J. Theory and Methodology in Respiratory Mechanics with Glossary of Symbols. In W. O. Fenn and H. Rahn (Eds.), *Handbook of Physiology, Section 3, Respiration,* Vol. 1. Washington, D.C.: The American Physiological Society, 1964. P. 363.
22. Milic-Emili, J., Orzalesi, M. M., Cook, C. D., and Turner, J. M. Respiratory thoraco-abdominal mechanics in man. *J. Appl. Physiol.* 19:217, 1964.
23. Naimark, A., and Cherniack, R. M. Compliance of the respiratory system and its components in health and obesity. *J. Appl. Physiol.* 15:377, 1960.
24. Otis, A. B., et al. Mechanical factors in distribution of pulmonary ventilation. *J. Appl. Physiol.* 8:427, 1956.
25. Otis, A. B. The Work of Breathing. In W. O. Fenn and H. Rahn (Eds.), *Handbook of Physiological, Section 3, Respiration,* Vol. 1. Washington, D.C.: The American Physiological Society, 1964. P. 463.
26. Turner, J. M., Mead, J., and Wohl, M. E. Elasticity of human lungs in relation to age. *J. Appl. Physiol.* 25:664, 1968.
27. Woolcock, A. J., Vincent, N. J., and Macklem, P. T. Frequency dependence of compliance as a test for obstruction in small airways. *J. Clin. Invest.* 48:1097, 1969.
28. Woolcock, A. J., and Macklem, P. T. Mechanical factors influencing collateral ventilation in human, dog and pig lungs. *J. Appl. Physiol.* 30:99, 1971.

THE BODY FLUIDS

IV

Kidney and Body Fluids Systems   **20**

Joel B. Mann

INTRODUCTION

The proper functioning of the various body cells depends upon the continuous maintenance of a favorable chemical environment, which is achieved primarily by the body
fluids system. This system consists of a multitude of processes and mechanisms, which
are both physicochemical and metabolic in nature, that act to maintain a biochemical
dynamic equilibrium. Chief among the operational units of the body fluid system
are the kidneys, aptly termed the "master chemists of the body," whose major functions and responsibilities include the regulation of volume and composition of the
body fluids. This unit deals with both the organization of the body fluid system as
well as the functions of the kidney and the regulatory mechanisms that maintain the
constancy of the internal environment.

In the average person, the kidneys receive 25% of the cardiac output or approximately 1.2 liters of blood per minute. Thus, over a 24-hour period, the kidneys are
exposed and reexposed to the entire blood volume 350 times, thereby enabling them
to alter readily the volume and composition of all the body fluids by their exposure
to a small subcompartment of the extracellular fluid space. The kidneys are uniquely
constructed to perform their function by the processes of filtration and reabsorption.

BIOLOGIC STRUCTURE

The kidneys normally weigh about 150 gm each and are situated behind the peritoneum
on opposite sides of the vertebral column (Fig. 20-1). They are surrounded by a protective environment of loose connective tissue and fat. Although each kidney usually
has only one renal artery, two or even three arteries are not uncommon. Urine is
excreted from the kidneys through the ureters into the bladder.

The *hilus* of the kidney, shown in sagittal section in Figure 20-2A, is a slit in the
middle third of the concave medial border of the kidney through which the ureter,
renal vein, and lymphatic vessels leave and the renal artery and nerves enter. The hilus
opens out into a space called the *renal sinus,* which is primarily taken up by the renal
pelvis with its extensions, the *major* and *minor calyces.*

Emerging from the renal parenchyma and protruding into each minor calyx are the
renal papillae. There are usually eight to ten papillae in each kidney, but the number
is quite variable. The papillae are cone-like, although flattened and with elliptical
bases. The papillae appear to be striated due to the *ducts of Bellini,* which converge
toward the papillary tips. The ducts of Bellini are themselves formed by the fusion
of many collecting ducts. Urine flowing out of the ducts of Bellini enters the minor

343

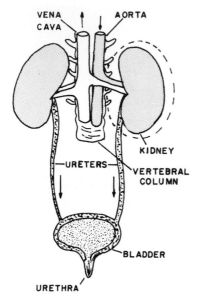

Figure 20-1
Location of the kidney.

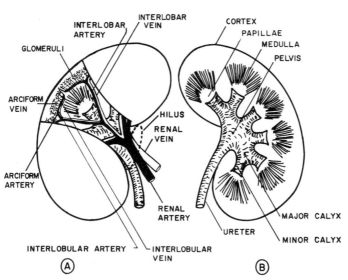

Figure 20-2
Sagittal section of the kidney.

calyx into which the papilla is thrust, then courses into the associated major calyx, the *renal pelvis,* and passes finally via the ureter into the bladder.

When cut in sagittal section, the kidney is readily divided by its appearance into an outer one-third, the *cortex,* and an inner portion, the *medulla* (Fig. 20-2B). Actually, the cortex dips down between the bases of the pyramids (the tips of which form the papillae) toward the renal sinus. The outer portion of the cortex has a granular appearance due to its content of convoluted proximal and distal tubules having been randomly cut in varying planes. The inner cortex is characterized by columns fanning outward from the pyramids. These radiating columns consist of mostly parallel, straight portions of the nephrons and blood vessels cut along their long axis.

As the renal artery enters the hilus, it divides into a number of branches, called the *interlobar arteries,* which course between the calyces and their contiguous pyramids (Fig. 2A). The interlobar arteries bend sharply over the bases of the pyramids to form the *arciform arteries,* which demarcate the junction of the cortex and medulla. The interlobular arteries arise at right angles from the arciform vessels and run radially toward the surface of the kidney, giving rise to short branches, the *afferent arterioles,* as shown in Figure 20-3. Each afferent arteriole supplies a single *glomerulus,* the ball-shaped mass of interconnected capillaries that forms the filtering mechanism of the kidney. The venous drainage corresponds roughly to the arterial supply.

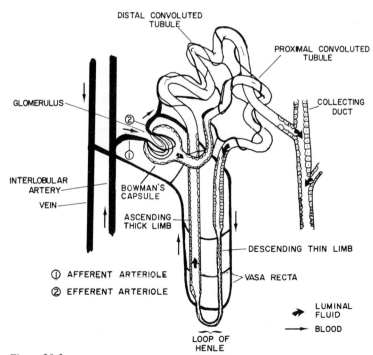

Figure 20-3
The nephron and its blood supply.

MICROSTRUCTURE OF THE KIDNEY

Each human kidney contains approximately one million fundamental units called *nephrons* (Fig. 20-3). Each nephron is composed of a renal corpuscle, a proximal convoluted tubule, a loop of Henle, and a distal convoluted tubule. Multiple nephrons drain into each collecting duct, which in turn coalesces with others to form the ducts of Bellini and ultimately drains into the renal pelvis.

The nephron, as defined morphologically, excludes the collecting duct because it has a different embryologic origin from the remainder of the tubule. Recently, the collecting duct has been found to play an active role in the formation of urine and is specifically involved in the concentration of urine, the regulation of acid-base balance, and the secretion and reabsorption of sodium and potassium. Although morphologically excluded from the nephron, the collecting duct must be included in any functional definition.

The *renal corpuscle* consists of a capillary tuft, or glomerulus, that is surrounded by the dilated blind end of the proximal tubule called *Bowman's capsule.* Within Bowman's capsule, the afferent arteriole ultimately branches into 20 to 40 capillary loops before reconverging to form the *efferent arteriole.* Both the afferent and efferent arterioles have smooth muscle fibers in their walls, and, by maintaining varying states of contraction, they are able to produce variable resistance to the flow of blood through the glomerulus.

Immediately before the afferent arteriole enters Bowman's capsule, there is a group of modified cells that form a cuff around the afferent arteriole known as the *polar cuff.* The distal convoluted tubule of that same nephron also has an area of specialized cells, called the *macula densa,* which is immediately adjacent to the polar cuff. These two structures together form the *juxtaglomerular apparatus.* The juxta-glomerular apparatus is a sensing mechanism for blood pressure, and, through an as yet unelucidated mechanism, it is responsible for the production of renin; the secretory rate of this is controlled by an autoregulatory feedback loop. *Renin* is an enzyme that initiates a series of reactions which ultimately result in the production of angiotensin, a very powerful vasoconstrictor.

After a short, straight connecting segment, the *proximal tubule* coils round on itself and then straightens out to enter a corticomedullary ray. It reaches the layers of the cortex next to the medulla and, in some instances, into the outer medulla itself. The proximal tubule is lined throughout its entire length by a single layer of cuboidal cells that rest on a basement membrane.

The portions of the cells facing the lumen tend to bulge into the lumen and nearly obliterate the space. The luminal surface of the cells is covered with filaments about one micron long that give the appearance of a brush border when they are viewed under the microscope. The presence of the brush border is a distinguishing feature of the proximal convoluted tubule, and it plays a critical role in the reabsorption of bicarbonate that has been filtered through the glomerulus.

Henle's loop is subdivided into several parts: the descending thick limb, the descending and ascending thin limbs, and the ascending thick limb which continues on to form the distal convoluted tubule. Those nephrons that originate in the outer two-thirds of the cortex have relatively short loops of Henle, whereas those that originate near the medulla (the juxtamedullary nephrons) have much longer loops. The ability to form a concentrated urine is directly related to the characteristic shape and total length of the loop of Henle.

The *distal tubule* arises initially as a straight segment from the ascending limb of the loop of Henle. It returns to its own renal corpuscle and there twists and turns on itself to form its convoluted portion; it ultimately continues on to join a collecting duct. The collecting duct is originally formed in the outer portion of the cortex by the union of several distal tubules. The duct increases in size by the addition of more distal tubules and ultimately by the fusion of several collecting ducts as it continues on its course to the renal papilla.

The cortical and juxtamedullary nephrons have quite different vascular supplies. In the cortex, the efferent glomerular arteriole immediately breaks up into a profuse anastomosis of capillaries surrounding the proximal and distal convoluted tubules. The loops of Henle and the immediately adjacent collecting ducts are also frequently included in this anastomosis. This capillary network ultimately reforms as venules, which coalesce to form the interlobular veins.

The vascular supply of the juxtamedullary nephrons is considerably more organized than that of the cortical nephrons (Fig. 20-3). The efferent glomerular arteriole of the juxtamedullary nephrons is of equal or larger diameter than the afferent glomerular arteriole. It also forms a network of anastomosing capillaries, but, in addition, it forms a series of straight capillary loops that parallel the descending and ascending limbs of loops of Henle. These vessels are called the *vasa recta,* and, as is so often the case with uniquely formed structures, they perform a special function in the production of a concentrated urine.

THE PROCESS OF BLOOD FILTRATION AND REABSORPTION

A schematic overview of the major functional aspects of the kidney is presented in Figure 20-4. The heavy block at the right side of the diagram represents the kidney, and each subdivision represents a functional unit of the kidney. Material is supplied to the kidney through the renal artery. The output of the kidneys consists of materials that are reabsorbed and appear in the renal vein, as well as urine, which is voided. Factors that regulate these input and output processes are shown in the blocks at the center and left of the figure. An increase in systemic blood pressure, for example, causes an increase in the glomerular hydrostatic pressure, which in turn increases the glomerular filtration rate.

Blood entering the kidney through the renal artery is partially filtered at the glomerulus. The unfiltered portion enters the peritubular capillary system and serves to provide substances for tubular secretion and to receive substances reabsorbed by tubular activity. Most of the filtered volume with its contained solutes is restored to the blood prior to its exit via the renal vein. As the filtrate courses through the proximal

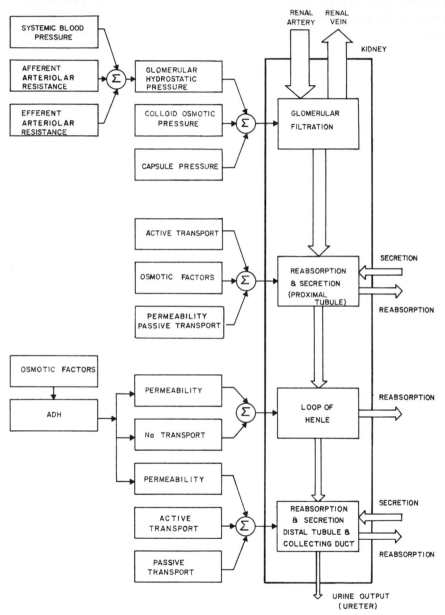

Figure 20-4
Functional operation of the kidney. (See text for explanation.)

convoluted tubule, the loop of Henle, the distal convoluted tubule, and the collecting duct, it is subject to various reabsorptive and secretory processes. These processes result in the ultimate production of the final urine, which is greatly changed in composition and reduced in volume as compared to the initial filtrate. The magnitude of these processes for some plasma constituents is listed in Table 20-1.

Table 20-1. Average Kidney Processing of Various Plasma Compounds in the Normal Human Adult

Substance	Per 24 Hours			
	Filtered	Reabsorbed	Secreted	Excreted
Water (ml)	180,000	179,000	–	1000
Sodium (mEq)	26,000	25,850	–	150
Chloride (mEq)	18,000	17,850	–	150
Bicarbonate (mEq)	4900	4900	–	–
Potassium (mEq)	900	900	100	100
Urea (mM)	870	460	–	410
Glucose (mM)	800	800	–	–
Uric acid (mM)	50	45	–	5
Creatinine (mM)	12	1	1	12

REFERENCES

1. Doyle, J. E. *Extracorporeal Hemodialysis Therapy in Blood Chemistry Disorders.* Springfield, Ill.: Thomas, 1962.

SELECTED READING

Atherson, J. C. Renal physiology. *Am. J. Med. Sci.* 263:335, 1972.
Ganong, W. F. *Review of Medical Physiology,* Ch. 38. Los Altos, Calif.. Lange Medical Publications, 1971.
Stacy, R. W., and Santolucito, J. A. *Modern College Physiology.* St. Louis: Mosby, 1966.

Renal Dynamics

21

Joel B. Mann

INTRODUCTION

Urine formation is the result of three separate processes: glomerular ultrafiltration, tubular reabsorption, and tubular secretion. An understanding of these processes and the methods by which they are measured is fundamental to the clear comprehension of the kidney's role in the maintenance of homeostasis.

This chapter deals with the mechanisms by which the nephron performs these processes and uses them to produce a concentrated urine. First, the nature of the filtration processes at the glomerulus and the corresponding filtration-rate measurement techniques are discussed. Next, the active and passive transport mechanisms involved in reabsorption from and secretion into the renal tubular lumen are examined. Finally, a mathematical treatment is employed to describe the processes of filtration, reabsorption, and secretion.

GLOMERULAR FILTRATION

CONCEPT OF CLEARANCE RATE

The term *clearance*, C, is a measure of the rate at which a substance is removed from the blood and excreted into the urine, and it is defined as the volume of plasma completely cleared of that substance per unit time. Alternatively, the clearance may be thought of as the least volume of plasma that would contain the amount of substance actually found to be excreted per unit time. For example, if, by actual measurement, we find that urine is being produced at the rate of 1 ml/min and the concentration of creatinine in that urine is 1 mg/ml, it is obvious that the kidneys are excreting creatinine at the rate of 1 mg/min. This may be expressed as:

$$\text{Excretion rate} = U_{creat} \; \dot{V} = (1 \text{ mg/ml}) \times (1 \text{ ml/min}) = 1 \text{ mg/min}$$

where C_{creat} = urine creatinine concentration (mg/ml)
$\dot{V}$ = urine flow rate (ml/min)

If the simultaneously determined plasma concentration of creatinine is 1 mg/100 ml, then at least 100 ml of plasma had to pass through the kidney in one minute to yield the 1 mg of creatinine that was actually excreted, and the creatinine clearance, C_{creat}, is therefore 100 ml/min.

Since it is entirely reasonable that all the creatinine may not be removed from each milliliter of plasma, there exists the possibility that a greater volume of blood was cleared of a smaller fraction of its creatinine. Since we do not know *a priori* which of these situations obtains, we can conclude only that at least the smaller volume had to pass through the kidney, and this volume, a virtual rather than a real volume, is therefore employed in the definition of the creatinine clearance rate.

The general clearance rate formula is:

$$\text{Clearance of } x = C_x = \frac{(c_x)_u \dot{V}}{(c_x)_{pl}}$$ (21-1)

where $(c_x)_u$ = concentration of x in urine (usually mg/ml)

$\dot{V}$ = urine flow rate (ml/min)

$(c_x)_{pl}$ = plasma concentration of x (mg/ml)

MEASUREMENT OF GLOMERULAR FILTRATION RATE

Any substance that is ideally suited for the determination of the glomerular filtration rate (GFR) must exhibit the characteristics shown in Table 21-1. Since such a substance would gain access to the urine solely by filtration in a concentration equal to that of plasma water and since it would be neither reabsorbed nor secreted by tubular

Table 21-1. Characteristics of Substances Used for the Determination of Glomerular Filtration Rate

Readily filterable (i.e., small molecular size)
Not reabsorbed or secreted by the kidney
Not stored in the kidney
Not metabolized
Does not become protein-bound
Nontoxic
Does not influence renal function

activity, its rate of filtration, $(\dot{Q}_x)_F$ (mg/min), must be equal to its rate of excretion, $(c_x)_u \dot{V}$, where $(c_x)_u$ is the urinary concentration of x (mg/ml) and $\dot{V}$ is the urine flow rate (ml/min). Furthermore, as for any filterable substance, the filtration rate of x equals the glomerular filtration rate times the plasma concentration of x, $(c_x)_{pl}$. Therefore,

$$(\dot{Q}_x)_F = (c_x)_u \dot{V} = \text{GFR} (c_x)_{pl}$$ (21-2)

and

$$\frac{(c_x)_u \dot{V}}{(c_x)_{pl}} = \text{GFR}$$

Since $(c_x)_u \dot{V}/(c_x)_{pl} = C_x$, then C_x = GFR, i.e., the clearance rate of this substance is equal to the glomerular filtration rate.

Inulin, a polysaccharide with a molecular weight of 52,000, is generally accepted as meeting all of the listed criteria in Table 21-1. Since inulin appears in the glomerular filtrate in a concentration equal to that of the plasma water and since it is neither reabsorbed from the tubules nor added to the tubular fluid by the process of tubular secretion, the amount appearing in the urine must be exactly equal to the amount filtered. If inulin then appears in the bladder at the rate of 125 mg/min and its plasma concentration is 1 mg/ml, then 125 ml of plasma must have been filtered through the glomeruli each minute. Given that $(c_{inulin})_u$ = 125 mg/ml, $\dot{V}$ = 1.0 ml/min, and $(c_{inulin})_{pl}$ = 1.0 mg/ml, then the calculation of C_{inulin} using Equation 21-1 yields a rate of 125 ml/min.

There are several lines of evidence supporting the use of inulin to determine the glomerular filtration rate: (1) the inulin molecule has a radius of 1.5 nm, as determined by the Einstein-Stokes equation, and would therefore be unlikely to back-diffuse passively through the tubular epithelium; (2) it is not metabolized, and any amount given can be completely recovered in the urine following parenteral administration; and (3) for any substance that is freely filterable and neither secreted nor reabsorbed, each increment in the amount filtered (or each increment in the plasma inulin concentration, assuming a constant filtration rate) must be paralleled by an exactly equal increment in the amount excreted. Therefore, the rate of excretion of such a substance must be directly proportional to and a linear function of its plasma concentration. It immediately follows that the clearance of such a substance is constant and is totally independent of its plasma concentration. Inulin behaves in exactly this fashion, as illustrated in Figure 21-1.

OTHER CLEARANCE PHENOMENA

Any substance that is both freely filtered at the glomeruli and actively secreted by the renal tubules should be excreted at a greater rate than inulin or any other substance

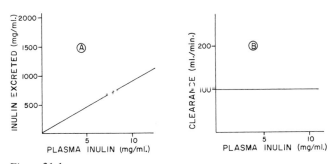

Figure 21-1
Inulin excretion and clearance as a function of plasma inulin concentration.

that gains access to the urine solely by glomerular filtration. Since the excretory rate, $(c_x)_u \dot{V}$, of such an agent is greater for any plasma concentration, $(c_x)_{pl}$, its clearance is also greater. Tubular secretory systems, however, become saturated readily, and increments in plasma concentration beyond the point of tubular saturation will produce increases in excretion due solely to incremental changes in the amount filtered. As the plasma concentration continues to increase, the amount filtered becomes an increasingly greater proportion of the total amount excreted, while the amount secreted becomes a continuously decreasing proportion. As the plasma level approache infinity, the clearance of any secreted substance will asymptotically approach the inulin clearance value (or the clearance value of any substance excreted solely by filtration). At an infinitely high plasma level, the contribution of tubular secretion would be infinitesimal, and the clearance would be equal to the rate of glomerular filtration. This phenomenon is known as the *self-depression of clearance* and is illustrated in Figure 21-2 with *p*-aminohippuric acid (PAH), a substance that is both filtered and secreted.

Exactly the opposite phenomenon occurs with substances that are known to be reabsorbed by the renal tubules. At low plasma levels, the tubules may completely reabsorb the substance in question (e.g., glucose), but at higher levels, as more and more is filtered, the tubular reabsorptive mechanism becomes saturated and the glucose begins to appear in the urine. Further increments in the filtered load that are produced by increasing the plasma level produce increasing excretion. The amount reabsorbed, which has reached a constant, maximum level, becomes an increasingly smaller fraction of the total amount excreted until, at infinitely high plasma glucose levels, the clearance of glucose is related solely to its rate of filtration, and the reduction in clearance due to tubular reabsorption becomes negligible. Again, under these conditions, the clearance of glucose should approach the inulin clearance value, just as the PAH clearance did, if inulin is a valid measure of filtration rate. Figure 21-3 shows that this is indeed the case.

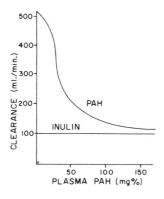

Figure 21-2
Self-depression of clearance.

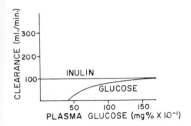

Figure 21-3
Glucose clearance phenomenon.

TUBULAR REABSORPTION

After the filtration procedure just described, the next step in renal processing is the tubular reabsorption of the bulk of the filtered water and solutes. *Reabsorption* involves the transfer of water and solutes from the tubular lumen across the tubular cell, into the interstitial fluid, and ultimately back into the blood. The importance and magnitude of reabsorption is made evident by examination of specific examples.

Of the 180 liters of filtrate produced each 24 hours, only 1 to 2 liters are actually excreted. Furthermore, of the 600 gm of sodium filtered, less than 0.3% is usually excreted, and the urine usually contains none of the approximately 200 gm of glucose, 100 gm of free amino acids, and varying quantities of other substances, which are totally reabsorbed.

Reabsorption is accomplished through a variety of transport mechanisms, both passive and active. These mechanisms and their operating principles form the topic of this section.

PASSIVE REABSORPTION

Passive reabsorption occurs whenever a substance moves from an area of high concentration to an area of lower concentration, i.e., down its electrochemical gradient. Water, urea, and probably chloride ions are passively reabsorbed within the renal tubule. Although no energy expenditure is involved directly in the reabsorption of these substances, the gradients down which they diffuse are established primarily by the energy expended on the active reabsorption of sodium.

When urine flows are low (less than 2 ml/min), the clearance rate of urea is very sensitive to changes in this flow, whereas at higher flow rates (more than 2 ml/min), the urea clearance becomes much less sensitive. This variation of urea clearance with urine flow is evidence for the passive reabsorption of urea. Further evidence for such passive reabsorption is provided by the observation that the fraction of filtered urea reabsorbed — which is expressed as:

$$\text{Fraction of urea reabsorbed} = 1 - \frac{c_{urea}}{c_{inulin}} \qquad (21\text{-}3)$$

where c_{urea} is the concentration of urea and c_{inulin} is the concentration of inulin in urine — remains unchanged with variations in filtered load exceeding two orders of magnitude. This strongly suggests that no limit for urea reabsorption exists, and that the reabsorption is achieved through passive diffusion rather than by energy-demanding mechanisms.

Reabsorption of chloride takes place passively in the proximal tubule and actively in the distal tubule. The passive transport that occurs in the proximal tubule, however, accounts for the greater portion of the reabsorption.

ACTIVE REABSORPTION

Active reabsorption is mediated through the action of discrete mechanisms. Many mechanisms may function to reabsorb more than one substance; the sharing of a single mechanism by a variety of substances is usually determined by certain chemical similarities among these substances. For example, hexose and pentose sugars such as glucose, fructose, and xylose share a common mechanism; saturation of this transport mechanism by any one of the sugars will result in the appearance of the other sugars in the urine. Similarly, the amino acids arginine, ornithine, lysine, and cystine have a common transport mechanism, and saturation of it by any one of these amino acids will result in the appearance of the others in the urine. Other basically discrete mechanisms may share one or two common steps in transport.

In addition to the simple reabsorption of a substance, exchange mechanisms also exist within the tubule, e.g., the exchange of H^+ and K^+ for Na^+ in the distal tubule. A transport mechanism is *active* if it transports a substance against an electrochemical gradient, i.e., against a gradient of electrical potential, of simple chemical concentration, or both. Any ion, for example K^+, will distribute itself across a membrane in relation to the potential difference that exists across that membrane according to the reasonably quantitative relation described by the simplified Nernst equation:

$$V = 61 \log \frac{[K^+]_{int}}{[K^+]_{ext}} \tag{21-4}$$

where $[K^+]_{int}$ and $[K^+]_{ext}$ are respectively the potassium concentrations on the internal and external sides of the membrane and V is the potential difference in millivolts. Ionic concentrations deviating from those predicted by the equation may be assumed to have been brought about by active transport mechanisms.

Active tubular reabsorption operates according to two separate mechanisms, one being transport-limited and the other gradient-time-limited. The transport-limited mechanisms are fixed in their reabsorptive capacity and can only transport a finite and limited amount of solute per unit time. This *maximum transport rate, T_{max},* can be measured when the tubular cells are presented with solute quantities greater than those they can transport. Under these circumstances, each increment in plasma concentration will produce a corresponding increment in urinary excretion.

The maximum transport rate is equal to the difference between the filtration rate and the excretion rate, or

$$T_{max} = \dot{Q}_F - \dot{Q}_E \tag{21-5}$$

where $\dot{Q}_F$ = filtration rate (mg/min)
$\dot{Q}_E$ = excretion rate (mg/min)

This concept is illustrated in Figure 21-4, where the reabsorption rate is shown to increase linearly and congruently with the filtration rate until the T_{max} level is reached. At this point, the tubule has reached its maximum transport capability, and no further increase in the reabsorption rate takes place despite an increase in the filtration rate.

Once the T_{max} is reached, the renal tubules can no longer increase their absorption rate, and any excess solute present will be excreted in the urine. The plasma concentration at which T_{max} is reached is called the *renal plasma threshold.* When this level is exceeded, each incremental increase of the substance under consideration in the blood produces a corresponding incremental increase in the urinary excretion of that substance. In actuality, the T_{max} point does not occur as the sharp break shown by the solid line in Figure 21-4, but it exhibits the rounded characteristic of the dashed lines shown. The smoothing out of this sharp transition is called the *splay* of the titration curve.

Although there are several explanations for the origin of this splay, the simplest is based on the concept of glomerulotubular balance. This concept states that for a variety of reasons, the volume rates of filtration at each glomerulus are not identical but are distributed about some statistical mean. Thus, each tubule will become saturated (T_{max}-limited) at a slightly different level of plasma concentration of the substance in question, and the overall effect will be the gradual transition shown by the dashed line in Figure 21-4.

Glucose, phosphate, and certain amino acids are a few of the substances that are reabsorbed by transport-limited mechanisms. With normal kidney function and normal plasma concentrations, most of these substances are completely or almost completely reabsorbed. Thus, the final urinary concentration of the substance is less than the

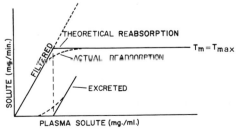

Figure 21-4
Transport-limited characteristic of reabsorption ($T_m = T_{max}$ = maximum transport rate).

plasma concentration, and the reabsorption is therefore active since it occurs against a concentration gradient.

The time-gradient mechanism limits transport according to the electrochemical gradient that can be established from the tubular lumen to the interstitial fluid during the solute transit period. This type of system does not transport substances completely, even for small quantities, and it does not exhibit the sharp transition characteristics of the transport-limited mechanism just discussed. The principal substances that are reabsorbed by gradient-time-limited mechanisms are sodium, bicarbonate, and, to a lesser extent, chloride [2, 3]. Hydrogen ion secretion is also gradient-time-limited.

TUBULAR SECRETION

Tubular transport may occur from the tubular lumen to the peritubular fluid as well as in the reverse direction. The former is termed *tubular reabsorption,* the latter, *tubular secretion.* The transport mechanisms operative in secretion are comparable to those involved in reabsorption and constitute passive secretion, maximum-transport-limited secretion, and gradient-time-limited secretion.

PASSIVE SECRETION

As with passive reabsorption, passive secretion consists of transport down an electrochemical gradient. In this case, however, the electrochemical gradient and the transport direction are from the peritubular fluid to the tubular lumen. The most significant substances that are passively secreted are potassium and several of the weak acids and bases.

Potassium is actively reabsorbed in the proximal tubules and is both passively secreted into and actively reabsorbed from the distal lumen. Nearly all of the 900 mEq of potassium that is filtered per day is thought to be reabsorbed by the proximal tubules, whereas the 100 mEq that is excreted daily into the urine is derived primarily from passive distal tubular secretion. Potassium balance is maintained, however, since the amount excreted is approximately equal to the normal daily intake, and the passive excretion rate will increase if the potassium intake is sustained above normal levels for a long time span. The details of potassium transport are discussed in Chapter 22.

Many weak bases, such as quinine, procaine, and ammonia, readily diffuse through the renal tubular cell and can easily move down a concentration gradient from the peritubular fluid to the tubular lumen. The excretion of these weak bases increases as the pH of the luminal fluid decreases by means of a mechanism called *ionic trapping* or *diffusion trapping.* Using ammonia as an example of a weak base, ionic trapping may be explained by the following analysis. The Henderson-Hasselbalch equation may be written in the form:

$$pH = pK_a + \log \frac{[\text{base}]}{[\text{acid}]} \qquad (21\text{-}6)$$

Since the pK_a of ammonia is 9.35, ammonia exists in the blood as an equilibrium mixture of ammonium ion, NH_4^+, and free ammonia, NH_3, in a ratio of approximately 100:1. As the pH of the luminal fluid is lowered by tubular secretion of H^+ to 5.35, for example, the ratio of NH_4^+ to NH_3 is increased to 10,000:1. The NH_4^+ is relatively lipid insoluble and cannot diffuse through the cell membrane nearly as readily as the free NH_3, and it hence becomes trapped in the tubular lumen. Thus, as NH_3 diffuses into the tubular lumen, it becomes converted to the ionic species, which results in the maintenance of a concentration gradient for free NH_3 that favors its continued, luminally directed diffusion. Although this mechanism is passive, energy must be supplied to transport H^+ actively into the luminal fluid, so that ionic trapping may continue. As with passive reabsorption, there is no limit to passive secretion, and, as the flow rate increases, secretion increases.

There is some evidence, on the other hand, that weak bases may be secreted by the proximal tubule, and that an alkaline urine in the distal tubule will increase the concentration of the nonionized molecular species and permit back-diffusion, perhaps even resulting in net reabsorption. If, however, an acid urine is present in the distal tubule, it will maintain a high concentration of the weakly diffusible ionized species, the effects of the secretion by the proximal tubule will be maintained, and net secretion will occur.

Ionic trapping of weak acids occurs in a similar fashion but in a reversed direction, i.e., acid excretion is increased with an alkaline pH of the luminal fluid and decreased with an acid pH. This phenomenon is of some clinical importance in the treatment of patients with certain drug intoxications, especially those due to salicylate and phenobarbital, since their excretion can be considerably enhanced by alkalinization of the urine.

TRANSPORT-LIMITED MECHANISMS

In contrast to the multiple maximum-transport-limited reabsorptive mechanisms, there are only three T_{max}-limited secretory mechanisms, which, in man, are primarily involved in the transport of foreign substances rather than naturally occurring substances. Since these mechanisms are identical, differing only with respect to the class of compound transported, the transport characteristics of p-aminohippurate (PAH), the best studied of this group, will be utilized in the description of this system.

Calculation of the Secretory Rate

Since substances secreted by the tubules are also filtered, the total rate of excretion, $(c_x)_u \dot{V}$, must equal the rate of filtration, $C_{inulin} (c_x)_{pl}$, plus the rate of tubular secretion, $(\dot{Q}_x)_S$. Rearranging,

$$(\dot{Q}_x)_S = (c_x)_u \dot{V} - (c_x)_{pl} C_{inulin} \tag{21-7}$$

Since many of these substances are variably bound to plasma proteins, and since the concentration found in the plasma ultrafiltrate is not equal to that in the plasma water but rather to the concentration of the unbound substance in plasma water, the value of the ratio $[(c_x)_{pl}]_F / [(c_x)_{pl}]_T$, or briefly, $c_{F/T}$ (the ratio of filterable to total plasma concentration), should be substituted for the value of $(c_x)_{pl}$ in the above equation. Furthermore, the plasma concentration should be corrected to the concentration in plasma water by dividing by the factor W, which is the fraction of plasma that is water and is normally 0.93 to 0.94.

Figure 21-5 depicts the general characteristics of the transport mechanism involved in the secretion of PAH. Over a range of plasma concentrations from 1 mg/100 ml to approximately 7 or 8 mg/100 ml of filterable (unbound) PAH, the amounts excreted, secreted, and filtered increase linearly in proportion to the increases in the plasma level. At these lower concentrations, for any given concentration of plasma PAH, the amount excreted equals the sum of the secreted PAH plus the filtered PAH. At plasma PAH levels above approximately 10 mg/100 ml, the T_{max} for PAH is obtained and there are no further increments in PAH secretion. At this point, the slope of the curve for excreted PAH decreases and becomes equal to that for filtered PAH. The curve for secretion intersects that of filtration at the plasma concentration level of approximately 65 mg/100 ml; at lower plasma levels, tubular secretion accounts for the major portion of urinary PAH, whereas at higher values, filtration provides the major portion.

If the plasma concentration of PAH is maintained at levels below the T_{max} — usually in the range of 4 to 6 mg/100 ml — then about 90% of the PAH is removed from the blood in one pass through the kidney, and its clearance can therefore be used to compute a reasonable approximation of renal plasma flow for clinical purposes.

Secretory Clearance

The concept of clearance applies not only to substances such as inulin, which are solely filtered, but also to substances that are secreted. The total clearance of such

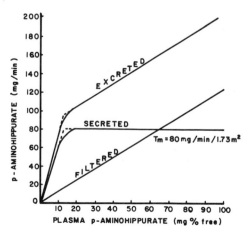

Figure 21-5
Transport-limited characteristic of secretion ($T_m = T_{max}$ = maximum transport rate).

substances is composed of two components: the glomerular clearance and the secretory clearance.

Except for the expanded scale, the framework of reference of Figure 21-2 is essentially identical to that of Figure 21-5. At plasma PAH levels up to 7 or 8 mg/100 ml, the PAH clearance is constant, with a rate of 660 ml/min. Since at these lower levels, the rate of excretion, $(c_{PAH})_u \dot{V}$, varies directly with the plasma concentration, $(c_{PAH})_{pl}$, then the total clearance, $[(c_{PAH})_u/(c_{PAH})_{pl}] \dot{V}$, is constant.

The glomerular clearance rate is equal to the rate of filtration of PAH — $C_{inulin} W c_{F/T} (c_{PAH})_{pl}$ — divided by the plasma concentration, $(c_{PAH})_{pl}$, and of course it is therefore constant at all plasma levels. At concentration levels below approximately 8 mg/100 ml, the tubular clearance rate remains constant, since the rate of secretion increases linearly with increases in plasma PAH (see Fig. 21-5). With plasma levels above approximately 10 mg/100 ml, the secretory transport system is saturated and the rate of secretion remains constant at the value of T_{max} (see Fig. 21-5). With a constant secretory rate, the tubular clearance rate must progressively decrease as the plasma concentration increases. Figure 21-2 shows the result of these changing relationships. As stated at the outset, the total clearance equals the glomerular clearance plus the tubular clearance:

$$\frac{(c_{PAH})_u \dot{V}}{(c_{PAH})_{pl}} = \frac{C_{inulin} W c_{F/T} (c_{PAH})_{pl}}{(c_{PAH})_{pl}} + \frac{(T_{PAH})_{max}}{(c_{PAH})_{pl}} \tag{21-8}$$

As $(c_{PAH})_{pl}$ increases without bound, the second term on the right side of the equation approaches zero since $(T_{PAH})_{max}$ is constant, and the total clearance rate approaches the glomerular clearance rate, $C_{inulin} W c_{F/T}$.

GRADIENT-TIME-LIMITED SECRETION

This tubular secretory mechanism is primarily involved in the secretion of hydrogen ions. Since all segments of the renal tubule are able to produce a tubular fluid with a lower pH than the surrounding interstitial fluid, all portions of the tubule are able to transport H^+ against an electrochemical gradient, i.e., engage in active transport. The proximal tubule, however, is able to maintain a minimum luminal pH of only 6.8, whereas the distal tubule is able to lower the luminal pH to approximately 5.0. If sufficient buffer is provided to prevent the pH of the luminal fluid from dropping below these limits and thereby keep the gradient within the capacity of the appropriate tubular cell, there is no absolute limit to the rate of H^+ secretion.

SUMMARY OF TUBULAR TRANSPORT RELATIONS

Figure 21-6 depicts the filtration, reabsorption, and secretion transport flow routes in the nephron. The following situations illustrate the contribution of these transport mechanisms to the final excretion rate.

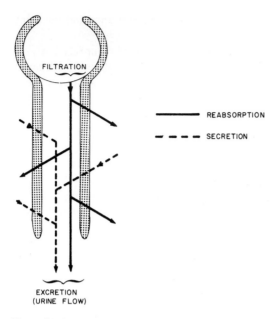

Figure 21-6
Reabsorption, secretion, and excretion flow paths in the nephron.

1. *No reabsorption or secretion.* In this case the excretion rate, $\dot{Q}_E$, is equal to the filtration rate, since none of the filtered substance is lost from or added to the tubular lumen. The following relations hold:

$$(\dot{Q}_x)_E = \text{GFR} = \frac{(c_x)_u \dot{V}}{(c_x)_{pl}} \tag{21-9}$$

where $(c_x)_u$ = urinary concentration of x (mg/ml)
$\quad\ \ (c_x)_{pl}$ = plasma concentration of x (mg/ml)
$\quad\ \ \dot{V}$ = urine flow rate (ml/min)

In this case, the rate of transport of substance x across the tubule walls is equal to zero. Inulin is a substance that exhibits these characteristics.

2. *Filtration and reabsorption; no secretion.* In this situation, the rate of excretion is equal to the filtration rate minus the reabsorption rate, $\dot{Q}_R$, or,

$$(\dot{Q}_x)_E = \text{GFR}\,(c_x)_{pl} - (\dot{Q}_x)_R \tag{21-10}$$

Glucose is a substance that is reabsorbed by the tubules but not secreted.

3. *Filtration and secretion; no reabsorption.* This case is similar to the above example except that now the transport is in the reverse direction, and the excretion rate becomes:

$$(\dot{Q}_x)_E = GFR\,(c_x)_{pl} + (\dot{Q}_x)_S \qquad\qquad (21\text{-}11)$$

where $\dot{Q}_S$ = the secretion rate.

4. *Filtration, reabsorption, and secretion.* With both transport mechanisms in operation, the excretion rate is:

$$(\dot{Q}_x)_E = GFR\,(c_x)_{pl} - (\dot{Q}_x)_R + (\dot{Q}_x)_S \qquad\qquad (21\text{-}12)$$

The two terms, $(\dot{Q}_x)_R$ and $(\dot{Q}_x)_S$, can be combined into one term, T_x, denoting the transport rate:

$$T_x = (\dot{Q}_x)_S - (\dot{Q}_x)_R \qquad\qquad (21\text{-}13)$$

Net reabsorption occurs when $(\dot{Q}_x)_R > (\dot{Q}_x)_S$ and $T_x < 0$; net secretion occurs when $(\dot{Q}_x)_S > (\dot{Q}_x)_R$ and $T_x > 0$. If $(\dot{Q}_x)_R = (\dot{Q}_x)_S$, the net tubular transport is zero, and the excretory mechanism cannot be distinguished from that of filtration alone, unless some chemical or physiologic manipulation is utilized to alter $(\dot{Q}_x)_R$ or $(\dot{Q}_x)_S$ separately, thereby unmasking the tubular component. Potassium excretion exemplifies this excretory mechanism [1].

REFERENCES

1. Giebisch, G., Klose, R. M., and Malnic, G. Renal Tubular Potassium Transport. *Proc. III Int. Cong. Nephrol.* Washington, D.C., 1966.
2. Rector, F. C., Jr., Bloomer, H. A., and Seldin, D. W. Effects of potassium deficiency on the reabsorption of bicarbonate in the proximal tubule of the rat. *J. Clin. Invest.* 43:1976, 1964.
3. Rector, F. C., Seldin, D. W., Roberts, A. D., Jr., and Smith, J. S. The role of plasma CO_2 tension and carbonic anhydrase activity in the renal reabsorption of bicarbonate. *J. Clin. Invest.* 39:1706, 1960.

SELECTED READING

Coelho, J. B., et al. Measurement of single nephron glomerular filtration rate without micropuncture. *Am. J. Physiol.* 223:832, 1972.
Gamble, J. L. *Extracellular Fluid.* Cambridge, Mass.: Harvard University Press, 1960.
Guyton, A. C. *Textbook of Medical Physiology,* 4th ed. Philadelphia: Saunders, 1971.

Materson, B. J. Measurement of glomerular filtration rate. *CRC Crit. Rev. Clin. Lab. Sci.* 2:1, 1971.

McSherry, E., et al. Renal tubular acidosis in infants. *J. Clin. Invest.* 51:499, 1972.

Pitts, R. F. *Physiology of the Kidney and Body Fluids.* Chicago: Year Book, 1968.

Russell, C. D. Response to bicarbonate in severe acidosis. *N. Engl. J. Med.* 289:755, 1973.

Smith, H. W. *The Kidney.* New York: Oxford University Press, 1951.

Winton, F. R. *Modern Views on the Secretion of Urine.* Boston: Little, Brown, 1956.

Tubular Transport of Water and Electrolytes

22

Joel B. Mann

INTRODUCTION

Although the mechanisms of electrolyte transport in the kidney have been studied extensively, there is still no universally accepted hypothesis to account for all the phenomena observed. The mechanisms are complicated by the fact that the various ions involved are frequently coupled to one another, the transport of water is also coupled to ion transport, and the transport activity of any given segment is influenced by the simultaneous operation of multiple hormonal, hydrodynamic, and chemical factors.

This chapter attempts to expand the concepts of tubular ion and water transport mechanisms that have previously been introduced. The following discussion is primarily didactic in nature and does not attempt to present a detailed explanation of the various discrepancies in transport phenomena that have been observed. Furthermore, significant variations have been found among different species; therefore, the extrapolation of these findings to man may not be entirely valid.

WATER TRANSPORT

Approximately 80% of the filtered water is obligatorily reabsorbed within the proximal tubule. This reabsorption is passive, following the osmotic gradient established by the reabsorption of solutes. Such obligatory reabsorption is not influenced by control mechanisms such as antidiuretic hormone (ADH) secretion, and it occurs spontaneously at a more or less constant rate and with no change in osmotic concentration. The remaining filtrate is subject to variable (facultative) reabsorption, it is influenced by multiple regulatory mechanisms as it progresses through the nephron, and it undergoes considerable processing before concentrated urine is ultimately produced. There are a number of mechanisms mediated by the kidney that are responsible for the regulation of the volume and concentration of the urine.

COUNTERCURRENT MULTIPLICATION AND EXCHANGE

The hairpin-shaped loop of Henle provides a means for the generation of a concentration gradient by utilizing a principle called *countercurrent multiplication.* The establishment and maintenance of this gradient is assisted by the presence of a *countercurrent exchange* process that takes place in the vasa recta (the small blood vessels paralleling the loop of Henle).

The process of countercurrent multiplication is initiated when sodium is actively reabsorbed from the ascending limb of Henle's loop and enters the interstitial space of the medulla (Fig. 22-1). Normally, water does not accompany the sodium due to the impermeability of the ascending limb; therefore, the luminal fluid becomes increasingly hypotonic as it flows through the ascending limb of the loop. The accumulation of sodium in the interstitial space surrounding the loops of Henle results in the establishment of a gradually increasing hypertonicity from the top to bottom of the loops, with an osmotic gradient extending from the nearly isotonic corticomedullary junction to the maximally hypertonic renal papillae.

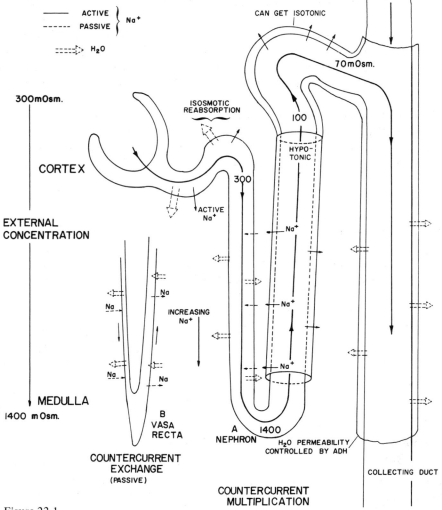

Figure 22-1
Countercurrent multiplication and exchange. *A.* Countercurrent multiplication. The double-layered wall of the ascending limb of Henle's loop denotes both its impermeability to water and its active Na^+ transport as well as serving to distinguish it from the thin-walled descending limb. *B.* Countercurrent exchange in the vasa recta.

The osmolality of the fluid in the descending limb increases progressively as it moves toward the tip, as shown in Figure 22-1. This increase is produced both by the passive movement of sodium from the hypertonic interstitium into the descending limb and by the passive movement of water down its concentration gradient from the descending limb into the hypertonic interstitium. This passive transfer results in the sequestration of sodium chloride in the renal medulla and the diversion of water away from the maximally hypertonic papillae. Recent evidence has indicated that urea is also of considerable importance in the operation of the countercurrent mechanism.

With the establishment of a concentration gradient, the crucial role of the distal tubule and the collecting duct in the formation of a concentrated urine comes into play. As previously discussed, the filtrate entering the distal convoluted tubule is hypotonic due to the active reabsorption of sodium in the ascending limb. Continued active reabsorption of sodium in the distal tubule tends to lower the osmolality of the intraluminal fluid even further. Consequently, if the distal tubule and collecting duct are impermeable to water, a very dilute urine may be formed with an osmolality as low as 70 mOsm/kg. However, if water permeability is high, the luminal fluid attains near equilibrium with the interstitial fluid, and an osmolality of approximately 300 mOsm/kg is found within the distal tubule. There is further progressive concentration due to simple outward passive diffusion of water down its osmotic gradient as the luminal fluid traverses the collecting duct and achieves osmotic equilibrium with the hypertonic interstitial fluid. When maximally concentrated, the urine osmolality in man may approach 1400 mOsm/kg. Thus, urine concentration can be controlled by adjusting the water permeability of the luminal walls.

The concentration of the urine is regulated by the circulating levels of antidiuretic hormone (ADH) in the blood, since the water permeability of the distal tubule and collecting duct is controlled by the concentration of ADH present in the plasma. Increasing quantities of ADH make these membranes increasingly permeable to water. Under normal circumstances, the ADH level is sufficient to adjust the membrane permeability so that the luminal fluid is nearly isosmotic by the time it enters the collecting duct. In addition, water reabsorption in the distal tubule results in a sizable fluid volume decrease between the loop of Henle and the collecting duct. As the remaining fluid flows toward the renal pelvis through the collecting duct, it enters areas of ever-increasing concentration that have been established by the countercurrent multiplication mechanism. Since the permeability to water of the collecting duct is maintained by the presence of ADH, urinary concentration is achieved by passive water diffusion out of the duct.

Although Henle's loop in its function as an active countercurrent multiplication system establishes and maintains the medullary hypertonicity required for the formation of a concentrated urine, this hypertonicity might quickly be dissipated were it not for the vasa recta functioning as a passive countercurrent exchange system to conserve the medullary hypertonicity and to decrease the total energy required to maintain it.

Figure 22-1B represents a capillary loop such as those of the vasa recta. As blood enters the top of the loop and descends into the ever-increasing hypertonicity of the medullary interstitium, it tends to approach osmotic equilibrium by the passive transfer of water out of the capillary loop and the equally passive transfer of osmotically active substances into the descending limb. It then reaches maximum hypertonicity at the tip. In the right-hand ascending limb of the loop, since the blood remains slightly hypertonic to the surrounding interstitium, water is osmotically transferred passively from the interstitium into the blood, and the osmotically active substances, e.g., sodium and urea, are similarly transferred down their concentration gradient into the interstitial fluid. Water is, in effect, shunted across the top of the loop, and, as shown in Figure 22-1B, the effective fluid flow is decreased with respect to the dissipation of the medullary hypertonicity that is established by the operation of the countercurrent multiplication system.

Measurement of Urinary Concentration and Dilution

For purposes of explanation only, let us assume that the production of a dilute urine is achieved by the addition of pure water to the isotonic glomerular filtrate and that the concentration of the urine is achieved by means of the removal of pure water from this isotonic glomerular filtrate. In other words, urinary concentration and dilution may be considered as deviations of volume in a negative or positive direction, respectively, from the volume required to produce an isotonic urine, i.e., the condition in which $O_u = O_{pl}$, where O_u is the urine osmolarity and O_{pl} is the plasma osmolarity (mOsm/ml).

The *osmolar clearance* may be defined, similarly to other clearances, in terms of the volume of plasma per minute completely cleared of osmotically active solutes and is expressed by the following equation:

$$C_{Osm} = \frac{O_u}{O_{pl}} \dot{V}_u \tag{22-1}$$

When the urine is isotonic, the ratio $O_u/O_{pl} = 1$, $C_{Osm} = \dot{V}_u$, and the entire urine volume is osmotically obligated in the elaboration of an isotonic urine.

If we now consider, for conceptual purposes only, that a hypotonic urine is produced by the addition of pure water to the isotonic filtrate, then the resultant volume is greater than the osmolar clearance and O_u is less than O_{pl}. This volume of water in excess of the volume required to produce an isotonic urine is called *free water*, and its rate of excretion, termed the *free water clearance*, C_{H_2O}, is the difference between the urine flow rate and the osmolar clearance rate:

$$C_{H_2O} = \dot{V}_u - C_{Osm} \tag{22-2}$$

For example, the urine flow may be increased to 15 to 20 ml/min after ingestion of a water load, and the urine osmolarity may drop to 40 to 50 mOsm/liter. The C_{H_2O} would then be:

$$C_{H_2O} = 16 \text{ ml/min} - \frac{(0.040 \text{ mOsm/ml}) (16 \text{ ml/min})}{0.285 \text{ mOsm/ml}} \tag{22-3}$$

C_{H_2O} = 16 ml/min − 2.2 ml/min = 13.8 ml/min

In this example, the osmolar clearance is 2.2 ml/min, i.e., the osmotically active constituents contained in 2.2 ml of plasma are excreted in 16 ml of urine in one minute, in effect removing 13.8 ml of pure water per minute.

With fluid restriction, the urine osmolarity can increase to as much as 1400 mOsm/liter, and the urine flow can be reduced to 0.5 ml/min. Under such circumstances, we might have the following results:

$$C_{H_2O} = 0.5 \text{ ml/min} - \frac{(1.280 \text{ mOsm/ml}) (0.5 \text{ ml/min})}{0.285 \text{ mOsm/ml}} \tag{22-4}$$

C_{H_2O} = 0.5 ml/min − 2.2 ml/min = −1.7 ml/min

Again, the term "free water" denotes that it is solute-free, and its arithmetic sign indicates that it is reabsorbed rather than excreted.

The rate at which the volume of solute-free water is reabsorbed is referred to as $(\dot{Q}_{H_2O})_R$ and is derived by rearranging Equation 22-2:

$$(\dot{Q}_{H_2O})_R = C_{Osm} - \dot{V}_u \tag{22-5}$$

HOMEOSTATIC REGULATORY AND CONTROL MECHANISMS

Isotonicity

The mechanism of maintenance of normal osmotic pressure of the body fluids can be characterized as a servomechanism with positive and negative feedback loops, as shown in the block diagram in Figure 22-2. A plus sign indicates that a positive increment at the location will produce an increase of activity in the function described in the following block. The transfer functions in the various blocks are not well-known, are highly nonlinear, and are not fully understood. Therefore, assigning numerical quantities or equations to these functions will only render inaccurate relationships. The relative changes, however, are known. For example, an increase in the osmolarity of the extracellular fluid will cause an increase in the activity of the osmoreceptors in the hypothalamus.

The osmolarity of the extracellular fluid, and, secondarily, of the total body water, is regulated almost exclusively through the control of water excretion. The osmolar regulatory mechanisms consist of the following components:

1. *Osmoreceptors.* These are the supraoptic nuclei located in the floor of the third ventricle of the brain. Changes in the osmolarity of extracellular fluid produce a shift of water across the membrane with an appropriate volume change, i.e., a decrease in extracellular osmolarity results in water movement into the osmoreceptor with a resultant increase in volume, and, conversely, an increase in extracellular fluid

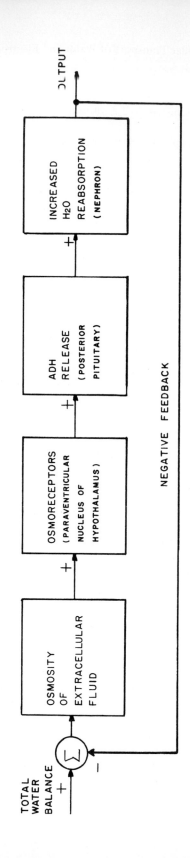

Figure 22-2
Osmolality regulation.

osmolarity initiates water movement out of the osmoreceptor and a concomitant decrease in volume. These receptors are responsive to changes of approximately ±2% in the osmolarity of the extracellular fluid.

2. *The integrative center.* Antidiuretic hormone is released in response to a wide variety of stimuli, such as pain, fear, nicotine, various drugs (e.g., epinephrine and histamine), changes in the osmotic activity of the body fluids, and, under some circumstances, the reduction of extracellular fluid volume. Conversely, ADH release is suppressed by the dilution of the body fluids, as might be expected, as well as after ingestion of alcohol or expansion of the body fluids, especially those that distend the right and left atria. In addition, there seems to be some degree of interplay between aldosterone secretion and renal vasomotor responses with ADH secretion. The proper control of these multiple stimuli implies the existence of some central integrating mechanism, although it has not yet been accurately identified.

3. *The ADH-releasing mechanism.* The integrative mechanism that reacts in response to the various stimuli causes ADH to be released from the posterior pituitary at a rate related to the needs of the moment.

4. *The effector mechanism.* The effector mechanism in this loop is, of course, the kidney, which responds to the level of circulating ADH by conserving or excreting free water. In addition, both dehydration and hypertonicity cause thirst and an increase in the ingestion of water. Presumably, the postulated central integrative mechanism must receive inputs from volume receptors and osmoreceptors, and, after appropriate processing of the signal, a sensation of thirst is produced.

In summary, dilution of the body fluids inhibits ADH secretion. The decrease in the circulating level of ADH causes the distal tubules and collecting ducts of the kidney to become impermeable to water, and the free water clearance is increased. Conversely, dehydration stimulates ADH secretion; in response to an increased circulating level of ADH, the distal tubules and collecting ducts become highly permeable to water, and free water is reabsorbed with a production of a concentrated urine. Thirst is also induced, and, as a result of increased water intake and decreased water excretion, the osmolarity is returned to normal.

Volume Control

Extracellular fluid volume is regulated by the control of both sodium and water excretion. Although the maintenance of body fluid volume is fundamental to the continued existence of the organism, and there are, no doubt, several interrelated mechanisms for the control of this volume, these mechanisms are still poorly defined and controversial. The identification of a supposed volume receptor by one group of investigators, for example, is frequently not confirmed by a second group of investigators. With this preface, we can say that those volume receptors that have been tentatively located in the head are supposedly sensitive to distension, those at the junction of the thyroid and carotid arteries are sensitive to pressure changes (and thereby function as baro-receptors), those in the left atrium are sensitive to the degree of distension, and, finally, those in the juxtaglomerular apparatus of the afferent renal arterioles function as a pressure-sensitive mechanism. The identification of the nature and locus of these variously described volume receptors as well as the location and function of an integrative network are still matters of considerable debate.

There are several renal effector mechanisms that are fairly well-defined and that are related primarily to aldosterone. Volume depletion, which is sensed either through a stretch receptor or baroreceptor located in the afferent arteriole in relation to the juxtaglomerular apparatus, causes this microorgan to release renin. The angiotensin-renin volume-regulating mechanism is shown in Figure 22-3. The renin reaching the bloodstream enzymatically initiates a two-step conversion of angiotensin precursors to angiotensin. The angiotensin thus formed has a dual function:

1. Stimulation of aldosterone release by the adrenal gland. This of course results in an increase in salt and water reabsorption by the kidney, with expansion of the extracellular fluid volume.
2. Vasoconstriction. This serves to increase blood pressure and renal perfusion.

In addition, the action of ADH on the renal tubules regulates the extracellular fluid volume by the pathway shown in Figure 22-4. Under normal or near-normal conditions, ADH has no role in volume regulation. With severe volume depletion, however, the ADH feedback loop illustrated in Figure 22-4 is activated and serves to augment the volume expansion effects of the aldosterone system. Volume receptors located in places such as the atrial wall of the heart influence the ADH output of the posterior pituitary. For example, a large decrease in the extracellular fluid volume has a positive effect on the release of ADH from the posterior pituitary gland. When the increased ADH concentration becomes effective at the site of the renal tubules, more water is reabsorbed and the extracellular fluid volume increases. This, however, is a secondary mechanism; the renin-angiotensin system is the primary mechanism for the maintenance of fluid volume.

TRANSPORT OF ELECTROLYTES

SODIUM ION TRANSPORT

Following filtration, sodium is actively reabsorbed in all portions of the nephron and collecting duct with the exception of the descending loop of Henle, which is impermeable to sodium. During the reabsorption process, electrical charge neutrality must be maintained, and consequently sodium reabsorption is associated with the transport of other ions. Thus, for every sodium ion reabsorbed, an oppositely charged ion (anion) must be reabsorbed or an ion of the same charge (cation) must be secreted

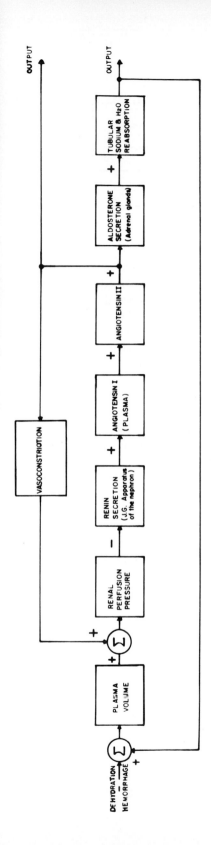

Figure 22-3
Angiotensin-renin volume-regulating mechanism.

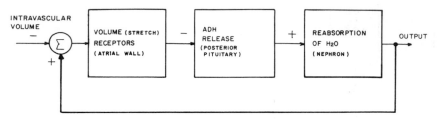

Figure 22-4
Role of ADH in fluid volume regulation.

in order to maintain this macroscopic charge neutrality. The ions most prominently involved in this exchange are the anions chloride and bicarbonate, which are reabsorbed, and the cations potassium and hydrogen, which are secreted.

A cellular structure model that is proposed to explain the reabsorption of sodium and other phenomena occurring in the proximal tubule is presented in Figure 22-5. Here, a tubular cell is shown with its luminal and peritubular membranes. An intercellular space separates adjacent cells and is probably continuous with basal infoldings. The intercellular space virtually disappears at the luminal border, where adjacent cells are connected by tight junctions. These junctions are much shorter in the proximal tubule than in the distal tubule, which perhaps accounts for the much greater permeability to ions and water found in the proximal tubule. Sodium ions, passively accompanied by chloride ions, are pumped into this extracellular, intercellular compartment, where they accumulate and produce a localized hyperosmolarity. This, in turn, results in water movement, probably both across the tight junction and through the cell. If the adjacent cell membranes are relatively inelastic, this local fluid accumulation should produce a small hydrostatic pressure gradient with a resultant movement of fluid from the luminal to the basal portions of this special compartment. As a final step, this fluid is moved across the contiguous capillary wall at a rate determined by the hydrostatic and oncotic pressure differences between the capillary lumen and the intercellular space.

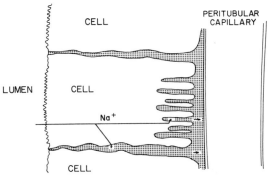

Figure 22-5
Tubular cell model.

There are three major arguments that support the concept of an intercellular compartment and its importance in proximal tubular ion and water reabsorption:

1. Utilizing a rather elaborate micropuncture technique, the electrical resistance across single tubular cell membranes has been measured and compared with the resistance across a segment of intact tubule. The transepithelial electrical resistance as measured between the tubular lumen and the peritubular surface of the intact tubule is several orders of magnitude less than the sum of the resistances of the individual cell membranes. These findings are best explained by a low-resistance shunt path in parallel with the cell membrane resistances, as shown in the electrical equivalent circuit of Figure 22-6.

In this illustration, the potential V_1 is the voltage difference across the peritubular membrane; E_1 represents the electromotive force (emf) generated by the concentration gradients for potassium, chloride, and sodium. The resistance R_1 is the specific ion resistance of the peritubular membrane. The voltage V_2 is the potential difference across the luminal membrane, E_2 represents the emf that is generated by the diffusion potentials across the luminal membrane and contributed to by chloride and sodium ions, and R_2 is the specific ion resistance of the luminal membrane. The voltage V_3 is the transtubular potential difference, and R_3 represents the low-resistance shunt path mentioned previously. The luminal transmembrane potential difference is thus provided to a significant extent by the peritubular emf generator via the extracellular shunt path, R_3. The luminal potential is therefore explained, at least in part, as a simple voltage drop across a shunting resistor.

2. The ionic selectivity of individual tubular cell membranes differs considerably from that of intact proximal tubules. The individual permeability characteristics of the luminal and peritubular cell membranes to a variety of ions can be measured by changes in the electrical potential that are induced across the respective membranes after experimental alteration of the luminal or peritubular ion concentrations. On the basis of such measurements, the tubular cell membranes are found to be quite selective; for example, the peritubular membrane of a single tubular cell is approximately 25 times more permeable to potassium ions than to chloride ions. In the

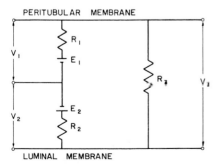

Figure 22-6
Electrical equivalent circuit of the peritubular and luminal membranes.

intact tubule, however, the permeabilities are entirely different, and the permeability ranking within a series of cations and anions is essentially that predicted from ion mobilities in free solution except for some reduction in anion mobility as compared to cation mobility. The transtubular sodium and potassium mobilities are quite similar, in contradistinction to their transmembrane mobilities. Since the intact proximal tubule fails to exhibit the pronounced ion selectivity of its component membranes, some structure other than the cell membranes must be the determining factor in ion movement across the proximal tubular epithelium. The previously described intercellular shunts could well be such an alternate pathway.

3. Depolarization of the peritubular membrane potential difference (see below), which may be induced by a sudden increase in the peritubular potassium concentration, is consistently accompanied by a depolarization of the luminal membrane; similarly, perfusion of the lumen with a sodium solution of low concentration will simultaneously increase both the luminal and peritubular potential differences. This rapid spreading of depolarization either from the peritubular membrane to the luminal membrane or from the luminal membrane to the peritubular membrane provides strong support for the hypothesis of the existence of a low-resistance shunting pathway between the peritubular and luminal surfaces.

Although studies have indicated the presence of active ion transport, a decision as to which ion is actively transported depends upon the identification of the direction and magnitude of the electrical potential difference across specific nephron segments. Utilizing small-diameter capillary electrodes, transtubular potential differences between the lumen and the peritubular fluid of 20 mV have been measured, with the potential at the lumen being negative in relation to that of the peritubular fluid. The peritubular membrane potential has been found to be approximately 70 mV, and the cell interior is negative with respect to the surrounding fluid.

This would seem to indicate that the movement of sodium against an electrical gradient across the proximal tubular epithelium in mammals is not clearly established, but evidence for such active transport may be obtained from stop-flow microperfusion experiments. Under these conditions, the electrochemical potential gradient of sodium can be measured under conditions of zero net flow. Then, Ussing's [2] work equation can be used to calculate the apparent emf of active sodium transport:

$$V_{Na^+} = \frac{RT}{2F} \ln \frac{c_o}{c_i} + V + \frac{RT}{2F} \frac{J_o}{J_i} \tag{22-6}$$

where c_i = concentration inside tubule
c_o = concentration outside tubule
J_i, J_o = respective unindirectional fluxes
V = electrical potential difference

Under experimental conditions of zero net flux and zero electrical gradient, only the first term of the right side of the equation remains, and, when solved, the result indicates an apparent emf of about 10 mV.

CHLORIDE ION TRANSPORT

The bulk of chloride reabsorption takes place passively in the proximal tubule. The potential inside the proximal tubule is about 20 mV and is negative with respect to the interstitial fluid. This electrical gradient allows passive diffusion from the lumen to the interstitial fluid whenever the tubular concentration of Cl^- is greater than 0.55 that of plasma. It may be necessary, however, to postulate the existence of active chloride transport under circumstances where no significant transtubular potential difference is present. Chloride transport, at least in the proximal tubule, may be coupled one-to-one with sodium ion transport by a single carrier.

In the distal tubule, the lumen-to-interstitial space potential difference is about 60 mV, which corresponds to the requirement that the luminal concentration of Cl^- be greater than 0.1 that of plasma in order for passive diffusion to occur.

At present, the exact nature of chloride reabsorption is not entirely clear. Although the chemical concentration gradient for chloride from the lumen to the cell favors chloride reabsorption, it is often too small to balance the intercellular electrical negativity. Possibly, the chloride ions may be transported by a carrier when combined with sodium ion as neutral sodium chloride. Active transport of chloride from the tubular lumen across the membrane into the cell has recently been demonstrated. On the other hand, chloride movement from the cell to the peritubular fluid occurs passively along an electrochemical potential gradient.

POTASSIUM TRANSPORT

Since essentially all the filtered potassium is reabsorbed in the proximal tubule, any potassium that appears in the urine must largely be the result of net secretion from the distal nephron. Although the potassium clearance is usually about 20% of the inulin clearance, various experimental procedures that are designed to enhance potassium excretion — such as potassium loading, urinary alkalinization, or infusions of salts of poorly reabsorbed anions (e.g., ferrocyanide or sulfate) — can increase the potassium/inulin ratio to 1.5, whereas measures designed to decrease potassium excretion can cause the reduction of this ratio to 0.05. Despite this 30-fold range of potassium excretion, proximal tubular potassium reabsorption tends to remain constant at about 85% to 90% of the filtered potassium, which thereby points to the distal nephron as the site that is responsible for the observed changes in urinary potassium excretion. The collecting duct is usually responsible for some potassium reabsorption, although under conditions designed to enhance potassium excretion, the net secretion of potassium from this source has also been observed.

Utilizing the Nernst equation and the average measured distal tubular transepithelial potential difference of 50 mV (the lumen being negative), the calculated equilibrium values for the potassium concentrations are regularly greater than the potassium concentrations actually measured, even at maximal potassium secretion rates. This immediately leads to three important conclusions:

1. Since the luminal potassium concentration is lower than that predicted on the basis of electrochemical potential gradients, there is no evidence for active potassium secretion in the distal tubule or collecting duct, and potassium secretion may be assumed to occur via passive transport.

2. Since under both stopped-flow and free-flow conditions, the luminal fluid potassium concentrations are less than predicted, then active reabsorptive transport must occur, presumably by means of a potassium pump located at the luminal membrane. In other words, purely passive distribution of potassium ion across the luminal membrane in the absence of any active reabsorptive process would result in intratubular potassium concentrations greater than those measured.

3. Net potassium transfer is determined in large part by the magnitude of the transepithelial potential difference. Experimentally, maneuvers designed to reduce intratubular electrical negativity regularly reduce intratubular potassium concentrations, whereas conversely, an increase in intratubular electrical negativity just as regularly increases the intratubular potassium concentration.

Distal tubular potassium transfer can therefore be considered to be regulated by a shift in the balance between a passive potassium leak into the lumen and the activity of the postulated reabsorptive pump. Factors that influence potassium transfer in the distal nephron do so more commonly by altering the passive transfer of potassium than by affecting the potassium pump. The three major factors controlling the electrochemical potential difference, and thereby the driving force for luminally directed passive potassium movement, are: (1) the electrical potential difference, (2) the concentration of potassium within the cells, and (3) the potassium conductance of the luminal cell membrane. These factors are of sufficient importance to warrant additional explanation.

A number of the experimental manipulations that were previously mentioned as producing a change in urinary potassium secretion may do so by altering the distal transtubular potential difference. For example, replacement of chloride ions by nonreabsorbable ions such as sulfate or ferrocyanide increases the excretion of potassium in relative proportion to the increase in the distal transtubular potential differences that they produce. Amiloride, a diuretic that reduces the distal transepithelial potential difference, also decreases potassium excretion. The changes in distal tubular transepithelial potential difference that are produced by changing the concentrations of sodium ion within the tubular lumen may provide the mechanism for the observed coupling between sodium reabsorption and potassium secretion. This possibility will be explored later in this chapter.

The electrochemical potential gradient that favors luminally directed potassium flux must quite obviously be greatly influenced by the effective intracellular potassium concentration. Experimentally, alterations in potassium transport have been produced by a number of factors known to change the effective intracellular potassium concentration. Alkalinization of the extracellular body fluids, for example, produces an increase in intracellular potassium, not only in the distal tubular cells, but throughout the body. Concomitantly with this increase in intracellular potassium, there is an increase in distal tubular potassium secretion that roughly parallels the degree of alkalinization. Conversely, extracellular acidosis is characterized by both a decrease in the cellular potassium concentration and a decrease in distal tubular potassium secretion. If, as in the proximal tubular cell, there is an active transport system that affects potassium uptake across the peritubular cell boundary, then changes in the transport rate would naturally affect the intracellular potassium concentration and, ultimately, the electrochemical potential gradient that is responsible for potassium movement across the luminal membrane. Cardiac glycosides, such as ouabain, that specifically inhibit active potassium uptake across the peritubular membrane also decrease distal tubular potassium secretion.

The *specific ion conductance* is defined as the current flowing across a barrier (i.e., membrane, cell, tubule, or whatever) for a given emf when that specific ion is the predominant current-carrying ion. Utilizing the micropuncture technique, sodium, potassium, and chloride conductances have been determined not only for the intact tubule, but also for the peritubular and luminal cell membranes as well. The ionic conductances of the peritubular and luminal membranes differ considerably, and this difference is of considerable importance in the production of the resting transtubular potential difference.

The peritubular membrane is highly potassium-selective, i.e., the value of potassium conductance for this membrane is high relative to that of other ions. This is in distinct contrast to the properties of the luminal membrane, which is much less selective. Several ions, most importantly sodium and potassium, contribute to the transluminal membrane ionic currents. Replacement of luminal sodium and potassium ions by choline, a cation with a very low conductance, results in a predictable increase in the potential difference across the luminal membrane and, as a consequence, a reduction in the transtubular potential difference. Conversely, the low chloride conductance of the distal tubular epithelium serves to decrease the electrical potential difference across the luminal membrane, and it thereby contributes importantly to the overall transepithelial potential difference. The high potassium and sodium conductances and the low chloride conductance serve to reduce the polarization of the luminal membrane, and this effect thereby enhances potassium movement from within the cell into the tubular lumen. The relative depolarization of the luminal cell membrane resulting from the above-described ionic conductance characteristics is of prime importance, since it is responsible for the asymmetrical cell polarization that is the genesis of the transepithelial potential difference with its important consequences on luminally directed potassium flux.

COUPLED ION TRANSPORT

Sodium-Potassium Coupling

The regulation of urinary potassium excretion is dependent upon changing reabsorptive or secretory activity of the distal nephron segment, since proximal tubular potassium reabsorption remains relatively constant under a wide variety of circumstances.

Urinary potassium excretion has long been known to vary directly with urinary sodium excretion, which thereby suggests that potassium secretion is in some way dependent upon the presence of sodium ions. Initially, a coupled-ion pump involved in a one-for-one sodium-for-potassium exchange was postulated. More recent evidence, however, suggests that the coupling between these two ions is less direct. Since coupling between sodium and potassium transport may occur at several sites and via several mechanisms, these will be discussed separately.

Under all experimental circumstances in which a decreased rate of sodium excretion is associated with a similar decrease in potassium excretion, there is invariably sufficient sodium throughout the distal tubule to permit a one-for-one sodium-potassium exchange rate that is greatly in excess of what is actually measured. Under conditions of sodium conservation, it is the collecting duct that produces the most impressive drop in sodium concentration. Furthermore, there is considerable evidence to indicate that increased reabsorption of potassium along the collecting duct is more responsible for a low rate of potassium excretion than is a decrease in distal tubular potassium secretion. These facts taken together indicate that a decreased distal tubular sodium load cannot per se be responsible for the observed relation between the urinary excretion of sodium and potassium, i.e., the existence of a coupled-ion exchange pump is unlikely.

Although a coupled-ion pump mechanism is unlikely, a less specific form of coupling is readily observed. As discussed in some detail, the transtubular electrical potential difference is directly dependent upon the luminal concentrations of sodium and chloride. As the tubular sodium concentration falls, the intratubular negativity is decreased, thereby diminishing one of the major forces responsible for the passive transfer of potassium ions into the tubular lumen and thus providing a mechanism for indirect ionic coupling.

In a variety of mammalian species, the collecting ducts are of only minimal importance in potassium secretion. A low sodium concentration in the luminal fluid, however, will reduce the intraluminal negativity within the collecting duct, just as it does within the distal tubule. Under these circumstances, the quantities of potassium in the luminal fluid that result from prior distal tubular secretory processes may back-diffuse out of the lumen into the papillary interstitial space, thereby reducing the urinary excretion rate of potassium. In the rabbit, unlike other mammals studied, there is active secretion of potassium in exchange for luminal sodium in the collecting duct. Teleologically, it is reasonable to postulate that in herbivorous animals such as the rabbit, adaptation to a chronic high-potassium diet results in active potassium secretion across the collecting duct, in addition to the mechanism that is operative in the distal convoluted tubule.

In contrast to the luminal membrane, there is believed to be a coupled sodium-for-potassium exchange mechanism in the peritubular membrane that is responsible for transfer of sodium from the intracellular fluid to the peritubular fluid. The operation of this coupled pump transport system, though one step removed, assumes the ultimate responsibility for net sodium reabsorption from the lumen as well as for the maintenance of a constant cell volume. There are a number of observations supporting this hypothesis, of which the major ones are: (1) sodium and potassium ions are found to move in opposite directions across the peritubular cell membrane, (2) net sodium excretion is dependent upon the potassium concentration in the peritubular fluid, and, conversely, (3) cellular potassium uptake depends upon the intracellular sodium concentration. Of prime importance in the support of this concept is the simultaneous inhibition of active transport of both sodium and potassium by ouabain-induced inhibition of sodium-potassium-activated renal adenosine triphosphatase. Furthermore, partial inhibition of sodium-potassium-activated renal adenosine triphosphatase is accompanied by a proportional inhibition of active sodium and potassium movement.

Recent evidence, however, supports the postulated existence of at least one additional sodium transport system of importance for the net tubular reabsorption of sodium and for the maintenance of a normal intracellular sodium concentration. The three major facets of this evidence are: (1) active sodium transfer from intracellular to peritubular fluid continues after potassium uptake and adenosine triphosphatase activity have been completely blocked by ouabain, (2) the sodium excretion that continues after adenosine triphosphatase inhibition is essentially unaffected by changing levels of the external potassium concentration, and (3) this system of sodium transport can be blocked by ethacrynic acid, a compound that does not inhibit adenosine triphosphatase in vivo. These two transport systems are depicted in Figure 22-7. Intracellular sodium ions are actively transported across the peritubular membrane by either pump A or pump B, as indicated in the illustration. Since pump A does not involve a sodium-potassium exchange, its operation contributes to intracellular electrical negativity, and it is therefore presumably of some importance in chloride transfer. Pump B is an ion-exchange pump that is sensitive not only to ouabain, but also to the extracellular potassium concentration. Theoretically, if the peritubular membrane were completely impermeable to potassium ions and pump B were completely inhibited, the continued operation of pump A should produce a sodium-potassium flux ratio of infinity. On

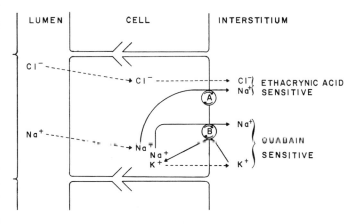

Figure 22-7
Sodium-potassium coupled-ion transport.

the other hand, complete inhibition of pump A with continued operation of pump B should produce a flux ratio of one. Experiments in perfused kidneys have tended to support these concepts.

Potassium-Hydrogen Coupling

A variety of experimental maneuvers designed to alkalinize the urine are frequently accompanied by an increase in potassium excretion. On the other hand, if potassium excretion is made the independent variable, potassium loading will raise the urine pH, and potassium depletion will enhance the excretion of hydrogen ions. This reciprocal relationship between potassium concentration and hydrogen ion excretion suggests, most simply, the existence of competition between potassium ions and hydrogen ions for a shared secretory mechanism located at the luminal membrane. Unfortunately, the problem is not so simple.

First, potassium depletion increases hydrogen ion secretion in the proximal convoluted tubule, a site of potassium reabsorption. This relationship, at least in the proximal tubule, cannot be postulated to be based on competition for some common secretory mechanism, since there is no potassium secretion in the proximal tubule. There is some evidence that potassium depletion increases the intracellular hydrogen ion concentration, thereby stimulating hydrogen ion secretion. Second, under conditions of bicarbonate loading, increased delivery of bicarbonate to the distal tubules has been shown to result in an increase in both hydrogen ion secretion and potassium secretion, which thereby demonstrates a parallel, rather than a reciprocal, change in secretory rates.

On the basis of these observations, the concept of potassium ions and hydrogen ions competing for a common secretory mechanism has been abandoned. It is more reasonable to postulate that intracellular pH changes somehow influence potassium secretion, and, conversely, changes in the intracellular potassium concentration somehow influence hydrogen ion secretion. In support of this, Giebisch [1] has recently shown that changes in the extracellular potassium level as well as in the intracellular pH will alter the active uptake of potassium across the peritubular membrane. Since the rate of peritubular potassium uptake is a major factor in determining the magnitude and concentration of the intracellular potassium pool that is available for secretion and the regulation of transmembrane potential differences, it must be the ultimate determinant of the potassium secretory rate. Factors affecting peritubular potassium transfer must therefore affect, directly or indirectly, distal tubular potassium secretion.

REFERENCES

1. Giebisch, G., Klose, R. M., and Malnic, G. Renal Tubular Potassium Transport. *Proc. III Int. Cong. Nephrol.* Washington, D.C., 1966.
2. Ussing, H. H. The Alkali Metal Ions in Biology. In Eichler, O., and Farah, A. (Eds.), *Handbuch der experimentellen Pharmakologie.* Göttingen: Springer Verlag, 1960.

SELECTED READING

Deane, N. *Kidney and Electrolytes.* Englewood Cliffs, N.J.: Prentice-Hall, 1966.

Giebisch, G., et al. Electrolyte transport in kidney tubule cells. *Annu. Rev. Physiol.* 12:141, 1972.

Maxwell, M. H., and Kleeman, C. R. *Clinical Disorders of Fluid and Electrolyte Metabolism.* New York: McGraw-Hill, 1962.

Schrier, R. W., et al. Tubular reabsorption of sodium ions, 1. *N. Engl. J. Med.* 285:1231, 1971.

Schrier, R. W., et al. Tubular reabsorption of sodium ions, 2. *N. Engl. J. Med.* 285:1292, 1971.

Acid-Base Balance **23**

Joel B. Mann

INTRODUCTION

The importance of acid-base balance in the body is evidenced by the fact that a stable pH of the body fluids is essential for life. Normal cell function requires an extracellular fluid pH value of very nearly 7.4, and this value is maintained within the narrow range of ±0.05 pH unit by the various regulatory mechanisms. This remarkable constancy is accomplished through the operation of buffers, ion-exchange mechanisms, and pulmonary and renal functions. An introduction to the major acid-base systems is presented in this chapter. The basic physicochemical properties related to these mechanisms were presented in Chapter 2.

REGULATION OF pH

BUFFERS

The most effective buffer in any chemical system is one with a pK_a approximately equal to the pH of the system in question. This fact may readily be seen from the Henderson-Hasselbalch equation (see Chap. 2, Eq. 2-38), which states that when the pH equals pK_a, the logarithm of $[A^-]/[HA]$ must equal zero, i.e., the concentrations of anion, $[A^-]$, and undissociated acid, $[HA]$, are equal, and therefore the system is able to resist a change in concentration in either direction on the addition of strong acids or bases.

The most important of the blood buffers is the bicarbonate system, even though its pK_a is 6.1. The reason why this system is effective despite its low pK_a is that it operates in a two-phase environment, rather than in the single phase of a glass beaker in a chemistry laboratory. The following equation describes the bicarbonate buffering system:

$$H_2O + CO_2 \rightleftharpoons H_2CO_3 \rightleftharpoons H^+ + HCO_3^- \tag{23-1}$$

The key to the efficiency of this system is the production of CO_2. With the addition of an acid, which might occur in metabolic acidosis, this reaction is shifted to the left and CO_2 is generated. Were the CO_2 to remain in solution, carbonic acid would accumulate, and there would be a rapid change in pH. The CO_2, however, is removed from the environment via the lungs, and thus a massive pH change is averted.

Conversely, if H^+ is removed from this system, which would occur in metabolic alkalosis, then the CO_2, which may be produced at the rate of about 10 mEq/min (expressed as carbonic acid), is excreted less rapidly by the lungs. This results in the addition of H^+ to the solution as the CO_2 retention forces the equilibrium to the right.

All proteins, including hemoglobin, exert their buffering activity via a series of titratable side-chain groups in the molecule. The pK_a values of the plasma proteins overlap, thereby producing a fairly efficient buffer system within the physiologic pH range. When titrated from pH 7.5 to pH 6.5, one gram of plasma protein will bind 0.110 mEq H^+, and one gram of oxyhemoglobin will bind 0.183 mEq H^+. Since 1 liter of blood contains approximately 150 gm of hemoglobin and 40 gm of plasma protein, the oxyhemoglobin in this amount of blood can buffer 27.5 mEq H^+ and the plasma protein can buffer about 4.4 mEq H^+.

ION EXCHANGE

This is a mechanism whereby the vast reservoirs of intracellular organic phosphate and protein are made available for the buffering of extracellular hydrogen ions. The mechanism operates primarily by means of an exchange across the cell membrane of extracellular Na^+ and H^+ for intracellular K^+. The K^+-H^+ exchange mechanism tends initially to elevate the serum K^+, and ultimately causes a depletion of total body K^+ since the K^+ is excreted by the kidneys.

There is, in addition, some evidence that Na^+ and Ca^{+2} from bone may be exchanged for H^+. The involvement of bone in the buffering defense mechanisms becomes much more important in chronic acidosis than in acute metabolic disturbances. With an acute acid load, which might be experimentally produced by the intravenous infusion of 150 mEq of dilute hydrochloric acid, approximately 15% to 20% of the load is neutralized by the intravascular buffers, 20% to 25% by interstitial buffers, and the remainder by intracellular buffers and bone.

PULMONARY REGULATION

The mechanisms of pulmonary regulation of acid-base balance are discussed in detail in Chapter 17. Briefly, the respiratory system assists in the regulation of pH either by removing the CO_2 generated via multiple metabolic pathways or by retaining CO_2 to generate H^+ (see Eq. 23-1).

RENAL REGULATION

Since the renal correction of changes in pH operates very slowly, requiring 48 to 96 hours to reach maximal efficiency, the immediate compensation for acid-base disturbances must be the responsibility of the chemical buffers of the body. The respiratory system acts as an important, rapidly acting, second line of defense. The correction of chronic acid-base disturbances produced by the accumulation or excessive loss of fixed acids or bases, however, must ultimately depend upon the renal excretion of hydrogen ions or bicarbonate.

CLINICAL CONCEPTS

RESPIRATORY ACIDOSIS

Respiratory acidosis is related to the inability of the lungs to remove carbon dioxide from the blood adequately. As a consequence of this, the PCO_2 and H_2CO_3 concentration rise, and, as predicted by the Henderson-Hasselbalch equation, the pH falls. Since the primary defect is pulmonary in origin, the initial compensation must result from the action of the buffer and ion-exchange mechanisms. If the pulmonary defect persists, renal compensation will ultimately come into play in the form of an increase in bicarbonate reabsorption and hydrogen ion excretion. As a result of respiratory acidosis, the serum bicarbonate level increases, the PCO_2 increases proportionately greater, and pH decreases, as indicated in Table 23-1.

Table 23-1. Blood Substance Levels During Alkalosis and Acidosis

	Metabolic Acidosis	Metabolic Alkalosis	Respiratory Acidosis	Respiratory Alkalosis
Blood pH	↓	↑	↓	↑
Blood $[HCO_3^-]$	↓	↑	↑	↓
PCO_2	↓	↑	↑	↓
$\dfrac{[HCO_3^-]}{PCO_2}$	↓	↑	↓	↑
$[Cl^-]$	↑↓ or 0	↓ or 0	↓ or 0	↑ or 0

RESPIRATORY ALKALOSIS

The defect involved in this disturbance is hyperventilation resulting in an increased removal of carbon dioxide from the blood. As Equation 23-1 demonstrates, this removal of CO_2 results in the consumption of H^+ as H_2CO_3 is dehydrated. Again, the immediate compensation must result from the action of the chemical buffers and ion-exchange mechanisms. Renal mechanisms are slowly activated and ultimately operate to lower the bicarbonate concentration, thereby lowering the pH, as dictated by the Henderson-Hasselbalch equation. During respiratory alkalosis, the CO_2 content is decreased, the PCO_2 is decreased proportionately, the ratio $[HCO_3^-]/PCO_2$ is thereby increased, and the pH is elevated (Table 23-1).

    In respiratory alkalosis, potassium ions are frequently lost in the urine, which results in hypokalemia and, ultimately, an intracellular deficit of potassium. If it is sufficiently severe, this condition can decrease the effectiveness of the renal compensatory mechanisms, with the resultant loss of additional hydrogen ion in the urine, increased bicarbonate reabsorption, and a worsening of the alkalosis.

METABOLIC ACIDOSIS

Metabolic acidosis results from the abnormal retention of fixed (i.e., nonvolatile) anions that are derived from the metabolism of organic and inorganic acids. The sulfur contained in ingested protein, for example, is oxidized by metabolic processes to H_2SO_4. This acid in turn reacts with the available buffers and decreases the buffer concentration. If the rate at which this occurs is faster than the rate at which the kidney is able to excrete the fixed anion and regenerate the consumed bicarbonate, then metabolic acidosis ensues. Ion-exchange mechanisms become operative in order to correct this defect, and the rate of respiration is increased. Carbon dioxide is then removed more rapidly, and, in accordance with the Henderson-Hasselbalch equation, the ratio of the buffer pair approaches its normal value and the pH moves toward normal. As a result of these multiple changes, the bicarbonate concentration is decreased. As the primary initiating event, the P_{CO_2} is decreased in partial compensation and the pH is decreased, but not to the extent that would obtain if there were no respiratory compensation. These effects are summarized in Table 23-1.

METABOLIC ALKALOSIS

This abnormality is seen less commonly than the conditions discussed previously, and it results from the loss of hydrogen ion or the administration of alkalinizing substances such as sodium bicarbonate or sodium lactate. The initial consequence of this problem is an elevation of bicarbonate levels, and, according to the Henderson-Hasselbalch equation, an increase in pH. The initial compensatory mechanisms are again mediated by buffers and ion exchange, but the effect of the latter may be limited by the renal K^+ loss that may occur in alkalosis and by the secondary movement of H^+ into cells, which is initiated by the K^+ loss. The rate of respiration is generally slowed so that some CO_2 retention ensues, which again tends to restore the buffer pair of the Henderson-Hasselbalch equation toward their normal relationship. The serum changes seen in metabolic alkalosis, as shown in Table 23-1, include an elevation of bicarbonate levels, an elevation of P_{CO_2}, and an elevation of pH.

It is important to note in Table 23-1 that in both respiratory acidosis and metabolic alkalosis, the bicarbonate concentration and P_{CO_2} are elevated; in respiratory alkalosis and metabolic acidosis they are decreased. Even though the bicarbonate concentration is the most easily measured of these parameters (it is usually determined as the closely related CO_2 content or the less valid CO_2 combining power), it is hazardous to make a clinical diagnosis of acid-base imbalance on the basis of this measurement alone. The actual state of acid-base balance is determined by the *ratio* of the various buffer pairs measured, which, in clinical medicine, is usually the bicarbonate-CO_2 system. Although, for example, the bicarbonate concentration is increased in respiratory acidosis, the partial pressure of CO_2 is increased even further, and the pH is, of course, lower.

SELECTED READING

Christensen, H. N. *Body Fluids and the Acid-Base Balance.* Philadelphia: Saunders, 1965.

Christensen, H. N. *pH and Dissociation.* Philadelphia: Saunders, 1964.

Crepeau, R., et al. Contribution of colon and kidney to acid base homeostasis in man. *Union Med. Can.* 101:735, 1972.

Davenport, H. W. *The ABC of Acid-Base Chemistry* (5th ed.). Chicago: University of Chicago Press, 1969.

Goldberger, E. *Water, Electrolyte and Acid-Base Syndromes.* Philadelphia: Lea and Febiger, 1960.

Papper, R. S. *Clinical Nephrology.* Boston: Little, Brown, 1971.

The Body Fluids $\qquad$ 24

Matthew B. Wolf

INTRODUCTION

It was recognized well over one hundred years ago by Claude Bernard that the proper functioning of the living organism depends critically upon the maintenance of near-normal chemical composition. In particular, it is the chemistry of the cell, the smallest viable unit, which must be maintained almost constant in the face of severe stresses to the organism.

Cells are found in a wide variety of shapes and sizes depending upon their function in the organism; however, they all have in common the processes of metabolism and subsequent energy production. These processes depend critically upon the maintenance of the internal chemical state of the cell. The energy produced is utilized partly for the self-protection of this chemical state and partly for the performance of the cell's specialized function in the entire organism.

The cell is adapted to protect its internal chemical composition by being enclosed in a membrane that is impermeable to the large protein molecules inside the cell, almost freely permeable to water, and selectively permeable to various other chemical species. In addition, part of the metabolic energy of the cell is used to "pump" certain chemical species in or out of the cell. These properties establish a unique internal chemistry whose volume and chemical composition are, to a degree, little affected by changes in the chemical environment of the cell.

This unique chemistry of cells allows them to develop and perform a variety of functions for the organism, including the mechanical contraction of muscle cells, the electrical signal conduction of nerve cells, the synthesis and secretion of hormones by gland cells, and the absorption of large quantities of water and other substances by cells of the kidney and the gastrointestinal system.

Just as for other cells, the maintenance of the ability of the red blood cell to carry oxygen depends upon the maintenance of its own chemical state by its internal processes and the constancy of the chemical composition of its external environment, the plasma.

The body fluids are maintained in a nearly constant state by a variety of processes and mechanisms, some of which are strictly physicochemical, some are related to metabolic events in the cells, and the remainder are those physiologic control systems that depend upon the functions of the renal and respiratory systems.

This chapter will focus upon the description of the state of the body fluid system and an explanation of how this state changes due to physicochemical and metabolic processes. The distribution of water, electrolytes, and other important chemical species contained in the body fluids of mammals is described, and methods of measurement of this distribution will be presented, including the tracer techniques used

391

to measure the volume of the body fluids. In addition, models of both equilibrium and transient water distribution between intracellular and extracellular fluids are developed.

WATER DISTRIBUTION

The fluids of the body are distributed in a nonuniform manner throughout every tissue and organ. Table 24-1 gives the percentages and amounts of water in various tissues and organs of a 70-kg man. The percentages range from 10% water for adipose (fat) tissue to 83% water for blood, with an average over the whole body of 60% water by weight. It is evident from the table that the water contained in skeletal muscle is about 50% of the total body water.

Table 24-1. Water Distribution in Organs and Tissues

Tissue	Percent Water	Percent of Body Weight	Liters of Water in 70-kg Man
Skin	72	18	9.0
Skeletal muscle	76	42	22.0
Bone	22	16	2.5
Blood	83	8	4.3
Adipose	10	variable	~1.0
Others	—	—	3.2
Total			42.0

Skeletal muscle is composed primarily of cells filled with fluid and many structured elements, all of which are enclosed by the cell membrane. The cell is obviously a nonhomogeneous structure, and the observed chemical properties of the cell depend upon where one is making measurements. Generally speaking, however, the concept of an overall intracellular fluid chemical composition is convenient. Other fluids in the muscle are the vascular fluid in the blood vessels and the interstitial fluid that fills the space between the blood vessels and the cells. An examination of any other organ or tissue reveals the same subdivisions of water distribution.

The first natural division of the body fluids is, then, between the intracellular fluid and the extracellular fluid, with the cell membrane forming the barrier between them. Of the intracellular fluids, that of the cellular part of the blood (red blood cells) is generally considered separately, since it has its own distinct chemical composition. The fluids of the remainder of the body cells (mainly of skeletal muscle cells) are conceptually lumped together and called the *intracellular fluid*, which, for the sake of convenience, is assumed to have a homogeneous chemical composition and a volume equivalent to the total volume of the body cells.

The *extracellular fluid* is also nonhomogeneous. The plasma and the interstitial fluid constitute two of its subdivisions. Fluids in supporting tissues, such as bone, and in connective tissues, such as cartilage and tendon, provide other subdivisions.

The remaining subdivision is a class of fluids called *transcellular fluids*. These are the digestive secretions, the cerebrospinal and intraocular fluids, the glandular secretions, and the fluids in the body cavities. The common property of transcellular fluids is that they are secreted into portions of the body that are separated from the main extracellular space by a continuous layer of epithelial cells. Even though the volume of these transcellular fluids is relatively small, their chemical composition is highly important for the proper functioning of the organism.

The distribution of water in the body may thus be defined as the volume of water in each of the individual categories of fluids. The measurement of these volumes cannot be directly performed in vivo, and thus the principles of indicator dilution are employed.

INDICATOR-DILUTION METHODS

Given a homogeneous fluid of unknown volume V, a small quantity m_I of some indicator or tracer is injected into the fluid. After uniform mixing is complete, a sample of the fluid is removed and analyzed for the concentration, c_I, of tracer. From mass balance principles,

$$m_I = c_I V \tag{24-1}$$

from which V can be calculated. Equation 24-1 is valid only if the indicator is neither lost from nor created in the volume to be measured and if the indicator itself does not alter this volume.

The volume of tracer distribution measured by this technique is usually referred to as a *compartmental volume*. Unfortunately, in the measurement of body fluid compartments, the conditions pertaining to Equation 24-1 are never completely satisfied. Usually the tracer substance leaves the compartment, either by excretion in the urine, by transfer into another fluid compartment, or by metabolism. In these cases, Equation 24-2 may be used:

$$V = \frac{m_I - m_E(t)}{c_I(t)} \tag{24-2}$$

where $m_E(t)$ is the cumulative amount of indicator excreted or lost from the compartment up to time t and $c_I(t)$ is the indicator concentration at that time, t, following the injection of indicator.

In general, one cannot directly determine how much of the tracer substance transfers from one body fluid compartment to another; however, an estimate can be made from the *plasma disappearance curve*. Figure 24-1 shows such a curve for inulin, a substance that is commonly used to measure extracellular fluid volume. The shape of the curve is assumed to be the result of a combination of the mixing transient distribution and the loss from the compartment. Plotting log c against the time from injection reveals

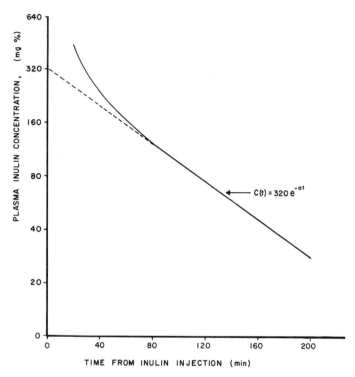

Figure 24-1
Plasma inulin disappearance curve.

a single exponential rate of disappearance after an 80-minute period. One then assumes that this is purely the rate of loss from the compartment. Extrapolation of the curve to the axis at zero injection time gives the theoretical concentration $c(0)$ that would be obtained if the injected substance were instantaneously distributed and no loss had yet occurred. Then Equation 24-2 is used, with $m_E(t)$ equal to zero and $c_I(t)$ equal to $c(0)$, to calculate V.

The distribution of the total body water in a normal, young adult male is shown in Table 24-2. This was determined using dilution techniques and individual tissue analyses. The total body water (TBW) was also measured using tracer-dilution methods; in this case, the tracer was tritiated water (HTO). The TBW averages about 60% of the body weight in young males, 50% in young females, and as high as 76% in babies. The variation is due to the variable fat content in these three groups. Fat is an almost water-free tissue, as shown in Table 24-1, and thus for a given body weight, the greater the amount of fat, the lower the percentage water, and vice versa.

In contrast to the tracer measurement of TBW, which yields results consistent with more direct measurements such as desiccation, the measurements of the volumes of extracellular water (ECW) and interstitial water (ISW) are much less reliable with respect to the relation between the measured volumes of distribution of tracers and

Table 24-2. Water Distribution in the Various Body Fluid Categories

Fluid	Water as % of Body Weight	Liters in 70-kg Man
Red blood cells	2.5	1.8
Plasma	4.5	3.0
Interstitial	12.0	8.4
Intracellular	31.0	21.8
Connective tissue	5.0	3.5
Transcellular	2.0	1.4
Bone	3.0	2.1
Total	60.0	42.0

the anatomic fluid spaces to be measured. The main difference is that in the case of measuring TBW, an isotope of water is used to trace the water distribution throughout the body, whereas in the case of ECW and ISW, there is no one substance that distributes uniformly just in the water inside or outside of the cells. As previously mentioned, the extracellular fluid is not homogeneous, and it is thus necessary to use various substances to measure the proportions of this fluid.

The evidence that chloride ion is almost exclusively localized extracellularly in skeletal muscle led to its use in the measurement of ECW. Subsequently, similar ions, such as bromide, sulfate, and so on, have also been used. The objections to using ions as tracers are that they all enter some cells to some extent and that their concentrations tend to be nonuniform throughout the extracellular space due to the Gibbs-Donnan distribution effect (see p. 396). These objections led to the use of saccharides, such as mannitol, sucrose, and inulin, to measure ECW; however, these substances have the disadvantage that they are rapidly excreted in the urine, and corrections must be made to account for this effect. A general discussion of these points can be found in the article by Edelman and Liebman [1].

In general, the saccharides can be assumed to measure the volume of water in plasma, interstitial fluid, and lymph, whereas ions such as chloride and bromide measure total ECW, the water in red blood cells, and a small portion of other cellular water. These generalizations are based on the relatively rapid equilibration (about one hour) of the saccharides within a particular volume, which is followed by a subsequent decline according to a single exponential function (see Fig. 24-1) over the next ten hours. These saccharides are presumably lost into a dense connective tissue space. Thus, the volume measured by back extrapolation of the disappearance curve has been referred to as the "rapidly equilibrating space." After two to three hours, the bromide or chloride plasma disappearance curves become almost horizontal, which indicates that their equilibration occurs in a space larger than that penetrated by the saccharides in the same time period.

To measure the ISW, the volume of water in the plasma (PW) must be subtracted from the ECW as measured by a saccharide disappearance technique. Thus, the plasma volume (PV) and the water content per volume of plasma must be determined. Plasma contains about 7% protein by volume, and thus the remaining 93% is water. Methods for estimating the PV are based either on tagging plasma proteins (albumin or globulin) with dyes (T-1824 or Evans blue) or isotopes (^{131}I or ^{125}I) or on using the measured red blood cell volume (RCV) and the blood hematocrit (the blood cell volume as a percentage of the total blood volume).

The plasma disappearance curve of tagged albumin shows a single exponential decline (due to loss into ISW) after a 10- to 15-minute mixing period. Back extrapolation of the curve and the use of Equation 24-2 yield the result that PW is 4.5% of the body weight.

The second method involves measuring the RCV by tagging a quantity of red blood cells with an isotope (^{51}Cr) and diluting these in the circulating red blood cell mass. Then, after a 10-minute mixing period, the concentration of tagged cells becomes constant and Equation 24-1 can be used to estimate the RCV. From this and the measurement of hematocrit, H, the PV can be computed:

$$PV = RCV \left(\frac{100}{H} - 1 \right) \tag{24-3}$$

If the hematocrit is measured from a sample of blood from a large blood vessel, then the PV calculated from Equation 24-3 will, in general, be about 10% to 15% lower than that estimated by using tagged plasma protein. This has led to the definition of a total body hematocrit, H_T, which is that calculated from the RCV and PV measured using tracers. The value of H_T is related to that of the large vessel hematocrit, H_{lv}, by a factor, f_{cell}, which is defined as the ratio of H_T to H_{lv}. The value of f_{cell} varies among animal species, as well as according to the state of the animal.

The volume of the water in cells, or intracellular water (ICW), cannot be directly measured. It can be estimated only by computing the difference between TBW and the measured volumes in the remainder of the body fluids. This estimate therefore conserves all the inaccuracies of the measurement methods discussed in this section.

ELECTROLYTE DISTRIBUTION

Each of the body fluids has its own distinct electrolyte pattern. In general, the extracellular fluids have high concentrations of sodium, Na^+, and chloride, Cl^-, and the concentration of bicarbonate, HCO_3^-, is also higher than in cellular fluids. In the intracellular fluid, potassium, K^+, is the predominant cation, while the phosphates, HPO_4^{-2} and $H_2PO_4^-$, are the inorganic anions in highest concentration. The proteins in the cell, though not present in extremely large amounts, are very highly negatively charged, and thus their distribution affects that of other chemical species. These general patterns for humans are shown in Figure 24-2. The data for dogs are listed in Table 24-3; however, with the exception of the red blood cell chemistry, these data could well apply to humans.

The electrolyte pattern in the extracellular fluids (i.e., plasma and interstitial fluid) is similar except that the cation concentrations are higher and the anion concentrations lower in plasma. This distribution is brought about by the negatively charged plasma proteins, as will be subsequently discussed.

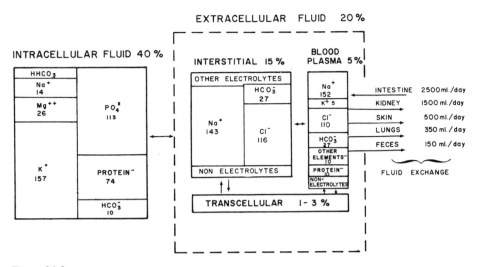

Figure 24-2
Human body fluid compartments and flow paths (electrolyte values are milliequivalents per liter of H_2O).

Table 24-3. Dog Body Fluid Electrolyte Concentrations[a]

Electrolytes	Red Blood Cells	Plasma	Interstitial Fluid	Intracellular Fluid
Sodium (Na^+)	134.0	153.0	146.0	10.0
Potassium (K^+)	10.0	4.8	4.6	160.0
Calcium (Ca^{+2})	3.4	3.0	2.7	0.2
Magnesium (Mg^{+2})	1.2	0.4	0.4	26.0
Total cations	148.6	161.2	153.7	196.2
Chloride (Cl^-)	88.0	112.0	118.0	3.0
Bicarbonate (HCO_3)	22.0	27.0	28.0	7.1
Phosphate (HPO_4^{-2})	0.5	0.7	0.8	100.0
Sulfate (SO_4^{-2})	0.4	0.6	0.7	20.0
Organic acids	1.5	1.9	2.0	0.1
Proteins	36.2	19.0	4.2	66.0
Total anions	148.6	161.2	153.7	196.2

[a]Concentration units in mEq/liter of H_2O.

The intracellular proteins produce a similar effect due to their electronegativity, which is reflected by the K^+ and Cl^- concentrations in the intracellular fluid; however, this effect is offset for ions such as Na^+ by mechanisms, or "pumps," inside the cell that tend to drive the Na^+ out. These will also be subsequently discussed.

Red blood cell chemistry is unique, and its electrolyte pattern depends upon the animal species. Dog red blood cell chemistry, as shown in Table 24-3, seems roughly similar to that of plasma, whereas the red blood cell chemistry of the human is more like the electrolyte pattern of the intracellular fluid.

A fact to note in Table 24-3 is that all the fluids are electrically neutral, i.e., the milliequivalents of total cations equal the milliequivalents of total anions. This observation leads to the *law of macroscopic electrical neutrality of solutions,* which holds both in vitro and in vivo. It is well known from physical principles that very small electrical charge imbalances produce large forces that tend to move ions toward bulk fluid electrical neutrality. This law does not pertain, however, to the observed local regions on the surfaces of biologic membranes where imbalances in electrical neutrality occur.

From total carcass analysis of ionic content, it is evident that not all the ions in the body are accounted for by the concentrations given in Table 24-3. Much of the electrolyte content of the body is in a bound or slowly exchanging form in the bone and dense connective tissue. This may contribute to the long-term electrolyte balance in the body, but it is usually not important in the day-to-day regulation of electrolytes.

MATHEMATICAL MODELS OF CHEMICAL EXCHANGE AMONG THE BODY FLUIDS

A MODEL OF EQUILIBRIUM WATER DISTRIBUTION BETWEEN INTRACELLULAR AND EXTRACELLULAR FLUIDS

It was hypothesized and experimentally verified by Wolf [3] that to a first approximation, the cells of the body acted as "perfect osmometers." Thus water, but not solute, is transferred between the intracellular and extracellular fluids in order to balance an osmotic pressure or osmolarity difference across the cell wall, thus bringing about a new equilibrium state.

This suggests a simple model to describe equilibrium water transfers between the body fluid spaces (Fig. 24-3). Consider a two-compartment body fluid system, with an intracellular compartment of volume V_{ic} and an extracellular compartment of volume V_{ec}. The total solute concentrations c in each compartment are assumed to be equal at equilibrium, and the solute content of the intracellular compartment is also assumed to remain constant. If a load of water (L_w) and solute (L_s) is added to

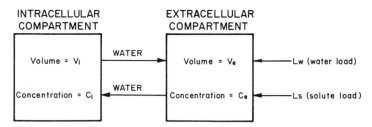

Figure 24-3

Two-compartment water distribution model. Initially, both compartments are at equilibrium and $c_i = c_e$. Addition of a load of water, L_w, and solute, L_s, to the extracellular compartment causes a water shift between the compartments so that the equality of solute concentrations is reestablished. The cell walls are assumed to be impermeable to solute.

the extracellular compartment and if it distributes uniformly, then water shifts will occur between the compartments in order to maintain the equality of solute concentrations. This is equivalent to saying that water will shift until the osmotic pressure difference, $\Delta\pi$, becomes zero. The equations that describe the equilibrium water distribution are:

$$c_{ic} = c_{ec} = c \tag{24-4}$$

and the mass-balance equations,

$$V_{ic} + V_{ec} = V_{ic}^{\,0} + V_{ec}^{\,0} + L_w \tag{24-5}$$

$$c(V_{ic} + V_{ec}) = c^0(V_{ic}^{\,0} + V_{ec}^{\,0}) + L_s \tag{24-6}$$

where the superscript zero refers to the initial values, i.e., those existing before water and solute addition. Solving Equations 24-5 and 24-6 for c yields:

$$c = \frac{c^0(V_{ic}^{\,0} + V_{ec}^{\,0}) + L_s}{V_{ic}^{\,0} + V_{ec}^{\,0} + L_w} \tag{24-7}$$

Also, since

$$c = \frac{c^0 V_{ec}^{\,0} + L_s}{V_{ec}} \tag{24-8}$$

then V_{ec} is

$$V_{ec} = \frac{(c^0 V_{ec}^{\,0} + L_s)(V_{ic}^{\,0} + V_{ec}^{\,0} + L_w)}{c^0(V_{ic}^{\,0} + V_{ec}^{\,0}) + L_s} \tag{24-9}$$

and V_{ic} can be calculated from Equation 24-5 using the result of Equation 24-9.

In order to examine the predictions of these equations, some simple cases of water and solute additions and withdrawals in a 70-kg normal adult male will be considered. From Table 24-2, his total body water $V_{ec}^{\,0} + V_{ic}^{\,0} = 42$ liters. Lumping the intracellular water with the red blood cell water content, $V_{ic}^{\,0} = 23.6$ liters and $V_{ec}^{\,0}$, the remainder, is then 18.4 liters. The initial solute concentration c^0 is taken as the normal osmolarity of the body fluids, or 0.3 Osm/liter.

First let us consider the loss of 10% of the total body water, which could occur in severe dehydration. Since no solute is gained or lost, then, from Equation 24-7,

$$c = \frac{(0.3 \text{ Osm/liter}) (42 \text{ liters})}{42 \text{ liters} - 4.2 \text{ liters}} = 0.333 \text{ Osm/liter} \tag{24-10}$$

From Equation 24-8, V_{ec} = 16.6 liters, and from Equation 24-5, V_{ic} = 21.2 liters. Note that the water is lost from each compartment in proportion to its initial volume.

If the individual loses solute along with the water, then the water shifts depend upon the solute concentration (osmolarity) of the fluid that is lost. Considering that this lost fluid also has a concentration of 0.3 Osm/liter, then $c = c^0$, V_{ec} = 14.2 liters, and V_{ic} would be unchanged from its initial value. Thus, if fluid of the same osmolarity as the body fluids is added to or subtracted from the body, then only the V_{ec} changes. This is what is implied by an isotonic fluid addition.

If the fluid that is lost is sweat, its composition is hypotonic, i.e., it has a lower osmolarity than the body fluids. Considering that its osmolarity is 0.15 Osm/liter, then $c = 0.317$ Osm/liter, V_{ec} = 15.4 liters, and V_{ic} = 22.4 liters.

A MODEL OF STEADY-STATE ION AND WATER DISTRIBUTION IN CELLS

In the previous model, it was assumed that the cell walls are impermeable to the solute. This assumption is consistent with the observed behavior of the cell volume in response to external changes in osmotic properties. However, after the introduction of isotopic tracer methods in the 1950s, the impermeability assumption was shown to be invalid, and thus another explanation for the volume stability of the cells had to be found. The finding that the maintenance of normal cation concentrations in cells depends upon metabolic processes led to the hypothesis that the active transport of these ions produces steady-state ionic gradients that affect the water distribution between the cells and their environment. A simple model is proposed that will illustrate this phenomenon.

Consider that our model system is composed of cells that contain Na^+, K^+, and Cl^- ions, which may permeate the membrane, and an impermeable anion, X^{-n}, in a concentration c_X and with a valence $z_X = -n$. The extracellular fluid, which has a large volume compared to that of the cell volume V_{ic}, contains only the permeable ions.

The electrochemical potential of an ion, as developed in Equation 5-1 (Chap. 5), may be expressed as:

$$\mu_i = \mu_i^0 + RT \ln a_i + z_i FV \tag{24-11}$$

where μ_i = electrochemical potential of ion i
$\quad \mu_i^0$ = partial molal free energy of ion i
$\quad z_i$ = number of charges carried by ion i
$\quad V$ = electrical potential
$\quad a_i$ = activity of ion i (a_i is proportional to the concentration, c_i)

The equilibrium condition for a solute distributed between two phases is:

$$(\mu_i)_1 = (\mu_i)_2 \tag{24-12}$$

where the subscripts refer to sides 1 and 2. Substitution from Equation 24-11 yields:

$$(\mu_i^0)_1 + RT \ln (a_i)_1 + z_i F V_1 = (\mu_i^0)_2 + RT \ln (a_i)_2 + z_i F V_2 \qquad (24\text{-}13)$$

Rearranging Equation 24-13 gives:

$$\ln \frac{(a_i)_1}{(a_i)_2} = \frac{(\mu_i^0)_2 - (\mu_i^0)_1}{RT} + \frac{z_i F}{RT} (V_2 - V_1) \qquad (24\text{-}14)$$

Using molar concentrations of sodium ion, the steady-state condition described by Equation 24-14 becomes:

$$\ln \frac{(c_{Na^+})_{ic}}{(c_{Na^+})_{ec}} = \frac{\Delta \mu_{Na^+}{}^0}{RT} + \frac{F}{RT} \Delta V \qquad (24\text{-}15)$$

The corresponding equations for potassium and chloride are:

$$\ln \frac{(c_{K^+})_{ic}}{(c_{K^+})_{ec}} = \frac{\Delta \mu_{K^+}{}^0}{RT} + \frac{F}{RT} \Delta V \qquad (24\text{-}16)$$

$$\ln \frac{(c_{Cl^-})_{ic}}{(c_{Cl^-})_{ec}} = \frac{\Delta \mu_{Cl^-}{}^0}{RT} - \frac{F}{RT} \Delta V \qquad (24\text{-}17)$$

Assuming that Cl^- is in electrochemical equilibrium, then $\Delta \mu_{Cl^-}{}^0 = 0$, and:

$$\frac{F \Delta V}{RT} = -\ln \frac{(c_{Cl^-})_{ic}}{(c_{Cl^-})_{ec}} \qquad (24\text{-}18)$$

Substituting Equation 24-18 into Equation 24-15 gives the steady-state condition that:

$$\frac{(c_{Na^+})_{ic} \, (c_{Ce^-})_{ic}}{(c_{Na^+})_{ec} \, (c_{Cl^-})_{ec}} = \exp \left(\frac{\Delta \mu_{Na^+}{}^0}{RT} \right) = A \qquad (24\text{-}19)$$

Similarly,

$$\frac{(c_{K^+})_{ic} \, (c_{Cl^-})_{ic}}{(c_{K^+})_{ec} \, (c_{Cl^-})_{ec}} = \exp \left(\frac{\Delta \mu_{K^+}{}^0}{RT} \right) = B \qquad (24\text{-}20)$$

The parameters A and B are assumed to remain constant and are therefore independent of concentration changes.

The condition of electrical neutrality inside the cell yields:

$$(c_{Na^+})_{ic} + (c_{K^+})_{ic} - (c_{Cl^-})_{ic} + \frac{z_X X}{V_{ic}} = 0 \qquad (24\text{-}21)$$

where X is the number of moles of the impermeable anion X^{-n}.

The osmotic pressure relationship developed in Equation 2-21 (Chap. 2) is expressed as:

$$\pi = cRT \qquad (24\text{-}22)$$

where π = osmotic pressure
   c = concentration
   T = absolute temperature

Thus, the difference in osmotic pressure across a membrane is:

$$\Delta\pi = RT\Delta c \qquad (24\text{-}23)$$

where $\Delta\pi$ = osmotic pressure difference
   Δc = difference in concentration of total impermeable dissolved solutes on both sides of the membrane

Assuming that the water is at equilibrium and that cells cannot support a hydrostatic pressure gradient, then from Equations 24-23 and 24-21:

$$\Delta\pi = (c_{Na^+})_{ic} + (c_{K^+})_{ic} - (c_{Cl^-})_{ic} + \frac{X}{V} - (c_{Na^+})_{ec} - (c_{K^+})_{ec} + (c_{Cl^-})_{ec} = 0 \qquad (24\text{-}24)$$

Adding Equation 24-21 and 24-24, dividing Equation 24-19 by 24-20 and then solving for $(c_{Na^+})_{ic}$

$$(c_{Na^+})_{ic} = \frac{\bar{c}_{ec} - (z_X + 1)\dfrac{X}{V_{ic}}}{2\left[1 + \dfrac{B(c_{K^+})_{ec}}{A(c_{Na^+})_{ec}}\right]} \qquad (24\text{-}25)$$

where $\bar{c}_{ec} = (c_{Na^+})_{ec} + (c_{K^+})_{ec} - (c_{Cl^-})_{ec}$. For known values of $z_X, A, B,$ and the extracellular concentrations, $(c_{Na^+})_{ic}$ can be found using the algebraic formula for the solution of quadratic equations. Then $(c_{Cl^-})_{ic}$ can be calculated from Equation 24-19, $(c_{K^+})_{ic}$ from Equation 24-20, and X/V_{ic} from Equation 24-21.

Using the data of Table 24-3 for the concentrations of the intracellular and extracellular fluids, $(c_{Na^+})_{ec} = 146$ mEq/liter, $(c_{K^+})_{ec} = 4.6$ mEq/liter, and, to have electrical neutrality, we take $(c_{Cl^-})_{ec} = 150.6$ mEq/liter; for the cell, $(c_{Na^+})_{ic} = 10$ mEq/liter, $(c_{K^+})_{ic} 160$ mEq/liter, $(c_{Cl^-})_{ic} = 3$ mEq/liter, and thus $z_X X/V_{ic} = 167$ mEq/liter. Assuming that $V_{ic} = 1$ liter, then from Equation 24-24, $X = 128.2$ and thus $z_X = -1.3$. Also, from Equation 24-19, we have $A = 0.00136$, and from Equation 24-20, $B = 0.693$.

Using these quantities, we can see the effect of a decrease in sodium pumping (i.e., an increase in A) on the intracellular ionic concentrations and volume (Table 24-4). Thus, as the effect of the Na^+ pump decreases, Na^+ and Cl^- enter the cell, K^+ leaves, and the cell swells simultaneously. This result is observed in cells whose metabolic pumps are "poisoned."

Table 24-4. Effect of Decrease in Sodium Pumping (Increase in A) on Intracellular Ionic Concentrations and Volume

A	$(c_{Na^+})_{ic}$ (mEq/L)	$(c_{K^+})_{ic}$ (mEq/L)	$(c_{Cl^-})_{ic}$ (mEq/L)	V_{ic} (liters)
0.00136	10.0	160	3.0	1.000
0.01	52.3	114	4.2	1.030
0.1	139	30.3	15.8	1.088
0.3	158	11.5	41.7	1.307
0.5	163	7.1	67.4	1.626

This analysis would not be completely valid in vivo, since the water and ionic shifts come from the extracellular space, thus changing its composition, whereas here we assumed the extracellular environment to be of constant composition.

A MODEL OF TRANSIENT EXCHANGES OF WATER AND SOLUTE AMONG THE BODY FLUIDS

The following analysis utilizes the principles of irreversible thermodynamics to form the basis for a model of water and solute exchanges among the body fluids. The analysis is presented in the form of an outline rather than a rigorous development, and it is only intended to acquaint the reader with the concepts involved in the model, so detailed derivations are omitted.

The body fluids are considered to exist in compartments that contain the fluids in the red blood cells, plasma, interstitial space, and other cells (intracellular fluid) (Fig. 24-4). For notational purposes, these will be denoted as compartments 1 through 4, respectively. Between any two compartments i and j, the volume flux may be expressed as:

$$(J_V)_{i,j} = (L_P)_{i,j} (\Delta P_{i,j} - \sigma_{i,j} \Delta \pi_{i,j})$$

(24-26)

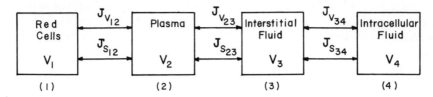

Figure 24-4
Compartmental model of solute and water exchange (J_V = volume flux; J_s = solute flux).

where $(J_V)_{i,j}$ = volume flux from compartment i to compartment j (liter/min)*
$\quad (L_P)_{i,j}$ = cross-coupling coefficient expressing the effect of pressure P_j in producing volume flux $(J_V)_{i,j}$ (liter/min/mm Hg)
$\quad \sigma_{i,j}$ = Staverman reflection coefficient
$\quad \Delta\pi_{i,j}$ = osmotic pressure difference between compartments i and j

A similar expression for solute flux is:

$$(J_s)_{i,j} = (\bar{c}_s)_{i,j} \, (1 - \sigma_{i,j}) \, (J_V)_{i,j} + \omega_{i,j} \, \Delta\pi_{i,j} \tag{24-27}$$

where $(\bar{c}_s)_{i,j} = \overline{X}_s/\overline{V}_s$, or the average molar concentration of solute across the membrane
$\quad \omega_{i,j}$ = solute mobility between compartments i and j

Conservation of water in each compartment requires that:

$$\dot{V}_i = -\Sigma_j (J_V)_{i,j}, \, i \neq j \tag{24-28}$$

and conservation of solute requires that:

$$\dot{m}_i = \dot{V_i(c_s)_i} = -\Sigma_j (J_s)_{i,j}, \, i \neq j \tag{24-29}$$

where m = mass of solute
$\quad c_s$ = solute concentration

    What we want to do with the model is to determine the volume of fluid in each compartment — V_1, V_2, V_3, and V_4 — and the concentration of solute in each — $(c_s)_1$, $(c_s)_2$, $(c_s)_3$, and $(c_s)_4$ — following some disruption of the equilibrium of the system; the equilibrium state is described by a set of initial conditions including c_s^0,

---

*Volume flux (J_V) is defined here in terms of liters per minute, although elsewhere in this book, it is defined as liters per minute per unit area of the partition separating each compartment. For the purposes of this model, however, these areas may be ignored. Similar considerations apply to solute flux (J_s).

V_1^0, V_2^0, V_3^0, and V_4^0. To do this, we will need to solve differential equations for $\dot{V}_i$ and $\dot{m}_i$. There will be one set of the conservation equations (24-28 and 24-29) for each of the four compartments, and one set of the flux equations (24-26 and 24-27) for the flows between each compartment. Thus, there possibly are eight differential equations to describe the conservation requirements of the distribution. We can simplify the set of conservation equations, however, by assuming that the solutes in question do not enter or leave the cells in the time span of interest. Thus,

$$\dot{m}_1 = \dot{m}_4 = 0 \qquad (24\text{-}30)$$

which leaves six differential equations in this set. Additional equations may also be eliminated. Assuming that the total volume of water V_T is known, then, for example, the differential equation for $\dot{V}_1$ can be eliminated since:

$$V_1 = V_T - V_2 - V_3 - V_4 \qquad (24\text{-}31)$$

and V_1 can be calculated without the need of the differential equation for $\dot{V}_1$. Finally, if the total mass of solute m_T is known, then

$$m_3 = m_T - m_1 - m_2 - m_4 \qquad (24\text{-}32)$$

from which, for example,

$$(c_s)_3 = \frac{m_3}{V_3} \qquad (24\text{-}33)$$

can be computed. Thus, only four differential equations are necessary.

The next task is to formulate the pressure changes in the system in terms of model variables. The assumption that cells do not support hydrostatic pressure gradients leads to:

$$\Delta P_{1,2} = \Delta P_{3,4} = 0 \qquad (24\text{-}34)$$

The osmotic pressure difference across membranes can be represented as a sum of two components. The crystalloid osmotic pressure due to the presence of solutes of low molecular weight can be written using the van't Hoff relation developed in Chapter 2 (Eq. 2-21) as:

$$\Delta(\pi_s)_{i,j} = RT \left[(c_s)_i - (c_s)_j \right] \qquad (24\text{-}35)$$

The colloidal osmotic pressure due to the presence of impermeable protein molecules can be written in general as:

$$(\pi_C)_i = f\left[(c_p)_i \right] \qquad (24\text{-}36)$$

where $(c_p)_i$ is the protein concentration in compartment i. The Staverman coefficient, σ, for protein is approximately one, and thus, in general,

$$\Delta\pi_{i,j} = \sigma_{i,j}RT\,[(c_s)_i - (c_s)_j] + (\pi_C)_i - (\pi_C)_j \tag{24-37}$$

The assumption that the body fluids have uniform solute concentrations or osmolarities, $(c_s)_i = (c_s)_j$, at steady state, coupled with the assumption of zero pressure gradient across the cell walls, leads, in the steady state $(J_V = 0)$, to the conclusion that:

$$[(\pi_C)_1 - (\pi_C)_2] = [(\pi_C)_3 - (\pi_C)_4] = 0 \tag{24-38}$$

We know that Equation 24-38 is not strictly true, since the protein content of cells, and thus the colloidal osmotic pressure, is much higher than that of the fluid outside the cells. It can be shown that cation pumping in cells may shift water across the cell walls. This effect is not explicitly included in this model. Instead, we have implicitly assumed that this effect is equal and opposite to the colloidal osmotic pressure effect so that Equation 24-38 holds across cell walls.

In plasma and interstitial fluids – i.e., compartments 2 and 3 – the colloidal osmotic pressure can be approximated by:

$$(\pi_C)_i = 2.1\,(c_p)_i + 0.16\,(c_p)_i^2 + 0.009\,(c_p)_i^3 \tag{24-39}$$

where $(c_p)_i$ is expressed as grams of protein per 100 ml of fluid. Conservation of the mass of protein in these compartments requires that:

$$(m_p{}^0)_i = (c_p)_i\,V_i \tag{24-40}$$

where $m_p{}^0$ is the initial mass of protein in grams.

The formulation of the hydrostatic pressure variation between the plasma and the interstitial compartments is a difficult problem. This is the pressure difference across the capillary wall where the exchange of water and solutes takes place. The capillary pressure normally varies from about 35 mm Hg at the arterial end to about 15 mm Hg at the venous end. There is fluid exchange along the entire length of the capillary; however, for present purposes, we wish to consider this as occurring at some point along the capillary length and driven by some mean capillary pressure, P_2. This pressure has been shown to be a function of the mean hydrostatic pressure of the blood, which in turn depends upon the blood volume, V_b. Therefore, we can formulate the changes in capillary pressure simply as:

$$P_2 = P_2{}^0 + K_2\,(V_b - V_b{}^0) \tag{24-41}$$

where $P_2{}^0$ is the mean capillary pressure at normal blood volume $V_b{}^0$ and K_2 is the inverse of the venous compliance.

The interstitial fluid pressure has been shown by Guyton et al. [2] to be a complex function of the interstitial fluid volume V_3. Data from his experiments fit the empirical relation:

$$P_3 = 2.0 + 2.67 \left(\frac{\Delta V_3}{W}\right) - 8.5 \exp\left[-36\left(\frac{\Delta V_3}{W}\right)\right] \tag{24-42}$$

where ΔV_3 is the change in volume from the initial or normal volume $V_3{}^0$ and W is the weight of the animal in kilograms.

We may now summarize the model equation set (see Fig. 24-4) as:

$$\dot{V}_2 = -(J_V)_{2,3} - (J_V)_{2,1} \tag{24-43}$$

$$\dot{V}_3 = -(J_V)_{3,2} - (J_V)_{3,4}$$

$$\dot{V}_4 = -(J_V)_{4,3}$$

$$(\dot{c}_s)_2 = -(J_s)_{2,3} - (c_s)_2 \, \dot{V}_2$$

where $(J_V)_{2,1} = -(L_P)_{2,1} \, \sigma_{2,1} RT \, [(c_s)_2 - (c_s)_1]$ $\tag{24-44}$

$$(J_V)_{2,3} = -(J_V)_{3,2} = (L_P)_{2,3} \left\{(P_2 - P_3) - (\pi_C)_2 + (\pi_C)_3 - \right.$$

$$\left. \sigma_{2,3} RT \, [(c_s)_2 - (c_s)_3] \right\}$$

$$(J_V)_{3,4} = -(J_V)_{4,3} = -(L_P)_{3,4} \, \sigma_{3,4} RT \, [(c_s)_3 - (c_s)_4]$$

$$(J_s)_{2,3} = -(J_s)_{3,2} = \bar{c}_{2,3} \, (1 - \sigma_{2,3}) \, (J_V)_{2,3} + RT\omega_{2,3} \, [(c_s)_2 - (c_s)_3]$$

$$\bar{c}_{2,3} \approx 0.5 \, [(c_s)_2 - (c_s)_3]$$

The only change in solute concentrations will occur in compartments 2 and 3 (see Eq. 24-30), so that solving for the concentration in either one of these compartments will be sufficient. Since we are contemplating an intervention in the solute concentration of compartment 2 (the plasma) in the "experiment" to follow (see pp. 409–410), we will use the differential equation for $(c_s)_2$ in our model.

Equation sets 24-43 and 24-44 can be solved on a computer to yield V_2, V_3, V_4, and $(c_s)_2$ as functions of time. Then V_1 may be obtained from Equation 24-31 and

$$V_T = V_1{}^0 + V_2{}^0 + V_3{}^0 + V_4{}^0 \tag{24-45}$$

where the superscript zero refers to initial conditions. The other concentrations can be obtained using equations analogous to 24-33 and the appropriate quantities. Thus,

$$m_1 = V_1{}^0 \, c_s{}^0 \tag{24-46}$$

$$m_3 = V_3{}^0 \, c_s{}^0$$

$$m_4 = V_4{}^0 \, c_s{}^0$$

where $c_s{}^0$ is the initial solute concentration of all the body fluids:

$$c_s{}^0 = (c_s{}^0)_1 = (c_s{}^0)_2 = (c_s{}^0)_3 = (c_s{}^0)_4 \tag{24-47}$$

The colloidal osmotic pressures $(\pi_C)_2$ and $(\pi_C)_3$ are determined from Equation 24-39, where

$$(c_p)_i = \frac{(m_p)_i}{V_i} \tag{24-48}$$

and

$$(m_p)_i = V_i{}^0 \, (c_p{}^0)_i \tag{24-49}$$

The hydrostatic pressure P_3 is determined from Equation 24-42. The capillary pressure P_2 is calculated from Equation 24-41, with $P_2{}^0$ being determined from the equilibrium relation,

$$P_2{}^0 - P_3{}^0 = (\pi_C{}^0)_2 - (\pi_C{}^0)_3 \tag{24-50}$$

A typical parameter set for a dog is:

$W = 20$ kilograms
$c_s{}^0 = 0.287$ Osm/liter
$V_1{}^0 = 0.652$ liter
$V_2{}^0 = 0.804$ liter
$V_3{}^0 = 3.2$ liters
$V_4{}^0 = 6.02$ liters
$RT \cdot (L_P)_{2,1} \cdot \sigma_{2,1} = 146$ liter^2/min/Osm
$RT \cdot (L_P)_{3,4} \cdot \sigma_{3,4} = 6$ liter^2/min/Osm
$(L_P)_{2,3} = 5 \times 10^{-3}$ liter/min/mm Hg
$\sigma_{2,3} = 0.05$
$RT \cdot \omega_{2,3} = 2.2$ liters/min
$(c_p{}^0)_2 = 6.24$ gm/100 ml
$(c_p{}^0)_3 = 2.0$ gm/100 ml
$K_2 = 20$ mm Hg/liter

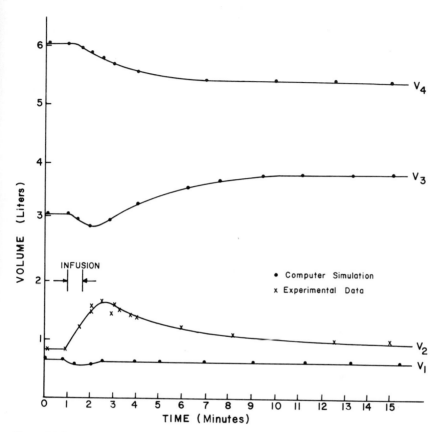

Figure 24-5
Computer simulation results of water distribution dynamics (V_1 = volume of fluid in red blood cells, V_2 = volume of plasma, V_3 – volume of interstitial fluid, and V_4 = intracellular fluid volume).

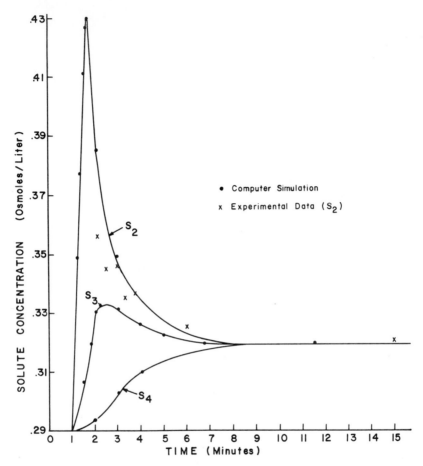

Figure 24-6
Computer simulation results of solute distribution dynamics (S_2 = concentration of solute in plasma, S_3 = concentration of solute in interstitial fluid, and S_4 = intracellular solute concentration).

Using this parameter set, the effects of a pulse infusion of solute and water on the distribution dynamics of water and solute between the various compartments may be calculated.

When the pulse is of 0.67 minute duration and consists of an infusion of 0.2 liter of a 2 Osm/liter solute into compartment 2, the dynamic changes of water volume and solute concentration during and after the infusion as predicted by the model equations are shown in Figures 24-5 and 24-6. In addition, some experimental data are shown for comparison. It should be noted that after a short time (9 to 10 minutes),

the characteristics of the curves of both Figures 24-5 and 24-6 are no longer indicative of a transient state, and the curves converge to steady-state values. The reasonable correlation between the theoretical and experimental data lends support to the
validity of the model.

REFERENCES

1. Edelman, I. S., and Liebman, J. Anatomy of body water and electrolytes. *Am. J. Med.* 27:256, 1959.
2. Guyton, A. C., Granger, H. J., and Taylor, A. E. Interstitial fluid pressure. *Physiol. Rev.* 51:527, 1971.
3. Wolf, A. V. Estimation of changes in plasma and extracellular fluid volume follow-
ing changes in body content of water and certain solutes, by means of an osmometric
equation. *Am. J. Physiol.* 153:499, 1948.

SELECTED READING

Curran, P. F., and Schultz, S. G. Transport Across Membranes. In M. B. Visscher (Ed.),
Handbook of Physiology, Section 6, Alimentary Canal, Vol. 3. Baltimore:
Williams and Wilkins, 1964.
Dick, D. A. T. *Cell Water.* Washington, D.C.: Butterworth, 1966.
Plonsey, R. *Bioelectric Phenomena.* New York: McGraw-Hill, 1969.
Ruch, T. C., and Patton, H. D. *Physiology and Biophysics.* Philadelphia: Saunders,
1965.
Snell, F. M., Shulman, S., Spencer, R. P., and Moos, C. *Biophysical Principles of
Structure and Function.* Reading, Mass.: Addison-Wesley, 1965.
Stacy, R. W., Williams, D. T., Wordon, R. E., and McMorris, R. C. *Essentials of Bio-
logical and Medical Physics.* New York: McGraw-Hill, 1955.
Strauss, M. B. *Body Water in Man.* Boston: Little, Brown, 1957.
Tosteson, D. C., and Hoffman, J. F. Regulation of cell volume by active cation trans-
port in high and low potassium sheep red cells. *J. Gen. Physiol.* 44:169, 1960.
Wiederhielm, C. A. Dynamics of transcapillary fluid exchange. *J. Gen. Physiol.*
52:29s, 1968.

THE NERVOUS SYSTEM

V

An Overview   **25**

Neil Schneiderman and Ronald S. Hosek

INTRODUCTION

Nervous systems are highly organized, interacting populations of living cells that are specialized for the acquisition, transmission, and storage of information. The processing of information within the nervous systems of higher organisms results in coordinated movements and complex behaviors. The cellular elements of the mammalian nervous system vary greatly in their structural properties, but all share, to some degree, the common property of their membranes being electrically active or excitable. It is the electrically responsive nature of nervous tissue that underlies the informational capabilities of the system, since neural information is coded in the form of time-varying transmembrane potentials.

The fundamental cellular element of the nervous system is the *neuron.* Neurons in the adult may vary greatly from a structural standpoint, but they typically have a clearly identifiable cell body, or *soma,* from which different types of branches or processes emanate (Fig. 25-1). The input elements of the neuron, known as *dendrites,* serve as receptors of information from other neurons. They often exist in great numbers and consist of arborized (tree-like) processes. The long process of the neuron is known as the *axon,* and it provides a means for the transmission of information over relatively long distances. Typically, one axon may branch into many *collateral axons,* each of which carries the same information as the original fiber. Many axons are coated with *myelin,* an insulating substance that consists of lipid molecules.

Nerve cells communicate with each other through junctions known as *synapses,* normally, a synapse occurs between the axon of one neuron and the dendrites or soma of another neuron. These synapses may be electrical or chemical in their mode of action. In the case of chemical synapses, which are more usual, a chemical substance is released by the axon terminal. This substance crosses the synapse and influences the membrane potential of the receiving cell. In this manner, chemical synapses provide a one-way conduction of information.

Structurally, neurons may be classified into several types. Thus, we may speak of *monopolar, bipolar,* and *multipolar* neurons as those with one, two, or more processes emanating from their cell bodies (Fig. 25-2). In the case of multipolar neurons, we may define as *isopolar* and *heteropolar* those neurons in which all the processes are or are not the same, respectively.

Estimates of the number of neurons in the human nervous system range from 2×10^{10} to 2×10^{11}. The entire nervous system may be characterized by the activities of reception, transmission, analysis, and command, and it should be noted that an individual neuron may also exhibit all of these activities. For this reason, the

415

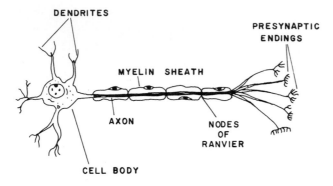

Figure 25-1
Vertebrate spinal motoneuron.

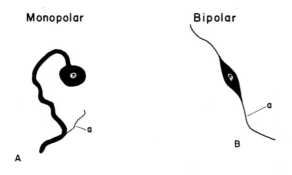

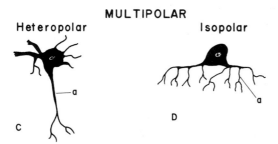

Figure 25-2
Types of neurons: *A.* Unipolar neuron from the dorsal root ganglion of an infant. *B.* Bipolar neuron from the nodose ganglion of an infant. *C.* Heteropolar multipolar motoneuron. *D.* Isopolar multipolar amacrine neuron in the retina. (*a* indicates axon.)

entire nervous system may be thought of as an extension or multiplication of a single neuron.

The nervous system in general is divided into the *central* and *peripheral nervous systems* (Fig. 25-3). The central nervous system (CNS) is made up of the *brain* and *spinal cord.* The peripheral nervous system (PNS) consists of the neural structures outside the CNS. The PNS includes the *cranial nerves,* which transmit information to and from the brain directly; the *peripheral nerves,* which carry information to and from the spinal cord; and the *autonomic nervous system,* which provides commands to the visceral muscles and glands. Although the distinction between the peripheral and central nervous systems is convenient, it should be noted that it is not definitive, since many afferent neurons terminate and many efferent neurons have their cell bodies within the CNS.

Generally, the PNS carries messages to and from the brain and spinal cord, whereas the CNS integrates incoming information and processes it for outgoing transmission.

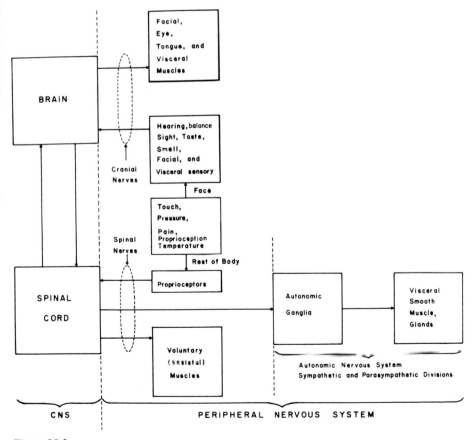

Figure 25-3
Block diagram representation of the nervous system.

Most of the data processing of the nervous system occurs within the CNS, although a certain amount is accomplished peripherally.

The informational input to the nervous system is provided by a special class of cells known as *receptors*. Neurons that transmit information from receptors to the CNS are known as *afferent* or sensory neurons. Afferent neurons provide information from receptors about the state of the internal organs as well as about the outside environment. Neurons within the CNS are referred to as internuncial neurons or *interneurons*. Output commands from the CNS to the muscles or glands (end organs) are transmitted via *efferent* neurons or *motoneurons*. Motoneurons are joined to muscles at special junctions, known as *motor end-plates*, that help transduce neural signals into muscular contraction; the motor end-plates also operate like chemical synapses.

Information is represented in the nervous system in the form of electrical potentials. Associated with all cellular membranes is a *resting potential* (see Chap. 5), which arises as a consequence of the ionic distributions surrounding the membranes. Superimposed on the resting potentials are time-variant potentials of various types that represent the biologically coded information. These potentials are usually identified in terms of the structure involved: associated with a receptor cell is a *generator potential*; the synapse is characterized by the *postsynaptic potential*; and characteristic of the motor end-plate is the *end-plate potential.* These particular potentials, though involving active membrane phenomena, may be classified as nonpropagating. The neuron has the unique ability to generate and sustain an impulse, known as the *action potential* (see Chap. 26), that can propagate undiminished along the entire length of an axon. In addition to the action potential, the neuron also exhibits different types of nonpropagating transmembrane potential changes.

Neurons are connected together for processing purposes into large families or *nets.* The communications within the nets occur in the form of combinations of the potentials described above. The general schema is that little gain or loss of information occurs during transmission along an axon. Processing takes place as a function of the convergences and divergences of different neurons at successive synapses. Thus, in the visual system where several synapses occur before the optic nerve enters the brain, considerable processing occurs in the periphery. In contrast, within the touch (somatosensory) system, little processing occurs in the periphery because the first synapse is located in the central nervous system.

In the remainder of this chapter, we shall examine the nervous system from an anatomic and organizational standpoint. This is done to acquaint the reader with how the various neural structures fit together, to familiarize him or her with some of the nomenclature, and to provide a general feeling of the extent and complexity of the system. In subsequent chapters, we shall examine the neuron and other related nervous system elements in greater detail. Engineering analysis and mathematical modeling techniques will be utilized to help elucidate the manner in which the the nervous system accomplishes the functions of reception, transmission, analysis, and command.

COMMON ANATOMIC TERMS

A number of common terms of orientation have been adopted to facilitate reasonably unambiguous descriptions of the relationships among anatomic structures. Several of these will be used in our description of the nervous system. *Ventral,* for example, refers to structures toward the belly, and *dorsal* to structures toward the back surface of the organism. *Cranial* or *anterior* refers to structures toward the head or cephalic end of the body, whereas *caudal* or *posterior* refers to structures toward the lower or tail end. In relating one structure to another, *superior* means above and *inferior* means below; *proximal* means near to and *distal* means away from the structure under consideration.

Any section dividing the body into left and right portions is a *sagittal* section. The perpendicular section running exactly down the midline is a *midsagittal* section. The relationship of structures toward or away from the midline is described by the terms *medial* (toward the midline) or *lateral* (away from the midline). *Ipsilateral* means on the same side; *contralateral* means on the opposite side.

A section dividing the ventral from the dorsal surface of the body would be termed a *frontal* section, and a section severing the head from the trunk would furnish a *horizontal* section. Brain structures are usually depicted as sagittal, frontal, or horizontal sections.

PERIPHERAL NERVOUS SYSTEM

The peripheral nervous system consists of cranial nerves and spinal nerves. Since nerves are bundles of neurons that travel together, they may be all afferent, all efferent, or mixed. The afferent fibers of nerves receive stimuli from receptors. Motor fibers are of two types: *somatic motor fibers* that terminate in the skeletal muscle and *autonomic fibers* that innervate the smooth muscle, cardiac muscle, and glands. Somatic motor fibers and autonomic fibers travel in separate bundles.

Most peripheral nerve fibers have a myelin sheath and a *neurilemma* (sheath of Schwann) as well. Some peripheral fibers, however, lack the myelin sheath but have a neurilemma. A completely severed peripheral nerve has some capacity to repair itself provided that the cell bodies have not been destroyed.

CRANIAL NERVES

There are 12 pairs of cranial nerves. These cranial nerves pass through foramina (openings) in the base of the cranium to reach their terminations. Most of them are mixed nerves having motor and sensory fibers, but a few, such as the olfactory and optic nerves, are only sensory. The numerical designation, name, and primary functions of the cranial nerves are summarized in Table 25-1.

Table 25-1. Summary of the Cranial Nerves

Number	Name	Primary Functions
I	Olfactory	Afferent for smell
II	Optic	Afferent for vision
III	Oculomotor	Afferent and efferent to all extraocular eye muscles except two Efferent to intrinsic eye muscles
IV	Trochlear	Afferent and efferent to one eye muscle
V	Trigeminal	Afferent from skin and mucous membranes of head as well as from chewing muscles Efferent to chewing muscles
VI	Abducent	Afferent and efferent to a single eye muscle
VII	Facial	Afferent from taste buds of anterior two-thirds of the tongue
VIII	Acoustic	Afferent for hearing (cochlear division) and equilibration (vestibular division)
IX	Glossopharyngeal	Afferent from throat and posterior one-third of the tongue
X	Vagus	Afferent from throat, larynx, and viscera Efferent to viscera, including lungs, heart, esophagus, stomach, liver, pancreas, small intestine, and large intestine
XI	Accessory	Afferent from baroreceptors Efferent to viscera along with vagus; efferent also to throat, larynx, and neck and shoulder muscles
XII	Hypoglossal	Afferent and efferent to tongue muscles

SOMATIC SPINAL NERVES

There are 31 pairs of spinal nerves. Each spinal nerve is attached to the spinal cord by a *dorsal* and a *ventral root.* Afferent impulses enter the cord via the dorsal root, whereas efferent impulses leave the cord by the ventral root. At a short distance from the cord, the roots join together to form the main trunk of the spinal nerve. The typical spinal nerve has (1) a *dorsal ramus* that supplies the skin and muscles of the back, (2) a *ventral ramus* that supplies the more ventral regions, and (3) a *communicating ramus* that contributes to the sympathetic division of the autonomic nervous system.

The spinal nerves and their associated dorsal roots have an anterior-to-posterior segmental arrangement. This means that the dorsal roots innervated by a particular level of the body come from a given level of the spinal cord. The skin region that provides nerve impulses to one dorsal root is called a *dermatome.* Most skin regions provide input to at least two dorsal roots; some regions share three dermatomes.

AUTONOMIC NERVOUS SYSTEM

The autonomic nervous system is an efferent neural system that innervates glands, cardiac muscle, and the smooth muscles of the viscera. These smooth visceral muscles also include those of the blood vessels. The autonomic nervous system thus supplies all of the effector organs of the body except the skeletal muscles, and it transmits motor commands to the circulatory, respiratory, digestive, urinary, reproductive, and endocrine systems.

The functions of the autonomic nervous system are exclusively motor inasmuch as its fibers only transmit efferent impulses to specific muscles and glands. These impulses tend to increase and decrease ongoing activities rather than initiate new actions. Thus, for example, the heart is intrinsically rhythmic and the autonomic nervous system modulates this rhythmicity.

The autonomic nervous system has two divisions. One of these, called the *sympathetic* (or thoracolumbar) division, originates from the thoracic and lumbar segments of the spinal cord (Fig. 25-4). In contrast, a *parasympathetic* (or craniosacral) division originates from the midbrain and medulla of the brain and from the sacral portion of the spinal cord in the lower back (Fig. 25-5).

In general, both the parasympathetic and sympathetic divisions of the autonomic nervous system innervate organs in a dual manner. In most instances, the influences of the two divisions are antagonistic. The two vagus nerves to the heart, for example, belong to the parasympathetic division and are responsible for slowing the heart rate. In contrast, the cardioaccelerator nerves to the heart belong to the sympathetic division and are influential in increasing heart rate and contractility.

Although in many instances the parasympathetic and sympathetic inputs to an organ act in an opposite manner, this is not invariably the case. The actions of the two divisions upon the salivary glands, for example, occur in the same direction with complementary effects. Since each of the two divisions of the autonomic nervous system can produce either inhibitory or excitatory effects, it is also possible for their actions either to oppose each other or to act synergistically. In the case of synergism, for example, the parasympathetic excitation of an organ may be accompanied by inhibition of the sympathetic input to the same organ.

In general, the autonomic pathways from the central nervous system to particular organs consist of two neuron chains. The cell body of the first-order neuron lies within the central nervous system, whereas the cell body of the second-order neuron lies within a ganglion or innervated organ. In the parasympathetic division, the synapse between the preganglionic and the postganglionic fibers is close to or within the innervated organ. In the sympathetic division, the synapse between the preganglionic and the postganglionic neuron is located in a sympathetic ganglion chain that runs outside of and parallel to the vertebral canal.

CENTRAL NERVOUS SYSTEM

The central nervous system consists of the brain and spinal cord. The brain is encased in the skull, or *cranium,* and is directly connected to the spinal cord through an

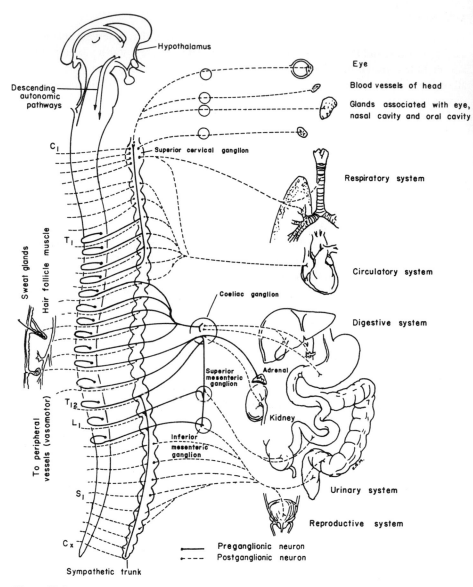

Figure 25-4
Sympathetic innervation of the autonomic nervous system. (From C. R. Noback. *The Human Nervous System: Basic Elements of Structure and Function.* Illustrated by Robert J. Demarest. Copyright © 1967 by McGraw-Hill, Inc. Used with permission of McGraw-Hill Book Company.)

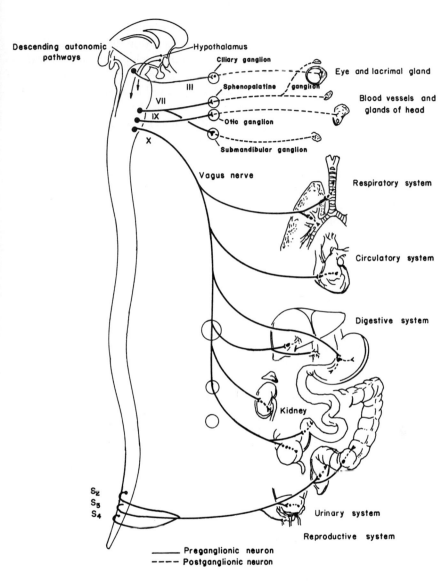

Figure 25-5
Parasympathetic division of the autonomic nervous system. (From C. R. Noback. *The Human Nervous System: Basic Elements of Structure and Function.* Illustrated by Robert J. Demarest. Copyright © 1967 by McGraw-Hill, Inc. Used with permission of McGraw-Hill Book Company.)

opening in the skull known as the *foramen magnum.* The spinal cord runs through a canal formed by openings in the *vertebrae,* the bones that form the spine. Both the brain and spinal cord are enclosed in three membranous envelopes called the *meninges.* If the brain or spinal cord were sliced open, the resulting section would reveal areas of *gray* and *white matter.* The gray matter includes neural cell bodies, whereas the white matter contains bundles of axons that appear white due to their myelin coating. The cell bodies are often clustered together to form structures known as *nuclei* or *ganglia.* Groups of axons often run together in a parallel fashion, forming cables known as *fiber tracts, lemnisci,* or *fasciculi.* In addition to neurons, the brain and spinal cord contain synapses and *glial cells.* The latter are actually much more numerous in the CNS than are neurons, and they are thought to serve the neurons in a supportive fashion, both metabolically and structurally.

From a structural standpoint, a preliminary overview of the central nervous system may best be conceptualized in terms of its embryologic development. During the course of embryonic development in mammals, the CNS begins to form as a tubular structure. The anterior portion of the neural tube develops into the brain, whereas the rest of the tube forms the spinal cord. Three conspicuous enlargements soon emerge in the anterior portion of the tube; these are the forebrain, midbrain, and hindbrain. As embryologic growth and differentiation continue, the forebrain and hindbrain subdivide, but the midbrain does not. The most anterior portion of the forebrain becomes the *telencephalon,* which consists of the cerebral hemispheres that form the outer mantle of the brain. The posterior portion of the forebrain becomes the *diencephalon,* which includes the thalamus and hypothalamus. In the hindbrain, the anterior portion forms the *pons* and *cerebellum,* and the posterior portion forms the *medulla* (Fig. 25-6). The tubular arrangement of the early stage of the embryo continues throughout development. As development progresses, the central cavity of the tube differentiates into the ventricles and canals of the brain and spinal cord.

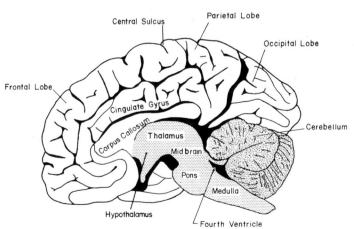

Figure 25-6
Midsagittal section of the human brain.

VENTRICULOMENINGEAL SYSTEM

The three meninges that enclose the brain and spinal cord are the *dura mater, arachnoid,* and *pia mater.* The dura mater is a tough, fibrous, inelastic membrane that forms the outermost covering of the brain. It is in two layers: the outer layer lines the bones of the cranial cavity; the inner layer surrounds the brain and spinal cord. Large folds of dura mater form partitions between various brain structures. Thus, for example, a structure called the *falx cerebri* separates the two cerebral hemispheres.

The arachnoid membrane is a delicate membrane just internal to the dura mater. Between the arachnoid and the deepest membrane (the pia mater) is the subarachnoid space, which is filled with cerebrospinal fluid. The arachnoid membrane may be visualized as a fluid-filled cushion separating the brain from the cranial bones. This fluid-filled cushion protects the brain by distributing the effects of abruptly applied forces. The fluid has a specific gravity of 1.0 so that the brain literally floats in it. The brain thus has a net weight of about 50 gm as a result of this buoyancy, whereas the actual weight of the brain is about 1400 gm.

The pia mater is the innermost membrane to cover the brain and spinal cord, and it is highly vascular. It closely invests the brain and dips into all the cerebral furrows. Invagination of the pia mater into the lateral ventricles of the cerebrum forms the *choroid plexuses,* which elaborate cerebrospinal fluid. *Cerebrospinal fluid* (CSF) is a clear, colorless liquid that contains proteins, glucose, urea, salts, and some leukocytes.

The brain itself has four major cavities that contain CSF. These cavities comprise the two *lateral ventricles* of the cerebrum, the *third ventricle* of the diencephalon, and the *fourth ventricle* in the lower brain stem. The ventricles communicate with each other and with the subarachnoid space to provide a continuous flow of CSF.

The two lateral ventricles lie within the lower, medial portion of the cerebrum. They are separated by a thin layer of tissue called the *septum pellucidum.* The lateral ventricles communicate with the third ventricle via the *intraventricular foramina,* and the third ventricle in turn communicates with the fourth via the *cerebral aqueduct.* A central canal in the lower medulla and spinal cord receives CSF from the fourth ventricle.

A secreting organ, the *choroid plexus,* is found within each ventricle. CSF arises within the choroid plexus and flows from the lateral ventricle downward toward the fourth ventricle. The fluid escapes through the foramina of the fourth ventricle into the subarachnoid space. This flow of CSF arises from the inequity between its production by the choroid plexus and its absorption by portions of the dura mater. The total volume of CSF in the adult is about 140 ml. Of this, about 100 ml is to be found in the subarachnoid space. The normal range of CSF pressure with the subject in the recumbent position is from 115 to 180 mm Hg. Approximately 50 to 100 ml of CSF is manufactured daily, but this amount can increase drastically under pathologic circumstances [3].

GLIAL CELLS

There are various kinds of supporting cells in the CNS that collectively are called *neuroglial cells,* or simply glial cells. These glial cells are perhaps ten times more numerous in the brain than are neurons. The processes of these glial cells weave between and around nerve cells. They are involved in producing the myelin sheaths that surround many axons in the CNS. There is some speculation that they may have important metabolic functions. Glial cells attach themselves to blood vessels as well as to neurons.

BLOOD-BRAIN BARRIERS

Many years ago, experimenters observed that when vital dyes are injected systemically into the bloodstream, most tissues of the body become stained, but the tissues of the brain and spinal cord do not. Subsequent research has shown that there are significant differences in the rates of transfer of various substances from the plasma to the CSF and from the plasma to brain cells. Some substances do not penetrate the walls of the cerebral capillaries at all, some do so slowly, and others penetrate very rapidly. The restricted diffusion of some molecules into the brain relative to their diffusion into other tissues is referred to as the *blood-brain barrier.* Because several variables influence the rate at which substances diffuse into the brain, the term "blood-brain barrier" is best defined in terms of rate-constants instead of in terms of an absolute barrier.

CEREBRAL HEMISPHERES

The cerebral hemispheres in humans envelop the anterior brain stem and form the largest portion of the brain. Roughly two-thirds of the neurons in the human brain are in the cerebrum. A deep, midsagittal cleft separates the two cerebral hemispheres from one another. Communication between the two hemispheres is carried out over distinct bands of fibers. The most prominent band crossing the midline is known as the *corpus callosum* (see Fig. 25-6).

The surface of the human cerebral hemispheres has a large number of convolutions and furrows. These convolutions are called *gyri,* whereas the furrows or crevices are called *sulci.* Deep furrows are sometimes called *fissures,* but the terms "fissure" and "sulcus" are occasionally used interchangeably.

Each cerebral hemisphere consists of three tiers. The outermost tier, called the *cerebral cortex,* has a grayish appearance because it contains many cell bodies. It is only about 2.5 mm thick, but its convolutions are so pronounced in humans that almost half of the cortex is hidden from view among the crevices. The phylogenetically newest portions of the cerebral cortex are referred to as the *neocortex.* In contrast, the more primitive portions of the cortex belong to a series of connected structures known as the *limbic system.* For the most part, these structures are found along the medial wall of the cerebrum and form the base of the cerebrum over the brain stem. The second tier of the cerebrum is whitish in appearance because of the heavy coating of myelin that covers the axons running to and from the cortex. The most internal of the three tiers of the cerebrum is the gray matter. This third tier consists of nuclear structures that are collectively called the *basal ganglia.*

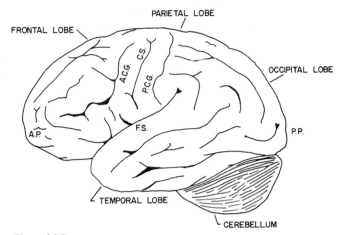

Figure 25-7
Lateral aspect of the brain showing major subdivisions and sulci. *ACG*: anterior (pre)central gyrus; *AP*: anterior pole; *CS*: central sulcus; *FS*: fissure of Sylvius; *PCG*: posterior central gyrus; *PP*: posterior pole.

The two hemispheres are divided by a deep crevice known as the longitudinal cerebral fissure or *sagittal sulcus.* Major furrows further divide each hemisphere into five lobes. These furrows are incomplete, however, so that complete separation of the lobes is, at least in part, arbitrary. The *frontal, parietal, temporal,* and *occipital* lobes are readily identifiable in the lateral view of the neocortex (Fig. 25-7); the fifth lobe, called the *insula,* is not. A crevice known as the central sulcus, or *fissure of Rolando,* separates the frontal lobe from the parietal lobe. Another cleft known as the lateral sulcus, or *fissure of Sylvius,* separates the temporal lobe from the frontal and parietal lobes. The insula is visible only when the sylvian fissure is retracted.

The Neocortex

On the basis of histologic, physiologic, and clinical evidence, the cortex has been divided into sensory, motor, and associational areas. Regions that receive impulses originating in the sense organs or receptors are designated as *sensory areas.* In contrast, regions that send output from the neocortex to the musculature are called *motor areas.* Most of the neocortex appears to be neither clearly sensory nor motor in function, and it is thus referred to as the *association cortex.*

The regions of the neocortex that receive primary sensory information are called *primary* sensory areas. The areas surrounding these primary sensory areas are called *secondary* sensory areas because they are more concerned with the organization of perceptions than they are with the reception of sensory input. The areas of the neocortex that control visual sensations and perceptions are located in the occipital lobe. The area for audition is located in the temporal lobe. An area of the neocortex that receives information about the body senses is located in the postcentral gyrus of the parietal lobe. This gyrus is situated immediately dorsal and posterior to the central sulcus and is concerned with the reception of touch, pressure, position sense, and skin temperature (proprioception).

Just anterior to the postcentral gyrus and across the central sulcus is the precentral gyrus. The motor area, which plays an important role in the regulation of body movement, lies in this precentral gyrus of the frontal lobe. Over 100 years ago, investigators demonstrated that electrical stimulation of the precentral gyrus induces movement in muscles on the contralateral (opposite) side of the body. This occurs because most of the neurons leaving the motor cortex cross the midline before synapsing in the spinal cord.

The body parts that receive projections from the motor cortex are topographically represented in the precentral gyrus. A similar representation for body parts that project somatosensory information can be drawn for the postcentral gyrus (Fig. 25-8). Although there are fundamental differences in the basic functions of the precentral and postcentral gyri, they are only relative; each area in part shares the functions of the other.

Immediately anterior to the primary motor area in the precentral gyrus is yet another area concerned with motor function. In contrast to the neurons of the precentral gyrus, which are responsible for individual voluntary motor acts, the neurons of the *premotor area* grade whole series of movements. Damage to the premotor area disrupts the timing of movements and coarsens higher motor organization.

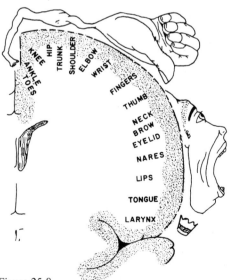

Figure 25-8
Frontal section of the left hemisphere through the postcentral gyrus (see Fig. 25-7) demonstrating the topographic specificity of the somatosensory cortex. Note that the sections of the body involved with fine control (i.e., the hands and face) have the greatest surface representation. (From W. Penfield and T. Rasmussen. *The Cerebral Cortex of Man.* New York: Macmillan, 1950. Reproduced with permission.)

In 1861, the neurologist Paul Broca related the loss of speech to the damage of an area in the left hemisphere that is approximately 2.5 cm in diameter [1]. Subsequent research has shown that the area of the neocortex that controls speech is more extensive than Broca thought, and it extends back into the parietal, occipital, and temporal lobes.

White Matter of the Cerebrum

The white matter of the cerebrum is internal to the cortex. It consists of myelinated *associational, commissural,* and *projectional* fibers. The associational fibers connect gyri within the same hemisphere. Commissural fibers cross the midline to connect structures at the same level. The largest commissure in the brain is the corpus callosum (Fig. 25-9); it connects the neocortex across the midline. Two other commissural bundles that connect forebrain structures across the midline are the *hippocampal commissure* and the *anterior commissure.* Unlike the commissural fibers, which join equivalent structures, the projectional fibers go from the cortex to lower structures, or vice versa. When projectional fibers cross the midline, they are said to *decussate.* Thus, for example, fibers originating in the motor cortex decussate in the medulla. Similarly, axons conveying somatosensory impulses decussate before ascending to the cortex.

Interruption of a decussation has immediate and observable results. If a motor pathway is severed, paresis (weakening) or paralysis occurs. If a sensory pathway is

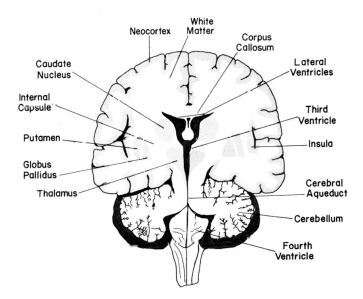

Figure 25-9
Frontal section through the cerebral hemispheres and brain stem showing the white matter and basal ganglia.

lesioned, there is a loss of sensation. In contrast to the effects caused by the interruption of a decussation, the effects of severing a commissure are more subtle and are not immediately apparent. This is because the organism continues to receive information in both hemispheres. Even when one eye is covered, for example, the other eye continues to send input to both halves of the brain. Experiments in which input is provided to only one side of the brain following section of the cerebral commissures, however, indicate that the human brain is asymmetrical. For most people, the left cerebral hemisphere appears to be crucial for verbal activity, whereas the right cerebral hemisphere appears to be superior for accomplishing nonverbal, perceptual tasks.

Basal Ganglia

The basal ganglia are a group of large nuclei lying deep within the cerebrum. The major nuclei of the basal ganglia include the *caudate nucleus,* the *putamen,* and the *globus pallidus.* Together the caudate nucleus and putamen make up the *corpus striatum.* The striatum plays an important role in the control of movement. Parkinson's disease, for example, which is characterized by tremor and delay in voluntary muscle action, is usually associated with lesions of the striatum.

DIENCEPHALON

The diencephalon is the most anterior portion of the brain stem. It includes the *epithalamus, thalamus, hypothalamus,* and *subthalamus.* Through its neural and vascular connections with the hypothalamus, the pituitary gland is intimately related to the diencephalon. In the center of the diencephalon is the third ventricle.

Epithalamus

The epithalamus consists of the *pineal body* and the *habenula.* The pineal body is a reddish-gray gland about 10 mm in length that is attached to the roof of the third ventricle (Fig. 25-10). Its function in mammals is to influence the activity of the reproductive system in response to circadian and seasonal variations in sunlight.

The manner in which light controls gonadal activity is complex [4]. Light-induced excitation influences the pineal gland in an indirect manner by way of activation within the sympathetic nervous system. Norepinephrine, which is released by the sympathetic nervous system, inhibits the activity of an enzyme, hydroxyindole-O-methyl transferase. This enzyme is needed by the pineal gland to synthesize melatonin. Since melatonin is a hormone that normally inhibits the gonads, prolonged exposure to light tends to release the gonads from inhibition. Tumors of the pineal gland are rare, but disorders leading to the hypersecretion of melatonin in children result in delayed sexual development, whereas hyposecretion results in precocious puberty.

The habenula is a triangular depression just anterior to the pineal body. Afferent inputs to the habenula come from olfactory nuclei and from portions of the limbic system. Efferent projections from the habenula ultimately synapse in the motor nuclei

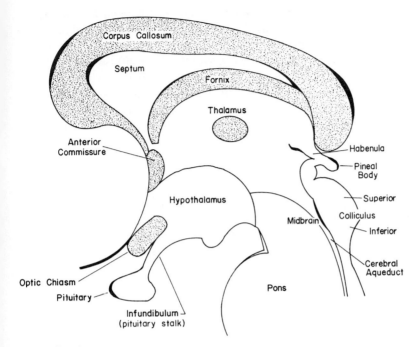

Figure 25-10
Midsagittal section depicting some important structures in the diencephalon and midbrain.

of some cranial nerves. Hence, the habenula appears to be involved in orienting organisms toward olfactory stimuli.

Thalamus

The thalamus constitutes the dorsal portion of the diencephalon. It consists of two, large, ovoid masses about 4 cm in length. Each thalamus is bounded medially by the third ventricle and laterally by the *internal capsule,* which is a large fiber pathway passing to and from the cortex.

One important role of the thalamus is in the processing of sensory information. Afferent fibers subserving all sensory modalities except olfaction synapse within the thalamus. The thalamic sensory relay nuclei include the *lateral geniculate nucleus,* the *medial geniculate nucleus,* and the *ventrobasal complex.* Fibers from the visual pathway are relayed from the lateral geniculate nucleus to the occipital cortex. Similarly, fibers from the auditory pathway are relayed from the medial geniculate nucleus to the temporal lobe, and fibers from somatosensory pathways are relayed from the ventrobasal complex to the parietal lobe.

In addition to the specific sensory relay nuclei just described, there are other cortical relay nuclei in the thalamus. The anterior nuclei of the thalamus appear to be part of a complex feedback system that involves both the hypothalamus and the most anterior portions of the frontal cortex. Similarly, the *ventrolateral nucleus* of the thalamus is part of a feedback system that involves both the cerebellum and motor cortex in the regulation of movement.

The thalamus includes a number of nuclear groups that have diffuse connections with the cortex, in contrast to the point-to-point connections that are characteristic of specific relay nuclei. These diffusely projecting nuclei play an important role in the regulation of the rhythmic electrical activity of the cortex, thereby modulating such states as attention, consciousness, and sleep. The perception of pain is also closely related to the functioning of the diffusely projecting regions of the thalamus.

Hypothalamus

The hypothalamus constitutes the major portion of the ventral region of the diencephalon and forms the floor of the third ventricle. A major landmark of the posterior hypothalamus is a pair of rounded structures known as the *mamillary bodies.* In front of the mamillary bodies lies the *tuber cinereum,* from which descends a hollow process known as the *infundibulum* or pituitary stalk. At the distal end of the infundibulum is the pituitary gland. Through its connections with the pituitary, the hypothalamus plays an important role in water, fat, and carbohydrate metabolism. It coordinates neural and endocrine activity, and it is the source of posterior pituitary hormones.

Subthalamus

The subthalamus forms the ventrolateral portion of the diencephalon. It is the diencephalic area at the junction of the thalamus and the midbrain. The subthalamus receives fibers from the striatum and functions in the regulation of motor control. In man, large isolated lesions of the subthalamus give rise to coarse, involuntary movements of the upper extremities. In the pathologic condition known as *hemiballismus,* this may be expressed as the continuous, uncontrolled flinging of an arm.

MIDBRAIN

The midbrain, or mesencephalon, is a short, constricted segment of the brain stem. It connects the diencephalon above with the pons and cerebellum below. The *cerebral aqueduct* passes through the midbrain and connects the third ventricle of the diencephalon with the fourth ventricle of the medulla.

The midbrain consists of a dorsal portion, called the *tectum* or roof, and a more massive, ventral portion known as the *cerebral peduncles.* The peduncles in turn consist of a dorsal portion, called the *tegmentum,* and a more ventral section, termed the *basis pedunculi.*

In the upper part of the midbrain the tectum is composed of the *superior colliculi* (Fig. 25-11), whereas more caudally, it is composed of the *inferior colliculi.* The superior colliculi are concerned with visual functions and the inferior colliculi with auditory ones.

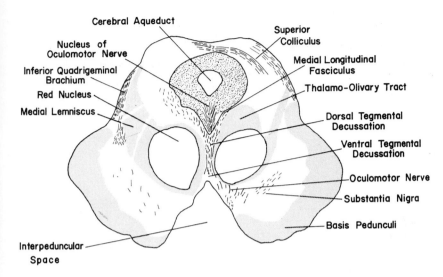

Figure 25-11
Section through the mesencephalon at the caudal borders of the superior colliculi.

Major somatosensory pathways ascend through the more lateral and dorsal portions of the midbrain tegmentum. These include the *medial lemniscus* and the *spinothalamic tract*. More medially, the tegmentum consists of complex polysynaptic
pathways and associated nuclei that make up the *reticular formation*. Ascending
portions of the reticular formation are involved in arousal. Descending portions are
importantly involved in motor functions. A prominent structure in the tegmental
reticular formation is the *red nucleus*. The afferent fibers projecting to the red nucleus
are principally derived from the motor cortex and cerebellum. The efferent projections
from the red nucleus are involved in the regulation of movement.

Whereas the red nucleus is part of an intricately organized feedback system that is
involved in the overall regulation of movement, more somatotopically organized motor
pathways originating in the motor cortex descend through the basis pedunculi. These
pathways descend through the pons to the medulla and spinal cord.

PONS

The midbrain continues below into a massive, rounded structure known as the pons
(Fig. 25-12). The pons lies directly above the medulla and ventral to the cerebellum.
It contains a tegmental, or dorsal, portion and a basilar, or ventral, portion.

The tegmental portion forms the upper part of the floor of the fourth ventricle.
It contains longitudinal fiber tracts that are continuous with those found in the medulla.
Many of these fiber tracts are sensory in origin.

The basilar portion of the pons contains both transverse and longitudinal fibers.
The transverse fibers lie in a thick layer on the ventral surface of the pons. They form

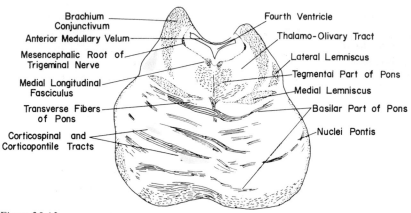

Figure 25-12
Transverse section through the brain stem showing the pons and the pathways leading to and from the cerebellum.

a bridge between the pons and the cerebellum. The longitudinal fibers in the basilar portion of the pons consist of three descending motor pathways. These are the *cortico-spinal tract,* the *corticobulbar tract,* and the *corticopontine tract.* Both the cortico-spinal and corticobulbar tracts continue their descent after leaving the pons. The corticospinal tract continues to the spinal cord, and the corticobulbar tract terminates in the motor nuclei of cranial nerves in the medulla. In contrast, the fibers of the corticopontine tract synapse in the pontine nuclei; these pontine nuclei send fibers to the cerebellum.

CEREBELLUM

The cerebellum lies above the medulla, dorsal to the pons, and inferior to the occipital lobes of the cerebral hemispheres. The cerebellum's three major connections are with the brain stem. A *superior cerebellar peduncle* connects the cerebellum with the mid-brain. The fiber connections between the pons and cerebellum consist of the *middle cerebellar peduncle.* A third fiber bundle that connects the cerebellum with the medulla is referred to as the *inferior cerebellar peduncle.* The fourth ventricle separates the pontine tegmentum from the cerebellum.

The human cerebellum consists of two hemispheres that are connected by a median strip called the *vermis cerebelli.* Numerous parallel folds traverse the hemispheres and the vermis. These folds, known as *folia,* give the surface of the cerebellum its charac-teristic corrugated appearance. As was the case for the cerebral hemispheres, the cere-bellum consists of a superficial mantle of gray matter, the *cortex,* and an internal mass of white matter in which are buried several gray nuclear masses. These nuclear masses within the cerebellum are collectively referred to as the *deep cerebellar nuclei.*

Afferent inputs arrive in the cerebellar cortex from numerous places. One main source is from receptors in muscles and glands, a second is from the vestibular system, and a third source of cerebellar afferent information takes the form of motor and sensory inputs from the cerebral cortex.

The efferent fibers from the cerebellum typically originate in the deep cerebellar nuclei. They project to the motor cortex, ventrolateral thalamus, red nucleus, vestibular nucleus, and inferior olive. The efferent outputs from the cerebellum contribute to several loops that provide feedback to the cerebellum. In this manner, the cerebellum appears to function as a servomechanism that exercises considerable control over the coordination of movement, muscle tone, and equilibrium [2].

MEDULLA OBLONGATA

The medulla or bulb is the lowermost portion of the brain. It lies below the pons and above the spinal cord (Fig. 25-13). The dorsal surface of the upper half of the medulla forms the floor of the fourth ventricle. As it proceeds caudally (away from the cortex), the ventricle narrows into a central canal that continues into the spinal cord. The demarcation between the medulla and the cord is the large opening in the skull that is called the foramen magnum. Since the pathways between the medulla and cord run uninterruptedly through the foramen, this demarcation between the brain and spinal cord is somewhat arbitrary.

A pair of enlargements called the *pyramids* form the ventral surface of the medulla. The pyramids consist of nerve fibers that decussate at the level of the pyramids, whereas the remaining fibers descend ipsilaterally to the cord. For this reason, lesions above the pyramids cause a greater weakening (paresis) of contralateral than ipsilateral motor function.

The dorsal portion of the lower medulla contains the dorsal columns and their associated nuclei, the *nuclei gracilis* and *cuneatus*. These tracts and nuclei play an important role in conveying somatosensory information toward the cortex. Many of the first-order neurons involved in discriminative touch ascend through the dorsal columns and synapse in the dorsal column nuclei. The second-order neurons in this system decussate in the medulla and ascend to the thalamus in a tract known as the *medial lemniscus*. Other somatosensory pathways ascend through the lateral region of the medulla. The inferior olive, which connects with the cerebellum, is also to be found in the lateral region of the upper medulla.

The internal portion of the medulla includes the reticular formation. Embedded within the reticular formation are nuclei and regions that play an important role in the control of vital functions such as blood pressure. The decussations of the pyramids and the medial lemniscus also constitute internal medullary regions.

SPINAL CORD

The spinal cord is the portion of the CNS that is located within the *vertebral canal*; it lies outside of the cranial cavity. Two major functions of the cord are: (1) it forms a conducting pathway to and from the brain and (2) it serves as a site of reflex

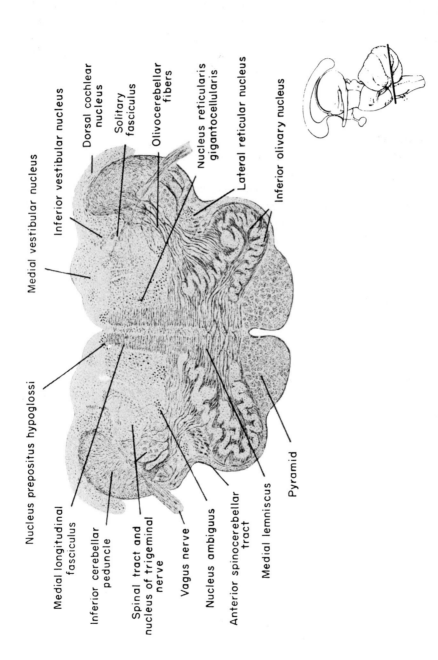

Nucleus prepositus hypoglossi

Medial vestibular nucleus

Inferior vestibular nucleus

Dorsal cochlear nucleus

Solitary fasciculus

Olivocerebellar fibers

Nucleus reticularis gigantocellularis

Lateral reticular nucleus

Inferior olivary nucleus

Medial longitudinal fasciculus

Inferior cerebellar peduncle

Spinal tract and nucleus of trigeminal nerve

Vagus nerve

Nucleus ambiguus

Anterior spinocerebellar tract

Medial lemniscus

Pyramid

Figure 25-13

Transverse section of the brain through the medulla at the level of the cerebellum. (From R. C. Truex and M. B. Carpenter. *Human Neuroanatomy* (6th ed.). Baltimore: Williams & Wilkins, 1969. Reproduced with permission of The Williams & Wilkins Co.)

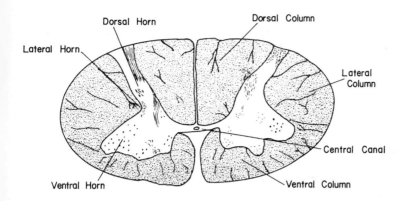

Figure 25-14
Cross section through the cervical region of the spinal cord.

mechanisms. The spinal cord, like the brain, may be divided into gray and white sections; the former contains mostly cell bodies, and the latter, myelinated axons or fibers.

The gray matter is conveniently divided into *ventral* and *dorsal horns*. In the upper portion of the cord (Fig. 25-14), *lateral horns* may also be distinguished. The ventral horns are occupied principally by the cell bodies of large motor neurons. The extensive destruction of these cell bodies, which occurs in poliomyelitis, induces paralysis. The dorsal horns contain the cell bodies of secondary or higher order sensory neurons. Lateral horn cells are efferent neurons that supply the sympathetic nervous system. Both the conducting pathways and the local spinal reflexes can involve synapses within the gray matter.

The white matter of the cord is divided by the gray matter into three general regions; these are referred to as the *dorsal, lateral,* and *ventral columns.* The dorsal columns are entirely afferent in nature, but both the lateral and the ventral columns contain discrete tracts of afferent and efferent fibers.

Some afferent fibers synapse in the gray matter of the cord. These may then give rise to interneurons that ascend to the brain via the white matter. In the same fashion, some efferent fibers descend through the white matter before synapsing in the gray matter of the cord. As previously indicated, however, not all afferent fibers ascend or give rise to fibers that ascend within the cord. The simple *spinal reflex arc,* for example, consists of (1) an afferent neuron that has its cell body outside of the cord, (2) an interneuron that lies within the gray matter, and (3) an efferent neuron that terminates upon a muscle but has its cell body within the gray matter of the cord.

In such a reflex, receptor stimulation gives rise to afferent activity that enters the spinal cord and causes efferent activity on the part of the spinal motoneuron. This efferent activity ultimately gives rise to muscle contraction; no higher decision centers are necessary.

REFERENCES

1. Broca, P. Remarques sur le siège de la faculté du langage articulé, suivies d'une observation d'aphémie (perte de la parole). *Bull. Soc. Anat. Paris*, 6:330, 1861.
2. Eccles, J. C., Ito, M., and Szentagothai, J. *The Cerebellum as a Neuronal Machine.* New York: Springer Verlag, 1967.
3. Minckler, J. Tissues of the Nervous System. In J. Minckler (Ed.), *Introduction to Neuroscience.* St. Louis: Mosby, 1972.
4. Wurtman, R. J., and Axelrod, J. The pineal gland. *Sci. Am.* 213:50, 1965.

SELECTED READING

Matzke, H. A., and Foltz, F. M. *Synopsis of Neuroanatomy.* New York: Oxford University Press, 1967.
Netter, F. H. *Nervous System.* New York: Ciba, 1962.
Penfield, W., and Rasmussen, T. *The Cerebral Cortex of Man.* New York: Macmillan, 1950.
Truex, R. C., and Carpenter, M. S. *Human Neuroanatomy.* Baltimore: Williams & Wilkins, 1969.

Neuronal and Synaptic Function   26

Leon A. Cuervo

INTRODUCTION

Although the neurons in the mammalian nervous system range over a wide spectrum structurally, all have characteristic types of electrical behavior in common. In speaking of the electrical behavior of a neuron, we are really talking about the behavior of the transmembrane potential as a function of time and the location in the cell with respect to the kinds of external perturbations or stimuli that are applied to it. The transmembrane potential arises mainly as a consequence of the ionic concentrations on either side of the membrane, and it is a function of the membrane permeabilities to the different ions.

In Chapter 5, the mechanisms underlying the generation and maintenance of the resting potential were discussed. In the present chapter, we shall first examine the passive behavior of the membrane that occurs when a nerve cell is stimulated in a subthreshold fashion; this involves the so-called *cable properties* of the neuron. Following this, the active membrane potentials will be discussed. And, finally, through an examination of synaptic transmission, we shall see how inputs from other cells are transformed into slowly varying potentials, which become integrated in the cell body and which can serve as an activity trigger for the integrating cell.

MEMBRANES AND MYELIN

The knowledge of the properties of biologic membranes is essential to an understanding of the biophysical basis of excitability in living cells. Unfortunately, however, we have at present only an incomplete understanding of the molecular structure of the membrane of any cell and almost no clues as to the differences in membrane structure that might account for the radically different response properties of such membranes. The general concept is that cell membranes are composed of various lipids that are associated with proteinaceous components.

A number of proposals based on chemical analyses and electron microscopic studies suggest that the lipid component probably consists of a layer two molecules thick. Substantiation of this model comes from studies with artificial membranes, which are constructed of lipid bilayers, that possess certain electrical and chemical properties similar to those of biologic membranes. Such bilayers are rather easily constructed in aqueous solution if one end of the lipid molecule has hydrophobic properties (i.e., repels water) and the other end has hydrophilic properties (i.e., attracts water). With such molecules, the hydrophobic ends turn inward and the hydrophilic ends project outward in contact with the aqueous phase. These "membranes" have properties of electrical resistance and capacitance that roughly resemble those of the membranes of living cells. Recently, several laboratories have successfully imitated more intricate properties of biologic membranes by adding various substances to these and other simple artificial membranes. The addition of a substance derived from bacterial cultures and certain antibiotics, for example, endows artificial membranes with the ability to produce gross imitations of the electrical changes seen in excitable cells, such as repetitively discharging action potentials. These chemical models may, in the near future, provide useful suggestions for the investigation of the molecular basis of the behavior of biologic membranes.

The electron microscope, although it cannot reveal any details of molecular structure, indicates that the outer cell membrane consists of alternating bands of dense and lighter materials. This apparent banded structure has been termed the *unit membrane* and is thought to be a basic property of all living cells. The appearance of a unit membrane in fixed and stained preparations is probably due to the configurational arrangement of the lipid and protein components, although its final interpretation must await an understanding of the chemical architecture of the membrane.

A predominant feature of many nervous system elements is the sheath that is formed by glial cells and a membranous structure, *myelin*. Myelin bears many ultrastructural similarities to cell membranes, and it is thought to be derived from the membrane of the Schwann cell as it spirals around nerve fibers to produce a multilayered, insulating, outer covering.

As the myelin is laid down around the axon, gaps are left at regular intervals. The axonal membrane within these gaps, which are known as the *nodes of Ranvier* (see Fig. 25-1), is probably directly exposed to the surrounding fluid, whereas the axonal surface between the nodes is well-insulated by the myelin. The transverse resistance of a unit area of the myelin sheath of, for example, frog myelinated fibers has been calculated to be 0.1 to 0.16 megohm·cm^2. As might be expected, the properties of the myelin sheath are significant in the conduction of nerve impulses along the axons.

CABLE PROPERTIES

Following the application of a stimulating current, the changes in the transmembrane potential of a nerve cell may be either active responses, which involve changes in certain membrane properties, or passive, linear responses. The type of response depends upon the direction or the intensity of the stimulating current, or both. Figure 26-1A shows both types of responses as recorded from an unmyelinated axon to which current pulses of constant duration but variable intensity and polarity were applied. The experimental system for determining these responses is illustrated in Figure 26-1B. Stimulating currents are applied using a glass microelectrode that is inserted in the cell at a short distance away from the point of recording, and the circuit is completed by a reference electrode that is immersed in the fluid bathing the axon. Small depolarizing pulses and hyperpolarizing pulses of any amplitude elicit responses that are merely a distortion of the applied waveform (records *a, a', b, b',* and *c'* in Fig. 26-1A).

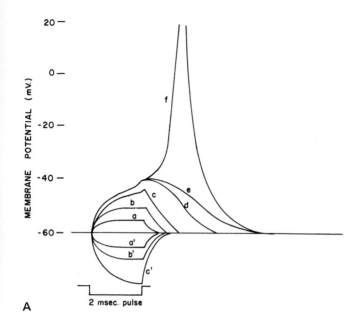

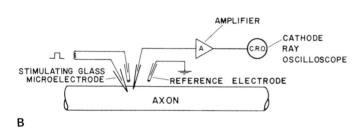

Figure 26-1
A. Response of an axon following the application of current pulses of constant duration but of varying intensity and polarity. *B.* Experimental arrangement for obtaining records shown in *A.*

The potential rises and decays exponentially instead of suddenly, as if the membrane contained a capacitor that is being charged by the applied current. These are passive responses, and analogous responses may be obtained from electrical circuits built from linear components (e.g., resistors and capacitors). Larger depolarizing pulses produce responses that increasingly deviate from this exponential behavior (records *c, d,* and *e* in Fig. 26-1A). Such responses can be simulated with electrical circuits, but these circuits must include active, nonlinear components, such as transistors.

When current pulses of long duration are applied, the intracellular potential varies exponentially, but it reaches a steady value that is maintained as long as the pulse lasts. By using several recording electrodes along the axon (Fig. 26-2A), it is possible to

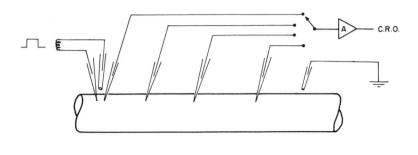

A

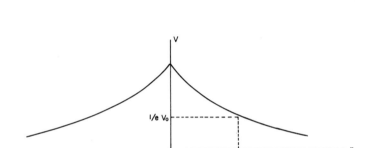

B

Figure 26-2
A. Recording setup for studying the longitudinal variation in potential following the application of current pulses of long duration. *B.* Plot of longitudinal variation of potential in the axon (λ is the length-constant).

determine the longitudinal spread of the perturbation. Figure 26-2B shows a plot of this steady potential level versus the distance from the point of current injection for a nonmyelinated axon. The curve shows an exponential decay on both sides of the stimulating electrode ($x = 0$), which is characterized by the length-constant λ. The magnitude of the length-constant increases with the diameter of the axon, reaching values of several millimeters for the largest fibers.

Experiments of this sort, together with measurements yielding membrane resistivity values that are several orders of magnitude higher than that of axoplasm, have suggested a model for the axon that represents it as a long, cylindrical conductor surrounded by an insulating sheath with capacitance and parallel leakage resistance, much like an electrical cable. An electrical equivalent circuit for this model is shown in Figure 26-3A. The electrical model represents a segment of an axon and accounts for its passive or "cable" properties. In this figure, the resistors R_o and R_i represent the resistances of the external fluid and the axoplasm, respectively, and the resistance-capacitance (RC) networks interconnecting them account for small areas of the membrane in the shape of a ring. The emf, which represents the resting potential, maintains the charge on

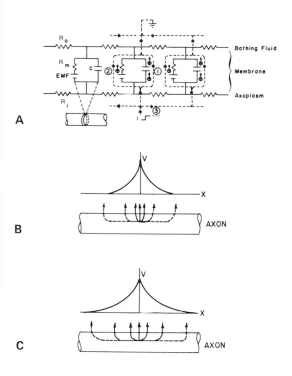

A

B

C

Figure 26-3
A. Equivalent circuit model of a biologic membrane that accounts for its passive or "cable" properties. B. Potential versus distance soon following a step input of current. C. Same as B but several milliseconds later.

the membrane capacitance, C. Ions moving across the membrane appear in the model as a current flowing through R_m. Ions moving toward the membrane or away from it, which would either create or neutralize fixed charges in its molecular structure, appear as a current flowing through the capacitor C. According to this model, the axon behaves as a leaky cable and should obey the equations of cable theory.

Before attempting the mathematical description of this system, it will be useful to gain a qualitative understanding of its behavior. Consider a long, nonmyelinated axon impaled by a microelectrode through which current is applied to the inside of the fiber. This current has three possible pathways to follow (see Fig. 26-3A):

1. Through the membrane capacitance, increasing or decreasing the charge accumulated in it, but bypassing the membrane resistance. This implies that no change occurs in the transmembrane ionic fluxes.
2. Through the membrane resistance, as a net transfer of charge across the membrane by the ionic fluxes. This can occur if the ionic concentrations, the permeabilities, or the potential difference deviate from their values at the resting state.
3. Along the axoplasm, driven by an electrical potential gradient along the inside of the fiber. This will occur only if two regions of the axon have different transmembrane potentials.

At the resting state, there is no net current flow through either of these pathways, but when a current pulse is applied to the axon via a microelectrode, a series of events begins to take place.

Initially, since the conditions for current flow through pathways 2 and 3 are not met, all the current will flow through the capacitance of the membrane area that is directly opposite the microelectrode. This surge of capacitive current alters the membrane potential at that point, thus creating conditions such that pathways 2 and 3 may begin conducting. At this time, a portion of the membrane current will be carried by the ionic fluxes through R_m, and some of the current injected, seeking a lower resistance pathway, will spread longitudinally to alter the charge of the membrane capacitance at points away from the electrode. At the point of current injection, the potential keeps changing at an increasingly slower rate, since less of the injected current flows through that portion of the membrane and the resistance R_m is taking part of this fraction of the total current.

The potential in this region does not change indefinitely, because eventually all of the membrane current is carried by transmembrane ionic fluxes and none goes to vary the charge at C. This state is reached shortly after the initiation of the current pulse and is represented in Figure 26-3B. At the point of current injection, the continuous arrow in the illustration indicates the current carried by ions through R_m, and, in the adjacent portions of the axon, the discontinuous arrows show the distribution of the mostly capacitive current in these regions. The corresponding plot of voltage versus distance along the axis at this point in time also follows the exponential decay form already shown in Figure 26-2B for the steady-state case.

A few milliseconds after the state shown in Figure 26-3B, we have the situation that is depicted in Figure 26-3C. Here, a larger portion of the membrane has reached the steady-state condition, and it now presents a relatively high resistance to the ionic current flow. Thus, more current will follow the lower resistance pathway of the axoplasm, and it will go into the membrane capacitance of more distant regions. Eventually, the membrane potential at different points in the fiber reaches values that induce ionic fluxes large enough to carry all the current crossing the membrane. Then the capacitive current becomes zero, and the distribution of membrane potential attains a steady state until the applied current pulse ceases.

In deriving the equation that describes the passive behavior of the axon, we will refer to Figure 26-4A. It represents an axon, surrounded by a thin film of salt solution, to which a current pulse is being applied through two electrodes placed opposite each other across the membrane. This system has axial symmetry, and, for the purpose of simplifying the problem, we will assume that the potentials throughout the axoplasm and bathing fluid are independent of radial distance and that their spatial dependence is only with respect to the distance along the x-axis. The length of the axon is so much greater than its diameter that it may be considered infinite. Current flowing from the external to the internal electrode spreads along the bathing fluid as the external current, I_o, crosses the membrane and returns along the axoplasm as the internal current, I_i. Both I_o and I_i are assumed to flow parallel to the axis. The equivalent circuit of Figure 26-4B will be considered to represent an axonal segment that is *not* affected by the distortions of this pattern of current flow that must occur

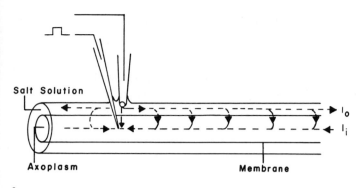

A

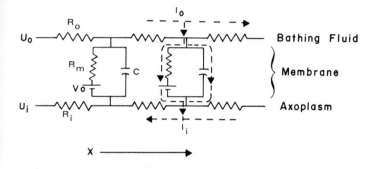

B

Figure 26-4
A. Axon surrounded by a thin film of salt solution with a current pulse being applied. *B.* Equivalent circuit model of the membrane showing current paths.

at the point of stimulation. At any point in this region, $I_o + I_i = 0$. The membrane current, I_m, is the sum of the resistive and capacitive components, and V_r is the emf that is responsible for the resting potential difference across the membrane.

The dimensions of the parameters for the model circuit are:

R_m = ohms·cm
R_o, R_i = ohms/cm of axonal length
I_i, I_o = amperes
I_m = amperes/cm of axonal length
C = farads/cm of axonal length

Let us define

$$V_m = V_i - V_o$$

as the difference between the internal and external potentials, and

$$V' = V_m - V_r$$

as the deviation of this potential difference from the resting potential difference of the cell. When no current is flowing, $V_m = V_r$ and $V' = 0$. We have:

$$\frac{\delta V'}{\delta x} = \frac{\delta V_m}{\delta x}; \quad \frac{\delta V'}{\delta t} = \frac{\delta V_m}{\delta t} \tag{26-1}$$

From Ohm's law,

$$\frac{\delta V_i}{\delta x} = -I_i R_i; \quad \frac{\delta V_o}{\delta x} = -I_o R_o \tag{26-2}$$

and

$$I_m = -\frac{\delta I_i}{\delta x} = \frac{\delta I_o}{\delta x} \tag{26-3}$$

since $I_i + I_o = 0$. Equation 26-3 means that any change in the external and internal longitudinal currents is due to leakage through the membrane. The choice of signs also defines the outward membrane current as positive. From the above definitions and Equations 26-1 and 26-2,

$$\frac{\delta V'}{\delta x} = \frac{\delta V_i}{\delta x} - \frac{\delta V_o}{\delta x} = -I_i R_i + I_o R_o \tag{26-4}$$

By further differentiation,

$$\frac{\delta^2 V'}{\delta x^2} = -R_i \left(\frac{\delta I_i}{\delta x}\right) + R_o \left(\frac{\delta I_o}{\delta x}\right) \tag{26-5}$$

$$\frac{\delta^2 V'}{\delta x^2} = (R_o + R_i) I_m \tag{27-6}$$

which is the *general cable equation.*

When the applied current is such that the membrane behavior is linear, the membrane current is the sum of the resistive and capacitive components. Thus,

$$\left(\frac{1}{R_i + R_o}\right)\left(\frac{\delta^2 V'}{\delta x^2}\right) = \frac{V'}{R_m} + C\frac{\delta V'}{\delta t} \tag{26-7}$$

where the left-hand side of the equation represents the total membrane current, I_m. The capacitive current is considered to represent the effect of the membrane dipoles, whereas the resistive (ionic) current is due to the movement of ions across the membrane. Solutions of this equation for particular cases can be found in the literature [9, 11].

When a direct current is applied to the axon, at the steady state Equation 26-7 becomes;

$$\frac{d^2 V'}{dx^2} = V' \frac{R_i + R_o}{R_m} \tag{26-8}$$

which has a solution

$$V' = A \exp(-x/\lambda) + B \exp(x/\lambda) \tag{26-9}$$

where

$$\lambda = \sqrt{R_m/(R_o + R_i)}$$

is the characteristic length or *length-constant* of the axon. For boundary conditions $V' = V_o$ at $x = 0$ and $V' = 0$ at $x = \infty$, $A = V_o$ and $B = 0$, and Equation 26-9 becomes:

$$V' = V_o \exp(-x/\lambda) \tag{26-10}$$

which shows that V' will decay exponentially along the axon from the point of application of the current (see Fig. 26-3B).

Equation 26-7 can also be written as:

$$-\lambda^2 \left(\frac{\delta^2 V'}{\delta x^2} \right) + \tau_m \left(\frac{\delta V'}{\delta t} \right) + V' = 0 \tag{26-11}$$

where

$$\tau_m = R_m C$$

is the *time-constant* of the membrane. The solution of Equation 26-11 will show, at each x, an exponential rise of V' with time following the application of the stimulating current (see Fig. 26-3B and C). Thus, an unmyelinated axon will distort an applied signal, as would an RC network, due to the combined effects of its membrane resistance and capacitance. Also, the signal will be attenuated as it is conducted along the axon. The magnitude of both effects is determined, respectively, by the time-constant and the length-constant.

The measurements of the impedance of axons performed by Cole and Curtis [1] and the analyses of axonal behavior under direct current pulses by Hodgkin and Rushton [9], and Lorente de No [11] have yielded values of the cable constants for giant unmyelinated fibers of several species (Table 26-1).

Table 26-1. Cable Constants of Unmyelinated Nerve Fibers

Species	Fiber Diameter	Length-Constant (mm)	Time-Constant (msec)	Membrane* Resistance (ohm·cm^2)	Membrane Capacity (farad·cm^{-2})
Squid	500	5	0.7	700	1
Lobster	75	2.5	2.0	2000	1
Crab	30	2.5	5.0	5000	1

*Resistance of a cm^2 of axonal membrane given by $2\pi r \cdot R_m$, where r is the radius of the axon.

For more complicated geometrical configurations, such as that of the soma of the nerve cell and its dendritic tree, the behavior of the potential cannot be expected to follow an equation as simple as Equation 26-7. Yet, the basic qualitative concepts of the spatial spread and time-dependence of the potential not only apply but are fundamental to the understanding of the mechanisms of neural integration at the neuronal level. These mechanisms are examined in detail in Chapter 29.

The membrane resistance and capacitance may be associated with structural features of the membrane. The bimolecular layer of lipid material that is thought to constitute a fundamental part of the membrane structure would present a high resistance to the passage of current and, at the same time, would have dielectric properties that match, nearly quantitatively, the requirements for the measured membrane capacitance. The way in which current-carrying ions traverse this lipid layer is not understood. A body of evidence is building up that suggests that the proteinaceous structures embedded in the lipid layer may offer pathways for the ions and possibly constitute the long-postulated ionic channels.

The presence of myelin around axons has a drastic effect on their cable properties. Because of the increase in thickness of the dielectric medium, the membrane resistance is increased a few orders of magnitude and the capacitance is greatly decreased. Both effects tend to decrease the membrane current to insignificant values except at the nodes of Ranvier, where the myelin's coverage is interrupted and the membrane recovers its normal resistance and capacitance. Thus, when a current is applied to the surface of a myelinated axon, the fraction that finds its way into the fiber will flow along the axoplasm, with only insignificant decrements occurring due to capacitive or ionic currents across the myelin internodes. Only at the nodes will the current leak out through the membrane and thereby produce significant attenuation and distortion of the signal.

NONLINEAR PHENOMENA

In the last section, it was noted that if small depolarizing current pulses are applied to an axon, the responses obtained are similar in shape but opposite in sign to those resulting from hyperpolarizing pulses of any amplitude. Figure 26-1A also illustrates responses to larger depolarizing pulses that deviate significantly from the passive behavior of a linear *RC* network. When the stimulus exceeds the threshold value, the most characteristic nonlinear response, the *action potential,* is elicited. These and other phenomena that involve some active component in the membrane are called *electrogenic responses.* It should be noted that the words "passive" and "active" are used here in the same sense as they are when a transistor is described as an active component and a capacitor as a passive component of a circuit, and they do not necessarily refer to a dependence on the metabolic activity of the cell.

The origin of this nonlinearity rests on the ability of the membrane to alter its permeability to one or more ions, which in turn causes a change in the transmembrane potential as predicted by the constant-field equation. Although the cable properties are probably similar in all regions of the cellular membrane, a neuron's ability to change its *permeability* characteristics varies widely from one region to another. At sites of synaptic attachment, the small patches of postsynaptic membrane undergo very specific changes in permeability under the influence of the chemical transmitter released by the incident axon. The potential variation that results is called a *post-synaptic potential.* Most of the rest of the membrane area of the dendrites is assumed to be purely passive in behavior. The soma, axon, and perhaps short, proximal portions of the dendrites are electrically excitable; i.e., an adequate electrical current will induce the changes in permeability that are characteristic of the action potential.

In order to elicit an action potential, the stimulating current should be depolarizing and of sufficient intensity to discharge the membrane capacitance to a value of the potential called the *threshold potential.* This current may be applied experimentally, or it may arise during the physiologic spread of depolarization from one region of the cell to another as a consequence of the cable properties of the cell. This phenomenon will be discussed in detail later when the propagation of the action potential is examined. The lowest threshold is found at the *initial segment* of the axon near the cell body, and the highest is found at the soma. Thus, depolarizations of sufficient magnitude of the dendritic tree will spread passively through the soma and on to the initial segment, initiating an action potential that propagates along the axon and back over the membrane of the cell body.

A better understanding of how changes in permeability result in changes in the transmembrane potential may be acquired by referring back to Figures 5-2 and 5-3. In these figures, each major ionic system is represented by an electromotive force and a variable resistor. The electromotive force, V_i, is the potential difference that would exactly balance the tendency of the ion to diffuse down its concentration gradient, i.e., the equilibrium potential given by the Nernst equation. Normal values of these equilibrium potentials for the squid axon at 10°C are $V_{Na} = 58$ mV, $V_K = -79$ mV, and $V_{Cl} = -62$ mV. The variable resistor represents the permeability of the membrane to the ion. The transmembrane potential, $V_m = V_i - V_o$, is the voltage difference measured across the membrane capacitance, C_m. As long as the resistors remain at a fixed value, the system is in a steady state and the capacitive current is zero. Therefore, the voltage difference across the capacitor is constant. The current through each ionic system is proportional to the difference between its emf and V_m:

$$I_{Na} = \frac{V_{Na} - V_m}{R_{Na}}; \qquad I_K = \frac{V_K - V_m}{R_K}; \qquad I_{Cl} = \frac{V_{Cl} - V_m}{R_{Cl}} \qquad (26\text{-}12)$$

Also, since $I_C = 0$, the only return path for I_K and I_{Cl} is through R_{Na}, which demands that $-I_{Na} = I_K + I_{Cl}$. If R_{Na} is now decreased, the absolute magnitude of I_{Na} will increase correspondingly while I_K and I_{Cl} remain unchanged, and $-I_{Na} > I_K + I_{Cl}$. Then the excess inward current will have to flow outwardly through the capacitor, accumulating positive charges inside and thus causing a depolarization of the membrane. This new value of V_m will reduce I_{Na} while increasing the magnitudes of I_K and I_{Cl} (Eq. 26-12), so that a new balance of the currents is attained, and again $-I_{Na} \cdot I_K + I_{Cl}$ and $I_C = 0$. If either R_K or R_{Cl} were decreased, the effect produced on the state of polarization of the membrane would be the opposite. Thus, an increase in inward Na^+ current will depolarize the cell, whereas an increase in outward K^+ currents, Cl^- currents, or both will hyperpolarize it. The reverse relationship, however, holds for currents that are applied to the membrane from an external source. In this case, an inward current will tend to charge the membrane capacitor further, whereas an outward current will have the contrary effect.

Both postsynaptic potentials and the depolarizations generated by many sense organs, or *generator potentials,* are *graded responses,* i.e., their amplitude is variable and depends on the magnitude of the responsible input. They may or may not give rise to an action potential, according to whether the threshold potential of the appropriate region of the cell is reached or not. The action potential, however, is not a graded response. Its *all-or-nothing* property resembles that of an explosive charge: if the triggering energy is not enough, nothing happens; but if the triggering energy is just sufficient, the full explosion occurs. The permeability changes that are responsible for the action potential are complex and will be described later in this chapter.

THE ACTION POTENTIAL

If a neuron is depolarized to the threshold potential, an action potential (AP) will be initiated. Although action potentials may vary slightly from cell to cell, all have several properties in common. Some of these shared characteristics will be described using the AP recorded from a squid giant axon as an illustration (Fig. 26-5). The initial slow rise, known as the *foot* of the AP, is followed by an increasingly faster depolarization. The potential overshoots the zero line, thus reversing the polarity of the membrane, and levels off at about 40 mV. The potential then begins to return slowly to the resting potential level. The duration of this sequence is about 1 or 2

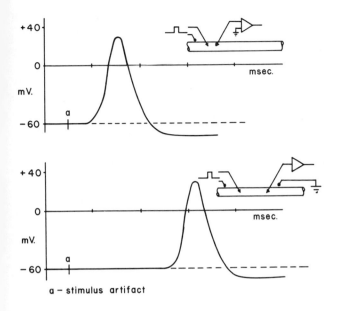

a – stimulus artifact

Figure 26-5
Action potential recorded from a squid giant axon.

milliseconds in nerve cells, and the amplitude (from resting potential to the tip of the AP) varies between 100 and 120 mV under normal conditions.

If the recording electrode is placed farther away from the point of stimulation, the AP is recorded somewhat later, but its magnitude and shape do not change, as is shown in the lower part of Figure 26-5. The time that elapses between the observation of the stimulus artifact and the foot of the AP is called the *latency* time. It has also been observed that the latency is proportional to the distance between the stimulating and the recording electrodes. These two observations indicate that the AP travels along the fiber with a constant velocity but without diminution or distortion. The velocity of propagation may be calculated from the latency.

Since axons are very poor conductors of passive electrical signals, the observed distortion-free propagation implies that the AP receives some sort of reinforcement as it travels. Otherwise, the decrement and distortion due to the length-constants and time-constants, as discussed in connection with cable properties, would be evident a few millimeters away from the stimulating electrode. The propagation of an AP is not a function of cable properties, however; rather, it depends upon the fact that each segment of membrane undergoes this characteristic voltage variation when it is stimulated by an AP occurring in an adjacent region of membrane. In this way, an AP that is originated at one region of membrane, either physiologically or experimentally, will trigger the portion of membrane just ahead to produce its own AP, and the continuation of this process is responsible for the propagation of the wave.

When a rectangular current pulse is applied to an axon, the capacitance of the small area of membrane that surrounds the stimulating electrode accumulates charge. The stronger the current, the faster the charge will accumulate, thereby giving rise to change in the transmembrane potential, V_m. Thus, a relatively weak current will require a longer time to reach the threshold potential than a stronger one. The maximum value that the change in potential may reach is given by the voltage drop, IR_m. Obviously, a current whose value is such that this voltage drop would be insufficient to bring V_m to the threshold value could never excite the cell. Figure 26-6 illustrates these points. This illustration shows the effects when rectangular current pulses of different durations and strengths are applied to an axon through a stimulating electrode that is placed very near to the recording electrode. Pulse a, which is of long duration and weak intensity, elicits only a passive response, whereas the other pulses are just large enough to elicit an AP 50% of the time. In the remaining 50% of the time, a *local response* is obtained. The local response, as the name implies, does not propagate, and it can be recorded only when the distance between the two electrodes is minimal. At longer distances, the local effects would not be recorded, and the AP would look like the one depicted in Figure 26-5, with its foot rising directly from the resting potential level. The action potentials b', c', and d' are identical in amplitude and shape, as would be any other impulses elicited from the given fiber under the same conditions, independently of the shape and intensity of the stimulus.

A plot of current intensity versus the time that it must be applied to reach threshold is called a *strength-duration* curve (Fig. 26-7). The shape of this curve depends on the cable equation for the tissue, and consequently such curves for all long, cylindrical

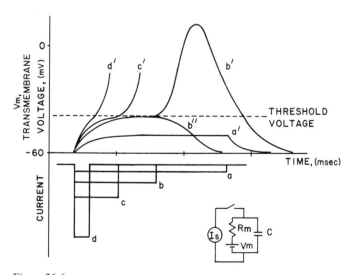

Figure 26-6
Membrane response to different current pulses. The equivalent circuit is based on the assumption that the current affects a region of membrane that is small enough to be adequately represented by a single RC circuit as shown.

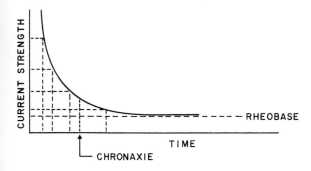

Figure 26-7
Strength-duration curve.

cells (nerve or muscle) exhibit the same shape. The curves for different cells vary
only with respect to their intensity and time scales, which in turn depend on the
values of the equation parameters. By assuming that the current affects a region of
membrane that is small enough to be adequately represented by a single RC circuit
(see Fig. 26-6) and that the threshold is reached at a fixed voltage, the threshold
potential V_{th}, it is easy to arrive at an equation that fits the data reasonably well.
It is to be noted that the first assumption neglects the spread of current along the
fiber. The change in potential produced by a current applied to this circuit is given by:

$$\Delta V_m = I_s R_m \left(1 - e^{-t/\tau}\right) \tag{26-13}$$

where I_s is the total applied current (which flows through the membrane capacitance
and partly through the membrane resistance) and τ is the membrane time-constant.
If we let $\Delta V_m = \Delta V_{th}$ represent a change in potential just large enough to reach
threshold and solve for I_s, we then have:

$$I_s = \left(\frac{\Delta V_{th}}{R_m}\right)\left(\frac{1}{1 - e^{-t/\tau}}\right) \tag{26-14}$$

which expresses the intensity of the current as a function of the time that it must be
applied in order to reach the threshold potential. For $t \ll \tau$, we have $1 - e^{-t/\tau} \approx t/\tau$,
yielding:

$$I_s t = \left(\frac{\Delta V_{th}}{R_m}\right)\tau \tag{26-15}$$

Note that the right-hand side of Equation 26-15 is a constant. This implies that for
shocks of very short duration, the amount of charge that has to be delivered to the
membrane is constant, i.e., all the current flows through the membrane capacitance.

Two important characteristics of strength-duration curves are the *rheobase, I_{rh}*, which is the current intensity that will bring the fiber asymptotically to threshold, and the *chronaxie, t_{ch}*, which is the application time required for a current that is twice the rheobase value to reach threshold. Since, after a long enough period of time, all the stimulating current will flow through the membrane resistance, then:

$$\Delta V_{th}/R_m = I_{rh}$$

Therefore, from Equation 26-14,

$$1 - \frac{I_{rh}}{I_s} = e^{-t/\tau} \tag{26-16}$$

From the definition of chronaxie and the above equation, we have:

$$t_{ch} = \tau \ln 2 \tag{26-17}$$

In other words, the value of the chronaxie is proportional to τ. When a strength-duration curve is determined by employing external electrodes on a whole nerve, the value of the chronaxie that is obtained yields a good estimate of the membrane time-constant of the axons, in spite of the complexity of the current flow patterns.

When two suprathreshold current shocks are applied in succession to a fiber, the effect of the second depends on the time interval between the two. If the two stimuli are 10 milliseconds apart, the axon responds to both with identical action potentials. As the interval is shortened, the second AP becomes smaller in amplitude, and the current intensity that is required to trigger it increases until the point that when the second stimulus falls on the repolarizing phase of the first AP, the second response cannot be obtained no matter how strong a stimulus is applied. At this point, the fiber is said to be *absolutely refractory*. Thus, the membrane properties that render an axon excitable are lost during an AP, but they are progressively recovered during the following several milliseconds until full excitability is restored. During this period of recovery, the fiber is said to be *relatively refractory*: its excitability has been partially recovered, and an accordingly small response may be elicited. Evidently, the frequency or firing rate of an axon is limited by its refractoriness. Under physiologic conditions, this limit is not usually reached, although firing rates in excess of 500 per second have been observed.

The excitability of nerve and muscle cells depends on the ability of their membranes to undergo specific permeability changes, which in turn give rise to transmembrane potential variations in a manner already discussed. The critical event in the generation of an AP is the rapid transient increase in the permeability coefficient for sodium ion, P_{Na}, upon depolarization. Around the threshold point, a regenerative process takes place between P_{Na} and the membrane potential: P_{Na} grows rapidly, allowing a rush of sodium ions into the fiber, which are driven by both the concentration and the electrical gradients; this inward sodium current further depolarizes the membrane, driving P_{Na} to even larger values. Near the peak of the AP, the value of P_{Na} has increased about 500-fold, and, for all practical purposes, the sodium ion concentration determines the potential. In experiments where the external sodium concentration is decreased, the peak of the AP reaches values that follow a slope similar to that predicted by the Nernst equation (see Chap. 5). The fast swing of the membrane potential toward positive values is checked by three factors: (1) as the potential approaches V_{Na}, the sodium current decreases; (2) the increase in P_{Na} is short-lived; thus, after reaching a high value and while the fiber is still depolarized, P_{Na} decreases spontaneously (a phenomenon called *inactivation*); and (3) the permeability coefficient for potassium ion, P_K, also increases with depolarization but considerably slower than does the P_{Na}; when the fiber is fully depolarized, the outward potassium current is favored by the electrical and the concentration gradients. This last factor not only contributes to stop the upward trend of the potential, but is also responsible for returning it to the resting level.

When an AP is traveling along an axon, each successive region of the membrane goes, in turn, through the sequence of events just described. The initial slow depolarization that corresponds to the foot of the AP is caused largely by the passive spread of current from the advancing wave, since at this point P_{Na} still has a very low value. Figure 26-8 represents an axon along which an AP is traveling to the right. The polarity of the active region with respect to its adjacent regions generates a longitudinal potential

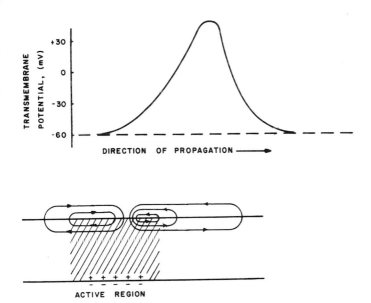

Figure 26-8
Membrane current distribution associated with a propagating action potential. (Adapted from B. Katz. *Nerve, Muscle and Synapse.* New York: McGraw-Hill, 1966.)

gradient that gives rise to longitudinal currents. The current pathway is completed through the membrane, and, as indicated in the figure, an outward current crosses the membrane just ahead of the traveling AP. This outward current will cause depolarization, since its origin is external to the area of membrane that it crosses. As this passive depolarization grows, the inward I_{Na} increases very slowly due to an increase in P_{Na}. Meanwhile, I_K and I_{Cl} are also increasing, since their electrochemical potentials are also increasing, and the system remains stable. Near the threshold point, the rise in P_{Na} becomes much steeper, causing I_{Na} to override the outward currents completely; at this point, the system becomes unstable and the potential goes steeply positive, driven by the regenerative relationship between P_{Na} and the membrane potential.

The mathematical description of the currents flowing during the conduction of the AP can be made before the changes in permeability that are associated with the AP are discussed quantitatively. The general cable equation (Eq. 26-6) does not imply a linear behavior of the membrane, and it hence may be used as a starting point.

Consider an axon of radius r. The longitudinal resistance of the axoplasm per unit length of axon, R_i, is $\rho/\pi r^2$ ohms/cm, where ρ is the resistivity of the axoplasm. The membrane current density or flux, J (amperes/cm^2) is defined as:

$$J = \frac{I_m}{2\pi r}$$

where I_m is, as in Equation 26-6, the current per unit length of the axon and $2\pi r$ is the circumference of the axon. The external resistance, R_o, in Equation 26-6 is relatively small, and it can be neglected when the axon is surrounded by a large volume of conducting fluid. Substitution for R_i and I_m in Equation 26-6 yields:

$$\frac{\delta^2 V'}{\delta x^2} = \frac{2\rho J}{r} \tag{26-18}$$

The action potential is a potential wave that propagates without diminution at a constant velocity θ. Then $x = \theta t$ and

$$\frac{\delta^2 V'}{\delta x^2} = \frac{1}{\theta^2}\left(\frac{\delta^2 V'}{\delta t^2}\right) = \frac{1}{\theta^2}\frac{2\rho J}{r} \tag{26-19}$$

Impedance measurements during the propagation of action potentials have shown that although the capacitance remains practically constant, the resistance undergoes complex variations [1]. Thus, the current density may be expressed in terms of its ionic and capacitive components by describing its complex behavior in terms of a nonlinear dependence of the resistance with respect to voltage and time. From Equation 26-7 and the definition of membrane current density,

$$J = \frac{V'}{2\pi r R_m} + \frac{C}{2\pi r}\left(\frac{dV'}{dt}\right) \tag{26-20}$$

where R_m and C have, respectively, the units of ohms·cm and farads/cm. Then, from Equations 26-19 and 26-20, the total membrane current density during the action potential is:

$$\frac{r}{2\rho\theta^2}\left(\frac{d^2V'}{dt^2}\right) = \frac{C}{2\pi r}\left(\frac{dV'}{dt}\right) + \frac{V'}{2\pi r R_m} \qquad (26\text{-}21\text{A})$$

Rearranging and solving for the currents, we have:

$$\frac{\pi r^2}{\rho\theta^2}\left(\frac{d^2V'}{dt^2}\right) - C\left(\frac{dV'}{dt}\right) - \frac{V'}{R_m} = 0 \qquad (26\text{-}21\text{B})$$

Or, since $R_i = \rho/\pi r^2$,

$$\frac{1}{R_i\theta^2}\left(\frac{d^2V'}{dt^2}\right) - C\left(\frac{dV'}{dt}\right) - \frac{V'}{R_m} = 0 \qquad (26\text{-}21\text{C})$$

Each term of this equation may be considered to be a current component:

$$I_m - I_C - I_i = 0 \qquad (26\text{-}21\text{D})$$

where I_m = total membrane current
I_C = capacitive current
I_i = ionic current

For the capacitive component, the current-carrying ions will accumulate at the membrane boundaries, thereby altering the charge and the state of polarization of the membrane dipoles. These ions may originally come from the same side of the membrane where they are accumulating, or from the opposite side. The resistive component, which is usually called *ionic current,* is carried by those ions that actually cross the membrane. Thus, a group of ions crossing the membrane will constitute an ionic current, and, if those same ions then accumulate at the membrane boundary, they constitute a capacitive current.

The curves in Figure 26-9 show the current distribution during an action potential. The equivalent circuits for each segment of membrane are separated from one another to emphasize that they represent the same region in space at different times, since we have derived the equations in the time domain. Nevertheless, due to the wave properties of the action potential, it is permissible to perform the same derivations for the distribution of currents in space at a given instant. The capacitive current curve, I_C, is proportional to the first derivative of the potential with respect to time, whereas the total membrane current curve, I_m, is proportional to the second derivative of the potential. The ionic current curve, I_i, is obtained as the difference of the other two.

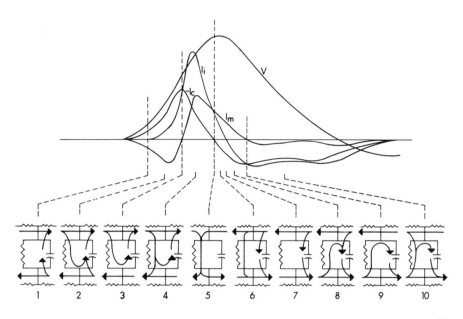

Figure 26-9
Current distribution during an action potential (see text for explanation). (Modified from A. L. Hodgkin and A. F. Huxley. A quantitative description of membrane current and its application to conduction and excitation in nerve. *J. Physiol.* 117:500, 1952.)

In addition to the ionic and capacitive currents, the diagrams account for the longitudinal currents that flow through the axoplasm and the external fluid. These pathways complete the local circuits through which the current flows during an action potential (see Fig. 26-9).

During the initial portion of an AP, there is no ionic current, but there are only small amounts of longitudinal current spreading along the axoplasm that leak outwardly through the membrane capacitance, thus depolarizing the membrane (step 1, Fig. 26-9). When this depolarization reaches the threshold point, the membrane resistance decreases rapidly (P_{Na} increases), and an inward ionic current flows through the resistance and joins that contributed by the longitudinal flow to depolarize the fiber further (step 2). When the total membrane current density is zero (step 3), the ionic and capacitive currents are equal in magnitude but opposite in direction. At this point, the AP has its fastest rate of rise. As the ionic current continues to increase while the rate of change of the potential decreases (i.e., the capacitive current decreases), some of the ionic current will begin to feed the longitudinal axoplasmic component (step 4), until, at the peak of the AP, the capacitive current becomes zero (no change in potential) and all the membrane current is ionic (step 5). At this point, the potential gradient along the fiber changes sign, and consequently the longitudinal currents will reverse their directions. Also, P_{Na} has begun to decrease spontaneously while P_K is increasing. This results in the rapid decrease of the absolute value of the ionic current (step 6), and, after its value becomes zero in step 7, the reversal of its direction (step 8). Here, there is still a net inward membrane current. In step 9, the outward ionic current has grown to equal the inward capacitive current, and in step 10, it has become larger since the net outward current is beginning to contribute to the external longitudinal current.

The phenomena underlying this whole process are the changes in P_{Na} and P_K, whereas the forces that drive it are the electrochemical potentials of sodium and potassium. The resulting inward sodium currents and outward potassium currents become overpowering at the regions represented by steps 4 and 10, respectively. In these regions, the excess ionic currents find their path down the longitudinal potential gradients, thus augmenting the longitudinal currents. These eventually find their way back through the membrane; in one case, they depolarize the fiber and are responsible for the propagation of the AP (steps 1 and 2), and, in the other case, they initiate the repolarization of the fiber (steps 6, 7, and 8).

The study of the complex kinetics associated with the changes in permeability during an action potential required some simplification of the system. This was achieved through the use of a very ingenious device that came to be called a *voltage clamp*. The voltage clamp was originally used only on squid axons, because their large size allowed certain manipulations that were impossible in thinner axons. Further developments of the technique, however, have made its application to smaller cells feasible. In essence, what the voltage clamp does is to suddenly impose a given membrane potential on a length of axon and maintain it at a constant value while the transmembrane currents are measured. Figure 26-10 shows a simplified version of the equipment and the experimental setup. A thin platinum wire, which is insulated except for the last 2 cm of its length, is introduced axially into the axon; a helical or sleeve electrode surrounds the outside of the axon. The low longitudinal resistance of the wire renders the length of axoplasm that is in electrical contact with it practically isopotential, thus eliminating the spatial derivative in Equation 26-11. Under these conditions, the AP is transmitted instantaneously along those 2 cm of axon; i.e., that section of the fiber will undergo the potential variations simultaneously. A command signal (V_c) is applied to one input of a voltage subtractor, whose output drives a current generator. The output of the generator is connected to an axial electrode that lies within a short length of axon, and the current return lead is connected to the sleeve electrode that surrounds the axon. The axon's transmembrane potential is recorded between a second axial wire and the external electrode, and, after amplification by amplifier 1, it is fed back to the second input of the subtracter.

Since the output of the subtractor is proportional to the difference $V_c - V_m$, it is clear that current will be passed continuously across the membrane in the manner required to maintain the equality $V_m - V_c$. Thus, V_m may be set at a particular value and maintained at that level, regardless of changes in the membrane's impedance. Moreover, the total transmembrane current (I_m) will be strictly radial, and it will, at all times, be equal to the applied current; consequently, it can be measured by simply recording the potential drop across a small resistance in series with the current electrode. Since V_m can be maintained constant at any desired value, the time derivative in Equation 26-20 also becomes zero; only at the on and off points of the command

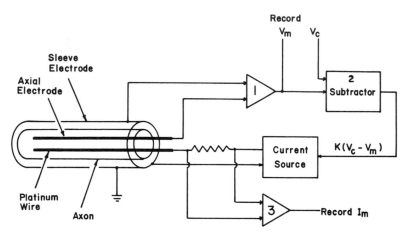

Figure 26-10
Voltage clamp procedure employed in the experiments of Hodgkin and Huxley [8]. The principles of operation are discussed in the text.

pulse, V_c, will a capacitive current flow. Consequently, all of the current flowing through the membrane during the pulse will necessarily be ionic current. The voltage drop is sensed by amplifier 3, which will provide an output voltage that is proportional to the current.

The outputs of amplifiers 1 and 3 (V_m and I_m, respectively) may be displayed on an oscilloscope screen and photographed. Figure 26-11 shows the current record obtained for a depolarization of 30 mV. During the first millisecond or so, the ionic current begins flowing inwardly, reaches a peak, and then declines until it becomes outward. It slowly grows in that direction and reaches a plateau that is maintained for as long as the command pulse lasts. If the sea water that bathes the axon is changed to a solution in which choline chloride is substituted for NaCl, the early occurring, inward component is no longer present (Fig. 26-11A). This is a strong indication that the initial current is carried by sodium ions. By varying the amplitude of the command pulse, a series of current curves is obtained (Fig. 26-11B).

The magnitude of the peak of the early component grows with the degree of depolarization up to membrane potentials of about −20 mV. For larger pulses, the initial current decreases, then becomes zero, and finally flows outwardly. This is better seen in Figure 26-12A, which is a plot of ionic current versus membrane potential. Curve 1 shows how the peak of the early component varies with respect to the membrane potential at which the axon is held by the command pulse. The zero current point occurs at V_m = 58 mV, which is the equilibrium potential of sodium. Since this zero current occurs at a time when the membrane permeability is large, it must mean that the driving force is zero, i.e., that the electrochemical potential for the ion that is carrying the current is zero. This condition exists only when the membrane potential exactly balances the effect of the concentration gradient, or, in other words, when

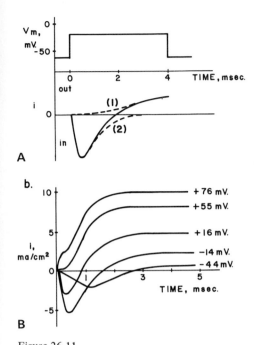

Figure 26-11

A. Current obtained during voltage clamp for a depolarization of 30 mV. Dashed curve 1 is the current recorded for the depolarization when external Na^+ is removed. Curve 2 is the difference between the normal current and curve 1. *B.* Curves of current versus time that are obtained when the amplitude of the command pulse is varied so that the membrane potential acquires the values indicated by the numbers at the right.

the membrane potential is at the equilibrium potential of the ion. This argument confirms the hypothesis that sodium ions carry the initial component of the current.

The late component, which reaches a steady state, always has an outward direction. Curve 2 in Figure 26-12A is a plot of the steady-state current values for different depolarizations. Early experiments suggested that this current is carried by potassium ions. This was later confirmed in experiments in which artificial solutions of varying potassium concentrations were substituted for the axoplasm. Under these conditions, the amplitude of the late component for a given depolarization is a linear function of the equilibrium potential for potassium.

In Figure 26-12A, both curves 1 and 2 are far from showing a linear relationship between current and voltage. The rectification or nonlinear character of the late component (curve 2) is an indication of the relationship between the potassium conductance (G_K) and the membrane potential. In experiments in which the potassium concentrations are constant $(V_K = \text{constant})$, the nonlinearity of the current curve shows that G_K varies with V_m, since:

$$G_K = \frac{1}{R_K} = I_K \frac{1}{V_m - V_K}$$

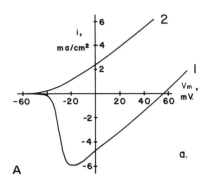

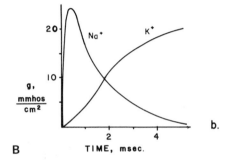

Figure 26-12

A. Record of the steady-state values of the early component (curve 1) and the late component (curve 2) of current density for different depolarizations. *B.* Time-course of ionic conductances.

The sodium current (curve 1) shows a more dramatic nonlinearity; in the region of potential from -60 mV to -20 mV, the current increases for larger depolarizations, even though a smaller driving force, $V_m - V_{Na}$, is present. This is due to the fact that in this region of potential, the increase in G_{Na} with depolarization has a greater effect on the current than the decrease in driving force.

It was already mentioned that by removing all the sodium ion from the external solution, it is possible to obtain records of the potassium current (see Fig. 26-11A, curve 1). By subtracting this from the whole ionic current, curve 2 in the same figure is obtained, which represents the sodium current. Since these curves are obtained at a constant V_m, the time-courses of the conductances and the currents are the same (Fig. 26-12B).

Detailed studies of permeability changes led Hodgkin and Huxley [8] to a mathematical description of the phenomena involved. In their model, which will be studied in more detail in Chapter 29, the conductances show a dependence on time and voltage, as indicated by the following equations:

$$G_{Na} = (G_{Na})_{max}\ m^3 h \tag{26-22}$$

$$G_K = (G_K)_{max} \, n^4 \tag{26-23}$$

where $(G_{Na})_{max}$ and $(G_K)_{max}$ are the maximum possible conductances for the corresponding ion and m, h, and n are parameters that are defined by expressions similar to the one written below for m:

$$m = m_\infty [1 - \exp(-t/\tau_m)] + m_0 \exp(-t/\tau_m) \tag{26-24}$$

in which the subscripts infinity and zero refer to time and τ_m is an instantaneous function of the membrane potential. The parameters m and n have values near zero at the resting potential and grow toward unity upon depolarization at a rate that is determined by τ_m and τ_n, respectively. The value of h is approximately 0.6 at the resting potential and decreases toward zero upon depolarization.

Based on their measurements under voltage clamp conditions, Hodgkin and Huxley were able to calculate the time-course of an action potential. Qualitatively, upon the initial depolarization, m increases rapidly, while n increases and h decreases much more slowly. In approximately 0.5 millisecond, G_{Na} has reached a value near $(G_{Na})_{max}$, and the sodium current is at its maximum. This further depolarizes the cell and thus accelerates the changes in m, h, and n. As h approaches zero, G_{Na} becomes negligible, as does I_{Na}. Meanwhile, n has been increasing, and the potassium current, driven now by a high electrochemical potential, starts flowing outwardly and the cell repolarizes. This repolarization tends to diminish the value of n until a steady state is reestablished at the resting potential.

The parameter h is a measure of the activation of the sodium conductance and is responsible for the refractoriness of the fiber. During the second half of the action potential, when h has a value near zero, an imposed depolarization may increase the value of m. However, G_{Na} will not be essentially affected, because of the low value of h, and the fiber will not respond with an action potential. After the fiber has repolarized completely, h, which has slow kinetics, gradually returns to the value of 0.6. During this period, the fiber is partially refractory, since a sufficiently large depolarization will increase m enough to overcome the partially depressed value of h and bring G_{Na} to a level such that the resultant I_{Na} will depolarize the fiber and produce a small action potential.

Due to the cable properties of the axon, a subthreshold depolarization will linger for some time after the pulse is ended. A second depolarizing pulse, applied while the residual depolarization lasts, should need less strength than normally to bring the fiber to threshold level (Fig. 26-13A).

Experimental results agree with this prediction as long as the first (conditioning) pulse has a duration of a fraction of a millisecond. For longer durations of the conditioning pulse, this facilitation becomes less pronounced, and, finally, for sufficiently long pulses, the strength of the second (test) pulse that is required to reach threshold becomes larger than normal. This phenomenon is called *accommodation* and can be explained on the basis of the Hodgkin-Huxley theory. Figure 26-13B represents the steady-state values of h as a function of the membrane potential for the squid axon.

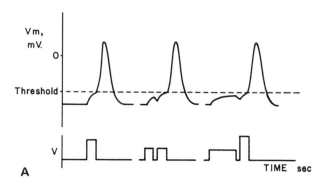

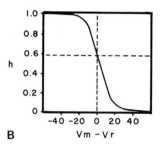

A

B

Figure 26-13
A. Effect of cable properties on the action potential when two depolarizing pulses are delivered in succession. *B.* Steady-state values of *h* as a function of membrane potential for the squid axon.

When a depolarizing pulse is applied, *h* begins to decrease slowly until it attains the value given by the curve for that potential. If the pulse is short, this process is interrupted before a significant change in *h* has occurred, and the summing effect of the test pulse with the residual depolarization of the conditioning pulse predominates. On the other hand, if the conditioning pulse is long, *h* has time to decrease sufficiently to inactivate the sodium conductance partially, and this effect will have more influence over the required strength of the test pulse than will the summation of the depolarizations. The shape of this curve for *h* also explains why long, hyperpolarizing, conditioning pulses promote larger sodium currents as well as action potentials of higher amplitude

THE PROPERTIES OF THE SYNAPSE

The site of functional interconnection between two neurons is known as a *synapse,* whereas the connection between a neuron and a muscle is referred to as a neuromuscular junction or *motor end-plate.* When an action potential arrives at an axon ending, the state of the postsynaptic cellular membrane is altered. There are two ways in which this may be accomplished: (1) in electrical junctions, the membranes of the two cells fuse, and their resistance is so low at the point of contact that the depolarization arriving at the presynaptic membrane is conducted with small loss to the postsynaptic membrane; (2) in chemical synapses, the depolarization arriving at the axon terminal causes the release of a chemical transmitter substance through the presynaptic membrane. This substance diffuses to and reacts with the postsynaptic membrane to alter its permeability characteristics. These changes in permeability result in decreased membrane resistance with consequent modifications in the postsynaptic membrane potential.

The concept of synapse, whether electrical or chemical, implies the one-way transmission of information. This unidirectional property determines the pathway for information flow in nervous systems from sensory receptors to central processors to effector organs. In this regard, we may define transmission pathways as being mono-, di-, or polysynaptic, depending upon the number of connections involved (Fig. 26-14). In studying the action of synapses, the systems most often employed have been the reflex mechanisms in the spinal cord; much of what is known about synapses has come from the work of Sir John Eccles, who studied the monosynaptic motoneuron reflex

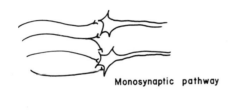

Monosynaptic pathway

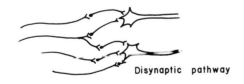

Disynaptic   pathway

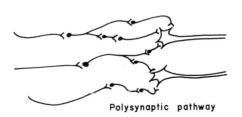

Polysynaptic   pathway

Figure 26-14
Monosynaptic, disynaptic, and polysynaptic pathways. The salient feature that differentiates these is the number of synapses interposed in series between the input and the output (i.e., the minimum number of synapses crossed). (From R. F. Thompson. *Foundations of Physiological Psychology.* Reproduced with permission. New York: Harper & Row, 1967.)

system in the cat. For this work, he shared the Nobel Prize in medicine with Hodgkin and Huxley in 1963 [5, 8].

CHEMICAL CONSIDERATIONS

Of the many types of synapses found in vertebrates, the vast majority may be classified as chemical. Morphologically, the chemical synapse may be thought of as a
minute, fixed junction between the presynaptic and postsynaptic cellular membranes.
The chemical synapse may be categorized further into type I or type II, depending
upon whether or not the junction involves a postsynaptic spine, respectively. The
spine is an outgrowth of membrane from the postsynaptic side. The presynaptic side
of the junction is formed by the terminus of an axon that is slightly enlarged into the
form of a small button or knob. Electron microscopy reveals that a definite gap 15 to
50 nm wide, the *synaptic cleft,* exists between the two apposed membranes (Fig. 26-15).
Also notable in electron micrographs are the *synaptic vesicles,* the small round or oval
structures that are contained within the synaptic knob. These vesicles are the site of
the synthesis, storage, or both, of the chemical transmitter substance.

    The events that are thought to occur at a chemical synapse during transmission are
outlined in Figure 26-16, which presents a simplified, functional picture [11]. With
the arrival of an action potential at the terminal of the nerve (event 1 in Fig. 26-16),
there is a transient synchronous release (2) of the transmitter T, which diffuses across
the cleft (3) and combines with special receptor sites in the postsynaptic membrane (4).
The transmitter and receptor molecules interact in such a way as to cause permeability

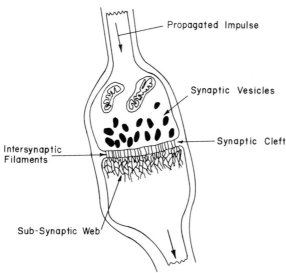

Figure 26-15
Chemical synapse.

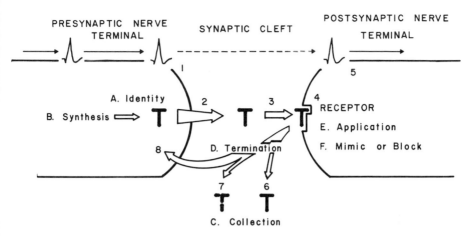

Figure 26-16
Hypothetical transmission processes at a synapse; T represents a hypothetical chemical transmitter. The numbers refer to the events taking place, and the letters refer to the criteria for a transmitter. (From R. H. Rech and K. E. Moore [Eds.]. *An Introduction to Psycho-pharmacology.* Reproduced with permission. New York: Raven Press, 1971.)

changes in the postsynaptic membrane, which ultimately lead to membrane hyper-polarizations or depolarizations (5) that are commonly known as postsynaptic potentials (PSP). Once the transmitter has exerted its desired effect, its action is terminated by its removal from the receptor sites either by diffusion (6), enzymatic destruction (7), or reuptake into the presynaptic terminals (8).

The neuromuscular junction or motor end-plate has been studied in great detail; since its mode of operation is quite similar to that of a chemical synapse, much can be learned from it about the operation of synapses. The function of the end-plate is that of transferring motor nerve impulses from motoneurons to muscles, and, as such, it behaves as an impedance-matching device between a high-resistance axon and a relatively low-resistance muscle fiber. For each volley in the motoneuron, there is a consequent impulse generated in the muscle, which leads to contraction.

In 1950 Fatt and Katz [6] observed that in "resting" muscle preparations, the end-plate region is the site of spontaneous miniature end-plate potentials of approximately 0.5 mV in amplitude that occur randomly at a nominal rate of about one per second. Except for their small sizes and random occurrence, these discharges are indistinguishable from the end-plate potentials that are produced by nerve impulses. It is assumed that they arise from the localized impacts of small quantities of the transmitter substance, acetylcholine, on the postsynaptic membrane. Analysis of the size of the miniature potentials shows that the smallest ones have a surprisingly uniform amplitude and that the larger ones are integral multiples of this basic quantal amplitude. These spontaneous discharges have been observed at all kinds of neuromuscular junctions and at synapses in the CNS, where the chemical transmitters have not even been identified. It is thought that incident nerve impulses intensify this release process momentarily, so that a few hundred of these events occur simultaneously (Fig. 26-17). Nerve impulses then may be considered as increasing the probability of this release. With the discovery of synaptic vesicles through electron microscopic techniques, it is now thought that each quantum may represent the transmitter that is contained in one or more vesicles.

Although it is now assumed that transmitters are synthesized in the axon terminals and stored in the local vesicles, it is interesting to note that very few transmitter substances have been unequivocally identified. A specific set of criteria must be satisfied to prove beyond doubt that a particular chemical is in fact the excitatory or inhibitory transmitter substance at a particular synapse. These criteria may be summarized as follows [14]:

1. Presynaptic terminals must contain the substance or an immediate chemical precursor.
2. Neural impulses in presynaptic terminals must result in liberation of the substance.
3. Injection of normal (i.e., physiologically compatible) amounts of the substance must cause the activation of postsynaptic cellular membranes.

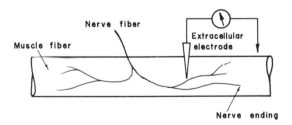

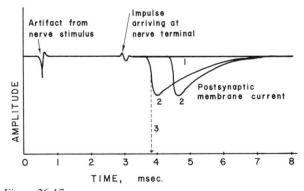

Figure 26-17
Focal surface recording from a neuromuscular junction. Under conditions of low calcium and raised magnesium concentrations, the postsynaptic deflection varies between failures (1), and quantal responses (2). Normally, a much larger deflection (3) leads to a muscle spike. (From B. Katz. *Nerve, Muscle and Synapse.* Reproduced with permission. New York: McGraw-Hill, 1966.)

4. An enzyme that destroys or inactivates the substance must be found in the vicinity of the synapse.
5. If drugs are known that block the action of the substance, they must block transmission at the synapse in question.

Although transmitters have been identified and confirmed in the peripheral nervous system, none has been conclusively determined within the CNS. The major reason for this is that many types of synapses exist in great abundance in CNS tissue, and they are packed in such a way that they cannot be easily isolated. It is difficult, therefore, to perfuse solutions selectively around them or to isolate and remove them for chemical analysis.

Only acetylcholine and norepinephrine have been completely confirmed as transmitter substances. Acetylcholine is released at the neuromuscular junctions of vertebrates, at the preganglionic nerve terminals of the autonomic nervous system, at the postganglionic endings of parasympathetic nerves, and at some sympathetic nerve endings. Norepinephrine is liberated at most sympathetic postganglionic terminals. Other partially confirmed transmitters are L-glutamate in the excitatory nerve-muscle synapse of insects and crustaceans, γ-aminobutyric acid (GABA) in the inhibitory nerve-muscle synapse of insects and crustaceans as well as in the inhibitory synapses of mammalian cortical cells, and glycine in the inhibitory synapses on motoneurons of the spinal cord.

Studies of the chemical nature of synaptic activity have resulted in a much more complete understanding of the dynamics of drug action within the nervous system. Traditionally, pharmacodynamic studies have been based on the assumption that compounds that are similar chemically would also be similar functionally. Often, however, substances that are vastly different chemically may produce the same behavioral effects. Strychnine and tetanus toxin, for example, will elicit similar seizure patterns but do not appear to be similar chemically. With the elucidation of the chemical action of synapses and neuromuscular junctions, it is now possible to explain many seemingly anomalous observations. Curare, for example, and a series of related compounds produce muscular paralysis and decreases in autonomic activity by blocking the normal transmitter effects of acetylcholine (ACh) at all ACh synapses. There are several ways in which such blocking may occur; these include (1) competitive inhibition in which the blocking agent competes for the receptor sites but does not activate the postsynaptic membrane, (2) long-lasting depolarization of the postsynaptic membrane so that the transmitter substance has no effect, and (3) the prevention of transmitter release from presynaptic terminals. Curare is believed to work by competitive inhibition, whereas botulin is thought to prevent transmitter release. Barbiturate anesthetics such as sodium pentobarbital are believed to act through the reduction of presynaptic activity. There are literally thousands of drugs that affect brain activity and behavior; although some act directly on cellular metabolism, it is likely that many exert their action by means of involvement at the synaptic level. Synapses provide the link, then, that enables external control of the nervous system through chemical means.

ELECTRICAL PROPERTIES

Although most synapses operate through chemical means, the direct electrotonic transmission of signals from one cell to another is possible, as stated previously, if certain requisites are fulfilled. Figure 26-18 shows the distribution of current flow around the junction when an action potential arrives at the terminus of an axon A that is adjacent to cell B. The incident depolarization produces transmembrane currents that flow into the cleft separating the two cells, and from there it follows two parallel pathways: (1) through R_L, out of the cleft, and into the open extracellular spaces, and (2) through the postsynaptic membrane (R_m) into the cell B (see Fig. 26-18). The relative values of R_L and R_m will determine whether the amount of current flowing into cell B will be sufficient to depolarize the nonsynaptic membrane of this cell to any significant extent. Several cases have been found in which these conditions are met. In cardiac muscle, in some smooth muscles, and in the segmental septa of invertebrate giant nerve fibers, the cells form electrical connections that are so effective that impulses travel from one cell to another with ease in either direction. Because two-way conduction is possible, such a connection, strictly speaking, is not a synapse. In other cases, such as the excitatory synapse of the crayfish abdominal nerve cord that was studied by Furshpan and Potter [7], the current will flow only in one direction. In this synapse, depolarizations of the presynaptic cell are transmitted, but hyperpolarizations are not. The rectifying property of this synapse resembles that of a single axonal membrane, and it is not completely understood.

Given that electrical synaptic connections are possible, it is of interest to speculate about the functional necessity for chemical synapses. One possible reason for this need seems to involve the quantal release mechanism of transmitter. The number of quanta released appears to be a function of the number of spikes arriving per unit time, the amplitude of the action potential, and the state of polarization of the presynaptic terminal. It is therefore possible to obtain different amounts of transmitter release that permit gradations in the level of the postsynaptic potential changes. This

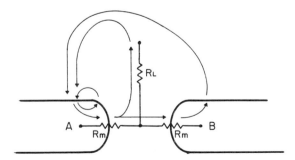

Figure 26-18
Distribution of the current that is presumed to exist around an electrical junction with the arrival of an action potential at axon terminal A.

would appear to be of some importance in the integrative properties of postsynaptic cells, and it could not be obtained as specifically if only strictly electrotonic transmission occurred. Another possible explanation appears to be that the synapse may serve as an impedance-matching structure that permits efficient transmission between a high-impedance nerve ending and the much lower impedance postsynaptic cell or muscle. It is interesting, therefore, to examine the motor end-plate to see if purely electrical, nonchemical transmission would be possible [10]. Consider the situation in the frog where a relatively large motoneuron (150 μ diameter) impinges upon a muscle fiber. To a current pulse such as an action potential, the muscle fiber presents a load impedance of approximately 50,000 ohms, which is several orders of magnitude less than that of the axon terminal. In order to excite the fiber, its membrane potential must be lowered from 90 mV to approximately 50 mV, which requires a current pulse of about 10^{-6} ampere. Interestingly, the effects of chemical transmission are such as to produce a current of approximately 3×10^{-6} ampere at many end-plates, thus providing an adequate safety factor for reliable transmission. The nerve ending itself has a contact area that may be roughly estimated as 2×10^{-5} cm^2. If the fiber produces a spike whose longitudinal current density is about 10^{-3} amp/cm^2, a total current of about 2×10^{-8} ampere is available. Even if this total current were to enter the muscle (which assumes the existence of protoplasmic continuity), a membrane potential change of only 2 mV would result. In the actual case, a 50-nm gap exists, which acts as a series resistance that would lower the current entering the fiber to 10^{-11} ampere; this would only be sufficient to produce a potential change of less than 1 μV.

Based upon such structural considerations, it is clear that the occurrence of purely electrical transmission is not likely. The problem is one of severe impedance mismatching, which implies that the end-plate serves as a matching network, as stated previously.

In order to study the potentials that are associated with synaptic action, microelectrode recording techniques are used in which the electrode is inserted directly into the postsynaptic cell. In neuromuscular junctions, the penetration has to occur near the point of contact of the membranes; otherwise, the leaky cable properties of the muscle fiber reduce the signal to undetectable levels (Fig. 26-19). The impaling of large neurons, such as the motoneurons of the vertebrate spinal cord and some ganglionic neurons of invertebrates, is sufficient for the detection of postsynaptic potentials from synapses at the proximal portions of dendrites and at the soma. To study the shape of postsynaptic potentials without interference from the resulting action potentials, several pharmacologic agents have been used. Curare reduces the size of the potentials at the vertebrate neuromuscular junction by competing with acetylcholine, whereas some anesthetics, such as barbiturates, have been used for studies of spinal motoneurons. Figure 26-20 shows the potentials evoked from a cat motoneuron by the stimulation of monosynaptically related afferents. The curves represent the average of many responses to three different strengths of stimulation, and they show the gradation of the postsynaptic potential. The fast rise and slow exponential decay exemplified in these curves are typical of postsynaptic potentials (see also Fig. 27-17).

The stimulation of some afferents to motoneurons produces a hyperpolarization instead of a depolarization (Fig. 26-21). If shortly after the hyperpolarizing stimulation

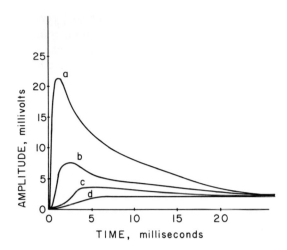

Figure 26-19
Time-course of the end-plate potential of curarized frog muscle recorded with an intracellular microelectrode at different distances from the synaptic area (*a*: 0 mm; *b*: 1 mm; *c*: 2 mm; *d*: 3 mm). (Modified from P. Fatt and B. Katz. Reproduced with permission. *J. Physiol.* [Lond.] 115:320, 1951.)

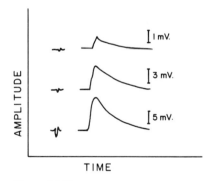

Figure 26-20
Excitatory postsynaptic potentials recorded with an intracellular microelectrode at a cat's moto-neuron. The small records at the left indicate the size of the synaptic input to the motoneuron, as recorded from the dorsal root. (Modified from J. S. Coombs, J. C. Eccles, and P. Fatt. *J. Physiol.* [Lond.] 130:347, 1955.)

a depolarizing one is delivered, then the intensity of the depolarizing input that is required to reach the threshold value in the postsynaptic cell is increased. In this way, some neuronal inputs tend to inhibit the firing of the cell. These two opposite postsynaptic potentials are known as the *excitatory postsynaptic potential* (EPSP) and the *inhibitory postsynaptic potential* (IPSP). Like the EPSP, the IPSP is a graded and nonpropagated response. It is much more limited in amplitude, however, since it rarely hyperpolarizes the cell beyond 75 mV, notwithstanding the intensity of the stimulus.

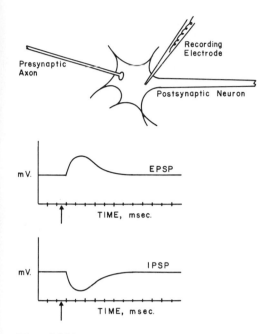

Figure 26-21
Synaptic inputs may be either excitatory or inhibitory (EPSP: excitatory postsynaptic potential;
IPSP: inhibitory postsynaptic potential).

Studies of the influence of the ionic environment on postsynaptic potentials have
revealed that lowering the external sodium concentration diminishes the size of the
EPSP, which thus indicates that the excitatory transmitter increases the permeability
of the membrane to sodium ions. In addition, when the resting potential of the cell
is decreased by current injection, the amplitude of the EPSP is diminished, and it
becomes zero when the resting potential is set at zero volts. Thus, the excitatory
transmitter induces permeability changes that tend to bring the membrane potential
to neutrality. These and other experimental results are best interpreted by assuming
that excitatory inputs cause a nonspecific increase in ionic permeability and create
a short circuit of the membrane at the synapse.

Experiments in which chloride ions are injected into motoneurons result in a reduction of the IPSP and even in the inversion of its polarity. Also, substitution of internal
potassium by sodium produces the same effect. If, under normal ionic conditions,
the cell is hyperpolarized by current injection, the IPSP is reduced and becomes
inverted in polarity at hyperpolarizations beyond −80 mV.

The interpretation of these results is that the inhibitory transmitter increases the
permeability of the membrane to K^+ and Cl^- ions, hence the membrane potential will
move toward a value between the equilibrium potentials of these ions. The interaction of opposing inhibitory and excitatory synaptic influences constitutes a fundamental aspect of the integrative action of single neurons.

When the depolarization of the postsynaptic soma reaches threshold, an action potential will be initiated at the axon hillock (i.e., the site of lowest threshold), and it will travel not only down the axon but also in the opposite direction, i.e., back into the soma and dendrites. This retrograde firing of the soma and dendrites serves to overwhelm all the synaptic input influences due to the large increase in P_K during repolarization, and it serves to return the membrane potential to the resting level. If the impinging excitatory inputs are sufficiently great at this point in time, the cell will rapidly reach its threshold potential again and fire another action potential. In this way, repetitive firing may be obtained. The above sequence of events during neuronal spike generation was deduced from intracellular recordings.

Figure 26-22 shows an action potential that was recorded from the cell body of a motoneuron. The initial slow depolarization is the EPSP which, at a level of approximately 10 mV of depolarization, leads to a propagating impulse. The lower curve is the derivative of the upper one and serves to accentuate the changes in slope that are proportional to transmembrane currents. The portion of the spike that recorded before the first dashed line in the figure is the action potential at the initial segment, which is reflected in the soma by electrotonic spread. Immediately, there is another change in slope that begins at about 30 mV of depolarization, which signals the beginning of the action potential in the soma. It is possible to block the soma-to-dendrite action potential by hyperpolarization of the soma and still elicit the axonal action potential. In this case, only the initial spike is observed.

The state of polarization as manifested by the axon hillock is the resultant potential due to the summation within the soma of all the excitatory and inhibitory inputs incident on the cell. This phenomenon, known as *spatial summation,* is described schematically in Figure 26-23. In Figure 26-23A, three EPSPs arrive simultaneously at the soma via different dendrites. The resultant potential at the cell body is much

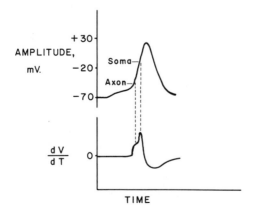

Figure 26-22
Action potential recorded from the cell body of a motoneuron. (Modified from J. S. Coombs, D. R. Curtis, and J. C. Eccles. *J. Physiol.* [Lond.] 139:232, 1957.)

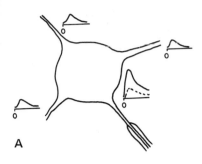

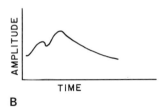

A

B

Figure 26-23
A. Effect of three action potentials arriving simultaneously; spatial summation occurs. *B.* Temporal summation following the arrival of two action potentials in a closely spaced fashion on the same fiber.

larger and more prolonged than any of the inputs taken individually. In Figure 26-23B, two EPSPs arrive in rapid succession through the same or different dendrites so that the soma receives the second EPSP before the effect of the first has vanished. This mechanism, termed *temporal summation,* accounts for reciprocal facilitation (i.e., one "helps" the other), and it is a consequence of the cable properties of the cell. Most excitatory synapses are located far from the cell body on very thin dendrites, and their electrotonic spread will exert only a slight influence on the membrane potential at the axon hillock. It is, then, the cooperative effect of many EPSPs that serves to bring the cell to its threshold. Inhibitory synapses, on the other hand, are most often located directly on the soma or even on the axon, and, as such, they exert a much more immediate influence.

Two general characteristics of synapses that are related to the processes of stimulus-secretion coupling and transmitter release and diffusion are (1) the *synaptic delay* and (2) the *frequency response* of synapses. The synaptic delay, or the interval between the arrival of an action potential at the axon terminal and the initiation of the response in the postsynaptic cell, has been measured in many preparations (Fig. 26-24). It has been found to have an average value of 0.8 msec in the CNS synapses of vertebrates and somewhat larger values in the ganglia and neuromuscular junctions. In most synapses, the response of the postsynaptic cell does not faithfully reproduce the firing pattern of

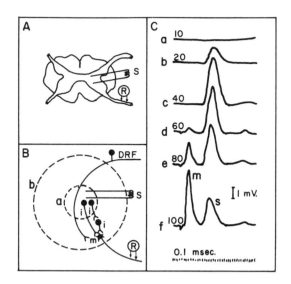

Figure 26-24
Measurement of synaptic delay in the spinal cord. *A.* Arrangement of stimulating (*S*) and recording
(*R*) electrodes. *B.* Interpretation of traces in *C. C.* Responses as the stimulus strength (indicated
by numbers above traces) was increased. In part *B,* only the dorsal root fibers (*DRF*) and interneurons (*i*) within the dashed circle *a* were excited by a weak stimulus; the shortest path to *R*
therefore included one synapse, and a delayed response, marked *s* in part *C,* resulted (traces *b–f*
in part *C*). With a strong stimulus, elements lying within dashed circle *b* were excited; these included
some motoneurons (*m*), whose discharge gave rise to *m* in part *C* (traces *d–f*). The difference in
latency between *m* and *s* (about 0.8 msec) is the approximate duration of synaptic delay. (From
B. Renshaw. Reproduced with permission. *J. Physiol.* [Lond.] 3:373, 1940.)

the innervating fiber. This phenomenon of temporal summation, or frequency response,
is not due to a prolonged refractory period, since the same cells are capable of repetitive
firing at frequencies as high as those of any axon when they are stimulated simultaneously through different inputs. In this case, the summation of the separate inputs in
the soma produces a large depolarization that elicits the train of action potentials. When
action potentials arrive in closely spaced fashion and temporal summation occurs, a
new postsynaptic potential may be generated before the former one can completely
decay (Fig. 26-25). Through this form of summation, it is possible to maintain a relatively steady level of depolarization or hyperpolarization within the postsynaptic cell.
When postsynaptic potentials are maintained in such a fashion, they are often termed
slow potentials. If the slow potential is maintained above the threshold value, the cell
will respond with a sustained burst of action potentials. Interestingly, when given cells
with appropriate properties, it is theoretically possible through spatial summation to
transfer different slow potentials from several cells to a single cell and thereby to perform simple arithmetic operations [14].

Not all the integrative neural functions occur postsynaptically; both facilitation
and inhibition may take place presynaptically. When two consecutive shocks are
delivered to the same presynaptic fiber, the synaptic potential resulting from the

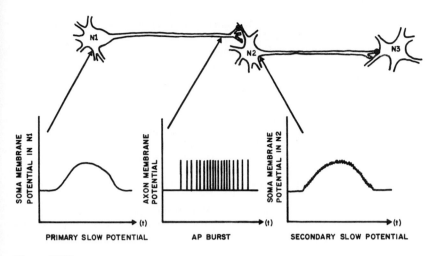

Figure 26-25
Generation of slow potentials through the temporal summation of incident impulses at a synapse. The transfer of slow potentials between cells is also indicated. (Modified from C. F. Stevens. *Neurophysiology, a Primer.* New York: Wiley, 1966.)

second is measurably increased if the interval between inputs is less than 20 msec and it is unaffected if the interval is lengthened. These effects have been traced to alterations in the amounts of transmitter being released. Another presynaptic effect is a form of inhibition that is observed in vertebrate nervous systems as well as in invertebrate neuromuscular junctions. It is mediated through an axoaxonic synapse. Frequently, in chains of neurons, a recurrent collateral fiber synapses at the terminus of an excitatory axon several neurons earlier in the chain. This synapse produces a depolarization of the presynaptic fiber, which, under these conditions, releases smaller amounts of transmitter upon the arrival of an action potential. A mathematical description of the generation of postsynaptic potentials may be found in Chapter 29.

REFERENCES

1. Cole, K. S., and Curtis, H. J. Electrical impedance of the squid giant axon during activity. *J. Gen. Physiol.* 22:649, 1939.
2. Coombs, J. S., Curtis, D. R., and Eccles, J. C. The generation of impulses in motoneurons. *J. Physiol.* (Lond.) 139:232, 1957.
3. Coombs, J. S., Eccles, J. C., and Fatt, P. Excitatory synaptic action in motoneurons. *J. Physiol.* (Lond.) 130:347, 1955.
4. Eccles, J. *Australian Academy of Sciences Year Book.* Sydney: Waite and Bull, 1963.
5. Eccles, J. C. *The Physiology of Synapses.* Berlin: Springer Verlag, 1964.
6. Fatt, P., and Katz, B. Analysis of the end plate potential recorded with an intracellular microelectrode. *J. Physiol.* (Lond.) 115:320, 1951.

7. Furshpan, E. J., and Potter, D. D. Mechanism of nerve impulse at a crayfish synapse. *Nature* (Lond.) 180:342, 1957.
8. Hodgkin, A. L., and Huxley, A. F. A quantitative description of membrane current and its application to conduction and excitation in nerve. *J. Physiol.* (Lond.) 117:500, 1952.
9. Hodgkin, A. L., and Rushton, W. A. The electrical constants of a crustacean nerve fiber. *Proc. R. Soc. Britain* 133:444, 1946.
10. Katz, B. *Nerve, Muscle and Synapse.* New York: McGraw-Hill, 1966.
11. Lorente de No, R. *Study of Nerve Physiology.* Studies of the Rockefeller Institute of Medical Research. Vols. 131–132. 1947.
12. Rech, R. H., and Moore, K. E. (Eds.) *An Introduction to Psychopharmacology.* New York: Raven Press, 1971.
13. Renshaw, B. Activity in the simplest spinal reflex pathways. *J. Physiol.* (Lond.) 3:373, 1940.
14. Stevens, C. F. *Neurophysiology, a Primer.* New York: Wiley, 1966.
15. Thompson, R. F. *Foundations of Physiological Psychology.* New York: Hoeber Med. Div., Harper & Row, 1967.

SELECTED READING

Leibovic, K. N. *Nervous System Theory.* New York: Academic, 1972.

Basic Sensorimotor Mechanisms   27

Richard E. Poppele

INTRODUCTION

The intent of this discussion of the bases of sensorimotor integration is to provide an introduction to concepts that deal with sensory processes and the control of movement. Simple neuronal circuits known as reflexes will serve as the focus of our discussion. The term *reflex,* meaning to turn back or to reflect, implies the occurrence of a response to a stimulus. The behavior of lower animals, which can be described primarily in terms of stimuli and responses, does depend largely on reflex circuits. Animals higher in the evolutionary scale exhibit behavior that cannot be defined simply by stimulus-response patterns, since their behavioral output is a complex function of the state of the organism and its history. Yet the basic reflex, which may however be internally modified, forms a primary component for nervous system function, and it is for this reason that we begin this study of the nervous system with the reflex.

A one-celled animal, such as the paramecium, is capable of changing its behavior in response to a stimulus. If the animal meets an obstacle or is poked with a small probe, its beating cilia, which propel it through the water, reverse their direction, and the animal swims away from the source of irritation. This illustrates the ability of this single cell both to recognize the presence of a stimulus and to produce an appropriate response. In higher animals, the two functions of recognition and response are specialized in separate classes of cells. *Receptor* cells are specialized for the recognition of particular changes in the environment, and *effectors* are specialized for the production of changes in behavior. A single receptor with its associated effector constitutes the simplest *reflex arc.* This combination leads to a stereotyped behavior in which responses are little altered from one stimulus to the next; changes in behavior result only from changes in stimulus parameters. In most cases, however, we find that the reflex arc includes a third interposed element, the *neuron,* which introduces a flexibility and a complexity that are necessary to help account for the behavior of higher animals.

The property of the neuron that lends these attributes to the reflex is its ability to sum or integrate inputs from various sources; this ability is a direct consequence of the cable properties of the neuronal membrane. Information, which is received by the neuron as voltage changes occurring at synapses that are distributed along its membrane, is summed, and it is a weighted sum of such individual events that determines the output of this element.

When there is a neuron in the reflex arc, the result of applying a stimulus to the receptor does not simply depend on stimulus parameters, but it also depends on the state of the neuron, which is a function of stimuli applied to other receptors and of information received from other neurons. Therefore, as an element of the reflex arc, the neuron can be considered in electronic terms as a *gate* or *gain control* (Fig. 27-1).

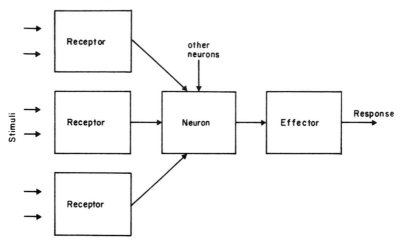

Figure 27-1
The relationship among the elements of the reflex arc.

The simplest reflex in man is the knee-jerk or tendon-tap reflex; it is the only one that has a single neuron in the reflex arc. Most reflexes, such as the withdrawal reflexes that produce responses to painful stimuli, involve many neurons between the receptor input and muscle output as well as many parallel pathways. Yet even the simple case of the knee jerk illustrates the variability introduced by the neuron. The force of the knee jerk in response to a tap is not dependent only on the strength or location of the tap. In fact, the clinical value of this reflex comes about because the response is a function of the state of the spinal motoneurons. Besides being a useful clinical tool, the reflex serves as an introduction to basic strategies employed by the nervous system. The properties and function of this particular reflex will be treated in some detail later in this chapter. First, however, we will examine a general case of the single-neuron reflex arc, with particular emphasis on the way in which the basic mechanisms that were discussed in the previous chapter are assembled as a sequence of events occurring between the stimulus and the response.

RECEPTOR MECHANISMS

The receptor is a transducer that interfaces the nervous system with its environment, thereby allowing the nervous system to sense various parameters of its environment, which may be either internal or external to the body. Receptors that monitor the internal environment are involved with control functions and usually provide a feedback that is incorporated into some control process. Receptors that sense the external environment may be involved either with control or with the acquisition of new information. In fact, a single group of receptors may serve both functions. The visual receptors, for example, provide feedback information to control processes that determine the pupil diameter and eye movement. They also provide information that is incorporated into the overall state of the nervous system, which we might call "awareness of the environment."

Each receptor type is sensitive primarily to only one form of incident energy. The structure of receptors often varies widely according to their particular functional specificity. Structural details of the receptor cells, their processes, and the surrounding tissue all contribute to specificity. In general, we can divide receptors into two broad classifications: *modified nerve endings* and *modified epithelial cells*. The epithelial cells primarily serve the special senses of vision, hearing, taste, smell, and balance. The modified nerve cells include temperature receptors and chemoreceptors, as well as a large class of receptors that are sensitive to various forms of mechanical energy. Receptors of the epithelial type are actually composed of two cells: an epithelial cell, which is some form of hair cell, and a nerve cell, which makes synaptic contact with it. The neuronal receptors, on the other hand, consist of a single cell with a cell body in or near the central nervous system and a long process to the periphery that is often associated with some special structure in the skin, muscle, tendon, joint, or internal organ.

Epithelial hair cells have a special structure called a *cilium*, which is an organelle having the same complex morphologic characteristics that are associated with locomotive cilia in single-celled animals [2]. All such cells have one or more cilia or some modification of a cilium. The outer segments of the rod and cone receptors of the retina, for example, have the same substructure as the kinocilia of the vestibular receptors of the semicircular canals of the ear. Each of the hair cell receptors, with the exception of the olfactory hair cell, makes synaptic contact with a receptor or an afferent neuron, which is usually a bipolar cell. (The olfactory unit, being less differentiated than the others, contains both the receptor and the neuron in the same cell.) In addition to their connection with receptor neurons, many motor or efferent nerve fibers make synaptic connection with these cells. Although the function of such motor control is largely unknown, it is widely observed, and it is believed to serve as a kind of feedback control for receptor input.

Neural receptors can be subdivided into at least four types. Simplest in appearance are the naked nerve terminals, which have no apparent specialization for receptor function. In fact, such endings are believed to serve more than one function, since areas like the cornea, which have only this type of receptor ending, are differentially sensitive to pain, temperature, and touch. A second type of neuronal receptor that functions as a touch receptor is associated with somatic hairs. A single hair may be innervated by 10 to 15 individual nerve fibers that form a collar parallel to the shaft of the hair. The whiskers of a cat are highly specialized sensory hairs that are each innervated by over 100 nerve endings. Hairless skin, such as that of the palms and soles, contains other sorts of sensory endings, which function similarly to the somatic hair receptors. These are called *Merkel's disks* and consist of a complex of nerve endings and epithelial cells. Finally, there are the corpuscular or encapsulated endings, in which the sensitive receptor endings are encased in a connective tissue capsule. Notable examples are the pacinian corpuscle, which is sensitive to pressure vibration the muscle spindle, which is sensitive to skeletal muscle stretch, and the Golgi tendon organs, which are sensitive to muscle tendon stretch.

Actually, the numbers and types of receptors that exist, particularly in the skin and internal organs, are quite large. The brief account given here is only to provide some basic orientation.

Receptors respond only to stimuli that occur within a well-defined physical domain. This domain is called the *receptive field* of the receptor. The resolution or grain of a sensory image that is conveyed to the CNS by a homogeneous population of receptors is a function of the receptive fields of the individual receptors. Where the fields are small — e.g., in the fingertips for touch receptors and in the fovea for visual receptors — resolution is best, and where fields are large — e.g., in the forearm or peripheral retina — resolution is poor.

Despite the diversity of receptor morphology, we can identify at least four functional parts that are common to most receptors (Fig. 27-2). The stimulus that forms the input is always some form of energy, either electromagnetic, mechanical, chemical, or thermal. It operates on a *filter*, which does not change the energy form, but modifies it to amplify or suppress certain attributes. The ear, for example, is an elaborate filter that enhances certain frequencies of sound and attenuates others. It is also an impedance-matching device with sufficient gain to overcome the losses that are associated with the transmission of sound from the air to the fluid medium of the internal ear, which contains the receptor hair cells. Similarly, the skin acts as a mechanical filter that alters the touch, pressure, or vibration that is detected by receptor endings embedded in the skin, and the eye redirects incident light to form an image on the retina.

The second, or *transducer*, stage converts the incident energy as modified by the filter into a modulation of the receptor membrane potential. Potential changes are then transmitted by some means of *coupling*, either a synapse or the electrical cable of the cell membrane, as an intracellular current to an *encoder*. Finally, the encoder converts the current into a modulated train of action potentials for transmission to the CNS.

Any of the compartments in this representation of the receptor can have properties that dominate the overall response. Many receptors, for example, have a tendency for all or part of their response to a maintained stimulus to decay with time. This property, called *adaptation*, can result from the behavior of any or all of the functional parts. The general observation is that receptor responses can be represented by two components: one that is proportional to the stimulus and one that is related to the rate of change of the stimulus. If we assume that the rate response decays

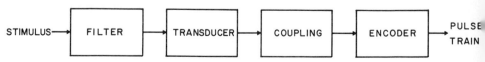

Figure 27-2
Functional compartmentalization of a receptor.

exponentially with time and that individual responses add linearly, then the total response at some time, t, is a function of the stimulus applied at t, $\alpha S(t)$, plus the sum of the residual responses due to stimuli acting earlier, which is given by the second term in Equation 27-1. Then the relation between stimulus intensity, $S(t)$, and response, $R(t)$, becomes:

$$R(t) = \alpha S(t) + \beta \int_0^t \dot{S}(t') \exp\left[-\gamma(t - t')\right] dt' \qquad (27\text{-}1)$$

If $\beta = 0$, then Equation 27-1 describes a *tonic* receptor whose response is always proportional to the stimulus. If $\alpha = 0$, then Equation 27-1 describes a *phasic* receptor that responds only when the stimulus changes. The rate of adaptation, which can be rapid or slow, is given by the rate constant γ. Notice that when α and β are > 0 and $S(t)$ is a step function whose derivative, $dS(t)/dt = \dot{S}$, is a Dirac delta function, the response is a step plus $\beta \exp -\gamma t$. Such a receptor displays both phasic and tonic properties. Typical receptor responses determined from Equation 27-1 are plotted in Figure 27-3 for step and ramp stimulus functions. It is possible to match the responses of most receptors qualitatively by suitable adjustment of the constants α, β, and γ.

The way in which the specific properties of a component of the receptor can lead to adaptation is illustrated by the following well-studied example of a receptor filter in the pacinian corpuscle. The mechanical properties of its onion-like capsule act as a high-pass filter that transmits only mechanical transients to the encased nerve ending. Figure 27-4 illustrates that when almost all the capsular lamellae are experimentally removed, the rapid adaptation to pressure stimuli that is normally displayed by these

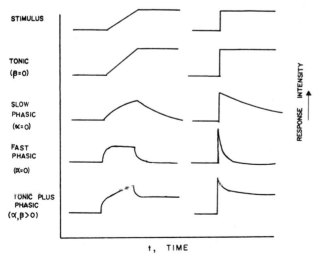

Figure 27-3
Receptor adaptation.

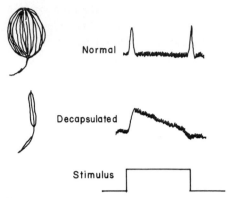

Figure 27-4
Decapsulation of the pacinian corpuscle. Receptor potential in response to a square-wave stimulus
is recorded in the intact, normal receptor before and after decapsulation. Response after decapsulation is much less phasic. (Reproduced with permission. From W. R. Loewenstein and
M. Mendelson. *J. Physiol.* [Lond.] 177:377, 1965.)

receptors is lost. The properties of the capsule have also been studied analytically by
means of a model of an equivalent cylindrical lamellar structure that treats each
lamella as an elastic element that is connected to the next lamella by a largely viscous
element. The frequency-response characteristics of this model were calculated, and
they are plotted in Figure 27-5. They correspond reasonably well to those of the intact
receptor, since there is no response to static pressure and the sensitivity is reduced to
vibrations below 200 Hz. Even so, the capsule is not the only source of adaptation in
this receptor. The transducer also fails to produce a prolonged response to a maintained
stimulus, even after the capsule is removed, as Figure 27-4 illustrates. The encoder,
too, is phasic, since it produces only a burst of action potentials when it is stimulated
with a steady current. In fact, in this particular example, it appears that each compartment contributes to the adaptive properties of the receptor. In contrast, all components of a tonic receptor respond proportionally to applied stimuli with little or no
adaptation.

The filter component may take on various forms and its operation may vary widely
from receptor to receptor, whereas the operation of the transducer is always the same:
to convert incident energy into a change in membrane potential. This change is
referred to as the *receptor potential,* or, in those receptors where it leads directly to
the generation of action potentials, it is often called a *generator potential.*

Although the process of transduction in receptors is only poorly understood, it is
generally believed to involve mechanisms that alter the permeability of the receptor
membrane. It was noted in Chapter 5 that the cell membrane is permeable to both
sodium and potassium ions and that neither is in thermodynamic equilibrium. In fact,
the membrane potential, in the absence of concentration changes, is a function of the
ratio of the sodium and potassium permeabilities (see Eq. 5-46). Any alteration of
that ratio acts as a valve that taps the potential energy stored in the concentration

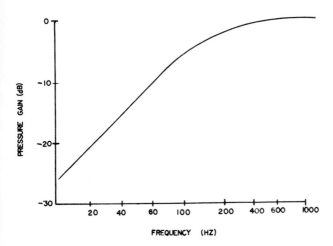

Figure 27-5
Frequency-response characteristics of a model of the pacinian corpuscle filter. The ratio of the applied pressure to the pressure transmitted to the nerve ending is plotted on the ordinate as gain and the modulating frequency is plotted on a logarithmic scale on the abscissa.

differences for these ions. If the ratio is increased, thus making sodium relatively more permeable, then sodium ions move inward and the membrane potential is driven toward the Nernst equilibrium potential for sodium. Transducer function is believed to be a consequence of the ability of a stimulus to alter the permeability ratio and thereby produce a potential change; this is termed the *ionic hypothesis* for transduction.

There is a body of evidence that supports the ionic hypothesis, but it has not been proved that the mechanism applies to all receptors. Perhaps the most confirming evidence comes from studies of receptors that can be removed from their normal environment and examined in isolation, such as the frog muscle spindle, the pacinian corpuscle, and the stretch receptor of the crayfish. A common finding in studies of these receptors is that when the sodium ions of the extracellular fluid are replaced with ions to which the cell membrane is impermeable, they produce an attenuated response to applied stimuli. Therefore, whatever the mechanism, it can be concluded that transduction depends, at least in part, on the presence of extracellular sodium.

However, if sodium were the only ion involved, we would expect that, as a consequence of its removal, the receptor response would be completely abolished rather than only attenuated. Hence, it has been suggested that a stimulus may cause a nonspecific change in receptor membrane permeability that allows other ions to move as well. This concept is supported by the finding that the membrane potential at which there is no longer a net driving force for the generation of a receptor potential is between 0 and −20 mV, rather than +60 mV as would be expected if the membrane became permeable only to sodium [14] (Fig. 27-6). Even so, if we accept that sodium and potassium are the ions most likely to be involved, it is clear that sodium permeability must increase more than potassium permeability in order to produce a

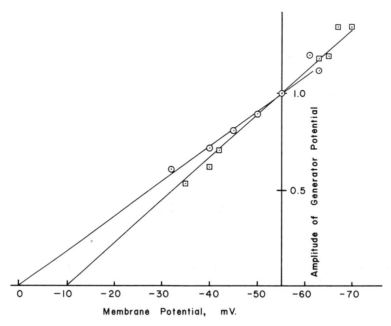

Figure 27-6
Equilibrium potential of the generator potential in the crayfish stretch receptor. The amplitude of a generator potential evoked by a constant amount of stretch (ordinate) is plotted as a function of membrane potential (abscissa). As the driving force ($V_{eq} - V_r$) is decreased by depolarization, the amplitude of the generator potential decreases. When extrapolated to zero amplitude, the curve shows that the V_{eq} is between 0 and -10 mV. (Reproduced with permission. From C. A. Terzuolo and Y. Washizu. *J. Neurophysiol.* 25:62, 1962.)

depolarization. This would generate a net inward current, and indeed such currents can be measured at the site of stimulation.

A notable exception to this basic scheme, but one that is still consistent with the general principles of the ionic hypothesis, is the vertebrate visual receptor. These cells have a resting potential of about -30 mV due to a high conductance to sodium ion at rest, i.e., in the dark. Light causes a hyperpolarizing receptor potential by reducing the sodium permeability, thereby driving the potential toward the potassium equilibrium potential.

One consequence of the ionic hypothesis for transducer action is that it implies a nonlinear relationship between the stimulus and the receptor potential. Experimentally, this relationship has been found to be linear over a relatively small range of stimulus intensities and logarithmic over larger ranges (see Fig. 27-8). As we shall see by examining a simplified version of the ionic permeability mechanism, these relationships are what we would expect.

Consider the electrical circuit in Figure 27-7A, which contains the passive membrane resistance (R_m) and capacitance (C_m) plus a conductance channel in series with a voltage source that represents the net driving force for ions through that channel. The

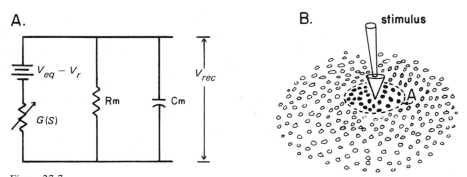

Figure 27-7
A. Equivalent electrical circuit of the transducer membrane (see text for details). B. Stimulus activates the conductance channels within an area A that is proportional to the stimulus intensity; channels outside this area are not activated.

driving force $(V_{eq} - V_r)$ is the difference between the equilibrium potential, at which there is no net movement of ions through the channel, and the resting membrane potential, where there is no net ionic current across the membrane in the absence of a stimulus. The conductance, $G(S)$, depends on the stimulus intensity, S, and when it is greater than zero, it leads to a time-varying voltage, V_{rec}, which is the receptor potential. If we neglect the intracellular resistance and thereby only allow voltage drops across the membrane, and further if we consider only the steady-state condition where $dV_{rec}/dt = 0$, then:

$$I = G(S) [V_{rec} - (V_{eq} - V_r)] = -\frac{V_{rec}}{R_m}$$

and

$$V_{rec} = (V_{eq} - V_r) \frac{R_m \, G(S)}{1 + R_m \, G(S)} \qquad (27\text{-}2)$$

where V_{eq} and V_r are constant emf's.

To get some idea of how G might depend on the stimulus intensity, suppose the membrane has a number of uniformly distributed sites that contain conductance channels which may be activated by a stimulus. Let the total value of the conductance be proportional to the fraction of those sites lying within a concentric area, A, around a point of stimulation. If we now assume that A is proportional to the stimulus intensity, S, then it follows that $G(S) = \alpha S$, where α is a constant of proportionality. Substituting this into Equation 27-2 and assuming that R_m and $(V_{eq} - V_r)$ are independent of the stimulus intensity, then:

$$V_{rec} = K \frac{bS}{1 + bS} \qquad (27\text{-}3)$$

where K and b are constants.

   Notice that for small stimuli, i.e., when $(1 + bS) \approx 1$, Equation 27-3 predicts a
nearly linear relation between the receptor potential and the applied stimulus intensity, which is consistent with the observed behavior of receptors noted above (Fig. 27-8.
As the stimulus becomes more intense, however, Equation 27-3 shows that the relation
deviates from linearity and that the receptor potential saturates for large values of S.
Since many receptors exhibit a logarithmic relationship between the stimulus and the
response for larger stimulus ranges, it would be interesting to see if Equation 27-3
could also predict that behavior. We will attempt to show that it does by substituting
into Equation 27-3 a variable, w, that represents the logarithm of the stimulus, and
observing the relationship between V_{rec} and w [10]. If we let w be the following linear
function of $\ln S$,

$$w = \frac{1}{2}(\ln b + \ln S)$$

then:

$$bS = \exp(2w)$$

Substituting this function into Equation 27-3 and multiplying both sides by 2,

$$2V_{rec} = K \frac{2\exp(2w)}{1 + \exp(2w)} = K\left[1 + \frac{\exp(w) - \exp(-w)}{\exp(w) + \exp(-w)}\right]$$

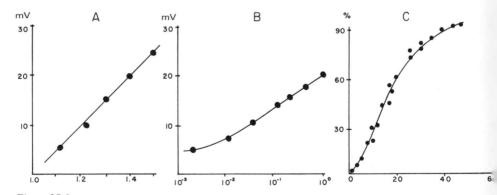

Figure 27-8
Relation between the receptor potential and the stimulus. *A*. Crayfish stretch receptor (abscissa:
multiples of rest length; ordinate: receptor potential in mV). (Reproduced with permission. From
C. A. Terzuolo and Y. Washizu. *J. Neurophysiol.* 25:62, 1962.) *B*. Limulus photoreceptor
(abscissa: light intensity in relative units, logarithmic scale; ordinate: receptor potential in mV).
(From M. G. F. Fuortes. *Am. J. Ophthalmol.* 46:210, 1962.) *C*. Pacinian corpuscle (abscissa:
amplitude of compression as a percentage of maximum; ordinate: receptor potential as a percentage
of maximum). (From W. R. Loewenstein. In W. R. Loewenstein [Ed.], *Handbook of Sensory Physiology*, Vol. 1. Berlin: Springer-Verlag, 1971. P. 269.)

therefore,

$$V_{rec} = K \left(\tfrac{1}{2} + \tfrac{1}{2} \tanh w\right)$$

If w is shifted an amount w_0, which is the value of w when $V_{rec}/K = \tfrac{1}{2}$, then:

$$\frac{V_{rec}}{K} = \tfrac{1}{2} + \tfrac{1}{2} \tanh (w - w_0) \qquad\qquad (27\text{-}4)$$

A graph of this function is plotted in Figure 27-9, and it can be seen that there is a range near w_0 where the curve is nearly linear. Since $(w - w_0) = c \ln (S) + d$, where c and d are constants, the linearity implies that the receptor potential is nearly proportional to the logarithm of the stimulus intensity in this range. If, on the other hand, we plot $\ln (V_{rec}/K)$ as a function of $(w - w_0)$, we find that the relationship is linear over an even greater range. In this range, therefore, $\ln V_{rec}$ is proportional to $\ln S$.

This is in agreement with the psychophysical observation that the relationship between the perceived magnitude of a stimulus and its physical intensity obeys a simple power law:

$$\Psi = K(S)^n$$

where Ψ is the perceived magnitude. This is not to say that the transducer mechanism is the only factor that contributes to the psychophysical relation between the stimulus and the perceived event, but it may be one of them.

Even though we started with a simple model based on the ionic hypothesis, we find that it can indeed predict both the linear and the logarithmic relationships that

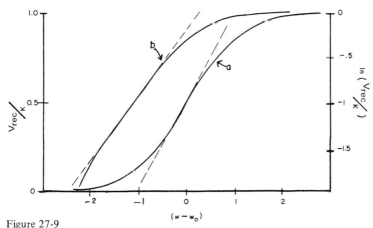

Figure 27-9
Curve a is a plot of V_{rec}/K as a function of $(w - w_0)$ from Equation 27-4 and represents the relation between the receptor potential and the logarithm of the stimulus. Curve b is a plot of $\ln (V_{rec}/K)$ as a function of $(w - w_0)$ and therefore represents the relation between the logarithm of the response and the logarithm of the stimulus.

hold between the stimulus intensity and the receptor potential amplitude, depending on the operating range and the biasing of the system. The same ionic hypothesis can be modeled to show how the transducer can introduce adaptation into the receptor response.

Suppose the instantaneous value of the receptor potential depends only on the membrane permeability for sodium ions and the driving force for that ion, $(V_{Na} - V_r)$. This supposition entails that we neglect the participation of other ions and the membrane capacitance. When the membrane permeability to sodium increases, sodium ions flow inward, where they may increase the concentration of that ion [5]. The magnitude of the concentration change depends on the magnitude and duration of the permeability increase, the area of the active membrane, and the volume of the intracellular compartment at the receptor ending. The net inward flow of sodium is given by:

$$\dot{Q}_{Na} = PA\,([Na]_o - [Na]_i)$$

where P is a permeability coefficient with dimensions of cm/sec and A is the membrane area. The rate of change of $[Na]_i$ as a result of this flow is equal to the flow divided by the volume of the receptor ending:

$$d[Na]_i/dt = \dot{Q}_{Na}/V = PA/V\,\{[Na]_o - [Na]_i(t)\} = k\,\{[Na]_o - k\,[Na]_i(t)\}$$

where $k = PA/V$. Applying the initial condition $[Na]_i(0) = c_i$ and letting $c_o = [Na]_o$, the above differential equation yields the solution:

$$[Na]_i(t) = c_i \exp(-kt) + c_o\,[1 - \exp(-kt)] \tag{27-5}$$

For a given step increase in permeability, the membrane potential will be determined by V_{Na}, which is approximately given by:

$$V_{Na} = 60 \log \frac{[Na]_o}{[Na]_i}$$

Substituting Equation 27-5 for $[Na]_i$ in the above, we have:

$$V_{Na}(t) = -60 \log\,[(c_i/c_o) \exp(-kt) + 1 - \exp(-kt)]$$

If we assume that the initial concentration ratio, c_i/c_o, is 0.1, then:

$$V_{Na}(t) = -60 \log\,[1 - 0.9 \exp(-kt)] \tag{27-6}$$

This function is plotted in Figure 27-10. It is a decaying exponential function with a single rate-constant, k, that is directly proportional to the sodium permeability

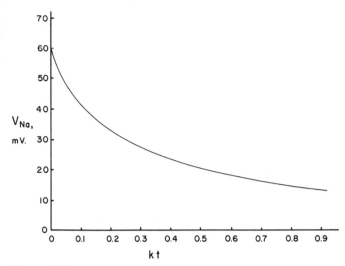

Figure 27-10
A plot of Equation 27-6 showing V_{Na} as a function of kt.

coefficient and the area of membrane excited and inversely proportional to the volume of the receptor ending.*

In this example, we have neglected the effect of a sodium pump, which would generate a flux of sodium ions out of the receptor ending. Unless the rate-constant of the pump were equal to or greater than k, however, the qualitative result would be the same, only instead of the equilibrium potential decaying to zero, it would decay to to a steady-state value given by the ratio of the rate-constants.

Changes in transmembrane potential generated at the receptor ending by the trans-ducer are propagated by a longitudinal current flow from the depolarized ending to the adjacent regions of the receptor cell. Since some receptors are connected to afferent neurons by synapses and do not produce action potentials themselves, it is assumed that the synaptic transmitter is released in quantities proportional to the magnitude of the receptor potential. The postsynaptic potential then becomes a generator poten-tial that leads to action potentials. In either case of coupling, whether of the electrical cable or the synapse type, the net effect is a flow of intracellular current toward the encoder or trigger zone of the cell. This is an area of excitable membrane with a low threshold for impulse initiation, and it is often associated with the first node of Ranvier, where the myelin sheath begins. Generator current flowing into that region depolarizes the membrane, and if the depolarization is sufficient, an action potential is generated.

The encoder region is the output stage of the receptor. Its function is to transform stimulus-modulated current into a coded train of action potentials. The operation of the encoder is analogous to that of a relaxation oscillator in which an input current, I_0,

*Note that k has the correct dimensions for a rate-constant since $k = PA/V = (cm \cdot sec^{-1})\ (cm^2)/ (cm^{-3}) = sec^{-1}$.

is linearly summed by a capacitor to a threshold voltage, V_{th}, in the time interval T:

$$V_{th} = \frac{1}{C} \int_0^T I_0\, dt = \left(\frac{I_0}{C}\right) T \tag{27-7}$$

If the integrator is reset each time it reaches threshold when a pulse is produced, then the pulse rate, f_0, is:

$$f_0 = \frac{1}{T} = \left(\frac{1}{V_{th}C}\right) I_0 \tag{27-8}$$

Thus, the frequency or rate of pulse generation is directly proportional to the applied current. This is, in fact, the relationship that is found for many actual receptor cells, as is shown in Figure 27-11, in which the current is supplied either by an intracellular electrode or by the transducer.

 The parameters of the model considered, however, are not a very realistic approximation of the parameters of the cell. A more appropriate model is one in which the integrator has a current leak, e.g., a parallel resistor-capacitor circuit, which would approximate the cable properties of the axon at the encoder site. Referring to the cable equation for membrane current (Chap. 26, Eq. 26-7), if we let $\delta V/\delta x = 0$, with all of the generator current flowing through the membrane at the encoder site, then:

$$I_m = I_0 = \frac{V(t)}{R_m} + C_m \frac{dV(t)}{dt} \tag{27-9}$$

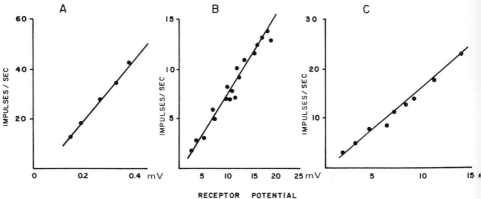

Figure 27-11
Relationship between the rate of impulse activity and the receptor potential. *A*. Frog muscle spindle. (Reproduced with permission. From D. Ottoson and G. M. Shepherd, *Acta Physiol. Scand.* 79:423, 1970). *B*. Limulus photoreceptor. (Reproduced with permission. From E. F. MacNichol, "Visual receptors as biological transducers." In R. G. Grenell and L. G. Mullins (Eds.), *Molecular Structure and Functional Activity of Nerve Cells.* Washington, D.C.: Am. Inst. of Biol. Sci., pp. 34-62, 1956.) *C*. Crayfish stretch receptor. (Reproduced with permission. From C. A. Terzuolo and Y. Washizu. *J. Neurophysiol.* 25:62, 1962.)

If this integrator operates to threshold, V_{th}, with the boundary conditions $V(0) = 0$ and $V(T) = V_{th}$, then:

$$V_{th} = I_0 R_m \left[1 - \exp\left(-T/R_m C_m\right)\right] \tag{27-10}$$

Letting $f_0 = 1/T$,

$$f_0 = \frac{-1/R_m C_m}{\ln\left[1 - (V_{th}/I_0 R_m)\right]} \tag{27-11}$$

which is no longer linear. In fact if the current I_0 is too small, there will be no output, since the real logarithm does not exist for $[1 - (V_{th}/I_0 R_m)] \leq 0$. For the membrane voltage to reach threshold, the generator current must be at least as large as V_{th}/R_m. This also turns out to be a basic property of sensory receptors. Unless the stimulus reaches a certain minimum value, there is no response from the receptor. A stimulus that reaches this minimum value is termed a *threshold stimulus*.

A plot of Equation 27-11 shows that there is a region over which the relation between f_0 and I_0 is nearly linear (Fig. 27-12). It can also be seen from the series expansion of the logarithm,

$$\ln\left(1 - a/x\right) = -a/x - a^2/2x^2 - \ldots - a^n/nx^n \ldots$$

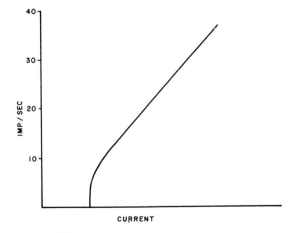

Figure 27-12
Relation between the rate of impulse activity and the input current for the leaky integrator encoder model.

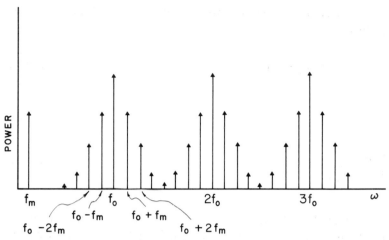

Figure 27-13
Power spectrum of a pulse-frequency modulated pulse train with a mean rate f_0 and a single modulation frequency f_m.

that:

$$f_o = \frac{-1/R_m C_m}{\ln\,[1 - (V_{th}/I_o R_m)]} \approx \left(\frac{1}{V_{th}C_m}\right) I_o \qquad (27\text{-}12)$$

for large I_o, i.e., $I_o R_m \gg V_{th}$. Therefore, a model that approximates the passive properties of the axon membrane behaves in a way that is at least qualitatively similar to the behavior of the receptor encoder.

The relationship between the generator current and the impulse rate is known as the *frequency code*. It implies that there is a modulation of the pulse rate by the generator current, which is called *pulse-frequency modulation* (PFM). Suppose that the pulses produced have a duration that is negligible compared to the interval between them and that the modulation is a sinusoidal current of the form, $I_0 + I_1 \cos (\omega t + \phi)$. It then follows that the pulse rate will increase and decrease sinusoidally about a mean rate as long as $I_0 - I_1 > V_{th}/R_m$ and that its power spectrum can be determined from a Fourier transformation of the time sequence. It is a discontinuous spectrum with discrete components at $0, f_0, 2f_0, 3f_0, \ldots$, each having sidebands at $nf_0 \pm m\omega/2\pi$, where $n \neq 0$ and $m = 1, 2, 3, \ldots$, and a single component at $\omega/2\pi$ [1]. From an inspection of this spectrum (Fig. 27-13), it is evident that as long as the modulation frequency, f_m, is much less than f_0, the modulation component can be separated from the rest of the pulse train, with no harmonic distortion, by low-pass filtering.

Demodulation of the signal that is generated by the receptor takes place in the central nervous system at the synaptic junction between the receptor axon and the second-order neuron. The operation performed at the synapse is equivalent to low-pass filtering. Information in the incoming signal is only about the intensity of the stimulus; there is nothing in the pulse modulation that specifies the modality of the stimulus or its origin. Such specificity is strictly a function of the connectivity of the network.

NEURONAL MECHANISMS

The integrative elements of the reflex arc are the neurons whose properties are tied closely to those of synaptic transmission. As stated in Chapter 26, the synapse provides one-way transmission of information from one cell to the next, usually without direct electrical coupling. Instead, the presynaptic cell secretes a chemical transmitter that induces a brief (2 millisecond) conductance change in the postsynaptic membrane; the resulting brief flow of transmembrane current causes a change in the membrane potential that spreads by electrotonic conduction. From this description, it may be observed that there are similarities between the response of a neuron to a synaptic stimulus and the response of a receptor to an environmental stimulus. In both cells, there is a unidirectional flow of information from an input transduction site, via some form of coupling, to an output stage that is the encoder. In the neuron, the synapse acts as an electrochemical transducer, the coupling is provided by the cable properties of the neuronal membrane, and the encoder, or trigger zone for impulse initiation, is located in the initial segment of its axon near the cell body or soma.

Many synapses are located in the neuronal dendrites, far from the trigger zone. Each synapse that is activated induces a small change in the membrane potential, and it must sum with many others before it can depolarize the cell sufficiently to reach the threshold value for an action potential. In addition, the response to a synaptic input that is located in the neuronal dendrites is attenuated by the cable properties of the cell membrane. Therefore, many neurons of the CNS require a considerable volume of synaptic input before they can respond with an action potential of their own.

In order to examine the effect of the membrane cable properties on the integration and flow of information in neurons, let us assume that the soma-dendritic complex can be represented by an equivalent uniform electrical cable (see Chap. 31). Then the time-course and amplitude of the postsynaptic potential (PSP), as measured at one end of the cable at the trigger zone, are a function of the distance of the synapse from that end. In fact, the synaptic potential is nearly equivalent to the response of the cable to an impulse applied at various points along its length. Although this problem is treated in detail in Chapter 31, we can observe some basic properties by making a few simplifying assumptions.

In the lumped equivalent circuit of Figure 27-14, assume that the individual longitudinal resistances, R_i, are large enough so that very little current passes from one segment to the next; i.e., the loading produced by each succeeding segment on the previous one may be neglected. Then the impulse response that is measured at the nth segment of the lumped cable can be approximated by:

$$h(t)^{*n} = [\exp(-t/R_m C_m)]^{*n} \qquad (27\text{-}13)$$

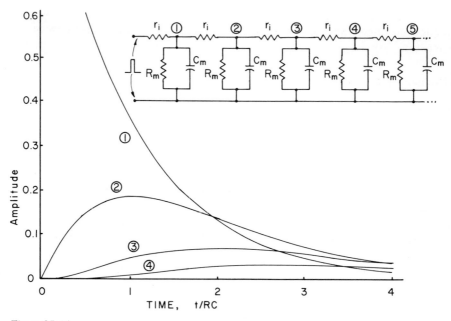

Figure 27-14
A lumped cable network that has properties similar to those of the neuronal membrane. The plots are the responses recorded at various nodes of the cable to an impulse applied at the left-hand end.

where $h(t)$ is the impulse response of one segment of the cable and the superscript $*n$ refers to the nth order convolution of $h(t)$ with itself. For this example, the response to an impulse as measured at the second segment would be:

$$h(t) * h(t) = \int_0^t h(\tau)h(t - \tau)d\tau$$

$$= t \exp(-t/R_m C_m) \qquad (27\text{-}14)$$

and at the third, it would be:

$$h(t) * h(t) * h(t) = \int_0^t \tau \exp(-\tau/R_m C_m)\, h(t - \tau)\, d\tau$$

$$= (t^2/2) \exp(-t/R_m C_m) \qquad [27\text{-}15)$$

and so on. Plots of the responses of the first four segments of this circuit are presented in Figure 27-14.

This simplified analysis of the cable leads to several conclusions that are applicable to synaptic transmission in neurons:

1. The amplitude of the PSP decreases with distance from the synapse.
2. The time-course and time-to-peak of the PSP increase with distance from the synapse.
3. The PSPs generated by synaptic inputs applied at different locations on the neuron and applied at different times sum linearly, as long as the synaptic current is proportional to the conductance change and the cable properties remain constant (see below).
4. The postsynaptic membrane acts as a low-pass filter and is therefore an effective demodulator for a frequency-modulated pulse train.

The property of filter demodulation is illustrated in Figure 27-15, which shows the result of applying a frequency-modulated pulse train to a synapse. The envelope of the PSPs measured some distance from the synapse is a copy of the original modulating signal. Very often, though, the PSP envelope is a smoother copy of the modulating signal than that which can simply be accounted for by the filtering process alone. This can happen when several nerve fibers carrying the same information converge on a single neuron. The summation of signals that takes place in the postsynaptic cell is an ensemble average, which enhances the common mode signal, the modulation. The pulse trains themselves are independently generated and are therefore uncorrelated, except for the modulation component that is common to them all. In terms of the

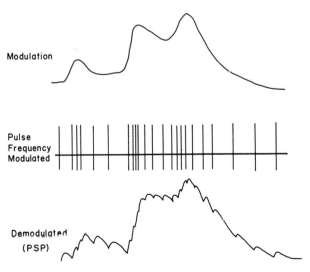

Figure 27-15
Demodulation of a pulse-frequency modulated pulse train by low-pass filtering (see text for details).

frequency spectrum of a pulse train, each pulse train has a different set of spectral components depending on its value of f_0, whereas each also contains an identical set of components due to the common modulation. In adding the spectra together, either before or after filtering, only the common terms will sum, and the other spectral terms will be relatively suppressed. It can be shown, in fact, that enhancement will occur, even if the f_0 of each train is the same, as long as the pulse trains are independent i.e., as long as pulses do not occur at the same time in each train.

The averaging mechanism for demodulation allows transmission to occur with relatively low pulse rates in several parallel channels without requiring the modulation components to be much slower than the pulse rate, as is the case when demodulation is carried out by filtering only. This may partly account for the apparent redundancy in neural pathways, where several independent parallel channels carry the same information. In all cases, though, it is the role of the impulse train to provide an undiminished data link between elements of the train, whereas the logic operations performed by a neuron consist of the appropriate summations of postsynaptic potentials.

In addition to the linear properties of synaptic transmission, there are also nonlinearities that should be mentioned. First, because the PSP is generated by a permeability change that allows ions to move along their electrochemical gradients, there is a nonlinear relation between voltage and conductance change, which is the same as that discussed previously for the receptor transducer. This nonlinearity is particularly notable for the inhibitory synapse. An inhibitory PSP is generated when the synaptic transmitter causes a selective increase in the permeability to potassium. Because the equilibrium potential for potassium is quite close to the resting potential, the driving force for potassium current is small. Thus, the range of linear behavior for the inhibitory PSP is also small, and it follows that the summation of inhibitory potentials is, in general, nonlinear. Furthermore, because the potentials are small, they have relatively little effect if they are generated far from the trigger zone of the neuron. It is interesting to note in this context that in two well-studied neurons — the spinal motoneurons and the cerebellar Purkinje cells — a major inhibitory synaptic input is found in the cell soma adjacent to the trigger zone.

A second nonlinearity is the result of current shunting due to the membrane conductance changes that accompany synaptic transmission. Since the local parameters of the cable change momentarily at the beginning of the PSP, summation during this period is not linear. The shunting effect is maximized for adjacent synapses, where a major portion of the synaptic current generated by one synapse is shunted through the membrane by the increased conductance caused by the other. The effect diminishes with distance, however, so that for synapses located on different dendrites, for example, simultaneous PSPs can sum linearly.

MUSCLE

The voluntary skeletal muscles are effectors that allow the nervous system to act on its environment. In fact, they provide the sole output to the external environment from the central nervous system, since it is the concerted activity of these muscles that constitutes the behavior of an animal. Their function is to change the electrical signals that are generated by the motoneurons of the nervous system into a mechanical force that is applied, usually across a joint, to help support a posture or to produce a movement. The mechanisms by which electromechanical transduction takes place are discussed in detail in Chapter 5. Here, we will only reemphasize some of the basic mechanical properties of muscle that are discussed in that chapter.

As a transducer for nervous system output, it would be desirable if muscle behavior depended only on that output with a flat frequency response. Since this is not the case, however, it presents the nervous system with specific problems that must be overcome by appropriate control mechanisms. The reflexes that are discussed below illustrate some of the strategies employed by the nervous system to idealize the response of muscles, so that their response is more nearly a copy of the command signal. Before we consider the reflexes, though, let us identify specifically what these nonideal properties are.

As noted in Chapter 5, the force that is generated by the contractile machinery of muscle is a function of the muscle length as well as of the input pulse rate. Actually, the tension that can be measured in the tendon of a contracting muscle has two components: a *passive component* due to the elasticity of the muscle cells, connective tissue, and tendon, and an *active component* that is generated by the contractile elements. The passive tension is almost linearly dependent on length, so that the passive stiffness is nearly constant. In addition, the passive dynamic stiffness has a nearly flat frequency response, indicating that under conditions of passive elongation, the elastic elements of the muscle prevail and the muscle behaves much like a spring. The total stiffness of a contracting muscle, however, which is a measure of both passive and active properties (see Chap. 31), is not constant, but rather it depends on both the muscle length and the rate of modulation of the input pulses. Let us consider these two properties — the steady-state behavior and the dynamics — separately.

The maximum steady-state tension was shown in Chapter 5 (see Fig. 5-9) to occur when the muscle is at or near its rest length. The total tension (Fig. 27-16) is the sum of the active and passive components. It shows that there is only a small operating range near the rest length where the total muscle force is relatively independent of the muscle length. Otherwise, the force produced by a given rate of motor nerve impulses varies over a wide range, depending on the length of the muscle.

As for the dynamics of the relation between input and force, the muscle has the properties of low-pass filter. In this case, synaptic transmission is not responsible for the behavior, because the nerve-muscle synapse acts as a one-to-one relay. Each motor nerve impulse leads to the production of a postsynaptic end-plate potential that is sufficiently large to generate a muscle action potential without the need for summation. Nevertheless, a single contraction that is caused by a single stimulus, called a *twitch,* may last for hundreds of milliseconds, which is many times the duration of the active state during which active tension is being generated. We might consider the twitch to be the impulse response of a muscle and a measure of the mechanical elasticity and viscosity engaged by a muscle contraction. When a second stimulus is presented to a muscle before the response to the first has decayed, the twitch responses

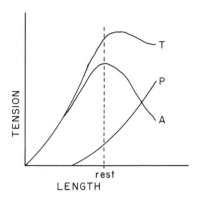

Figure 27-16
Muscle length-tension characteristics. *A*: active tension generated by contraction; *P*: passive tension due to the elastic properties of the muscle; *T*: total measured tension.

are added (see Fig. 5-10). If stimuli are presented in rapid succession, the summation builds up to an almost steady tension, called *tetanus* or *tetanic tension*. The properties just described are similar to those of the electrical cable, which operates as an integrator or a low-pass filter. These properties are also evident from the frequency-response characteristics of muscle that are obtained by sinusoidally varying the rate of the action potentials stimulating a muscle. Data obtained from such experiments are shown and discussed in Chapter 29.

In summary, there are two basic properties of muscle that lead to the nonideal behavior of this output transducer: a nonlinear dependence of output force on length, and a severely attenuated response to a rapidly changing input drive.

THE STRETCH REFLEX

When a normally innervated muscle is stretched, it contracts. This response, which is known as the *stretch reflex,* is the basis of the tendon-tap reflex mentioned previously. Rapping on the patellar tendon, for example, induces a quick stretch of the thigh muscle that excites a group of stretch receptors known as *muscle spindles.* These receptors make excitatory synaptic connection with the motoneurons that control that same muscle, causing it to contract.

The muscle spindles, which were described briefly earlier, are highly specialized mechanoreceptors consisting of an encapsulated bundle of muscle fibers, called *intra-fusal muscle,* which are innervated by two types of receptor endings. These are the *primary endings,* formed by the large, myelinated, group Ia nerve fibers that make monosynaptic connections with motoneurons, and the *secondary endings* formed by smaller, myelinated, group II nerve fibers that are apparently not involved in simple reflex behavior. The two parts of the spindle receptor respond to different parameters of muscle stretch (Fig. 27-17). Primary endings are basically phasic receptors and are therefore mostly concerned with changes in muscle length, whereas secondary endings

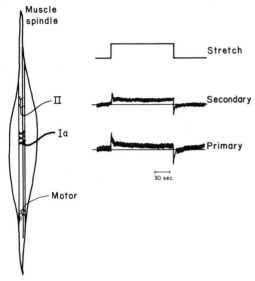

Figure 27-17
Responses of the primary and secondary muscle spindle endings to a step change in length.
Secondary endings give rise to group II nerve fibers. Primary endings, located at the center of
the intrafusal muscle, give rise to group Ia nerve fibers. The secondary response is tonic with a
rapid early adaptation. The primary response is phasic with a slow adaptation. The motor inputs
that contract the intrafusal muscle will stretch the sensory endings, since only the ends of the
muscle contract. (Reproduced with permission. From R. E. Poppele and R. J. Bowman. *J. Neurophysiol.* 33:59, 1970.)

are basically tonic receptors. In terms of Equation 27-1, the response of primary
endings can be characterized by a large β and a small α factor and two components
of adaptation, a rapid one and one with a time-constant of around 25 seconds. The
response of the secondary ending, on the other hand, is equivalent to having a large
α factor, a small β factor, and a rapid adaptation. Figure 27-17 compares the responses
of these receptors to step changes in length.

Although these are basically stretch receptors, they provide measurements of muscle
length and the rate of change of length, but not of tension, because they lie in parallel
with the contractile elements of the muscle, the *extrafusal* fibers. There is another
group of stretch receptors that serve as tension receptors, the *Golgi tendon organs,*
which are found in the muscle tendon (Fig. 27-18). In that position, they are sensitive to any force that generates tension in the muscle and thereby stretches the tendon.
When a muscle contracts, therefore, the tendon organs respond with increased activity,
whereas muscle spindle activity is decreased by the shortening of the muscle. In contrast, a passive stretch of the muscle and tendon excites both types of receptor. The
activity of a Golgi tendon organ influences a motoneuron via an interneuron, and it is
therefore part of a two-neuron or disynaptic reflex arc. The role of this reflex, which
inhibits muscle contraction in opposition to the stretch reflex, will be discussed below.
First, however, we will consider the details of the stretch reflex.

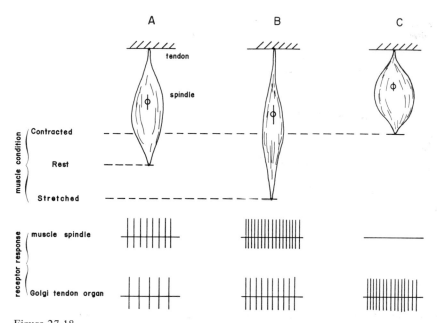

Figure 27-18
Response of the muscle spindles and the Golgi tendon organs to three muscle conditions.
A. Muscle at rest; both receptors active. *B.* Muscle stretch; activity of both receptors increased.
C. Muscle contracted; spindle is silent and the tendon organ maximally active.

The number of muscle spindles found in a muscle seems to depend on the specific role of the muscle. Those involved in fine discriminatory movements generally have more spindles than those involved in gross movements, even though the latter muscles are often larger. Nerve fibers from the group of spindles within a muscle form a part of the peripheral nerve that innervates that particular muscle. The cell bodies of the receptors are located in the dorsal root ganglia (see Chap. 25). The nerve fibers enter the spinal cord via the dorsal roots, and they then branch, sending processes to several locations. One set of branches goes to the ventral horn, where it synapses with the motoneurons (Fig. 27-19). Each single motoneuron, in turn, innervates a group of muscle fibers that are scattered throughout the target muscle. The functional unit, consisting of a motoneuron and the muscle fibers it innervates, is referred to as a *motor unit.* Graded contractions of a muscle are accomplished by variations in the number of active motor units as well as by the rate of impulse activity generated by each motoneuron. It is seldom, if ever, that all the motor units of a muscle are simultaneously active.

The sequence of events that constitute the stretch reflex may be summarized as follows. The stimulus that starts the sequence is a pull on the muscle tendon (such as might result by tapping the tendon) that lengthens the muscle. Muscle spindles within the muscle respond to this stimulus with increased impulse activity. Because of the specific connectivity of the group Ia fibers from the primary endings of the spindle,

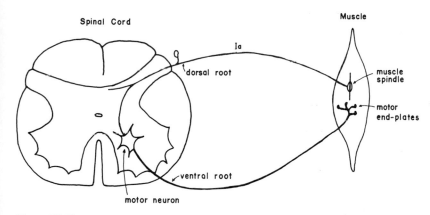

Figure 27-19
Stretch reflex (see text for description).

the level of excitatory synaptic activity in the group of motoneurons controlling the muscle that was stretched is then increased. The activity is normally sufficient to decrease the membrane potential of a certain percentage of these neurons sufficiently to either increase their rate of impulse activity or initiate impulse activity in previously silent neurons. A number of neurons, however, that were previously silent will remain so, because the net depolarization produced by this new input is still inadequate to reach threshold. The final result will be a recruitment of previously silent motor units and an increase in the rate of activity in other motor units that were previously active. Both factors will cause an increase in motor force, which, in the case of the tendon tap at the knee, will result in a movement of the leg.

It turns out that the force generated by the stretch reflex has a surprisingly flat frequency response with respect to the applied stretch. For a step change in length, the peak force is generated more quickly than would be expected from the discussion of muscle dynamics presented previously. It is true that a part of the force is passive and that it is expected to be rapid, but even the contractile force, which can be assessed by observing the response before and after cutting the muscle nerve in order to abolish the reflex, rises quite rapidly, following a short latency period due to conduction times along the reflex pathway. The reason for this rapid response is found in the compensatory dynamics of the muscle spindle receptor [13]. The spindle has a frequency response that just complements that of the muscle. In operational terms, the spindle acts as a differentiator, whereas the muscle is an integrator; combining the two elements in series produces a proportional operator.

Figure 27-20 illustrates how this complementary interaction operates when a step change in length is applied as an input to the receptor. The overshoot of the receptor response leads to a rapid initial buildup of muscle force. This interaction is perhaps even more evident when it is analyzed in the frequency domain. If the muscle is stretched sinusoidally, it is observed that the tension produced is proportional to and in phase with the applied stretch up to around 10 Hz. This results from compensation

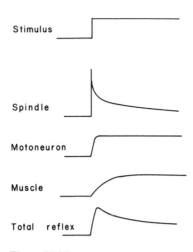

Figure 27-20
Response of each element of the reflex arc to a step function stimulus. Spindle response to a step increase in length; motoneuron response to a step input of current applied across its membrane; muscle response to a constant rate of nerve stimulation that is suddenly applied; total reflex is the muscle response to a sudden stretch with the reflex arc intact.

for the low-pass dynamics of the muscle. Further details about this dynamic compensation and the frequency analysis of the reflex are presented in Chapter 31.

This example illustrates a basic strategy employed by the nervous system to overcome an inherent limitation that is imposed by the output machinery. Other limitations, like the nonlinear tension production by muscle, are dealt with by other reflex mechanisms that will be discussed below.

The stretch reflex represents an elementary level of muscle control that functions simply to help maintain a nearly constant muscle length in spite of externally applied forces. Since the muscle usually operates across a joint, it becomes a reflex that helps to maintain a given joint angle. Because a muscle can generate force in only one direction, however, it takes at least two muscles to control the movement of a joint: one to flex the joint, and the other to extend it. Simultaneous contraction of these muscles allows positive positional stabilization of the joint as well. Two such muscles are referred to as *antagonists*. Movement requires a coordinated control of antagonistic muscles, which can be illustrated even at the level of the stretch reflex.

In addition to their excitatory connection with motoneurons of the *agonist* muscle, the group Ia fibers make an inhibitory connection with antagonist motoneurons via an interneuron. Therefore, when a joint is flexed, muscle spindles in the extensor muscles are stretched, and they induce the muscle to contract and oppose the flexion. At the same time, motoneurons of the flexor muscle are inhibited and that muscle is relaxed. The circuitry illustrated in Figure 27-21 is referred to as the *reciprocal innervation* of antagonistic muscles. The same principle of antagonistic control is found again and again throughout the motor control system.

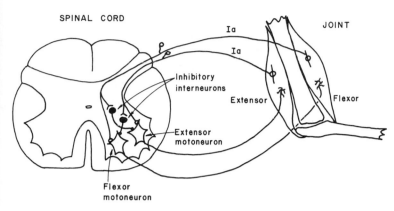

SPINAL CORD

Ia

Ia

JOINT

Inhibitory
interneurons

Extensor

Flexor

Extensor
motoneuron

Flexor
motoneuron

Figure 27-21
Reciprocal innervation of antagonistic muscles (see text for details).

From what has been stated about the properties of the elements of this reflex, it should be obvious that the implied role of feedback stabilization can only be effective at some optimum angle of the joint. If the limb is extended, for example, by voluntary contraction of the extensor muscle, the spindles in the extensor muscle fall silent and are no longer responsive to stretch, while the flexor spindles become active so as to induce a flexor contraction. Therefore, if this were the only reflex mechanism operating, voluntary movements would become very difficult, being always opposed by the stretch reflex. Recall, however, that the spindle consists of a special muscle to which the receptor is attached. The attachment is such that if that muscle contracts, it also stretches the receptor (see Fig. 27-17).

Now suppose that when a motoneuron becomes active, it causes both the extrafusal and intrafusal muscles to contract. Let us then reconsider the example given above. Contraction of the extensor muscle will now cause a simultaneous shortening and stretching of the spindle, which, if properly balanced, will have little effect on the rate of spindle activity. Likewise, the passive lengthening of the flexor muscle leads to a stretch of flexor spindles, which is counteracted by a relaxation of the intrafusal muscle. With this arrangement, the reflex would tend to assist, rather than oppose, a voluntary movement. Furthermore, if the external load on the moving limb were to be suddenly increased, thus causing the movement to lag behind the expected movement for a particular force, there could be a discrepancy between the shortening of the spindle due to extrafusal contraction and the stretching due to intrafusal contraction, which would cause the spindle activity to accelerate. This is because the intrafusal muscle shortens independently from the extrafusal muscle as well as independently from the external load. The reflex response to the added load, therefore, is an added contractile force that helps to overcome the load. Thus, the system acts as a servoassist mechanism for load compensation. As described, however, it is basically a positive feedback system, and it is therefore likely to become unstable. The actual circuitry, which involves a separate class of motoneurons, appears to

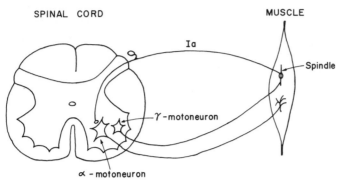

Figure 27-22
Motor innervation of muscle spindles by the gamma motoneurons.

provide the same tendency for load compensation, but without the positive feed-back.

The intrafusal muscles are actually innervated by separate motoneurons, which are termed the *gamma motoneurons* to distinguish them from the motoneurons that innervate the extrafusal muscles, the *alpha motoneurons* (Fig. 27-22). The contraction of intrafusal muscles results from the activity generated by gamma motoneurons, but only alpha motoneurons receive the afferent feedback directly from the spindle receptors, so that positive feedback is avoided. The control of intrafusal muscle and its effect on muscle spindle behavior is actually quite elaborate and is beyond the scope of this chapter. Suffice it to say that the gamma motoneuron innervation alters the sensitivity of the spindle receptors, and the phasic and tonic sensitivities can be adjusted independently.

The role of the gamma motoneuron in the task of motor control is poorly understood at the present time. It has been suggested that the alpha-gamma motoneuron system could act as a servomechanical follow-up device in which the spindle activity represents an error signal; negative feedback, which cancels the error, then results when a muscle shortens. In this scheme, movements could be initiated by gamma motoneuron activity alone, which would represent a certain muscle length. Spindles responding to that signal would continue to be active until the muscle shortened sufficiently to silence them. The amount of shortening would correspond to the length signaled by the gamma motoneurons. Such a mechanism requires a rather high loop-gain to be effective. Analyses such as those reviewed in Chapter 31, however, have shown that the loop-gain in this system is probably too low for this to be a likely mechanism for the initiation or control of movement. On the other hand, recent experimental evidence suggests that voluntary movements are generated by means of a different strategy, since it has been found that alpha and gamma motoneurons are usually activated simultaneously. This would imply that the role of the gamma motoneuron system is more like that described above, namely, to maintain the activity of spindles and to provide load compensation in a contracting muscle.

In summary, the stretch reflex is an elementary neuronal mechanism that illustrates the principle of dynamic compensation and the closely coupled control of antagonistic muscles. The central nervous system can apparently utilize the reflex in different ways, depending on the central commands to the alpha and gamma motoneurons. It should be emphasized, though, that many other reflexes and control strategies are involved in the total depiction of the control of skeletal muscle.

OTHER SPINAL REFLEXES

To extend the picture of reflex control of muscle contraction somewhat, we will consider the reflex arc that contains the Golgi tendon organ. As explained previously, it is disynaptic pathway that is inhibitory for a given muscle (Fig. 27-23). The tendon organs are particularly sensitive to tension caused by muscle contraction; therefore, the greater their contractile tension, the greater the reflex inhibition on the motoneurons responsible for the contraction. Such behavior, called *autogenic inhibition,* or, sometimes the inverse stretch reflex, is believed to help make muscle force less dependent on muscle length. In theory, it functions as a negative-feedback automatic gain control. When the gain is high, such as at rest length where the active tension is maximal, this feedback maximally suppresses the motoneuron output. Conversely, when the gain is low, the feedback is minimal and this output is not suppressed. The result is that the gain, or force delivered for some given state of the motoneurons, is less dependent on muscle length than it is without the operation of such a reflex. The fact that there is a neuron other than the motoneuron in the reflex arc suggests that the operation of the reflex is not always the same. Indeed, experimental evidence has shown that under certain conditions, the reflex is suppressed by inhibition from higher motor centers, thereby reducing or eliminating the automatic gain control.

So far, the reflexes that we have considered operate on a single muscle or an antagonistic pair of muscles. Reflex motor control, however, is much more complex than would be suggested by these few simple reflexes, and it would therefore be misleading not to at least mention other reflexes that may operate on entire limbs and pairs of limbs. Such reflexes may incorporate some of the same receptors that we have already

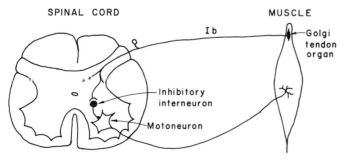

Figure 27-23
Golgi tendon organ reflex (see text for details).

discussed; however, many others — such as the joint receptors, various kinds of touch and pressure receptors, and receptors of noxious stimuli — are involved as well. Reflex arcs tend to be polysynaptic, involving many interneurons and considerable divergence, so that a local stimulus may evoke a response that involves many muscles. The scratch reflex, for example, that is exhibited by many animals is a case in which a very localized irritation activates the majority of the musculature of a single limb into a complex pattern of activity directed at the source of irritation. Therefore, even though there is often considerable divergence of activity, it is not unorganized divergency, but rather it is highly specific and purposeful.

Another example of highly directed reflexes is provided by the *withdrawal* or *flexion reflexes.* The stimulus may be any noxious input that might be potentially damaging to the body, and the response is a withdrawal of the threatened part, which is usually induced by the contraction of flexor muscles. However, in this case, too, the action is quite specifically dependent on the location and intensity of the stimulus, and, moreover, it does not always result in the activation of flexor muscles. If, for example, you were to sit accidentally on a sharp tack, you would probably extend rather than flex your legs in order to avoid the painful stimulus. On the other hand, if you were to step on that same tack, there would be a flexion of the appropriate leg. But simply flexing the leg that happened to step on the tack is not enough, for if the opposite leg were not simultaneously extended, you might fall. This reaction, called the *crossed extensor reflex,* is also a spinal cord reflex. A basic circuit such as crossed extension, which causes one limb to extend when the other flexes and vice versa, is obviously useful for locomotion as well. In fact, if there is a higher motor center to control locomotion, it could operate by appropriately biasing reflexes such as this. Thus, the basic circuitry of a reflex may serve more than one function; it may be evoked consequent to activity in other reflex pathways, or it may be influenced by higher centers.

In general, we might say that the reflex forms a basic functional building block upon which the function of more complex systems can be superimposed. At every neuron in a reflex arc, the specific function of the reflex can be modified by integrating reflex information with data formulated elsewhere in the system. The generation of a behavioral output may then be looked upon, in part, as the result of the continuous interaction among a large number of reflexes, which are in turn being manipulated by the more complex systems of the brain.

REFERENCES

1. Bayly, E. J. Spectral analysis of pulse frequency modulation in the nervous system. *IEEE Trans. Biomed. Eng.* 15:227, 1968.
2. Flock, Å. Electron microscopic and electrophysiological studies on the lateral line canal organ. *Acta Otolaryngol.* [Suppl.] (Stockh.) 199:1, 1965.
3. Fuortes, M. G. F. Electric activity of cells in the eye of Limulus. *Am. J. Ophthalmol.* 46:210, 1962.
4. Fuortes, M. G. F. Generation of Responses in Receptor. In W. R. Loewenstein (Ed.), *Handbook of Sensory Physiology,* Vol. 1. Berlin: Springer-Verlag, 1971. P. 243.

5. Houk, J. C. Rate sensitivity of mechanoreceptors. *Ann. N.Y. Acad. Sci.* 156:901, 1969.
6. Loewenstein, W. R. Mechano-electric Transduction in the Pacinian Corpuscle. Initiation of Sensory Impulses in Mechanoreceptors. In W. R. Loewenstein (Ed.), *Handbook of Sensory Physiology,* Vol. 1. Berlin: Springer-Verlag, 1971. P. 269.
7. Loewenstein, W. R., and Mendelson, M. Components of receptor adaptation in a pacinian corpuscle. *J. Physiol.* (Lond.) 177:377, 1965.
8. MacNichol, E. F. Visual Receptors as Biological Transducers. In R. G. Grenell and L. J. Mulling (Eds.), *Molecular Structure and Functional Activity of Nerve Cells.* Washington, D.C.: Am. Inst. Biol. Sci., 1956.
9. Munger, B. L. Patterns of Organization of Peripheral Sensory Receptors. In W. R. Loewenstein (Ed.), *Handbook of Sensory Physiology,* Vol. 1. Berlin: Springer-Verlag, 1971. P. 523.
10. Naka, K. I., and Rushton, W. A. H. S-potentials from colour units in the retina of fish (Cyprinidae). *J. Physiol.* (Lond.) 185:536, 1966.
11. Ottoson, D., and Shepherd, G. M. Steps in impulse generation in the isolated muscle spindle. *Acta Physiol. Scand.* 79:423, 1970.
12. Poppele, R. E., and Bowman, R. J. Quantitative description of the linear behavior of mammalian muscle spindle. *J. Neurophysiol.* 33:59, 1970.
13. Poppele, R. E., and Terzuolo, C. A. The myotatic reflex: Input-output relationship. *Science* 159:743, 1968.
14. Terzuolo, C. A., and Washizu, Y. Relation between stimulus strength, generator potential and impulse frequency in stretch receptor of Crustacea. *J. Neurophysiol.* 25:62, 1962.

SELECTED READING

Cold Spring Harbor Symposium on Quantitative Biology. Vol. 33, 1965.
Loewenstein, W. R. (Ed.). *Handbook of Sensory Physiology,* Vol. 1. Principles of Receptor Physiology. Berlin: Springer-Verlag, 1971.
Matthews, P. B. C. *Mammalian Muscle Receptors and Their Central Actions.* Baltimore: Williams & Wilkins, 1972.
Montcastle, V. B. (Ed). *Medical Physiology,* Vol. 1 (13th ed.). St. Louis: Mosby, 1974.

Neural Coding in Sensory Systems   28

Ray W. Winters

INTRODUCTION

The various sensory systems provide the link between physical energies in the external and internal environments and the neural systems that give rise to sensations and perceptions. The various sensory modalities of vision, audition, taste, olfaction, balance, tactility, temperature, and pain have associated with them particular receptor elements that often are located in specific organs such as the eye, ear, tongue, nose, and so on. These receptor elements, which typically behave as electrochemical or electromechanical transducers, provide afferent impulses to particular segments of the peripheral and central nervous systems, which ultimately elicit sensations corresponding to the appropriate stimuli.

A complete account of sensation and perception must include both a description of psychophysical relationships (i.e., the functional relationships between energy fluctuations in the environment and sensations) and the neural mechanisms that underlie them. As S. S. Stevens [32] put it, " . . . psychophysics defines the challenge; it tells what the organism can do and it asks those who are inspired by such mysteries to try with scalpel, electrode, and test tube, to advance our understanding of how such wonders are performed."

The developing use of single-cell recordings as an electrophysiologic method has added a new dimension to the study of sensory mechanisms. To a large degree, it has enabled the neurophysiologist to meet many of the "challenges" offered by psychophysics. In allowing the investigator to examine directly the activity of single neurons in sensory pathways, microelectrode recording techniques permit him to delineate how sensory information is processed.

This chapter mainly addresses itself to single-cell *coding* (or processing) of information at the initial stages of the various sensory pathways. In accordance with the request expressed by Stevens, an attempt will be made, when possible, to relate these data to those of psychophysical experiments.

CHARACTERISTICS OF SENSORY MECHANISMS

ENERGY SELECTIVITY

Sensory systems exhibit energy selectivity in three ways. First, a particular sensory system is most sensitive to a particular type of energy. The visual system, for example, responds best to electromagnetic radiation, and the tactile system, to mechanical stimulation. The type of energy that is most effective for a particular system is referred to as the *adequate stimulus* for that modality. It is sometimes possible to activate a sensory system by a stimulus that is not its adequate stimulus — e.g., mechanical pressure on the eyeball will generate visual sensations — but thresholds are much higher for these stimuli.

Second, sensory systems are selective in that they only respond to a limited portion of an energy spectrum. The human auditory system, for example, responds only to sound pressure changes with frequencies between 20 and 20,000 Hz. In a similar manner, the human visual system is sensitive only to electromagnetic radiation with wavelengths between 400 and 700 nanometers.

Some sensory modalities show a third type of limitation in that they are not equally sensitive to all energy fluctuations within their stimulus domain. A photic flash at a wavelength of 550 nm, for example, will appear brighter to a human subject than a 620-nm flash of the same energy. Human auditory thresholds for a 3000 Hz tone are lower (and hence sensitivity is higher) than for a 15,000 Hz tone.

RECEPTORS

Energy fluctuations in the environment lead to activity in receptors. The first step in the processing of sensory information is *transduction,* a mechanism whereby physical energy (stimulus) is converted to neural activity in first-order neurons. The initial neural event is graded, but this activity eventually leads to all-or-none activity in the afferent axons leading to the CNS. Some receptors are similar anatomically in that many have cilia-like structures at their receiving end and synapse-like junctions at their sending end [12] (see also Chap. 27).

NEURAL PATHWAYS

All sensory systems make connections with the central nervous system. The axons that emerge from the sensory organ constitute, in part, the peripheral nerves. For vision, audition, and other senses in which the sensory organ is located in the head, the cranial nerves are involved. The spinal nerves carry somatic sensory information from the body.

Upon entering the CNS, sensory information for a particular system can be carried to several areas within the brain or spinal cord. The number of nuclei and tracts in sensory pathways varies, but all systems, except that for olfaction, synapse in the thalamus. The axons of cell bodies that are located in the thalamus connect to primary sensory projection regions in the cerebral cortex. Each sensory modality has a cortical projection area, which, to a large extent, is spatially segregated from other sensory projection areas. It is presumed by most investigators that the electrochemical events in the cerebral cortex underlie the conscious experiences elicited by sensory stimuli.

ADAPTATION

Most sensory modalities show adaptation, i.e., a change in response strength, to stimuli of long duration. The mechanism for adaptation for some senses, such as touch, lies in the processes of the receptors. Mechanical characteristics of the tissue that surrounds many tactile receptors cause them to respond at the onset and termination of a stimulus, but not to a sustained stimulus. Adaptation in some systems, such as that for vision, is the result of neural interactions occurring beyond the receptors. Recurrent inhibition, for example, would be such a neural mechanism of adaptation.

SENSORY CODING: TOPOGRAPHIC AND NONTOPOGRAPHIC MODALITIES

Topographic modalities [1] are modalities in which the stimulus dimension is coded spatially at the receptor and is represented spatially in the neural pathway [8]. The pitch (frequency) modality in audition would be a topographic modality. Receptors at the base of the cochlea of the ear are maximally sensitive to high-frequency tones, whereas receptors at the apex of the cochlea respond best to low-frequency tones. Auditory neurons in the CNS are also spatially segregated according to their responses to frequency; neurons that are most sensitive to high frequencies are spatially separate from neurons with a high sensitivity to low-frequency tones.

Visual position is also a topographic modality. Different points in visual space are given topographically to different points on the retina of the eye, and this topography is preserved in visual structures in the CNS.

A nontopographic modality is a modality in which the stimulus dimension is not given spatially at the receptor surface and hence is not represented spatially in the CNS. Visual color and gustatory quality (e.g., sweet, sour, and bitter) are examples of nontopographic modalities.

ACROSS-NEURON PATTERN THEORY

Several modalities share the same neurons, so changes along any one of several stimulus dimensions can change the neural activity in a single receptor or afferent neuron. In the visual system, for example, a change in stimulus position, wavelength, or intensity could change the firing rate of a single axon in the visual pathway. It has been suggested by Ericson [8] that stimulus coding must therefore be viewed in terms of the relative amount of neural activity in the total group of cells activated by a stimulus. Qualitative aspects of the stimulus, such as taste quality or visual color, are represented neurally by an *across-neuron pattern.*

This is the neural pattern that is produced by many parallel neurons leading from one part of a sensory pathway to another. Each stimulus would give rise to a unique across-neuron pattern, and this pattern would be the neural representation of the stimulus. Stimuli that give rise to similar perceptions produce similar across-neuron patterns. According to Ericson [8], stimulus intensity — e.g., brightness in vision and loudness in audition — is represented neurally by the total amount of neural activity, independently of the shape of the across-neuron profile. A stimulus that produces changes in both the across-neuron pattern and the total amount of neural activity would produce both a qualitative and quantitative change in sensation.

To summarize, there are three major mechanisms involved in afferent coding: (1) neural topography, (2) across-neuron patterns, and (3) the total amount of neural activity. Topographic coding is involved in those modalities — such as visual position, somatesthetic position, and auditory frequency — in which the stimulus dimension is given topographically at the sense organ. Almost all stimuli presented to an organism activate many neurons, and each neuron responds to changes along several stimulus dimensions. Thus, coding for both topographic and nontopographic modalities must be viewed in terms of the total neural activity elicited by many parallel neurons. Qualitative aspects of the stimulus are coded by the across-neuron pattern elicited, and quantitative features are coded by the total amount of neural activity.

SOMATIC SENSORY SYSTEM

BODY

The major receptors for the somatic sensory system are Meissner's corpuscles, tactile disks, pacinian corpuscular receptors in muscle spindles. Ruffini's endings, bulbs of Krause, and Golgi tendon organs. Information from these receptors, which are located in the body, travels along the processes of unipolar neurons, whose cell bodies are located in the dorsal root ganglion (Fig. 28-1). These fibers enter the spinal cord through the dorsal root. The medial portion of the dorsal root contains heavily myelinated fibers (*A fibers*) that connect to those receptors that carry tactile and proprioceptive information (Meissner's corpuscles, tactile disks, pacinian corpuscular receptors in muscle spindles, and Ruffini's endings are the main types). A lateral bundle of the root is composed of thinly myelinated and unmyelinated processes (*C fibers*) that come mostly from free nerve endings, the bulbs of Krause, and probably some other encapsulated endings. The fibers of the lateral bundle are responsive to pain, temperature, and light touch.

The axons of the medial bundle bifurcate near the dorsal horn; a long process ascends the dorsal column as a part of either the *fasciculus gracilis* (for the lower body) or the *fasciculus cuneatus* (for the upper body), and short collaterals enter the gray matter of the dorsal horn (see Fig. 28-1).

The longer processes making up the dorsal white column synapse in the medulla in either the *nucleus cuneatus* or the *nucleus gracilus*. The axons of the cells of these nuclei cross the midline and travel up the brain stem as the *medial lemniscus*. Their axons end in the *ventrolateral nucleus* (which is also called the ventrobasal nucleus) of the thalamus. The final stage of the pathway involves the axons of the thalamic neurons. They travel to the somatic sensory cortex (the *postcentral gyrus*) via the internal capsule.

The thinly myelinated and unmyelinated axons of the dorsal-root lateral bundle synapse in various nuclei of the dorsal horn, including the *substantia gelatinosa*. Cells of the substantia gelatinosa can send axons to adjacent dorsal horn nuclei (such as the nucleus proprius, for example) or to the substantia gelatinosa at other

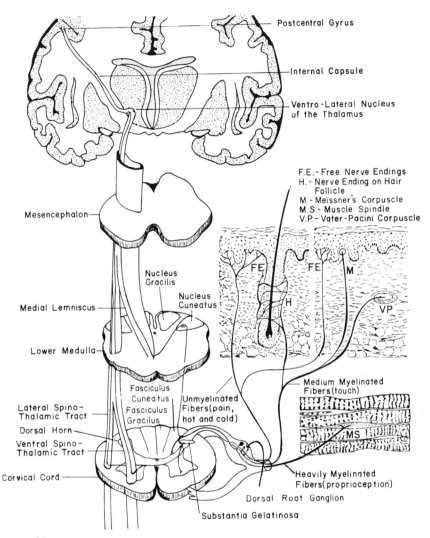

Figure 28-1
Anatomy of the somatic sensory system. The fibers ascending the dorsal white column are mainly concerned with touch and proprioception, the fibers of the anterior spinothalamic tract carry light touch information, and those of the lateral spinothalamic tract are concerned with pain and temperature. (After Frank H. Netter, M.D. From The CIBA Collection of Medical Illustrations, Reproduced with permission. © Copyright 1953, 1972, CIBA Pharmaceutical Company, Division of CIBA-GEIGY Corporation.)

levels of the cord. These latter connections are made by a small tract, called the *tract of Lissauer.*

Fibers from the dorsal horn cross the midline to form the anterior (light touch) and lateral (pain and temperature) *spinothalamic tracts.* These axons connect to the ventrolateral or ventrobasal nucleus of the thalamus. The neurons of this nucleus send axons to the postcentral gyrus of the cerebral cortex.

HEAD

In most ways, the organization of the somatic sensory system in the head region parallels that which is found for the body. Pain and temperature areas remain separate from touch and proprioceptive areas. Somatic sensory information reaches the arcuate nucleus of the thalamus via a path that runs parallel to the medial lemniscus. The arcuate nucleus projects to the postcentral gyrus. Most of the sensory information reaches the CNS through the trigeminal nerves.

SENSORY CODING: TOUCH AND PROPRIOCEPTION

SPATIAL ORGANIZATION OF PROJECTIONS

Most sensory systems show topographic coding in one form or another. In the somatic sensory system, there is an orderly arrangement of projections to the various nuclei and to the somatic sensory cortex such that stimulation in one region of the body gives rise to neural activity in a restricted region of the sensory pathway. Poggio and Mountcastle [26] and Mountcastle et al. [20] have also demonstrated that this principle of spatial representation applies to submodalities within the somatic sensory system; that is, within a given block of cells (in the ventrobasal thalamus, for example) that is activated by a particular portion of the body surface, submodalities like light touch, deep pressure, and joint movement are spatially segregated. In the somatic sensory cortex, the modalities are arranged in vertical columns, so that penetration with an electrode perpendicular to the cortex will detect cells that respond to the same modality as it is advanced from the surface to the white matter [19, 21].

It is generally accepted that the relative extent of the somatic sensory projections that are devoted to a particular region of the head or body is proportional to the use and sensitivity of that region.

RECEPTIVE FIELD STUDIES OF TACTILE NEURONS

The *receptive field* of a single tactile neuron is defined as the region of the skin which, when mechanically stimulated, leads to a change in the firing rate of the neuron. Tactile neurons have receptive fields that show spatial antagonism [21], which is such that the stimulation of one skin area causes excitation of the neuron and the stimulation of a nearby region causes inhibition. This neural mechanism is referred to as *lateral inhibition.* Figure 28-2 shows an example of a cortical neuron with a

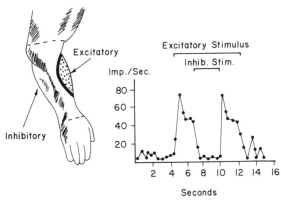

Figure 28-2
Receptive field of a single tactile neuron in the somatic sensory cortex. The receptive field has an excitatory central region and an inhibitory surrounding region. Response measure is the number of nerve impulses per second. (From V. B. Mountcastle and T. P. S. Powell. Central neural mechanisms subserving position sense and kinesthesis. *Bull. Johns Hopkins Hosp.* 105:173, 1959. Copyright The Johns Hopkins University Press, used with permission.)

receptive field that exhibits lateral inhibition. This unit shows an increase in firing rate when a restricted region of the skin on the preaxial side of the arm is stimulated, which is shown as the excitatory area in the figure. Stimulation of the skin within a much larger surrounding area, however, causes a cessation of the unit's activity. This type of receptive field is called an *on-center, off-surround* field. Figure 28-2 also shows, in the plot on the right of the figure, how an inhibitory stimulus can modulate the response to a steady stimulus in the excitatory region. At the onset of the excitatory stimulus, the firing increases. When an inhibitory stimulus is also present, the firing is completely suppressed, but the firing rate returns to a high level when the inhibitory stimulus is removed.

Mountcastle and Powell [21] argue that the spatially antagonistic field organization that they found at all levels of the somatic sensory pathway enhances the discrimination of two tactile stimuli located near each other on the skin. In Figure 28-3, three across-neuron patterns are shown. The upper one is for a single point stimulus on the skin; the other two traces are the patterns elicited by two points on the skin. A single tactile stimulus would fall into a number of receptive fields of single afferent neurons. For some cells, the stimulus would be in the surrounding area of the receptive field, whereas for others, the stimulus would be in the receptive field center. All points on the contours above the base line in Figure 28-3 represent firing rates for those cells in which the stimulus fell into the center of their receptive fields; points below the base line represent the firing rates for those cells in which the stimulus fell into the inhibitory surrounding area. The lowest two across-neuron patterns in Figure 28-3 show the importance of lateral inhibition to sensory discrimination. When there is no lateral inhibition, the across-neuron pattern that is elicited by two adjacent points on the skin is very flat and discrimination would be poor. With lateral inhibition, the contour is sharpened and hence discrimination would be more acute.

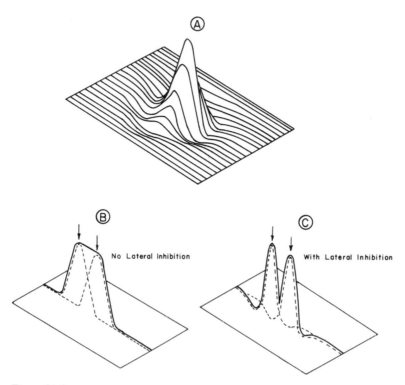

Figure 28-3
Across-neuron patterns in somatic sensory central nervous system structure. Each graph plots
the position, e.g., in the ventrobasal thalamus, (x and z axis) against the firing rate (y axis).
A. The contour for a point stimulus on the skin. *B.* The contour of the across-neuron pattern
that would be elicited in a system without lateral inhibition (i.e., without an inhibitory surrounding area). *C.* The pattern for a system with lateral inhibition. (*A* from Vernon B. Mountcastle
and Ian Darian-Smith: Neural Mechanisms in Somesthesia. In Vernon B. Mountcastle, editor,
Medical Physiology, ed. 12, St. Louis, 1968. The C. V. Mosby Co. *B* and *C* modified from this
publication. Used with permission.)

CODING OF STIMULUS INTENSITY

Mountcastle et al. [20] and Werner and Mountcastle [37] have conducted the most
comprehensive analysis to date of intensity-coding mechanisms in the somatic sensory
system. They have examined the firing rate changes in single proprioceptive and
tactile cells in response to variations in stimulus strength.

   In their studies of the tactile modality, they recorded the activity of single saphenous
nerve fibers in monkeys and cats while they varied the amount of skin indentation.
The saphenous nerve fibers that were studied ended in encapsulated mechanoreceptors,
called *Iggo receptors,* which are found in hairy skin areas. The mechanical properties
of these receptors are thought to underlie the relatively fixed neural response patterns
found in the saphenous nerve fibers that they studied. At stimulus onset, one component

of the response is a rapid-discharge transient. It lasts about 100 msec and then declines exponentially to a steady firing level. This latter component, which they call the "steady state," occurs during the remaining time that the stimulus is applied.

Stimulus (amount of indentation) versus response (firing rate) functions were found to be negatively accelerated, and a power function, $R = kS^n$, was determined to be the mathematical relationship that best described the data. The value of exponent for the function was usually about 0.5 for any response-observation time during the transient or steady-state components of the response.

The studies of Mountcastle et al. [20] on proprioception involved recordings from single cells in the ventrobasal complex of the monkey. In these studies, the stimulus intensity was defined as the amount of rotation in a joint of a peripheral limb. The neural response patterns, as in the case of the tactile system, had an early transient component that was followed by a steady-state component. Once again, the stimulus-response relationships were best described by a power function with an exponent of about 0.5.

A behavioral prediction that one might make on the basis of the data from the studies on both the saphenous nerve and the ventrobasal thalamus is that tactile and proprioception discrimination would be better at the lower end of the stimulus scale than at the higher end. Werner and Mountcastle [37] argue that this may not be true, however. They have also observed that there is a direct relationship between the inter-spike interval length (the inverse of the firing rate) and the variability of spike activity as measured by the standard deviation of an averaged response; i.e., the higher the firing rate, the lower the variability. They argue that therefore the higher $\Delta R/\Delta S$ ratio at the lower end of the stimulus scale may be offset by the lower variability at the higher end of the stimulus scale, thereby permitting equal or nearly equal discrimination throughout the entire stimulus range of a nonlinear system (Fig. 28-4). The main assumption of the statistical model shown on the right of Figure 28-4 is that a discrimination is made between two tactile stimuli of different intensities located at the same point on the skin when the area under the curve formed by the overlapping distributions (shaded areas in the figure) reaches a small criterion value. At the lower end of the stimulus scale, the area of overlap is relatively small because the slope of the function is high. At the higher end of the stimulus scale, the slope of the function is low but the variability is also low, as is indicated by the steepness of the Gaussian distribution curves. Thus, discrimination would also be good at the higher intensities.

It was proposed at the outset of this chapter that the stimulus intensity was neurally represented by the total amount of neural activity that is generated by all the neurons responding to a peripheral stimulus. The analysis by Mountcastle et al. [20] of intensity coding was made only for individual cells, but it is possible to apply their model to the total amount of neural activity without making any substantial changes.

SENSORY CODING OF PAIN

Perhaps the most curious of the senses is that of pain. Because of its clinical significance, it has been one of the most intensively studied skin senses, yet it is still perhaps

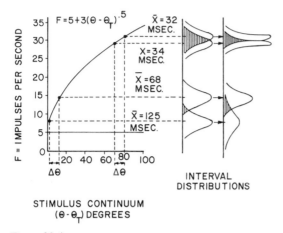

STIMULUS CONTINUUM
$(\theta - \theta_T)$ DEGREES

Figure 28-4
Werner and Mountcastle model for discrimination in the tactile and proprioceptive system. The value of θ represents either the angle of a joint or the amount of depression on the skin. It is possible to achieve equal discrimination with a negatively accelerated intensity-response function because response variability is higher at the lower end of the response scale than at the higher end (see the Gaussian distributions at far right). If the variability were the same (Gaussian distributions nearest the graph), then discrimination would be best at the lower end of the scale. (From G. Werner and V. B. Mountcastle. *J. Neurophysiol.* 26:958, 1963. Used with permission.)

the one that is least understood. Most of the controversy over the past 100 years concerning pain mechanisms has been centered around two opposing theories: (1) the *specificity theory,* which argues that pain, like other senses, is a specific modality with its own receptors, peripheral pathways, and central nuclei and tracts, and (2) the *pattern theory,* which holds the opposite viewpoint: that there are no specific receptors or pain pathways. Melzack and Wall [15] have, however, presented a model of pain perception that accounts for the clinical and experimental data that cannot be adequately accounted for by either previous theory.

Melzack and Wall argue that pain perception is the result of an interacting gate control and action system located in the spinal cord, but it is monitored by central efferents. They contend that the substantia gelatinosa of the spinal cord acts as a gate control system that modulates the activity of cells, which they call "T cells," that are located in the adjacent portion of the dorsal horn. The axons of the T cells presumably cross the spinal cord and become the lateral spinothalamic tract. Figure 28-5 shows a schematic diagram of the Melzack and Wall model. Input to the system is thought to come from two sources: large, fast-conducting A fibers (L in the diagram) and small, slow-conducting, unmyelinated C fibers (S in the diagram). A branch of the A fibers also makes up the dorsal column tract that mediates touch and proprioception (see Fig. 28-1). Both the A and C fibers make connections to the cells of the substantia gelatinosa and the T cells. The branches of the A and C fibers that synapse on the T cells are thought to be excitatory, whereas the A fibers are excitatory and the C fibers inhibitory to the substantia gelatinosa cells. The

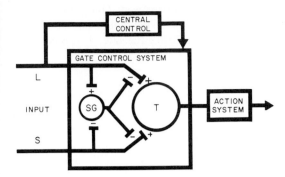

Figure 28-5
Gate control model of pain perception. The T cell is the neuron of the dorsal horn nucleus (like
the nucleus proprius). Its axons cross the midline, become a part of the lateral spinothalamic
tract, and connect to a pain "action system" at some higher level in the CNS. *L* refers to large,
myelinated A fibers; *S* refers to small, unmyelinated C fibers. The branch of *L* leading to the
central control would be axons ascending the dorsal column of the spinal cord. *SG* is the substantia gelatinosa. (From R. Melzack and P. D. Wall. *Science* 150:971, 1965. Reproduced with
permission of the American Association for the Advancement of Science. Copyright 1965.)

substantia gelatinosa cells feed back onto the afferent terminals of the A and C fibers
and exert presynaptic inhibition. In this way, they can modulate the excitatory
effect of these axons on the T cell. Central control of the gating mechanism is also
possible according to this model.

Iggo [11] has shown that about 30% of the C fibers respond only to noxious
stimuli (i.e., stimuli that are strong enough to produce pain) in comparison to about
2% for the A fibers. This observation, coupled with the fact that A-fiber responses
are phasic (i.e., the response adapts rapidly to a prolonged stimulus) and C-fiber
responses are tonic (i.e., slow to adapt), has led some investigators to argue that
tactile information, which shows fast adaptation perceptually, is mediated by the
A fibers, and pain information, which does not show adaptation, is mediated by the
C fibers. Melzack and Wall have argued for an interaction between the two fiber
systems to account for pain perception [15], but they would probably agree that
tactile information is carried mostly by the A fibers up the dorsal column.

Because the spontaneous activity level for C fibers is higher than that for A fibers,
Melzack and Wall maintain that in the absence of stimulation, the gate remains in a
relatively open position [15]. Since A fibers show very little spontaneous activity,
light pressure to the skin would produce a greater relative increase in their activity
than it would in C fiber activity. The gate would close somewhat or, at best, remain
in the same position, and no pain would be perceived. Tactile information would
still ascend the dorsal column, of course, and there would be tactile sensations elicited
by the stimulus. If the intensity of the stimulus is increased to the point of being
noxious, relatively more C fibers than A fibers will be recruited, because a higher
percentage of them respond only to noxious stimuli, and, if the stimulation is prolonged, the A fibers will adapt. Both of these changes would cause the gate to open,

and, if the T cell activity increases enough, pain will be perceived. One could counter-act the effect by rapidly changing the stimulus (because of the rapid adaptation of the A fibers), thereby increasing the A fiber activity and closing the gate. If it is closed sufficiently, the pain will be allayed.

THE AUDITORY SYSTEM

STRUCTURE

The ear can be partitioned into three compartments: the outer, middle, and inner ear. Most of the structural characteristics of the outer and middle ear are designed to minimize the energy losses that normally occur when vibrations from an air medium are transferred to the liquid medium of the inner ear. The structures of the inner ear transform vibrations there to nerve impulses, which are then sent to the brain (Fig. 28-6)

The two major components of the outer ear are the *pinna* and the *external auditory meatus* (canal). The former structure probably has little or no function in man, but in animals such as the cat, the pinna reflexively rotates toward a sound source and hence improves auditory acuity. The external auditory meatus serves as a protection for the delicate *tympanic membrane* (eardrum) of the middle ear. The eardrum obviously would be more subject to damage if it were placed unprotected at the surface of the body. The resonant characteristics of the external auditory canal serve to amplify frequencies between 2000 and 5500 Hz and hence, in part, account for the heightened sensitivity of human hearing to frequencies in this range.

The middle ear includes the tympanic membrane and the *ossicles:* the *malleus* (hammer), *incus* (anvil), and *stapes* (stirrup). Functionally, there are two major pur-poses of the middle ear. First, it serves as mechanical transducer for the transfer of energy from an air medium to the fluid medium of the inner ear. The motion of the ear drum is transmitted to the *oval window* (the initial structure of the inner ear) via

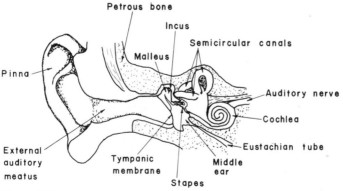

Figure 28-6
The ear and the vestibular system.

the ossicles. A large possible energy loss is reduced because the area of the tympanic membrane is much greater than the area of the oval window, thus increasing the pressure transmitted to the oval window. Were it not for this mechanism, about 99% of the energy reaching the inner ear would be lost by reflection.

A second function of the middle ear is to regulate the intensity of sound waves reaching the inner ear. This is accomplished by the movement of two muscles, the *tensor tympani* and *tensor stapedius,* that control the spatial arrangement of the ossicles. High-intensity sounds cause the muscles to contract (through feedback from the CNS), which can lead to an attenuation of the sound by as much as 20 decibels.

The organ for hearing in the inner ear is called the *cochlea.* This snail-shaped structure is divided into three fluid-filled chambers (Fig. 28-7): the *scala vestibuli* (the upper chamber), the *scala media* (the middle chamber; also called the *cochlear duct*), and the *scala tympani* (the lower chamber). The scala vestibuli and scala tympani are connected at the apex of the cochlea by a small opening, the *helicotrema.* The upper and middle chambers are separated by *Reissner's membrane*; the *basilar membrane* forms a boundary between the scala media and scala tympani. It is the motion of the basilar membrane that activates the receptors for hearing, the hair cells (see Fig. 28-7), which are found in the *organ of Corti.* These receptors are embedded in a gelatinous structure known as the *tectorial membrane.*

Vibrations of the ossicles are transmitted to the cochlea via a membrane in the scala vestibuli, the oval window. Except for a second membrane in the scala tympani, the *round window,* the cochlea is completely enclosed by bone. The fluid of the cochlea is incompressible, so that displacements at the oval window result in fluid movement out through the round window with little distortion.

The adequate stimulus for hearing is a shearing force that results from the movement of the basilar membrane. The shearing force occurs because the basilar membrane and the tectorial membrane are attached at different places (the tectorial membrane, in fact, is only attached at one point). When a low-frequency tone (below 20 Hz) reaches the cochlea, the fluid of the scala vestibuli leaks through the helicotrema into

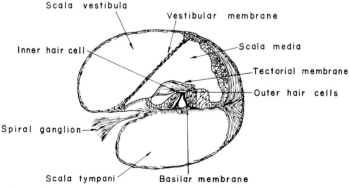

Figure 28-7
Cross section of the cochlea.

the scala tympani. At somewhat higher frequencies, the fluid cannot move through the helicotrema as fast as the displacements at the oval window. Since the basilar membrane lies between the oval and round windows, it is set in motion by these vibrations. A traveling wave moves from the apex to the base of the basilar membrane (Fig. 28-8). The basilar membrane is thinnest and widest, and thus is not as stiff, at its base. At low frequencies, the traveling wave causes maximal displacement at the base of the basilar membrane. High-frequency traveling waves do not reach the base of the membrane. As the wave accelerates, back pressure is incurred, and it becomes easier to displace the stiffest part of the membrane (the apex) than to counteract the inertia of the fluid.

ANATOMY OF THE AUDITORY PATHWAY

Pressure changes in the scala media that result from sound waves cause a shearing force on the tectorial membrane. This shearing force activates the hair cells of the organ of Corti. The hair cells are innervated by the dendritic processes of the first-order neurons of the auditory system. These neurons are bipolar cells whose cell bodies are located in the spiral ganglion (Fig. 28-9). One branch of these bipolar cells projects to the medulla, where it can synapse in either the dorsal or ventral cochlear nucleus.

There are three major fiber bundles, which are not shown in Figure 28-9, that leave the dorsal and ventral cochlear nuclei. First, a dorsal bundle, whose cell bodies lie in the dorsal cochlear nucleus, crosses the midline and enters the lateral lemniscus of the opposite side of the brain. A second, intermediate bundle, which originates in the ventral cochlear nucleus, also crosses the midline and enters the contralateral lateral lemniscus. The third group, the ventral bundle, also arises from the ventral cochlear nucleus. Many of the fibers of the ventral bundle terminate in the ipsilateral or contralateral superior olivary complex. The fibers that cross the midline form a

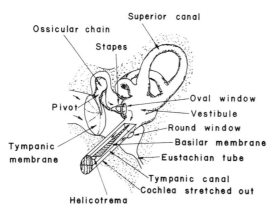

Figure 28-8
The middle and inner ear showing the cochlea uncoiled.

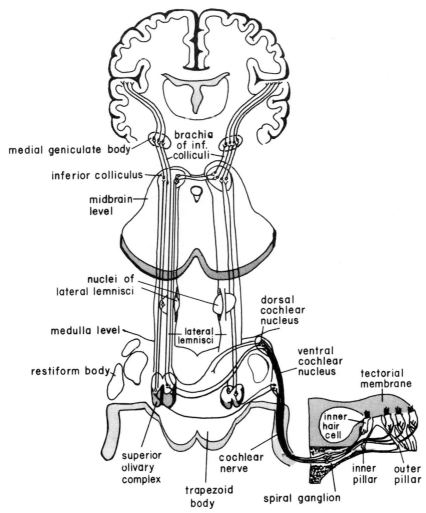

medial geniculate body

brachia of inf. colliculi

inferior colliculus

midbrain level

nuclei of lateral lemnisci

dorsal cochlear nucleus

medulla level

lateral lemnisci

ventral cochlear nucleus

restiform body

tectorial membrane

inner hair cell

superior olivary complex

cochlear nerve

inner pillar

outer pillar

trapezoid body

spiral ganglion

Figure 28-9
Anatomy of the auditory system. (After Frank H. Netter, M.D. From The CIBA Collection of Medical Illustrations. Reproduced with permission of the CIBA Pharmaceutical Company, Division of CIBA-GEIGY Corporation. © Copyright 1953, 1972. All rights reserved.)

conspicuous tract called the *trapezoid body*. The nuclei that are scattered among these fibers are called the *nuclei of the trapezoid body*. Many of the ventral bundle fibers also synapse here. Some axons from the ventral bundle go directly to the lateral lemniscus of the opposite side without synapsing in the trapezoid nuclei or the superior olivary complex.

The major ascending auditory pathway of the brain stem is the *lateral lemniscus*. It is composed of axons coming from the contralateral ventral and dorsal cochlear nuclei and the ipsilateral superior olivary and trapezoid nuclei. Most of the axons of the lateral lemniscus terminate in the ipsilateral *inferior colliculus,* which is located in the midbrain (see Fig. 28-9). Some axons, however, connect to the *nucleus of the lateral lemniscus,* which in turn sends axons back to the lateral lemniscus. Also, some lateral lemniscus fibers project directly to the *medial geniculate body* of the thalamus. Information can reach the opposite side of the brain through either the commissure of the lateral lemniscus or the commissure of the inferior colliculus. The medial geniculate body, which receives most of its input from the inferior colliculus via the brachium of the inferior colliculus, sends axons to the auditory cortex.

CODING OF FREQUENCY

Von Bekesy's classic work [35] has shown that movement of the stapes causes a wave to travel from the base to the apex of the basilar membrane. The location of the peak deflection of the wave is dependent upon the frequency of the stimulus. Since the stiffness of the basilar membrane is greater at the apex than at the base, low frequencies cause a relatively greater deflection at the base. With higher frequencies, the back pressure incurred is of a high enough magnitude to prevent the traveling wave from reaching the base. The most obvious conclusion to be drawn from these findings is that frequency, at least at the receptor level, is a topographic modality; that is, low frequencies cause a displacement of the basilar membrane at the base, thereby activating one subset of receptors, whereas high frequencies cause a displacement at the apex of the membrane, thereby activating another subset of receptors. A problem with such a scheme, however, arises with frequencies below 4000 Hz. In this range, the entire basilar membrane vibrates, and the difference in deflection magnitude between the apex and base is not great enough to account for the rather acute pitch discriminations that can be made in this frequency range. Since the membrane displacements that are associated with high frequencies are more restricted, a topographic code seems more appropriate for this range of frequencies.

There is also evidence for topographic coding in single auditory neurons. Most cells in the auditory pathway respond only to a restricted portion of the sound spectrum. This response area is usually defined by a *tuning curve,* which is a plot of tonal frequency against the threshold of firing for a single cell (Fig. 28-10). Each tuning curve has a single frequency, called the *best frequency,* at which the threshold is lowest. Most tuning curves have a fairly sharp high-frequency cutoff, but they respond to a much wider range of low frequencies. In general, the tuning curves become progressively sharper at higher levels of the auditory pathway up to the medial geniculate nucleus (see Fig. 28-10). Cortical neurons have rather broad tuning curves. It is also known that auditory cortex lesions and lesions of the brachium of the inferior colliculus have little effect on pitch discrimination [2].

Auditory neurons are topographically arranged according to the best frequency. This *tonotopic organization,* as it has been called, is well illustrated in a study of the inferior colliculus by Rose et al. [30]. They found that there is an orderly sequence of best frequencies from high to low in the external nucleus of the inferior colliculus,

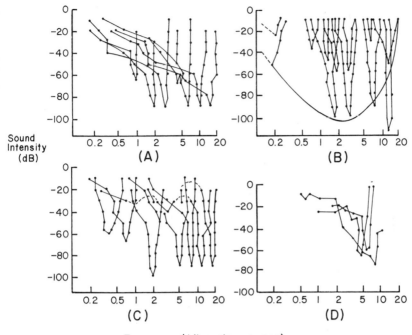

Frequency (kilocycles per sec)

Figure 28-10
Tuning curves for single auditory neurons at different levels in the auditory pathway of the cat.
A. Auditory nerve; *B.* Inferior colliculus; *C.* Trapezoid body; *D.* Medial geniculate body.
Most curves show a high-frequency cutoff at all levels. Each curve represents the tuning curve
of one neuron; for example, the tuning curves of seven neurons are shown in part *C.* The tuning
curves become sharper up to the medial geniculate body, then they become wider. (From
Y. Katsuki. In W. A. Rosenblith [Ed.], *Sensory Communication.* New York: Wiley, 1961.
P. 561. Used with permission.)

and that the reverse sequence, i.e., from low to high, occurs in the central nucleus
(Fig. 28-11). A tonotopic arrangement of neurons also has been reported for the
cochlear nuclei [29], the superior olivary complex [26], the medial geniculate
nucleus [9], and the auditory cortex [40].

It seems evident from the studies mentioned in the last few paragraphs that tonal
frequency is, at least in part, coded topographically by single auditory neurons. Rose
and colleagues [28] have also shown, however, that the temporal pattern and the
frequency of firing of single neurons are important in the coding of this stimulus
dimension. They recorded from single auditory nerve fibers in the squirrel monkey
while they varied the tonal frequency of low-frequency (below 5000 Hz) stimuli.
The neurons that were examined had best frequencies below 5000 Hz. One of the
major means of displaying their data was to plot interval histograms. In this type of
histogram, the interval (in microseconds) between adjacent action potentials is plotted
against the frequency of occurrence of each interval. The stimuli were pure tones,
which lasted for one second and were repeated ten times. The counts on the ordinate

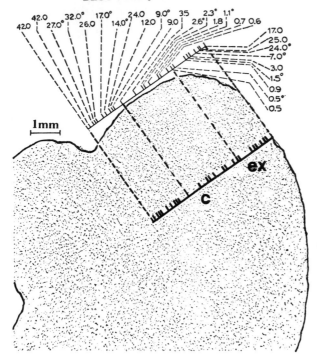

BEST FREQUENCY KC

Figure 28-11
Photomicrograph of the coronal section in the inferior colliculus of the cat. The solid, thick line shows the location of the electrode tract. The external nucleus is nearest to surface; the central nucleus is deeper. The best frequencies of the units found in the nuclei are indicated above. (From J. E. Rose et al. *J. Neurophysiol.* 26:294, 1963. Used with permission.)

of the histogram thus represent the sum of ten stimulus trials. As can be seen from Figure 28-12, each frequency generates a number of partial distributions. The peak of each distribution was usually found to be an integral multiple of the period of the stimulating tone. The modal value of the first peak for a 1000 Hz tone, for example, would be 1000 μsec, the second peak 2000 μsec, and the third 3000 μsec. An obvious limitation of such a period-time code comes about for high-frequency tones, where the period of the sine wave is less than 1000 μsec. Since the mammalian neuron cannot fire much faster than once every millisecond, modal values of less than 1000 μsec are not observed for these frequencies. The first peak for these frequencies is usually at a value near 1000 μsec, and other peaks are integral multiples of the period of the sine wave.

Interval histograms similar to the ones shown in Figure 28-12 were found for all stimulating frequencies that activated a cell, regardless of such attributes as best frequency, stimulus duration, or intensity. The interval histogram data suggest that the action potentials usually occur during a restricted segment of the cycle of the stimulating

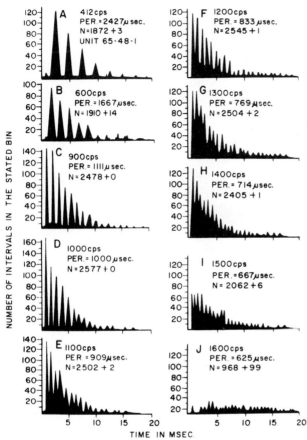

Figure 28-12
Interval histograms for a single neuron in the squirrel monkey auditory nerve. Each graph is for
a different tonal frequency. The period of each frequency is also indicated; N refers to the number of intervals plus the number of intervals longer than 20 mscc. (From J. E. Rose et al. *J. Neurophysiol.* 30:769, 1967. Used with permission.)

tone, a phenomenon that Rose et al. [28] refer to as *phase locking*. The folding histogram (Fig. 28-13) graphically depicts the degree to which the response is locked to the
stimulus cycle. For these histograms, the data are recycled through a computer every
time a stimulus cycle is completed, so that all nerve spikes are represented as if they had
occurred during one cycle of the sine-wave stimulus. This allows one to determine when,
on the average, spikes are occurring during the stimulus period. In Figure 28-13, folding histograms are shown for several frequencies delivered to a unit that has a best
frequency of 4000 Hz. The extent of phase locking is quantitated for each histogram
by measuring the percentage of spikes during the most efficient half-cycle of the tone,
i.e., the half-cycle where more than 50% of the spikes occur. This value, which Rose
et al. [28] called the *coefficient of synchrony* (the value of S in Fig. 28-13), can then

UNIT 66-86-16

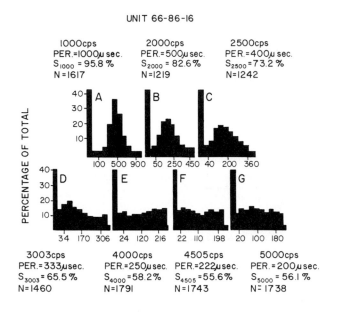

Figure 28-13
Folding histograms for a single auditory neuron in the squirrel monkey. The symbols are the same as in Figure 28-12 except that the S percentages indicate the coefficient of synchrony. (From J. E. Rose et al. *J. Neurophysiol.* 30:769, 1967. Used with permission.)

be statistically analyzed by a chi-square test to ascertain if it differs substantially from a chance variation. Percentage values greater than 56% were statistically significant. As can be seen from Figure 28-13, the best phase locking for this unit occurred for a 1000 Hz tone. Spikes tended to distribute themselves randomly at frequencies approaching 5000 Hz. Phase locking was readily demonstrated for frequencies below 5000 Hz; it is not systematically altered by varying the stimulus intensity or the frequency within the response area of the cell.

To summarize, it appears that topographic variables are extremely important in receptor coding of tones with frequencies above 4000 Hz. Below 4000 Hz, the basilar membrane displacement is relatively uniform from base to apex, and topography becomes less important in coding. Topographic coding for single auditory neurons can be demonstrated throughout the auditory pathway. Most cells have best frequencies, and there is an orderly arrangement of cells, according to the best frequency, in nuclei at various points in the auditory pathway. Spiral ganglion cells have been shown to transmit tonal frequency information by producing discharges at intervals grouped around integral multiples of the stimulus period. The period-time code is particularly important for tones with best frequencies below 5000 Hz.

CODING OF INTENSITY

Intensity-response functions for single auditory neurons are dependent upon the frequency of the stimulating tone. For any given unit, the firing rate may show either a mathematically monotonic relationship to the intensity or a relationship characterized by an inverted "U" function. In the latter case, the firing rate increases up to a point (usually about two logarithmic units above threshold intensity); then further increases in the intensity cause a decrease in firing rate.

Intensity-response functions for auditory neurons change as one ascends the auditory pathway. For auditory nerve and trapezoid body cells, the stimulus-response functions have a steep slope, and maximum firing rates reach as high as 800 spikes per second. For the cortex and the medial geniculate nucleus, however, the slopes are more gradual, and the maximum firing rates are diminished. Lesion studies are consistent with these results, since they indicate that intensity discrimination is reduced only slightly after lesions are made in the auditory cortex and the brachium of the inferior colliculus.

AUDITORY LOCALIZATION

Psychophysics

Psychophysical experiments clearly demonstrate that the localization of a sound in space is determined by two binaural cues. First, there is a binaural phase or time differential for any stimulus that is not located in the median plane, i.e., a plane through the midline of the body. Second, there is a "sound shadow" cast as sound waves bend around the head, so that the stimulus that reaches the distal ear is weaker than the one that arrives at the ear closest to the sound. Both of these cues are based on the very obvious geometrical fact that the ears are not located in the same place. For any given *direction angle* – i.e., the angle between the midsagittal plane and the sound source – one can calculate the additional distance that the sound must travel in order to reach the more distant ear. It is then relatively straightforward to calculate Δt, the binaural time difference, and ΔS, the binaural intensity difference, for any stimulus. For example, a stimulus in the median plane will have Δt and ΔS values equal to zero. For a low-frequency tone near the head with direction angles of $25°$, $50°$, and $90°$, the Δt values will be 0.222, 0.444, and 0.799 μsec, respectively, for humans.

There are several limitations involved with these two binaural cues. One limitation is that there are many places in auditory space that will produce the same binaural input to the organism. For each direction angle, therefore, there is a "cone of confusion" in which the subject is not able to discriminate the sound location from other regions with the same direction angle. As an example, with a direction angle of $0°$, a blindfolded subject would not be able to determine if a sound is directly in front, behind, or above him, since Δt and ΔS are equal to zero for all stimuli along the midsagittal plane. Another limitation is that the binaural time-difference cue is limited to low frequencies. For humans, the upper limit is 1000 Hz. Also, binaural intensity difference cues operate best at high frequencies; for humans, this range is above 3000 Hz [39].

TASTE

The receptors for gustation, the taste buds, group together and form small elevations on the tongue, larynx, and pharynx; the elevations are called *papillae*. Each taste bud is innervated by a number of neural processes and contains several supporting cells (Fig. 28-14). The neural processes that enter the receptor areas are dendrites of unipolar cell bodies located in the *geniculate, petrosal,* or *nodose ganglia.* Most taste information from the anterior two-thirds of the tongue travels along the chorda tympani, a branch of the seventh cranial nerve, to the geniculate ganglion; some of the fibers from the anterior portion of the tongue follow a more indirect route through the greater superficial petrosal nerve to reach the geniculate ganglion. Information from the posterior one-third of the tongue travels along the ninth cranial nerve to the petrosal ganglion; cell bodies of the nodose ganglion send many of their peripheral fibers (fibers of the tenth nerve) to the pharynx and larynx. The central processes of the cell bodies that are located in the geniculate, petrosal, and nodose ganglia enter the medulla and synapse in the nucleus of the solitary tract. Axons from the cells of the nucleus of the solitary tract cross the midline and ascend the brain stem in association with the trigeminal lemniscus. These axons synapse in the arcuate nucleus of the thalamus. The cortical projection area for taste is located in the lower part of the postcentral gyrus.

PERIPHERAL CODING OF TASTE QUALITY

It is generally accepted that there are four taste qualities: sour, salty, sweet, and bitter. The tongue has been shown to have regions that are differentially sensitive to chemicals associated with these sensations. The anterior portion of the tongue is more sensitive to chemicals that give rise to sweet and bitter tastes, whereas the posterior portion of the tongue is more concerned with chemicals that produce sour and salty sensations. Von Bekesy [34] has shown that chemical and electrical stimulation of individual taste papillae in humans elicits only one of the four basic taste qualities. The spatial arrangement of taste receptors is, therefore, evidence for a topographic code for taste quality at the receptor level.

The coding picture is much more complicated with respect to single afferent neurons in the gustatory system. Single cells have very broad chemical sensitivity curves, and, although these cells are usually more sensitive to one class of chemicals than others, it has not been possible to categorize these cells according to the four basic taste qualities.

Pfaffman [24] and Ericson and Doetsch [4, 7] have done the most comprehensive work in the area of single-cell coding in the gustatory system. They recorded from single cells in the chorda tympani and in the nucleus and tractus solitarius of the rat

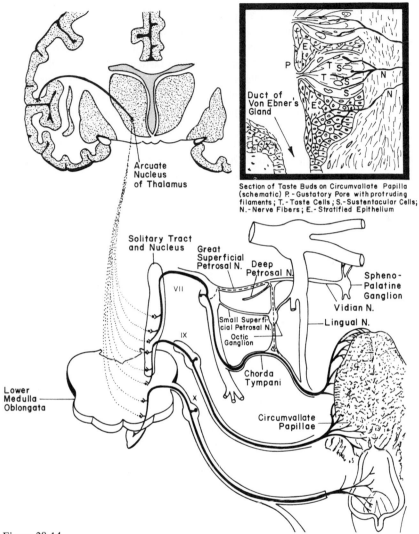

Section of Taste Buds on Circumvallate Papilla (schematic) P.-Gustatory Pore with protruding filaments; T.-Taste Cells; S.-Sustentacular Cells; N.-Nerve Fibers; E.-Stratified Epithelium

Figure 28-14
Anatomy of the gustatory system. Receptors are shown in the upper right of the figure. (After Frank H. Netter, M.D. From The CIBA Collection of Medical Illustrations. © Copyright 1953, 1972, the CIBA Pharmaceutical Company, Division of CIBA-GEIGY Corporation. All rights reserved.)

while they chemically stimulated the anterior portion of the tongue. Chemical stimulation was achieved by using a flow chamber in which the tongue was inserted. Each stimulus flowed over the tongue for about 5 seconds, and, after a delay of 10 seconds, the tongue was rinsed with distilled water for about 30 seconds. Microelectrode recordings were made using conventional techniques. As mentioned earlier, the units that they studied in the chorda tympani (CT) and the nucleus and tractus solitarius (NTS) did not show chemical specificity, but instead these units were differentially sensitive to a rather broad range of stimuli. A simple topographic code seemed unlikely in the face of these data, so these investigators emphasized the importance of across-neuron response patterns in coding taste quality. Any chemical stimulus excites a large number of cells to various degrees, producing an across-neuron pattern of activity that is unique to that stimulus and constitutes the neural representation of the stimulus. The across-neuron patterns for chorda tympani and nucleus and tractus solitarius cells in response to three chemicals are shown in Figure 28-15. The cells represented on the abscissa are arbitrarily arranged in order of their decreasing sensitivity to one of the chemicals used, NaCl. The upper graph shows the response pattern for NTS neurons, and the lower graph depicts the response for CT cells.

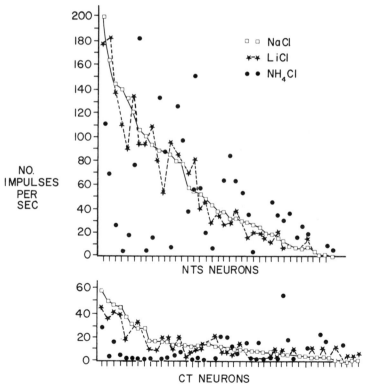

Figure 28-15
Across-neuron patterns for nucleus and tractus solitarius (NTS) neurons (*above*) and chorda tympani (CT) neurons (*below*). The patterns for three chemicals are shown. The abscissa indicates types of neurons; about 40 NTS and 40 CT neurons are shown. The ordinate shows the firing rate of a particular neuron in response to the specified chemical. Technically, this is not a function and should be drawn as a histogram. The neuron units on the abscissa are arbitrarily arranged in order of decreasing sensitivity to NaCl. (From G. S. Doetsch and R. P. Ericson. *J. Neurophysiol.* 33:490, 1970. Reproduced with permission of the American Physiological Society.)

Doetsch and Ericson [4] argue that since it is the neural pattern that determines the taste quality, it follows that chemicals that have similar across-neuron response patterns should give rise to similar taste sensations, i.e., should be difficult to discriminate. In an attempt to quantitate the similarities and dissimilarities among response patterns, Doetsch and Ericson computed Pearson product-moment correlation coefficients for the response patterns for all possible pairs of chemicals used in their CT and NTS experiments.

The response-pattern correlation coefficients for various pairs of chemicals allows one to make predictions about taste discrimination at the behavioral level. Chemicals with correlation coefficients greater than 0.90, such as NaCl and LiCl, for example, should be difficult for the rat to discriminate; ones with low correlation coefficients, such as KCl and sucrose, should be easier to discriminate. These predictions are, for the most part, consistent with the results of behavioral taste experiments with the rat [18, 23].

Doetsch and Ericson also attempted to describe the neural transformations that take place at the NTS, so they made a quantitative comparison between their NTS data and the CT data. They found several major differences between the input to the NTS and its output. First, as Figure 28-15 shows, the firing rates for NTS units are, on the average, higher (by a factor of 4.3) than those for CT cells. Also, the Pearson product-moment correlation coefficients for pairs of chemicals were usually higher for NTS cells than for CT units. The relative position of the correlation coefficient for a pair of chemicals, however, remains relatively constant; that is, chemical pairs with relatively low correlation coefficients for CT cells also had relatively low correlation coefficients for NTS, and ones with relatively high CT correlation coefficients had relatively high NTS coefficients.

Behavioral experiments (as well as common experience) demonstrate that the recognition of a taste quality develops relatively slowly over time, and, in some cases, a taste quality can actually change with time; e.g., the taste of saccharin is initially sweet but it can take on a bitter taste with time. It is therefore of primary interest to know how the response pattern of single afferent gustatory neurons varies over time. Doetsch and Ericson did a correlation analysis on the across-neuron response pattern during the first, second, and third seconds of the response. In general, they found that the stability of the taste message varied from chemical to chemical. For example, the NaCl correlation coefficients for CT cells between the first and second, second and third, and first and third seconds were 0.86, 0.93, and 0.90, respectively. For NH_4Cl, these values were 0.70, 0.82, and 0.57, respectively. As a rule, the correlation coefficients were higher for the cells at the NTS, which thus indicates that the taste message is more stable there. It would be interesting to see if these correlations are meaningful from a behavioral standpoint. One might ask, for example, what correlation coefficient is required to lead to a change in taste sensations over time?

In similarity to the units reported in other sensory systems, gustatory cells have a spike discharge pattern that includes an early transient phase followed by a steady level of firing. Doetsch and Ericson contend that the taste message is contained in the later portion of the response and that the transient component is a period in which the message is very unstable. The ratio of the transient-component firing rate to the steady-state firing rate is about 2.5 to 1 for NTS cells and 5 to 1 for CT units. If one can presume that this is a measure of the stability of the response, then these data, they contend, confirm the above-mentioned data in showing that the response is more stable at the NTS level than at the CT level.

PERIPHERAL CODING OF INTENSITY

Intensity (concentration) versus response (firing rate) functions for single afferent gustatory neurons are suggestive of those reported for single auditory neurons. Many cells show mathematically monotonic relationships between the logarithm of the concentration of a number of chemicals and the firing rate, whereas for others, the functions are shaped like an inverted "U," i.e., there is an initial rise in spike frequency over about two logarithmic units of concentration, but a decrease in firing rate occurs for increases in concentration beyond this point [13]. In a general way, as in other sensory systems, it appears that the stimulus intensity (concentration) is coded by the total firing rate of all the cells activated by a chemical.

Since a single gustatory cell can change its firing rate with respect to a change in either the concentration or the chemical composition of a stimulus, it is worthwhile to know how the coding mechanisms for taste quality are related to those for stimulus intensity. Or, stated in terms of an across-neuron pattern code, how stable is the across-neuron pattern as the intensity is changed? If the pattern is unaltered as the intensity is varied, then the perceptual quality should remain constant, but the magnitude of the sensation should vary. If the across-neuron pattern is altered when the intensity is varied, then both the qualitative and quantitative components of the sensation should change.

Ganchow and Ericson [9a] studied the effects of intensity variations on the across-neuron patterns of gustatory neurons. They found that for some chemicals (HCl and NH_4Cl), changes in chemical concentration altered the height of the across-neuron profiles but did not substantially affect the shape of the profiles. It can be inferred that when the concentration is varied, taste quality remains constant for these chemicals. Some stimuli, such as KCl, NaCl, and sodium saccharin, showed variations in their across-neuron patterns as the intensity was changed. It was inferred that for rats, KCl tastes bitter at a 0.03 M concentration, but it becomes salty tasting as the concentration increases. Sodium saccharin is sweet at weak concentrations, but it becomes more similar in taste to NaCl and NH_4Cl with increases in concentration. These expectations are consistent with the available behavioral data [18, 23, 25].

VISION

OPTICS OF THE EYE

A light ray is bent when it passes from one transparent medium to another with a different index of refraction. This bending is referred to as the *refraction* of light rays. If a refracting surface is appropriately curved, the rays may meet in the same plane and form an image of the object. For example, the light rays from the point on the left side of Figure 28-16 are focused by the convex lens so as to produce an inverted image on the right side of the lens. The image, in effect, is a surface of bright points, each of which has been focused in the same plane.

At least from an optical standpoint, there is a close resemblance between the eye and a camera. In both cases, light rays are brought to a focus upon a light-sensitive surface: in the case of a camera, the photographic film, and in the case of eye, the *retina*. The refracting medium of the eye, of course, is much more complicated than the lenses of a camera. The major refracting surface of the eye is the clear outer surface, the *cornea* (Fig. 28-17). The crystalline lens of the eye also plays an important role in forming an image; light rays are also refracted, to a lesser extent, by the aqueous humor and by the vitreous humor of the eye.

The eye is approximately spherical in shape. Most of the outer surface is a tough, white substance called the *sclera*. The sclera serves to protect the delicate structures on the inside of the eye; it also maintains the eye's spherical shape. Light rays are admitted through the clear anterior surface, the cornea. The rays are brought to a focus on the retina after passing through the aqueous humor of the anterior chamber, the crystalline lens, and the vitreous humor. The shape of the crystalline lens is controlled by the ciliary muscles. When the object to be focused is close to the eye, the crystalline lens is rounder than when the object is more distant.

The *iris,* a muscular membrane that is located anterior to the crystalline lens, controls the amount of light admitted; the opening that is formed by the iris is called the *pupil.* In addition to compensating for changes in illumination, the pupil also helps to improve the depth of focus of the lens.

Light rays activate the photoreceptors of the retina, which are classified as *rods* and *cones*. These elements transform electromagnetic radiation into graded electrochemical activity, and they hence begin the processing of visual information. There are more rods than cones, and the two types of receptors are not distributed in the same manner across the retina. The cones, which mediate color vision and are essential to acute form vision, are concentrated in and around the *fovea*. Maximal visual

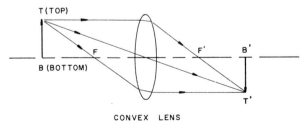

CONVEX LENS

Figure 28-16

Image construction by principal rays. The path of a ray that passes through the center of the convex lens is unaltered. A ray that is parallel to the axis when it strikes the lens passes through the focal point of the lens (*F'*), and a ray that passes through the focal point (*F*) before entering the lens is parallel to the axis when it leaves the lens.

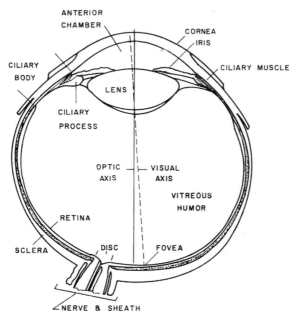

Figure 28-17
Horizontal section of the human eye.

acuity and color discrimination occur at the fovea, and, under normal viewing conditions, the position of the eye is adjusted by the extraocular muscles so as to bring the image of the attended object to focus on the fovea. The inner layers of the retina (see next section) and the superficial blood vessels are not found in the fovea.

The rods, which are not found in the fovea, have their highest concentration in the more peripheral regions of the retina. These receptors, as a system, have been developed to detect low levels of illumination. There are no receptor elements at the point in the retina where the optic nerve leaves the eye. This optic disc produces a blind spot in vision.

RETINAL ANATOMY

Anatomists have defined ten layers in the vertebrate retina. Starting from vitreous humor and working toward the sclera, there is the internal limiting membrane, the optic nerve fiber layer, the ganglion cell layer, the inner plexiform layer, the inner nuclear layer, the outer plexiform layer, the outer nuclear layer, the external limiting membrane, the layer of rods and cones, and the pigment epithelium (Fig. 28-18; the optic nerve layer and pigment epithelium are not shown in the figure). Using both electron microscopic and light microscopic techniques, Dowling and Werblin [5, 6, 36] have extensively examined vertebrate retinal fine structure, particularly the synaptic connections in the inner and outer plexiform layers. The synapses in the outer layers are found to occur in invaginations in the receptor terminals.

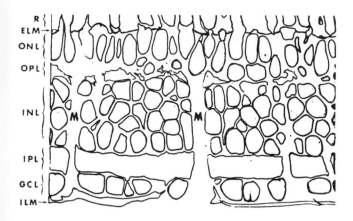

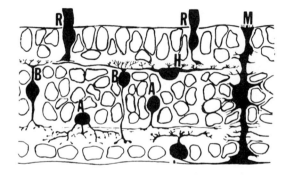

Figure 28-18
Layers of the vertebrate retina. The upper figure shows eight layers: the receptor layer (R), the external limiting membrane (ELM), the outer nuclear layer (ONL), the outer plexiform layer (OPL), the inner nuclear layer (INL), the inner plexiform layer (IPL), the ganglion cell layer (GCL), and the internal limiting membrane (ILM); M refers to Mueller cells (glia). The lower figure shows the cells located in the various layers. The cell bodies of the amacrine (A), bipolar (B), and horizontal (H) cells are found in the inner nuclear layer; the outer nuclear layer contains the cell bodies of the photoreceptors (R). Synapsing among the horizontal cells, bipolar cells, and receptors occurs in the outer plexiform layer; synapsing among the ganglion cells, bipolar cells, and amacrine cells occurs in the inner plexiform layer. (From J. E. Dowling. *Invest. Ophthalmol.* 9:655, 1970. Used with permission.)

The anatomy of the central visual pathways has been found to be species-dependent. For most higher mammals, the retinal ganglion cell axons project to the lateral geniculate nucleus (LGN) of the thalamus. These axons are referred to collectively as the *optic nerve* when they lie outside of the central nervous system. Upon entering the brain, they become the optic tract until they reach the LGN. At the transition point between the nerve and the tract is the *optic chiasm*. This represents a decussation of the optic nerve, and it is found at the base of the hypothalamus. Axons from the nasal hemiretina cross the midline to become a part of the contralateral optic tract; axons of the temporal hemiretina, at least in man, enter the ipsilateral optic tract. In man, about 50% of the fibers cross at the optic chiasm (Fig. 28-19).

Lateral geniculate axons, which are called the *optic radiations,* project to the visual cortex of the occipital cortex (Brodmann's area) [13]. From here, connections are made to several areas, including the temporal lobe, the frontal lobe, the adjacent occipital lobe areas, and several subcortical structures such as the superior colliculus. In some lower animals, there is no visual cortex, or, when it is present, it is poorly developed. In these species, most of the optic nerve fibers project to the optic tectum, a midbrain structure that is homologous to the mammalian superior colliculus.

RECEPTIVE FIELD STUDIES OF SINGLE VISUAL CELLS

The *receptive field* of a single visual neuron is defined as that region of the retina or visual field which, when presented with a photic stimulus, will lead to a change in the

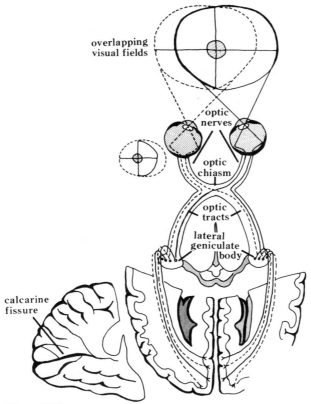

Figure 28-19
Anatomy of the visual system. (Structures covering the thalamus have been removed for sake of illustration.) (After Frank H. Netter, M.D. From The CIBA Collection of Medical Illustrations. © Copyright 1953, 1972. Reproduced with permission of the CIBA Pharmaceutical Company, Division of CIBA-GEIGY Corporation. All rights reserved.)

neural activity (usually the action potentials) of the cell. Receptive field organization varies according to the animal species and the level of the visual system studied. In general, however, it is possible to define two broad classes of receptive fields [5]. The first group, called *simple fields,* has two concentric, antagonistic zones. Stationary spots of light flashed in one zone excites the cell at "light on," whereas a static spot flashed in an antagonistic zone inhibits neural activity at "light on" and excites at "light off." The second class is referred to as *specialized fields.* These usually require more than the flashing of static spots to determine the receptive field organization adequately. Visual cells with specialized receptive fields often respond best to slits (white rectangles), bars (black rectangles), or edges (stimuli with adjacent areas of differing brightness). In contrast to simple fields, stationary spots of light in many cases are not effective stimuli in specialized fields. Targets must be moved, often in a specific direction, through the receptive field in order to activate the cell.

Michael [17] has attempted to account for the species differences that have been found in the receptive field organization of many animals on anatomic grounds. Animals such as the cat and monkey, he argues, have optic nerves that project mainly to the LGN. LGN cells then project to what is usually a well-developed visual cortex. The superior colliculus in these animals is fed mainly by visual cortex neurons. Lower animals, such as the frog and pigeon, have either no visual cortex or a very poorly developed one. The optic nerve fibers project directly to the optic tectum (Fig. 28-20). Retinal ganglion cells with simple receptive fields would, according to Michael, follow the geniculate-visual cortex-superior colliculus path, whereas cells with specialized fields would project directly to the optic tectum. There is some experimental evidence

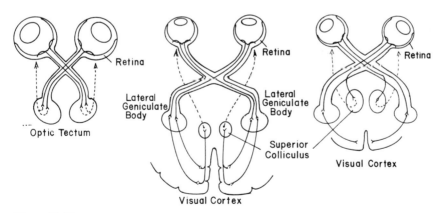

Figure 28-20
Comparison of the anatomy of the visual pathway in several species. Left figure shows the pathway for the frog and pigeon, which have no visual cortex. The retinal ganglion cells project directly to the optic tectum. The middle figure shows the visual system for the cat and primates. The retinal ganglion cells project to the lateral geniculate body; then the lateral geniculate cells project to the visual cortex. Visual cortex cells project to the superior colliculus. The figure on the right shows the pathway for the ground squirrel. Some retinal ganglion cells project to the superior colliculus and some go to the lateral geniculate nucleus. (From C. R. Michael. *Sci. Am.* 220:105, 1969. Reproduced with permission.)

to support this position. The ground squirrel, which Michael has studied extensively [16], has a rather poorly developed visual cortex, but it has both simple and
specialized retinal receptive fields. He has found that in this species, the optic nerve
fibers with simple receptive fields project to the LGN, whereas those with specialized
fields project directly to the optic tectum. This is also apparently the case for
rabbits [14, 33].

In the cat, Wickelgren and Sterling [38] have showed that the receptive field
characteristics of the superior colliculus neurons were dependent upon the integrity
of the visual cortex. With the visual cortex intact, specialized receptive fields, much
like those reported by Hubel and Wiesel in the cortex of the cat [10], were observed
in the superior colliculus. Lesioning the cortex, however, considerably altered these
characteristics. The statements of Hubel and Wiesel [10] summarize the present
thinking on the subject:

At first glance it may seem astonishing that the complexity of third order neurons
in the frog's visual system should be equalled only by that of sixth-order neurons in
the geniculo-cortical pathway of the cat. Yet this is less surprising if one notes the
great anatomical difference in the two animals, especially the lack in the frog of any
cortex or dorsal lateral geniculate body. There is undoubtedly a parallel difference
in the use each animal makes of its visual system: the frog's visual apparatus is presumably specialized to recognize a limited number of stereotyped patterns or situations, compared with the high acuity and versatility found in the cat. Probably it is
not so unreasonable to find that in the cat the specialization of cells for complex
operations is postponed to a higher level, and that when it does occur, it is carried
out by a vast number of cells, and in great detail.

NEUROANATOMIC BASIS OF RETINAL RECEPTIVE FIELD ORGANIZATION

In order to determine the neuroanatomic connections that underlie retinal receptive
field organization, it is necessary to make single-cell recordings from each of the
five retinal elements and correlate the results of such experiments with electron
microscopic and light microscopic studies of the retinal fine structure. Dowling and
Werblin [6, 36] have been able to accomplish both types of studies for the retinal
cells of the mudpuppy. These cells are large relative to those of most animals, which
thus makes intracellular recordings possible that have been unsuccessful with other
vertebrates. The mudpuppy offers another advantage for those interested in studying
receptive field organization in that some of its retinal ganglion cells have simple
receptive fields and others have specialized fields.

Figure 28-21 shows the intracellular responses of single cells in the mudpuppy
retina. Three stimuli were used in the experiments: a 100-micron photic spot
flashed in the center of the receptive field, a 250-micron annulus, and a 500-micron
annulus.

Photoreceptors always give a strong hyperpolarizing response to spot illumination
and a weak hyperpolarizing response to annuli. The responses elicited by the annuli
are thought to be the result of scattered light (i.e., light that is scattered back to the

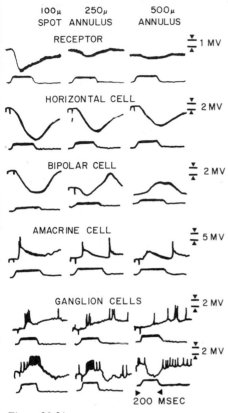

Figure 28-21
Intracellular responses from the mudpuppy retina. Hyperpolarizing responses are downward; depolarizing responses are upward. The stimulus marker is shown below each response. Receptors, horizontal cells, and bipolar cells show only graded responses. Amacrine cells and ganglion cells show action potentials as well. The bottom two traces are from the two types of ganglion cells mentioned in the text. The upper one is for the "on-off" type of ganglion cell (movement-sensitive). The lower one is for an "on-center, off-surround" ganglion cell. (From F. S. Werblin and J. E. Dowling. *J. Neurophysiol.* 32:339, 1969. Reproduced with permission.)

center of the receptive field). The receptive field of a photoreceptor, therefore, is very small and according to Dowling, probably only as large as the receptor itself. Receptors do not make lateral connections with other receptors, so these interpretations are consistent with the anatomic data.

Horizontal cells also give hyperpolarizing responses, but their receptive fields are much larger than the ones reported for photoreceptors. They give strong hyperpolarizing responses to both spots and annuli, and, when presented together, the effects of spots and annuli are additive. As the anatomic characteristics suggest, horizontal cells receive receptor input over a wide field.

Two types of bipolar cells have been observed: ones that depolarize to spot illumination and ones that hyperpolarize to spot illumination. In both cases, there is an antagonistic "surround" in the receptive field, so that simultaneous stimulation with spot and annulus stimuli leads to a cancellation of the response. It is at the bipolar cell, therefore, that the first simple receptive fields are seen with a simple (*center-surround*) organization.

Amacrine cells are the first unit to give spike activity. The spikes are superimposed upon a depolarization response. The neural activity of the amacrine cell is phasic, and it usually occurs at "light on" and "light off," i.e., it is an on-off response. The response is dependent upon the stimulus configuration, and there is a great deal of variability from cell to cell. Werblin and Dowling [36] showed that these cells are sensitive to movement. Thus, they are probably the first retinal element to show the characteristics that are associated with specialized receptive fields. They probably receive inputs from both types of bipolar cells as well as from other amacrine cells.

Retinal ganglion cells also exhibit spike activity that is superimposed on a depolarization response. There are two types of ganglion cells: ones with a sustained response and a center-surround organization, and ones with phasic, on-off responses that are movement-sensitive. The receptive field of each type of ganglion cell closely resembles those found for the bipolar cell. As was the case for bipolar cells, the receptive fields were of the simple concentric types, with center and surrounds showing spatial antagonism. These two subjects are thought to be fed inputs by the on-center and the off-center bipolar cells, respectively.

The phasic type of ganglion cell has been referred to as a specialized cell. It responds well to moving targets, and many of these cells show directionally selective responses. They are fed inputs, according to Dowling [5], mainly by amacrine cells.

To summarize, it appears that the simple type of receptive field with center-surround antagonism is formed in the outer plexiform layer by receptor-bipolar-horizontal cell interactions. Figure 28-22 gives an anatomic scheme that proposes the neural connections that might underlie the behavior of a simple retinal ganglion cell. Receptors are thought to have an excitatory effect on "on-center, off-surround" bipolar cells and an inhibitory effect on "off-center, on-surround" bipolar neurons. Peripheral antagonism in both cases is thought to be mediated by horizontal cells. These effects could occur either at horizontal-bipolar cell synapses or possibly within receptor invaginations (not shown in the diagram). For "on-center" bipolar cells, the effect of the horizontal cell would be inhibitory, and for the "off-center" bipolar cell, this effect would be excitatory. Excitation here would lead to a hyperpolarization of the bipolar cell, and inhibition would lead to depolarization of the bipolar cell. As the diagram shows, the two mechanisms are spatially overlapping. According to this model, "on-center" retinal ganglion cells are fed by "on-center" bipolar cells.

There are probably a number of different types of specialized retinal ganglion cells in the mudpuppy retina. Figure 28-23 shows Dowling's scheme for the type that is sensitive to the direction of movement [5]. For this ganglion cell, a target moving from right to left evokes a strong response, whereas a target moving from left to right, the null direction, does not. In this model, bipolar cells provide the input to amacrine cells, and the main input to the ganglion cell is from the amacrine cell. Directional

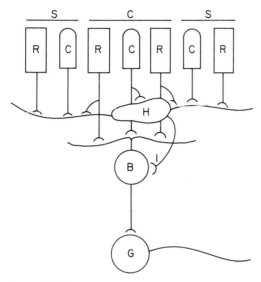

Figure 28-22
Anatomic schema for a simple "on-center, off-surround" cell. At the top, C and S refer to center ("on") and surround ("off"). H indicates a horizontal cell; R and C, the rod and cone photoreceptors; B, a bipolar cell; and G, a ganglion cell. The photoreceptors are excitatory to the bipolar cell, and the horizontal cell is inhibitory to the bipolar cell (I). The bipolar cell is excitatory to the ganglion cell.

selectivity comes about by inhibitory interactions between amacrine cells that are connected in series.

Comparative studies of retinal fine structure, coupled with the results of intracellular recordings discussed above, indicate that there are several synaptic pathways through which visual information flows and these routes determine the type of receptive field observed at the ganglion cell. In all retinas, the path through the outer plexiform layer is nearly the same; there are very few species differences in the anatomy of this layer. When the anatomy of the inner plexiform layer of various vertebrates is compared, however, it is found that animals with primarily simple receptive fields, such as the cat and monkey, have mostly amacrine-ganglion cell dyad pairings, a relatively low number of serial synapses between amacrine cells, and, in general, fewer amacrine synapses per unit area. On the other hand, animals such as the frog and pigeon, which have specialized receptive fields, have dyad pairings consisting of two amacrine cells; there are a large number of serial synapses between amacrine cells and a larger number of amacrine cell synapses per unit area. Mixed retinas that have an equal number of both types of receptive fields show approximately equal numbers of amacrine-ganglion dyads and amacrine-amacrine dyads.

It appears, then, that the outer plexiform layer of the vertebrate retina is concerned with the spatial distribution of illumination; it is in this layer that concentric receptive fields are formed. On the other hand, the inner plexiform layer is concerned with the

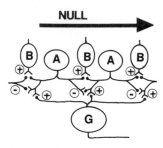

Figure 28-23
Neural scheme for a directionally selective, retinal ganglion cell. The pluses indicate excitatory synapses; minuses are inhibitory synapses. Bipolar cells (B) synapse mainly with amacrine cells (A). The ganglion cells (G) receive their primary input from the amacrine cells. Serial synapses between the amacrine cells are also shown. (From J. E. Dowling. *Invest. Ophthalmol.* 9:655, 1970. Reproduced with permission.)

dynamic characteristics of the visual target; it is in this layer that specialized receptive fields are formed.

SUMMARY

In an attempt to understand single-cell coding, the neurophysiologist has emphasized the specialized function of individual cells in sensory neural structures. The neurophysiologist who studies vision, for example, reports "color coded" neurons and "movement detectors." Investigators studying audition discuss binaural "phase detectors" and cells that are tuned to low-frequency tones. Although it is very convenient to examine cell specificity, it must be kept in mind that a stimulus on the receptor surface activates many neurons and that several modalities often share the same neurons. A single visual cell can change its firing as a result of a change in stimulus intensity, spectral composition, or the spatial distribution of the stimulus. An auditory neuron could be affected in the same way by either a change in the location of the stimulus in space or a change in the tonal frequency of the stimulus. Thus, it is important to consider the collective output of many parallel neurons in a sensory pathway; the neural representation of a stimulus is an across-neuron pattern.

A number of sensory modalities — such as visual position, somatic sensory position, and auditory tonal frequency — are given topographically at the receptors, and, to a large extent, a stimulus characteristic can be determined according to which neurons in the sensory pathway are firing. Even in these topographic modalities, however, the across-neuron patterns elicited by stimuli must be considered. It is a relatively straightforward matter for the organism to discriminate by neurotopography between a tactile stimulus located on the back and a tactile stimulus located on the hand. However, if the two tactile stimuli are located near each other on the hand, the shape of the across-neuron contour that is elicited must be considered. In a similar way, a tone of 18,000 Hz and one of 2000 Hz can be differentiated topographically, but a comparison of across-neuron patterns must be made when a 3300 Hz tone and a 3400 Hz tone are compared.

Perhaps the most important neural mechanism used in sensory systems is lateral inhibition. As was demonstrated previously, lateral inhibition serves to sharpen the across-neuron pattern elicited by stimuli that are similar. Lateral inhibition can be shown to sharpen wavelength discrimination (color), discrimination of tonal frequency (pitch), two-point tactile discrimination, and visual acuity.

REFERENCES

1. Barlow, H. E., Hill, R. M., and Levick, W. R. Retinal ganglion cells responding selectively to direction and speed of image motion in a rabbit retina. *J. Physiol.* (Lond.) 173:377, 1964.
2. Butler, R. A. Discrimination learning by rhesus monkeys to auditory incentives. *J. Comp. Physiol. Psychol.* 50:239, 1957.
3. Corbit, J. Behavioral regulation of hypothalamic temperature. *Science* 166:256, 1970.
4. Doetsch, G. S., and Ericson, R. P. Synaptic processing of taste-quality information in the nucleus tractus solitarius of the rat. *J. Neurophysiol.* 33:490, 1970.
5. Dowling, J. E. Organization of vertebrate retinas. *Invest. Ophthalmol.* 9:655, 1970.
6. Dowling, J. E., and Werblin, F. S. Organization of retina of the mudpuppy, *Necturus maculosus.* I. Synaptic structure. *J. Neurophysiol.* 32:315, 1969.
7. Ericson, R. P. Sensory Neural Patterns and Gustation. In Y. Zotterman (Ed.), *Olfaction and Taste.* Oxford: Pergamon, 1963. Pp. 205–213.
8. Ericson, R. P. Stimulus coding in topographic and nontopographic afferent modalities: On the significance of the activity of individual sensory neurons. *Psychol. Rev.* 75:447, 1968.
9. Galambos, R. Microelectrode studies on medial geniculate body of cat. III. Response to pure tones. *J. Neurophysiol.* 15:301, 1952.
9a. Ganchow, J. R., and Ericson, R. P. Neural correlates of gustatory intensity and quality. *J. Neurophysiol.* 33:768, 1970.
10. Hubel, D. H., and Wiesel, T. N. Receptive fields, binocular interactions, and functional architecture in the cat's visual cortex. *J. Physiol.* (Lond.) 160:106, 1962.
11. Iggo, A. Cutaneous heat and cold receptors with C afferent fibers. *Q. J. Exp. Physiol.* 44:362, 1959.
12. Johnson, J. I., Hatton, G. I., and Goy, R. W. The Physiological Analysis of Animal Behavior. In E. S. E. Hafez (Ed.), *The Behavior of Domestic Animals* (2nd ed.). London: Balliere, Tindall and Cassell, 1969.
13. Makous, W., Nord, S., Oakley, B., and Pfaffmann, C. The Gustatory Relay in the Medulla. In Y. Zotterman (Ed.), *Olfaction and Taste.* Oxford: Pergamon, 1963. Pp 381 393.
14. Masland, R. H., Chow, K. L., and Stewart, R. H. Receptive field characteristics of superior colliculus neurons in the rabbit. *J. Neurophysiol.* 34:148, 1971.
15. Melzack, R., and Wall, P. D. Pain mechanisms: A new theory. *Science* 150:971, 1965.
16. Michael, C. R. Receptive field of single optic nerve fibers in a mammal with an all-cone retina. *J. Neurophysiol.* 31:249, 1968.
17. Michael, C. R. Retinal processing of visual images. *Sci. Am.* 220:105, 1969.

18. Morrison, G. R. Behavioral response patterns of salt stimuli in the rat. *Can. J. Psychol.* 21:141, 1967.
19. Mountcastle, V. B. Modality and topographic properties of single neurons of cat's somatic sensory cortex. *J. Neurophysiol.* 20:408, 1957.
20. Mountcastle, V. B., Poggio, G. F., and Werner, G. The relationship of thalamic cell response to peripheral stimuli over an intensive continuum. *J. Neurophysiol.* 26:807, 1963.
21. Mountcastle, V. B., and Powell, T. P. S. Central neural mechanisms subserving position sense and kinesthesis. *Bull. Johns Hopkins Hosp.* 105:173, 1959.
22. Myers, R. D. Temperature Regulation. In W. Haymaker, E. Anderson, and W. J. H. Nauta (Eds.), *The Hypothalamus.* Springfield, Ill.: Thomas, 1969.
23. Nachman, M. Learned aversion to the taste of lithium chloride and generalization to other salts. *J. Comp. Physiol. Psychol.* 56:343, 1963.
24. Pfaffman, C. The Sense of Taste. In Field, J., Magoun, H. W., and Hall, V. E. (Eds.) *Handbook of Physiology, Section 1, Neurophysiology,* Vol. 1. Washington, D. C.: The American Physiological Society, 1959. Pp. 507–533.
25. Pfaffman, C. Taste Stimulation and Preference Behavior. In Y. Zotterman (Ed.), *Olfaction and Taste.* Oxford: Pergamon, 1963. Pp. 257–273.
26. Poggio, G. F., and Mountcastle, V. B. The functional properties of ventrobasal thalamic neurons studied in unanesthetized monkeys. *J. Neurophysiol.* 26:775, 1963.
27. Ranson, S. W. Regulation of body temperature. *Res. Publ. Assoc. Res. Nerv. Ment. Dis.* 20:342, 1940.
28. Rose, J. E., Brugge, J. F., Anderson, D. J., and Hind, J. E. Phase-locked response to low-frequency tones in single auditory nerve fibers of the squirrel monkey. *J. Neurophysiol.* 30:769, 1967.
29. Rose, J. E., Galambos, R., and Hughes, J. R. Microelectrode studies of the cochlear nuclei of the cat. *Bull. Johns Hopkins Hosp.* 104:211, 1959.
30. Rose, J. E., Greenwood, D. D., Goldberg, J. M., and Hind, J. E. Some discharge characteristics of single neurons in the inferior colliculus of the cat. I. Tonotopical organization, relation of spike-counts to tone intensity and firing patterns of single elements. *J. Neurophysiol.* 26:294, 1963.
31. Satinoff, E., and Rutstein, J. Behavioral thermoregulation in rats with anterior hypothalamic lesions. *J. Comp. Physiol. Psychol.* 71:77, 1970.
32. Stevens, S. S. The Psychophysics of Sensory Function. In W. A. Rosenblith (Ed.), *Sensory Communication.* New York: Wiley, 1961.
33. Stewart, D. L., Chow, K. L., and Masland, R. H. Receptive field characteristics of lateral geniculate neurons in the rabbit. *J. Neurophysiol.* 34:139, 1971.
34. Von Bekesy, G. Taste theories and the chemical stimulation of single papillae. *J. Appl. Physiol.* 21:1, 1966.
35. Von Bekesy, G., and Rosenblith, W. A. The Mechanical Properties of the Ear. In S. S. Stevens (Ed.), *Handbook of Experimental Psychology.* New York: Wiley, 1951.
36. Werblin, F. S., and Dowling, J. E. Organization of the retina of the mudpuppy, *Necturus maculosus.* II. Intracellular recording. *J. Neurophysiol.* 32:339, 1969.
37. Werner, G., and Mountcastle, V. B. Neural activity in mechanoreceptive cutaneous afferents: Stimulus-response relations, Weber functions and information transmission. *J. Neurophysiol.* 28:359, 1965.
38. Wickelgren, B. G., and Sterling, P. Influence of visual cortex on receptive fields in the superior colliculus of the cat. *J. Neurophysiol.* 32:16, 1969.
39. Woodworth, R. S., and Schlosberg, H. *Experimental Psychology.* New York: Holt, 1960.
40. Woolsey, C. N. Organization of Cortical Auditory System. In W. A. Rosenblith (Ed.), *Sensory Communication.* New York: Wiley, 1961. Pp. 233–257.

SELECTED READING

DeValois, R. L. Neural Processing of Visual Information. In R. W. Russel (Ed.),
    Frontiers in Physiological Psychology. New York: Academic, 1966. Pp. 31–92.
Geisler, C. D., Rhode, W. S., and Hazelton, D. W. Responses of inferior colliculus
    neurons in the cat to binaural acoustic stimuli having wide-band spectra. *J. Neuro-
    physiol.* 32:960, 1969.
Goldberg, J. M., and Brown, P. B. Response of binaural neurons of dog superior olivary
    to dichotic tonal stimuli. Some physiological mechanisms of sound localization.
    J. Neurophysiol. 32:613, 1969.
Mountcastle, V. B., and Darian-Smith, I. Central Nervous Mechanisms in Somesthesia.
    In V. B. Mountcastle (Ed.), *Medical Physiology,* Vol. 2 (12th ed.). St. Louis:
    Mosby, 1968. P. 400.
Mountcastle, V. B., and Powell, T. P. S. Neural mechanisms subserving cutaneous
    sensibility with special references to the role of afferent inhibition in sensory
    perception and discrimination. *Bull. Johns Hopkins Hosp.* 105:201, 1959.
Rose, J. E., Gross, N. B., Geisler, C. D., and Hind, J. E. Some neural mechanisms in
    the inferior colliculus of the cat which may be relevant to localization of a sound
    source. *J. Neurophysiol.* 29:288, 1966.
Werner, B., and Mountcastle, V. B. The variability of central neural activity in a sensory
    system, and its implications for the central reflection of sensory events. *J. Neuro-
    physiol.* 26:958, 1963.

The Organization of Movement by the Brain   29

Michael Gimpl, Neil Schneiderman, and Ronald S. Hosek

INTRODUCTION

The neural bases of motor activity ultimately depend upon the discharges of spinal motoneurons and the motor nuclei of cranial nerves. Three interacting systems in the brain influence the motor outflow: the pyramidal, extrapyramidal, and cerebellar systems. Together, they help to integrate (1) skilled, voluntary movement, (2) the coordination of groups of muscles to make movements smooth and precise, and (3) the adjustment of body posture. The systems in the brain that aid in organizing movement interact with local reflex systems. Most of these reflex systems, which are discussed in Chapter 27 (pp. 500–508), have their synapses within the spinal cord. The organization of movement by the brain is also influenced by sensory feedback to the central nervous system (CNS) from muscle.

SENSORY FEEDBACK FROM MUSCLE

In Chapter 27, the three types of sensory stretch receptors — groups Ia, Ib, and II — which are located in skeletal muscle and which form the basis for feedback information about muscle length and tension, were introduced and discussed (pp. 500–507). The classification of fiber types, along with their origin, is summarized in Table 29-1.

Primary spindle endings (group Ia fibers) are most sensitive to the velocity or rate of change of muscle length; secondary spindle endings (group II fibers) respond proportionately to the instantaneous muscle length. Golgi tendon organs (group Ib fibers), because of their location in muscle-tendon junctions, discharge proportionately to the tension existing in the extrafusal fibers. The group Ia, Ib, and II fibers from the muscle receptors synapse upon motoneurons that control the activities of both synergistic and antagonistic muscles. It is at this level that sensory feedback from the muscles and tendons is integrated into control systems for the regulation of muscle length and tension.

The motoneurons that directly innervate the extrafusal muscle fibers are called *alpha motoneurons.* Lloyd [11] demonstrated the monosynaptic excitatory effect of group Ia fiber activity on alpha motoneurons that innervate the same muscle. This forms the basis for a facilitatory feedback effect from group Ia fibers via collaterals to the motoneurons of direct synergists to the original muscle. The group Ia fibers provide a feedback signal that is proportional to the velocity of shortening or lengthening in the extrafusal muscles. Reflexive compensation occurs via the synergistic alpha motoneurons to maintain a given muscle length. Activity in the group Ia fibers not only facilitates the alpha motoneuron activity of synergistic muscles, but it also inhibits

Table 29-1. Function and Diameter of the Afferent Fibers of Mammalian Muscle Nerves

Group	Diameter (μ)	Conduction Velocity (m/sec)	Sensory Endings (Origins)
Ia	12–20	72–120	Primary endings on muscle spindles
Ib	12–20	72–120	Golgi tendon organs
II	4–12	24–72	Secondary endings on muscle spindles
III	1–4	6–24	Pressure or pain receptors
IV	Nonmyelinated fibers		Pain receptors

the alpha motoneuron activity of antagonistic muscles. This allows length compensation to occur through the combined contraction and relaxation of appropriate flexor and extensor muscles. Alpha motor activity, then, is influenced by the feedback from muscle to the CNS; it is also affected by gamma motoneurons originating within the CNS.

The *gamma motoneurons* influence the output of alpha motoneurons by adjusting the sensitivity of the spindle receptors (see Fig. 27-22). The manner in which gamma efferents can lead to subsequent alpha efferent activity is as follows. Gamma efferent activity produces contraction of the intrafusal fibers of the muscle spindle. The spindle receptors, in turn, activate afferent fibers that enter the dorsal horn of the spinal cord. These afferent neurons synapse either directly or via interneurons upon the alpha motoneurons. The afferent neurons also provide input to the cerebellum via the spinal (spinocerebellar) tracts. Although gamma efferent activity can influence subsequent motoneuron activity, the most usual case is for alpha and gamma efferent activity to act concomitantly. The reasons for this are discussed in Chapter 27 (pp. 505–507).

The gamma efferent neurons form part of a system in which muscle spindle organs can be biased so that the output from the spindle reflects (1) the actual degree of stretch placed upon it and (2) the input reaching it from the CNS via gamma motoneurons. These gamma efferent neurons, in turn, are influenced by more central neurons in the brain. The gamma efferents, then, appear to be involved in a mechanism by which the brain anticipates possible adjustments in the lengths of various muscles and accordingly biases the sensitivity of the spindle organs in preparation.

The presence of interneurons in two of the feedback circuits for length and tension (i.e., those of the secondary muscle spindle endings and Golgi tendon organs) provides for the distribution of incoming sensory impulses from the muscle to the appropriate motoneurons. This allows for neural *convergence* – the integration of sensory input prior to output – as well as *divergence* – the activation of a greater number of output channels from a lesser number of input channels. To this end, the spinal interneurons perform several important functions. The most important of these are the gating and modulating of signals obtained by allowing or preventing incoming signals from reaching motoneurons. Other important functions include (1) signal amplification, either in intensity or in time, and (2) the inversion of incoming signals from excitatory to inhibitory.

Thus, the activity that arises in the attached receptors may or may not reach the motoneurons in an intact system.

Granit [7] has suggested that the impact of sensory activity in those feedback circuits possessing interneurons may well depend upon how the motoneurons and interneurons are biased from other converging inputs. Those cortical pyramidal cells in primates that do form monosynaptic connections with motoneurons, combined with the presence of interneurons in some of the feedback systems, apparently provide a means of "driving" the skeletal musculature as well as overriding the feedback systems during voluntary movement.

In addition to the reflexive distribution of sensory information from muscle spindles and Golgi tendon organs at the spinal level, the discharges of these muscle and tendon receptors are fed to the cerebellum. There is little evidence for any differential pathway of ascent for the sensory input, since the group Ia, Ib, and II fibers all make connections with both the anterior and posterior spinocerebellar tracts. The tract, however, is somatotopically organized in its course and cerebellar termination. Present evidence seems to indicate that the ascending sensory information from different body parts is used in the fine coordination of posture and movement.

THE PYRAMIDAL SYSTEM

ANATOMY

The pyramidal system is a tract of fibers whose cell bodies are located in the cerebral motor cortex (Fig. 29-1). The axons of these cells descend through the brain stem, where some of them terminate. Many axons, however, continue on to the medullary pyramids, where most of them decussate. The axons ultimately terminate on motoneurons and interneurons at all levels of the spinal cord.

Historically, the cells of origin of the pyramidal tract were thought to be the large Betz's cells of the fifth cortical layer. More recently, this view has been modified, since there are approximately 1,000,000 axons in the medullary pyramids and only about 34,000 cells of Betz in the motor cortex. Additionally, most of the pyramidal tract fibers are small and not large as one would assume if they arose from large cell bodies. Thus, the cells of origin of the pyramidal tract consist not only of Betz's cells, but also of other cells that are widely distributed throughout the cortex.

FUNCTION

By delivering brief stimulating currents to the motor cortex by means of surface electrodes, the contraction of specific groups of muscles at various body parts can be elicited. This technique has yielded information about the manner in which the cortical areas that give rise to the pyramidal fibers are organized. The motor cortex is somatotopically organized; that is, innervation of the pathways leading to different body parts arises from fairly specific regions of the motor cortex. With the exception

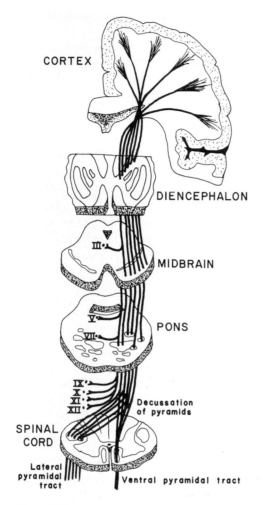

Figure 29-1
The pyramidal tract from the left motor cortex. Most of the fibers decussate in the medulla and descend in the dorsolateral column of the spinal cord of the opposite side.

of the face, the various body parts are unilaterally arranged, and each cortical motor area controls the musculature on the opposite side of the body. These mapping experiments have revealed an important principle in the organization of movement by the brain: the cortical representation of body parts is proportional in size to the degree of fine control required to perform a particular voluntary movement. Thus, in man, the muscles involved in speech and skillful hand movements are well represented in the cortex, since these are both highly coordinated activities.

   The motor cortex of the precentral gyrus is not the only source of activity in the pyramidal system. Woolsey and Chang [21] delivered electrical stimulation to the medullary pyramids and recorded antidromic pyramidal activation at the cortex. They found that a sizable area in front of the precentral gyrus as well as some areas of the postcentral gyrus also contribute fibers to the pyramidal system.

Both lesioning and extracellular recording experiments also indicate that the pyramidal system provides the main pathway for the production of skilled voluntary movements. Following destruction of the pyramids in cats, for example, three main deficits were observed [19]. First, there was a definite loss or impairment of specific responses dealing with locomotion; the extent of the deficit was comparable to that observed following removal of the cortex. Second, there was a pronounced deficit in such normal movements as pawing and clawing. Third, spinal flexor activity exhibited a significantly higher threshold of activation. When pyramidal sectioning was carried out on monkeys, similar results were obtained [18]. In the case of the monkey, sectioning the pyramids in only one side resulted in an almost complete failure to use voluntarily the extremities on the opposite side of the body. The affected side of the body, however, was used in gross motor activities, such as climbing and positional righting, particularly if the unaffected side was restricted from maximal use. Therefore, the influence of the pyramidal tract innervation seems to be important in the execution of fine, voluntary movements. Even more significant is the finding that following pyramidal section, the monkeys lost the ability to make fine, discrete use of the digits on the extremities of the affected side.

The microelectrode recording experiments of Evarts [4] and Evarts and Thach [5] further support the view that the pyramidal system plays an important role in the execution of voluntary movement. In these studies, microelectrodes were placed in the motor cortex and were used to record the patterns of discharge of single pyramidal tract neurons. The monkeys were trained to perform a number of different motor tasks, for which they were reinforced. Evarts found activity in pyramidal tract neurons that correlated well with arm movement when a monkey reached for food. Some of these neurons showed a brief burst of discharge with movement, and others showed a tonic discharge rate even in the absence of movement.

In an attempt to determine the temporal relationship of neuronal activity in the pyramidal tract to movement, Evarts [4] trained monkeys to release a key whenever they were presented with a light flash. The latency of the motor cortex response to a light flash is about 30 msec, but the pyramidal cells did not discharge until 100 msec after the flash. The response of releasing the key occurred at 180 to 200 msec. Thus, the activity of the pyramidal tract neurons preceded the movement, but it occurred at a latent period, which suggests the existence of additional routing of information from the light flash prior to the appearance of a corresponding discharge in the pyramidal cells.

Evarts also recorded neuronal activity during flexion or extension and when a posture of flexion or extension of the wrist was held (Fig. 29-2). Somewhat surprisingly, force, and especially rapidly changing force, rather than displacement or position, correlated most closely to the activity of the pyramidal tract neurons. The recordings of neuronal activity in Figure 29-2 show the discharge patterns of two pyramidal cells during flexion and extension of the wrist. The larger of the two action potentials (spikes) is from a

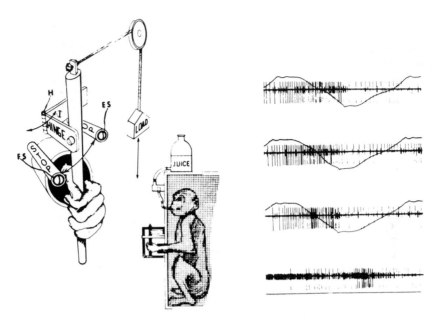

Figure 29-2
Pyramidal cell discharges during movements. (From E. V. Evarts. Representation of Movements and Muscles by Pyramidal Tract Neurons of the Precentral Motor Cortex. In M. D. Yahr and D. P. Papura [Eds.], *Neurophysiological Basis of Normal and Abnormal Motor Activity.* Hewlett, N.Y.: Raven Press, 1967. Reproduced with permission.)

cell close to the recording electrode, whereas the smaller is from a more distant cell. It can be seen in Figure 29-2 that both cells show a high rate of discharge during flexion (downward movement of the superimposed tracing) and a low rate of discharge during extension (upward movement of the tracing). Therefore, during wrist flexion, these two pyramidal tract neurons are well correlated with the specific movement. The lowermost tracing in the figure shows the discharge pattern of these same two cells during a different movement, and it can be seen that they are no longer related to each other in discharge pattern. It appears, then, that the discharge of pyramidal cells is closely related to the production of specific voluntary movements, and that the particular movement produced is determined in large part by the combination of pyramidal cells in activation at that time.

Pyramidal outflow influences movement by interacting with neural systems at the spinal level. It seems likely that pyramidal impulses generate movement by the simultaneous activation of alpha and gamma motoneurons. In so doing, the specification of a change in force is transmitted to the muscles, and the servomechanical characteristics of the intrafusal muscle adjust in synchrony with the change in force. The results are a shift in position of the attached body part and a sensory response that is returned to the brain via the afferents. In summary, the pyramidal tract provides an avenue by which nerve impulses from the cortex exert an influence on spinal mechanisms, thereby providing one aspect of the organization of movement by the brain.

THE EXTRAPYRAMIDAL SYSTEM

The extrapyramidal system provides a second descending pathway for the brain to influence movement by interacting with the local neuronal systems at the spinal level. The extrapyramidal system is a much more complicated neural network than is the pyramidal system, and the specific relationships of many of its components to movement have not yet been determined.

ANATOMY

The extrapyramidal system originates from efferent projection fibers of the cortex. The cortical localization of the cells of origin for these fibers is the same as that of the pyramidal system. For this reason, somatotopic organization is characteristic of the extrapyramidal system as well. The extrapyramidal projection fibers from the motor cortex end in the basal ganglia, the red nuclei, the substantia nigra, the brainstem reticular formation, and several other subcortical nuclei and structures (Fig. 29-3). The major descending pathway emerging from these structures arises in the reticular formation and is referred to as the *reticulospinal tract*. A secondary pathway, the *rubrospinal tract,* is nearly absent in man, although it is well-defined in lower animals.

Each of the aforementioned nuclei has direct or indirect connections with each other, as well as with the cerebellum, thalamus, and cortex. It is this complexity of interconnections and the feedback loops formed by them that has complicated the understanding of the role played by many components of the extrapyramidal system. The extrapyramidal components may, in general, be discussed in terms of three major subdivisions: (1) the cortical components, (2) the basal ganglia, and (3) the descending reticular system.

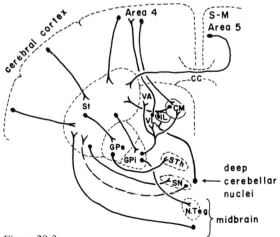

Figure 29-3
Some of the important anatomic connections of the extrapyramidal system. *S-M*: somatomotor cortex, *St*: neostriatum, *GP*: globus pallidus, *STh*: subthalamus, *SN*: substantia nigra, *N.Teg*: tegmental nuclei of reticular formation, *VA, VL, IL*, and *CM*: thalamic nuclei. (Adapted with permission. From J. M. Kemp and T. P. S. Powell. *Philos. Trans. R. Soc. Lond.* [Biol. Sci.] 262:441, 1971.)

CORTICAL COMPONENTS

Experiments designed to elucidate the functions of the pyramidal system have also provided valuable information concerning extrapyramidal involvement in movement. Tower [17] stimulated the motor cortex before and after pyramidal destruction. Prior to pyramidal destruction, stimulation of the cortex produced phasic movement with a short latency period. Following destruction of the pyramids, the movement elicited by cortical stimulation was slow to appear, and it tended to outlast the stimulus. In addition, the stimulation of the cortex during a reflexively produced movement resulted in attenuation of that movement. This suggests that inhibitory functions of the cerebral cortex over spinal mechanisms occur via extrapyramidal pathways.

Subsequent experiments by Hongo and Jankowski [9] indicated that the extra-pyramidal system mediates excitatory as well as inhibitory effects. Hongo and Jankowski stimulated the sensory motor cortex of the cat following destruction of the pyramids. They then recorded the effect of cortical stimulation on spinal moto-neurons. Excitatory as well as inhibitory postsynaptic potentials appeared in the motoneurons leading to extensor muscles, while predominately excitatory effects appeared in the motoneurons of flexor muscles.

Although anatomically the extrapyramidal and pyramidal systems are difficult or impossible to separate at the cortical level, it is clear that a second, slower acting, descending pathway from the cortex to the spinal motoneurons does exist.

BASAL GANGLIA

The basal ganglia are made up of two major parts. One part consists of the caudate nucleus and the putamen, which are collectively known as the *neostriatum.* The other part is the *globus pallidus.* Afferent connections to the neostriatum arise in the motor cortex. The globus pallidus is made up of large motor cells and is the major efferent mechanism of the basal ganglia. The globus pallidus projects direct connections to three nuclear masses in the midbrain and lower diencephalon. These are the subthalamic nucleus, the substantia nigra, and the red nucleus. Direct connec-tions also exist between the globus pallidus and the descending reticular formation of the brain stem.

Research utilizing lesion and stimulation techniques on the basal ganglia has yielded little conclusive information regarding their role in the control of movement. However, one reliable effect has been noted. Stimulation of the caudate nucleus leads to the cessation of movement and a general arrest reaction. This is consistent with the con-cept that the cortical inhibition of spinal reflexes occurs over extrapyramidal pathways. Early investigations by Rioch [14] of the basal ganglia employed comparisons of the behavioral effects of decortication with those produced by a combination of decorti-cation and removal of the basal ganglia. These experiments essentially showed no differences between the effects of the two procedures. More recently, however, Wang and Akert [20] have shown that lesions of the basal ganglia in combination with decortication produce more severe deficits than decortication alone. Thus, the basal ganglia may play a role in the modulation of descending cortical activity.

Clinical studies of diseases of the basal ganglia have also shed some light on the role of these ganglia in the regulation of movement. The movement disorders associated with diseases of the basal ganglia are of two types: hyperkinetic and hypokinetic. Hyperkinetic disorders, in which there is excessive or abnormal movements such as twitching or uncontrolled jerking, have been related to (1) degeneration of the caudate nucleus (chorea), (2) lesions of the globus pallidus (athetosis), and (3) damage to the subthalamic nuclei (ballismus). The major hypokinetic disorder related to basal ganglia disease is parkinsonism. Here the general lack of movement, rigidity, and resting tremor have been closely related to pathologic changes in the substantia nigra. Most notably, there is a decrease in the content of the neurotransmitter substance, dopamine, in the basal ganglia.

In general, the basal ganglia appear to modulate corticospinal outflow in order to prevent afterdischarge and oscillation in the extrapyramidal system.

THE DESCENDING RETICULAR FORMATION

The brain-stem reticular formation has been shown by Magoun and Rhines [13] to exert both inhibitory and facilitatory effects on the stretch reflexes of skeletal muscles. Two general types of effects have been observed. Stimulation of the ventral and medial portions of the medulla inhibits knee jerk and flexion reflexes as well as the movements elicited by electrical stimulation of the motor cortex, whereas stimulation of the lateral portions results in a facilitation of these same responses.

Granit and Kaada [8] have recorded the discharge activity of the spinal gamma efferent neurons in response to stimulation of the reticular formation. They found that stimulation of the anterior portions of the reticular formation resulted in increased gamma efferent discharges, whereas stimulation of the caudal region of the reticular formation led to decreased discharges. Since many of the components of the extrapyramidal system have connections with the reticular formation, a large part of the inhibition and facilitation of local spinal interactions, as well as the cortical outflow, appears to be affected by the reticular formation's control of the gamma motoneuron system.

In general, it appears that it is the extrapyramidal system which is primarily responsible for regulating the local spinal events that maintain position and posture. Additionally, the extrapyramidal system is closely affiliated with the cerebellum in providing a basis for the feedback control of movement and posture by the brain.

THE CEREBELLUM

The role of the cerebellum in the regulation of movement was poorly understood until recently. As more has become known about the morphology, physiology, and anatomic connections of the cerebellum, it has come to be seen as critical for the fine control of both posture and voluntary movement. The cerebellum is unique among the three major brain components that regulate movement in that it has no direct pathways of its own to the spinal motoneurons. Instead, its influence is exerted via the extensive connections it has with the basal ganglia and other nuclei of the extrapyramidal system, as well as with the vestibular system, cerebral cortex, and pyramidal system.

ANATOMY

The general structure of the cerebellum is similar to the cerebrum in that it consists of a highly convoluted outer cellular layer, an intermediate layer of white matter, and several deep nuclei [1, 6]. The convolutions in the cerebellum, however, are more extensive than those in the cerebrum. Although the cerebellum weighs approximately 10% as much as the cerebrum, its surface area is about 75% that of the cerebral cortex. Also, unlike the cerebral cortex which contains just two hemispheres, the cerebellum contains two hemispheres plus a medial portion called the *vermis cerebelli* or the vermiform process of the cerebellum. Several fissures divide the vermis and the hemispheres into lobes and lobules.

The input and output relations of the cerebellum are relatively simple. Two types of afferent nerve fibers enter the cerebellum and project to the cerebellar cortex; these are called *climbing fibers* and *mossy fibers*. They convey information from muscle, the tendon receptors, the vestibular system, and other sensory modalities, as well as from the pyramidal and extrapyramidal systems. After processing this information in the cerebellar cortex, efferents project to the deep cerebellar nuclei. Whereas two types of fibers — mossy and climbing — project to the cerebellar cortex, only one type of fiber projects from it. This is the Purkinje neuron; its output is inhibitory. Because the cerebellar cortex has only two inputs and a single output, it has provided an ideal system for studying the neuronal processing of information. An understanding of this processing depends upon a knowledge of the synaptic connections that occur in the cerebellar cortex.

SYNAPTIC ORGANIZATION OF THE CEREBELLAR CORTEX

The cerebellar cortex consists of three discrete layers: an outer molecular layer, a Purkinje cell layer, and an inner granular cell layer. The middle layer of the cerebellar cortex contains the cell bodies of Purkinje neurons (Fig. 29-4). Each Purkinje neuron has a large and elaborate dendritic tree that arborizes in fan-like fashion along the sagittal plane. This dendritic tree extends into the outer molecular layer of the cerebellar cortex. The climbing fibers entering the cerebellar cortex synapse on the dendrites of the Purkinje cells. These synapses are excitatory [3]. Each climbing fiber possesses an extensive all-or-none excitatory connection with the dendrites of the Purkinje cell. Thus, whenever the climbing fiber discharges, the Purkinje cell also discharges. The action of the climbing fiber is excitatory, whereas the action of the Purkinje neuron is to inhibit cells of the deep cerebellar nuclei.

The second afferent projection to the cerebellar cortex occurs via the mossy fibers. These fibers do not ascend directly to the outermost or molecular layer of the cerebellar cortex, but instead they synapse on small granule cells in the inner granular layer of the cortex. The synapse between the mossy fiber and the granule cell is also excitatory.

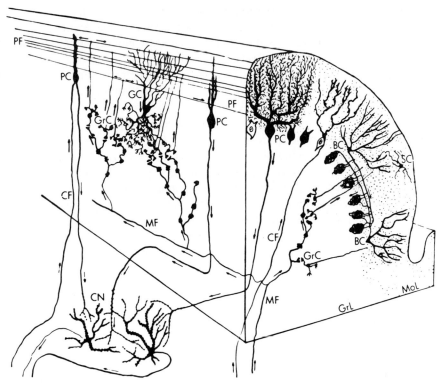

Figure 29-4
Segment of a cerebellar folium. *CN*: cerebellar nuclei, *CF*: climbing fiber, *MF*: mossy fiber, *PC*: Purkinje cell, *GC*: Golgi cell, *PF*: parallel fibers, *SC*: stellate cells, *BC*: basket cells, *GrC*: granule cells. (From C. A. Fox. *Correlative Anatomy of the Nervous System.* New York: Macmillan, 1962. P. 193. Reproduced with permission.)

The axon of the granule cell ascends to the outer molecular layer of the cerebellar cortex, where it bifurcates to form parallel fibers (see Fig. 29-4). These parallel fibers travel perpendicularly to the dendritic expansion of the Purkinje cells, traversing the Purkinje cell dendrites much like telephone wires strung along the branches of a tree. The number of possible synaptic connections is immense. In the monkey, for example, each Purkinje cell has about 9×10^4 dendritic spines, with approximately 3×10^4 parallel fibers passing through the dendritic tree. The numbers of neurons themselves are staggering. In humans, for instance, there are about 3×10^{10} granule cells, 3×10^7 Purkinje cells, and 2×10^8 basket and stellate cells in the cerebellar cortex.

The parallel fibers of the granule cells form excitatory synapses upon Purkinje neurons as well as upon basket and Golgi cells. Whereas activation of Purkinje cells by the parallel fibers contributes to the inhibition of deep cerebellar nuclei, the activation of the basket and Golgi cells has an opposite effect, because the latter neurons inhibit the Purkinje cells. Thus, although the parallel fibers excite some Purkinje cells, they inhibit other Purkinje cells via the activation of basket cells.

In the case of the Golgi neuron, the situation is somewhat more complex. Purkinje cells have recurrent collaterals that form axosomatic synapses with Golgi neurons. The Golgi neurons, however, receive their primary excitatory input from the parallel fibers of the granule cell. These Golgi neurons synapse between the mossy fiber and the granule cell plexus (Fig. 29-5). Since the Golgi cell is inhibitory, whereas the mossy fiber is excitatory, the Golgi neuron functions primarily to provide a negative feedback to the mossy fiber-granule cell relay [3].

The sole output from the cerebellar cortex is from the Purkinje cells, whose function is to inhibit the deep cerebellar nuclei. Because of this, the cerebellar cortex has the greatest predominance of inhibition of any portion of the brain [2]. All the neurons originating in the cerebellar cortex are inhibitory except the granule cells. This dominance of inhibition is of considerable value in providing for accuracy in computation by the cerebellum. Since all the inputs to the cerebellum are transformed into inhibitory action within two synapses, the possibility of prolonged chattering in chains of excitatory neurons via collaterals is eliminated. Consequently, within 100 msec after a computation, that portion of the cerebellar cortex is "clear" and ready for the next computation. This, of course, is invaluable for the reliable performance of quick, skilled movements.

CONNECTIONS OF THE CEREBELLUM

Three peduncles — the inferior, middle, and superior cerebellar peduncles — connect the cerebellum with the rest of the brain. They connect the cerebellum with the medulla, pons, and midbrain, respectively. Most afferent fibers enter the cerebellum from the inferior and middle cerebellar peduncles, although a small number enter through the superior peduncles. There are nearly three times as many afferent cerebellar fibers as efferent ones.

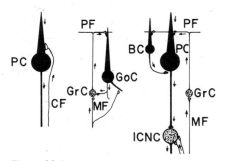

Figure 29-5
Synaptic connections of the major cells in the cerebellar cortex. Arrows show the lines of operation; inhibitory cells are shown in black. *PC*: Purkinje cell, *CF*: climbing fiber, *GrC*: granule cell, *PF*: parallel fiber, *GoC*: Golgi cell, *MF*: mossy fiber, *BC*: basket cell, *ICNC*: intra-cerebellar nuclear cell. (From J. C. Eccles. *The Understanding of the Brain.* New York: McGraw-Hill, 1973. P. 126, Fig. 4-12. Reproduced with permission of McGraw-Hill Book Co.)

The afferent cerebellar pathways convey impulses from a wide variety of sources. Most important of these are the proprioceptive impulses from the vestibular system and the impulses from the stretch receptors in muscles and tendons. Auditory, visual, and tactile impulses also reach well-localized areas of the cerebellum. The tactile impulses are somatotopically organized [16]. In addition to receiving input from several sensory systems, the cerebellum also receives indirect inputs from the cerebral motor cortex (Fig. 29-6).

The efferent connections of the cerebellum are best conceived of as feedback loops. Axons originating in the deep cerebellar nuclei transmit impulses to the pyramidal system, the extrapyramidal system, and the vestibular system, and, in turn, they receive projections from them to the cerebellar cortex. Thus, for example, cerebellar efferents project to the contralateral motor cortex via thalamic nuclei. The motor cortex in turn gives rise to pyramidal fibers, whose collaterals convey impulses back to the cerebellar cortex via the pontine nuclei.

Another feedback loop involves the efferent cerebellar projection to the red nucleus of the extrapyramidal system. The red nucleus projects to the inferior olive, then back

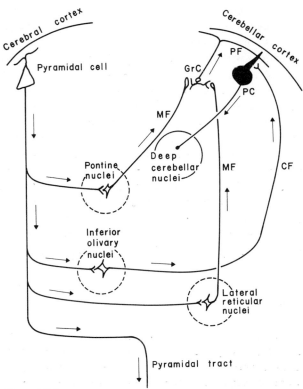

Figure 29-6
Pathways from the cerebral cortex to the cerebellum. *PF:* parallel fiber, *GrC:* granule cell, *PC:* Purkinje cell, *MF:* mossy fiber, *CF:* climbing fiber.

to the cerebellar cortex. Other deep cerebellar efferents project to the vestibular nuclei, which in turn send impulses back to the cerebellum. In this manner, the connections of the cerebellum make it an ideal servomechanism for posture and movement control.

SPECIFIC CEREBELLAR INFLUENCES ON MUSCLE AND MOVEMENT

Cerebellar output via the various extrapyramidal nuclei has been shown to alter the excitability of alpha and gamma spinal motoneurons [8]. Stimulation of one part of the anterior lobe of the cerebellum produces a decrease, whereas stimulation of another part of the lobe produces an increase, in the afferent discharge of muscle spindle receptors. The changes in spindle afferent discharge are accompanied by corresponding increases or decreases in muscle tone.

Damage to portions of the cerebellum may lead to deficits that are quite predictable from the results of stimulation studies. Damage to the medial anterior lobe, for example, produces an increase in extensor muscle tone, whereas damage to the more lateral regions of the anterior lobe leads to a decrease in this tone. In summary, stimulation and lesioning of regions of the anterior cerebellar lobes produce effects that are consistent with the concept that this region is importantly involved in the regulation and maintenance of muscle tone. This is not surprising in view of the fact that the anterior lobe also has many sensory connections with the muscle and tendon receptors through the spinocerebellar tracts. Thus, the anterior lobe of the cerebellum may be the locus of control for the regulation of alpha-gamma motoneuron linkage during movement and the control of posture. The destruction of various regions of the cerebellum annihilates the normal balance of activity in alpha and gamma motoneurons. This destruction of the alpha-gamma linkage would also result in a faulty and inaccurate feedback signal from the muscles to the brain, and it could explain some of the movement deficits that appear following cerebellar damage in man. For example, damage to the pars intermedia and the hemispheres in man often leads to deficits involving the force, rate, and steadiness of voluntary movements. Movements like touching the nose with one's finger often show deficits in accuracy, overshooting of targets, and, occasionally, tremor. These are precisely the kind of deficits one would expect to occur if a servomechanism based on gamma afferent and alpha efferent interaction were damaged.

SENSORIMOTOR INTEGRATION

A full understanding of the manner in which the brain organizes movement cannot be obtained from a detailed analysis of motor systems alone. The organization of movement and posture arises from the interactions of sensory input and motor output. Several examples of sensorimotor interactions have been isolated and may serve as illustrations of the different types and locations of integration that occur in the organization of movement and posture.

THE VESTIBULAR SYSTEM

The vestibular system is a sensory system, which, like the muscle spindles and Golgi tendon organs, is closely linked to the regulation of movement and posture. It provides information to the CNS about the position and movement of the head, and it has important influences on posture and several reflex responses.

The vestibular apparatus consists of a set of sense organs located in the bony portions of the inner ear. The discharge of some of these receptors is governed by the position of the head in relation to gravitational vectors, whereas others are related to the direction and amount of either rotary or linear acceleration of the head.

The vestibular nerve fibers that connect the sensory apparatus with the vestibular nuclei make up part of the eighth cranial nerve. There are five different vestibular nuclei in the CNS that relay information from the vestibular apparatus to different portions of the CNS. Projections exist between one vestibular nucleus and the cerebellum. Four other vestibular nuclei are located in the brain stem, three of which relay information to ascending systems, while one relays information downward to the spinal cord via the vestibulospinal tract.

The vestibulospinal tract terminates on spinal interneurons that regulate the excitability of both alpha and gamma motoneurons. Increases as well as decreases in the activity of the gamma system can be produced by stimulation of the vestibular nuclei. Thus, the central regulation of movement and posture incorporates sensory information from the vestibular system into the final output to the muscles.

The ascending pathways from the vestibular nuclei regulate reflexes involving the muscles of the eye and neck. The vestibuloocular reflex consists of compensatory eye movement in response to head rotation. This serves to prevent images from sweeping too quickly across the retina.

From the signal flow diagram shown in Figure 29-7, it may be seen that the semicircular canals of the inner ear are stimulated by the angular acceleration of the head that is incurred by positional changes. To insure the correct eye position in the head, the brain must be able to sense the head position dynamically, which implies that two integrations of the acceleration signal must take place. It is known that one such integration takes place in the semicircular canals, so that the vestibular nuclei are supplied with velocity information. Based upon studies of neurons of the sixth cranial nerve (nervus abducens), Skavenski and Robinson [15] have demonstrated that the second integration arises as a combination of brain-stem and extraocular muscular and mechanical effects. It is their contention that the signal arriving at the oculomotor nuclei obeys an equation of the form:

$$\frac{dR(t)}{dt} = T\frac{dR_v(t)}{dt} + R_v(t) \qquad (29\text{-}1)$$

where $R(t)$ = the signal arriving at the oculomotor nuclei
$R_v(t)$ = the signal arriving at the vestibular nuclei
T = a system constant

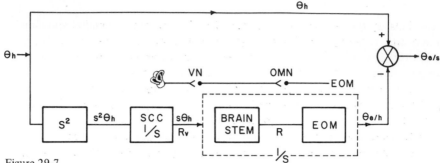

Figure 29-7
Signal flow diagram for the vestibuloocular reflex. The second derivative of head position, $s^2\theta_h$, is integrated in the semicircular canals (SCC) to provide head velocity information, R_v, to the vestibular nucleus (VN) of the brain stem. The signals from the vestibular nucleus impinge upon the oculomotor nucleus (OMN), which provides a positioning signal, $\theta_{e/h}$, to be summed with θ_h to control. The extraocular muscles (EOM) thus provide the appropriate eye position, $\theta_{e/s}$. (Modified from A. A. Skavenski and D. A. Robinson. *J. Neurophysiol.* 36:724, 1973.)

If a Laplacian transformation is performed on this equation, the transfer function of the vestibuloocular system becomes:

$$\frac{R}{R_v}(s) = T + 1/s = \frac{sT + 1}{s} \tag{29-2}$$

The Laplace transformation transforms an integro-differential equation in the time domain, t, such as Equation 29-1 into an algebraic equation in the s-domain such as Equation 29-2. The Laplace transformation, often called the *transform,* is mathematically defined as:

$$L[f(t)] = F(s) = \int_0^\infty f(t)e^{-st}\,dt$$

The complex Laplace variable $s = \sigma + j\omega$, σ and ω being in units of reciprocal time. The necessary initial conditions are introduced in the transformed equation, so that the solution proceeds directly to the desired result.

If the output of this system is applied to the oculomotor system, which has a transfer function [20] approximated by:

$$\frac{\theta_{e/h}}{R}(s) = \frac{1}{sT + 1} \tag{29-3}$$

where $\theta_{e/h}(s)$ is the eye position as a function of head position, then the total system transfer function becomes:

$$\frac{\theta_{e/h}}{R}(s) \cdot \frac{R}{R_v}(s) = \frac{sT + 1}{s} \cdot \frac{1}{sT + 1} = \frac{1}{s} \tag{29-4}$$

thus providing the desired integration.

The presumed neurologic mechanism for obtaining this transfer characteristic is shown in Figure 29-8. In this diagram, a brain-stem neural integrator is postulated to act in parallel with a direct vestibuloocular pathway of gain T, which is provided by the medial longitudinal fasciculus.

Presumably, the high-frequency oscillations (above 1 Hz) that are associated with rapid head movements are shunted by the medial longitudinal fasciculus and are integrated by the orbital mechanics. Low-frequency movements (below 1 Hz) are then integrated by the neural integrator. The feedforward path for velocity information through the fasciculus probably accounts for the fact that the system works well up to at least 5 Hz. Without this path, the mechanical properties of the orbital tissues would cause the eye position to lag far behind any rapid changes in head position.

In addition to the apparent integration by the brain of sensory information from the vestibular system, increasing attention has recently been paid to the interaction of the more classic sensory systems (i.e., visual and auditory) with portions of the various motor systems. For example, Maekowa and Simpson [12] reported a close relationship between the visual system and that portion of the cerebellum receiving projections from the vestibular nuclei (the vestibulocerebellum). Visual activity in the accessory optic tract results in a pronounced activation, via climbing fibers, of Purkinje cells in the vestibulocerebellum. Stimulation of this portion of the cerebellum leads to the inhibition of the vestibuloocular reflex. Apparently, then, motor output from the cerebellum depends upon the integration of input from visual and vestibular sources.

SENSORIMOTOR INTEGRATION AND THE MOTOR CORTEX

The motor cortex, as well as the cerebellum, appears to be involved in sensorimotor integration. It has long been known, for example, that the motor cortex receives

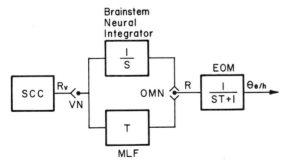

Figure 29-8
Position of the presumed brain-stem neural integrator in parallel with the direct pathways provided by the medial longitudinal fasciculus (*MLF*) for processing velocity information from the vestibular nucleus (*VN*). SCC: semicircular canal, *OMN*: oculomotor nucleus, *EOM*: extraocular muscles. (See text for further discussion.) (Modified from A. A. Skavenski and D. A. Robinson. *J. Neurophysiol.* 36:724, 1973.)

somatesthetic, kinesthetic, auditory, and visual information from the thalamus. The interaction of sensory input with motor outflow has been increasingly studied in recent years.

We previously saw how sensory information in the muscle spindle and Golgi tendon receptors is capable of influencing motor activity at the spinal level. The muscle spindle, however, both receives and projects information to supraspinal levels. Foci of these interactions include the pyramidal, extrapyramidal, and cerebellar systems. Although the most direct route for muscle spindle afferents to reach the cerebellum is via the spinocerebellar pathways, an indirect pathway involves the motor cortex. Sensory information from the receptors projects indirectly to the pyramidal neurons of the motor cortex, and the motor cortex, in turn, sends impulses to the cerebellum via cerebrocerebellar projections.

The central initiation of voluntary movement may also involve both the spino-cerebellar and cerebrocerebellar pathways, which, in this case, would function as a comparator system. Collaterals from the pyramidal system to the cerebellum would provide information just prior to and during a movement. The initiation of movement also causes impulses to be sent from the motor cortex to the target muscles. Sensory feedback from the muscle, in turn, projects back to the cerebellum. Thus, two separate pathways originating in the motor cortex may provide the cerebellum with a comparator function based upon the two input systems. In this manner, the cerebellum would provide a corrective signal to the extrapyramidal system that would make the movement smooth and precise.

This line of reasoning is consistent with that of Eccles [2], who views the cerebellum as a computer that assimilates essentially all the sensory inputs and arrives at an appropriate output signal for fine motor control. The volitional act of placing a glass of water on a table top, for example, requires this sort of fine control, and it involves proprioceptive and visual cues as well as velocity feedback. Presumably, this act would be initiated by the pyramidal system, which would mobilize the appropriate large muscle groups in the arm. Inhibitory control signals would then be sent to these muscle groups from the cerebellum to insure that the act is completed successfully, i.e., that the glass reaches the table top — not a level above or below it — with the proper velocity.

REFERENCES

1. Barker, D. *Zoology and Medical Research.* Durham, England: University of Durham Press, 1963.
2. Eccles, J. C. *The Understanding of the Brain.* New York: McGraw-Hill, 1973.
3. Eccles, J. C., Ito, M., and Szentagothai, J. *The Cerebellum as a Neuronal Machine.* Berlin: Springer-Verlag, 1967.
4. Evarts, E. V. Relation of discharge frequency to conduction velocity in pyramidal tract neurons. *J. Neurophysiol.* 28:216, 1965.
5. Evarts, E. V., and Thach, W. T. Motor mechanism of the CNS: Cerebrocerebellar interrelations. *Annu. Rev. Physiol.* 31:451, 1969.

6. Fox, C. A. *Correlative Anatomy of the Nervous System*. New York: Macmillan, 1962. P. 193.
7. Granit, R. *The Basis of Motor Control*. New York: Academic, 1970.
8. Granit, R., and Kaada, B. E. Influence of stimulation of central nervous structures on muscle spindles in cat. *Acta Physiol. Scand.* 27:161, 1942.
9. Hongo, T., and Jankowski. E. Effects from the sensorimotor cortex on the spinal cord in cats with transected pyramids. *Exp. Brain Res.* 3:117, 1967.
10. Kemp, J. M., and Powell, T. P. S. The connexions of the striatum and globus pallidus: Synthesis and speculation. *Philos. Trans. R. Soc. Lond.* [Biol. Sci.] 262:441, 1971.
11. Lloyd, D. P. C. Conduction and synaptic transmission in the reflex response to stretch in spinal cats. *J. Neurophysiol.* 6:317, 1943.
12. Maekowa, K., and Simpson, J. Climbing fiber responses evoked in vestibulo cerebellum of rabbit from visual system. *J. Neurophysiol.* 36:649, 1973.
13. Magoun, H. W., and Rhines, R. An inhibitory mechanism in the bulbar reticular formation. *J. Neurophysiol.* 9:165, 1946.
14. Rioch, D. M. Functions of the brainstem in preparations with extensive lesions of the neocortex. *Res. Publ. Assoc. Res. Nerv. Ment. Dis.* 21:133, 1942.
15. Skavenski, A. A., and Robinson, D. A. Role of abducens neurons in vestibulo ocular reflex. *J. Neurophysiol.* 36:724, 1973.
16. Snider, R. S. Recent contributions to the anatomy and physiology of the cerebellum. *Arch. Neurol. Psychiat.* 64:196, 1950.
17. Tower, S. S. Extrapyramidal action from the cat's cerebral cortex: Motor and inhibitory. *Brain* 59:408, 1956.
18. Tower, S. S. Pyramidal lesion in the monkey. *Brain* 63:36, 1940.
19. Tower, S. S. The dissociation of cortical excitation from cortical inhibition by pyramid section and the syndrome of that lesion in the cat. *Brain* 58:238, 1935.
20. Wang, G. H., and Akert, K. Behavior and reflexes of chronic striatal cats. *Arch. Ital. Biol.* 100:48, 1962.
21. Woolsey, C. N., and Chang, H. T. Activation of the cerebral cortex by anti-dromic volleys in the pyramidal tract. *Res. Publ. Assoc. Res. Nerv. Ment. Dis.* 27:146, 1947.

SELECTED READING

Aidley, D. J. *The Physiology of Excitable Cells*. London: Cambridge University Press, 1971. P. 360, Table 16-2.
Aseltine, J. A. *Transform Method in Linear System Analysis*. McGraw-Hill Electrical and Electronic Engineering Series. F. E. Terman, W. W. Harman, and J. G. Truxal (Consulting Eds.). New York: McGraw-Hill, 1958.
Evarts, E. V. Representation of Movements and Muscles by Pyramidal Tract Neurons of the Precentral Motor Cortex. In M. D. Yahr and D. P. Papura (Eds.), *Neurophysiological Basis of Normal and Abnormal Motor Activity*. Hewlett, N.Y.: Raven Press, 1967 Pp. 215–253.
Robinson, D. A. Oculomotion for unit behavior in monkey. *J. Neurophysiol.* 33:393, 1970.

Neural Substrates of Behavior 30

Neil Schneiderman

INTRODUCTION

One of the major goals of medicine and the life sciences is to understand the biologic functioning of the behaving individual. The purpose of this chapter is to review some of our knowledge concerning the neural integration of consciousness, emotion, and motivated behavior.

MODULATION OF CONSCIOUSNESS

ASCENDING RETICULAR ACTIVATING SYSTEM

The reticular formation is a midventral core of neural tissue that extends from the spinal cord to the thalamus. It is surrounded by the long ascending fibers of the classic somatosensory pathways and by long descending motor pathways.

The reticular activating system receives sensory input from most modalities. The organization of the system, however, is complex, and the amount of synaptic convergence within it abolishes most modal specificity. For the most part, reticular neurons are activated with equal facility by different sensory stimuli. Few sensory neurons can be driven by only a single sensory mode, and few, if any, neurons in the reticular formation are influenced by all modalities of stimulation. The general nonspecificity of reticular-formation neurons in response to stimulation from various sensory modalities suggests that the general level of sensory activity in the reticular system is abstracted from the various specific sensory inputs feeding into it. Many neurons in the reticular formation, however, do not respond to stimulation from any sensory modality.

Moruzzi and Magoun [25] found that electrical stimulation of the reticular core of the brain stem in lightly anesthetized cats produced an electroencephalographic (EEG) pattern of arousal in the cerebral cortex. Their discovery prompted interest in many areas of brain function, ranging from those of attention and wakefulness to coma and sleep. The initial studies of the reticular activating system provided, at least in part, a link between an understanding of specific sensory systems and that of behavioral phenomena such as attention, arousal, and sleep.

In the experiment of Moruzzi and Magoun, electrical stimulation of the reticular formation induced EEG arousal. Subsequent investigation showed that electrical stimulation of the ascending reticular formation induced behavioral arousal. Additional evidence that the ascending reticular formation played a key role in wakefulness was provided by lesion studies. In these experiments, either the reticular formation or the classic sensory pathways were lesioned in different cats. Animals with lesions that interrupted the classic sensory pathways displayed normal sleeping-waking cycles, with appropriate EEG activity. In contrast, those animals with reticular lesions were stuporous and showed EEG patterns typical of sleep. The initial stimulation and lesion studies thus suggested that cortical EEG activity and behavioral arousal are mediated by the ascending reticular formation.

Numerous lesion studies have subsequently modified our conclusions about the ascending reticular activating system. Although a massive one-stage bilateral lesion of the midbrain reticular formation produces severe coma and death, a similar sized lesion made in multiple stages permits considerable recovery. Smaller one-stage lesions sparing the most medial portions of the midbrain produce transitory coma with good recovery, and similar lesions made in several stages produce little behavioral impairment. When one considers that a large number of regulatory nuclei and projection fibers are imbedded in the brain-stem reticular formation, it is hardly surprising that animals die from massive one-stage lesions. These lesions may therefore have relatively little to do with the mechanisms of wakefulness. Carefully performed lesion experiments, however, have allowed a distinction to be made between the portions of the reticular formation involved in the electroencephalographic and behavioral manifestations of arousal. Lesions in the posterior hypothalamic outflow of the reticular formation result in a behavioral somnolence that persists even when stimulation of the midbrain reticular formation produces an EEG pattern of wakefulness. In contrast, cats with lesions of the midbrain reticular formation are not completely somnolent, but show an EEG pattern characteristic of wakefulness.

THE ELECTROENCEPHALOGRAM

The gross electrical activities of different parts of the brain can be recorded electrographically. As early as 1875, evidence of spontaneous electrical activity was detected in the brains of animals. Observing that sensory stimulation transformed the characteristics of the ongoing electrical activity, the early pioneers of electrophysiology concluded that what they were recording was somehow related to cerebral function. By 1930, sensitive string galvanometers were being used to record the electrical activity from the human brain through the intact skull.

The recording of gross electrical activity from the brain through the intact skull provides an *electroencephalogram* (EEG). When the skull is trephined and the electrodes are placed either on the meninges covering the brain or within the cortex itself, the recordings constitute an *electrocorticogram* ("electrocorticogram" is abbreviated ECoG so as not to confuse it with the English abbreviation for "electrocardiogram"). In a third recording procedure, the electrodes penetrate deep within the brain, and the readings obtained are called *depth recordings.*

To record voltage changes through the skin covering the skull, either from the surface of the cortex or from deep within the brain, the electrodes may be arranged for either *bipolar* or *monopolar* (a misnomer) recording. Typically, bipolar recordings are used to assess the fluctuations in potential between two electrodes that are relatively close to one another. When recordings are made from electrodes deep within the brain, the electrode tips are frequently less than 0.5 mm apart. The recordings provide the algebraic resultant of the potentials under each electrode.

In contrast to bipolar recording, in which both leads are in proximity to the brain structures supplying the bioelectrical activity, the technique of monopolar recording attempts to provide a record of the potential differences between an active cortical or subcortical electrode and an indifferent electrode that is situated at a location relatively distant from the brain, such as the ear. For the most part, the active lead picks up the brain potentials that are immediately near it, but the interpretation of results is sometimes difficult. This is particularly true of human EEG recordings, in which both electrodes are actually distant from the brain. For this reason, a common reference electrode is sometimes affixed to the skull, and readings between this lead and various other placements are compared. Although the term "monopolar" is inexact and misleading, it continues to be widely used in both research and clinical practice.

It was not until after the turn of this century that observations were made indicating that the normal oscillation found in the recordings of brain activity could be interrupted by the stimulation of an afferent nerve. By 1930, it was known that the EEG was dominated by large-amplitude, slow-wave fluctuations during sleep, while the recordings showed considerably less amplitude and less obvious periodicity during wakefulness.

The EEG reveals a number of distinct patterns that are usually correlated with sleep, wakefulness, and attention (Fig. 30-1). In the resting human with his eyes closed, *alpha waves* are typically recorded from the occipital cortex. The alpha rhythm consists of high-amplitude, well-synchronized sinusoidal waves having a frequency of 8 to 12 Hz. When the person opens his eyes, the synchronized alpha rhythm gives way to a desynchronized, lower-amplitude pattern — *beta waves* — in which some high-frequency responses can be detected by computer analysis. Desynchronization is not only characteristic of the transition between the closing and the opening of the eyes, but also of the most alert, attentive, or excited states.

As a person becomes drowsy and drifts off to sleep, relatively high-voltage, slow waves appear in the EEG record. These slow waves, with a frequency of 2 to 4 Hz, are called *delta waves*. Also, during drowsiness and in light or moderate sleep, *spindle bursts* occur that are interspersed between periods of delta activity. These bursts occur at a frequency of from 12 to 15 Hz, and they frequently have a spindle form, from which they derive their name. Although fairly deep sleep is usually accompanied by high-voltage delta activity and spindles, another deep stage of sleep is accompanied by EEG episodes that resemble those of wakefulness. At first glance, the desynchronized EEG of deep sleep looks so much like the EEG of wakefulness and so unlike the EEG of light or moderate sleep that the term "paradoxical sleep" has been applied to the deep sleep state.

REGULATION OF SLEEP

Transection of the brain at the level of the midbrain makes an animal comatose with an accompanying EEG pattern that is characteristic of sleep. At one time, the effect was thought to be due to the interruption of the long sensory tracts. These were believed to maintain wakefulness by constant bombardment of the cortex. This contributed to the view that sleep was essentially a passive phenomenon due to an

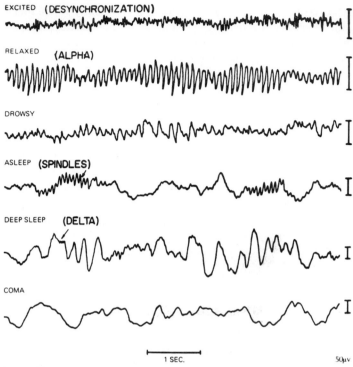

Figure 30-1
Characteristic patterns of the electroencephalogram. (Adapted from W. Penfield and H. Jasper. *Epilepsy and the Functional Anatomy of the Human Brain.* Boston: Little, Brown, 1954. Reproduced with permission.)

absence of sensory stimuli impinging upon the cortex. The experiments of Moruzzi and Magoun [25] and their subsequent collaborators, however, shifted the locus of the waking mechanism from the specific sensory pathways to the reticular formation.

The passive theory of sleep was challenged by the experiments of W. R. Hess [12]. These studies indicated that an electrical stimulation at 5 to 6 Hz of the midline structures of the thalamus induced a normal, awake cat to show all of the signs of sleep. Upon receiving this stimulation, the animal would search for a comfortable place, curl up, and go to sleep.

Even more convincing demonstrations of an active sleep-inducing system were provided by experiments in which it was found that midpontine transections produced increases in the behavioral and electroencephalographic manifestations of wakefulness. These findings suggested that the sleep-inducing structures were located in the lower brain stem.

The most comprehensive analysis of sleep states has been provided by Jouvet [16]. According to Jouvet, mammalian sleep may be characterized by two different states that can be recognized easily by means of polygraphic analysis. In one of these states, which is referred to as *slow-wave sleep,* the electrical activity of the cortex includes delta activity and spindle bursts. In the second state, which is referred to as *paradoxical sleep,* the cortical activity resembles that of the waking state, but, behaviorally, the organism is in a state of deep sleep.

During paradoxical sleep, there is an almost total absence of electromyographic activity in the antigravitational jaw and neck muscles. This state appears to represent a true paralysis that is induced by tonic inhibitory influences upon motor output. The blood pressure drops, but it is accompanied by rapid oscillations. Rapid eye movements also occur and are quite pronounced. A sleeping person who is awakened during paradoxical sleep is likely to report dream activity. Under normal circumstances approximately 80% of a night's sleep is spent in slow-wave sleep, whereas about 20% is spent in paradoxical sleep.

Several findings suggest that serotonin may be involved in slow-wave sleep, whereas the catecholamines such as norepinephrine and dopamine may be involved in paradoxical sleep. In one experiment, Jouvet found that reserpine, which produces decreases in both serotonin and catecholamine concentrations in the brain, suppresses both slow-wave and paradoxical sleep. In some animals, secondary injections of 5-hydroxytryptophan were given in order to restore normal concentrations of serotonin in the brain. In other animals, secondary injections of dihydroxyphenylalanine (DOPA) were given in order to restore normal brain concentrations of the catecholamines. The injections of 5-hydroxytryptophan that increased the concentrations of serotonin in the brain immediately restored slow-wave sleep, and injections of DOPA that restored the normal brain concentrations of catecholamines restored paradoxical sleep. The concentrations of serotonin in the brain can be decreased without altering the concentrations of the catecholamines by means of *p*-chlorophenylalanine injections. Jouvet found that injections of this agent induced insomnia, whereas injection of 5-hydroxytryptophan returned the animal to normal sleep. A significant correlation was thus found to exist between the decrease in slow-wave sleep and the diminution in cerebral concentrations of serotonin.

Histofluorescence techniques have made it possible to study the distribution of monaminergic neurons in the brain. Serotonergic neurons display a yellowish fluorescence, whereas catecholamine-containing neurons have a greenish fluorescence. These studies have indicated that serotonin-containing neurons are prevalent in the raphe (dorsal midline) system of the lower midbrain, pons, and medulla. In contrast, norepinephrine-containing neurons are located in the lateral part of the pontine tegmentum (locus coeruleus), as well as in many other places. The terminals of the serotonergic and noradrenergic neurons are widely distributed throughout the brain stem, diencephalon, and neocortex.

Jouvet found that bilateral lesions of the locus coeruleus of the pontine tegmentum suppressed paradoxical sleep without altering slow-wave sleep. In contrast, lesions of the raphe system produced insomnia. Since paradoxical sleep appears only when slow-wave sleep reaches a daily threshold of about 15%, Jouvet proposed that paradoxical sleep depends upon (1) serotonergic "priming" mechanisms located medially in the raphe system and (2) a noradrenergic "triggering" system located laterally in the pontine tegmentum. Jouvet has also found evidence that acetylcholine may play a role in activating the noradrenergic triggering mechanism.

Although Jouvet's experiments have strongly implicated midbrain and pontine structures in the triggering of both behavioral and electroencephalographic sleep patterns, there is considerable evidence that rostral structures play an important role in the EEG aspects of slow-wave sleep. Coagulation of the midline structures of the thalamus, for example, abolishes the cortical spindle patterns that occur at the onset of sleep. The removal of the neocortex eliminates slow-wave activity in the thalamus and the reticular formation for several months. Further evidence for a descending cortical synchronizing influence upon the brain stem may be inferred from the finding that following the section of the brain stem at the junction of the midbrain and diencephalon, spindle activity persists above, but not below, the point of section.

The relationship between dream activity and the states of sleep has been studied. Rapid eye movements (REM) typically occur during low-voltage, desynchronized sleep [1]. When human subjects are awakened from so-called REM or paradoxical sleep, they almost invariably report dreaming. In contrast, when they are awakened from slow-wave sleep, subjects rarely report dreams. A brief dream state normally occurs after about 90 minutes of sleep. Increasingly longer episodes recur approximately every 1 to 1-1/2 hours during a night's sleep. Typically, when human subjects are systematically deprived of REM sleep for prolonged periods of time, they report psychologic discomfort. Subsequently, when the subjects are permitted to sleep uninterruptedly, a greater proportion of time is spent in a state of paradoxical sleep. The biologic utility of the different sleep states is unknown at present.

CENTRAL REGULATORY MECHANISMS AND MOTIVATIONAL STATES

During the course of evolution, important neuronal and hormonal control systems have developed in complex animals with regard to temperature regulation, water regulation, feeding, drinking, and sexual behavior. In addition, general approach and avoidance systems were developed. The hypothalamus and limbic system play an important role in the regulaton of all these activities.

TEMPERATURE REGULATION

The maintenance of a relatively constant body temperature in birds and mammals depends upon a relatively large number of regulatory functions. Thus, when the body temperature falls, a number of reactions occur that lead to both heat conservation and increased heat production. Conversely, rises in the body temperature above normal levels cause various heat-dissipating mechanisms — such as panting, sweating, and vasodilation — to be activated.

A fall in body temperature will usually lead to a constriction of cutaneous blood vessels and an inhibition of perspiration, which contribute to heat conservation. Increased heat production is concomitantly brought about by shivering, augmentations in thyroid and adrenal secretions, and an increase in cardiac activity. The augmented thyroid secretion increases the rate of metabolism; augmented adrenal activity leads to increased sugar release into the blood. In addition, various behavioral processes, ranging from foot stamping to shelter building or migration, may occur.

The hypothalamus plays a prominent role in thermoregulation. Apparently, central receptors in this structure receive thermal inputs from the blood supply. These inputs are then transmitted to regulatory cells that make the adjustments necessary to maintain a genetically-imposed temperature set-point. As early as 1940, Ranson [32] demonstrated that lesions of the anterior hypothalamus cause a pronounced decrease in an animal's ability to counteract overheating when subjected to high temperatures. Subsequent experiments have shown that electrical stimulation of the anterior hypothalamus can lead to a suppression of shivering and hypothermia. Evidence of thermal receptors in the anterior hypothalamus has been provided by microelectrode studies that have recorded single-unit activity in that structure during changes in hypothalamic temperature. Heat-sensitive units have been found in the hypothalamus that increase their discharge rate as the hypothalamic temperature is increased. Single units have also been observed in this same structure that discharge differentially in response to heating and cooling.

The experimental evidence suggests that the anterior hypothalamus contains the principal thermostat of the body. Experiments in which the temperature of animals could not be driven beyond certain limits by regional heating and cooling, however, have led to the conclusion that the anterior hypothalamus may not be the only site that provides inputs to thermoregulatory effector mechanisms. Satinoff and Rutstein [3], for example, found that although rats with lesions in the anterior hypothalamus could not maintain their body temperature at normal levels in a cold environment, they would artificially obtain thermoregulation in the cold if they were permitted to press a bar repeatedly in order to activate a heat lamp. In another experiment, the interaction between the responses to skin temperature and hypothalamic temperature changes was examined, and the responses were found to be additive [4]. In general, when appropriate behavioral responses are available, organisms will attempt to escape from ambient temperatures requiring physiologic adjustments that have a high energy cost.

Although the anterior hypothalamus appears to behave in a thermostatic fashion, the posterior hypothalamus has been implicated in the physiologic effector responses that contribute to heat production. The contention that the posterior hypothalamus does not have thermostatic functions rests upon the experimental evidence that heating or cooling of the posterior hypothalamus does not lead to compensatory physiologic changes, and that pyrogens that raise the body temperature when administered to the anterior hypothalamus do not have this effect upon the posterior hypothalamus.

The observation that the posterior hypothalamus plays a role in temperature regulation, however, dates back to the early years of this century, when it was found that animals with lesions in part of the hypothalamus lose several of their responses to cooling, such as shivering. Subsequent lesion and stimulation experiments have indicated that the dorsomedial portion of the posterior hypothalamus is essential for shivering, whereas the ventrolateral portion of the posterior hypothalamus is responsible for the inhibition of heat production. The physical relationship of the hypothalamus and the pituitary glands makes the hypothalamic control over the endocrine system self-evident.

WATER REGULATION

Living organisms depend upon water. As discussed in Chapter 24, water accounts for approximately 70% of the mammalian body weight. In addition, the secretion and excretion of water takes place as part of the processes of bodily waste removal and temperature regulation. Water intake takes place by drinking and by consuming food with water content; its loss occurs through respiratory evaporation, perspiration, urination, and defecation.

One of the primary effects of water intake is to decrease the concentration of solids in the blood. This dilution of the plasma causes a decrease in osmotic pressure. If the intracellular and extracellular fluids have the same osmotic pressure, net fluid transmission across the membrane is essentially zero. The solution is then said to be *isotonic*. Considerable evidence has now been accumulated that some cells within the hypothalamus can detect departures from isotonicity. These cells are referred to as *osmoreceptors*.

The importance of the brain in water regulation was suggested by experiments in which hypertonic saline was injected into the systemic circulation [45]. Injections into the carotid artery, but not into other arteries that are more removed from the brain, produced a rapid and pronounced increase in antidiuretic hormone (ADH) secretion. The neuroendocrine mechanisms relating to ADH are discussed in detail in Chapters 22 and 43.

The dependence of ADH secretion on the hypothalamo-hypophysial system has been demonstrated by both stimulation and lesion experiments. Electrical stimulation by means of electrodes implanted in the medial eminence of the hypothalamo-hypophysial tract increases ADH secretion, and injection of hypertonic saline into the same location reduces urinary flow. Conversely, injury to the hypothalamo-hypophysial system at any point in the supraoptic nucleus, the tract itself, or the posterior pituitary leads to diabetes insipidus (see Chap. 33).

Peck and Novin [28] have provided evidence that the osmoreceptors that mediate drinking in rabbits are located in the lateral preoptic area of the hypothalamus. In order to demonstrate this, they had to show that a decrease in cell volume was a necessary and sufficient condition for eliciting drinking behavior. One way to decrease cell volume is to inject hypertonic saline into the region. Saline, however, could cause nonspecific excitation of neural tissue due to the ionic properties of sodium. Thus, on the basis of saline injection alone, it would not be possible to distinguish whether the drinking was elicited by the decreased cell volume or by the excitation of individual neurons due to the ionic properties of the sodium.

In contrast to hypertonic saline, which is ionic, hypertonic sucrose is electrolytically inert but it may cause a decrease in cell volume. The effects of both hypertonic sucrose and hypertonic saline upon drinking behavior were therefore examined. As a further check, the effects of hypertonic urea were also investigated. Whereas hypertonic sucrose and hypertonic saline produce a decrease in cell volume inasmuch as they are excluded from cells, urea enters both the intracellular and extracellular compartments.

Hypertonic saline, but not hypertonic sucrose or urea, elicited drinking, eating, and gnawing in some locations such as the lateral and dorsal hypothalamus. Apparently, the saline injections nonspecifically stimulated circuits that are involved in drinking as well as other behaviors. In contrast to the findings following injection into the lateral and dorsal hypothalamus, injections of hypertonic saline and hypertonic sucrose, but not hypertonic urea, into the lateral preoptic nucleus elicited drinking. These injections into the lateral preoptic nuclei were therefore effective only to the extent that they dehydrated cells.

Although the osmoreceptors for drinking appear to be restricted to the lateral preoptic area, many other brain structures within and closely related to the hypothalamus appear to be involved in the circuits for drinking behavior. Stimulation and lesion studies have helped to delineate these pathways. Electrical stimulation of the lateral hypothalamus, for example, leads to profuse drinking. In contrast, bilateral lesions in the lateral hypothalamus lead to adipsia (nondrinking) as well as a refusal to accept food [41]. The aphagia or refusal to eat disappears slowly, but the adipsia lasts much longer or even permanently.

CENTRAL CONTROL OF EATING BEHAVIOR

It has long been known that a tumor in the region of the hypothalamus can lead to excessive weight gain. Obesity in hypothalamically lesioned rats is due to overeating rather than to a change in metabolic functions. Following ventromedial hypothalamic lesions, rats go through two phases. During the first stage, they may eat as much as two to three times their normal amount of food and gain weight rapidly. During the second stage, they reach a plateau, but they maintain their excess weight. Interestingly, animals with lesions of the ventromedial hypothalamus are more influenced by the taste of food than are normal animals.

On the basis of these lesion experiments, it has been suggested that the ventromedial hypothalamus is a satiation center that normally inhibits eating when an adequate food intake has been attained. Presumably, the satiation mechanism breaks down after the lesion, at least until some other tissue can take over. In contrast to the lesions of the ventromedial hypothalamus that lead to hyperphagia, stimulation of the ventromedial hypothalamus leads to a cessation of eating. Since stimulation of this structure is also aversive to the animal, however, it is difficult to draw conclusions about the effects of stimulation of this structure upon eating behavior.

Whereas lesions of the ventromedial hypothalamus lead to hyperphagia and obesity, lesions of the lateral hypothalamus lead to aphagia as well as adipsia. If animals with lesions in the lateral hypothalamus are not force-fed, they soon die. If they are force-fed, the aphagia slowly disappears [41]. In contrast to the effects of lesions in the lateral hypothalamus, electrical stimulation leads to eating behavior. The results of lesioning or stimulating the ventromedial or lateral hypothalamus are shown in Figure 30-2.

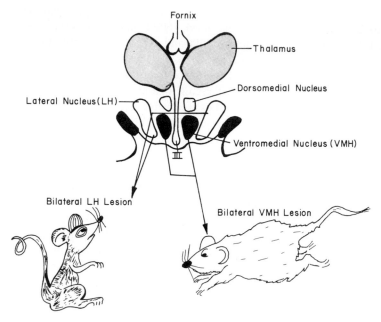

Figure 30-2
Bilateral destruction of the lateral hypothalamus (*LH*) leads to aphagia and adipsia. Bilateral destruction of the ventromedial hypothalamus (*VMH*) leads to hyperphagia and obesity.

   Although the reciprocal relationship between the ventromedial and the lateral hypothalamus suggests the existence of adjoining satiation and feeding centers, the situation appears to be actually more complex. Pronounced increases in eating, for example, have been elicited not only by electrical stimulation of the lateral hypothalamus, but also by stimulation of the posterior hypothalamus and ventral midbrain.
   Recent evidence has suggested that two different types of adrenergic receptors in the brain may be involved in eating and drinking behavior. A reciprocal inhibitory relationship has been proposed to exist between two different adrenergic hunger-and-thirst regulating systems in the rat [19]. This may help to provide a neurochemical explanation for the ability of animals to maintain food and water consumption at a constant ratio.
   The distinction between alpha-adrenergic and beta-adrenergic receptors was originally proposed on the basis of the different autonomic nervous system reactions that occur to agonist drugs (in this case, sympathomimetic), such as norepinephrine or phenylephrine, which are predominantly alpha-adrenergic, and isoproterenol, which is primarily beta adrenergic.
   In one experiment, drugs that had been demonstrated to be agonists or antagonists in the peripheral autonomic system (Table 30-1) were injected directly into the perifornical region of the hypothalamus of rats through permanently implanted cannulas [18].

Table 30-1. Innervation of the Cardiovascular System

Structure (Division)	Transmitter	Receptor	Agonist	Antagonist
Heart				
parasympathetic	Acetylcholine	Muscarinic	Acetylcholine	Atropine
sympathetic	Norepinephrine	Beta	Isoproterenol	Propranolol
Arterioles of skin and mucosa (sympathetic)	Norepinephrine	Alpha	Norepinephrine Phenylephrine	Phentolamine

The alpha-agonist norepinephrine stimulated eating, whereas the beta-agonist isoproterenol suppressed food intake. Conversely, the alpha-antagonist phentolamine suppressed eating, but the beta-antagonist propranolol enhanced eating.

In another investigation, opposite effects upon water intake were found when cannulas were placed primarily in the lateral hypothalamus [19]. The alpha-agonist suppressed thirst, whereas the beta-agonist stimulated thirst. Considered together, the two intracranial chemical stimulation studies suggest that at least in the rat (1) an alpha-adrenergic "hunger" system opposes a beta-adrenergic "food-satiety" system, (2) a beta-adrenergic "thirst" system opposes an alpha-adrenergic "water-satiety" system, and (3) a reciprocal inhibitory relationship exists between the hunger and thirst regulating systems (Table 30-2). The alpha-receptors appear to be predominantly located in the ventromedial hypothalamus, whereas the beta-receptors have been reported to be primarily located in the lateral hypothalamus.

Table 30-2. Reciprocal Effects of Intracranial Injections of Alpha- and Beta-Adrenergic Agonists and Antagonists upon Eating and Drinking in Rats

Substance	Function
Alpha-adrenergic	
Agonist (norepinephrine)	Enhances eating Suppresses thirst
Antagonist (phentolamine)	Enhances drinking Suppresses eating
Beta-adrenergic	
Agonist (isoproterenol)	Enhances drinking Suppresses eating
Antagonist (propranolol)	Enhances eating Suppresses drinking

SEXUAL BEHAVIOR

The mechanisms controlling sexual behavior are complex. Almost all levels of the central nervous system — from cortex to cord — are involved in its integration. Moreover, the endocrine system (see Chap. 35) also plays a crucial role in the elaboration of sexual behavior. Pivotal to the regulation of sexual behavior is the hypothalamus.

The hypothalamus exercises a direct control as a neural mechanism and a relatively more indirect control through its influences upon the pituitary.

Most of the reflexes involved in copulatory behavior are located in the lumbar and sacral segments of the spinal cord. Erection and ejaculation are still possible after spinal transection. Since lesions in the hypothalamus can permanently abolish sexual excitability, it appears that there is an inhibitory system in the brain stem that is capable of antagonizing the spinal reflexes for sexual behavior.

Male and female cats or guinea pigs with lesions in the anterior hypothalamus do not mate. The deficit is not due to hypothalamic control over the endocrine system, because treatment with estrogen or androgen does not restore the sexual behavior. In a classic experiment, Sawyer and Robinson [35] demonstrated the existence of dual hypothalamic control mechanisms governing sexual behavior in female cats and rabbits. Confirming previous experiments, they found that lesions of the anterior hypothalamus abolished sexual behavior and that the behavior could not be restored by the injection of estrogen. In contrast, however, lesions of the tuberal hypothalamus, including the ventromedial nucleus or the mammillary region, abolished mating behavior and caused atrophy of the ovaries, but both of these effects could be prevented by estrogen replacement. A similar dual mechanism also appears to govern male sexual behavior, since lesions in the ventromedial hypothalamus of the male rat abolish mating behavior, but this can be restored by testosterone therapy.

Electrical stimulation studies of the hypothalamus have provided further evidence that this structure plays a role in the organization of sexual behavior. Stimulation of the anterior hypothalamus in the males of several species, including monkeys, opossums, and rats, has elicited pronounced sexual behavior. These effects have included continuous erection, attempts to copulate with inanimate objects, and repeated ejaculations. One heroic rat is reported to have had some 20 ejaculations within a single hour of stimulation [44].

Chemical stimulation of the hypothalamus has also been found to induce sexual behavior, although presumptive transmitters, such as acetylcholine and norepinephrine, have not been effective. Minute injections of testosterone, however, into the lateral preoptic region of the anterior hypothalamus in male rats have been found to elicit increased sexual behavior. Similarly, the implantation of a solid estrogen pellet into the mammillary region in ovariectomized female cats has produced continuous receptivity lasting for several months [11]. Evidence of the presence of estrogen-sensitive neurons in the hypothalamus has been obtained histologically and by the use of autoradiography. Gross EEG activity recorded from the hypothalamus also appears to correlate well with the biochemical evidence, thereby implicating this structure in the regulation of sexual behavior. Following vaginal stimulation in the estrous cat, for example, behavioral orgasm is accompanied by high-amplitude slow waves in the EEG recorded from the anterior hypothalamus [31].

Although evidence derived from several lines of research shows that the hypothalamus plays an important role in the regulation of sexual behavior, the exact relationship of different hypothalamic structures in the neurohormonal integration of sexual behavior is not thoroughly understood. It does appear, however, that the tuberal region of the hypothalamus exercises control over the secretion of gonadotrophic hormones from the pituitary. These, in turn, affect the secretion of gonadal hormones. The increased levels of these gonadal hormones in the blood possibly trigger the activity of chemically sensitive cells in the anterior hypothalamus, which then elicit sexual behavior. The manner in which pathways to the tuberal region are stimulated to initiate control over the hormones has not yet been adequately delineated.

EMOTIONAL BEHAVIOR

Integrated emotional behavior patterns require an intact hypothalamus. Cats with all the brain tissue above the midbrain removed show only fragmented components of the rage reaction; in contrast, cats with all the brain tissue removed rostral to the hypothalamus display fully developed rage reactions.

Electrical stimulation of the hypothalamus by means of chronically implanted electrodes produces various emotional behaviors, including well-directed attack behavior. Experiments employing brain stimulation have, to a limited extent, distinguished among flight, rage, and stalking behavior patterns. Flight responses have readily been evoked by stimulation of the lateral regions of the anterior hypothalamus, rage responses characterized by hissing and snarling have been elicited from the ventromedial hypothalamus, and stalking behavior followed by killing has been obtained following electrical stimulation of the lateral hypothalamus. Such distinctions have not, however, always been clear-cut, in part because various experimenters have used different species, procedures, and stimulation parameter values in their studies.

In one experiment that was aimed at elucidating the differences among the aggressive responses that can be elicited by stimulating the lateral or the ventromedial hypothalamus of the rat, cholinergic and anticholinergic substances were injected into the two structures [36]. The investigators found that injection of a cholinergic substance into the lateral, but not the ventromedial, hypothalamus, caused rats that had never killed spontaneously to stalk and kill mice. Conversely, injections of the anticholinergic drug, methylatropine, into the lateral hypothalamus blocked the killing response in rats who killed naturally.

Intracranial stimulation has been used to investigate other emotional behaviors besides flight and aggression. Olds and Milner [27], for example, found that electrical stimulation of the septal region in rats could serve as a reward. In their studies, the animals were trained to press a lever in order to receive a short train of electrical stimulation. Previously, many investigators had trained rats to press a similar bar at high rates to receive food or water, but nobody had guessed that animals could be trained to press a bar to receive electrical stimulation of the brain.

In the same year that Olds and Milner reported their experiment, Delgado, Roberts, and Miller [5] reported the first experimental proof that electrical stimulation of the brain can motivate avoidance. Delgado et al. first trained cats to rotate a wheel in response to a buzzer in order to avoid a painful shock. They then paired the same buzzer sound with brain stimulation delivered to the midbrain, lateral mass of the thalamus, or hippocampus. In each case, the cat learned to rotate the wheel to avoid this stimulation.

Subsequent work confirmed and extended the findings of both Olds and Milner and Delgado et al. These experiments have shown that the medial forebrain bundle that travels through the lateral hypothalamus plays a central role in the mediation of reward, and that the periventricular system (i.e., the structures surrounding the ventricles of the telencephalon, diencephalon, and midbrain) plays a major role in the mediation of punishment. In general, the structures involved in reward, such as the medial forebrain bundle, may be considered to be part of a "go" mechanism that facilitates ongoing behavior; in contrast, the periventricular structures involved in punishment appear to be part of a "stop" mechanism that inhibits ongoing behavior.

The effects of drugs upon electrical self-stimulation of the brain have been widely studied. In general, it appears that the self-stimulation reward system is adrenergic. Thresholds of bar-pressing for self-stimulation are lowered by sympathomimetic (adrenergic) drugs, such as amphetamine, and are elevated by chlorpromazine and other drugs that interfere with adrenergic transmission. Interestingly, norepinephrine-containing terminals have been located in the medial forebrain bundle by means of histochemical fluorescence techniques. Stein and Wise [39] have further shown that norepinephrine is released following the electrical stimulation of the medial forebrain bundle at the same parameter values that lead to high rates of bar pressing.

In the experiment of Stein and Wise, the hypothalamus and the amygdala of unanesthetized rats were perfused with Locke's solution through a permanently implanted push-pull cannula after the intraventricular injection of radioactively labelled norepinephrine. The push-pull cannula consisted of two concentric tubes mounted in a nylon screw. Ringer-Locke's solution ran down the inner tube into the brain and was collected through the outer tube. Rewarding electrical stimulation of the medial forebrain bundle caused a pronounced increase in the radioactivity of the perfusate, which indicated that there was an increased release of norepinephrine.

The mapping of the brain in self-stimulation experiments has taken on increased significance since mounting evidence has strongly implied that the concentrations of monoamines in the brain are related to mood and emotion. Tranquilizing drugs, such as reserpine and chlorpromazine, decrease the availability of norepinephrine at the central synapses; in contrast, antidepressants, such as imipramine and the monoamine oxidase inhibitors, increase its availability at central synapses.

ROLE OF THE HYPOTHALAMUS IN MOTIVATION

Electrical stimulation of the hypothalamus has been used to elicit a number of behavioral patterns. These have included eating, drinking, hoarding, stalk-attack, and sexual behavior. Each has been considered to reflect motivated behavior rather than a stereotyped motor act for several reasons. First, the stimulated animal does not engage in the behavior unless appropriate goal objects are present. Second, the animal from whom the behavior is elicited can be taught to perform some learned task, such as a bar-press maneuver, in order to obtain the goal object. Third, the animal will usually tolerate a moderate amount of electrical shock or other aversive stimulation in order to obtain the object.

Upon initial investigation, some of the behaviors that are induced by intracranial stimulation seemed to be elicited from reasonably specific places. For this reason, the common belief developed that discrete hypothalamic regions are involved in the regulation of behavior patterns underlying specific biologic needs. Thus, for example, the lateral hypothalamus was conceived to be a center that is responsible for the integration of eating and drinking behavior. The results of several experiments, however, have provided evidence that the behavior elicited by hypothalamic stimulation is subject to change as a function of alterations in the environment.

In the experimental situation described by Valenstein, Cox, and Kakolewski [42], for example, a food- and water-satiated rat was placed in an enclosure with food, water, and a stick. At first, the intracranial stimulation elicited eating behavior. When the food and water were removed, however, the animal began gnawing the stick. Subsequently, when the food and water were again made available, the stimulation continued to elicit the gnawing behavior.

Not only oral behaviors, but other very different kinds of behaviors can be interchanged by altering environmental conditions. Behaviors as different as clucking, preening, and wing-scratching in fowl have been elicited by stimulation through the same electrodes at different times using identical parameter values [46]. More recently, it was found that male sexual behavior or eating behavior could be elicited in the rat by stimulating the posterior hypothalamus using the same parameter values of stimulation [3].

Valenstein et al. [42] have suggested that rather than motivating a specific behavior, hypothalamic stimulation elicits species-specific response patterns under environmental conditions that normally elicit them in aroused animals. They have proposed that electrical stimulation of the hypothalamus does not in itself create hunger, thirst, or sexual appetite, but it instead excites part of the neural substrate that underlies well-established responses. Environmental conditions then interact with internal factors, such as the effects of food deprivation, to determine which response is chosen.

THE LIMBIC SYSTEM

At one time, many of the structures that are now included in the limbic system were thought to be involved in the sense of smell. For this reason, the structures forming a medial border at the junction between the diencephalon and the telencephalon were referred to as the *rhinencephalon* (Fig. 30-3). At present, the equivalent of an olfactory projection area seems to include only the olfactory tubercle, the prepyriform cortex (the uncus in man), and the corticomedial nuclei of the amygdala. Other structures historically included in the rhinencephalon — such as the septal region, hippocampus, cingulate gyrus, and the basolateral portion of the amygdala — apparently have little if anything to do with olfaction. Experiments have shown, for example, that animals

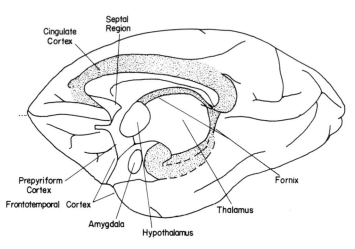

Figure 30-3
The structures of the limbic system form a medial border at the junction between the diencephalon and the telencephalon.

can learn olfactory discriminations after important rhinencephalic structures, such as the hippocampus, have been extirpated. Consequently, MacLean [21] popularized the term *limbic system* to avoid the olfactory connotations of "rhinencephalon."

An important experiment linking the limbic system with the elaboration of emotion was reported by Klüver and Bucy [17]. They performed bilateral, anterior temporal lobectomies upon rhesus monkeys and observed striking changes of behavior. The lesions destroyed the amygdala, entorhinal cortex, ventral hippocampus, and portions of the temporal neocortex. The resulting symptoms included docility, loss of fear, hypersexuality, psychic blindness, and compulsive oral tendencies. The monkeys appeared to be unable to discriminate between food and potentially dangerous objects, and they placed all kinds of things in their mouths.

Subsequent to the Klüver and Bucy experiment, a large number of investigators became interested in determining the anatomic bases of the different behaviors involved in what has become known as the Klüver-Bucy syndrome. In spite of the fact that none of these studies has been entirely successful and that many paradoxical results have been obtained, several generalizations have emerged. The taming effects definitely appear to be due to the bilateral destruction of the amygdala. The psychic blindness, or inability to recognize the meaning of things, is apparently due to the damaging of the temporal neocortex, although the hippocampus may also be involved. In part at least, the hypersexuality seems to be due to the damage to the pyriform cortex.

The striking findings of Klüver and Bucy and the subsequent difficulties investigators have had in disentangling the anatomic bases of the symptoms have continued to stimulate investigations in this area. These studies have, for the most part, increasingly focused upon the functions of individual structures and upon the interrelationships between closely related structures. It is difficult to say whether this strategy reflects the difficulty of dealing with a large, complex, comprehensive system at one time, or whether it reflects a growing conviction among investigators that many of the structures of the limbic system may not be functionally related to one another. In any event, most investigators have focused upon either a single structure or upon the relationship of a particular structure with the hypothalamus.

THE AMYGDALA

The amygdala lies deep within the base of the temporal lobe. The corticomedial portion of the amygdala receives direct input from the olfactory bulb, but the basolateral portion does not. Other major inputs to the amygdala are from the reticular formation, the pyriform cortex, and the hippocampus. The amygdala also has connections with the orbitoinsulotemporal cortex. The major efferent outflow is to the septal region and the hypothalamus by way of the stria terminalis. Another connection between the amygdala and the hypothalamus is through a ventral amygdalofugal pathway provided by fibers lying just above the optic tract.

Stimulation of the amygdala produces a variety of visceral and somatic responses, which include cardiovascular, respiratory, and pupillary changes similar to those elicited by hypothalamic stimulation. Stimulation studies have been used to plot the extensive circuit involved in defensive reactions. The structures involved include the amygdala, stria terminalis, hypothalamus, and central gray matter of the midbrain. Although section of the stria terminalis does not abolish the defensive reactions elicited by stimulation of the amygdala, the subsequent interruption of the ventral amygdalofugal pathway does [13].

Electrophysiologic data have shown that evoked potentials can be elicited from the hypothalamus following stimulation of the amygdala. Impulses from the amygdala have been shown to be capable of inhibiting or facilitating individual neurons in the hypothalamus [7]. Behavioral evidence also shows that stimulation of the amygdala can facilitate or inhibit the attack behavior elicited by hypothalamic stimulation [40].

Stimulation of the amygdala in cats produces a pronounced defense reaction, which includes arching of the back, hissing, and bared teeth. Interestingly, stimulation of the amygdala never arouses the quiet, stalking response that can be elicited from parts of the hypothalamus. Furthermore, stimulation of the amygdala suppresses rat-killing behavior in cats that are usually motivated to engage in this behavior by electrical stimulation of the hypothalamus.

Lesions of the amygdala have a pronounced effect upon behavior. A survey of 43 studies dealing with the behavioral effects of amygdalectomy found that taming and docility were obtained even from such animals as the lynx and wild rat; whereas 10 studies noted significant rage reactions following amygdalectomy [8]. In one experiment, investigators obtained either rage or taming [10]. Animals showing rage behavior also had damage to the hippocampus and developed epileptic seizures.

The amygdala appears to play a role in maintaining social dominance. Following bilateral amygdalectomy, monkeys show a decrease in social dominance if they are kept in a group situation [33]. In humans, stereotactic lesions in the amygdala reduce emotional excitability and aggressiveness, and they tend to normalize social behavior in patients with severe antisocial behavior disorders [26].

THE SEPTAL REGION

The septal region is a midline structure just anterior to the preoptic region of the hypothalamus. Although several distinct nuclear groups may be identified in the septal region, it is usually divided into medial and lateral septal areas. The septal region forms the medial wall of the anterior horn of the lateral ventricles. It is a much more pronounced structure in lower mammals than in higher primates. The septal region has reciprocal connections with several structures. It sends and receives many fibers from the hippocampus by way of the fornix. Fibers to and from the midbrain and hypothalamus connect with the septal region by way of the medial forebrain bundle and the periventricular pathway.

Septal stimulation in cats and rabbits can lower the heart rate and blood pressure, reduce respiratory rate and cortically induced movement, and constrict the pupils. These findings are consistent with reports that an important effect of septal stimulation is to induce motor inhibition [23].

Lesions of the septal region appear to interfere with the inhibitory influences that are normally exerted by this area. Lesions of the septal region, for example, have been observed to produce increases in the reactivity to sensory stimuli and have induced vicious behavior and hyperemotionality [3]. Most of these symptoms gradually disappear during the course of several weeks if the animals are frequently handled.

The medial septal region contains rhythmically firing cells that pace the hippocampal theta rhythm. This theta rhythm is a characteristic EEG pattern recorded from the dorsal hippocampus; it has a frequency of 4 to 8 Hz and consists of high-amplitude sinusoidal waves. Stimulation of the medial septal region can facilitate theta activity, whereas lesions of the same region abolish it.

THE HIPPOCAMPUS

The hippocampus is a structure of considerable size. In some mammals, it occupies up to one-quarter of the forebrain. Its motor and sensory cells are neatly segregated in layers, and it contains a unique type of cell called the *double pyramid cell.*

The entire hippocampal formation includes the entorhinal cortex, the subiculum hippocampi, and a combination of pyramidal and dentate areas referred to as *Ammon's horn* (Fig. 30-4). Generally, the term "hippocampus" is meant to exclude the entorhinal cortex and the subiculum. Afferent fibers are believed to reach the hippocampus from the medial septal region and the entorhinal cortex. More indirectly, afferents appear to reach the hippocampus from the midbrain, the midline nuclei of the thalamus, and the cingulate gyrus. The major outflow of the hippocampus is through the fornix, with efferent fibers terminating in the septal region and throughout the hypothalamus.

Although the hippocampus has always been classified as part of the rhinencephalon, it clearly has no olfactory function. Removal of the hippocampus interferes with

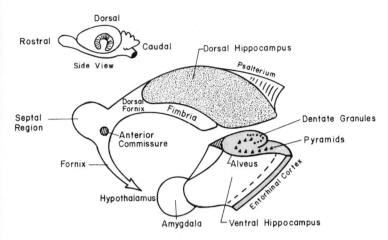

Figure 30-4
Side view of the hippocampus in the rat, with a section through the middle portion. (From
R. J. Douglas. *Psychol. Bull.* 67:416, 1967. Copyright 1967 by the American Psychological
Association. Reproduced with permission.)

neither olfactory discriminations nor conditioning to olfactory cues. In addition,
the hippocampus is well developed in porpoises and whales, both of which appear
to be anosmic.

Considerable evidence indicates that the hippocampus is concerned with the storage of recent memories. Although hippocampectomized human patients can recall
information for seconds or even a few minutes after it is presented, other postopera-
tive events cannot be recalled. Ability to recall events occurring before the operation,
however, is unimpaired.

The hippocampus has been implicated in the elaboration of other behaviors besides
the storage of recent memories. Grastyan [9] and his collaborators provided evidence
that the hippocampal theta rhythm may be inhibitory and that one of the functions
of the hippocampus may be to prevent the occurrence of orienting responses to insignificant stimuli. Grastyan observed that during the first few presentations of sensory
stimuli in a conditioning experiment, desynchronization occurred in the absence of
theta activity. A reciprocal relationship between theta activity and desynchronized
EEG activity then developed after several trials. In other experiments, Grastyan and
his co-workers observed that cats who received electrical stimulation of the dorsal
hippocampus stopped all ongoing behavior and oriented toward a moving object.

Additional support for the hypothesis that the hippocampus may have important
inhibitory functions comes from a number of lines of inquiry. One of these is from
studies in which hippocampectomized and normal animals have been trained to avoid
shock. These studies have indicated that hippocampectomy inhibits passive-avoidance
behavior. In the passive-avoidance situation, a warning signal indicates that the animal's
drinking dish or feeding dish is electrified for a limited period of time. An animal displays a passive-avoidance response by refraining from eating or drinking during the

time that the signal is operative. The impaired performance in this situation that is observed following hippocampectomy suggests that the hippocampus is important for the inhibition of approach responses.

Another line of inquiry that implicates the hippocampus in the inhibition of responding has been reported by Douglas [6]. He observed that mice younger than 24 to 26 days that have not been operated upon can readily learn an active-avoidance response, but they cannot learn a passive-avoidance response. In the active-avoidance situation, the animal must make an appropriate response, such as a lever-press, in the presence of a signal in order to avoid a noxious shock. Douglas has suggested that the failure of his animals to learn the passive-avoidance task occurred because the complete organization of the hippocampus in young mice does not develop until after 24 days. Evidence of hippocampal immaturity in mice until at least three weeks after birth has been provided by autoradiographic studies in which tritiated thymidine has been used to radioactively mark neuroblasts. One way of testing Douglas' hypothesis concerning hippocampal influences upon inhibition would be to x-irradiate the hippocampus shortly after birth. Since x-irradiation would selectively damage the neuroblasts, the hippocampus in irradiated animals should remain immature and the animals should show deficits in performance on a passive-avoidance task even after several months.

A final hypothesis about hippocampal function that has been proposed is based upon observations that theta rhythms recorded from the entorhinal cortex or hippocampus may precede or accompany goal-directed motor acts. Theta waves recorded from the entorhinal cortex have been reported to appear just before cats begin to approach a goal box [14]. Hippocampal theta waves have also been reputed to accompany voluntary phasic responses in rats, such as bar-pressing or running [43]. In the experiment with rats, the theta rhythm did not accompany voluntary immobility nor involuntary, instinctive, and consummatory responses such as eating, drinking, grooming, or yawning.

THE NEOCORTEX: NONSENSORY, NONMOTOR FUNCTIONS

FRONTAL-LOBE SYNDROME

For many years, the frontal lobes were commonly believed to be the seat of intelligence. Examination of human patients suffering from frontal-lobe injury, however, indicates that such patients score normally on most standard intelligence tests. Instead of being considered as the basic integrating center for intelligence, the frontal lobes are better conceptualized as being involved in complex motivational processes.

In the late 1930s, Jacobsen examined the role of the anterior frontal cortex in primates [15]. He found that lesions in the anterior frontal cortex produced pronounced impairment of delayed-response tasks. In a typical delayed-response experiment, a monkey was shown a piece of food, which was then placed under one of two or more identical cups. After a delay, the monkey was permitted to choose one of the cups. Jacobsen's "frontal" monkeys did well on a variety of discrimination tasks, but they did very poorly whenever a delay was introduced. Jacobsen also trained chimpanzees to use sticks to reach for food. After ablation of the anterior frontal cortex, the chimpanzees could perform the response only if the food and stick were simultaneously in their visual field. Jacobsen concluded that the anterior lobes are necessary for immediate memory.

Subsequent experiments have indicated that the delayed-response deficit is not due to a loss of memory, but it is due, at least in part, to the hyperexcitability and distractibility caused by the lesion. "Frontal" monkeys, for instance, could perform delayed-response tasks successfully if they were kept in the dark during the delay intervals [22]. Monkeys tested while under barbiturates also did well after anterior frontal lobe ablation [47]. Interestingly, the delayed-response deficit does not appear in humans when they are tested on simple tasks. Presumably, this is because of the ability of humans to use verbal mediation to bridge the temporal delays.

In addition to their hyperexcitability and distractibility, animals with frontal lobe damage tend to relapse to perseverational behavior once they are distracted. When "frontal" monkeys are allowed to make a single, rewarded response before the delay, they show no deficit. Also, if the monkey has only to determine whether or not a cup is baited rather than which of two cups to choose, the "frontal" animal usually succeeds.

The perseverational nature of the behavioral deficit following frontal lobectomy can also be shown by means of delayed-alternation and reversal-learning tasks. In the alternation task, for example, the animal must learn that the food reward is under the cup that was *not* baited on the previous trial. Unlike the deficit on the choice-delay tasks previously described, deficits on the delayed-alternation task show up following lesions of structures other than the anterior frontal cortex, including the cingulate cortex, the hippocampus, and the head of the caudate nucleus. These structures, however, are closely related to the frontal cortex.

Human patients with damaged anterior frontal lobes have difficulty in performing a card-sorting task [24]. In this task, the patient has to change his or her set whenever the experimenter pronounces the previous mode of responding incorrectly. Thus, if "clubs" are the correct suite and the experimenter switches to "diamonds," the subject has to learn to recognize that "clubs" are now incorrect and that "diamonds" are correct. Patients with damaged anterior frontal lobes have a great deal of difficulty on this task, although they show little or no deficit either in short-term memory or in their performance on a standard IQ test. Data such as these suggest that the anterior frontal cortex functions to institute flexibility in well-organized subcortical motivational systems.

SPEECH, LANGUAGE, AND THE LATERALIZATION OF FUNCTION

In most people, disturbances of speech, termed *dysphasia,* follow damage to the lateral surface of the left cerebral hemisphere. The most vulnerable areas are those in the third frontal convolution, called *Broca's area,* and in the posterior temporoparietal region, known as *Wernicke's area* (Fig. 30-5).

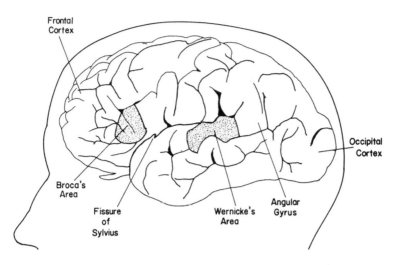

Figure 30-5
Lateral view of the human cortex showing Broca's area and Wernicke's area.

In general, frontal lesions tend to produce deficits in speech production that are known as *expressive dysphasias,* whereas lesions in Wernicke's area tend to produce deficits in language comprehension known as *receptive dysphasias.* The distinctions that allocate different types of dysphasia to various brain areas, however, are far from clear-cut. Penfield and Roberts [30], for example, found that dysphasias resulting from electrical stimulation of the cortex were often quite specific within a particular patient, but that there was very little tendency for one type to appear exclusively in a particular area among different patients. A few impairments of speech ability do seem to be fairly well localized. Luria's [20] "dynamic aphasia," in which the patient has reduced spontaneous speech and poor word fluency, is caused by injury to the frontal lobe on the side that is dominant for speech. In general, however, lesions restricted to one of the classic speech areas leave only very mild residual dysphasic symptoms.

In right-handed people, the left hemisphere is dominant for speech; this is also true for most left-handed and ambidextrous people. Evidence is beginning to accumulate, however, that indicates that the lateralization of speech is less complete in left-handed people. One of the interesting characteristics of the associative areas of the neocortex is their functional asymmetry. Whereas the perception and learning of verbal material appears to take place largely in the left cerebral hemisphere, the perception and retention of music and geometrical forms appear to take place largely in the right cerebral hemisphere. In recent years, experiments involving section of the corpus callosum, which is the major communications channel between the two hemispheres, have provided important information about functional asymmetry.

The so-called *split-brain operation* is performed to prevent massive epileptic seizures in one hemisphere from invading the other. In order to accomplish this, the corpus callosum, the anterior commissure, and the hippocampal commissure are sectioned.

The operation produces patients whose hemispheres have been isolated from each other except through the lower brain stem. Because of the organization of the visual system, which provides input from each eye to both hemispheres, little if any behavioral deficit is ordinarily observed after the split-brain operation.

In an experimental situation, however, Sperry and Gazzaniga [37] restricted the visual input to the different eye fields of split-brain patients in order to examine the consequences of (1) the right hemisphere having no speech mechanism and (2) factual information from the left hand going only to the right hemisphere. Sperry and Gazzaniga found that subjects could not identify an object placed in their left hand, but they could identify it when it was placed in their right hand. Similarly, if an image was projected to the right occipital cortex, the subjects would report seeing a flash, but they could not identify what they saw. One method that Sperry and Gazzaniga used to demonstrate that the right hemisphere did receive visual input was to flash pictures of a nude girl. In this case, the patient's emotional response was unmistakable, but he could not tell the experimenter what caused it.

Sperry and Gazzaniga also found that patients who could not identify an object placed in the left hand could still use this hand to select a writing instrument from an array of objects when instructed to do so. Thus, the right cerebral hemisphere can preside over the performance of simple tasks involving speech comprehension. This is hardly surprising when it is considered that nonhumans lacking speech can also do these tasks. What is puzzling, however, is why patients with intact commissures, but with lesions restricted to the left cerebral hemisphere, show receptive dysphasias or asphasias for simple speech or writing. One possibility is that the intact portions of the damaged left hemisphere may inhibit the right hemisphere.

REFERENCES

1. Aserinsky, E., and Kleitman, N. Regularly occurring periods of eye motility and concomitant phenomena during sleep. *Science* 118:273, 1953.
2. Brady, J. V., and Nauta, W. J. H. Subcortical mechanisms in emotional behavior: The duration of effective changes following septal and habenular lesions in the albino rat. *J. Comp. Physiol. Psychol.* 48:412, 1955.
3. Cagguila, A. R. Analysis of the copulation-reward properties of posterior hypothalamic stimulation in male rats. *J. Comp. Physiol. Psychol.* 70:399, 1970.
4. Corbit, J. Behavioral regulation of hypothalamic temperature. *Science* 166:256, 1970.
5. Delgado, J. M. R., Roberts, W. W., and Miller, N. E. Learning motivated by electrical stimulation of the brain. *Am. J. Physiol.* 179:587, 1954.
6. Douglas, R. J. The hippocampus and behavior. *Psychol. Bull.* 67:416, 1967.
7. Driefuss, J. J., Murphy, J. T., and Gloor, P. Contrasting effects of two identified amygdaloid efferent pathways on single hypothalamic neurons. *J. Neurophysiol.* 31:237, 1968.
8. Goddard, G. V. Functions of the amygdala. *Psychol. Bull.* 62:89, 1964.
9. Grastyan, E. The Hippocampus and Higher Nervous Activity. In M. A. B. Brazier (Ed.), *Central Nervous System and Behavior.* Washington, D.C.: Josiah Macy Foundation, 1959. Pp. 119–205.

10. Green, J. D., Clemente, C. D., and deGroot, J. Rhinencephalic lesions and behavior in cats. *J. Comp. Neurol.* 108:505, 1957.
11. Harris, G. W., and Michael, R. P. The activation of sexual behavior by hypothalamic implants of oestroegen. *J. Physiol.* (Lond.) 171:275, 1964.
12. Hess, W. R. Le sommeil. *C. R. Soc. Biol.* (Paris) 107:1333, 1931.
13. Hilton, S. M., and Zbrozyna, A. W. Amygdaloid region for defense reactions and its efferent pathway to the brain stem. *J. Physiol.* (Lond.) 165:160, 1963.
14. Holmes, J. E., and Adey, W. R. Electrical activity of the entorhinal cortex during conditional behavior. *Am. J. Physiol.* 199:741, 1960.
15. Jacobsen, C. F. Studies of cerebral function in primates: I. The function of the frontal association areas in monkeys. *Comp. Psychol. Monogr.* 13:3, 1936.
16. Jouvet, M. Biogenic amines and the states of sleep. *Science* 163:32, 1969.
17. Klüver, H., and Bucy, P. C. Preliminary analysis of functions of the temporal lobes in monkeys. *Arch. Neurol. Psychiatr.* 42:979, 1939.
18. Leibowitz, S. F. Reciprocal hunger-regulating circuits involving alpha- and beta-adrenergic receptors located respectively, in the ventromedial and lateral hypothalamus. *Proc. Natl. Acad. Sci. U.S.A.* 67:1063, 1970.
19. Leibowitz, S. F. Hypothalamic alpha- and beta-adrenergic systems regulating both thirst and hunger in the rat. *Proc. Natl. Acad. Sci. U.S.A.* 68:332, 1971.
20. Luria, A. R. Factors and Forms of Aphasia. In A. V. S. de Reuck and M. O'Connor (Eds.), *CIBA Foundation Symposium on Disorders of Language.* London: Churchill, 1964. Pp. 143–167.
21. MacLean, P. D. Psychosomatics. In J. Field, H. W. Magoun, and V. E. Hall (Eds.), *Handbook of Physiology, Section 1, Neurophysiology,* Vol. 3. Washington, D.C.: The American Physiological Society, 1958. Pp. 1723–1744.
22. Malmo, R. B. Interference factors in delayed response in monkeys after removal of frontal lobes. *J. Neurophysiol.* 5:295, 1942.
23. McCleary, R. A. Response specificity in the behavioral effects of limbic system lesions in the cat. *J. Comp. Physiol. Psychol.* 54:605, 1961.
24. Milner, B. Effects of different brain lesions on card sorting. *Arch. Neurol.* 9:90, 1963.
25. Moruzzi, G., and Magoun, H. W. Brain stem reticular formation and activation of the EEG. *Electroencephalogr. Clin. Neurophysiol.* 1:455, 1949.
26. Narabayashi, H., Nagao, T., Saito, Y., Yoshida, M., and Nagahata, M. Stereotaxis amygdalotomy for behavior disorders. *Arch. Neurol.* 9:1, 1963.
27. Olds, J., and Milner, P. Positive reinforcement produced by electrical stimulation of septal area and other regions of rat brain. *J. Comp. Physiol. Psychol.* 47:419, 1954.
28. Peck, J. W., and Novin, D. Evidence that osmoreceptors mediating drinking in rabbits are in the internal preoptic area. *J. Comp. Physiol. Psychol.* 74:134, 1971.
29. Penfield, W., and Jasper, H. *Epilepsy and the Functional Anatomy of the Human Brain.* Boston: Little, Brown, 1954.
30. Penfield, W., and Roberts, L. *Speech and Brain Mechanisms.* Princeton, N.J.: Princeton University Press, 1959.
31. Porter, R. W., Cavanaugh, E. G., Critchlow, B. V., and Sawyer, C. H. Localized changes in electrical activity of the hypothalamus in estrous cats following vaginal stimulation. *Am. J. Physiol.* 189:145, 1957.
32. Ranson, S. W. Regulation of body temperature. *Res. Publ. Assoc. Res. Nerv. Ment. Dis.* 20:342, 1940.
33. Rosvold, H. E., Mirsky, A. F., and Pribram, K. H. Influence of amygdalectomy on social interaction in a monkey group. *J. Comp. Physiol. Psychol.* 47:173, 1954.

34. Satinoff, E., and Rutstein, J. Behavioral thermoregulation in rats with anterior hypothalamic lesions. *J. Comp. Physiol. Psychol.* 71:77, 1970.
35. Sawyer, C. H., and Robinson, B. Separate hypothalamic areas controlling pituitary gonadotrophic function and mating behavior in female cats and rabbits. *J. Clin. Endocrinol.* 16:914, 1956.
36. Smith, D. E., King, M. B., and Hoebel, B. G. Killing: Cholinergic control in the lateral hypothalamus. *Proc. 77th Annu. Conv. Am. Psychol. Assoc.* 4:875, 1967.
37. Sperry, R. W., and Gazzaniga, M. S. Language Following Disconnection of the Hemispheres. In F. L. Darley (Ed.), *Brain Mechanisms Underlying Speech and Language.* New York: Grune & Stratton, 1967.
38. Stein, L. Reciprocal Action of Reward and Punishment Mechanisms. In J. Heath (Ed.), *The Role of Pleasure in Behavior.* New York: Hoeber Med. Div., Harper & Row, 1964. P. 113.
39. Stein, L., and Wise, C. D. Release of norepinephrine from hypothalamus and amygdala by rewarding stimulation of the medial forebrain bundle. *J. Comp. Physiol. Psychol.* 67:189, 1969.
40. Stokman, C. L. J., and Glusman, M. Amygdaloid modulation of hypothalamic flight in cats. *J. Comp. Physiol. Psychol.* 71:365, 1970.
41. Teitelbaum, P., and Epstein, A. N. The lateral hypothalamic syndrome: Recovery of feeding and drinking after lateral hypothalamic lesions. *Psychol. Rev.* 69:74, 1962.
42. Valenstein, E. S., Cox, V. C., and Kakolewski, J. W. The role of the hypothalamus. *Psychol. Rev.* 77:16, 1970.
43. Vanderwolf, C. H. Hippocampal Electrical Activity and Voluntary Movement in the Rat. *Tech. Report No. 17.* MacMaster University, Hamilton, Ontario, Canada, 1968.
44. Vaughan, E., and Fisher, A. E. Male sexual behavior induced by intracranial electrical stimulation. *Science* 137:758, 1962.
45. Verney, E. B. The antidiuretic hormone and the factors which determine its release. *Proc. R. Soc. Lond.* [Biol.] 135:25, 1947.
46. Von Holst, E., and Von Saint Paul, U. On the functional organization of drives. *Anim. Behav.* 11:1, 1963.
47. Wade, M. The effect of sedatives upon delayed response in monkeys following removal of the prefrontal lobes. *J. Neurophysiol.* 10:57, 1947.

Membranes and Neuronal Models 31

Donald R. Humphrey

INTRODUCTION

A *model* of a biologic system is itself a physical or a mathematical system that has been constructed in such a way that the behavior of certain of its variables will quantitatively parallel that of *selected* variables within the biologic prototype. As one may assemble an analog computing network, for example, in which the output voltage mimics the displacement behavior of a mechanical system, so might one construct mathematical or electrical analogs whose major parameters, inputs, and outputs correspond *quantitatively* to those of a particular biologic system. It is the purpose of this and the subsequent chapter to discuss briefly the general functions of such models, to indicate the general levels of analysis within neurophysiology at which modeling attempts are carried out, and to discuss in some detail a small number of contemporary models that have been of major importance in promoting an understanding of neuronal input-output relations.

In the attempt to understand the basic properties of nervous systems and how they process information, experimental and theoretical studies have been carried out on at least four different levels. These are the subcellular, single-unit, network, and systems levels. The present chapter will be concerned with the first two of these, and Chapter 32 will be addressed to the latter two levels.

At the first, or *subcellular,* level are those studies that are concerned with the basic properties of neuronal membranes (see Chaps. 5 and 26) and their mathematical representations. Attention has focused in the past upon the mechanisms that are responsible for maintaining a cell's resting potential, upon the changes in a membrane's ionic conductance that are produced by synaptic transmitters, and upon the ionic mechanisms that underlie the action potential. Many of the contemporary models at this level are based upon the experimental work of Fatt and Katz [8], of Eccles and his colleagues [5, 7], and of Hodgkin and Huxley [14d, 16]. They include Rall's mathematical formulation for the electrical behavior of postsynaptic membranes [29], Hodgkin's and Huxley's mathematical model of the action potential [15], and the electronic implementation of this model developed and explored by Lewis [20, 21].

At the second, or *single-unit,* level are those studies that are concerned primarily with the input-output functions of single neurons (see Chap. 32). The principal goal is to obtain a quantitative description of the major input-output transformations performed by a neuron, so that it might be conveniently treated as an element in studies of the behavior of interconnected cells. Because it is often impossible to measure accurately or to control all the inputs to a real neuron, much of our knowledge of neural input-output relations has come from simulation experiments with mathematical or electronic model neurons. Contemporary neuron models that have yielded important insights, and which are thus of particular interest, are the lumped-parameter models of Harmon [14], Lewis [22], and Perkel et al. [26] as well as the distributed-parameter model developed by Rall [29, 30], which takes into account a cell's dendritic tree and the distribution of synapses over its surface. In addition to these models, there have been a number of recent, noteworthy attempts to quantify the input-output relations of single sensory receptors with the use of systems analysis techniques [14c, 16a, 27a, 35].

MODELS OF PASSIVE, SUBSYNAPTIC, AND EXCITABLE PATCHES OF MEMBRANE

The schematic drawing of a neuron shown in Figure 31-1 may be used to point out some important general properties of mammalian cells. In the absence of a synaptic input, a typical neuron is assumed to have a resting transmembrane potential that is constant over its entire surface. Evidence indicates, however, that the different membrane regions of a single cell may differ significantly with respect to their excitability properties. Studies of spinal motoneurons and other cells, for example, have shown that the initial segment has a lower threshold for generating an action potential than does the cell body, and these studies have suggested that the dendrites may be electrically inexcitable [12, 23]. In addition, synaptic terminals are distributed over a neuron's soma and its dendrites, and transmitters may act on each small patch of subsynaptic membrane to produce specific changes in its ionic conductance.

A single neuron is composed, then, of three spatially distributed and electically distinct regions of membrane. The first, or *passive,* region is made up of these soma-dendritic areas that are not covered by synaptic terminals and that do not generate action potentials; for modeling purposes, the values of its resistance and capacitance per unit area are usually considered constant. The second, or *synaptic,* membrane region is both anatomically intermingled and electrically in parallel with the first; it consists of the collection of subsynaptic membrane sites that may undergo increases in ionic conductance. In modeling the electrical behavior of this type of membrane,

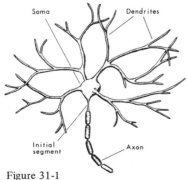

Figure 31-1
Schematic drawing of a neuron.

its conductance must be considered to be a function of the synaptic input, and the equilibrium potentials of the excitatory postsynaptic potential (EPSP) and the inhibitory postsynaptic potential (IPSP) must be taken into account. The third, or *excitable,* membrane region is made up of those areas of the cell, generally the soma and the initial segment, that may generate action potentials. By analogy with studies on the squid axon, the ionic conductances in this region will be functions of membrane potential as well as of time.

In order to simulate accurately the electrical behavior of an entire cell, it is therefore necessary to have an electrical model for each of these types of membrane. Let us turn now to an examination of the major models that have been used for this purpose. Unless indicated otherwise, the circuits and equations used in this section will be valid only for a patch of membrane that is sufficiently small that the instantaneous potentials and the transverse current density are constant over its surface.

EQUIVALENT CIRCUITS FOR PASSIVE AND RESTING MEMBRANE

In Chapter 5, Figure 5-2B shows an equivalent circuit for a small patch of neural membrane that can be derived from a consideration of the membrane's dielectric properties and the passive electrochemical forces that act to produce transmembrane ionic currents or fluxes [32]. The voltage sources (V_{Na}, V_K, V_{Cl}) represent the equilibrium potentials for sodium, potassium, and chloride ions, respectively, and the series resistances ($R_{Na} = 1/G_{Na}$, $R_K = 1/G_K$, $R_{Cl} = 1/G_{Cl}$) represent the *resting* membrane's resistance to various ionic currents.

Figure 5-2C shows the equivalent circuit for a patch of passive membrane that has been derived from numerous experimental studies of the subthreshold electrical behavior of nerve cells and axons. The voltage source V_r represents the resting potential, and R_r (= $1/G_r$) is the transverse resistance of a unit area of passive membrane. Both circuits have been used for modeling purposes, and it will be appropriate to write down the equations that describe them and to show that the two circuits are consistent.

Using circuit B (Fig. 5-2), the net transmembrane current density (J_m) will be equal to the sum of the capacitative (J_C) and ionic components (J_{Na}, J_K, J_{Cl}):

$$J_m = J_C + J_{Na} + J_K + J_{Cl} \tag{31-1}$$

Or, equivalently,

$$J_m = C_m \frac{\delta V_m}{\delta t} + G_{Na}(V_m - V_{Na}) + G_K(V_m - V_K) + G_{Cl}(V_m - V_{Cl}) \tag{31-2}$$

where C_m is the membrane capacitance per unit area of membrane, G is the conductance per unit area, $V_m = V_{int} - V_{ext}$ is the transmembrane potential difference, and

J_m is positive for outwardly directed current. Equation 31-2 is easily rearranged to obtain:

$$J_m = C_m \frac{\delta V_m}{\delta t} + (G_{Na} + G_K + G_{Cl}) \left(V_m - \frac{G_{Na} V_{Na} + G_K V_K + G_{Cl} V_{Cl}}{G_{Na} + G_K + G_{Cl}} \right) \qquad (31\text{-}3)$$

If we now define:

$$G_r = G_{Na} + G_K + G_{Cl} \qquad (31\text{-}4)$$

and:

$$V_r = \frac{G_{Na} V_{Na} + G_K V_K + G_{Cl} V_{Cl}}{G_{Na} + G_K + G_{Cl}} \qquad (31\text{-}5)$$

then Equation 31-3 reduces to:

$$J_m = C_m \frac{\delta V_m}{\delta t} + G_r(V_m - V_r) \qquad (31\text{-}6)$$

which is the equation for the transmembrane current density given by circuit C in Figure 5-2. Note that under steady-state conditions and in the absence of applied currents, $J_m = \delta V_m / \delta t = 0$, and Equation 31-6 reduces to $V_m = V_r$, as is required. In deriving Equation 31-6, in which G_r and V_r are constants, we have, of course, assumed that the ionic conductances are constant. It is also generally assumed that C_m is constant over the entire surface of the neuron and is thus independent of regional variations in membrane excitability. Experimental studies have shown that Equation 31-6 may be applied to excitable as well as to strictly passive regions of the neural membrane, provided that attention is confined to very small, subthreshold changes in V_m.

EQUIVALENT CIRCUITS FOR SUBSYNAPTIC MEMBRANE

Beginning with the important work of Fatt and Katz [8] and of Coombs et al. [5], experimental studies have outlined the major electrical events involved in postsynaptic excitation and inhibition [7, 25, 33]. When invaded by an impulse, a presynaptic terminal releases a small quantity of transmitter substance, which acts upon the post-synaptic or subsynaptic membrane to produce a very localized and specific increase in its ionic conductance. EPSPs appear to result from transient increases in G_{Na} and G_K, which thus allow a net ionic current to flow inward across the subsynaptic membrane and to neutralize some of the negative charge on the inner surface of adjacent regions of the membrane. IPSPs, on the other hand, appear to result from transient increases in G_K and G_{Cl}, and a net flow of membrane current in the opposite direction. Each type of PSP reverses its direction (from depolarizing to hyperpolarizing, or vice versa) at a particular background level of V_m, which is referred to as its *equilibrium potential.*

The exact relationship between the magnitude of a conductance increase and the transmitter concentration is not known; however, the two variables are positively correlated, and it is often assumed that they are proportional. In addition, studies have shown that a brief presynaptic volley produces a conductance transient in the postsynaptic membrane that consists of an abrupt increase in conductance followed by an exponential decay [7]. On the assumption of proportionality, this transient would reflect the abrupt release and subsequent removal or enzymatic breakdown of the transmitter substance. Regardless of the interpretation, however, these results indicate that the conductance transients evoked by a single presynaptic spike can be approximated by simple, decaying exponential functions.* If the time-constant of the postsynaptic membrane is long compared to the duration of transmitter action, the transient can in fact be approximated by a brief pulse, or a Dirac delta function, with little resultant error.

An equivalent ionic circuit for a small patch of subsynaptic membrane is shown in Figure 31-2A; it is similar to the circuit of Figure 5-2B, but in this case the ionic conductances are considered to be functions of the transmitter concentration or of the presynaptic firing rate and time. It will again be appropriate to show how simplified representations of excitatory and inhibitory subsynaptic membranes can be derived from the general ionic model in Figure 31-2A. To do so, let us first equate each ionic conductance function to the sum of its constant resting value, G', and the time-varying increment from this value, $\Delta G(t)$, that is produced by synaptic bombardment:

$$G_{Na}(t) = G_{Na}' + \Delta G_{Na}(t) \tag{31-7}$$

$$G_K(t) = G_K' + \Delta G_K(t) \tag{31-8}$$

$$G_{Cl}(t) = G_{Cl}' + \Delta G_{Cl}(t) \tag{31-9}$$

Consider now the subsynaptic membrane at an *excitatory* synapse. From the circuit of Figure 31-2A and the definitions given above, the expression for the transmembrane current density becomes:

$$J_m = \left[C_m \frac{\delta V_m}{\delta t} + G_{Na}'(V_m - V_{Na}) + G_K'(V_m - V_K) + G_{Cl}'(V_m - V_{Cl}) \right] \tag{31-10}$$

$$+ \left[\Delta G_{Na}(V_m - V_{Na}) + \Delta G_K(V_m - V_K) \right]$$

*The amplitudes of the conductance transients evoked by successive spikes need not, however, be constant. By confining our attention to the postsynaptic membrane, we have overlooked the fact that the amount of transmitter released following a spike will depend upon the presynaptic terminal's past history or firing rate. This dependence is evidenced by such phenomena as presynaptic facilitation, depression, and posttetanic potentiation [7, 33].

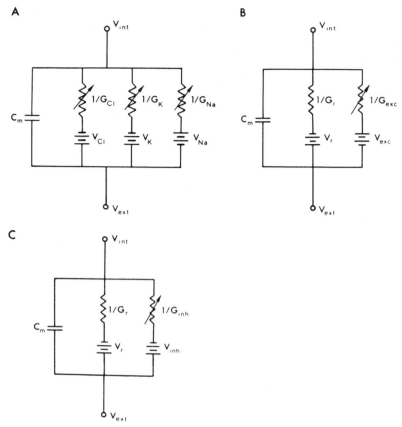

Figure 31-2

Equivalent circuit representations for a unit area of subsynaptic membrane. *A.* Representation in terms of variable ionic conductances, whose values depend upon the subsynaptic transmitter concentration and type. *B.* Representation of the membrane at an excitatory synapse (G_r = the resting membrane conductance; G_{exc} = a variable, depolarizing conductance; V_{exc} = the equilibrium potential for the EPSP). *C.* A similar circuit for the membrane at an inhibitory synapse (V_{inh} = the equilibrium potential for the IPSP). The mathematical relationships between the various representations are discussed in the text.

where the constant and incremental conductance terms have been separated. The first expression in brackets on the right represents the fraction of the membrane current density (J_r) that is attributable to the membrane capacity and the resting values of the ionic conductances; the second expression represents the synaptic current density (J_{syn}) that results from a transmitter-induced "opening" or increase in the sodium and potassium conductance channels. From Equations 31-2 to 31-6, it can be seen that:

$$J_r = C_m \frac{\delta V_m}{\delta t} + G_r(V_m - V_r) \tag{31-11A}$$

The second bracketed expression in Equation 31-10 can be simplified by noting that,

$$J_{syn} = (\Delta G_{Na} + \Delta G_K) \left(V_m - \frac{\Delta G_{Na} V_{Na} + \Delta G_K V_K}{\Delta G_{Na} + \Delta G_K} \right) \tag{31-11B}$$

and by defining:

$$\Delta G_{exc} = \Delta G_{Na} + \Delta G_K \tag{31-12A}$$

which is the increase in membrane conductance evoked at each excitatory synapse; and

$$V_{exc} = \frac{\Delta G_{Na} V_{Na} + \Delta G_K V_K}{\Delta G_{Na} + \Delta G_K}$$

which is the reversal or equilibrium potential for the EPSP. With these definitions, Equation 31-11B reduces to

$$J_{syn} = \Delta G_{exc}(V_m - V_{exc}) \tag{31-12B}$$

and Equation 31-10 can be placed in the simple form:

$$J_m = C_m \frac{\delta V_m}{\delta t} + G_r(V_m - V_r) + \Delta G_{exc}(V_m - V_{exc}) \tag{31-13}$$

Thus, the simple equivalent circuit shown in Figure 31-1B can be used to represent the subsynaptic membrane at an excitatory synapse.

In a similar manner, it is possible to define the membrane conductance increment at an inhibitory synapse and the reversal potential for the IPSP by the expressions

$$\Delta G_{inh} = \Delta G_K + \Delta G_{Cl} \tag{31-14}$$

$$V_{inh} = \frac{\Delta G_K V_K + \Delta G_{Cl} V_{Cl}}{\Delta G_K + \Delta G_{Cl}}$$

and arrive at the following equation for the subsynaptic membrane current density at each *inhibitory* synapse:

$$J_m = C_m \frac{\delta V_m}{\delta t} + G_r(V_m - V_r) + \Delta G_{inh}(V_m - V_{inh}) \tag{31-15}$$

The equivalent circuit described by this equation is shown in Figure 31-2C.

Although they are quite simple and are applicable only to a membrane patch or certain simple neuronal geometries,* Equations 31-13 and 31-15 compactly summarize a number of important properties of PSPs that have been observed experimentally. For example, in the absence of applied currents, $J_m = 0$, and at the peak of a PSP, $\delta V_m/\delta t = 0$; consequently Equation 31-15 can be rearranged to obtain:

$$\Delta V_m = \frac{\Delta G_{inh}(V_{inh} - V_r)}{\Delta G_{inh} + G_r} \tag{31-16}$$

where $\Delta V_m = V_m - V_r$ and, in this particular case, represents the peak amplitude of the IPSP when measured from the value of V_r. An analogous equation can be derived from Equation 31-13 for the excitatory synapse under the same conditions. A number of important conclusions can be drawn from an examination of Equation 31-16:

1. The occurrence of an IPSP and its direction (i.e., whether depolarizing or hyper-polarizing) depend upon the relationship between V_{inh} and V_r. In mammalian motoneurons, $V_{inh} \approx -80$ mV and $V_r \approx -70$ mV; consequently, ΔV_m is in the hyperpolarizing direction, but this need not be true for all cells.
2. The magnitude of ΔV_m is linearly related to the difference between the equilibrium potential for the PSP and the background or resting level of V_m, i.e., V_r. In most well-studied neurons, $V_e \approx 0$ mV; consequently, the absolute value of $(V_{exc} - V_r)$ is much larger than that of $(V_{inh} - V_r)$, and the amplitude of the EPSP evoked by an excitatory conductance increment will be much larger than that evoked by an equal increment in inhibitory conductance.
3. The amplitude of a PSP is *not* linearly related to the amplitude of the underlying conductance increase, as can be seen from the expression:

$$\frac{\delta(\Delta V_m)}{\delta(\Delta G_{inh})} = \frac{G_r}{(\Delta G_{inh} + G_r)^2}(V_{inh} - V_r)$$

Note, however, that the error introduced by assuming proportionality is minimal when the conductance increments are small compared with G_r.

RALL'S EQUIVALENT CIRCUIT FOR A UNIT AREA OF DENDRITIC MEMBRANE

At this point, it will be appropriate to show how the various circuits described above can be combined into an equivalent circuit for a unit area of *inexcitable* dendritic

---

*As will be shown in Chapter 32, the "membrane patch" equations are valid also for spherically shaped neurons without dendrites and other structures in which the transmembrane current flow is not associated with potential gradients along the internal and external surfaces of the membrane.

membrane that is covered partially by excitatory and partially by inhibitory synaptic terminals. To do so, let us first define:

$\alpha(t)$ = the proportion of a unit area of membrane that is covered at any time, t, by active excitatory terminals

$\beta(t)$ = the proportion covered at any time, t, by active inhibitory terminals

$(1-\alpha-\beta)$ = the proportion that is not covered by active terminals and is therefore strictly passive

The net transmembrane current density (J_m) will now be equal to the sum of the current densities across the three subareas of membrane:

$$J_m = (1 - \alpha - \beta)J_{pass} + \alpha J_{exc} + \beta J_{inh} \tag{31-17}$$

where J_{pass} is given by the expression for J_m in Equation 31-6, J_{exc} by that in Equation 31-13, and J_{inh} by that in Equation 31-15. If we now insert the expressions for the various current densities and, for compactness, define $\alpha(t)\Delta G_{exc}$ as G_{exc} and $\beta(t)\Delta G_{inh}$ as G_{inh}, we obtain:

$$J_m = C_m \frac{\delta V_m}{\delta t} + G_r(V_m - V_r) + G_{exc}(V_m - V_{exc}) + G_{inh}(V_m - V_{inh}) \tag{31-18}$$

Note that C_m, G_r, V_r, V_{exc}, and V_{inh} are constants, whereas G_{exc} and G_{inh} are functions of the synaptic input density and time. Note also that in the absence of synaptic input, $G_{exc} = G_{inh} = 0$, and Equation 31-18 reduces to the equation for a passive patch of membrane (Eq. 31-6), as is required.

The equivalent electrical circuit representation of Equation 31-19 is shown in Figure 31-3. Although G_r, G_{exc}, and G_{inh} were not defined explicitly in terms of ionic conductances (as has been done above), this circuit was first presented and used by Rall in his theoretical studies of PSP distributions in model neurons with dendritic trees [29, 30]. It will be instructive to briefly examine the responses of this model to simple forcing functions, since it can also be used to approximate the subthreshold behavior of an entire neuron with either a nonexistent or a negligible dendritic tree.

In the absence of extraneous sources, any current that flows through the membrane as a result of increases in G_{exc} or G_{inh} must return through G_r and C_m; consequently, $J_m = 0$, and Equation 31-18 may be divided by G_r to obtain:

$$0 = \tau_m \frac{\delta V_m}{\delta t} + (V_m - V_r) + \xi(V_m - V_{exc}) + \zeta(V_m - V_{inh}) \tag{31-19}$$

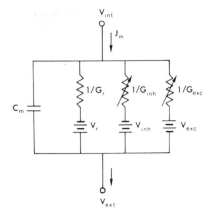

Figure 31-3
Rall's equivalent circuit for a unit area of dendritic membrane that contains both excitatory and inhibitory subsynaptic sites.

where $\tau_m = C_m/G_r$, the time-constant of resting or passive membrane   (31-20)
$\xi = G_{exc}/G_r$
$\zeta = G_{inh}/G_r$

If we now introduce the new variables,

$$T = t/\tau_m \tag{31-21}$$

and

$$V = \Delta V_m = V_m - V_r \tag{31-22}$$

then Equation 31-19 may be expressed in the convenient form:

$$\frac{\delta V}{\delta T} + (1 + \xi + \zeta)V = \xi(V_{exc} - V_r) + \zeta(V_{inh} - V_r) \tag{31-23}$$

For simple delta-function conductance transients $\xi = K_{exc}\,\Delta(T)$ and $\zeta = K_{inh}\,\Delta(T)$, where K_{exc} and K_{inh} are constants, and for the initial condition $V(0) = 0$, the solution to Equation 31-23 may be expressed as:

$$V(T) = [K_{exc}(V_{exc} - V_r) + K_{inh}(V_{inh} - V_r)]\, u(T) \exp(-T - K_{exc} - K_{inh}) \tag{31-24}$$

where $u(T)$ is the unit step function and is defined here as zero for $T \leqslant 0$.

   Plots of Equation 31-24 for various combinations of K_{exc} and K_{inh} are shown in Figure 31-4A. Note that despite the simplified conductance transients, which produce

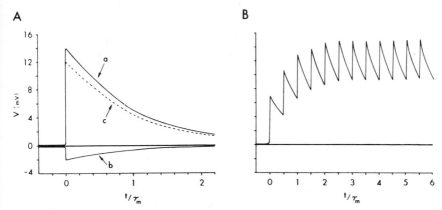

Figure 31-4
Theoretical postsynaptic potentials produced by delta-function conductance inputs to Rall's dendritic membrane model. *A.* Theoretical EPSPs (curve *a*) and IPSPs (curve *b*) computed from Equation 31-24 with the following parameter values: $V_r = -70$ mV; $V_{exc} = 0$ mV; $V_{inh} = -80$ mV; $\xi = 0.2\Delta(T)$ for curve *a* and $\zeta = 0.2\Delta(T)$ for curve *b*. Curve *c* represents the solution obtained when both ξ and ζ occur simultaneously. *B.* Temporal summation of the EPSPs evoked by a simulated step increase in the presynaptic firing rate. Each presynaptic spike was assumed to produce a delta-function conductance increase of intensity $\zeta = 0.1\Delta(T)$.

a rising phase that is too rapid, the PSPs generated by this simple model are not unlike those observed in real neurons. The EPSP produced by a given increment in the model's excitatory conductance (G_{exc}) is considerably larger than that produced by an equal increment in G_{inh} for reasons that have been discussed above; consequently, simultaneous and equal increases in the two conductances lead to a depolarizing potential that differs only slightly from that produced by an increment in G_{exc} alone. This type of amplitude discrepancy has been observed experimentally, and it is often impossible for an investigator to determine whether a depolarizing potential recorded from a real neuron is a pure EPSP or whether it reflects a mixture of excitatory and inhibitory bombardment.

    If the conductance transient evoked by each presynaptic spike is again approximated by a delta function, the response of the model to a presynaptic spike train, $P(T)$, can be obtained by simply convolving the spike train with Equation 31-24. For a spike train that produces excitatory conductance changes only, the appropriate expression becomes:

$$V(T) = K_{exc}(V_{exc} - V_r) \int_0^T u(t') \exp(-t' - K_{exc}) P(T - t') \, dt' \qquad (31\text{-}25\text{A})$$

For a step increase in the presynaptic firing frequency, f, from $f = 0$ at $T = 0$ to $f = 2/\tau_m$, the pulse train is given by:

$$P(T) = \sum_{n=0}^{\infty} \Delta(T - n/2) \qquad (31\text{-}25\text{B})$$

and the solution to Equation 31-25 that is illustrated in Figure 31-4B is easily obtained.
Note that the theoretical PSPs exhibit temporal summation, a phenomenon that is
seen, of course, in real neurons. Thus, even this simple model of a single patch of
dendritic membrane exhibits electrical phenomena that are qualitatively similar to
those observed in real cells; we shall return to it when we discuss the techniques employed by various investigators in modeling the electrical properties of an entire neuron.

THE HODGKIN-HUXLEY MODEL OF EXCITABLE MEMBRANE

The quantitative model of the action potential developed by Hodgkin and Huxley is
perhaps the most well-known, experimentally tested, and widely accepted neural
model in existence; moreover, their studies constitute one of the most elegant examples in physiology of a successful interplay between experiment and theory [14d–16].
It will therefore be appropriate to discuss some of the mathematical reasoning behind
their experiments, and the techniques they employed, as well as the pertinent features
of their model.

By 1950, a number of studies had shown that the general cable equation (see
Chap. 26) could be used to describe the subthreshold electrical behavior of unmyelinated axons, and that a propagating action potential was associated with a propagated
increase in membrane conductance [4, 6]. It thus appeared that the same general
equation could be applied in the case of an action potential, provided that the membrane conductance (G_m) was considered to be a variable and a function of V_m, distance
(x), and time (t). If so, the following relations would be expected to hold:

$$J_m(x,t) = \frac{-r_a}{2} \frac{\delta J_{lo}}{\delta x} = \frac{-r_a}{2} \frac{\delta}{\delta x} \left(-\frac{1}{\rho_{int}} \frac{\delta V_{int}}{\delta x} \right) \tag{31-26}$$

$$J_m(x,t) = C_m \frac{\delta V_m}{\delta t} + G_m(V_m,x,t)(V_m - V_r) \tag{31-27}$$

$$\frac{r_a}{2\rho_{int}} \frac{\delta^2 V_m}{\delta x^2} = C_m \frac{\delta V_m}{\delta t} + G_m(V_m,x,t)(V_m - V_r) \tag{31-28}$$

where J_m = membrane current density (ampere/cm^2)
   J_{lo} = longitudinal axoplasmic current density (ampere/cm^2)
   V_{int} = the potential within the axoplasm
   ρ_{int} = specific resistivity of the axoplasm (ohm·cm)
   C_m = membrane capacitance per unit area (farad/cm^2)
   G_m = membrane conductance per unit area (mhos/cm^2)
   r_a = the axon's radius
   x = the distance along the axon in the direction of propagation

Equation 31-26 assumes that the axoplasmic current is strictly longitudinal and that the membrane current is radial, whereas Equation 31-28 assumes that V_{ext} is negligible in comparison with V_{int}, so that $V_m \approx V_{int}$.

As a result of preliminary studies, Hodgkin et al. [16] had proposed that the action potential resulted from transient, propagated increases in G_{Na} and G_K. In order to study these transients, however, three major steps were necessary. First, a method was required for eliminating the dependence of J_m, V_m, and G_m upon distance, so that Equation 31-27 could be reduced to the simpler form:

$$J_m(t) = C_m \frac{\delta V_m}{\delta t} + G_m(V_m,t)(V_m - V_r) \tag{31-29}$$

Second, a method was required for measuring or controlling the variables other than G_m, so that its dependence upon V_m and time might be determined. Finally, a method was required for resolving G_m into its component ionic conductances.

The first requirement was met through use of the giant, unmyelinated axon of the squid as an experimental model. Because of the axon's large diameter (up to 0.8 mm), a thin, uninsulated wire can be inserted into the axoplasm and along its interior; if a short length of axon is used, it can also be encased in an uninsulated, external sleeve electrode, as is shown in Figure 26-10. The two electrodes act to short-circuit the longitudinal resistances along the interior and exterior of the axon, thus ensuring iso-potentiality along its axis. If the entire length of axon is then excited simultaneously by the same or an additional pair of electrodes, a membrane action potential will occur in which there are no longitudinal variations in V_m, G_m, and J_m. Consequently, the entire structure will behave as a uniform patch of membrane, and Equation 31-29 can be applied.

The second requirement was met by employing a *voltage-clamp technique* that had been developed previously by other researchers [3]. The principle of the method is also illustrated in Figure 26-10 and is discussed in detail in Chapter 26. The advantages offered by this method can be illustrated with the aid of Equation 31-29. Suppose, for example, that V_m is suddenly displaced from its resting value to a more depolarized level (V_C) and maintained there. After a brief surge of capacitative current, $\delta V_m/\delta t = 0$, and Equation 31-29 will reduce to:

$$I_m/A = J_m = G_m(V_C,t)(V_C - V_r) \tag{31-30}$$

where A is the axon's surface area, and I_m is the total membrane or applied current. Thus, J_m will be strictly ionic, and, because V_C is constant, it will be directly propor-tional to the membrane conductance transient.

In their initial studies, Hodgkin et al. [14d] found that a stepwise depolarization of the membrane produces a transient, inward membrane current, that is followed by a slowly rising and sustained outward current (see Fig. 26-11A). The first component reverses its direction when the driving potential for sodium current $(V_C - V_{Na})$ reverses its sign and is absent in sodium-free media; it is therefore clearly carried by sodium ions.

Their evidence also indicated that the delayed outward current is independent of the Na^+ current and is carried primarily by K^+ ions. As a result of these findings, they adopted the equivalent circuit for a patch of excitable membrane that is shown in Figure 31-5. In this model, the sodium and potassium currents are additive and their associated conductances are functions of V_m and time; an additional, constant leakage conductance (G_L) was included in order to account for a small current density that is associated with J_K, but that is apparently carried by other ions. Having adopted this circuit, the following equation could then be written for the transmembrane current density under voltage-clamp conditions:

$$I_m/A = J_m = G_{Na}(V_C,t)\,(V_C - V_{Na}) + G_K(V_C,t)\,(V_C - V_K) \qquad (31\text{-}31)$$

$$+ G_L(V_C - V_L)$$

Using this model as a guide, Hodgkin and Huxley were then able to resolve G_m into its ionic components in the following way. By varying V_C, a set of curves was obtained that related the total ionic current density $(J_{Na} + J_K + J_L)$ to time, with V_m as a parameter; an example of one of these measurements is shown in Figure 26-11B. The procedures were then repeated with the axon in a solution in which the sodium had been replaced with impermeable choline ions. This maneuver yielded a set of curves in which the total ionic current density was equal to $J_K + J_L$. The J_{Na} transients were then estimated by simply subtracting the two sets of measurements (see Fig. 26-11A). It was then a simple matter to ignore the small leakage current and to obtain a family of sodium and potassium conductance curves from the expressions:

$$G_{Na}(V_C,t) = \frac{J_{Na}(V_C,t)}{(V_C - V_{Na})} \qquad (31\text{-}32)$$

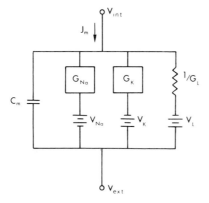

Figure 31-5
The general membrane equivalent circuit diagram adopted by Hodgkin and Huxley in their studies of the ionic bases of the action potential. The conductances G_{Na} and G_K were found to be functions of both membrane potential and time.

and,

$$G_K(V_C,t) = \frac{J_K(V_C,t)}{(V_C - V_K)} \tag{31-33}$$

Examples of these conductance transients are shown in Figure 26-12B.

Having computed the conductance curves, Hodgkin and Huxley then formulated a set of *empirical* equations that would relate them to V_m and time. They found that a given G_K curve was well fit by the equation:

$$G_K = (G_K)_{max}\, n(t)^4 \tag{31-34}$$

where $(G_K)_{max}$ is the constant, maximum value of the potassium conductance and $n(t)$ is a solution of the equation:*

$$\frac{dn}{dt} = \alpha_n(1 - n) - \beta_n n \tag{31-35}$$

Furthermore, by defining the rate constants α_n and β_n by means of the empirical formulas,

$$\alpha_n = \frac{0.01\,(V + 10)}{\exp\,[(V + 10)/10] - 1} \tag{31-36}$$

$$\beta_n = 0.125\,\exp\,(V/80) \tag{31-37}$$

where $V = V_m - V_r$, it was possible to use Equation 31-34 to describe all of the experimentally obtained G_K curves.

Because of its transient nature, Hodgkin and Huxley found it convenient to describe the sodium conductance curve as the product of two hypothetical state variables:

$$G_{Na} = (G_{Na})_{max}\, m^3 h \tag{31-38}$$

The variable m is given by the solution to

$$\frac{dm}{dt} = \alpha_m(1 - m) - \beta_m m \tag{31-39}$$

---

*Although they had no direct supporting evidence, Hodgkin and Huxley noted that Equations 31-34 and 31-35 could be given a simple physical interpretation. Assume, for example, that in order for a microscopic potassium conductance channel to open, four charged particles must simultaneously occupy particular sites on the inner surface of the membrane. If the probability of each particle being in position is n and if their movements are independent, then the probability of all sites being occupied is n^4. Moreover, if the movement of the particles is governed by first-order kinetics, then n will be given by the solution to Equation 31-35, where $\alpha_n(1 - n)$ is the rate at which particles move into position and $\beta_n n$ is the rate at which they move out of position.

and m^3 describes a hypothetical process that activates or increases G_{Na}. The variable h is given by the solution to

$$\frac{dh}{dt} = \alpha_h(1 - h) - \beta_h h \qquad (31\text{-}40)$$

and describes a process that inactivates or decreases G_{Na}. As before, $(G_{Na})_{max}$ represents the maximum, constant value of the sodium conductance, and the rate-constants α and β, which are functions of membrane potential but not of time, are defined by the formulas:

$$\alpha_m = \frac{0.1\,(V + 25)}{\exp\,[(V + 25)/10] - 1} \qquad (31\text{-}41)$$

$$\beta_m = 4\,\exp\,(V/18) \qquad (31\text{-}42)$$

$$\alpha_h = 0.07\,\exp\,(V/20) \qquad (31\text{-}43)$$

$$\beta_h = \frac{1}{\exp\,[(V + 30)/10] + 1} \qquad (31\text{-}44)$$

Having obtained expressions for G_{Na} and G_K as functions of V_m and time, it was then possible to write the following equation for the transmembrane current density in an excitable patch of membrane:

$$J_m = C_m \frac{\delta V_m}{\delta t} + (G_{Na})_{max}\,m^3 h(V_m - V_{Na}) + (G_K)_{max}\,n^4(V_m - V_K) \qquad (31\text{-}45)$$

$$+ G_L(V_m - V_L)$$

As indicated previously, this equation also applies to an (unmyelinated) axon in which axial variations in V_m are abolished, and, in both cases, J_m will be equal to the applied current density, J_{appl}. Consequently, J_{appl} can be made a hypothetical stimulating pulse, and Equations 31-34 *through* 31-45 can be solved numerically to obtain a theoretical membrane action potential. One of the membrane action potentials computed by Hodgkin and Huxley is shown in Figure 31-6 along with one recorded experimentally from a segment of squid axon. As can be seen, the general agreement is excellent.

In addition to these computations, Hodgkin and Huxley obtained solutions for the more realistic case of a propagating action potential. On the basis of previous studies, it was known that the action potential propagates in an unmyelinated fiber with both a constant velocity and a constant shape. Mathematically, this implies that the action potential satisfies the one-dimensional wave equation,

$$\frac{\delta^2 V_m}{\delta x^2} = \frac{1}{\theta^2}\frac{\delta^2 V_m}{\delta t^2} \qquad (31\text{-}46)$$

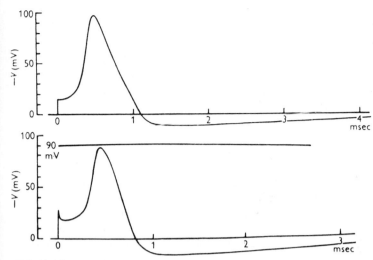

Figure 31-6
Theoretical and experimentally recorded membrane action potentials. *Upper curve*: Solution of Equation 31-45 for an initial depolarizing pulse of 15 mV and a temperature of 18.5°C. (Although not indicated in the text, the theoretical rate-constants in the Hodgkin-Huxley model are assumed to be functions of temperature.) *Lower curve*: Tracing of a membrane action potential recorded from a squid axon at a temperature of 20.5°C. The time scales differ by a factor that is appropriate for the temperature difference. (From A. L. Hodgkin and A. F. Huxley. *J. Physiol.* [Lond.] 117:500, 1952. Reproduced with permission.)

where θ is the velocity of propagation. In view of this relation, it was possible to place Equation 31-28 in the form,

$$\frac{r_a}{2\rho_{int}\theta^2}\frac{d^2 V_m}{dt^2} = C_m \frac{dV_m}{dt} + (G_{Na})_{max} m^3 h(V_m - V_{Na}) \tag{31-47}$$

$$+ (G_K)_{max} n^4 (V_m - V_K) + G_L(V_m - V_L)$$

and to solve it numerically. An example of the solution is shown in Figure 5-3 along with the theoretical G_{Na} and G_K transients.

In addition to predicting the shape of the action potential, the model of Hodgkin and Huxley accurately predicted (1) the shape of the membrane impedance changes associated with an action potential, (2) the existence of the threshold phenomenon, (3) refractory periods, (4) oscillations in V_m produced by subthreshold stimulating currents, (5) accommodation phenomena, and (6) anodal-break excitation. Although it was formulated some twenty years ago, the model has withstood the test of time, and it has been extended to cover conduction in myelinated axons [11] and a number of additional phenomena [24]. There are, of course, other models that deal with the same phenomena; of particular note among these are the general models explored by

Fitzhugh [10], Hoyt's alternative formulation of the Hodgkin-Huxley equations [17], and the alternative view of the nature of membrane excitation presented by Tasaki and Singer [34]. Nevertheless, the Hodgkin-Huxley model stands today as the most widely accepted *macroscopic* view of the electrical properties of excitable membranes. Consequently, although it may not be necessary to use so detailed a model in studying certain aspects of neuronal input-output behavior, the models that are employed must be consistent with its general properties.

THE SIMULATION STUDIES OF LEWIS: AN EXTENSION OF THE ECCLES AND HODGKIN-HUXLEY MODELS

Studies have shown that single neurons may differ widely in terms of the spontaneous fluctuations in V_m that they exhibit in the absence of external stimuli, as well as in their responses to synaptic input. Many cells, for example, will remain essentially quiescent in the absence of synaptic input, whereas others may exhibit a continually drifting membrane potential that erupts periodically in rhythmic bursts of spikes. It has in fact been noted that the spontaneous fluctuation in V_m may be nearly sinusoidal in some cells, whereas in others, it consists of a series of sawtooth or ramp-like depolarizations, each of which is terminated by a spike or a transient resetting of the membrane potential (e.g., Fig. 31-7A) [2, 13]. In addition, the EPSPs evoked by a high-frequency train of presynaptic spikes may be of near-constant amplitude in some cells, whereas in others, each successive PSP in the train is either larger or smaller than the preceding one. The latter two cases are referred to as synaptic *facilitation* and *antifacilitation,* respectively, and, on occasion, both phenomena may be observed

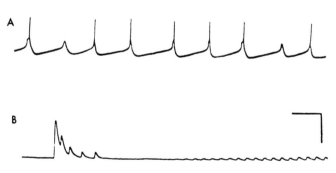

Figure 31-7
A. Spontaneous pacemaker potentials recorded intracellularly from a neuron in the cardiac ganglion of the lobster. Note that action potentials are associated with most of the rhythmic depolarizations, but they are not necessary for their occurrence. *B.* Antifacilitation (*left*) and facilitation (*right*) of evoked EPSPs in a cardiac ganglion cell. Note the progressive decline in EPSP amplitude on the left and the gradual increase in size on the right. The two sets of PSPs were evoked by trains of constant-intensity electrical pulses that were delivered to two separate presynaptic pathways (*horizontal bar*: 400 msec in part *A*, 100 msec in part *B*; *vertical bar*: 10 mV in part *A*, 20 mV in part *B*). The action potentials are small in part *A* because the recording electrode was at some distance from the site of spike generation. (Adapted from T. H. Bullock. *Science* 129:997, 1959. Reproduced with permission.)

within the same cell following the stimulation of different presynaptic pathways (Fig. 31-7B).

It is not clear from experimental studies, however, whether or not these various forms of neuronal behavior can all be accounted for in terms of the Eccles and Hodgkin-Huxley models of subsynaptic and excitable membranes. Moreover, the associated mathematical formulations are sufficiently complicated that digital computer simulation studies of such general problems are economically unfeasible. Recognizing these facts, Lewis [20–22] designed a set of solid-state circuits that simulated accurately the behavior of the variables contained in the Hodgkin-Huxley and Eccles hypotheses. The circuits thus constituted an economical and special-purpose analog computer, that could be used to study the types of neuronal behavior to be expected from varying the parameters of the Eccles and Hodgkin-Huxley models.

The basic membrane equivalent circuit that was simulated by Lewis is shown in Figure 31-8. The portion to the left of the dashed line represents a patch of sub-synaptic membrane with three major conductance channels. The first conductance is that underlying the EPSP; it has no series voltage source, because the equilibrium potential for this PSP is approximately 0 mV. The next two branches represent G_{Cl} and G_K, with their associated equilibrium potentials, and are the pathways for inhibitory synaptic currents. The portion to the right of the dashed line is the general Hodgkin-Huxley model for a patch of excitable membrane that either lies adjacent to, or is coextensive with, the subsynaptic region.

The techniques used by Lewis in simulating the conductance elements varied as his circuits were simplified and refined, but the general principles may be summarized as follows.

The general methods used to simulate *subsynaptic conductances* may be illustrated with reference to G_K. A voltage pulse is first applied to the simple filter shown in

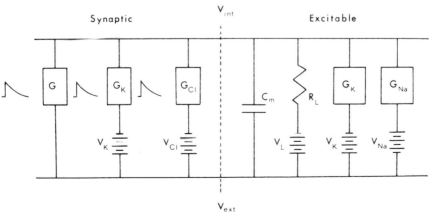

Figure 31-8
The general membrane model simulated electronically by Lewis [22]. (A brief description of the circuit is given in the text.)

Figure 31-9A, which produces an output proportional to $\exp(-t/\tau_s)$. This signal represents a decaying transmitter concentration or conductance transient, $G_K(t)$, and its amplitude and decay rate (τ_s) are controlled by means of potentiometers R_1 and R_2. The signal is then applied to one input of a multiplier, and the voltage difference $V_m - V_K$ is applied to its second input. The output of the multiplier, which is a current that is proportional to the product $G_K(V_m - V_K)$, is then applied

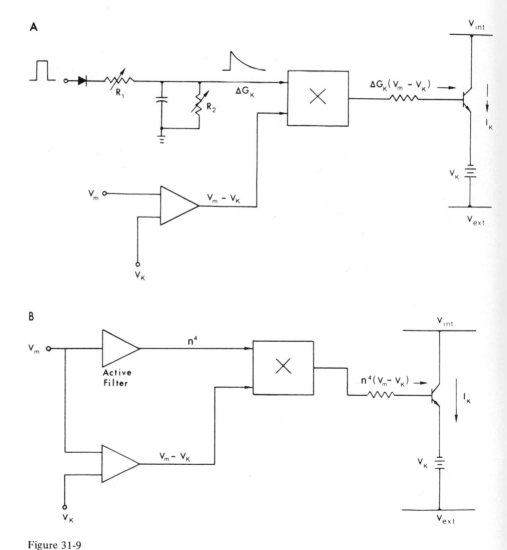

Figure 31-9
A. General features of the circuit employed by Lewis [20] in simulating synaptically evoked increases in potassium conductance (see text for details). B. The general method employed by Lewis in simulating the voltage-dependent potassium conductance of excitable membrane.

to the base of a transistor, whose collector and emitter are connected to the "inner surface" of the model membrane and a voltage source that represents V_K. This input produces a current, I_K, across the "conductance" transistor and, consequently, across the model membrane, that is proportional to the product $G_K(V_m - V_K)$; this is, of course, the desired result. The methods used to simulate I_{Cl} and the excitatory synaptic conductance current are similar in principle, with the "conductance" transistors and their connections being those required to produce the appropriate directions of current flow.

The methods used to simulate *excitable membrane conductances* may again be illustrated with reference to G_K (Fig. 31-9B). In this case, the transmembrane potential is applied to an *active* filter, which produces an output that simulates the potassium conductance variable, $n^4(V_m,t)$, of the Hodgkin-Huxley model; the filter is constructed so that its steady-state output matches the dependence of G_K on V_m that was found in the experimental data reported by Hodgkin and Huxley. The output of the filter is fed to one input of a multiplier, and the difference signal, $V_m - V_K$, is applied to its second input. The output of the multiplier is then applied to a "conductance" transistor, as described above, which produces a simulated transmembrane current that is proportional to the product $n^4(V_m,t)(V_m - V_K)$, as is required. The method used to model the excitable membrane's sodium current was similar, but, in this case, two active filters were necessary to simulate the variables $m^3(V_m,t)$ and $h(V_m,t)$ of the Hodgkin-Huxley model; in addition, two multipliers were required to form the successive products m^3h and $m^3h(V_m - V_{Na})$.

The circuits employed by Lewis were constructed in such a way that the major parameters of the Eccles and Hodgkin-Huxley models could be varied independently. With the parameters set in accordance with known values for motoneurons and squid axons, the circuits generated realistically shaped PSPs and spikes, as can be seen in Figures 31-10 and 31-11. The most significant information was obtained, however, when the parameters were varied.

Consider first, for example, the responses to simulated synaptic inputs. If the feedback loop representing the voltage dependence of G_{Na} is disconnected, a train of conductance transients, which may be evoked by applying pulses to either the EPSP conductance or directly to the sodium conductance, will produce a train of EPSPs of near-constant amplitude (Fig. 31-10A). In this case, of course, the model membrane is completely passive.* With the feedback loop closed, however, the train of EPSPs exhibits either facilitation or antifacilitation, as is shown in Figure 31-10B and C. The occurrence of these phenomena was found to depend upon (1) the non-linear dependence of G_{Na} upon V_m, and (2) the relationship of the time-constant of transmitter decay (τ_s) to the time-constant of sodium conductance inactivation (τ_h). For example, over a limited range, the derivatives $\delta G_{Na}/\delta\Delta V_m$ and $\delta(dG_{Na}/dt)/\delta\Delta V_m$ are increasing functions of membrane depolarization ($\Delta V_m = V_m - V_r$) in both the Hodgkin-Huxley and electronic models. Consequently, the increment in voltage-dependent or *active* G_{Na} that results from a synaptically induced depolarization will

---

*In this particular simulation experiment, the potassium conductance was also passive [20].

A

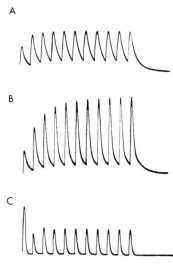

B

C

Figure 31-10

Examples of the simulated EPSPs generated by Lewis' electronic model. To obtain each trace, a train of pulses was applied to the model's excitatory synaptic conductance input, and the resulting voltage change was measured across the simulated membrane. *A.* EPSPs of near-constant amplitude are evoked when the feedback loop representing the voltage-dependence of G_{Na} is open. *B.* Facilitation occurs when the feedback loop is closed and the time-constant of simulated transmitter decay (τ_s) is longer than the interpulse interval. *C.* Antifacilitation occurs when τ_s is shorter than the interpulse interval; although the trace is faint, note that the first EPSP in the train is considerably larger than the rest. (Adapted from E. R. Lewis. In R. Reiss [Ed.], *Neural Theory and Modeling.* Stanford, Calif.: Stanford University Press, 1964. Reprinted with permission.)

be enhanced if the membrane is slightly depolarized when the synaptic input occurs, and if inactivation is not in progress.

Consider next the sequence of events during the first two subthreshold EPSPs of a train. The first conductance transient will produce a synaptic current that depolarizes the membrane. As a result, the active G_{Na} will increase slightly, and more current will flow inward across the membrane. At this time, however, both G_K and inactivation of G_{Na} are increasing, thus producing an increase in the ratio of outward-to-inward current. In addition, the synaptic conductance transient is declining, and, as a result, V_m begins to return toward the resting level.

Consider now three separate cases. If the interval between synaptic inputs is greater than τ_h but less than τ_s, then G_{Na} inactivation will be negligible, and the residual transmitter will maintain a slight depolarization of the membrane. Consequently, the second synaptic input will produce a depolarization that adds to that remaining, and it will thus produce a greater increment in active G_{Na} than that which occurred during the first EPSP. As a result, the second EPSP will be larger than the first; i.e., *facilitation* will occur. If the interval between synaptic inputs is greater than τ_s and less than τ_h, however, the increment in active G_{Na} that is produced by the second input will be less than that which was produced by the first. In this case, the second EPSP will be

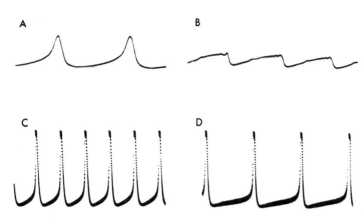

Figure 31-11
Spontaneous potentials generated by varying the simulated membrane conductance parameters in Lewis' electronic model. *A* and *B*. Simulated subthreshold pacemaker potentials. Note the similarity of the waveforms to the neuronal potentials shown in Figure 31-6. *C*. Rhythmic action potentials produced by increasing the average level of G_{Na}. *D*. Reduction in spontaneous firing rate produced by increasing the time-constant of recovery from G_{Na} inactivation. (Adapted from E. R. Lewis. In R. Reiss [Ed.], *Neural Theory and Modeling*. Stanford, Calif.: Stanford University Press, 1964; and E. R. Lewis. *Kybernetik* 5:30, 1968.)

smaller, and *antifacilitation* will be exhibited. Finally, it is clear that with the appropriate relationship between input intervals, τ_s and τ_h, a train of EPSPs of essentially constant amplitude could result. Thus, Lewis' results suggested that a patch of membrane with a voltage-dependent sodium conductance could exhibit any one of three types of EPSP summation, depending upon the dynamics of the presynaptic spike train and the conductance parameters. They also suggested that a synaptic input over one pathway could facilitate the EPSPs evoked in a cell by a second pathway; the basic requirement is that both inputs depolarize a common area of membrane that contains a voltage-dependent sodium conductance. This type of mutual facilitation was, in fact, observed subsequently in the nervous system of *Aplysia,* but it was attributed, for various reasons, to presynaptic mechanisms [19].

In addition to these results, Lewis found that by varying the Hodgkin-Huxley parameters, a variety of "spontaneous" fluctuations in membrane potential could be produced that were in many cases similar to those observed in real cells. For example, upon (1) slightly increasing the resting level of G_{Na}, (2) decreasing the voltage-dependence of G_{Na}, and (3) slightly increasing the voltage-dependence of G_K, subthreshold oscillations were produced that varied in shape from quasisinusoidal to the ramp-like depolarizations that are characteristic of pacemaker neurons (Fig. 31-11A and B) [13]. In addition, by simply increasing the equilibrium level of G_{Na}, it was possible to produce rhythmic, spontaneous firing (Fig. 31-11C). The spontaneous firing frequency could then be modified by altering two parameters that control the duration of refractoriness: the time-constants of the decline in G_K and in G_{Na} inactivation (Fig. 31-11D). By combining a steady level of G_{Na} that is sufficient to produce firing with conditions (2) and (3) above, it was also possible to generate "spontaneous" bursts of spikes with interspike interval distributions that resemble those generated by cells in the mammalian central nervous system [1].

The similarities between these results and the physiologic phenomena described above do not prove, of course, that the explanations obtained from the model are correct. However, Lewis' studies have shown that with minor parameter adjustments, the combined Hodgkin-Huxley and Eccles model can provide explanations for a much wider range of phenomena than had previously been anticipated. These studies have also shown the value of employing carefully constructed hardware models to illuminate and explore the ramifications of mathematically complex theories. For these reasons, the results obtained by Lewis are a significant contribution to our current understanding of the behavior of neuronal membranes.

WHOLE NEURON MODELS

In the previous section, we reviewed the major electrical models of different types or regions of neuronal membrane. In this section, a number of whole neuron models will be described that have been used in quantitative studies of neuronal input-output behavior. As will become apparent, the models differ considerably in complexity, ranging from simple, lumped-parameter representations of major neural properties to complex, distributed-parameter models that incorporate not only the membrane models discussed above, but also the neuron's major geometrical features. Each of these models was developed, however, for a specific purpose, and its utility cannot be judged solely in terms of apparent realism or complexity; each has, in fact, been useful in revealing important aspects of neuronal behavior. The descriptions are collected here simply for convenience and to provide an overview of the types of modeling techniques that have been employed.

LUMPED-PARAMETER MODELS

The well-known models of Perkel et al. [26], Harmon [14], and Lewis [22] are all of the lumped-parameter type. In each case, the neuron is represented, either explicitly or implicitly, by a lumped circuit of the type described previously for an isopotential patch of membrane. Thus, the cable properties of a cell's dendrites are not considered, and no provision is made for studying the variations in neuronal output that result from different spatial distributions of synaptic input. The amount of error introduced by this simplification depends, however, upon the intended use of the model. The error will be small, for example, in the case of a spherical neuron with no dendrites, because the interior of its soma will be roughly isopotential at all times [28]. Certain cells within the nervous system of *Aplysia* are of this type, and Perkel's model was constructed specifically for theoretical studies of their input-output spike trains. The approximation is also reasonable if one wishes to model only the behavior that is produced by the synaptic inputs to a cell's soma; such is the stated purpose of the neuron models employed by Harmon and by Lewis.

Thus, each of the models described below has been quite suitable in view of the restricted range of applications for which it was designed. The reader should realize, however, that a lumped-parameter model is *not suitable for detailed studies* of the input-output behavior of neurons with extensive dendritic trees and synaptic inputs [30].

Perkel's Software Model

The software "neuron" employed in the simulation studies of Perkel et al. [26, 27] is an example of a useful minimum-parameter model. From intracellular recordings, it appeared to these investigators that the major state variables that influenced the spike train of an *Aplysia* neuron were (1) the relationship between the cell's membrane potential (V_m) and its threshold value (V_{th}), (2) the amplitudes and time-constants of the cell's PSPs, and (3) the cell's postspike refractory periods. Consequently, a software model neuron was designed around these basic variables.

The features of the model are summarized in Figure 31-12. Each input produces a PSP that consists of a stepwise increment (or decrement) in V_m, followed by an exponential decay. In simulating a cell with no spontaneous discharge, the subthreshold value of V_m is determined simply by a linear superposition of PSPs, provided that a recent spike has not occurred. With a sufficient temporal summation of PSPs, however, V_m reaches a threshold value, and a spike is generated. Each spike is followed by an absolute refractory period, after which V_m at first jumps to a

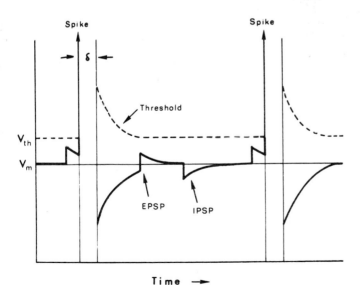

Figure 31-12
Summary of the major features of Perkel's software model neuron [26]. The time interval δ represents a cell's absolute refractory period, V_m its membrane potential, and V_{th} its threshold voltage. (Further details are given in the text.)

hyperpolarized value and then decays exponentially to its resting level. The model's postspike threshold is also elevated, and it then returns exponentially to its prespike value. These transient changes in V_m and the threshold value thus define a relative refractory period.

To simulate the behavior of a spontaneously firing, pacemaker neuron, the model's asymptotic threshold voltage is set at a more hyperpolarized level than its hypothetical resting potential. Consequently, the model cell discharges continuously even in the absence of synaptic input, with the interspike interval being the time required for V_m to return from its postspike value to V_{th}.

Although it contains only a few basic properties, Perkel's model has been subjected to reasonably realistic inputs, and it has generated theoretical spike trains that are similar to those recorded from *Aplysia* neurons; some of these results are described in Chapter 32.

Harmon's Neuromime

The term *neuromime* is often used to refer to an electronic circuit that has been designed to mimic or simulate the electrical behavior of a neuron. One of the simplest but historically most useful neuromimes was that constructed and employed in the important studies of Harmon [14, 14a, 14b]. Much like Perkel's software model (which it preceded), Harmon's neuromime was not based explicitly on the membrane equivalent circuits that had been derived from electrophysiologic studies. Instead, Harmon attempted to simulate only those general properties that appeared to be most directly involved in determining a neuron's input-output behavior.* A list of these simulated properties is given in Table 31-1, and the general features of the neuromime are shown in Figure 31-13A. Its principles of operation may be summarized as follows.

The property of *temporal summation* is simulated by using the capacitor C (see Fig. 31-13) to integrate the various input or presynaptic pulse trains. The model

Table 31-1. Properties of Harmon's Neuromime

Simulated Neuronal Property	Model Parameters
Resting threshold	−1.5 volt for a steady, transducer-like input; −3.0 volt for a spike input
Temporal summation or synaptic time-constant	0.5 to 1.2 msec, depending upon input configuration
Input-output delay	0.1 to 1.0 msec, depending upon input configuration
Spike output	0.8 to 1.2 msec pulse of −10 volt amplitude
Refractory periods	Absolute: ≈ 1.2 msec Relative: 2.7 msec time-constant
Firing frequency range	0−600 pulses/sec
Inhibition	Produced by graded subtraction from excitation

*The neuromime was designed primarily for neural network simulation studies.

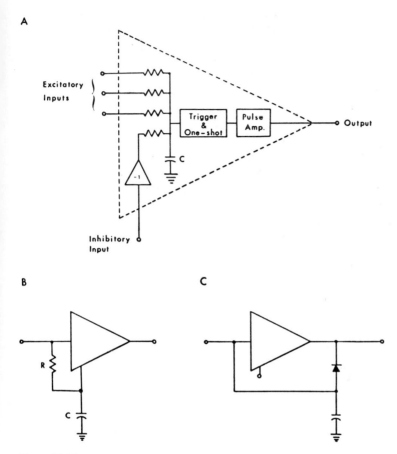

Figure 31-13
A. General features of Harmon's neuromime [14]. *B.* A method for producing adaptation in the neuromime's firing rate. The input pulse train is low-pass filtered and applied to the model's inhibitory input, thus producing an exponentially increasing voltage across the capacitor that subtracts from the integrated excitatory input. *C.* Method for producing sustained or spontaneous firing. The neuromime's output is applied to its excitatory input terminal, thereby producing a regenerative, positive-feedback network.

contains one input channel with an inverting amplifier, so that standardized input pulses may be used to produce both *excitatory* and *inhibitory potentials*; the two types of potentials are summed algebraically at *C.* The integrated voltage is then applied to a combined Schmitt trigger and multivibrator (one-shot) circuit, whose triggering voltage simulates a neuron's *threshold.* When the triggering voltage is exceeded, the one-shot circuit produces a simulated output *spike* that consists of a negative pulse. The output pulse is then fed to an amplifier with a low output impedance, so that it might be used as an input to other neuromimes or to various external circuits. The pulse duration is approximately one millisecond, thus providing an *absolute refractory period* of the same duration. Each output pulse is integrated and fed back to the circuit in such a way that its triggering threshold at first jumps to a high level and then decays exponentially with a time-constant of 2.7 msec. This transient increase in the threshold value thus simulates a neuron's *relative refractory period.*

The neuromime thus exhibits a number of neuron-like properties, which include, in addition to those listed above, a stimulus strength-duration curve similar to that exhibited by many nerve cells and axons, and a near-logarithmic dependence of pulse output frequency upon the level of steady input voltage. Moreover, its capabilities can be extended considerably by simply adding external circuitry to its inputs.

By appending the circuit of Figure 31-13B, for example, a single input can produce both excitatory and inhibitory transients. By adjusting R and C in this circuit so that the time-constant of inhibition is longer than that of excitation ($>$ 3 msec), the neuromime can simulate the processes of *accommodation* and *adaptation.* In the former case, the apparent threshold to an input voltage (V) is an inverse function of dV/dt, which is precisely the defining property for accommodation. In the latter case, the inhibitory time-constant is made comparatively long; the neuromime will then respond to a step increase in input voltage with a train of spikes whose frequency declines exponentially over time. This type of firing rate adaptation is characteristic not only of many single neurons and sensory receptors, but also of recurrent inhibitory networks. Spontaneous or self-sustained firing can also be produced by connecting several neuromimes together in a positive feedback loop or by using the simple feedback circuit shown in Figure 31-13C.

Harmon's model is thus quite flexible, and it may be used to simulate a fairly wide range of neuronal behavior. Because of this flexibility and its comparatively simple design, the neuromime is well suited for simulation studies of small neuronal systems and networks; an example of this type of application is given in Chapter 32.

Lewis' Model Neuron

As can be seen in Figure 31-7, the general membrane model simulated by Lewis [21, 22] is made up of two parallel circuits: the portion to the left of the dashed line represents a region of passive, subsynaptic membrane, and the portion to the right, a region of excitable membrane. It is possible to separate the circuits that simulate these two regions and then recouple them with resistors placed in series with the branches that represent the internal and external surfaces of the membrane. The resulting *two-compartment* model can then be used to simulate the behavior of a passive neuron soma, which receives synaptic inputs, and a nearby initial segment or trigger zone. The coupling resistors thus represent the lumped resistances of the intracellular and extracellular current pathways that link these two regions. The Lewis model has in fact been used in this way in theoretical studies of the behavior of simple neuronal networks; some of the results of these studies are discussed in Chapter 32.

DISTRIBUTED-PARAMETER MODELS

Rall's Equivalent Cylinder and Compartmental Models

Because of the high dendritic core resistance,* a synaptic input at the periphery of a cell's dendritic tree will change the membrane potential at its trigger zone significantly less than will a similar input at the soma. The effects will not be negligible, however, and it is therefore necessary to consider the dendritic locus of a synaptic input in attempting to quantify its effects on neuronal input-output behavior. Although this fact had long been recognized by a number of neurophysiologists, there was, for many years, no clear method available for taking it into account. It was well known, of course, that the classic cable equation can be used to estimate the potential distribution along a single, cylindrical core conductor (see Chap. 26). It was not until the important work of Rall [30], however, that a method became available for extending the general cable equation to a complex and repeatedly branching dendritic tree. The major features of Rall's method for accomplishing this may be summarized as follows.

Consider first the dendritic tree of Figure 31-14, in which the number (n) of equal branches at any distance (x) from the soma and their common radius (r) are both step-wise functions of x. At any given value of x, the total dendritic core current, I_{lo}, will be equal to the sum of the core currents in the n branches:

$$I_{lo} = n(x) \left(\frac{-\pi r^2(x)}{\rho_{int}} \cdot \frac{\delta V_{int}}{\delta x} \right) \tag{31-48}$$

where ρ_{int} is the specific resistivity (ohm·cm) of the dendritic core. In addition, the membrane current density, J_m, will be given by:

$$J_m = \frac{-1}{2\pi nr} \frac{\delta I_{lo}}{\delta x} \tag{31-49}$$

where $2\pi nr$ is the sum of the circumferences of the n cylindrical branches. The product $J_m R_m$ may therefore be expressed as:

$$J_m R_m = \frac{R_m}{2\rho_{int} nr} \left[nr^2 \frac{\delta^2 V_{int}}{\delta x^2} + \frac{\delta V_{int}}{\delta x} \frac{d}{dx} (nr^2) \right] \tag{31-50A}$$

*The core resistance of a 50-micron length of cylindrical dendrite, which is 5 microns in diameter, is in the order of 2.5 megohms.

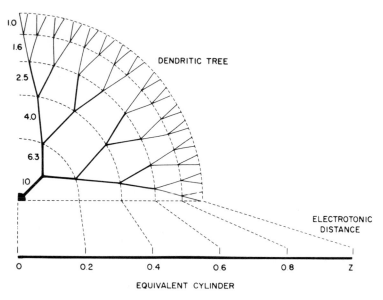

Figure 31-14
Geometrical illustration of Rall's method for transforming the cable equations for a branching dendritic tree into the equation for a single-membrane cylinder. As shown by the dotted lines, equal increments in axial distance along the equivalent cylinder correspond to equal increments in dendritic surface area rather than length. In order for this transformation to hold, a cell's dendritic tree must conform approximately to the mathematical rules outlined in the text; the dendritic tree shown here was constructed according to one of these rules. The numbers to the left indicate the radii of the dendritic branches (in microns) at various distances from the model cell's soma. (From W. Rall. In R. Reiss [Ed.], *Neural Theory and Modeling*. Stanford, Calif.: Stanford University Press. Reprinted with permission.)

Or, equivalently, as:

$$J_m R_m = \frac{R_m r}{2\rho_{int}} \left[\frac{\delta^2 V_{int}}{\delta x^2} + \frac{\delta V_{int}}{\delta x} \frac{d}{dx} \ln (nr^2) \right] \tag{31-50B}$$

where R_m is the transverse resistance of a unit area of passive or resting membrane ($R_m = 1/G_r$).

For an *unbranched*, cylindrical dendrite of constant radius r_0, the *electrotonic length-constant*, λ, is defined by:

$$\lambda = \left(\frac{R_m r_0}{2\rho_{int}} \right)^{1/2} \tag{31-51}$$

Thus, by considering λ to be a function of x, Equation 31-50B may be written as:

$$J_m R_m = \lambda^2 (x) \left[\frac{\delta^2 V_{int}}{\delta x^2} + \frac{\delta V_{int}}{\delta x} \frac{d}{dx} \ln (nr^2) \right] \tag{31-52}$$

We wish now to find a *new distance variable, $Z(x)$,* that will reduce Equation 31-52 to a simpler form. This may be accomplished in the following way. If we use the relations,

$$\frac{\delta V_{int}}{\delta x} = \frac{\delta V_{int}}{\delta Z}\frac{dZ}{dx}$$

and,

$$\frac{\delta^2 V_{int}}{\delta x^2} = \frac{\delta}{\delta Z}\left(\frac{\delta V_{int}}{\delta Z}\frac{dZ}{dx}\right)\frac{dZ}{dx}$$

then a bit of algebra will show that $J_m R_m$ can be expressed in terms of the new variable Z as follows:

$$J_m R_m = \lambda^2(x)\left(\frac{dZ}{dx}\right)^2\left[\frac{\delta^2 V_{int}}{\delta Z^2} + \frac{\delta V_{int}}{\delta Z}\left(\frac{dZ}{dx}\right)^{-1}\frac{d}{dx}\ln\left(nr^2\frac{dZ}{dx}\right)\right] \qquad (31\text{-}53)$$

Note that if $Z(x)$ is defined as:

$$Z(x) = \int_0^x \frac{1}{\lambda(x')}dx' \qquad (31\text{-}54)$$

then

$$\frac{dZ}{dx} = \frac{1}{\lambda(x)} \qquad (31\text{-}55)$$

and the coefficient of the bracketed expression on the right of Equation 31-53 is equal to one. In addition, if

$$nr^2\frac{dZ}{dx} = \text{constant} \qquad (31\text{-}56)$$

then the coefficient of $\delta V_{int}/\delta Z$ is zero; note that from the definitions of dZ/dx and $\lambda(x)$, this latter condition implies that:

$$n(x)\, r(x)^{3/2} = \text{constant} \qquad (31\text{-}57)$$

Thus, with Z defined by Equation 31-54 and the dendritic branching pattern by Equation 31-57, Equation 31-53 can be reduced to:

$$J_m R_m = \frac{\delta^2 V_{int}}{\delta Z^2} \qquad (31\text{-}58)$$

The significance of this result should not be underestimated. To illustrate its importance, it may first be noted that a surprisingly large number of dendritic branching patterns can be approximated by Equation 31-57. Second, it is to be noted that for a single, unbranched cylinder of constant radius, the product $J_m R_m$ is given by:

$$J_m R_m = \frac{\delta^2 V_{int}}{\delta X^2}$$

where $X = x/\lambda_0$ and λ_0 is constant (see Chap. 26). *Thus, if the branching pattern can be approximated by Equation 31-57, then the equation for the transmembrane current density in a complex dendritic tree can be reduced to that for a single equivalent cylinder, where distance is measured in terms of the variable, Z.* *

A reduction of this type is illustrated in Figure 31-14, where the dendritic tree was actually constructed in accordance with Equation 31-57. It should be noted that increments in Z are *not* proportional to increments in the distance x along successive dendritic branches; rather, they are proportional to increments in dendritic surface area A, as can be seen from the relations:

$$\frac{dA}{dx} \propto nr$$

$$\frac{dZ}{dx} \propto (r)^{-1/2}$$

and, as implied by Equation 31-57,

$$nr \propto (r)^{-1/2}$$

This proportionality is illustrated in Figure 31-14 by the dashed lines, which mark off equal increments in dendritic surface area and in Z.

Given this transformation of distance variables, an equivalent cable equation can be derived by (a) noting that the potential gradients over a cell's surface are, in general, much smaller than those internally, so that $\delta^2 V_{int}/\delta Z^2 \approx \delta^2 V_m/\delta Z^2$, and (b) by using Rall's equivalent circuit for the dendritic membrane to obtain a second expression for $J_m R_m$ (see Eq. 31-19). Using the variable $T = t/\tau_m$, the resulting equation becomes:

$$\frac{\delta^2 V_m}{\delta Z^2} = \frac{\delta V_m}{\delta T} + (V_m - V_r) + \xi(V_m - V_{exc}) + \zeta(V_m - V_{inh}) \tag{31-59}$$

---

*As noted by Rall, a much wider class of branching patterns can be handled in a similar way by requiring that the coefficient of $\delta V_{int}/\delta Z$ in Equation 31-53 be a nonzero constant. In this case, the equation for the dendritic tree reduces to that for a cylinder of nonconstant radius.

where the variables $\xi(Z,T)$ and $\zeta(Z,T)$ and the constants V_r, V_{exc}, and V_{inh} are defined as previously (Eq. 31-21).

With the appropriate boundary and initial conditions, Equation 31-59 may be solved analytically for a voltage, a current, or a synaptic input at a particular distance along the equivalent cylinder. Solutions have been obtained, for example, for the electrotonic spread of potentials from a cell's soma into a passive dendritic tree [18] and for step or delta-function inputs at a particular value of Z [29]. For all but the simplest cases, however, Equation 31-59 must be solved numerically. Moreover, the steps involved in obtaining a numerical solution are virtually identical to those that would be followed if the dendritic tree were simply represented by a series of lumped compartments, as is shown schematically by the connected circles in Figure 31-15. Here, each compartment consists of an equivalent circuit that is identical to that for a unit area of dendritic membrane (see Fig. 31-3); each represents, however, the lumped membrane properties of all of the dendritic branches within a length, ΔZ. In addition, each compartment is coupled to its neighbor by a resistance, $R_d = 1/G_d$, that is equal to the combined dendritic core resistance over the length, ΔZ. For most purposes, then, it is reasonable to use simply a compartmental model instead of a continuous cylinder representation. The mathematical methods employed in such a model may be summarized as follows.

By noting that the difference between the core current that enters the kth compartment and that which leaves it must equal the compartment's transmembrane current, $I_{m,k}$, the following equation is easily derived.

$$I_{m,k} = G_d \Delta^2 V_k = \frac{\Delta A}{R_m} \left[\frac{\delta V_k}{\delta T} + (V_k - V_r) + \xi_k(V_k - V_{exc}) + \zeta_k(V_k - V_{inh}) \right] \quad (31\text{-}60\text{A})$$

Here, V_k is the value of V_m at compartment k, ΔA is the surface area of the dendritic region represented by k, and

$$\Delta^2 V_k = (V_{k-1} - V_k) - (V_k - V_{k+1}) \quad (31\text{-}60\text{B})$$

It can also be shown that the compartmental coupling conductance, G_d, is given by:

$$G_d = \frac{\Delta A}{R_m(\Delta Z)^2} \quad (31\text{-}60\text{C})$$

Consequently, Equation 31-60A may be placed in the form:*

$$\Delta^2 V_k = (\Delta Z)^2 \left[\frac{\delta V_k}{\delta T} + (V_k - V_r) + \xi_k(V_k - V_{exc}) + \zeta_k(V_k - V_{inh}) \right] \quad (31\text{-}61)$$

*Note that Equation 31-61 reduces, as limits are taken, to Equation 31-59, thus proving the consistency of the compartmental and continuous cylinder representations.

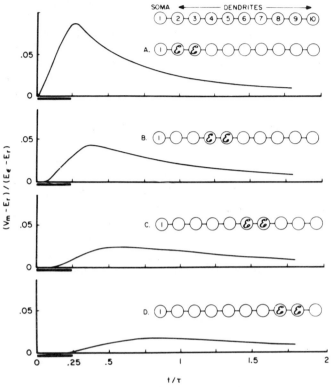

Figure 31-15
Theoretical EPSPs generated by synaptic inputs to Rall's compartmental model. As shown in the insets, a chain of ten compartments is used to represent a cell's soma and its dendritic tree. The PSPs shown are those generated at the soma by excitatory conductance pulses at the compartments indicated by the symbol ξ. The duration of the pulses is indicated by the horizontal bar under each curve. Note that the amplitude and time-course of each PSP is expressed in terms of the dimensionless variables $(V_m - V_r)/(V_{exc} - V_r)$ and $T = t/\tau_m$. To convert to real time, a τ_m value of 5 msec is appropriate for mammalian motoneurons. (From W. Rall. In R. Reiss [Ed.], *Neural Theory and Modeling.* Stanford, Calif.: Stanford University Press. Reprinted with permission.)

By writing a similar equation for each of the N compartments, a system of first-order equations is obtained that can be solved simultaneously for N values of the dendritic potential distribution, $V_m(Z,T)$, at successive points in time. In addition, the conductance functions ξ and ζ can be continuously updated at each compartment, so that complex patterns of synaptic input can be simulated.

Some examples of the simulated PSPs computed with Rall's model are shown in Figure 31-15. In this case, nine compartments were used to simulate a dendritic tree, and these were connected to a tenth compartment representing a cell's soma. The simulated tree was 1.8 λ long, which is on the order of the estimated electrotonic length for the dendrites of spinal motoneurons. The EPSPs shown are those evoked

at the soma by single excitatory conductance pulses at various dendritic compartments. As can be seen, the PSPs evoked by distal dendritic inputs rise and decay
slowly, but are not negligible. In fact, the slow time-course of these PSPs suggests
that distal dendritic synapses would be particularly effective in converting noisy and
irregular presynaptic spike trains into smoothly graded and sustained changes in
postsynaptic firing frequency. Thus, even these simple results illustrate the possible
importance of the dendritic locus of the synaptic input in studies of spike-train processing by single cells.

The important features of Rall's models may thus be summarized as follows. By
using a generalized distance variable, the models allow a wide range of complexly
branching dendritic trees to be represented electrically either by an unbranched
equivalent cylinder or by a series of interconnected compartments. Each compartment is described by a first-order differential equation, which relates its membrane
potential to the excitatory and inhibitory synaptic conductances and to the membrane potentials of adjacent compartments. Although it was not discussed above,
Rall and Shepherd [31] have also shown that a set of nonlinear equations in ξ, ζ, and
V_m can be used to generate theoretical action potentials. Consequently, one or two
compartments can be used to simulate a cell's soma and its initial segment or trigger
zone. Thus, Rall's model takes into account not only the different electrical properties of a cell's membrane, but also the cell's geometrical configuration and the spatial
properties of its synaptic input. At present, there are no other integrated neuron
models of equal predictive power and generality. Consequently, Rall's theoretical
studies must surely be recognized as a major development in the history of neural
modeling.

An Electronic Version of Rall's Compartmental Model

Because of its complexity, it is not economically feasible to use Rall's model for
digital-computer studies of neuronal spike trains. The computation of only a few
PSPs and spikes, for example, may require several minutes of computer processing
time. Because of its predictive power and realistic properties, however, a version of
Rall's model is needed that can be used for this purpose and that can also be made
available to neurophysiologists who have no access to a large computing facility.
For these reasons, Pottala et al. [27b] have developed an electronic analog of Rall's
model that is consistent with its major mathematical properties. An earlier, similar
model was also developed by Fernald [9].

The model neuron that is simulated by Pottala's analog is shown in Figure 31-16.
An equivalent dendritic cylinder is represented by a chain of four compartments,
and a fifth compartment represents the neuron's soma and initial segment. The
dendritic cylinder and the soma are assigned anatomic dimensions, and physiologic
extimates of specific membrane resistance ($R_m \approx 1600$ ohm·cm^2), capacitance
($C_m \approx 3 \, \mu\mathrm{F/cm^2}$), and intracellular resistivity ($\rho_{int} \approx 100$ ohm·cm) are used to
compute the appropriate values for the lumped compartmental membrane impedances

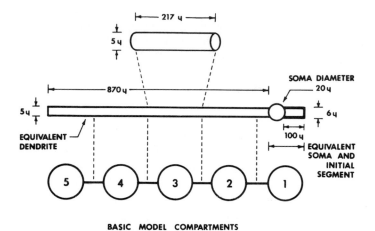

BASIC MODEL COMPARTMENTS

Figure 31-16

Geometrical features of the model neuron simulated electronically by Pottala et al. Four lumped compartments are used to represent an equivalent dendritic cylinder, while a fifth compartment represents the cell's soma and initial segment. (From E. W. Pottala et al. *IEEE Trans. Biomed. Eng.* 20:132, 1973. Reproduced with permission.)

and coupling resistors.* These values are then scaled so as to reduce noise and measurement problems, but in such a way that the simulated dendritic length-constant (λ_d = 435 microns) and the membrane time-constant (τ_m = 5 msec) are unchanged.

The basic circuit that is used for the soma and initial segment compartment is shown in Figure 31-17. The portion labelled "synaptic region" represents the combined passive and synaptic regions of soma membrane and is identical to the circuit used for each dendritic compartment. Each synaptic conductance (G_{exc} and G_{inh}) is simulated by applying a pulse to the gate of an appropriately connected field-effect transistor (FET), thus altering its transconductance. The PSPs generated in this way prove to be of realistic shape and agree well with those computed previously by Rall (Fig. 31-18). The portion labelled "trigger zone" in Figure 31-17 represents the spike-generating regions of the soma and initial segment. The two FETs are connected to reference voltages that are proportional to the equilibrium potentials, V_{Na} and V_K. By applying appropriate voltage functions to their gates, depolarizing and hyperpolarizing conductance increases (G_d and G_h) are generated, with time-courses similar to those of the G_{Na} and G_K transients described by Hodgkin and Huxley [19]. As a result, a realistically shaped action potential occurs, that is followed by a period of hyperpolarization (Fig. 31-19A).

The overall operation of the model is illustrated in the block diagram of Figure 31-20. Rectangular conductance pulses are used to generate PSPs at various

---

*For a 217-micron length of dendrite, for example, the lumped membrane resistance, R, is given by: $R = R_m/(\text{surface area}) = 1600 \text{ ohm·cm}^2/[\pi(5 \times 10^{-4})(2.17 \times 10^{-2})] \text{ cm}^2 \approx 50$ megohms.

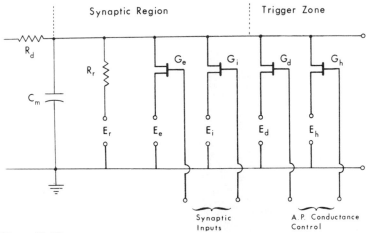

Figure 31-17
The circuit used by Pottala et al. in simulating the electrical behavior of a neuron's soma and initial segment. Synaptic conductance changes (G_e, G_i) are simulated by gating field-effect transistors connected to voltage sources that represent the PSP equilibrium potentials (E_e, E_i). Action potentials are generated by applying shaped, gating signals to field-effect transistors that represent the spike-associated hyperpolarizing (G_h) and depolarizing (G_d) conductance channels. The voltage sources E_d and E_h represent the approximate sodium and potassium equilibrium potentials. The resting membrane is simulated by the components E_r, R_r, and C_m. The compartmental coupling resistor R_d represents the lumped core resistance between the center of the model cell's soma and the adjacent segment of dendrite. (From E. W. Pottala et al. *IEEE Trans. Biomed. Eng.* 20:132, 1973. Reproduced with permission.)

compartments, and the resulting electrotonic potential at the soma is continuously recorded and compared with a threshold voltage. When the threshold value is exceeded, a Schmitt trigger circuit discharges into a high-pass filter circuit, which is used to simulate the membrane process of accommodation. If the output of the filter is sufficient, a one-shot circuit is triggered, and its pulse is fed to two active filters. The outputs of the filters are then used to generate the transient conductance increases at the soma that produce the action potential.

The voltage waveforms that are generated by Pottala's analog model are in good agreement with the PSPs and action potentials recorded from spinal motoneurons and other cells; examples of the simulated potentials are shown in Figures 31-18 and 31-19. Although this method for simulating an active region of membrane is less detailed and realistic than that used by Lewis, the basic conductance-transient phenomena are retained. Consequently, it is possible to study the interactions between dendritic PSPs and somatic action potentials in the presence of the shunting membrane impedance changes that are known to occur in real cells. In addition to generating realistically shaped potentials, Pottala's analog model exhibits a number of other neuron-like properties, including accommodation and firing-rate adaptation (see Fig. 31-19B). It promises to be of considerable value for theoretical studies of the input-output behavior of anatomically realistic cells.

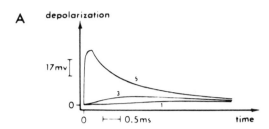

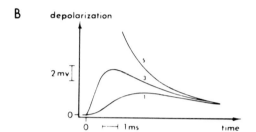

Figure 31-18
Examples of the simulated EPSPs generated by Pottala's model neuron. *A.* EPSPs generated at compartments 1, 3, and 5 by a brief excitatory conductance pulse at compartment 5. *B.* EPSPs as in part *A*, but recorded at a higher gain and slower sweep speed to show the shapes of the electrotonically spread potentials. (From E. W. Pottala et al. *IEEE Trans. Biomed. Eng.* 20:132, 1973. Reproduced with permission.)

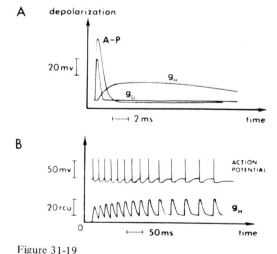

Figure 31-19
A. Time-courses of the conductance changes (g_d, g_h) used in Pottala's electronic model to produce simulated action potentials (*A-P*). *B.* An example of firing-rate adaptation. A step increase in excitatory conductance was produced at compartment 5, and the resulting firing pattern was recorded at the soma compartment. The gradual decline in firing rate was found to be due to temporal summation of the hyperpolarizing conductance transients (g_h) associated with each spike, as can be seen in the lower trace. The magnitude of the conductance change is measured in arbitrary (but linear) relative conductance units (rcu). (From E. W. Pottala et al. *IEEE Trans. Biomed. Eng.* 20:132, 1973. Reproduced with permission.)

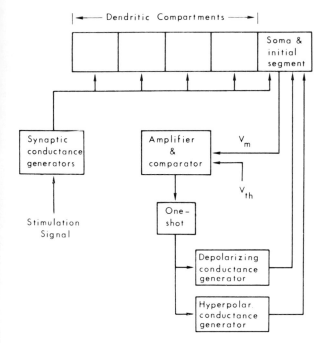

Figure 31-20
Summary of the general operation of Pottala's electronic analog. (From E. W. Pottala, et al. *IEEE Trans. Biomed. Eng.* 20:133, 1973. Reproduced with permission.)

REFERENCES

1. Bishop, P. O., Levick, W. R., and Williams, W. O. Statistical analysis of the dark discharge of lateral geniculate neurons. *J. Physiol.* (Lond.) 170:598, 1964.
2. Bullock, T. H. Parameters of integrative action of the nervous system at the neuronal level. *Exp. Cell Res.* (Suppl.) 5:323, 1958.
3. Cole, K. S. Dynamic electrical characteristics of the squid axon membrane. *Arch. Sci. Physiol.* (Paris) 3:253, 1949.
4. Cole, K. S., and Curtis, H. J. Electric impedance of the squid giant axon during activity. *J. Gen. Physiol.* 22:649, 1939.
5. Coombs, J. S., Eccles, J. C., and Fatt, P. Electrical properties of the motoneuron membrane. *J. Physiol.* (Lond.) 130:291, 1955.
6. Davis, L., Jr., and Lorente de No, R. Contribution to the Mathematical Theory of Electrotonus. In *A Study of Nerve Physiology*. New York: Rockefeller Institute for Medical Research, 1947.
7. Eccles, J. C. *The Physiology of Synapses*. New York: Academic, 1964. Pp. 37–53, 152–188.
8. Fatt, P., and Katz, B. An analysis of the end-plate potential recorded with an intra-cellular electrode. *J. Physiol.* (Lond.) 115:320, 1951.
9. Fernald, R. D. A neuron model with spatially distributed synaptic input. *Biophys. J.* 11:323, 1971.

10. Fitzhugh, R. Impulses and physiological states in theoretical model of nerve membrane. *Biophys. J.* 1:446, 1961.

11. Fitzhugh, R. Computation of impulse initiation and saltatory conduction in a myelinated nerve fiber. *Biophys. J.* 2:11, 1962.

12. Fuortes, M. G. F., Frank, K., and Becker, M. C. Steps in the production of motor-neuron spikes. *J. Gen. Physiol.* 40:735, 1957.

13. Hagiwara, S., and Bullock, T. H. Intracellular potentials in pacemaker and integrative neurons of the lobster cardiac ganglion. *J. Cell Comp. Physiol.* 50:25, 1957.

14. Harmon, L. D. Studies with artificial neurons. I. Properties and functions of an artificial neuron. *Kybernetik* 1:89, 1961.

14a. Harmon, L. D. Neuromimes: Action of a reciprocally inhibitory pair. *Science* 146:1323, 1964.

14b. Harmon, L. D., Levinson, J., and Van Bergeijk, W. A. Studies with artificial neurons. IV. Binaural temporal resolution of clicks. *J. Acoust. Soc. Am.* 35:1924, 1963.

14c. Hasan, Z., and Houk, J. C. Transition in sensitivity of spindle receptors that occurs when muscle is stretched more than a fraction of a millimeter. *J. Neurophysiol.* 38:673, 1975.

14d. Hodgkin, A. L., and Huxley, A. F. Currents carried by sodium and potassium ions through the membrane of the giant axon of *Loligo*. *J. Physiol.* (Lond.) 116:449, 1952.

15. Hodgkin, A. L., and Huxley, A. F. A quantitative description of membrane current and its application to conduction and excitation in nerve. *J. Physiol.* (Lond.) 117:500, 1952.

16. Hodgkin, A. L., Huxley, A. F., and Katz, B. Measurement of current-voltage relations in the membrane of the giant axon of *Loligo*. *J. Physiol.* (Lond.) 116:424, 1952.

16a. Houk, J. C., and Simon, W. Responses of Golgi tendon organs to forces applied to muscle tendon. *J. Neurophysiol.* 30:1466, 1967.

17. Hoyt, R. C. The squid giant axon: Mathematical models. *Biophys. J.* 3:399, 1963.

18. Humphrey, D. R. Re-analysis of the antidromic cortical response. II. On the contribution of cell discharge and PSPs to the evoked potentials. *Electroencephalogr. Clin. Neurophysiol.* 25:421, 1968.

19. Kandel, E. R., and Tauc, L. Mechanism of prolonged heterosynaptic facilitation. *Nature* (Lond.) 202:145, 1964.

20. Lewis, E. R. An Electronic Model of the Neuron Based on the Dynamics of Potassium and Sodium Fluxes. In R. Reiss (Ed.), *Neural Theory and Modeling.* Stanford, Calif.: Stanford University Press, 1964.

21. Lewis, E. R. An electronic model of neuroelectric point processes. *Kybernetik* 5:30, 1968.

22. Lewis, E. R. Using electronic circuits to model simple neuroelectric interactions. *Proc. IEEE* 56:931, 1968.

23. Nelson, P. G., and Frank, K. Extracellular potential fields of single spinal motor-neurons. *J. Neurophysiol.* 27:913, 1964.

24. Noble, D. Application of Hodgkin-Huxley equations to excitable tissue. *Physiol. Rev.* 46:1, 1966.

25. Patton, H. D. Spinal Reflexes and Synaptic Transmission. In T. C. Ruch and H. D. Patton (Eds.), *Physiology and Biophysics.* Philadelphia: Saunders, 1965. Pp. 153–180.

26. Perkel, D. H., Moore, G. P., and Segundo, J. P. Continuous time simulation of ganglion nerve cells in *Aplysia*. In F. Alt (Ed.), *Biomedical Sciences Instrumentation*, Vol. 1. New York: Plenum, 1963. Pp. 347–357.
27. Perkel, D. H., et al. Pacemaker neurons: Effects of regularly spaced synaptic input. *Science* 145:61, 1964.
27a. Poppele, R., and Bowman, R. Quantitative description of the linear behavior of mammalian muscle spindles. *J. Neurophysiol.* 33:59, 1970.
27b. Pottala, E. W., Colburn, T. R., and Humphrey, D. R. A dendritic compartment model neuron. *IEEE Trans. Biomed. Eng.* 20:132, 1973.
28. Rall, W. Branching dendritic trees and motorneuron resistivity. *Exp. Neurol.* 1:491, 1959.
29. Rall, W. Theory of physiological properties of dendrites. *Ann. N. Y. Acad. Sci.* 96:1071, 1962.
30. Rall, W. Dendritic Trees and Neuronal Input-Output Relations. In R. Reiss (Ed.), *Neural Theory and Modeling*. Stanford, Calif.: Stanford University Press, 1964.
31. Rall, W., and Shepherd, G. M. Theoretical reconstruction of field potentials and dendrodendritic synaptic interactions in olfactory bulb. *J. Neurophysiol.* 31:884, 1968.
32. Stevens, C. F. *Neurophysiology: A Primer.* New York: Wiley, 1966. Chap. 10.
33. Stevens, C. F. Synaptic physiology. *Proc. IEEE* 56:916, 1968.
34. Tasaki, I., and Singer, I. Membrane macromolecules and nerve excitability: A physico-chemical interpretation of excitation in squid giant axons. *Ann. N.Y. Acad. Sci.* 137:792, 1966.
35. Terzuolo, C. A., and Bayly, E. J. Data transmission between neurons. *Kybernetik* 5:83, 1968.

SELECTED READING

Bullock, T. H. Neuron doctrine and electrophysiology. *Science* 129:997, 1959.
Harmon, L. D., and Lewis, E. R. Neural modeling. *Physiol. Rev.* 46:513, 1966.
Hodgkin, A. L. *The Conduction of the Nervous Impulse.* Liverpool: Liverpool University Press, 1964.
Moore, J. Specifications for nerve membrane models. *Proc. IEEE* 56:895, 1968.
Rall, W. Electrophysiology of a dendritic neuron model. *Biophys. J.* 2:145, 1962.
Stevens, C. F. Synaptic physiology. *Proc. IEEE* 56:916, 1968.

Neural Networks and Systems Modeling **32**

Donald R. Humphrey

In the preceding chapter, equations and equivalent circuits were presented that describe
the electrical behavior of different types of neural membranes. It was also shown how
these may be incorporated into a model of an entire cell. In this chapter, a number of
studies will be described in which models have been used to interpret certain aspects
of neuronal behavior at the network and systems levels.

Network level studies focus upon the behavior of a population or a group of interconnected cells. The major variable of interest is the pattern of synaptic connectivity
within the population (i.e., its internal "wiring diagram"), and the manner in which
this pattern determines the behavior of the group's single units, and its input-output
relations. It is important to note that the many theoretical studies at this level differ
considerably in their approaches, and can be placed along a continuum in terms of the
apparent biologic applicability of the models they have employed and the results that
have been obtained.

At one end of the continuum are those studies, by far the most common, in which
computer techniques are used to explore the behavior of large arrays of neuron-like
elements that are connected either randomly or with arbitrarily specified patterns.
The "elements" are, in general, highly simplified representations of known neural
properties, which is perhaps in keeping with an emphasis upon studying the behavior
that is due primarily to network connections. Studies of this type are often quite
mathematically sophisticated, and they have yielded a number of interesting hypotheses about the types of connections necessary for network stability or instability,
as well as various types of hypothetical "learning" processes. In general, however,
the hypotheses have not been amenable to experimental verification, and in some
cases the biologic significance of the models employed has not been clear; for these
reasons, they will not be discussed further here.

At the opposite end of the continuum are those studies in which carefully constructed mathematical or electronic model neurons are used to simulate the behavior
of a known neural subsystem or network [e.g., 17, 18]. Studies of this type have
often provided reasonable explanations for otherwise puzzling, experimentally observed
phenomena, and they are thus valuable adjuncts to biologic research into the properties
of small neuronal networks. Examples of this type of approach will be given in the
present chapter.

The *systems level of analysis* involves studies in which a reflex arc component, a
reflex arc, or an entire neuromuscular control system is treated as a single "black box";
control systems techniques are then used to derive transfer functions that will quantitatively link a particular set of input-output measurements under a given set of

experimental conditions. In many of these studies, the input is an external stimulus (e.g., a light, a tone, or an external force applied to a limb or joint) and the output is some measure of the behavioral response that it evokes. An example would be a study of the reflex contraction evoked in a group of muscles by a controlled stretching of their spindles; in this case, the input variable might be the angular position of the joint spanned by the muscle group, and the output variable the torque generated by their reflex contraction. Although such measurements are distinctly "nonneural," they can indicate a great deal about the information-processing properties of the neural systems that link them. Moreover, theoretical studies at this level help to establish quantitative ties between behavioral and nervous system variables; for this reason, they must be viewed as an important part of the neural modeling enterprise. The studies by Poppele and Terzuolo [25] and by Roberts et al. [28] on the input-output properties of the stretch reflex and Stark's impressive studies [32] of the hand-movement control system in man provide excellent examples of this type of approach.

The examples described in this chapter represent only a small cross section of the neural modeling studies that have been done at the network and systems levels. These examples are sufficient, however, to illustrate a number of useful modeling applications. As will be seen, a model has been used in each case to integrate and explain a number of experimental observations or to explore a hypothesis about neuronal input-output behavior that could not be fully explored experimentally.

THEORETICAL STUDIES WITH MODEL NEURONS

MONOSYNAPTIC INPUTS TO SPINAL MOTONEURONS

Intracellular recordings from spinal motoneurons have shown that the muscle spindle afferents that connect synaptically with the cell can produce excitatory postsynaptic potentials (EPSPs) with a variety of time-courses and shapes. This is true even in the case of miniature EPSPs, where each appears to be due to a solitary spike in a separate, single, presynaptic fiber; some examples are shown in Figure 32-1. The different time-courses are thus apparently not due to differences in the pattern of presynaptic discharge.

In such studies, evidence has suggested that the intracellular electrode is usually located in the cell's soma; thus, the different types of EPSPs are all observed at a single cellular location. Anatomic studies have shown, however, that each motoneuron does in fact have an extensive dendritic tree, and that synaptic terminals are distributed over its entire surface.*

Given these facts, at least two explanations can be advanced to account for the observed differences in EPSP shape. The first, which we shall call *hypothesis A,* is

---

*The synapses on each motoneuron are derived, however, from a number of incoming, ascending and descending spinal fiber tracts. Thus, the anatomic studies do not indicate whether a particular pathway impinges primarily upon a cell's dendrites, its soma, or both.

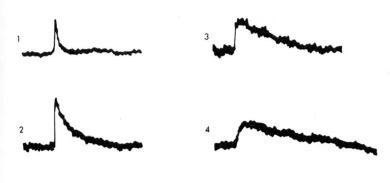

Figure 32-1
Miniature EPSPs recorded with intracellular microelectrodes from cat spinal motoneurons. Each EPSP is apparently evoked by the synchronous release of transmitter substance from the terminals of a single, presynaptic axon. The EPSP amplitudes range from 0.75 to 1.5 mV, and the horizontal bar represents 2 msec. Note the comparatively wide range of rise and decay times. (Adapted from R. E. Burke. *J. Neurophysiol.* 30:1114, 1967.)

based upon a traditional view that is held by a number of neurophysiologists concerning the primary location of the physiologically effective synapses [3]. It may be summarized as follows:

A. Because of the high dendritic core resistance, synaptic inputs to a cell's dendrites will be relatively ineffective in producing changes in the somatic membrane potential and firing rate. Consequently, the recording of significant EPSPs upon stimulation of spindle afferents indicates that they terminate primarily on or near the cell's soma. The different EPSP time-courses must therefore be due to differences in the duration of transmitter action at various subsynaptic sites.

For a number of reasons, however, Rall and others have proposed a second hypothesis, *hypothesis B,* which may be summarized as follows [26a]:

B. Anatomic measurements suggest that the dendrites of many motoneurons are not *electrotonically* longer than 1.5 to 2.0 λ, where λ is the dendrite's electrotonic length constant. Consequently, dendritic synapses may contribute significantly to the somatic membrane potentials and firing rates. This would be particularly true if the terminals on a number of branches discharged simultaneously; in this case, the electrotonic PSPs from each branch would sum linearly at the soma. Moreover, differences in the EPSP shape could simply reflect differences in the location of active synapses: even with a constant duration of transmitter action, dendritic inputs would produce slower PSPs at the soma than would somatic synapses, because of electrotonic distortion.

To test these two hypotheses, Rall [26a] therefore compared the recordings of motoneuron EPSPs obtained by Smith et al. [31] and Burke [1] with those generated by the simulated inputs to his model cell. The time-course of the synaptic conductance transient (i.e., the duration of transmitter action) and the location of the input were varied independently.

As expected, Rall found that increasing the duration of a conductance change, $\xi(t)$, that was evoked simultaneously in all compartments,* increased both the time-to-peak and the duration of the resulting EPSP. Both measurements were *also* increased by holding $\xi(t)$ constant and increasing the *distance* between the soma compartment and the *single* dendritic compartment in which it occurred. The two cases could be clearly distinguished, however, by plotting two EPSP measurements against one another. These were (1) the interval between the points on the rising and falling phases of each PSP at which it attained one-half of its peak amplitude, and (2) the time from the onset of the PSP to its peak. A plot of this type is shown in Figure 32-2. The EPSP measurements obtained by increasing the duration of ξ were all found to fall along the solid line in the figure. On the other hand, the measurements obtained by varying the dendritic location of ξ were all found to fall along the dashed curve.

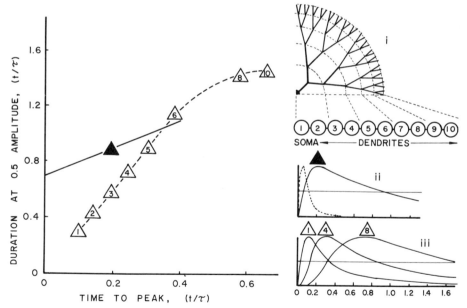

Figure 32-2
Relationships between the duration and the rise time of intrasomatic EPSPs when the duration of assumed transmitter action is varied (curve ▲) and when the dendritic location of the synaptic input is changed (curve △). The insets to the right show (*i*) the simulated dendritic tree and compartmental model used in computing the theoretical PSPs, (*ii*) the EPSP produced by an increase in excitatory conductance (dotted curve) that occurs simultaneously in all compartments, and (*iii*) the EPSPs generated at the soma by brief conductance increases in compartments 1, 4, and 8. The time scales for the inset graphs are in units of $T = t/\tau_m$. (From W. Rall et al. *J. Neurophysiol.* 30:1169, 1967. Reproduced with permission.)

*This condition is equivalent to using the one-compartment soma model that is implied by hypothesis A.

A plot of similar measurements from motoneuron EPSPs was found to agree much more closely with the dashed curve in Figure 32-2 than with the solid line. Thus, support was obtained for hypothesis B. In addition, Rall's model predicted that if hypothesis B were correct, one would expect the following experimental results: (1) because of electrotonic losses, the changes in membrane conductance measured by injecting current at a cell's soma will be smaller during long-duration (dendritic) EPSPs than during brief (somatic) EPSPs, and (2) because the areas of membrane conductance change do not overlap, the EPSPs evoked by spatially distinct synaptic inputs will sum linearly when they are evoked together. The experiments of Smith et al. [31] and of Burke [1] provided clear and direct support for both predictions. It was therefore concluded that the spindle afferent terminals are distributed over both the soma and the dendrites of a motoneuron, and, perhaps most important, that the synaptic inputs to a cell's dendrites may be quite effective in modifying its input-output behavior.

Thus, a theoretical model served in these studies not only to integrate and explain a number of observations, but also to suggest experiments that would allow an intelligent choice between two fundamentally different and competing hypotheses about the effectiveness of a neuron's synaptic inputs. This type of result illustrates, of course, one of the most important uses of a quantitative model.

INPUT-OUTPUT RELATIONS IN INVERTEBRATE NEURONS*

Because its cells are geometrically simple, easily penetrated by microelectrodes, and connected in known ways, the abdominal ganglion of the sea slug *Aplysia californicus* is often used for studies of neuronal input-output behavior. The ganglion contains a number of pacemaker neurons, which discharge spontaneously with a "natural" or characteristic interspike interval. Each pacemaker receives excitatory and inhibitory synaptic inputs from surrounding cells, however, and when an EPSP or IPSP is evoked, the interspike interval is shortened or lengthened (Fig. 32-3).

In intracellular recordings from pacemaker cells, a stable relationship has been observed between a cell's spikes and rhythmic IPSPs that are apparently evoked by inputs from another pacemaker. Each spike or pair of spikes, for example, may be followed with a constant phase lag or latency by an IPSP (Fig. 32-3B); in other cells, the ratio is reversed. with two or more IPSPs occurring between consecutive spikes. The coupling can be disrupted by stimulating the cell electrically or through one of its afferent inputs, but it is reestablished when the stimulus is removed. This type of stable, *phase-locked* behavior on the part of a group of cells is of major theoretical interest, because in this case no closed-loop feedback connections are known to exist between them.

To explore the phenomenon further, Perkel's software model (see Chap. 31) was used to determine whether or not such behavior could be expected from a pacemaker cell when it was subjected, under known open-loop conditions, to various frequencies

*See also the systems analysis approach to this problem that was taken by Terzuolo and Bayly [34].

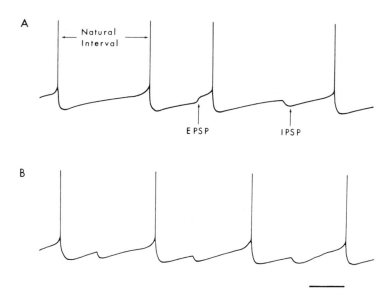

Figure 32-3
A. An example of the changes in interspike interval that are produced by EPSPs and IPSPs in a spontaneously firing, pacemaker cell. *B.* An example of the stable phase-locking between cell discharge and evoked IPSPs that is often observed in *Aplysia* pacemaker neurons. The horizontal bar represents approximately 100 msec. The records are simulated, but are quite similar to those obtained experimentally by a number of investigators.

of excitatory and inhibitory synaptic inputs [24]. Perhaps surprisingly, the simulation results predicted that a stable phase-locking should occur at certain PSP input frequencies. The results for various IPSP input frequencies, for example, are shown in Figure 32-4A, where the dashed lines indicate the segments on the input-output curve for which phase-locking was predicted. In addition, the results predicted that over a number of limited ranges, the cell's firing frequency would show an anomalous increase as the frequency of the inhibitory input was increased.*

It was, of course, important to test these predictions experimentally. For a number of reasons, however, it was not possible to control the PSP input frequency over a sufficient range in *Aplysia* neurons. Consequently, the stretch receptor neuron of the crayfish was used as an experimental model. The short dendrites of this neuron are imbedded in muscle, and they are depolarized when the muscle is stretched. In addition, the cell has direct, inhibitory fiber inputs that are easily isolated and stimulated. It was therefore possible to set up a steady depolarization at the cell's trigger zone by stretching the muscle, and thus produce a sustained, repetitive discharge much

---

*The explanation for this result is complex, but it stems from the fact that during locking, the model's discharge frequency depends *inversely* upon the phase lag between a spike and the ensuing input IPSP. With a slight increase in IPSP input frequency, the phase lag can be slightly decreased. This decrease is more than enough to offset the slowing effects of the increment in IPSP input rate, and, as a result, the cell's discharge frequency will rise.

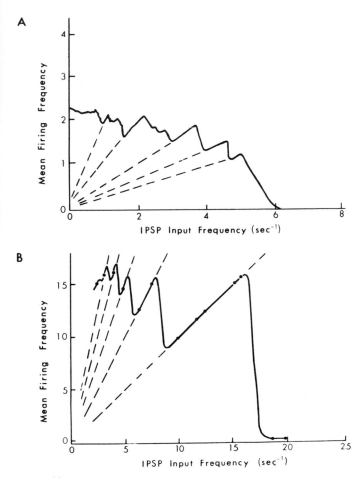

Figure 32-4
A. Theoretical relationship between mean discharge frequency and mean IPSP input frequency for a spontaneously firing, pacemaker model neuron. B. Experimentally observed relationship between firing frequency and mean IPSP input frequency in a rhythmically discharging crayfish stretch receptor (additional details are discussed in the text). (Adapted from D. H. Perkel et al. *Science* 145:61, 1964.)

like that of a pacemaker neuron. It was also possible to vary the IPSP input frequency over a wide range and observe its effects on the discharge properties of the "spontaneously" firing cell.

The major results of this experiment are shown in Figure 32-4B. The dots in this figure represent the experimental data, and the solid curve that connects them was derived (for input frequencies less than 15 pulses per second) from Perkel's model. The dashed lines again represent the zones over which input-output phase locking occurred. As can be seen, the agreement between experiment and theory was excellent, and the predicted increases in firing rate with increases in IPSP input frequency were indeed observed. Although the general correlation between input and output would be opposite to that shown, Perkel's model also predicted that such phenomena would occur as the EPSP input frequency is increased.

The most significant result of these studies, then, is the concept that the activity of a group of cells might become temporally locked in time, with stable phase relationships, even without connections of the reciprocal or closed-loop type. Although it has been verified only for certain invertebrate neurons, this concept may prove to be important for understanding the phase-locked behavior of mammalian cells that occurs in different regions of the brain during rhythmic movements and during sleep. It should also be noted that the concept would perhaps not have been derived, nor even explored, without the aid of an explicit, quantitative model.

MUTUAL EXCITATION AND INHIBITION IN SIMPLE NEURONAL NETWORKS

Examples of rhythmic, bursting behavior may be found in many neuronal systems, ranging phylogenetically from those that control movement and flight in common insects [35] to those that control respiration in man [2, 30]. Many of these systems are characterized by the presence of two reciprocally active groups of cells. In some cases, the groups consist of the separate sets of motoneurons that control reciprocally acting or antagonistic muscles; in others, they correspond to separate groups of higher-order cells, which in turn control antagonistic motoneuron pools. During rhythmic movements, the cells within each group tend to discharge in synchrony, but their collective activities alternate in a cyclic, on-off fashion (Fig. 32-5). If a pathway that excites both groups is stimulated, the average discharge rate of the cells within each

Cell 1

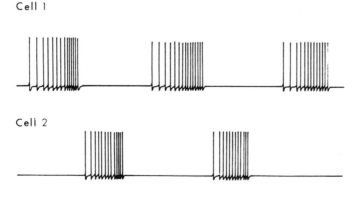

Cell 2

Time ⟶

Figure 32-5

An example of the alternating bursts of spikes that are displayed by members of reciprocally active populations of cells in many invertebrate and mammalian nervous systems. The records are theoretical, but they are similar to the extracellular recordings that have been obtained, for example, from the brain stem neurons that control the inspiratory and expiratory phases of respiration.

group may increase or the burst duration may change, but the cyclic, on-off behavior usually persists. Moreover, any input to one of the groups affects the behavior of both. The groups are thus interconnected in some type of closed-loop system.

A commonly accepted explanation for such rhythmic, reciprocal behavior is based on the concept of *mutual inhibition*. If, for example, the outputs of two groups of cells inhibit one another, they will tend to discharge in reciprocal bursts even when excited by a common, constant-frequency, driving source. The reciprocal behavior is then easily explained on the basis of adaptation. If one group is initially dominant, its cells will adapt over time and exert less inhibition on the second group; the second group will then begin to discharge in response to the common excitatory input and will in turn inhibit the first. The discharges of the two groups will thus alternate in a cyclic, on-off manner. The general validity of this explanation has, in fact, been demonstrated in a number of model neuron simulation studies [9, 18, 23a].

In attempting to use this general model to interpret neural recordings from the flight control system of the locust, however, Wilson and Waldron [36] encountered a number of difficulties. The simulation experiments that they undertook to resolve these difficulties are of particular interest, for they illustrated not only some interesting properties of simple networks, but also the value of having an explicit model available for checking intuitive explanations of neural behavior.

The basic experimental observations of Wilson and Waldron are summarized in Figure 32-6. Their recordings were obtained from two sets of motoneurons, which act to produce reciprocal, upward and downward movements of the locust's wings. When they were excited weakly by a common and maintained input, the two groups of cells were found to fire reciprocally with an average of one or two spikes per cell per cycle (Fig. 32-6A). When the input excitation was increased, two major changes were observed: first, the frequency of on-off alternation increased; second, each unit now exhibited a burst of spikes, with a firing rate that *accelerated* from the beginning to the end of each burst (Fig. 32-6B). It was also observed that the duration of the burst in one group did not necessarily correspond with the related period of silence in the other.

Upon attempting to simulate this behavior with a pair of mutually inhibiting neuromimes, however, Wilson found that a number of details could not be reproduced. The neuromimes used in this case were the two-patch models developed by Lewis that have been described in Chapter 31. The neuromimes were connected so that the output spikes from each were fed back to the inhibitory synaptic input circuit of the other. When both units were driven by a common input, reciprocal bursts indeed occurred, as is shown in Figure 32-7A. The burst duration was also increased, as in the locust, when the input intensity was raised (Fig. 32-7B). Unlike the locust pattern, however, an increasing input was found to produce (1) a *decrease* in burst alternation frequency, (2) a constant or decelerating discharge rate during each burst, and (3) an interburst interval that was proportional to the duration of the burst in the opposing unit. It was therefore clear that the simple mutual inhibition model was insufficient. Both Lewis [18] and Wilson [36] had observed, however, that accelerating bursts

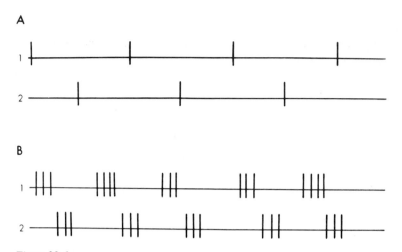

Figure 32-6
Patterns of reciprocal discharge exhibited by locust flight-control motoneurons. The two traces
in parts *A* and *B* represent typical extracellular recordings obtained simultaneously from the
motoneurons that control the upward and downward phases of wing movement in the locust.
In each trace, the abscissa indicates time, and each vertical pulse represents an action potential.
A. Responses of two antagonistic motoneurons to low-intensity stimulation of an afferent pathway that excites both populations. The members of each cell group respond with only one or
two spikes per cycle, and the activity of the two groups alternates in a regular, cyclic fashion.
It is important to note that the frequency of alternation is not directly related to the afferent
stimulation frequency. *B.* Responses to the stimulation of the afferent pathway at a higher
intensity. Note that the cells in each group now exhibit a greater number of spikes per burst,
and that the alternation frequency is increased (further discussion is given in the text). (Adapted
from D. M. Wilson and I. Waldron. *Proc. IEEE* 56:1058, 1968. Reproduced with permission.)

of spikes could be produced by coupling neuromimes in an *excitatory* manner. In
this case, a positive feedback loop was formed, and a weak input was sufficient to
cause a regenerative, accelerating discharge on the part of each unit. As the discharge
progressed, however, the factors that are responsible for refractoriness were also building up; consequently, the discharge would eventually terminate, and a period of
silence would occur until the recovery was sufficient to allow both units to begin to
discharge again. An example of this type of behavior on the part of a single neuromime is shown in Figure 32-8.

In such cases of excitatory coupling, Wilson found that an increase in input intensity produced an increase in burst duration and repetition rate that was very similar
to that observed in the locust. Moreover, by first forming two separate groups of
mutually exciting units and *then* cross-coupling them with inhibitory connections,
Wilson was able to simulate all the major features of the locust motoneuron behavior.
In this case, however, the inhibitory coupling served only to ensure a proper phasing
between the bursts of activity in each group; the properties of each burst depended
primarily upon each group's intrinsic connections. Thus, contrary to the original
hypothesis, Wilson found that the most accurate model for the locust system was one
in which *excitatory* feedback connections played a major role.

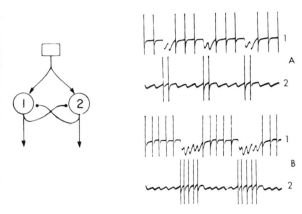

Figure 32-7
Discharge patterns produced by a common excitatory input to two mutually inhibiting neuromimes. The neuromimes employed were the two-compartment models developed by Lewis [18]
and described in Chapter 31. A common excitatory stimulus was applied to the EPSP conductance
channel of each neuromime, causing them to discharge. The output of each unit was then
coupled to the inhibitory synaptic input channel of the other, as shown in the inset. The
resulting discharge patterns are shown in *A* and *B*, where the traces are the simulated membrane
potentials recorded at each neuromime's trigger region. *A*. Responses produced by comparatively
low-intensity stimulation. *B*. Responses during stimulation at a higher intensity. Note that the
burst-alternation frequency is *lower* in part *B* than in *A*. (Adapted from D. M. Wilson and
I. Waldron. *Proc. IEEE* 56:1058, 1968. Reproduced with permission.)

Two general and important conclusions may be derived from even the few studies
and examples mentioned here. First, simulation studies of this type have made it
clear that rhythmic, bursting behavior can occur in a group of cells simply because
of the connections that exist between them. This fact has been overlooked by some
neurophysiologists, who have attempted to explain the rhythmic behavior observed
in various parts of mammalian nervous systems on the basis of spontaneous "pacemaker" cells, with unique membrane properties, such as those found in *Aplysia*.
Second, the general pattern of motoneuron behavior that was simulated by Wilson is
very similar to that observed in the respiratory systems of many mammals, including
man [2, 30], and his results may therefore be of more general significance. It is quite
clear, then, that important concepts and hypotheses can be derived from studies of
even the simplest of nervous systems.

FREQUENCY-RESPONSE PROPERTIES OF THE MAMMALIAN RETINA

If the intensity of a light is sinusoidally modulated, it will appear, at lower frequencies, to "flicker." If the modulation frequency is increased, however, a point will
eventually be reached at which the light appears to glow with a steady intensity.*

---

*This phenomenon accounts for our tolerance of fluorescent light bulbs, which actually "flicker"
at 120 Hz, or some 60 Hz higher than the average flicker detection threshold.

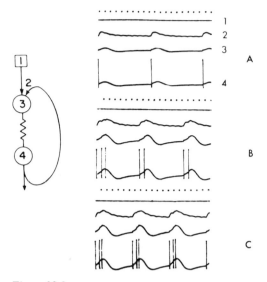

Figure 32-8
Simulation of a mutually excitatory network by a single neuromime with a positive feedback loop. The neuromime contains two simulated synaptic regions and a spike trigger zone. In the schematic network, 1 is a simulated presynaptic input source, 2 is the synaptic region to which the model's output is connected, 3 is the synaptic region to which the input is connected, and 4 is the model's trigger zone; regions 2, 3, and 4 are connected electrotonically by coupling resistors. Panels $A-C$ show the simulated membrane potentials at the four regions as monitored with an oscilloscope. The intensity of the presynaptic input (trace 1) is gradually increased from part A to part C. Note the resulting increase in the number of spikes per burst, the increased burst duration, and the accelerating discharge frequency during each burst. These results are much like those observed in the locust flight system. (Adapted from D. M. Wilson and I. Waldron. *Proc. IEEE* 56:1058, 1968. Reproduced with permission.)

The frequency at which this occurs is termed the *flicker-fusion frequency* (f_s), and its value is a function of both the average intensity of the light (I) and the modulation amplitude (i_0). If i_0 is increased, for example, then f_s is also increased. Conversely, if i_0 is held constant and I is increased, then fusion occurs at a lower frequency. The *detectability* of a flickering light thus depends upon the relative modulation amplitude, i_0/I; the greater this amplitude, the greater the range over which the mammalian visual system can sense a changing light intensity.

Although flicker fusion had been studied in various experimental contexts for over a century, no general neurophysiologic theory had been advanced for it prior to 1961. Upon reviewing studies of fusion, however, Levinson and Harmon [17] noted that the properties described above would be those that would be expected if the peripheral visual system contained an equivalent low-pass filter stage, followed by a set of neurons that acted, by virtue of their nonlinear, threshold properties, as flicker "detectors." Detection would occur, they reasoned, as long as the cells discharged with a distinct burst during each light intensity cycle, thus sending a series of repetitive, distinct signals to higher levels of the nervous system. Fusion would occur, on the other hand, when the output of the filter elements was not modulated sufficiently to cause the cells to discharge in temporally distinct bursts.

This general idea was not without support. By replotting the fusion data from previous psychologic experiments, Levinson and Harmon obtained the series of curves relating i_o and f_s at various values of I that are shown in Figure 32-9. Note, as they did, that the curves tend to merge at frequencies greater than 10 Hz, with an envelope that resembles the gain-frequency plot of a low-pass filter. In addition, previous studies of the responses of cat retinal ganglion cells to a flashing light [4] suggested that these cells might function as flicker detectors. Tracings of the responses

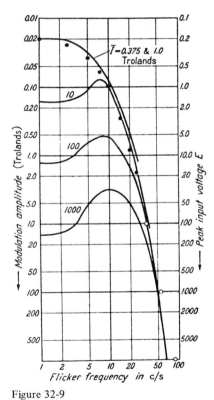

Figure 32-9
Frequency characteristics of the human "flicker-fusion" system. Each solid curve shows the modulation amplitude at which apparent fusion occurs as a function of the frequency of light-intensity modulation; note that increasing modulation amplitudes are plotted in a downward direction. The different curves were obtained with different mean light intensities ($\bar{I}$ = 10 to 1000 trolands). Light intensity is expressed in trolands, which is a measure that is proportional to both the area and the intensity of *retinal* illumination. Note that progressively greater modulation amplitudes are needed to *avoid* apparent fusion as the modulation frequency is raised, which suggests that a low-pass filtering of the signal occurs within the observer's visual system. The voltage scale on the right shows the amplitude in millivolts of the simulated light input to the electronic flicker-fusion model; the series of dots superimposed on the human frequency-response curve was obtained from the model. (Adapted from J. Levinson and L. D. Harmon. *Kybernetik* 1:107, 1961.)

of a single ganglion cell are shown, for example, in Figure 32-10. As can be seen in this figure, the number of spikes in each burst declines and the apparent phase lag (dashed lines) increases as the flash frequency is raised; both results suggest an attenuation of the cell's *input* at higher frequencies. Since the ganglion cell is only two synapses removed from the retina's photoreceptors, it was proposed, in accordance with others [4], that the filtering effects occurred in the first two cellular stages. Flicker fusion was therefore viewed as a retinal phenomenon.

To test these ideas quantitatively, Levinson and Harmon used the simple three-component model shown in Figure 32-11. It was assumed that the first stage in the light transduction process would pass a signal without distortion; its output was therefore simulated by a series of rectangular voltage pulses similar to the assumed light intensity input. The pulses were then applied to a five-stage, low-pass filter, which represented the lumped frequency characteristics of the retina's first two cellular layers. The filter converted its pulsatile input into a sinusoidal output, which was then applied to a model ganglion cell consisting of a single neuromime. Because of the neuromime's threshold characteristics, it fired only during the supra-threshold phase of each sinusoid.

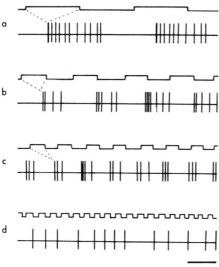

Figure 32-10   100 msec

Responses of a retinal ganglion cell to a flashing light. The photoreceptors of an anesthetized cat's retina were stimulated with a flashing light, and extracellular microelectrode recordings were obtained from one of its activated ganglion cells. The upper traces in *a–d* represent the stimulus signal, and an upward deflection indicates the period during which the light is on. The lower lines are tracings of the simultaneously obtained microelectrode recordings, with each upward deflection representing a spike. The apparent phase lag of the cell's response is indicated by dashed lines, which connect the first evoked spike to the onset and termination of the associated flash. The flash frequency is gradually increased in *a* through *d*. Note that the apparent phase lag increases as the frequency is raised, and that the cell's responses become less vigorous. At the highest frequencies shown, the cell fails to respond to each flash. (Adapted from C. Enroth. *Acta Physiol. Scand.* 27:Suppl. 100, 1952.)

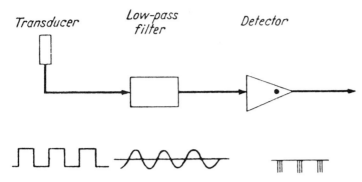

Figure 32-11
The general model used by Levinson and Harmon in simulating the frequency-response characteristics of the retina. The first stages of the retinal transduction process were assumed to convert
a square-wave light stimulus into a square-wave physiologic signal. Subsequent stages introduce
low-pass filtering, which converts the square waves into sinusoidal waves. The ganglion cells then
act as detectors, responding with a burst of spikes only during the suprathreshold phases of the
sinusoidal input signal. (Adapted from J. Levinson and L. D. Harmon. *Kybernetik* 1:107, 1961.)

Some of the model's responses to various input frequencies are shown in Figure 32-12.
Two response properties similar to those of ganglion cells are obvious: the apparent
phase lag is increased and the number of spikes per cycles is decreased as the input frequency is raised. In this example, the phase-locked discharge of the neuromime was
abolished at 25 Hz, thus defining a "fusion" frequency for the particular input intensity employed.

By determining the model's fusion frequency for various input amplitudes, Levinson
and Harmon were then able to reproduce the envelope of the curves obtained in
previous psychophysical studies; their theoretical results are illustrated by the series
of dots that are shown in Figure 32-9. They stressed, however, that their model
was only a first approximation, and that the low-frequency portions of the curves
might also have been reproduced had they incorporated frequency-dependent or
intensity-dependent feedback. Such feedback could represent, for example, the lateral
inhibitory connections that are known to exist in the retinas of mammals and other
species.

Although they were only a first approximation, the results obtained by these
investigators were nonetheless in reasonably good agreement with existing experimental data. In addition, Levinson and Harmon were able to derive predictions from
their model that were verified in subsequent psychophysical experiments. It is also
significant that later studies of single photoreceptors and ganglion cells have provided
evidence in support of the general type of filtering operation that they postulated [19].
Their model was therefore phenomenologically correct, and it served the important
function of drawing together and offering explanations for apparently diverse types
of experimental data.

In addition to the application described here, Harmon's model neuron has been
used to study the behavior of simple networks [9] and the mechanisms of signal

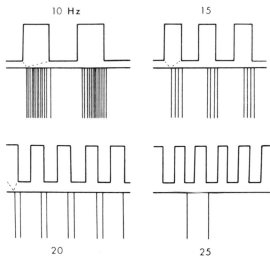

Figure 32-12
Responses of the fusion model to simulated light flashes. The upper traces represent the light-flash signal, which is delivered at the indicated frequency. The lower traces show the simulated output spikes generated by the model ganglion cell. Note the decrease in burst duration and firing frequency and the increasing phase lag (dashed lines) as the flash frequency is raised. Fusion occurs at 25 Hz when the model fails to respond with a discrete burst of spikes to each simulated light flash. (Adapted from J. Levinson and L. D. Harmon. *Kybernetik* 1:107, 1961.)

processing in the mammalian auditory system [10]. In each of these cases, the biologic problem selected for study has been a significant one, and the simulation results have been illuminating. The serious student can learn much from a careful reading of these well-written and important papers.

USING MODEL NEURONS TO INTERPRET EXTRACELLULAR RECORDINGS

During a localized, excitatory increase in membrane conductance, current will flow into a cell across the short-circuited area of the membrane and along its interior; it will then flow outward with varying density across the surrounding regions of membrane, and it will return to the site of the conductance change through a series of divergent, looping pathways in the extracellular medium (Fig. 32-13). With a hyperpolarizing increase in the inhibitory membrane conductance, the flow of current will be in the opposite direction. In both cases, however, the flow of charge across the resistive extracellular medium will generate a potential field, whose temporal variations will be in synchrony with those of the cell's transmembrane current distribution. The extracellular current density and voltage gradient will be greatest in the vicinity of the cell, but current may flow throughout the surrounding medium. A time-varying voltage may therefore be recorded by placing a microelectrode in the vicinity of the cell and a reference lead at some other, usually remote, location in the same volume conductor.

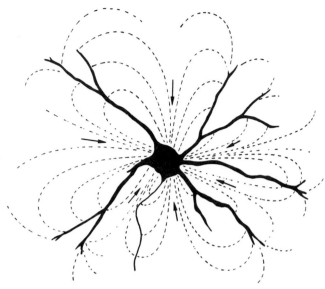

Figure 32-13
General directions of extracellular current flow around a stellate-shaped neuron during a depolarizing increase in membrane conductance at its soma. In this case, the general direction is radially
inward, toward the soma. Although not shown, the current loops are completed by an oppositely
directed flow along the interior of each dendrite.

The extracellular potentials that are generated and recorded in this way have provided neurophysiologists with much of their present information about the properties
and behavior of central nervous system (CNS) cells. It is, of course, possible to penetrate CNS cells and to record their transmembrane potential variations directly. Electrode stability and cell injury are always problems in such cases, however, because of the
mechanical vibrations arising from nearby, pulsating blood vessels and from the animal's
respiratory movements. In addition, stable recordings are usually obtained only when
the electrode tip penetrates the cell's soma, where the surface area is comparatively
large and slight movements produce less damage than would occur in the case of the
penetration of a slender and fragile dendrite.

There are certain questions of importance, however, that cannot be answered from
transmembrane potential recordings at a single cellular locus. A question of this type
that is of major theoretical interest concerns the excitability properties of a cell's
dendrites.* If a neuron's dendrites are excitable, for example, it will respond with an
action potential to a suprathreshold, synaptic input at any point on its surface, much
like a simple trigger element or switch. If the dendrites are passive, however, the cell

*For purposes of discussion, it was assumed in the section on membrane patch models that the
dendrites of a typical neuron are inexcitable. This is not universally true, however, and the
question of dendritic excitability has not yet been answered for many types of cells in the mammalian CNS.

is capable of more complex forms of spatial and temporal summation; the inputs to dendritic regions will affect the cell's output less than those near the trigger zone, and it will behave more like a spatially distributed integrator in series with a switch. Clearly, then, the question of dendritic excitability is important for understanding input-output relations at the single-neuron level; it cannot be answered, however, by recording only a cell's somatic potential. In an attempt to resolve this problem, neurophysiologists have therefore recorded the *extracellular* potentials generated by a single active neuron or by a group of synchronously discharging cells [5, 6, 21]. In each case, it has been assumed that the potential waveforms and their distribution around a cell with active dendrites would differ significantly from those around a cell with passive dendrites.

One of the first and technically best of these studies was that conducted by Fatt [5] . By carefully dissecting its axon free from an adjacent muscle nerve, Fatt was able to antidromically* activate a single motoneuron and to record the potentials that it generated in the spinal cord of an anesthetized cat; some of the waveforms that he recorded are shown in Figure 32-14B. Having obtained these data, however, he was faced with a very difficult problem in interpreting them. Although it was known that most motoneurons have a shape similar to that shown in Figure 32-14A, it was not actually possible to visualize the cell that was being recorded from; consequently, there was no way of precisely correlating the observed potentials with the cell's dendritic geometry. Moreover, no quantitative model existed that would allow an estimate of the types of potentials to be expected from a cell with excitable dendrites, as opposed to one with passive dendrites. Fatt was therefore forced to rely primarily upon intuitive interpretations. He noted, for example, that each recorded waveform had a prominent inflection on its rising phase, which suggested that it was the sum of two separately generated potentials, A and B. He was able to verify this by showing that only the A component was evoked by the second of two closely spaced stimuli, apparently because the area responsible for the B component was refractory. He also noted that the potentials were primarily negative, even though their small amplitude suggested that they were recorded from the periphery of the cell's dendritic tree. Since negative potentials had been recorded most often in the past when an electrode was near a region of active membrane, this finding suggested that the dendritic terminals were actively depolarized. Fatt also noticed that the latency of the B peak increased as the electrode was withdrawn from the apparent region of the cell body, which suggested the existence of a process that propagated at a finite velocity. He therefore concluded that the A component arose from an action potential in the cell's soma and the B component from a propagating action potential in its dendrites.

It was subsequently pointed out by Rall [26], however, that waveforms similar to those recorded by Fatt would also be expected from a motoneuron with *inexcitable*

---

*If a cell's axon is stimulated electrically, one action potential is sent toward the axon terminals and another toward the cell body. If the cell's soma and dendrites are excitable, both will be invaded by an action potential. This type of activation is referred to as *antidromic* in order to distinguish it from the normal sequence, in which a cell is activated synaptically and the action potential propagates from the cell body toward the axon terminals.

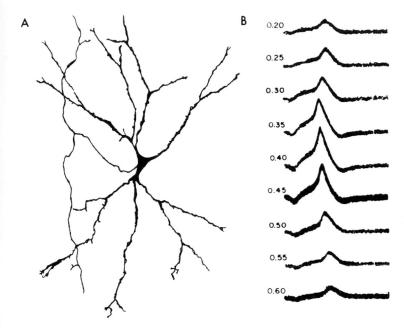

Figure 32-14
Extracellular potentials generated by the antidromically evoked discharge of a single motoneuron.
A. Camera-lucida tracing of a Golgi-stained, spinal motoneuron showing the general shape of the
type of cell from which the recordings were made. Five or six dendritic trunks and a bifurcating
axon can be seen to emerge radially from a central soma. *B.* Extracellular potentials recorded
along a single electrode track in the vicinity of an antidromically excited motoneuron. The numbers to the left indicate the distances along the track in millimeters from an arbitrary zero point
at which the recordings were obtained. Each trace is 2 msec in duration and consists of a number
of photographically superimposed responses. The upward direction indicates negativity of the
recording electrode in all traces. The largest spike shown is approximately 0.6 mV in amplitude.
(Part *B* adapted from P. Fatt. *J. Neurophysiol.* 20:27, 1957. Reproduced with permission.)

dendrites. Using a simple model cell with passive dendritic trunks, Rall computed a
theoretical, extracellular action potential that was quite similar to those observed by
Fatt. Subsequent studies have tended to support Rall's interpretation [26], and it is
now believed that motoneuron dendrites are passive, and that the *A* and *B* components
represent the discharge of the cell's initial segment and soma. In spite of Fatt's im-
peccable logic, then, his intuitively based conclusions were apparently wrong.

Because of the widespread use of extracellular recording methods in neurophysiology,
many interpretive problems of this general type still exist and will no doubt continue to
arise; it will therefore be appropriate to show how a model neuron and a digital computer can often be used to solve them. The general method has been outlined previously
by Humphrey [15]. In the example to be given here, which both confirms and extends
Rall's results, a model neuron was used to estimate the extracellular action potentials that would be generated by a cell with passive dendrites and by one with active
dendrites.

A theoretical motoneuron was first constructed using Rall's three-halves power
dendritic branching rule (see Chap. 31; Eq. 31-57). The cell consisted of a central
soma and six dendritic trunks that divided into 12 secondary and 24 terminal branches;
a scale drawing is shown in Figure 32-15. The cell was constructed so that it could be
represented mathematically by a ten-compartment model; its geometrical dimensions
and the regions represented by each compartment are shown in Table 32-1.

Two general sets of computations were then performed. In the first, an intracellular action potential similar to that recorded by Fatt was assumed to occur in the
soma (see Fig. 32-15). The compartmental equations were then used to compute the
intracellular voltages that would occur as the action potential spread electrotonically
into a chain of *passive* dendritic compartments. In the second case, the dendrites
were assumed to be excitable, and the action potential was allowed to propagate
along them with a velocity that was proportional to the dendritic branch diameter.
Both sets of computations were performed for each 0.05 msec increment in time
over a span of 2 msec.

Each of the instantaneous voltage distributions was then used in Equation 31-60A,
with ξ and ζ set equal to zero, to compute a *membrane current-density* distribution.
These distributions were then mapped onto the model cell's soma and dendrites. The

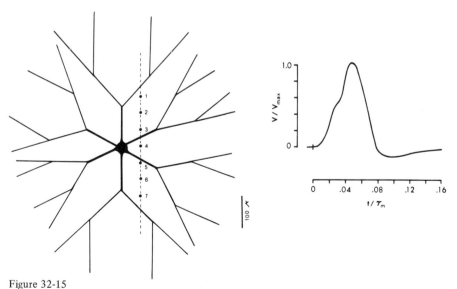

Figure 32-15
Left: Scale drawing of the model cell used in computing theoretical, extracellularly recorded
action potentials. Six branching dendrites emerge from a central, spherical soma, four along the
horizontal plane and two vertically. The dashed line represents a hypothetical electrode track
that travels vertically between the two dendritic trunks on the right. The numbered dots desig-
nate "recording" points along the track for which theoretical waveforms were computed (additional discussion is given in the text). *Right:* Time-course of the action potential that was
assumed to occur in the model cell's soma. The waveform is very similar to that recorded intracellularly by Fatt [5].

Table 32-1. Geometrical Features of the Model Neuron Used in Computing Theoretical, Extra-cellularly Recorded Action Potentials[a]

Region	Diameter (microns)	Model Compartment
Soma (spherical)	34.0	1
Dendrites:		
Primary trunks	3.0	2, 3
Secondary branches	1.9	4, 5, 6
Tertiary branches	1.2	7, 8, 9, 10

[a]In addition to these dimensions, the total electrotonic length of the cell was 1.8 λ, and the length-constant of the dendritic trunks was 335 microns. The total radial extent of the dendritic tree was 460 microns.

center of the soma was located at the origin of a three-dimensional coordinate system, so that the coordinates of each point on the cell's surface were known. The mapping procedure thus yielded a set of current source-sink distributions, each representing the distributions of inwardly and outwardly directed membrane currents over the surface of the cell at a single point in time.

An equation that had been derived previously by Stevens [33] and by Geselowitz [7] was then used to estimate the potentials that would be generated by each of the membrane current distributions if the cell were located in a homogeneous and isotropic conductor.* The equation is an approximate one, and it has the form:

$$\phi(p,t) \approx \frac{1}{4\pi\sigma} \int_A \frac{J_n(p',t)}{r(p',p)} dA \tag{32-1}$$

where $\phi(p,t)$ = the potential at the extracellular point, p, at time t

$\quad$ $J_n(p',t)$ = the magnitude of the normal component of the membrane current-density vector, at point p' on the surface of the cell and at time t

$\quad$ $r(p',p)$ = the linear distance between the point at which ϕ is evaluated and the point at which J_n is evaluated

$\quad$ dA = a differential element of membrane surface area

$\quad$ σ = the electrical conductivity of the extracellular volume

As indicated by the subscript A, the integral in Equation 32-1 is to be evaluated over the coordinates corresponding to the surface of the cell. It should also be noted that J_n is positive for an outwardly directed membrane current and negative where the current flows into the cell.

*Neither the brain nor the spinal cord is an isotropic, homogeneous conductor; a number of useful and instructive theoretical results can be obtained, however, by assuming them to be so. For a general, quantitative discussion of potentials in a volume conductor, as well as a number of related theoretical studies, the reader should consult Plonsey's well-written and authoritative book, *Bioelectric Phenomena* (see Selected Reading).

By evaluating Equation 32-1 numerically at successive points in time, a number of theoretical, extracellular action potentials were computed. The waveforms were those that would have been "recorded" at equally spaced points along the hypothetical electrode track shown by the dashed line in Figure 32-15. The potentials were evaluated at the same extracellular points for both cases of dendritic excitability.

The results of these computations are summarized in the two columns of theoretical, extracellular action potentials shown in Figure 32-16. By comparing these waveforms to those recorded by Fatt (see Fig. 32-14B), it can be seen that his data correspond most closely with those that would be expected from a cell with passive or *inexcitable* dendrites. Note that all of the major features of his data are reproduced theoretically, including the negativity of the potential and the increasing latency to its peak at remote regions of the cell's dendritic tree. Moreover, from the model and Equation 32-1, the polarities of both the observed and the theoretical potentials are easily explained. Because the dendritic membrane currents are distributed over a wide area, the potentials at all points more than a few microns from each dendrite are determined primarily by the time-course and the direction of the concentrated membrane current flow that

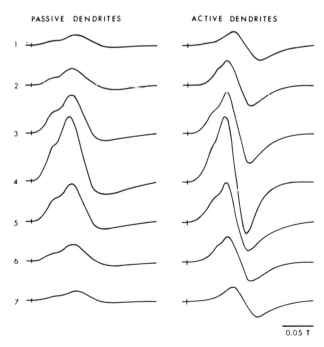

Figure 32-16
Theoretical extracellular potentials generated by the discharge of a model stellate cell with passive dendrites (*left*) and with excitable dendrites (*right*). The potentials shown in each column are those computed for the hypothetical recording positions shown in Figure 32-15. Negativity is indicated by the upward deflection in each trace. The time scale is in units of $T = t/\tau_m$; to convert to real time, a τ_m value of 5 msec is appropriate. Note the excellent agreement between the waveforms shown on the left and those recorded experimentally by Fatt (see Fig. 32-14B).

occurs at the soma. During an action potential at the soma, the membrane current flow is inward, and the potential field is negative with respect to a distant reference point. If the dendritic tree is actively invaded by the action potential, however, the current at the soma will suddenly reverse from inward to outward as this region becomes a source for the rapidly depolarizing dendrites; in this case, the field potentials will exhibit first a negative and then a positive peak, as shown in Figure 32-16.

It should be clear from this example that the extracellular waveforms that are generated by a cell will depend not only upon its regional membrane properties and the nature of the membrane conductance change, but also upon the cell's geometrical configuration. The potentials shown in Figure 32-16, for example, differ considerably from those reported previously by Humphrey [15] for a pyramidally shaped model neuron, even though the assumed dendritic properties and intracellular events were identical. For these reasons, it is clear that estimates of the cellular events underlying a set of extracellularly recorded waveforms are best obtained, when possible, with the aid of an explicit, quantitative model.

Before concluding this section, two related modeling applications may be noted. In the first, Humphrey [15] used a model cortical neuron to estimate the extracellular fields that would be generated by intrasomatic action potentials and PSPs. From his results, he was able to derive new theoretical explanations for a previously known but poorly understood fact: namely, that the EEG potentials recorded at the surface of the brain appear to arise more from low-amplitude cellular PSPs than from larger amplitude, synchronous action potentials. These results have been viewed by several authors as important for our current understanding of the cellular bases of the electro-encephalogram; the original article by Humphrey may be consulted for additional details

The second application was reported by Rall and Shepherd [27]. Using simple, cylindrical model neurons, these investigators attempted to reconstruct the evoked extracellular potentials that they had recorded from the olfactory bulb of the rabbit. They were able to do so, however, only by assuming that there were direct synaptic connections between the *dendrites* of the bulb's mitral and granule neurons. Although such dendrodendritic synapses were almost unheard of at the time, they have since been found by electron microscopists not only in the area postulated by Rall and Shepherd, but also in widespread regions of the mammalian brain. The biological significance of this type of modeling result needs no further emphasis.

There are many additional problems in neurophysiology that can benefit from theoretic modeling applications of the general type described here. These applications will be profitable, however, only in those cases in which the extracellular potentials that are simulated are those that have been recorded from anatomic regions with known cellular geometries and orientations. It is therefore necessary for the theoretician to collaborate with the experimental physiologist or to obtain his own experimental data.

THE SYSTEMS APPROACH IN NEURAL MODELING

The modeling applications discussed above have been primarily at the single-neuron level. From studies of this general type, it is quite clear that the input-output relations of any single neuron are distinctly nonlinear. Each cell exhibits, for example, a definite threshold and a saturation level in the curve that relates its input and output firing frequencies; moreover, the studies of Perkel [24] and others have shown that a cell's spontaneous rhythms may introduce nonlinear input-output irregularities even within intermediate firing-rate ranges. Indeed, it is the difficulty in handling such behavior mathematically that has prompted the development of a number of electronic model neurons.

An important property of the mammalian nervous system, however, is that many of the nonlinearities that are evident at the single-cell level tend to become functionally less important when one considers the input-output behavior of a cellular *population*. The fibers that connect synaptically with CNS cell groups, for example, tend to branch repeatedly and to diverge, so that each cell receives terminals from a large number of separate, but functionally related, axons. Because of these converging terminals and the low-pass filtering introduced by synaptic transmission, the effective input to each cell's trigger zone will consist of a weighted, highly smoothed sum of the firing frequencies in a large number of presynaptic terminals. Moreover, this type of mixing and temporal smoothing of inputs occurs at each synaptic relay. For these reasons, short-range irregularities in the input-output relations of individual cells (e.g., as shown in Fig. 32-4) tend to become less important when one considers the flow of information through neuronal populations.

In addition to this smoothing or averaging property, mechanisms exist within cellular populations for extending the dynamic input-output ranges considerably beyond those exhibited by single neurons. Because of differences in the density of synaptic inputs, for example, various cells within a CNS population will be recruited into activity at different levels of afferent input intensity.* Consequently, as the input intensity is raised, an increasing *number* of cells can be brought into play, thus compensating for the firing-rate saturation that is occurring in cells recruited at lower input levels. In this way, the population's collective output can be graded over a much wider range than would be anticipated from a knowledge of single-cell input-output relations alone. This type of range extension is also found at the receptor level in various sensory systems and at the motoneuron and muscle fiber levels in the mammalian motor system [1a, 8, 25].

Perhaps because of these mechanisms and others that are yet to be clarified, a number of investigators have found that a linear systems approach can be profitably employed in studying the input-output behavior of neurologic subsystems. Among the early and most important contributions in this area are the studies by Stark and his colleagues of the systems that control pupillary diameter, eye movements, and hand movements in man. Since this work has recently been summarized by Stark [32], however, it will not be discussed further here. We will instead concentrate upon a small number of more recent studies, in which systems techniques have been used to define the input-output relations of reflex arc components and of an entire reflex system.

---

*In the present context, *input intensity* is defined as the total number of impulses per unit time in a group of functionally related presynaptic fibers; it thus depends upon both the number of active fibers and their respective firing frequencies.

SYSTEMS REPRESENTATION OF THE STRETCH AND
TENDON-ORGAN REFLEXES (SEE ALSO CHAPTER 27)

The basic stretch and tendon-organ reflex arcs that act to control the overall length and tension of each skeletal muscle are shown schematically in Figure 32-17A.

Consider first the *stretch reflex* system. A given skeletal muscle will contain from 30 to 500 muscle spindles, all arranged in *parallel* with its contractile fibers (see Chaps. 5 and 27). Each spindle consists of one to three elongated nuclear bag fibers and four to six parallel nuclear chain fibers, surrounded by a single connective tissue sheath.* Each spindle is also innervated by three types of axons, two of which are sensory and one motor in function. The pattern of innervation is as follows:

1. *Sensory.* A single group IA axon pierces each spindle's sheath and divides into a number of branches. A separate branch then wraps around the central portion of each bag and chain fiber, thus forming a number of *annulospiral* endings. Because each of these terminals is depolarized by the stretching of its associated fiber, the firing pattern in the parent IA axon will be a function of the mechanical properties of both the bag and the chain fibers. These fibers differ somewhat in their distributed viscoelastic properties, with the net result that each group IA axon tends to discharge at a rate that is related to both the degree and the velocity of spindle stretch. Each spindle is also innervated by a number of smaller, group II axons, which terminate largely upon chain fibers. Their central connections are different than those of the group IA fibers, however, and their exact functions are unknown.

2. *Motor.* In addition to having centrally located, sensory terminals, each spindle fiber is innervated at its polar ends by small *gamma* motor axons, whose cell bodies lie within the spinal cord. When these cells discharge, a contraction is produced at the ends of

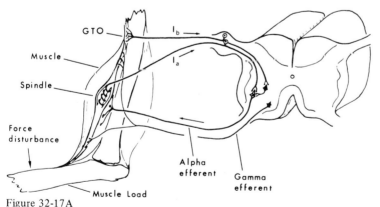

Figure 32-17A
Stretch and tendon organ reflex arcs, showing the orientation of the muscle's receptors, the afferent pathways, central synaptic connections and efferent pathway. The black neuron within the schematic cross section of the spinal cord represents a population of inhibitory interneurons. *Arrows* indicate the direction of "information flow" during a reflex contraction that is produced by applying a force disturbance to the muscle's attachments or its mechanical load. GTO = Golgi Tendon Organ.

*The nuclear bag and chain fibers are so named because of the geometrical distribution of their multiple nuclei; their sensory innervations and mechanical properties differ in important ways. For a comprehensive and up-to-date discussion of the muscle spindle and its role in the control of movement and posture, the reader may consult the excellent book by Matthews [20].

each spindle fiber, thus imparting a stretch to its central portion. As a result, the sensory terminals are depolarized and are caused to discharge. By way of synaptic inputs to gamma motoneurons, then, various supraspinal motor centers can affect the firing rates of spindle afferents and thus the pattern of sensory information that they provide to the CNS.

Consider now the flow of information around the reflex loop when a muscle is stretched by means of an external force that is applied to its attached limb; for convenience, let us refer to both the anatomic diagram in Figure 32-17A and the equivalent systems diagram shown in Figure 32-17B. As a result of the imposed stretch, a fraction of the muscle's spindle afferents will begin to discharge at rates that are functions of the magnitude of the stretch and the derivatives of this variable with respect to time. Within the spindle afferent population, however, there is a distribution of apparent stretch thresholds; consequently, the collective signal sent to the CNS will depend not only upon single-fiber firing rates, but also upon the number of axons that are active at a given amplitude and velocity of stretch. This signal will then be transmitted through monosynaptic connections to the population of alpha motoneurons that innervate the *same* muscle from which the afferent input arose. The motoneurons will therefore discharge, both in increasing numbers and with increasing rates, as the intensity of afferent bombardment is raised. As a result, the muscle will begin to contract, thus generating a force that opposes the initial disturbance. The magnitude of this force will depend not only upon the muscle's neural input, however, but also upon its inherent mechanical properties: at shorter lengths and greater velocities of shortening, its force or tension output will be reduced. This dependence of force output upon a muscle's passive mechanical properties is indicated by a separate loop in Figure 32-17B.

From this brief description and the related control diagram, it should therefore be clear that the stretch reflex arc is actually part of a closed-loop, negative feedback system, that acts to regulate a muscle's length.

Let us turn now to the *tendon-organ reflex* arc. Each tendon is innervated by large-diameter, group IB axons, whose terminals form, in conjunction with the tendon's fascicles, a number of mechanically sensitive Golgi tendon organs (GTOs). Because these mechanoreceptors are geometrically in *series* with the muscle's contractile fibers, they will discharge in response to the forces generated by muscular contraction as well as to those applied externally. Recent studies have, in fact, suggested that over the normal range of in vivo changes in muscle length, the effective force input to the GTO population arises primarily from active muscular contraction [13].

Consider now the activity within this reflex arc during a neurally evoked contraction. As a result of the stretching forces imposed by the contracting muscle, its tendon afferents will discharge and excite a set of inhibitory interneurons within the cord; the discharge of these cells will, in turn, reduce the output of the motoneurons that excite the contracting muscle. Thus, the tendon reflex arc is actually part of a negative feedback system that regulates a muscle's contractile force output. As can be seen in the control diagram, this system opposes the stretch reflex and reduces its overall gain. It has been proposed that this opposition tends to linearize the stretch reflex and to reduce the dependence of its gain upon mean muscle length and fatigue.

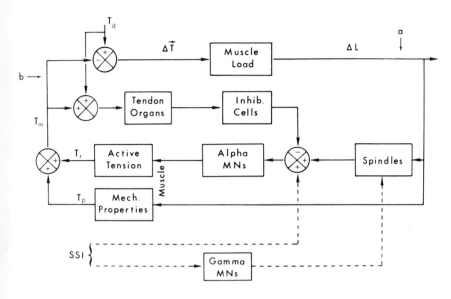

Figure 32-17B
An equivalent control or information-flow diagram for the stretch and tendon organ reflexes.
T_d = the tension or force disturbance commonicated to the muscle's tendon; ΔT = the net tension increment produced at the muscle's tendon by applied and reflex-generated muscle tensions; ΔL = the resultant change in muscle length or angular joint position; T_r = the reflexly or neurally generated component of muscle tension; T_p = the passive component of muscle tension; T_m = the net tension generated by the muscle; SSI = neural inputs from supraspinal structures. The arrows a and b indicate the points at which the overall control loop is broken in order to study its input-output properties.

The behavior of each skeletal muscle if regulated, then, by a basic *unit* control system of the type summarized in Figure 32-17B. In addition, the more complex systems that control limb position and movement each contain a number of these basic units, interconnected by both excitatory and inhibitory pathways [12]. A knowledge of the unit's dynamic properties is therefore essential for an understanding not only of simple reflex behavior, but also of the neural control of posture and voluntary movement. For this reason, a number of investigators have, in recent years, used control systems techniques to study the behavior of the stretch and tendon-organ reflexes and of their components. An important goal has been that of obtaining a set of transfer functions that might be used in modeling studies of the more complex systems in which these components and reflexes are embedded.

DYNAMIC PROPERTIES OF THE STRETCH REFLEX*

Recognizing the need for a quantitative approach in motor systems physiology, a number of investigators have employed the methods of control systems analysis in

*As can be seen in Figure 32-17, the tendon arc is an integral part of the overall stretch reflex; it has been treated as a separate reflex system in the preceding discussion only for convenience.

studying the behavior of stretch reflex arc *components* [11, 14, 22, 24a]. Houk et al. [14], for example, have developed models of the in situ spindle and tendon organ that have aided considerably in understanding their dynamic behavior; in addition, Partridge [22] has used frequency-analysis methods to characterize the relationship between motor nerve impulse frequency and evoked muscle tension. By far the most complete quantitative analyses of a stretch reflex system reported to date, however, are those performed by Poppele and Terzuolo [25] and by Rosenthal et al. [29]. Because these studies both confirm and integrate the findings of many previous authors we will concentrate in the present section upon a summary of their methods and major results.

To study the input-output behavior of the stretch reflex and its components, these workers chose the gastrocnemius and soleus muscles of the cat, which act to extend its hind foot. The basic experimental arrangement that they employed is summarized schematically in Figure 32-18. After anesthetizing the animal and immobilizing its knee, the tendons of one or more muscles were dissected free and coupled by means of a low-compliance string to the offset shaft of a feedback-controlled DC motor. A photocell was used to measure and control the linear displacement of the shaft and the attached muscle; a strain gage on the shaft also measured the tension exerted by the muscle and its attachments. With the use of this apparatus, the experimenters were able to stretch the muscle sinusoidally over a full range of physiologically important frequencies (0.05 to 20 Hz) and measure its net tension output. Note in the control diagram of Figure 32-17B that this procedure is equivalent to removing the muscle's load, applying a muscle-length forcing function or input at point *a*, and measuring the net tension output at point *b*. Note also that because the animal is anesthetized, the descending inputs to alpha and gamma motoneurons are presumably held constant.

As can be seen in the preceding control diagram, the increment in muscle tension (ΔT) that is produced by an increment in length (ΔL) will consist of two components: (1) a reflexly-generated component (ΔT_r), and (2) a component (ΔT_p) due to the

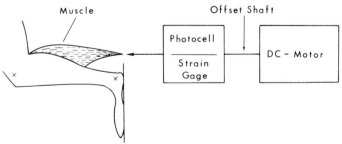

Figure 32-18
Simplified diagram of the procedure used in studying the input-output behavior of a stretch reflex system. The *x*s indicate the points at which the animal's hindlimb was immobilized (additional details are given in the text). (Adapted from R. Poppele and C. A. Terzuolo. *Science* 159:743, 1968.)

passive compliance properties of the muscle and its attachments. Because these components may differ in phase as well as in amplitude, they are best viewed as vectors (as indicated by bold-face type). If it is further assumed that they are additive, a *dynamic stiffness vector* (**S**) may then be defined by the equation,

$$S = \frac{\Delta T}{\Delta L} = \frac{\Delta T_r}{\Delta L} + \frac{\Delta T_p}{\Delta L} \tag{32-2}$$

Note now that if the dorsal roots that contain the muscle's spindle and tendon afferents are cut, the reflex loop is opened and ΔT_p can be studied in isolation. Thus, although the reflex component cannot be measured directly, it can be estimated by means of subtraction techniques from the measurements obtained under closed-loop and open-loop conditions.

Having observed these relationships, Rosenthal et al. [29] first studied the linearity of the system under both closed-loop and open-loop conditions. With the reflex loop intact, the mean output tension was varied by stretching the muscle about different mean lengths; with the loop open, the mean tension was adjusted by stimulating the muscle's cut nerve electrically at a particular intensity and pulse frequency. In both cases, the gain of the system (gain = ΔL) was found to increase as the mean output tension rose, which indicated an overall nonlinearity. The *shapes* of the gain and phase plots were independent, however, of the mean output level, and over limited ranges, both the muscle and the intact system behaved in a near-linear manner; a piecewise linear systems approach was therefore justified.

The experimental frequency-response curves obtained by Rosenthal et al. are shown in Figure 32-19 for both the muscle and the intact reflex system. Also shown are the curves for the reflex component alone ($\Delta T_r/\Delta L$ and ϕ_r versus frequency) that were estimated by Roberts et al. [28] from Rosenthal's data. The closed-loop and open-loop curves can be seen to differ somewhat, particularly with regard to the phase angle; moreover, the reflex component exhibits a significant reduction in gain and an increasing phase lag at frequencies greater than 8 Hz. The most significant feature of the results, however, is the near-flatness of the reflex gain and phase-angle curves between 0.2 and 10 Hz, a frequency span that corresponds to the physiologically important range in both postural and voluntary motor control systems.

In view of the known slow dynamics of a muscle's tension response to a neural input [22], it was clear that a system frequency response of this type could occur only if a frequency compensation was performed by some other reflex component. Because of the rate-sensitivity of muscle receptors, it appeared that their dynamic behavior might provide such a compensation [23]. In order to explore this point further, frequency-response curves were therefore obtained for each major component of the stretch reflex system. The basic measurements included (1) the changes in firing rate exhibited by single-spindle and tendon-organ afferents during sinusoidal muscle stretching, (2) the reflex changes in firing rate displayed by single motoneurons during muscle stretching, and (3) the tension responses of the muscle to stimulation of its cut nerve with sinusoidally modulated pulse amplitudes and frequencies. This last

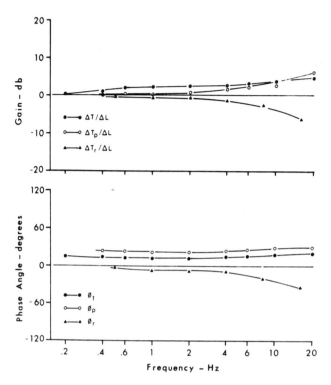

Figure 32-19
Frequency-response characteristics of the stretch reflex under closed-loop and open-loop conditions. The closed-loop curves (●) were obtained from the gastrocnemius muscle with the reflex
loop intact. The open-loop curves (○) were obtained from the same muscle after transecting its
afferent innervation. The curves representing the frequency characteristics of the reflex system
alone (▲) were obtained by a subtraction technique (additional information is given in the text).
(Adapted from N. P. Rosenthal et al. *J. Neurophysiol.* 33:713, 1970.)

procedure mimics sinusoidal changes in both motoneuron recruitment and firing
rate.

   The major results of these experiments are summarized in Figure 32-20. Three
points should be noted. First, both the spindle and the GTO afferents display a gain
and leading phase angle that are increasing functions of frequency. Second, the gain
and phase-angle curves describing the response of alpha motoneurons to muscle
stretching are strikingly similar to those exhibited by the muscle's receptors; thus,
over the range of frequencies explored, no major temporal changes were introduced
by the synaptic transfer of information from the muscle's receptors through its
motoneuron pool. Finally, note that the gain and phase-angle changes exhibited by
the muscle during a simulated input from its motoneurons are essentially a mirror
image of those exhibited by its receptor afferents. Thus, it is quite clear that the
dynamic behavior of the muscle's receptors compensates significantly for the phase

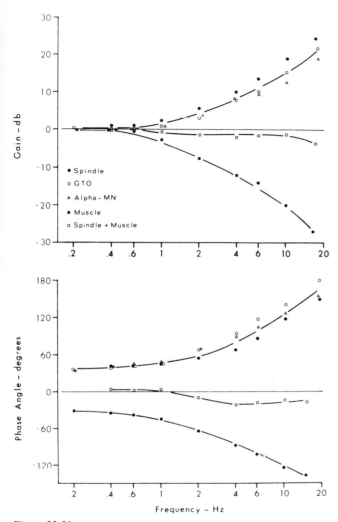

Figure 32-20
Frequency-response characteristics of stretch reflex arc components. For the "spindle," "GTO,"
and "alpha-MN" curves, the gain measurement consisted of the ratio $\Delta f_r/\Delta L$, where Δf_r is the
increment in single-unit firing rate produced by a given increment in muscle length (ΔL). For the
"muscle" curve, the gain measurement consisted of the ratio $\Delta T_i/\Delta S$, where ΔT_i is the increment
in isometric muscle tension produced by an increment ΔS in the intensity or frequency of motor
nerve stimulation. The symbol definitions given for the gain curves also apply to the phase-angle
plots. (Adapted from N. P. Rosenthal et al. *J. Neurophysiol.* 33:713, 1970.)

lag and loss of gain introduced by its tension-generating machinery.* As noted by Poppele and Terzuolo [25], this compensation no doubt contributes to the stabilization of the stretch reflex when it is viewed as a feedback control system.

In addition to demonstrating this important reflex property, Rosenthal et al. [29] were able to show that the linear input-output range for the overall reflex system was considerably greater than that exhibited by any single receptor element or neuron; their explanation for this difference is similar to the argument regarding threshold distribution that was presented above. Moreover, by estimating transfer functions from the component frequency-response curves, Roberts et al. [28] were able to construct a multicomponent model that predicted the frequency behavior of the overall reflex system with excellent accuracy. The success of the modeling procedure suggested that the component transfer functions had been accurately estimated and that no major system dynamics had been overlooked. The original article by Roberts et al. may be consulted for additional details.

Although these studies have provided no new information about the basic properties of reflex arc components, they are of major importance in the field of motor systems physiology for at least three reasons. First, when combined with the work of others [13], they provide a quantitative description of the stretch reflex system that will be of great value in estimating the role of various CNS structures in the control of movement and posture. The results summarized above, for example, have been used by Humphrey [16] in interpreting the significance of the neuronal firing patterns observed within the motor cortex of the primate during a voluntary, muscular contraction. Second, by viewing the spindle and tendon organ as parts of a reflex control system, these studies have provided an explanation for the biologic utility of their rate-sensitive behavior that has long escaped the notice of more traditionally oriented physiologists. Finally, by viewing the motoneuron pool as part of a control loop, they have shown that the dynamics of neuronal behavior may, in many cases, be negligible when compared with those arising from the mechanical time-constants of the muscle and its receptors. This point is of major importance for the design of future experiments, and it has been overlooked by many neurophysiologists who assume that the temporal behavior of neuronal networks is of central importance for understanding the dynamics of sensory and motor systems. Clearly, then, the use of a systems approach can be of value in providing quantitative results that are of use to other investigators and in placing the observations of the behavior of a system's components in proper perspective. Additional research of this type is needed not only in neurophysiology, but in other biologic disciplines as well.

---

*This point is well illustrated by the essentially flat curves in Figure 32-20 that represent the sums of the spindle and muscle-tension gain and phase curves. Note that because a logarithmic plot is used for the gain data, this summation is equivalent to *multiplying* the transfer functions of the two components and obtaining the composite gain and phase curves from the product.

REFERENCES

1. Burke, R. E. Composite nature of the monosynaptic excitatory postsynaptic potential. *J. Neurophysiol.* 30:1114, 1967.
1a. Carpenter, D. O., and Henneman, E. A relation between the threshold of stretch receptors in skeletal muscle and the diameter of their axons. *J. Neurophysiol.* 22:353, 1966.
2. Cohen, M. I. Discharge patterns of brain-stem respiratory neurons in relation to carbon dioxide tension. *J. Neurophysiol.* 31:142, 1968.
3. Eccles, J. C. The Excitatory Responses of Spinal Motorneurons. In J. C. Eccles and J. P. Schade (Eds.), *Progress in Brain Research.* New York: Elsevier, 1964.
4. Enroth, C. Mechanism of flicker fusion studied on single retinal elements in the dark-adapted eye of the cat. *Acta Physiol. Scand.* [Suppl.] 27:100, 1952.
5. Fatt, P. Electric potentials occurring around a neuron during its antidromic activation. *J. Neurophysiol.* 20:27, 1957.
6. Freygang, W. H., Jr. An analysis of extracellular potentials from single neurons in the lateral geniculate nucleus of the cat. *J. Gen. Physiol.* 41:543, 1958.
7. Geselowitz, D. B. On bioelectric potentials in an inhomogeneous volume conductor. *Biophys. J.* 7:1, 1967.
8. Granit, R. *Receptors and Sensory Perception.* New Haven, Conn.: Yale University Press, 1955.
9. Harmon, L. D. Neuromimes: Action of a reciprocally inhibitory pair. *Science* 146:1323, 1964.
10. Harmon, L. D., Levinson, J., and Van Bergeijk, W. A. Studies with artificial neurons. IV. Binaural temporal resolution of clicks. *J. Acoust. Soc. Am.* 35:1924, 1963.
11. Houk, J. C., Cornew, R. W., and Stark, L. A model of adaptation in amphibian spindle receptors. *J. Theor. Biol.* 12:196, 1966.
12. Houk, J. C., and Henneman, E. Feedback Control of Movement and Posture. In V. B. Mountcastle (Ed.), *Medical Physiology.* St. Louis: Mosby, 1968. Pp. 1681-1696.
13. Houk, J. C., Singer, J. J., and Goldman, M. R. An evaluation of length and force feedback to soleus muscles of decerebrate cats. *J. Neurophysiol.* 33:784, 1970.
14. Houk, J. C., Singer, J. J., and Henneman, E. Adequate stimulus for tendon organs with observations on mechanics of ankle joint. *J. Neurophysiol.* 34:1051, 1971.
15. Humphrey, D. R. Re-analysis of the antidromic cortical response. II. On the contribution of cell discharge and PSPs to the evoked potentials. *Electroencephalogr. Clin. Neurophysiol.* 25:421, 1968.
16. Humphrey, D. R. Relating motor cortex spike trains to measures of motor performance. *Brain Res.* 40:7, 1972.
17. Levinson, J., and Harmon, L. D. Studies with artificial neurons. III. Mechanisms of flicker fushion. *Kybernetik* 1:107, 1961.
18. Lewis, E. R. Using electronic circuits to model simple neuroelectric interactions. *Proc. IEEE* 56:931, 1968.
19. Maffei, L., Cervetto, L., and Fiorentini, H. Transfer characteristics of excitation and inhibition in cat retinal ganglion cells. *J. Neurophysiol.* 33:276, 1970.
20. Matthews, P. B. C. *Mammalian Muscle Receptors and Their Central Actions.* Baltimore: Williams & Wilkins, 1972.
21. Nelson, P. G., and Frank, K. Extracellular potential fields of single spinal motorneurons. *J. Neurophysiol.* 27:913, 1964.
22. Partridge, L. D. Modifications of neural output signals by muscles: A frequency response study. *J. Appl. Physiol.* 20:150, 1965.
23. Partridge, L. D., and Glaser, G. H. Adaptation in regulation of movement and

posture. A study of stretch responses in spastic animals. *J. Neurophysiol.* 23:25 1960.
23a. Perkel, D. H., and Mulloney, B. Motor pattern production in reciprocally inhibitory neurons exhibiting postinhibitory rebound. *Science* 185:181, 1974.
24. Perkel, D. H., et al. Pacemaker neurons. Effects of regularly spaced synaptic input. *Science* 145:61, 1964.
24a. Poppele, R., and Bowman, R. Quantitative description of the linear behavior of mammalian muscle spindles. *J. Neurophysiol.* 33:59, 1970.
25. Poppele, R., and Terzuolo, C. A. Myotatic reflex: Its input-output relations. *Science* 159:743, 1968.
26. Rall, W. Electrophysiology of a dendritic neuron model. *Biophys. J.* 2:145, 196.
26a. Rall, W. Distinguishing theoretical synaptic potentials for different soma-dendriti distributions of synaptic input. *J. Neurophysiol.* 30:1138, 1967.
27. Rall, W., and Shepherd, G. M. Theoretical reconstruction of field potentials and dendrodendritic synaptic interactions in olfactory bulb. *J. Neurophysiol.* 31:884, 1968.
28. Roberts, W. J., Rosenthal, N. P., and Terzuolo, C. A. A control model of the stretch reflex. *J. Neurophysiol.* 34:620, 1971.
29. Rosenthal, N. P., McKean, T. A., Roberts, W. J., and Terzuolo, C. A. Frequency analysis of stretch reflex and its main subsystems in triceps surae muscles of the cat. *J. Neurophysiol.* 33:713, 1970.
30. Salmoiraghi, G. C., and von Baumgarten, R. Intracellular potentials from respiratory neurons in brain stem of cat and mechanism of rhythmic respiration. *J. Neurophysiol.* 24:203, 1961.
31. Smith, T. G., Wuerker, R. B., and Frank, K. Membrane impedance changes during synaptic transmission in cat spinal motorneurons. *J. Neurophysiol.* 30:1072, 196.
32. Stark, L. *Neurological Control Systems: Studies in Bioengineering.* New York: Plenum, 1968. Pp. 297–367.
33. Stevens, C. F. *Neurophysiology: A Primer.* New York: Wiley, 1966. Chap. 10.
34. Terzuolo, C. A., and Bayly, E. J. Data transmission between neurons. *Kybernetik* 5:83, 1968.
35. Wilson, D. M. The Origin of the Flight Motor Command in Grasshoppers. In R. Reiss (Ed.), *Neural Theory and Modeling.* Stanford, Calif.: Stanford University Press, 1964. Pp. 331–345.
36. Wilson, D. M., and Waldron, I. Models for the generation of the motor output pattern in flying locusts. *Proc. IEEE* 56:1058, 1968.

SELECTED READING

Henneman, E. Peripheral Mechanisms Involved in the Control of Muscle. In V. B. Mountcastle (Ed.), *Medical Physiology.* St. Louis: Mosby, 1974. Chap. 22.
Houk, J., and Henneman, E. Feedback Control of Muscle: Introductory Concepts. In V. B. Mountcastle (Ed.), *Medical Physiology.* St. Louis: Mosby, 1974. Chap. 21.
Perkel, D. H. Spike Trains as Carriers of Information. In F. O. Schmitt (Ed.), *The Neurosciences: Second Study Program.* New York: Rockefeller University Press, 1970. Pp. 587–596.
Plonsey, R. *Bioelectric Phenomena.* New York: McGraw-Hill, 1969. Chap. 5.
Rall, W. Electrophysiology of a dendritic neuron model. *Biophys. J.* 2:145, 1962.
Terzuolo, C. A. Data Transmission by Spike Trains. In F. O. Schmitt (Ed.), *The Neurosciences: Second Study Program.* New York: Rockefeller University Press, 1970. Pp. 661–670.

THE ENDOCRINE SYSTEM VI

Harry Lipner

INTRODUCTION

The complexity of the metazoan organism, which is exemplified by the mammal, requires that it maintain its internal environment in a steady state; this condition of stability is termed *homeostasis*. It must occur despite the continuous activity of the cells of the organism and their constant exchange of materials with the extracellular fluid that composes the internal environment. The maintenance of a steady state — e.g., of the concentration of specific constituents — is brought about by the combined activity of the nervous and endocrine systems. The cells of the nervous system communicate among themselves and with effector organs by specific neurotransmitter substances. Neurotransmitter substances are low-molecular-weight, water-soluble compounds that bridge the minute gap between neurons or between neurons and effector cells. The endocrine glands or cells communicate with the cells of effector organs by means of specific secretions called *hormones* (from the Greek, *hormaein,* to arouse or excite). Hormones are organic substances that vary in complexity from simple compounds to large, high-molecular-weight proteins. They are secreted into the circulation and may travel great distances before exerting an effect. Since some cells show no response to certain of the hormones and others show highly specific responses, it is assumed that highly specific receptor sites exist on the various cells.

TRANSDUCTION MECHANISMS

Many cells exhibit the ability to transform one type of stimulus energy into another. Sensory cells detect a large variety of stimulus energies and transduce these to electrical signals; other cells detect stimuli and transduce these into chemical signals. Examples of the latter variety of cells are those associated with the regulation of plasma glucose or calcium levels. In both cases, the sensing cells are chemodetectors as well as hormone secretors or *chemoendocrine cells*. Similarly, cells that, when subjected to neurotransmitter excitation, release hormonal substances into the blood that act on distant specific target cells are termed *neuroendocrine cells* (Table 33-1).

CHARACTERISTICS OF HORMONES

The secretion of hormones may be *tonic* (continuous) or *phasic* (in surges), or this secretion may show a surge superimposed on a tonic pattern. Hormones do not act directly on the cells that produce them. Hormones may act specifically on a single

Table 33-1. Neuroendocrine Transducer Cells

Cell Type	Input Signal	Hormonal Output Signal	Feedback Signal
Hypothalamic releasing hormone cells	Catecholamines (dopamine and norepinephrine)	Releasing hormones	Short loop via pituitary hormone (primary) Long loop via pituitary target hormone (secondary)
Supraoptic nucleus cells of hypothalamus	Acetylcholine or norepinephrine	Vasopressin (ADH)	Osmolality of extracellular fluid of brain Osmoreceptors in liver; blood pressure in major arteries
Adrenal medulla	Acetylcholine	Epinephrine	High concentrations of glucocorticoids Neural signals from CNS
Pineal gland	Norepinephrine	Melatonin	Estradiol (in mammals) causes secretion
Juxtaglomerular cells of nephron	Norepinephrine	Renin	Sodium ion concentration contributes to regulation of secretion Sympathetic inflow; renal pressure and resistance and vascular volume
β Cells of pancreatic islets	Acetylcholine	Insulin	Glucose enhances secretion of insulin Hypoglycemia causes secretion to stop
α₂ Cells of pancreatic islets	Norepinephrine	Glucagon	Hyperglycemia suppresses secretion

Modified from R. J. Wurtman. *Fed. Proc.* 32:1769, 1973.

organ or they may exert general metabolic effects. The regulation of the secretion of hormones is generally under negative feedback control, and it is a function of the rate of disappearance of the hormone from the blood. Hormone concentrations determine the intensity of the cellular response. The magnitude of the acute response is a function of the capacity of the cellular response. The effects generated by some hormones persist as long as the hormone is present (e.g., the effects of insulin or aldosterone), whereas the effects of others persist long after detectable quantities have disappeared (e.g., the effects of thyroid hormone or estrogen). Hormones may be derivatives of the amino acid tyrosine — e.g., the catecholamines (dopamine, norepinephrine, and epinephrine) and the thyroid hormones (thyroxine and triiodothyronine). They may also be derived from cholesterol — e.g., progesterone, estradiol-17, estrone, estriol, testosterone, and the adrenal cortical steroids — or they may be synthesized from amino acids and appear as polypeptides or complex proteins.

SOURCE OF THE SECRETIONS

The hormones may be secreted by nerves, by clearly delineated endocrine glandular organs, or by discrete cells in organs, e.g., the hypothalamus, the thyroid, and the pancreatic islets of Langerhans. The endocrine glands consist of a variety of organs with pronounced variations in structure. The common feature possessed by all is an absence of ducts and an exceedingly well-developed blood supply. The specific endocrine glands of man are the adrenals, gonads, hypophysis, islets of Langerhans, parathyroid, and thyroid glands. Anatomically undefined secretory areas are exemplified by the hypothalamus and the intestinal mucosa.

The hypothalamus is a major source of neurohormones that regulate hypophysial function. The mucosa and submucosa of the pyloric antrum of the stomach and of the duodenum are the sources of hormones that promote or inhibit gastric juice secretion and gastric motility; these hormones also stimulate the secretion of pancreatic juice as well as of bile. Table 33-2 lists the endocrine organs, their secretions, their targets, and their effects. This unit will discuss the role and contribution of these hormones to the function and homeostasis of the organism. The emphasis of the presentation will be on functional, not biochemical, aspects (for complete biochemical analyses, see reference [2]).

There is a growing list of substances that are apparently ubiquitous to all cells and that are released by a variety of stimuli. These affect cellular activity locally. Because these materials may appear in the blood, some workers call them "hormones." Among these substances are the kinins and prostaglandins. The former are generally decapeptides or octapeptides, and the latter constitute a unique group of 20-carbon atom, lipid-soluble, unsaturated hydroxy acids.

Table 33-2. The Vertebrate Hormones: Classes, Targets, and Effects

Hormone	Target	Effects
I. Peptides and protein hormones		
1. Pituitary gland (hypophysis)		
A. Adenohypophysis (anterior lobe)		
a. Pars distalis		
1. Long-chain polypeptides		
Growth hormone (GH, somatotropin)	All tissues	Growth of tissues (easily seen in long bones, metabolism of pro e n_ and mobilization of fat) Lipolysis
Prolactin	Mammary glands (alveolar cell) Crop gland of pigeons	Milk production in prepared gland Crop gland "milk" production
Adrenocorticotropin (ACTH, corticotropin)	Adrenal cortex	Synthesis and release of glucocorticoids
2. Glycoproteins		
Thyroid-stimulating hormone (TSH, thyrotropin)	Adipose tissue Thyroid gland	Lipolysis Synthesis and secretion of thyroxine and triiodothyronine
Male:		
Follicle-stimulating hormone (FSH)	Seminiferous tubules	Production of sperm
Interstitial cell—stimulating hormones (ICSH) luteinizing hormone (LH)	Testes	Synthesis and secretion of androgens
Female:		
FSH	Ovary (follicles)	Follicle maturation
LH	Ovary (interstitial cells)	Final maturation of follicle; estrogen secretion; ovulation; corpus luteum formation; progesterone secretion
b. Pars intermedia		
Melanocyte-stimulating hormones (α- and β-MSH, intermedin)	Melanophores	Pigment dispersal in melanophores (darkening of skin)

Hormone	Target	Action
B. Neurohypophysis (posterior lobe)		
Oxytocin (letdown factor, milk-ejection factor)	Uterus	Contraction of smooth muscle; milk ejection
	Mammary glands	
Vasopressin (antidiuretic hormone, ADH)	Kidney	Reabsorption of water from urine
	Arteries	Contraction of smooth muscle
2. Pancreas		
Insulin	All cells	Carbohydrate, fat, and protein metabolism; hypoglycemia
Glucagon	Liver	Hyperglycemia
3. Ovary		
Relaxin	Pelvic ligaments	Separation of pelvic bones
4. Thyroid		
Thyrocalcitonin (calcitonin)	Bones, kidney	Excretion of calcium and phosphorus; inhibited calcium release from bones; decreased blood calcium levels
5. Parathyroid		
Parathyroid hormone	Bones, kidney	Elevated blood calcium and phosphorus levels; mobilization of calcium from bone; inhibited calcium excretion from kidney
6. Kidney[a]		
Erythropoietin	Bone marrow	Increased erythrocyte production
Renin	Adrenal cortex	Aldosterone synthesis and secretion
7. Stomach and duodenum		
Gastrin	Stomach	Acid secretion; gastric motility
Cholecystokinin (CCK)	Gall bladder	Contraction of the gall bladder
Secretin	Pancreas	Secretion of water and salts; inhibits gastric secretion and motility and the secretion of digestive enzymes
Pancreozymin (identical to CCK)	Pancreas	
II. Amino acid derivatives		
1. Thyroid		
Thyroxine	Most cells	Increased metabolic rate, growth, and development
Triiodothyronine		
2. Adrenal medulla		
Norepinephrine	Most cells	Increased cardiac activity; elevated blood pressure; glycolysis; hyperglycemia
Epinephrine		

[a]The renal hormones appear to be enzymes that activate plasma substrates, which act on the target organ.

Table 33-2 (Continued). The Vertebrate Hormones: Classes, Targets, and Effects

Hormone	Target	Effects
II. Amino acid derivatives (Continued)		
3. Pineal gland		
Melatonin	Melanophores	Dispersion of melanin
4. Argentaffin cells, platelets, nerves		
Serotonin (5-hydroxytryptamine)	Arterioles, central nervous system	Vasoconstriction
III. Steroids and lipids		
1. Testes		
Androgen (testosterone)	Most cells	Development and maintenance of masculine characteristics and behavior
2. Ovary		
Estrogen (17β-estradiol)	Most cells	Development and maintenance of feminine characteristics and behavior
3. Corpus luteum		
Progesterone	Uterus, mammary glands	Maintenance of uterine endometrium and stimulation of mammary duct formation
4. Adrenal cortex		
Hydrocortisone Cortisone	Most cells	Balanced carbohydrate, protein, and fat metabolism; antiinflammatory action
Aldosterone	Kidney	Reabsorption of Na^+ from urine
5. Prostate, seminal vesicles, brain, nerves		
Prostaglandins	Uterus, rabbit duodenum	Contraction of smooth muscle

FUNCTIONS OF HORMONES

REGULATION

In the developing fetus, hormones regulate the differentiation of sexual structures, and in amphibians, they are responsible for metamorphosis. They are necessary for normal growth in the growing organism and, at adulthood, for the cessation of growth.

INTEGRATION

Hormones are responsible for the regulation of the estrous cycle in lower vertebrates and the menstrual cycle in primates. They are responsible for the orderly progression of growth and the regression of the endometrium of the uterus as well as for the development of the ovarian follicle. The integrative role may be parallel to, supplement, or be part of a neural pathway. This role is illustrated in the process of lactation and the release of milk.

HOMEOSTASIS

At times, the homeostatic role of hormones is indistinguishable from the integrative one. The concept of homeostasis entails that at steady-state conditions, the concentration of many substances in the body fluids is apparently constant. The regulation of blood sugar, for example, is illustrated in Figure 33-1. Other examples of hormonal regulation, including feedback mechanisms, will be presented later in a similar manner.

In the diagrams of Figure 33-1, variables are treated as signals, e.g., the concentration of glucagon to the input of block 5, the liver. The output signal from a block may be directed to any number of other blocks without regard for the conservation of the signal. Each block is described by its own steady-state characteristic curve that defines the functional relationship between the output and input in graphic form. This relationship may also be expressed mathematically [3].

Negative feedback in a system is indicated by a resultant negative slope in a sequence of a series of blocks. There must be an odd number of blocks that exhibit sign reversal, that is, the overall function must have a negative slope. In Figure 33-1, negative feedback is exhibited in the system by α-cell function in the pancreas (block 4). For effective feedback, at least one component should exhibit unilateral coupling. This is necessary in order to maintain low sensitivity to disturbances.

Homeostasis is continually being disturbed, and a pure steady-state situation is never reached. Instead, the systems undergo a continuing sequence of responses to these disturbances. Therefore, it is necessary that the dynamic response, as well as the steady-state circumstances, be understood when studying feedback system behavior.

In Figure 33-1, block 1 represents the resultant effect of the combined actions of insulin from block 2 (the hypoglycemic factor) and glucagon from block 4 (the hyperglycemic factor) on blood glucose. The solid line in block 1 represents the usual postabsorptive sugar level (80 mg/100 ml of blood). There are two feedback loops shown in Figure 33-1: the first path controls glucose uptake and includes blocks 1, 2, and 3;

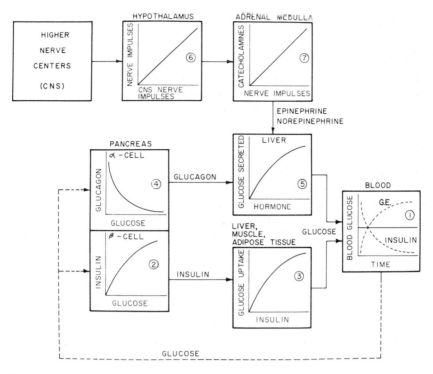

Figure 33-1
The regulating mechanism for blood glucose.

the second path controls glucose secretion and includes blocks 1, 4, and 5. These two paths work together in keeping the blood glucose level at a constant value.

The concentration of glucose in the blood is monitored by the cells in the islets of Langerhans in the pancreas. The α and β cells secrete the hormones that regulate blood glucose; each cell contains the receptor transducer and the effector structures. This regulating mechanism responds dynamically to a change of blood glucose concentration. Its elevation would induce an increase in the secretion of insulin (block 2), which in turn would enhance the uptake of glucose by the liver, muscles, and adipose tissues (block 3). The reduction of the blood glucose concentration would generate an increase in the secretion of glucagon (block 4), which would cause the liver to increase the secretion of glucose by stimulation of glycogenolysis (block 5). The action of the central nervous system via the adrenal medulla (blocks 6 and 7) is directed at the stimulation of hepatic glycogenolysis under conditions of emotional or physical stress.

PERMISSIVENESS

The ability of cells to respond optimally to a hormonal stimulus may require the presence of another hormone that acts to maintain a particular metabolic level of the original hormone. This is evidenced by the requirement of the ovarian follicle for estrogen so that luteinizing hormone may cause ovulation and the need for thyroid hormone if growth hormone is to be effective.

COMPARISON OF THE NERVOUS SYSTEM AND THE ENDOCRINE SYSTEM

Higher organisms utilize two integrative systems: the nervous system and the endocrine system. These interact to make exquisitely fine control of many body functions possible. They are very similar in their overall organization.

A broad comparison between the nervous and the endocrine systems is shown in Table 33-3. A relationship that is analogous to the monosynaptic reflex is seen in the endocrine system when the parathyroid gland cells are stimulated to release parathyroid hormone due to a reduction of the calcium concentration in the blood. In the endocrine system, the analog of the interneuron may be all the glands that interact regardless of their distance from each other; the secretion of several glands may be introduced between the initial stimulus and the response in the effector cell in a manner similar to the action of the many interneurons that are introduced between the afferent neuron and the effector cell.

MECHANICS OF HORMONE-RECEPTOR INTERACTION

If the administration of a hormone to an animal or an isolated tissue results in a measurable response, the occurrence of an interaction is implied between the hormone and its receptors, which are specific structures on or in the component cells in the animal or tissue. Hormone-receptor interactions are complex, and they are therefore most appropriately analyzed in experiments with isolated tissues or, better, with a homogeneous population of cells. Hormone-receptor interactions have been the subject of extensive studies; the following discussion relies heavily upon the analysis by Ariens et al. [1], which exemplifies such studies.

Hormone receptors in general remain uncharacterized; they may be single molecules, parts of molecules, or multimolecular complexes. They may be free in the cytosol or part of the organized structure of the cell, and they are organ-specific. A number of assumptions regarding hormone receptors have been deemed necessary: the receptor sites are homogeneous and noninteracting, one hormone molecule interacts with one receptor molecule, and no cooperative interactions occur. The hormone-receptor interaction may be transient with a large, negative free energy of dissociation, or the two moieties may be reversibly or irreversibly covalently bonded. In this discussion, the complex that is formed by hormone-receptor interaction will be assumed to be reversibly bonded, although it need not be. Upon the formation of the hormone-receptor complex, the hormone may either be inactivated or remain unchanged, but in any case, the complex formation initiates a stimulus that is proportional to the response. Ariens et al. [1] assume that linear relationships exist between the stimulus

Table 33-3. Comparison Between the Nervous System and the Endocrine System

Generic Designation	Components	Stimulus	Central Organization	Efferent Limb	Efferent Structures
Monosynaptic pathway	Afferent–efferent neuron pathway	Muscle stretch	Spinal cord	Motor cell	Muscles
Monohormonal pathway	Single-gland system	Change in blood composition (glucose, Ca^{+2}, NaCl)	Single gland α and β cells of islets of Langerhans Light cell of thyroid and parathyroid glands Supraoptic nucleus of hypothalamus	Hormones Glucagon and insulin Thyrocalcitonin and parathyroid hormone Vasopressin (antidiuretic hormone)	Responding tissues Liver and muscles Bones and kidney Kidney
Disynaptic pathway	Afferent neuron–interneuron–efferent neuron pathway	Pain	Spinal cord	Motor cell	Muscles
Dihormonal pathway	Two-gland system	Inhibition of tonic secretion	Hypothalamus, pituitary GH-RH[a] P-RIH MSH-RIH	Hormones Growth hormone Prolactin Melanocyte stimulating hormone	Responding tissues Body cells Mammary glands Melanocytes
Polysynaptic pathway	Afferent neuron–several interneurons (in sequence)–efferent neuron pathway	Most sensory inputs	Spinal cord	Motor cell	Muscles
Polyhormonal pathway	Several glands in sequence	External factors (light, temperature, changes in blood concentration of stimulating compounds)	Secretory cell: Hypothalamus / Pituitary / Target gland TSH-RH[a] / TSH / Thyroid gland LH-RH / LH / Corpus luteum FSH-RH / FSH / Ovarian follicle C-RH / ACTH / Adrenal cortex	Hormones Thyroid hormones Progesterone Estrogen Cortisol	Responding tissues Body cells Uterus and Body cells

[a]Abbreviations: GH-RH = growth-hormone releasing hormone; P-RIH = prolactin release-inhibiting hormone; MSH-RIH = melanocyte-stimulating hormone inhibiting hormone; TSH-RH = thyroid-stimulating hormone releasing hormone; LH-RH = luteinizing-hormone releasing hormone (interstitial cell-stimulating hormone); FSH-RH = follicle-stimulating hormone releasing hormone; C-RH = adrenocorticotropic releasing hormone.

and the response as well as between the stimulus and the relative concentration of the hormone-receptor complex. The proportionality constant in the latter relationship is a measure of the *intrinsic activity* of the hormone or its analogs. In summary, the activity of the hormone is a function of its intrinsic activity and its affinity for the receptor.

The mathematical analysis of the hormone-receptor interaction is based on the application of the law of mass action. The special application of the latter results in the Langmuir adsorption isotherm relationship and the Michaelis-Menten description of enzyme-substrate interaction. The interaction between molecules and specific sites on a surface may be described by a relationship that is analogous to the Langmuir adsorption isotherm. The reversible formation of the hormone-receptor complex is describable as:

$$[H] + [R] \underset{k_{-1}}{\overset{k_1}{\rightleftharpoons}} [HR] \tag{33-1}$$

where $[H]$ = concentration of free hormone
$[HR]$ = concentration of bound hormone
$[R]$ = concentration of free receptors

The concentration of the total number of receptors, $[R_T]$, equals $[HR] + [R]$. The rate of complex formation is $k_1 [H] \cdot [R]$, and the rate of dissociation is $k_{-1} [HR]$. At equilibrium, $k_1 [H] \cdot [R] = k_{-1} [HR]$. The equation for the Langmuir adsorption isotherm has the form of Equation 33-2A when it is assumed that the fraction of hormone molecules touching the receptors approaches unity. If the fraction of bound receptors is $[HR]/[R_T]$, then the number of free receptors is $1 - [HR]/[R_T]$, and the appropriate form of the Langmuir adsorption isotherm is:

$$k_1 [H] \left(1 - \frac{[HR]}{[R_T]}\right) = k_{-1} \frac{[HR]}{[R_T]} \tag{33-2A}$$

By rearrangement,

$$\frac{[HR]}{[R_T]} = \frac{[H]}{K_A + [H]} \tag{33-2B}$$

where $K_A = k_{-1}/k_1$.

The relationship described in Equation 33-2B can also be obtained by means of the Michaelis-Menten analysis of substrate-enzyme interaction:

$$[E] + [S] \underset{k_{-1}}{\overset{k_1}{\rightleftharpoons}} [ES] \overset{k_2}{\longrightarrow} [E] + [P] \tag{33-3}$$

Equation 33-3 may be written so as to represent the hormone-receptor relationship and its generation of a response:

$$[H] + [R] \underset{k_{-1}}{\overset{k_1}{\rightleftharpoons}} [HR] \xrightarrow{k_2} \text{Response} + [H] + [R] \tag{33-4}$$

At equilibrium, the application of the law of mass action gives:

$$K_H = \frac{k_{-1}}{k_1} = \frac{[H] \cdot [R]}{[HR]} \tag{33-5}$$

where K_H = the equilibrium constant
$\quad\quad k_1$ = the rate-constant of the formation of the hormone-receptor complex
$\quad\quad k_{-1}$ = the rate-constant of the dissociation of the hormone-receptor complex

Since the concentration of free receptor sites, $[R]$, is not known or easily determined, it must be expressed in terms of the concentration of bound sites, $[R_T] - [HR]$:

$$K_H = \frac{[H]\,([R_T] - [HR])}{[HR]} \tag{33-6}$$

The concentration of bound receptors can be determined by solving Equation 33-6 for $[HR]$ when radioactive hormones are used but not when the analysis is based on bioassay. By rearranging Equation 33-6, the value of $[HR]$ is given by:

$$[HR] = \frac{[R_T] \cdot [H]}{K_H + [H]} \tag{33-7}$$

The fraction of the total receptors occupied by hormone is obtained by dividing by $[R_T]$ (Eq. 33-8). The ratio of bound receptors to total receptors increases with an increase in the concentration of hormone, $[H]$, and decreases as the equilibrium constant, K_H, becomes larger.

$$\frac{[HR]}{[R_T]} = \frac{1}{1 + K_H/[H]} = \frac{[H]}{[H] + K_H} \tag{33-8}$$

Equations 33-2B and 33-8 are therefore equivalent expressions of the hormone-receptor interaction. The equilibrium constants, K_A and K_H, are the same in both equations, and this constant, which is the reciprocal of the dissociation constant, is what Ariens et al. [1] call the *affinity constant*.

Equation 33-7 cannot be used in analysis based on bioassay, because the concentration of the hormone-receptor complex, $[HR]$, cannot be accurately evaluated using this method. However, if the effect due to the hormone, E_H, is assumed to be proportional to the concentration of bound hormone [1], then $E_H = a[HR]$, and substitution into Equation 33-7 gives:

$$E_H = \frac{(1/a)\,[R_T]\,[H]}{K_H + [H]} = \frac{(1/a)\,[R_T]}{1 + K_H/[H]} \tag{33-9}$$

At high hormone concentrations, the number of bound receptors, HR, approaches the total number of receptors present, R_T, and the hormone-generated response, E_H, approaches the maximum response, $(E_H)_{max}$. The maximum response is proportional to the total number of receptors present, or:

$$(E_H)_{max} = a'[R_T] \tag{33-10}$$

From Equation 33-9 and the definition of *intrinsic activity*, α, as $1/(aa')$, the following relationship is obtained:

$$\frac{E_H}{(E_H)_{max}} = \alpha\,\frac{[HR]}{[R_T]} = \frac{\alpha}{1 + K_H/[H]} \tag{33-11}$$

The determination of the values for the intrinsic activity of the hormone, α, and the affinity constant, K_A, is now possible. A plot of the ratio of $E_H/(E_H)_{max}$ versus $[H]$ results in a curve such as that illustrated in Figure 33-2A. The intrinsic activity of the hormone, α, is inversely proportional to the number of receptors that have to be occupied to induce a stimulus and generate a response. The intrinsic activity of a hormone may be determined from the fact that at high doses, the response that the hormone induces approximates the maximum response that can be elicited in the tissue, and the ratio $E_H/(E_H)_{max} \approx \alpha \approx 1$. Thus, if this ratio for a specific hormone is arbitrarily given the value of unity, then the activity of analogs of the hormone may be expressed in relation to the parent hormone, e.g., as fractions of the maximum activity of the hormone.

The plot in Figure 33-2A indicates that the value of the dissociation constant (K) equals the value of the concentration of the hormone, $[H]$, when the response is equal to one-half the maximum response, i.e., when $E_H = 0.5\,(E_H)_{max}$. In other words, at $0.5\,(E_H)_{max}$, the value of the *affinity* constant, K_A, is equal to the reciprocal of the value of the hormone concentration, $1/[H]$.

Two different types of receptor sites have been described: a high-affinity, low-capacity site, and a low-affinity, high-capacity site. With no assumptions other than that the law of mass action applies, a plot of the ratio of bound-to-unbound hormone versus bound hormone, $[H_B/H_U]$ versus $[H_B]$, displays a linear relationship (Fig. 33-2B).

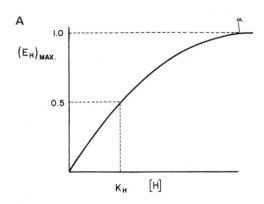

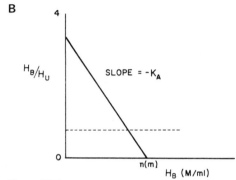

Figure 33-2
A. Plot of the ratio of hormone-generated response to maximum response versus hormone concentration, $[H]$. B. A Scatchard plot of the ratio of bound (H_B) to unbound (H_U) hormone.

In this plot, called a Scatchard plot, the intercept on the abscissa indicates the molar concentration of binding sites, $n(M)$, and the slope of the straight line is the negative of the affinity (association) constant, $-K_A$. This analysis is independent of the effector response, and it defines the receptor site in terms of molar concentration and the affinity constant. The experimental data are obtained by equilibrating the receptor protein with the radioactively labeled hormone and then determining the amount of bound and unbound hormone present. The analysis may be extended by determining the hormone specificity through comparison with analogs of the hormone as well as with dissimilar hormones. Studies of the cytosol progesterone-binding protein of the uterine myometrium, for example, indicate it to be a 4S protein; this protein is not entirely steroid-specific, since it binds glucocorticoids almost as well as progesterone.

MECHANISMS OF HORMONE ACTION

In the final analysis, the effects caused by hormones are the gross expressions of the biochemical regulations that are imposed on the cell. Hormones augment the regulation

of ongoing biochemical events at the cellular level, and they generally bring about
the integration of organ systems by means of the feedback signals that are elicited from
controlled target organs.

Hormones may exert their effects at a number of cellular sites. They may affect
membrane permeability by altering the availability of organic substrates or electrolytes
(Table 33-4). Also, they may trigger a second messenger at the cell membrane that
can cause a cascade effect that amplifies the signal generated by the primary messenger
(Table 33-5). A model of such a mechanism is described in Figure 33-3A. In this case,
it is assumed that the hormone interacts with a receptor protein and generates a change
in adenylate cyclase activity, with the subsequent formation of 3',5'-cyclic adenosine
monophosphate (cAMP). The results of such action are the conversion of nonactivated
to activated enzymes and the formation of substrates from cellular storage depots
(Fig. 33-3B). Hormones may also act at the nucleus of the cell, either directly or after
intracellular conversion to an active form. At the nuclear level, hormones regulate
protein synthesis (Figs. 33-4 and 33-5).

The process of activation and deactivation of glycogen phosphorylation is illustrated
stepwise in Figure 33-3B. The steps are: (1) hormones acting on the membrane affect
the activity of adenylate cyclase, thus inducing the formation of cAMP; the latter has
been termed an *intracellular* or *second messenger*; (2) cAMP activates the formation
of protein kinases, which are proteins composed of regulatory and catalytic subunits;
(3) phosphorylation by cAMP causes the activation of the protein kinase that, in the
liver or muscle cell, may cause activation of phosphorylase kinase; (4) activation of
phosphorylase b to phosphorylase a occurs; and (5) there is a breakdown of 1,4 bonds
in glycogen, with the formation of glucose 1-phosphate. The net result in the liver is
the release of glucose, and in muscle, the result is enhanced glycolysis and the release
of lactic acid. These effects are stimulated by epinephrine and glucagon. These hormones also activate, via reaction 6 (Fig. 33-3B), the cAMP mechanism. Synthetase
kinase I then converts (via reaction 7) synthetase I (independent) to the D (dependent)
form of the enzyme and thus decreases the availability of synthetase I for glycogen
synthesis (reaction 8). Insulin blocks reaction 6, the activation of synthetase kinase I,
which thus allows synthetase to remain in the active, I form of the enzyme, thereby
maintaining the rate of glycogen synthesis. In adipose cells, reaction 1 occurs, which
activates lipase and causes the release of free fatty acids.

The hormones that have been implicated in protein synthesis are listed in Table 33-6.
These hormones have been shown to induce protein synthesis in both in vivo and
in vitro systems. Several of these hormones have also been implicated in the synthesis
of specific proteins (Table 33-7).

The time of onset of action by hormones is variable, ranging from seconds in the
case of the catecholamines to days in the case of thyroxine. Equally variable is the
necessity of hormones for animal survival. In the case of some hormones (gonadal
hormones are a case in point), the survival of the individual is unaffected, whereas in
the case of parathyroid hormone or aldosterone, for example, survival depends on the
replacement of the former with calcium salts and the latter with sodium salts if these
hormones are ablated (Table 33-8).

Table 33-4. Hormone-Influenced Substrate Transport

Hormone	Substrate	Effect	Tissue
Thyroid hormones	L-Tyrosine L-Tryptophan L-Phenylalanine L-Methionine L-Histidine L-Leucine	Increased transport	Activate L-site and cell membranes of responding organs; cartilage, skeletal muscle
	Glycine L-Valine L-Isoleucine L-Glutamine L-Asparagine L-Hydroxyproline L-Norleucine	Inhibited transport	Activate L-site and cell membranes of responding organs; cartilage, skeletal muscle
Thyroid-stimulating hormone (TSH)	I^-	Increased transport into cells	Thyroid
Luteinizing hormone (LH); follicle-stimulating hormone (FSH); human chorionic gonodotropin (HCG)	Glucose	Increased transport into cells	Ovaries
Adrenocorticotropin (ACTH)	Ascorbic acid	Decreased or inhibited transport into cells	Adrenal cortex
Aldosterone and mineral corticoids	Na^+	Increased transport into cells	Toad bladder
	K^+	Decreased transport into cells	Renal nephron
Parathyroid hormone	Ca^{+2}	Increased transport across membranes	Intestine Bone Kidney
Calcitonin	Ca^{+2}	Decreased transport across membranes	Bone
Epinephrine	Amino acids	Increased transport into cells	Liver
Norepinephrine	Glucose	Increased transport out of cells	Adipose tissue Cardiac muscle Skeletal muscle
	Na^+		Cardiac muscle
	K^+		Intestinal smooth muscle

Table 33-5. Some Hormone Actions Mediated by Changes in Cyclic AMP

Hormone	Tissue	Effect
	Increased Cyclic AMP Levels	
Adrenocorticotropic hormone	Adrenal cortex	↑ Steroidogenesis
	Fat (rat)	↑ Lipolysis
Luteinizing hormone	Corpus luteum, ovary, testis	↑ Steroidogenesis
	Fat	↑ Lipolysis
Catecholamines	Fat	↑ Lipolysis
	Liver	↑ Glycogenolysis, ↑ gluconeogenesis
	Skeletal muscle	↑ Glycogenolysis
	Heart	↑ Inotropic effect
	Salivary gland	↑ Amylase secretion
	Uterus	Relaxation
Glucagon	Liver	↑ Glycogenolysis, ↑ gluconeogenesis, ↑ induction of enzymes
	Fat	↑ Lipolysis
	Pancreatic alpha cells	↑ Insulin release
	Heart	↑ Inotropic effect
Thyroid-stimulating hormone	Thyroid	↑ Thyroid hormone release
	Fat	↑ Lipolysis
Melanocyte-stimulating hormone	Dorsal frog skin	↑ Darkening
Parathyroid hormone	Kidney	↑ Phosphaturia
	Bone	↑ Ca^{+2} resorption
Vasopressin	Toad bladder, renal medulla	↑ Permeability
Hypothalamic releasing factors	Adenohypophysis	↑ Release of tropic hormones
Prostaglandins	Platelets	↓ Aggregation
	Thyroid	↑ Thyroid hormone release
	Adenohypophysis	↑ Release of tropic hormones
	Decreased Cyclic AMP Levels	
Insulin	Fat	↓ Lipolysis
	Liver	↓ Glycogenolysis, ↓ gluconeogenesis
Prostaglandins	Fat	↓ Lipolysis
	Toad bladder	↓ Permeability
Catecholamines (β-adrenergic stimuli)	Frog skin	↓ Darkening
	Pancreas	↓ Insulin release
	Platelets	↓ Aggregation
Melatonin	Frog skin	↓ Darkening

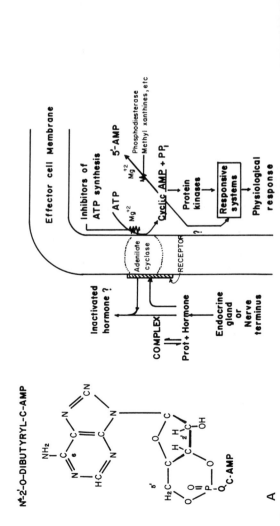

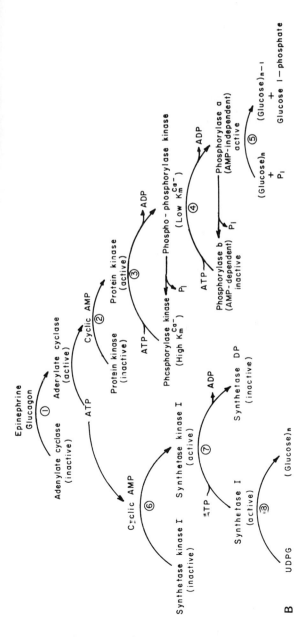

Figure 33-3

A. Model of the hormone target-cell receptor interaction and the generation of the "second messenger." (From E. W. Sutherland. *Science* 177:401, 1972. Reproduced with permission of the Nobel Foundation.) B. The cascade of activation and deactivation of glycogen phosphorylase. At each lower level of the activating cascade, the number of molecules involved increases by at least one order of magnitude.

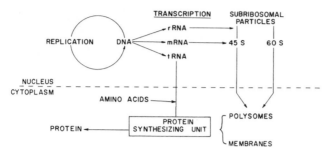

Figure 33-4

A simplified scheme showing the biosynthetic relationships between the nucleic acids (DNA and RNA) and proteins that determine information flow and protein synthesis in nucleated cells. As is discussed in more detail in the text, the transcription and translation processes are the logical sites of hormone action. It has been further proposed that the type of growth or development would be determined genetically via mRNA coding for specific proteins, but that its expression requires the generation of ribosomes and their topographic segregation by firm attachment to membranes. (Modified from J. R. Tata. *Nature* [Lond.] 219:331, 1968. Reproduced with permission.)

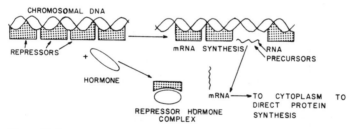

Figure 33-5

Induction of the synthesis of a new protein by a hormone acting as a derepressor. (Modified from P. Karlson. *Perspect. Biol. Med.* 6:203, 1963. Reproduced with permission of the University of Chicago Press.)

Table 33-6. Hormonal Stimulation of Protein Synthesis In Vivo and In Vitro

Hormone	Species	In Vivo	In Vitro
Growth hormone	Rat, man	Liver, kidney, heart, spleen, adrenal fat depot, muscle	Liver slices, cell-free system, adrenal quarters
ACTH	Rat	Adrenal	Adrenal quarters, cell-free system
TSH	Chicken, rat	Thyroid	Thyroid slices
Thyroxine, triiodothyronine	Rat, man	Liver, kidney, heart	Liver mitochondria, liver microsomes or ribosomes
	Frog, tadpole		Liver slices
Testosterone	Rat, mouse, guinea pig	Seminal vesicle, prostate, kidney	Slices, homogenates, muscle minces, cell-free preparations from seminal vesicle, prostate, kidney, muscle
Estrogen	Rat, hen	Rat uterus, hen oviduct	Slices, homogenates, cell-free preparations of rat uterus and hen oviduct
Cortisone	Rat	Liver	Liver microsomes
Insulin	Rat	Liver, diaphragm, skeletal muscle	Diaphragm, liver slices, cell-free preparations from liver and heart

Table 33-7. Hormonal Stimulation of Specific Protein Synthesis

Hormone	Specific Protein Synthesized
Thyroxine, triiodothyronine	Cytochrome C, isocitrate dehydrogenase, respiratory components and membranes
Estrogen	Phosvitin
Cortisone	Tryptophan pyrrolase, transaminase(s), gluconeogenic enzymes
Insulin	Glucokinase, pyruvate kinase, malic enzyme

From J. Larner. *Intermediary Metabolism and Its Regulation.* Englewood Cliffs, N.J.: Prentice-Hall, 1971. Reproduced with permission.

Table 33-8. Hormone Classification in Terms of Time of Onset of Action, General Type(s) of Effects, and Requirement for Survival

Hormone	Time of Onset of Action	General Type(s) of Hormone Effects	Survival Time After Ablation
Epinephrine and other catecholamines	Seconds–minutes	Metabolic	Unaffected
Glucagon	Seconds–minutes	Metabolic and growth	Unaffected
Insulin	Minutes–hours	Metabolic and growth	Greatly reduced
ACTH	Minutes–hours	Metabolic and growth	Greatly reduced
Cortisone	Minutes–hours	Metabolic and growth	Greatly reduced
Aldosterone	Minutes	Metabolic	Days
Testosterone	Hours–days	Growth	Unaffected
Estrogen	Hours–days	Growth	Unaffected
Growth hormone	Hours–days	Growth	Unaffected
TSH	Hours–days	Growth	Unaffected
Eodysone	?	Growth	Unaffected
Thyroid hormones	Days	Growth	?
Parathyroid hormone	Minutes–hours	Metabolic and growth	Hours
Calcitonin	Minutes	Metabolic	Unaffected

REFERENCES

1. Ariens, E. J., Simmonis, A. M., and Van Rossum, J. M. *Molecular Pharmacology.* New York: Academic, 1964.
2. Frieden, E., and Lipner, H. *Biochemical Endocrinology of the Vertebrates.* Englewood Cliffs, N.J.: Prentice-Hall, 1971.
3. Jones, R. W. *Principles of Biological Regulation.* New York: Academic, 1973.

SELECTED READING

Ekins, R. P. Basic Concepts of Saturation Analysis Techniques. In L. Martini and V. H. T. James (Eds.), *Current Topics in Experimental Endocrinology,* Vol. 1. New York: Academic, 1971. P. 1.
Karim, S. M. M. (Ed.). *The Prostaglandins.* New York: Wiley-Interscience, 1972.
Litwack, G. (Ed.). *Biochemical Actions of Hormones.* New York: Academic, 1970.

The Hypothalamo-Hypophysial Axis   34

Harry Lipner

THE HYPOPHYSIS

The pituitary gland or *hypophysis,* which is composed of a neural and a glandular part, is regulated by the neural and glandular activity of the hypothalamus that overlies it. The hypophysis is contained in a depression (the sella turcica) in the basisphenoid bone located in the floor of the cranium. In the human, it weighs about 0.5 gm. The hypophysis is composed of the *adenohypophysis,* or glandular portion, and the *neurohypophysis,* a nonglandular portion (Fig. 34-1). The adeno-hypophysis is subdivided into the anterior lobe (*pars distalis*), a tuberal part (*pars tuberalis*), and the intermediate part (*pars intermedia*). The neurohypophysis is composed of the neural lobe (*pars nervosa*), the neural stalk (*infundibulum*), and the *median eminence.* The hypophysis hangs below the brain and is connected to it by the neural stalk, which forms the link between the hypophysis and the median eminence.

The neural lobe is composed of pituicytes, which have no secretory function. They are derived from connective tissue elements of the nervous system, the neuroglia. The hypothalamus projects a hypothalamo-hypophysial tract — a nerve bundle originating in two of the hypothalamic nuclei: the supraoptic and paraventricular nuclei — into the hypophysis. The tract terminates around the capillaries in the neural lobe.

The adenohypophysis is the source of at least seven and possibly eight hormones, and it contains at least seven clearly demarcated cell types. The secretions, the cell types, their specific names, and granular size are listed in Table 34-1. Perhaps one of the most challenging problems whose resolution has appeared in the past 20 years was that of the regulation of secretion of the hypophysial hormones.

Hormonal control of the pituitary gland is made possible by its complex blood supply. In the human, the internal carotid artery provides the inferior hypophysial arteries to the neural lobe and, to a small degree, the sinusoids of the adenohypophysis. The internal carotid artery and the posterior communicating branch of the arterial circle of Willis give rise to several superior hypophysial arteries. These vessels enter the median eminence and the base of the pituitary stalk, where they arborize profusely. They are then gathered together at the surface into veins that pass down the stalk to supply the sinusoids of the adenohypophysis. The arrangement of two sets of capillaries separated by an efferent artery or vein constitutes a portal system. In this case, the connection of capillaries in the median eminence of the hypothalamus and the hypophysis gives rise to the hypothalamo-hypophysial portal system. The flow of blood is from the median eminence to the adenohypophysis. There are also long and short portal vessels, which make possible a higher degree of specificity of control. The primary control, however, is brought about by the secretion of neurohormones

697

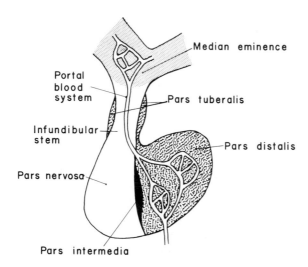

Median eminence

Portal
blood
system

Pars tuberalis

Infundibular
stem

Pars distalis

Pars nervosa

Pars intermedia

Figure 34-1
Sagittal section through the hypophysis. Its anatomic and functional interrelations are indicated
in the scheme below:

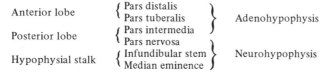

Anterior lobe $\left\{\begin{array}{l}\text{Pars distalis}\\ \text{Pars tuberalis}\end{array}\right\}$

Posterior lobe $\left\{\begin{array}{l}\text{Pars intermedia}\\ \text{Pars nervosa}\end{array}\right.$ Adenohypophysis

Hypophysial stalk $\left\{\begin{array}{l}\text{Infundibular stem}\\ \text{Median eminence}\end{array}\right\}$ Neurohypophysis

in the median eminence in the vicinity of the first capillary bed and the delivery of
these neurohormones to the adenohypophysial cells that surround the sinusoids of
the second capillary bed.

The function of this portal system became apparent when Harris [1] and his associates demonstrated that coitally induced ovulation in rabbits in which the hypophysial
stalk had been sectioned could occur if the portal system was allowed to regenerate. If,
however, a barrier (e.g., aluminum foil or waxed paper) was inserted into the site of
the hypophysial stalk section in order to prevent blood vessel regeneration, then ovulation did not occur. These early experiments led to an enormous proliferation of
research, which established that the control of hypophysial function resided, in large
part, in the *hypothalamus,* the area of the diencephalon immediately overlying the
pituitary gland.

THE HYPOTHALAMIC HORMONES

The hypothalamus is the source of an undetermined number of hormones whose functions include the regulation of the biosynthetic and secretory activities of the adenohypophysial cells. Initial studies suggested that each adenohypophysial cell was

Table 34-1. Adenohypophysial Hormones, Cell Type, and Granule Size

Hormone	Cell Type		Size of Granule (microns)
	General	Specific	
Growth hormone (GH, somatotropin)	Acidophil[a]	Somatotrope	350
Prolactin (luteotropin, mammo-tropin)	Acidophil	Mammotrope	600
Thyroid-stimulating hormone (TSH, thyrotropin)	Basophil	Thytrotrope	140
Follicle-stimulating hormone (FSH)	Basophil	Gonadotrope	200
Luteinizing hormone (LH)	Basophil	Gonadotrope	200
Adrenocorticotropic hormone (ACTH, corticotropin)	Chromophobe	Corticotrope	200
Melanocyte-stimulating hormone (α-MSH, β-MSH)	Basophil	–	–
Lipotropin (β-LPH, γ-LPH)	–	–	–

[a]Ehrlich divided stains into those with affinities for nuclear constituents and those with affinities for cytoplasmic constituents. Later workers showed that basic (cationic) dyes – for example, methylene blue or azure A and B – stain structures composed of nucleic acids, sulfated poly-saccharides, sialic and uronic polysaccharides, or negatively charged proteins, whereas the acidic (anionic) dyes – for example, eosin Y and B or orange G – react with positively charged proteins. The terms *basophil, acidophil,* and *chromophobe* were assigned, respectively, to the cells with cytoplasmic granules that stain with basic dyes, acidic dyes, or neutral dyes.

regulated by a specific neurohormone whose function was to stimulate the secretion of preformed and stored hormone; hence the name *releasing hormones* was adopted. Subsequent studies suggested that certain pituitary hormones (prolactin and melano-cyte-stimulating hormones) were tonically inhibited from being secreted; the hormones causing this were called *release-inhibiting hormones.* It is currently unclear whether the original concept of the high specificity of these neurohormones is correct. The same releasing hormone that causes LH secretion, for example, also causes FSH secre-tion. Similarly, the same releasing hormone that causes TSH secretion also causes prolactin secretion. Despite these limitations, however, it is still possible to describe the control of adenohypophysial hormone secretion and the relative specificity that exists in terms of the function of hypothalamic hormones.

Table 34-2 gives the names that have been assigned to each of the hypothalamic hormones and their commonly used abbreviations. In all cases, the primary target is assumed to be the *only* target of these regulating hormones. The concentrations of the regulating hormones in the hypothalamo-hypophysial portal blood are so small

Table 34-2. Hypothalamic Hormones

Hypothalamic Hormone	Abbreviation	Primary Target	Secondary Target	Hormonal Control	Feedback
Corticotropin (ACTH) releasing hormone	C-RH or CRF	Corticotrope	Adrenal cortex	Glucocorticoids	$(-)^a$
Thyrotropin (TSH) releasing hormone	TSH-RH, TRH, or TRF	Thyrotrope	Thyroid	Thyroid hormones	$(-)$
Luteinizing-hormone (LH) releasing hormone	LH-RH or LH-RF	Gonadotrope	Interstitial cells of testis and ovary Granulosa cells of ovarian follicle Corpus luteum	Sex hormones	$(+ \text{ or } -)$
Follicle-stimulating hormone (FSH) releasing hormone	FSH-RH or FSH-RF	Gonadotrope	Spermatogonia Ovarian follicle cells	Sex hormones	$(+ \text{ or } -)$
Growth-hormone (GH) releasing hormone	GH-RH or GH-RF	Somatotrope	Somatic cells	None	Metabolite regulation; glucose $(-)$, plasma free fatty acids $(-)$
Growth-hormone (GH) release-inhibiting hormone (somatostatin)	GH-RIH or GIF	Somatotrope	Somatic cells	None	—
Prolactin releasing hormone	P-RH or PRF (TSH-RH?)	Mammotrope	Mammary tissue	Estrogen Progesterone	Neural stimulus (suckling) $(+)$
Prolactin release-inhibiting hormone	P-RIH or PIF	Mammotrope	Mammary tissue	Estrogen Progesterone	Neural stimulus (suckling) $(-)$
Melanocyte-stimulating hormone (MSH) releasing hormone	MSH-RH, MRH, or MRF	Intermediate cells	Melanophores	None	Neural stimulus; visual cues $(+ \text{ or } -)$
Melanocyte-stimulating hormone (MSH) release-inhibiting hormone	MSH-RIH, MRIH, or MIF	Intermediate cells	Melanophores	None	—

[a]Positive feedback: $(+)$; negative feedback: $(-)$.

(picogram quantities) that these hormones would be greatly reduced in effectiveness once they were diluted in the systemic blood. The secondary targets are the structures that respond to *pituitary* hormone stimulation. These hormones are secreted in greater concentrations, and they appear in the systemic circulation in nanogram quantities. In effect, then, an amplification of the neurohormonal signal occurs. The hormones that are secreted by the pituitary as a result of releasing hormone activity are released in a pulsatile manner. Pulsatile secretion probably applies to all the adenohypophysial hormones, although it has been demonstrated only for GH, LH, and FSH.

The structures of these hormones are shown in Table 34-3. The amino acids, which are represented by their conventional three-letter abbreviations (Try, Cys, and so on), are connected by peptide bonds. The amino terminal amino acid is shown at the left. Sometimes, instead of a free amino group, an acetyl group (Ac) has been added, as in the structure given for C-RH. The *pyro* form of glutamic acid, with its internally formed ring structure, is found in TSH-RH and LH-RH. One interesting property of this class of hormones is that in the carboxyl terminal amino acid, the carboxyl group is often replaced by an amide group. For instance, the structure of the terminal amino acid of C-RH, glycinamide, is $-NHCH_2CONH_2$.

The hypothalamus plays a key role in the feedback control of many endocrine systems, and it serves as the transducer that converts neural signals into endocrine secretions (Fig. 34-2). The input signal to the hypothalamus may generate either a positive or a negative response. Exteroceptive stimuli generate neural signals that spread through the nervous system either into the brain via the spinal cord or directly into the higher centers of the brain. The neural signal eventually activates the hypothalamus and elicits the secretion of releasing or release-inhibiting hormones via the tuberoinfundibular tract. These relatively specific secretions then cause specific adenohypophysial cells to secrete pituitary hormones.

An example of a positive feedback response is the response to suckling in the lactating mother. Suckling by the infant stimulates receptors in the nipple that cause two different secretory patterns, both of which are reflex in nature and positive in direction. The first response is the release of oxytocin and the phenomenon of milk ejection or what dairymen call "letdown." This involves the excitation of the paraventricular nucleus in the hypothalamus and the secretion of oxytocin into the capillaries of the neurohypophysis. Simultaneously, the secretion of prolactin release-inhibiting hormone (P-RIH) stops, and there probably begins the secretion of prolactin releasing hormone (P-RH), the existence of which is subject to question. Prolactin, however, then begins to appear in the systemic blood. The net result is that the oxytocin causes the myoepithelial cells to contract, thus causing the milk in the alveoli of the mammary gland to move into the ducts and to be ejected. The prolactin stimulates the alveoli to form more milk.

Exteroceptive stimuli that represent stressful conditions may modulate the secretion of pituitary hormone. The secretion of adrenocorticotropin (ACTH) in response to stress is a positive feedback response. This same stress may cause a depression of TSH secretion, i.e., it induces a negative feedback response with regard to this hormone.

Table 34-3. Structure of Hypothalamic Hormones

Hormone	Structure
Corticotropin (ACTH) releasing hormone (C-RH)	Ac–Ser–Try–Cys–Phe–His–(Asp–NH, Glu–NH)–Cys–(Pro, Val)–Lys–Gly–NH$_2$
Thyrotropin (TSH) releasing hormone (TSH-RH)	(*pyro*)Glu–His–Pro–NH$_2$
Luteinizing-hormone releasing hormone (LH-RH)	(*pyro*)Glu–His–Trp–Ser–Tyr–Gly–Leu–Arg–Pro–Gly–NH$_2$
Follicle-stimulating hormone releasing hormone (FSH-RH)	
Growth-hormone releasing hormone (GH-RH)	(Not isolated)
Growth-hormone release-inhibiting hormone (somatostatin) (GH-RIH)	H-Ala–Gly–Cys–Lys–Asn–Phe–Phe–Trp–Lys–Thr–Phe–HO–Cys–Ser–Thr
Prolactin releasing hormone (P-RH)	(Probably TSH-RH)
Prolactin release-inhibiting hormone (P-RIH)	(Not isolated; possibly dopamine)
Melanocyte-stimulating hormone releasing hormone (MSH-RH)	Pro–Leu–Gly–NH$_2$ or Cys–Tyr–Ile–Gln–Asn–Cys–OH[a]
Melanocyte-stimulating hormone release-inhibiting hormone (MSH-RIH)	H-Cys–Try–Ile–Gln–Asn–Cys–OH[a]

[a]Not identified in hypothalamic tissue.

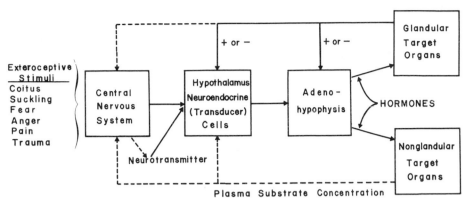

Figure 34-2
Generalized scheme of the neuroendocrine pathway for the regulation of tropic hormone secretion. Feedback controls probably act on the central nervous system and on the hypothalamus. Target hormones are also able to modulate the response of the adenohypophysial cells as well as the hypothalamus. The effect of the hormones may be positive (+) or negative (−) feedback control of secretion. Those adenohypophysial hormones that act on nonglandular target organs are under either neural control or metabolic substrate control. Dashed lines show negative feedback mechanisms.

Humoral stimuli arising within the organism may be of two varieties. They may be hormonal substances produced by target glands, or they may be osmotically active or chemical substances that exert effects by virtue of their osmotic properties or specific composition. Estrogens, when administered chronically in small amounts followed by a single large dose, cause a surge of LH secretion that persists for about two hours; the result is a positive feedback response. Ablation of the primary target glands results in the elevation of the plasma concentration of certain pituitary hormones, e.g., TSH. Depression of circulating TSH concentrations is readily achieved by the administration of thyroid hormone (T_4). Gonadotropin concentrations in castrated animals may be reduced to those found in intact animals by the administration of the appropriate gonadal steroids. These are examples of negative feedback controls that are induced by target-gland hormones.

Nonhormonal controls have also been demonstrated. Excessive consumption of water causes a reduction in plasma osmolality, with a reduction in antidiuretic hormone (ADH) secretion and a consequent increased loss of a dilute urine; this mechanism constitutes a negative feedback control. Growth hormone secretion has been linked with plasma free fatty acid concentrations. Hypolipemia causes an increase in GH secretion, which in turn causes an increase in the concentration of free fatty acids in the plasma. This is an example of the negative feedback control of a primary glandular secretion by nutrients contained in the plasma.

REFERENCES

1. Harris, G. W., and Donovan, B. T. *The Pituitary,* Vol. 3. Berkeley, Calif.: University of California Press, 1966.

SELECTED READING

Dannies, P. S., and Tashjian, A. H., Jr. Effects of thyrotropin-releasing hormone and hydrocortisone on synthesis and degradation of prolactin in a rat pituitary cell strain. *J. Biol. Chem.* 248:6174, 1973.

Gual, C., and Rosenberg, E. (Eds.). *Hypothalamic Hypophysiotropic Hormones.* Amsterdam: Excerpta Medica, 1973.

Guillemin, R., Burgus, R., and Vale, W. The hypothalamic hypophysiotropic thyrotropin-releasing factor. *Vitam. Horm.* 29:1, 1971.

Hinkle, P. M., and Tashjian, A. H., Jr. Receptors for thyrotropin-releasing hormone in prolactin-producing rat pituitary cells in culture. *J. Biol. Chem.* 248:6180, 1973.

Martini, L., and Ganong, W. F. *Neuroendocrinology,* Vols. 1 and 2. New York: Academic, 1966.

Schally, A. V., Arimura, A., and Kastin, A. J. Hypothalamic regulatory hormones. *Science* 179:341, 1973.

Harry Lipner

THE ADENOHYPOPHYSIAL HORMONES

As early as 1920, with the successful ablation of the pituitary gland in tadpoles, dogs, and rats, it was apparent that the pituitary is the source of specific hormones that control the growth and the reproductive and metabolic processes of vertebrate organisms. These functions could be recovered by the implantation of pituitary tissue or the administration of pituitary extracts.

Of the eight hormones that have been isolated from adenohypophysial tissue, only six have been extensively studied. The roles of the melanocyte-stimulating hormones and of the lipotropins in the economy of the mammalian organism have not been established. The mechanism of action of the adenohypophysial hormones is under intensive study. The molecular structures of these substances have been determined. The adenohypophysial hormones and their principal functions are listed in Table 35-1.

GROWTH HORMONE

Growth hormone (somatotropin) secretion is provoked under a number of conditions that vary in such a way as to defy the formulation of any simple rules for its prediction. Sleep, exercise, surgical trauma, hemorrhage, ether anesthesia, and cold stress all are stimulatory to GH secretion. Secretion is stimulated by lowering the blood sugar level by insulin or by simulating this effect with 2-deoxy-D-glucose, amino acids, and lysine vasopressin. Only a small number of conditions have been described that are inhibitory to GH secretion. GH is inhibitory to itself in that it exerts a "short-loop" inhibition, as do glucose (by means of hyperglycemia) and somatostatin.

The feedback control of GH secretion is summarized in Figure 35-1. The central nervous system (CNS, block 1) contains the structures that transduce chemical and mechanical stimuli to neural signals. These signals are conducted into the ventromedial and arcuate nuclei of the hypothalamus by neurons whose transmitter substance is a catecholamine (block 2). In primates, the receptor sites in the medial basal hypothalamus are α-adrenergic receptors. Blockade with phentolamine (an α-adrenergic blocking agent) prevents the secretion of GH. The controls of somatostatin secretion are not known, and those that are indicated (block 4) are hypothetical. It is known, however, that the administration of glucose or GH and the elevation of plasma free fatty acids are associated with decreased GH secretion (blocks 6 and 7). Unlike other hormones, there is no target gland hormone to exert a negative feedback control for GH. Regulation appears to depend on the concentration of tissue substrates. It should be noted, however, that GH and insulin are present in the liver in an approximately inverse relationship (block 8). By influencing the concentration of the nutrient substrate

Table 35-1. Hormones of the Adenohypophysis

Group	Hormone	Principal function
I	1. Adrenocorticotropin (ACTH)	Production and secretion of adrenal cortical hormones (stimulates the adrenal cortex to produce such steroid hormones as cortisol and corticosterone). Sheep: Isolated by Li, Geschwind, Levy, Harris, Dixon, Pon, and Porath, 1953. Structure determined by Li, Geschwind, Cole, Raacke, Harris, and Dixon, 1955. Pig: Isolated and structure determined by Bell, Shepherd, and co-workers, 1954–55.
	2 and 3. Melanotropin (α-MSH and β-MSH) (melanocyte-stimulating hormone)	Darkening of skin (pigmentation). β-MSH: Isolated and structure determined by Geschwind, Li, and Barnafi, 1956; Harris and Roos 1956. α-MSH: Isolated and structure determined by Harris and Lerner, 1957.
	4 and 5. Lipotropin (β-LPH and γ-LPH)	Release of lipid from adipose tissue. β-LPH: Isolated and structure determined by Li, Barnafi, Chretien, and Chung, 1965. γ-LPH: Isolated and structure determined by Chretien and Li, 1967.
	6. Somatotropin (growth hormone, GH)	General body growth; anabolic effect on various tissues of the body; effects on fat, carbohydrate, and protein metabolism. Bovine: Isolated and purified by Li and Evans, 1944 (bGH). (MW = 45,000.) Human: Isolated and purified by Li and Papkoff, 1956 (hGH). Structure determined by Li, Liu, and Dixon, 1966. (MW = 21,500.)
II	7. Prolactin (lactogenic hormone, Pr)	Growth, development, and lactation of the mammary gland. Isolated and purified by Li, Lyons, and Evans, 1942; White, Bonsnes, and Long, 1942. Structure determined by Li, Dixon, Lo, Schmidt, and Pankov, 1969. (MW = 23,300.)
	8. Interstitial cell-stimulating hormone (ICSH) (also called luteinizing hormone, LH)	In a woman – development of interstitial cells in the ovaries. Affects reproduction. In a man – development of interstitial cells in the testes. Affects reproduction. Isolated and purified by Li, Simpson, and Evans, 1940. Structure determined by Papkoff, Sairam, and Li, 1971.
III	9. Follicle-stimulating hormone (FSH)	In a woman – development of follicles in the ovaries. Affects reproduction. In a man – seminiferous tubules in the testes. Affects reproduction. Isolated and purified by Papkoff, Candiotti, and Li, 1964.
	10. Thyrotropic hormone (TSH)	Production and secretion of thyroid hormones. Isolated and purified by Pierce, Condliffe, and their co-workers, 1960. Structure determined by Liao and Pierce, 1971.

Modified from C. H. Li: Proc. Am. Phil. Soc. 116:365, 1972.

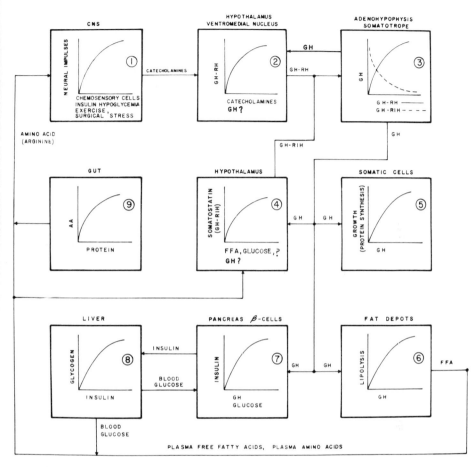

Figure 35-1
Control of the secretion of growth hormones (GH).

in the blood, a negative feedback control is exerted (block 9). Further control is exerted via the short-loop negative feedback due to GH acting on the hypothalamus (blocks 2 and 4).

PROLACTIN

Prolactin (lactogenic hormone, mammotropin) has a different pattern of secretion than does growth hormone. Human pituitaries synthesize and secrete prolactin primarily during pregnancy and the postpartum period. In large part, prolactin is responsible for the anabolic processes in the alveolar cells, of which milk formation is the final result.

The control of prolactin secretion in humans is diagrammed in Figure 35-2. The effects of exteroceptive stimuli (suckling, stress, and exercise) are assumed to excite the hypothalamus via other areas in the central nervous system (CNS, block 1). Since

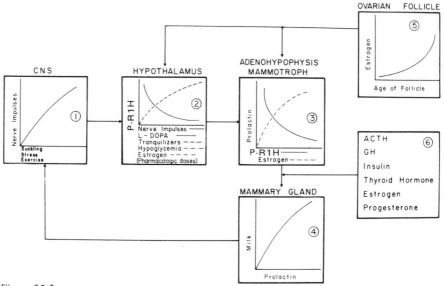

Figure 35-2
Control of prolactin secretion.

this hypothalamic input ultimately causes an increased plasma prolactin concentration, it is assumed that it decreases the secretion of prolactin release-inhibiting hormone (P-RIH) (block 2). It should be noted that this effect could also be induced by an increased secretion of a prolactin releasing hormone (P-RH). The hypothalamus is probably the site at which tranquilizers act (see block 2), but it need not be. L-Dihydroxyphenylalanine (L-DOPA) penetrates the blood-brain barrier and is probably converted to one of the active catecholamines; in any case, it acts in the same manner as an increased neural input does, namely, to increase P-RIH secretion and depress prolactin secretion. Hypoglycemia and estrogen both act on the hypothalamus to increase prolactin secretion (block 2).

Estrogen can act directly on the pituitary to increase prolactin secretion. A diagram indicating an ovarian follicular source of estrogen is included in Figure 35-2 (block 5), but it should be noted that in humans, no increased prolactin secretion has been observed at the time of peak gonadotropin secretion, i.e., at midmenstrual cycle. Figure 35-3 shows the plasma prolactin concentration in women who are nursing. In a 30-minute nursing session, the peak values are exhibited at the termination of nursing.

Prolactin may also play a role in the economy of the organism well beyond its role in lactation. Prolactin secretion in nonlactating humans is described in Chapter 36.

ADRENOCORTICOTROPIN

The control of adrenocorticotropin (ACTH) secretion from the adenohypophysis is multifaceted. ACTH secretion is directly under control of the adrenocorticotropin-

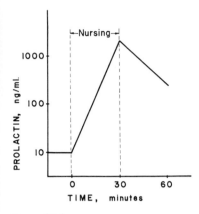

Figure 35-3
Plasma prolactin concentration in women during nursing.

releasing hormone (C-RH). C-RH secretion is controlled by ACTH via a short-loop negative feedback circuit and by a longer negative feedback loop involving gluco-corticoid secretion that is activated by ACTH. The glucocorticoids also act directly on the adenohypophysis to modulate the secretory activity of the corticotropic cells. Emotional and environmental stress as well as noxious stimuli and physical and surgical trauma may cause increased ACTH secretion. The administration of norepinephrine or L-dopamine is also a potent stimulator of ACTH secretion.

The interactions of the various factors contributing to the control of ACTH secretion are illustrated in Figure 35-4A and B. Physical and emotional stresses are potent stimuli for ACTH secretion. It is assumed that all physical and emotional stimuli funnel into the hypothalamus from the central nervous system (block 1). The neural inputs from the CNS act on the hypothalamus to either increase or decrease C-RH secretion (block 2). Hormonal signals also regulate C-RH secretion, which is increased by catecholamines and depressed by glucocorticoids and ACTH. The C-RH that is released from the hypothalamus causes the adenohypophysis to release ACTH (block 3). Figure 35-4B includes some of the targets of ACTH and the glucocorticoids. ACTH acting on the adrenal cortex increases glucocorticoid secretion (block 4); ACTH acting on fat depots causes the mobilization of free fatty acids (FFA) (block 5). The gluco-corticoids that are secreted as a result of the action of ACTH on the adrenal cortex act on the hypothalamus to cause negative feedback inhibition of C-RH secretion (Fig. 35-4A). The glucocorticoids acting on muscle cause a decreased synthesis of protein and an increased release of amino acids (Fig. 35-4B, block 6). They also act on the liver to cause an increase in the synthesis of glucose (gluconeogenesis) due to the increased concentrations of amino acids brought via the blood (block 7). The glucocorticoids act on both the liver and muscle to increase their stores of glycogen (block 8). With rising concentrations of ACTH and glucocorticoids, increased concentrations of glucose and free fatty acids appear in the blood. The ACTH that is secreted

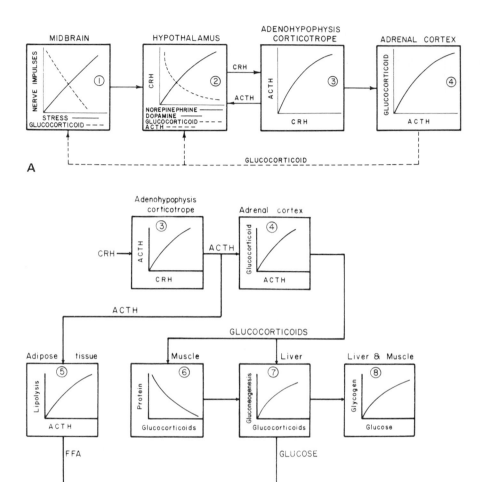

Figure 35-4
A. Control of adrenocorticotropin (ACTH) secretion. *B.* Effects of ACTH on free fatty acid (FFA) and glucose concentrations in the blood.

by the adenohypophysis acts by means of short-loop feedback control to inhibit C-RH secretion in the hypothalamus (Fig. 35-4A, blocks 2 and 3).

MELANOCYTE-STIMULATING HORMONE

The secretion of melanocyte-stimulating hormone (MSH) is controlled by the hypothalamus through a tonic inhibition effect that is exerted by the melanocyte-stimulating hormone release-inhibiting hormone (MSH-RIH) (Fig. 35-5, block 2). The secretion of

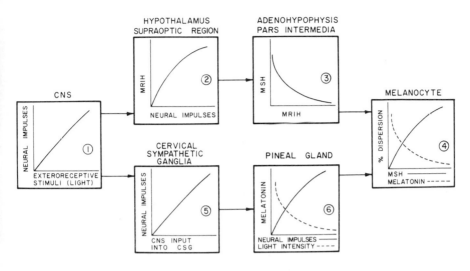

Figure 35-5
Control of melanocyte-stimulating hormone (MSH) secretion.

MSH-RIH is modulated by the influx of light into the eyes (block 1). The primary source of MSH-RIH is the supraoptic nucleus of the hypothalamus, which lies above the optic chiasm. The existence of a possible input from the optic chiasm into the overlying nuclei has been postulated. Light modulates the output of melatonin from the pineal gland (block 6). This mechanism involves the activation of the superior cervical ganglion and its discharge via the internal carotid plexus to activate the pineal gland (block 5).

The normal hue of the frog skin is correlated with the mean level of MSH and melatonin secretion, with extremes being represented by darkened or lightened skin. In higher vertebrates, the pineal gland may exert a control on the secretion of luteinizing hormone. Two substances have been isolated that are capable of suppressing MSH secretion by the pituitary. One is a tripeptide, Pro—Leu—Gly-amide, which is derived from the C-terminal end of oxytocin, and the other is a pentapeptide with the sequence Pro—His—Phe—Arg—Gly—NH_2.

Another neurohormone has been hypothetically postulated to increase pituitary MSH secretion. This substance has been termed melanocyte-stimulating hormone releasing hormone (MSH-RH). It may be a pentapeptide with the sequence Cys—Tyr—Ile—Gln—Asn—Cys—OH, which supposedly causes MSH activity to appear in the blood. Most recently, Schally's group [5b] noted that in rats with hypothalamic lesions involving the hypophysiotropic area but with no infarcts in the median eminence or the pituitary, MSH activity appeared in the blood. Since there was no intact connection between the hypothalamus and the pituitary, these investigators have suggested that there is only a release-inhibiting hormone but no specific releasing hormone for MSH.

THYROTROPIN

The thyrotrope in the adenohypophysis is the source of thyrotropin or thyroid-stimulating hormone (TSH). The thyroid gland is its usual target, and TSH in large excess affects fatty tissues by stimulating the secretion of fatty acids.

The control of TSH secretion is perhaps among the simplest of control mechanisms (Fig. 35-6). The thyroid hormones (T_3 and T_4) exert a direct inhibitory or negative feedback control on the adenohypophysis, whereas the hypophysiotropic neurohormone TSH-RH is responsible for direct stimulation (Fig. 35-6, blocks 2, 3, and 4). The relationship between plasma thyroxine (T_4) concentrations and plasma levels of TSH has been explored by Reichlin et al. [5a]. As the dose of thyroxine that was administered to hypothyroid patients was increased, the plasma TSH concentration declined, and the curve of a plot of the two variables leveled out at the normal range of T_4 (Fig. 35-7). In studies with rats, similar data were obtained. The role of the hypothalamus was in part delineated when it became apparent that rats with lesions in the hypothalamus could still respond to substances that prevent the secretion of thyroid hormones (i.e., goitrogens) with an increased TSH secretion. This elevation in TSH secretion, however, was less than that seen in the intact rats. In Reichlin's analysis of the control system, the pituitary is assigned the role of the "thyrostat," which is under direct negative feedback control by thyroid hormone, and the hypothalamus is assumed to modulate the responsiveness of the pituitary and to provide the "set point."

Electrical stimulation of the hypothalamus in the area that has been designated as the *thyrotropic area* causes a prompt increase in the plasma TSH level. Associated with the electrical stimulation of the thyrotropic area are demonstrable increases in TSH-RH in the portal blood. The secretion of TSH-RH is also affected by environmental temperature. Acutely increased plasma TSH levels can be demonstrated by the direct cooling of the hypothalamus with a double-barreled probe through which cold fluid is circulated. Chronically cold conditions might not directly influence hypothalamic function; however, there is probably an increased utilization of thyroid

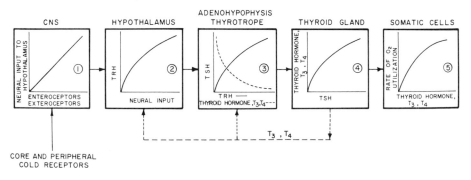

Figure 35-6
Control of thyroid-stimulating hormone (TSH) secretion.

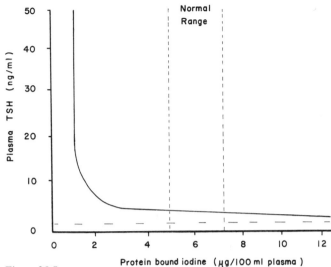

Figure 35-7
The effect of replacing successively larger doses of thyroxine at approximately 10-day intervals in six hypothyroid patients. (From S. Reichlin et al. *Recent Prog. Horm. Res.* 28:229, 1972. Reproduced with permission.)

hormones, and thus the rate of secretion of hormones may be changed without affecting their blood concentrations.

GONADOTROPINS: FOLLICLE-STIMULATING HORMONE AND LUTEINIZING HORMONE (INTERSTITIAL CELL-STIMULATING HORMONE)

The follicle-stimulating hormone (FSH) is perhaps the least understood of the adenohypophysial hormones. It is necessary for the maturation of ovarian follicles, but its precise role is undefined. In the male, it is necessary for spermatogenesis, but here, too, its precise role is not known inasmuch as it exerts an effect only in the presence of luteinizing hormone. The analysis of its function is complicated by the difficulty in demonstrating, in the intact, mature animal, an effect on ovarian cycling in the presence of FSH antiserum. There are conflicting data on whether FSH can induce steroid synthesis; however, it is generally believed that FSH is not steroidogenic.

The conditions involved in the regulation of FSH secretion are apparently similar to those for the regulation of luteinizing hormone and will be considered with this latter hormone.

In both males and females, luteinizing hormone (LH) or interstitial cell-stimulating hormone (ICSH) is unquestionably steroidogenic. LH and ICSH are the same hormone, but it is frequently referred to as ICSH in males. In the male, the target cells are the interstitial cells in the testis, and the result of the hormonal action is the secretion of testosterone. In the female, the target cells may be the interstitial cells in the ovary, but the primary target is the granulosa (or follicular) cells of the ovarian follicle.

The result of the action of the hormone is to induce the luteinization of the cells as well as steroidogenesis with secretion of progesterone.

The regulation of gonadotropin secretion is dependent on negative feedback control that is exerted primarily on the hypothalamus and secondarily on the adenohypophysis. The steroid that exerts the chief effect in the female is estrogen. This hormone reaches its peak level prior to the onset of the surge of the two gonadotropins, which, in women, occurs at about the midpoint of the menstrual cycle. Plasma progesterone concentration rises concurrently with the levels of the gonadotropins, and is dependent on these for its appearance (Fig. 35-8).

The neurons in the hypophysiotropic area of Halasz are influenced by neural inputs from other neural areas, and their activity is increased by L-dopa. It is not clear whether L-Dopa is the active factor or whether it must be converted to norepinephrine. L-Dopa, however, does penetrate the blood-brain barrier, and it exerts an effect on gonadotropin secretion. Melatonin, which is produced in the pineal gland, delays puberty in female rats and the development of sperm in the testes of pubescent male rats. Melatonin thus plays a transient role in sexual development, but it may not be significant in the adult.

The secretion of gonadotropins in the male is continuous, i.e., it is tonic. The male shows no cyclic rhythmicity in gonadotropin secretion, and in primates and domesticated

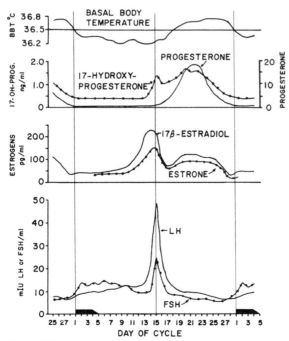

Figure 35-8
Representative basal body temperatures and daily serum concentrations of gonadotropins and gonadal steroids that might be obtained during a normal, 28-day, human menstrual cycle.

species such as dogs, cats, rats, sheep, swine, cattle, the male can mate at any time. Among mammalian species in the wild, however, the male generally is competent only when the female is in estrus (heat). Figure 35-9 illustrates the control of gonado-tropin secretion in males. The controls of secretion are oriented, in man and domesti-cated animals, around the maintenance of constant levels of activity. The central nervous system causes increased androgen (e.g., testosterone) secretion in the presence of females or as a result of copulatory activity (Fig. 35-9, block 2). Castration results in the atrophy of accessory sexual structures such as the prostate, seminal vesicles, and penis, the cessation of growth of pubic and facial hair, the redistribution of body fat according to the female pattern, and decreased muscle size and strength. Blood titers of LH and FSH rise to extremely high levels. The administration of an androgen (testosterone is the most potent of the androgens) reverses all these effects of castra-tion. In boys with pineal tumors, precocious puberty occurs, which suggests a pineal-gland control of gonadotropin secretion (block 1). The secretion of androgens plays a significant role in the negative feedback control of the secretion of gonadotropins; this effect is exerted at both the hypothalamic and adenohypophysial levels (blocks 3 and 4).

The feedback control mechanism for LH and FSH secretions by the gonadotropic cells of the adenohypophysis in females is shown in Figure 35-10. The LH and FSH levels are presented as a single curve (block 3) even though the onset of the secretion of FSH begins earlier than that for LH (block 2) and the secretion persists for a longer time. The steroid hormones (estrogen and progesterone) modulate the responsiveness of the gonadotropins to LH-RH and FSH-RH (block 3). The contribution of the central nervous system is manifested by the effects of coitus on the increased secretion of gonadotropins (block 1). Other neural areas impinge on the hypothalamus to regulate

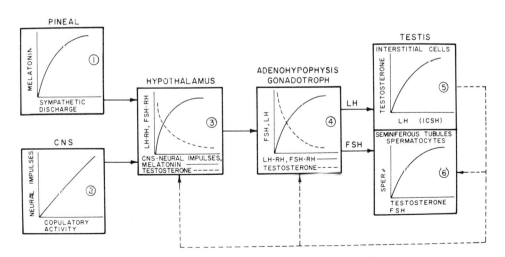

Figure 35-9
Control of the secretion of gonadotropins in males.

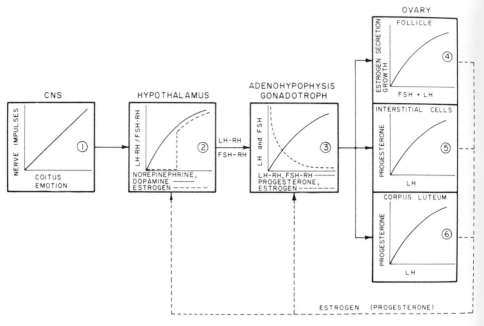

Figure 35-10
Control of the secretion of gonadotropins in females.

the secretion of the releasing hormones. The neuroactive substances acting on the hypophysiotropic area are dopamine and norepinephrine (block 2). These are released by the neurons that terminate in the hypophysiotropic area.

THE NEUROHYPOPHYSIAL HORMONES

The association of the neurohypophysis with endocrine function was first suggested in 1895 when Oliver and Schaefer [4] demonstrated increases in blood pressure (pressor effects) in dogs following the administration of posterior pituitary extracts. In 1909 Dale [1] noted that such extracts induced contractions in the rabbit uterus (oxytocic effect). Von den Velden [8] in 1913 observed the inhibition of urine excretion by posterior pituitary extracts (antidiuretic effect). Ott and Scott [5] in 1910 showed that these extracts caused the release of milk ejection (letdown effect).

Thus, all the actions of the hormones of neurohypophysis were well identified before 1920. Kamm et al. [3] in 1928 made the long leap to the conclusion that there are two such hormones: *oxytocin,* with oxytocic and milk ejection properties, and *vasopressin,* with pressor and antidiuretic properties. Although there has been some confusion as to whether a single, large molecule contains both hormones, the present consensus suggests that the two hormones are each attached to or are part of a large molecule, termed *neurophysin.* Each hormone is assumed to be synthesized in the cell bodies that compose the different hypothalamic nuclei. They are conveyed down the axons of neurosecretory neurons from their origin in the hypothalamic nuclei to their destination, the nerve terminals in the neurohypophysis.

Electrolytic lesions in the area of the supraoptic nucleus of the hypothalamus in rats cause the development of a condition that is similar to that seen in humans with damage to that region of the hypothalamus; this disease is *diabetes insipidus*. In the full-blown disease in man it is characterized by the excretion of large volumes, 4 to 11 liters, of dilute urine (specific gravity 1.002 to 1.006) in 24 hours (polyuria) and the consumption of large volumes of water (polydipsia); the urine produced has an extremely low specific gravity. Electrolytic lesions in the area of the paraventricular nucleus of the hypothalamus in pregnant rats result in difficulty in the delivery of young, and when lesions are made in lactating rats, the suckling pups fail to obtain milk, because milk ejection does not occur.

The amino acid composition of oxytocin was established by du Vigneaud and others in 1953 [2] and that of vasopressin or antidiuretic hormone (ADH) was determined by Turner et al. in 1951 [6]. In 1953 du Vigneaud et al. synthesized oxytocin and vasopressin [2]. These small peptide hormones, oxytocin and arginine-vasopressin, are found to be the same in the posterior pituitaries of all mammals except the hog, which has lysine-vasopressin. The neurohypophysial hormones of most vertebrates have been identified, and their amino acid sequences are shown in Figure 35-11. The classes and some orders of animals in which they are usually found are listed in Table 35-2.

Amino acid number (position)	1 2 3 4 5 6 7 8 9
Parent molecule:	Cys−Tyr−. . . , . . .−Asn−Cys−Pro−. . .−Gly−(NH$_2$)
Oxytocin:	Cys−Tyr−Ile−Gln−Asn−Cys−Pro−Leu−Gly−(NH$_2$)
Arginine-vasopressin:	Cys−Tyr−Phe−Gln−Asn−Cys−Pro−Arg−Gly−(NH$_2$)
Lysine-vasopressin:	Cys−Tyr−Phe−Gln−Asn−Cys−Pro−Lys−Gly (NH$_2$)
Arginine-vasotocin:	Cys−Tyr−Ile−Gln−Asn−Cys−Pro−Arg−Gly−(NH$_2$)
Ichthyotocin:	Cys−Tyr−Ile−Ser−Asn−Cys−Pro−Ile−Gly−(NH$_2$)
Mesotocin:	Cys−Tyr−Ile−Gln−Asn−Cys−Pro−Ile−Gly−(NH$_2$)
Glumitocin:	Cys−Tyr−Ile−Ser−Asn−Cys−Pro−Gln−Gly−(NH$_2$)

Figure 35-11
The molecular structure of the seven known natural neurohypophysial hormones and the probable parent molecule from which they originated. One gene duplication and a series of subsequent single substitutions in position 3, 4, or 8 could produce two molecular "lines." For example, substitution of isoleucine for glutamine in position 8 transforms glumitocin into ichthyotocin, and substitution of glutamine for serine in position 4 transforms ichthyotocin into mesotocin. The mammalian hormone oxytocin appears with the substitution of leucine for arginine in position 8 of vasotocin. Arginine-vasopressin appears by the substitution of phenylalanine for isoleucine in position 3 of arginine-vasotocin.

Table 35-2. Phylogenetic Distribution of the Neurohypophysial Hormones

Class or Order	Oxytocin Principle	Vasopressin Principle
Mammalia	Oxytocin	Arginine-vasopressin[a]
Aves	Oxytocin	Arginine-vasotocin
Reptilia	Mesotocin	Arginine-vasotocin (chromatographic and pharmacologic evidence only)
Amphibia	Mesotocin	Arginine-vasotocin
Pisces		
Teleosts	Isotocin	Arginine-vasotocin
Ray-finned fish	Glumitocin	Arginine-vasotocin(?)
Lungfish	Mesotocin (perhaps oxytocin also)	Arginine-vasotocin
Cyclostomes	–	Arginine-vasotocin

[a]An exception is the hog, which has lysine-vasopressin.

REGULATION OF VASOPRESSIN SECRETION

The regulation of vasopressin or antidiuretic hormone (ADH) secretion revolves around the homeostasis of body water. The classic observation is that of Verney [7], who showed that an increase in the osmotic pressure of blood in the internal carotid artery is followed by the secretion of ADH. Intracarotid injections of hypertonic saline solutions also cause the secretion of oxytocin, which is accompanied by milk ejection from lactating mammary glands and the contraction of the uterus. Expansion of the blood volume, which may be simulated by distension of the left atrium, causes a decrease of ADH secretion and a consequent diuresis. Hemorrhage with a reduction of the blood volume causes the secretion of ADH and induces the retention of urinary water. Alcohol is a most potent inhibitor of ADH secretion, and this effect results in diuresis. A detailed discussion of the role of ADH in the regulation of water metabolism is contained in Chapter 36.

REGULATION OF OXYTOCIN SECRETION

Oxytocin, apart from its role in lactation and parturition, has no known function in the organism (Fig. 35-12); many roles, however, have been suggested. These have been that it acts as a releasing hormone, that it causes sperm transport in males in preparation for subsequent ejaculation, that it aids in sperm transport in females by increasing uterine motility after coitus, that it acts as a natriuretic hormone (i.e., causes the specific excretion of sodium ion), and that it causes the increased uterine contractility at the termination of pregnancy.

The factors that are most usually associated with oxytocin secretion are parturition and suckling. In both situations, a reflex arc is involved, the afferent limb of which is neural and the efferent limb is hormonal. Distension of the cervix and vagina in pregnant ewes causes the secretion of oxytocin. During parturition in cattle and during the second stage of labor in women, the blood concentrations of oxytocin increase sharply.

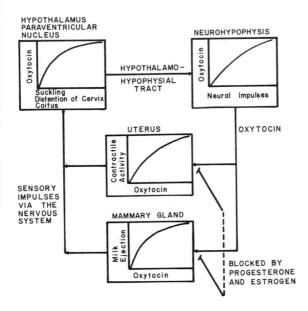

Figure 35-12
Control of oxytocin secretion.

In lactating females, the ejection of milk requires the secretion of oxytocin. Milk ejection can be induced either by the administration of oxytocin or by the electrical stimulation of the paraventricular nucleus in the hypothalamus or of any point along the hypothalamo-hypophysial tract. Milk ejection is prevented by anesthetizing the suckled nipple, by denervating the mammary gland, or by sectioning the infundibular stalk. Other factors that may cause the secretion of oxytocin are coitus and sexual excitation.

Progesterone and estrogen cause an inhibition of lactation. The mechanism appears to be due to a direct action of the steroids on the mammary glands, which makes them refractory to the hormones that are responsible for inducing the synthesis of milk. It has been suggested that the onset of parturition is subsequent to the declining production of progesterone by the aging placenta and the consequent release of inhibition of oxytocin secretion. Once initiated, the process becomes a positive feedback system, since the distention of the cervix and vagina induces further secretion of oxytocin. Suckling also acts as a positive feedback stimulus for the secretion of oxytocin, as does coitus.

REFERENCES

1. Dale, H. H. The action of extracts of the pituitary body. *Biochem. J.* 4:427, 1909.
2. du Vigneaud, V. Trail of sulfur: From insulin to oxytocin. *Science* 123:967, 1956.
3. Kamm, O., Aldrich, T. B., Grote, I. W., Rowe, L. W., and Bugbee, E. P. The active principles of the posterior lobe of the pituitary gland. *J. Am. Chem. Soc.* 50:573, 1928.
4. Oliver, G., and Schaefer, E. A. On the physiological action of extracts of pituitary body and certain other glandular organs. *J. Physiol.* (Lond.) 18:277, 1895.
5. Ott, I., and Scott, J. C. The action of infundibulum upon the mammary secretion. *Proc. Soc. Exp. Biol.* 8:48, 1910.
5a. Reichlin, S., et al. Tissue thyroid hormone concentration of rat and man determined by radio immunoassay. Biologic significance. *Mt. Sinai J. Med. N.Y.* 40:502, 1973.
5b. Schally, A. V., Arimura, A., and Kastin, A. J. Hypothalamic regulatory hormones. *Science* 26:179, 1973.
6. Turner, R. A., Pierce, J. G., and du Vigneaud, V. The purification and the amino acid content of vasopressin preparations. *J. Biol. Chem.* 191:21, 1957.
7. Verney, E. B. The antidiuretic hormone and the factors that determine its release. *Proc. R. Soc. Lond.* [*Biol.*] 135:27, 1947.
8. Von den Velden, R. Die Nierenwirkung von Hypophysenextrakten beim Menschen. *Berl. Klin. Wochenschr.* 50:2083, 1913.

SELECTED READING

Bishop, W. H., and Ryan, R. J. Human luteinizing hormone and its subunits. Physical and chemical characterization. *Biochemistry* 12:3076, 1973.
Frantz, A. G., Kleinberg, D. L., and Noel, G. L. Studies on prolactin in man. *Recent Prog. Horm. Res.* 28:527, 1972.
Jaffe, R. B., Yuen, B. H., Keye, W. R., Jr., and Midgley, A. R., Jr. Physiologic and pathologic profiles of circulating human prolactin. *Am. J. Obstet. Gynecol.* 117:757, 1973.
Li, C. H. Hormones of the adenohypophysis. *Proc. Am. Philos. Soc.* 116:365, 1972.
Li, C. H., and Dixon, J. S. Human pituitary growth hormone. XXXII. The primary structure of the hormone: Revision. *Arch. Biochem. Biophys.* 146:233, 1971.
Papkoff, H., and Ekblad, M. Ovine follicle stimulating hormone: Preparation and characterization of its subunits. *Biochem. Biophys. Res. Commun.* 40:614, 1970.
Saxena, B. N. Protein-polypeptide hormones of the human placenta. *Vitam. Horm.* 29:96, 1971.

The Adrenal Cortex 36

Harry Lipner

STRUCTURE OF THE ADRENAL CORTEX

The precursor cells of the adrenal cortex arise from the coelomic mesoderm on the posterior abdominal wall in the vicinity of the anterior margin of the mesonephros. The adrenal cortex eventually differentiates into three cell layers, called the *zona glomerulosa, zona fasciculata,* and the *zona reticularis.* The zona glomerulosa is located just beneath the adrenal capsule and is the source of aldosterone. The zona fasciculata is the widest of the three cell layers and is the primary source of gluco-corticoids (cortisol, cortisone, and corticosterone). The zona reticularis borders and abuts upon the medulla and is thought to be the source of gonadal steroids.

The secretory activity of the zona glomerulosa is controlled by several different factors, such as the action of angiotensin II (which in turn is induced by renin), the plasma sodium ion and potassium ion concentrations, and, minimally, the influence of adrenocorticotropic hormone (ACTH). The zona fasciculata and zona reticularis are under the primary control of ACTH, and the relationship is that of a negative feedback control system (see Chap. 35, Fig. 35-4A).

ROLE OF THE ADRENAL CORTICAL STEROIDS

Glucocorticoids are involved in the regulation of many aspects of cellular metabolism, as are all classes of steroids. The basis of action for all these classes is, in general, similar. The steroid molecule diffuses across cell membranes with ease. Within the cell, it binds with high affinity and stereospecificity to cytoplasmic receptor proteins. Changes of conformation in the steroid-receptor protein complex allow it to enter the cellular nucleus and interact with chromatin. The functions of the second complex are the regulation of transcription in specific parts of the genome and the induction of the synthetic events that become observed biologic phenomena.

The glucocorticoids — e.g., cortisol, cortisone, or corticosterone — enhance the conversion of amino acids to glucose (gluconeogenesis) and to lipids (lipogenesis), which is accompanied by the excretion of nitrogen, generally in the form of urea. These steroids induce the operation of amino transferases, the enzymes involved in this conversion. In adrenalectomized animals, for example, liver glycogen stores are rapidly depleted and hypoglycemia readily occurs. The administration of cortisol, however, quickly restores the carbohydrate levels in the liver and blood and simul-taneously causes the excretion of nitrogen.

Glucocorticoids when present in excess prevent the utilization of amino acids in muscle, with a consequent depletion of muscle protein. The loss of the protein is primarily due to protein turnover with inadequate concurrent synthesis. Protein synthesis in the liver is enhanced by excess glucocorticoids; however, the proteins thus formed are the enzymes that are involved in gluconeogenesis and lipogenesis.

Excess glucocorticoids also cause the mobilization of fat from some fat depots and its deposition in others. The individual on glucocorticoid therapy may develop fat pads in the cheeks, giving him a rounded face, as well as a large mound of fat between the shoulders ("buffalo hump").

The glucocorticoids are valuable in clinical medicine because of several other properties they possess. They are antiinflammatory in that they decrease hyperemia, fluid exudation, cellular migration and infiltration, as well as histamine and prostaglandin production. A consequence is that healing is facilitated. Steroids of this class also have an antipyrogenic effect by preventing the release of granulocytic pyrogens. The glucocorticoids have an antiallergic action by preventing the formation of antibodies in response to challenging antigens. They cause regression of lymphatic tissue, and they have been used to retard tissue rejection after organ transplantation.

The secretion of ACTH, which is under the control of the hypothalamic releasing hormone C-RH, results in the tropic stimulation of the adrenal cortex and the secretion of glucocorticoids. The resulting glucocorticoids exert a negative feedback control on ACTH secretion (see Chap. 35, Fig. 35-4A).

CONTROL OF WATER AND SALT METABOLISM

Homeostasis of salt and water is maintained by the functioning of aldosterone, antidiuretic hormone (ADH), and prolactin. Aldosterone regulates the renal reabsorption of sodium ions at the expense of potassium ions. ADH controls the reabsorption of water by altering the permeability to water of the distal tubule and the collecting duct of the nephron. Prolactin is a third humoral factor involved in homeostasis of body salt and water; its administration causes antidiuresis as well as the retention of salt and, with it, water.

The diagrams in Figure 36-1 summarize the control mechanisms of water and salt metabolism. The diagram is divided into three portions in order to allow each of the three generally recognized pathways to be identified and evaluated. In the intact human, all three pathways are operational, and each contributes to the regulation that is encompassed by the term *homeostasis of body fluids.* In this series of diagrams, the solid lines represent the input pathways and the dotted lines, the feedback pathways; the neural routes are indicated by thin solid lines and the vascular routes, by thick solid lines.

The regulation of water excretion by ADH is described in Figure 36-1A. The ingestion of water causes a decrease and the ingestion of salt causes an increase in the osmolality of the blood leaving the gut (block 1). All blood that enters the vascular supply of the stomach and small intestine leaves by the hepatoportal vein and passes through the liver (block 2). Osmoreceptors in the liver discharge to the brain (block 3).

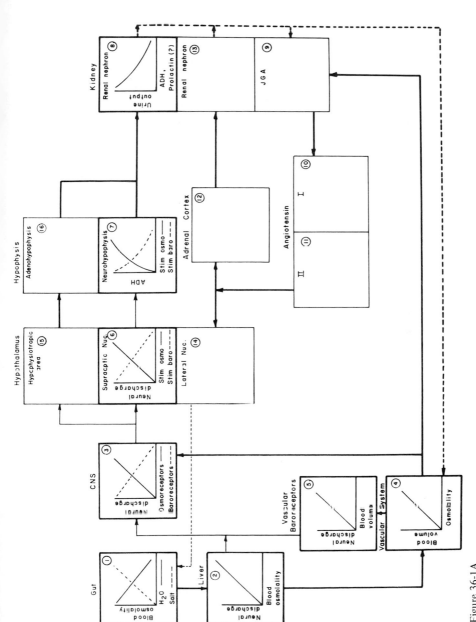

Figure 36-1A
Control of water and salt metabolism. Antidiuretic hormone (ADH) regulation of water excretion.

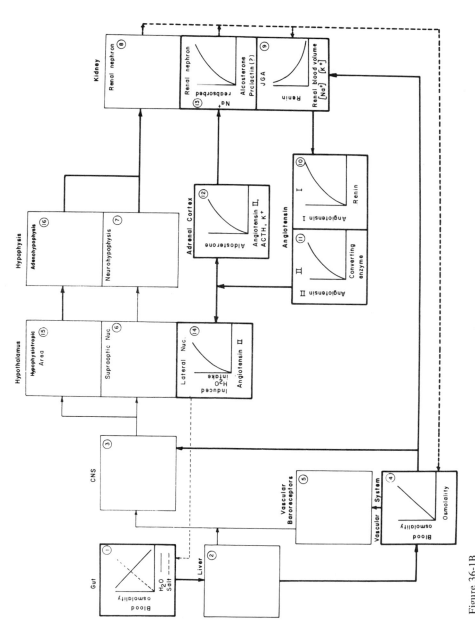

Figure 36-1B
Control of water and salt metabolism. Renin-angiotensin-aldosterone mechanism.

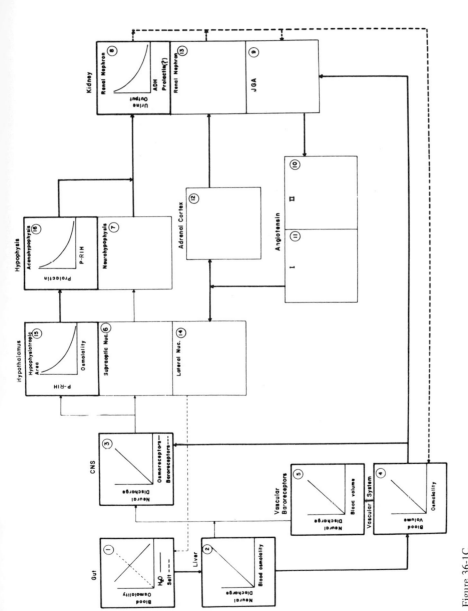

Figure 36-1C
Control of water and salt metabolism. Prolactin regulation of the osmolarity of blood.

The blood in the vascular compartment (block 4) increases in volume when its osmolality is raised, and it decreases in volume quite rapidly when its osmolality falls.

The osmolality of the blood in the vascular compartment also acts on the osmoreceptors in the CNS (block 3). The combined action of the neural input from the liver and the direct effect of the osmolality of the blood on the CNS is to stimulate the supraoptic nucleus of the hypothalamus (block 6) to secrete ADH by activating neural discharges down the hypothalamo-hypophysial tract that terminates in the neurohypophysis (block 7); excitation of the tract results in neurosecretion. The ingestion of salt would result in the increased secretion of ADH, with a resultant increased retention of water due to increasing its reabsorption in the distal tubule and collecting ducts of the renal nephron (block 8). Excessive ingestion of water or of dilute fluids (e.g., beer) results in the reduced secretion of ADH and an increase in the loss of water from the kidney through an increased production of urine.

The regulation of salt and water excretion by the renin-angiotensin-aldosterone mechanism is illustrated in Figure 36-1B. The ingestion of salt or water (block 1) affects the sodium concentration and the osmolality of the blood (block 4). The concentrations of sodium and potassium ions influence the chemoreceptors and baroreceptors in the juxtaglomerular apparatus (JGA, block 9) with a consequent effect on renin secretion. Increased concentrations of sodium or potassium ions and the expansion of the renal blood volume or the elevation of renal blood pressure result in decreased renin secretion. The secretion of renin into the vascular compartment causes the conversion of angiotensinogen to angiotensin I (block 10), which is converted to angiotensin II (block 11). Angiotensin II stimulates the secretion of aldosterone from the zona glomerulosa in the adrenal cortex (block 12). The aldosterone acting on the distal tubule of the renal nephron induces the reabsorption of sodium (block 13). A low intake of sodium or the excessive ingestion of dilute fluids results in increased aldosterone secretion, which is a mechanism that protects the organism from the loss of sodium. Elevated levels of renin secretion with enhanced angiotensin II production may induce drinking behavior by acting on the lateral nuclei of the hypothalamus (block 14).

Recently, the role of prolactin in the regulation of salt and water metabolism has been reported; this is diagrammed in Figure 36-1C. The ingestion of salt or water (block 1) affects the osmolality of the blood; this in turn affects the neural discharge from the liver (block 2). Neural discharges from the liver may act directly on the CNS (block 3), and other neural discharges arise from the vascular baroreceptors (block 5). The sum of these inputs to the CNS is to modulate prolactin release-inhibiting hormone (P-RIH) secretion from the hypothalamus (block 15). Decreased blood osmolality enhances P-RIH secretion, and increased blood osmolality depresses P-RIH secretion. The secretion of prolactin by the mammotropic cells of the adenohypophysis (block 16) may act on the renal nephron to promote the retention of water when the blood osmolality is elevated.

The operation of these mechanisms may be better understood if the control system is considered to be subject to a stress. The consumption of a large volume of water would result in a decreased blood osmolality and a consequent decrease in ADH secretion with a resultant loss of water via the urine. Concurrently, prolactin secretion is also depressed because more P-RIH is secreted, and increased aldosterone secretion prevents the excessive loss of sodium ions. The consumption of large amounts of salt, on the other hand, results in an increase in the blood volume with a corresponding increase in blood pressure and an elevation in the blood osmolality. The resultant excitation of the vascular baroreceptors inhibits ADH secretion; however, the increasing blood osmolality incites osmoreceptor discharge, counteracts the depression of ADH secretion, and increases prolactin secretion. The combined result is the retention of sufficient water to facilitate the return of blood osmolality to normal values, and the net effect is an expansion of total body fluid volume. The increased sodium ion concentration causes a depression of the production of aldosterone, as does the increased renal blood volume and pressure. Decreased aldosterone secretion is accompanied by the excretion of sodium and a slow return of the volume of body fluids to the state that existed before the stress was applied.

SELECTED READING

Bentley, P. J. *Endocrines and Osmoregulation.* New York: Springer-Verlag, 1971.
Bransome, E. D., Jr. Adrenal cortex. *Annu. Rev. Physiol.* 30:171, 1968.
Buckman, M. T., and Peake, G. T. Osmolar control of prolactin secretion in man. *Science* 181:755, 1973.
Davis, J. O. The control of renin release. *Am. J. Med.* 55:333, 1973.
McKerns, K. W. (Ed.). *Functions of the Adrenal Cortex,* Vols. 1 and 2. New York: Appleton-Century-Crofts, 1968.
Relkin, R. Effects of alterations in serum osmolality on pituitary and plasma prolactin levels in the rat. *Neuroendocrinology* 14:61, 1974.
Williams, R. H. (Ed.). *Textbook of Endocrinology* (5th ed.). Philadelphia: Saunders, 1974.

The Thyroid Gland 37

Harry Lipner

THE DEVELOPMENT OF THE THYROID

The thyroid gland originates in the one-month-old human embryo tissue from the primitive pharynx and grows posteriorly, carrying along a duct (the thyroglossal duct) until it reaches the developing trachea, to which it attaches itself. The thyroid becomes bilobate and attaches to the lateral aspects of the trachea just below the thyroid cartilage. The thyroglossal duct normally becomes obliterated. When obliteration fails, cyst formation occurs along its tract. The only sign of the thyroid's site of origin is a dimple on the posterior third of the tongue, the foramen cecum. The lower portion of the duct gives rise to the pyramidal lobe of the thyroid gland.

The cells composing the lobes undergo two rearrangements: first, they align themselves into cords, and then they are transformed into tube-like structures. The tubes then break down into follicles, each surrounded by a dense, basket-like arrangement of capillaries and larger blood vessels. Each follicle is lined by a single layer of cuboidal cells. The cells are the source of thyroglobulin, which fills the follicular space. The proteolysis of thyroglobulin gives rise to thyroid hormones. Interspersed among the follicles are the parafollicular cells, which are thought to be the source of a hormone (calcitonin or thyrocalcitonin) that is concerned with calcium metabolism and causes a reduction in plasma calcium concentrations.

METABOLISM OF IODIDE

The healthy human in a euthyroid state excretes as much iodide as he ingests. His needs are remarkably small, being in the range of 10 to 30 ng per day. Most of the iodine in the organism (5 mg) is contained in the thyroid gland. Of this, less than one percent is inorganic; the rest of the iodine is covalently bound to tyrosine molecules. Iodine is lost in two forms: it may be incorporated into thyroid hormones, which appear in the bile as both sulfates and glucuronides, or it may be lost as iodide that appears in the urine. The pathways for iodine metabolism are illustrated in Figure 37-1.

THE SYNTHESIS OF THYROID HORMONES

The synthesis of thyroid hormones has been studied for many years, and now, for research as well as pedagogic purposes, it may be divided into several stages. These

729

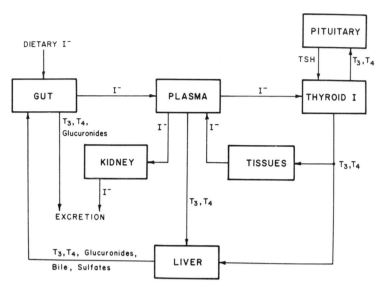

Figure 37-1
The metabolism of inorganic and organic iodine in humans. Dietary iodide maintains the iodide pool in the organism and provides the thyroid with the iodide that it needs for the synthesis of thyroid hormone. The thyroid hormone goes through an enterohepatic circulation involving the liver and the intestine. The glucuronides and sulfates formed in the liver are split in the intestine and reabsorbed from the intestine. This enterohepatic circulation contributes to the maintenance of the thyroid hormone pool.

have been termed iodide accumulation, iodide oxidation and organification, the coupling of iodinated tyrosine, and storage and release (Fig. 37-2).

IODIDE ACCUMULATION

Uptake of iodide by the thyroid epithelial cell is an active process that results in an accumulation of iodide that greatly exceeds its concentration in the plasma. The system responsible for iodide accumulation is called the *iodide trap* or *iodide pump*. The activity of this pump may be examined independently of the other phases of iodide utilization by poisoning the subsequent steps in the biosynthesis of thyroid hormone. The compounds used are sulfhydryl-containing goitrogens such as 6-propyl-thiouracil (PTU) or methimazole (thiamazole). One hour after treatment with PTU, the administration of a tracer dose of ^{131}I results in a ratio of thyroid ^{131}I concentration to serum ^{131}I concentration that ranges between 25 and 100. The accumulation of iodide must be an active process, since the iodide crosses the epithelial cell membranes into the cell against an electrochemical gradient. Furthermore, the fact that the pump is energy-dependent is indicated by its being subject to inhibition by anoxia, cyanide, 2,4-dinitrophenol, or hypothermia. The activity of the pump is increased by

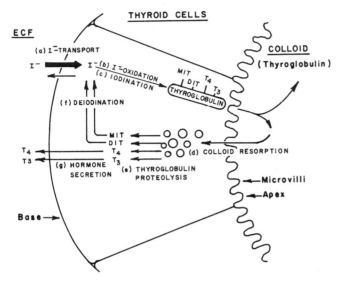

Figure 37-2
The pathways of iodine metabolism in the thyroid. (From W. Tong. In S. C. Werner and
S. H. Ingbar [Eds.], *The Thyroid: A Fundamental and Clinical Text* [3rd ed.]. New York:
Harper & Row, 1971. Reproduced with permission.)

thyroid-stimulating hormone (TSH) and can be blocked competitively by thiocyanate
and perchlorate anions.

IODIDE OXIDATION AND INCORPORATION

The iodide that is accumulated by the thyroid is rapidly incorporated into thyroglobulin. In reactions of iodide ion and casein in vitro, thyroxine is formed if an
oxidizing enzyme or hydrogen peroxide (H_2O_2) is added. Furthermore, the reaction
may be prevented by substances that reduce iodine to iodide. The assumption has
therefore been made that in vivo oxidation of iodide also must occur. The oxidation-
reduction potential of $I_2 \rightleftharpoons I^-$ is as high as +0.535 volt. In biologic systems, the substances with redox potentials that are sufficiently high to oxidize the iodide ion are
O_2 and H_2O_2, with redox potentials of +0.82 and +1.3 volts, respectively. Any direct
evidence that H_2O_2 is clearly involved in the oxidation, however, has not been reported.
It could be that H_2O_2 is generated from the autoxidation of flavine oxidases (Fig. 37-3).
The peroxide might then oxidize the iodide ion to a highly reactive molecular form.
The latter could react with the tyrosyl units that are already incorporated in the
thyroglobulin molecule. In any case, the result of the oxidation of iodide is the
appearance of molecules of monoiodotyrosine (MIT), diiodotyrosine (DIT), triiodothyronine (T_3), and thyroxine (T_4), which are obtained on the hydrolysis of thyroglobulin from thyroid tissue.

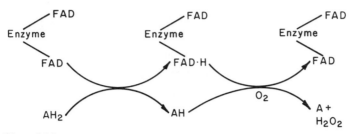

Figure 37-3
The autoxidation of the flavine adenine dinucleotide cofactor (FAD) of a flavine oxidase with
the resulting oxidation (loss of hydrogens) from a substrate (AH_2) and the formation of peroxide.
This type of reaction must occur at a slow rate to generate sufficient peroxide to catalyze the
oxidation of iodine to a reactive form.

The iodination of thyroglobulin occurs in the apex of the epithelial cells in the
thyroid follicles (see Fig. 37-2). Recent evidence suggests that iodination may occur
on the microvilli at the junction of the thyroid cells with the thyroglobulin, which
occurs in the colloid of the follicles. Time-course studies of the appearance of the
various iodinated molecules indicate that MIT appears most rapidly, reaching a peak
in 24 hours, and its appearance is followed by that of DIT at a slightly slower rate.
The formation of thyroxine is quite slow, reaching its maximum over a period of days.
The formation of thyroxine is thought to require the coupling of two molecules of
DIT, whereas the formation of triiodothyronine requires the coupling of one molecule
of MIT and one of DIT. Several hypotheses have been suggested to explain the
coupling reaction. Perhaps the most feasible of these is the one based on the observation that the incubation of purified thyroglobulin, stable iodide, radioactive iodide
(^{131}I), an oxidase from thyroid microsomes, and an H_2O_2-generating system (glucose
plus glucose oxidase) resulted in the formation of ^{131}I-iodothyronines (T_3 and T_4). It
therefore appears that an oxidase or oxidases contained in thyroid microsomes can,
in the presence of a peroxide-generating system, iodinate tyrosyl units in thyroglobulin
and couple them to form iodothyronines.

SECRETION OF THYROID HORMONES

The iodothyronines are secreted by the thyroid gland in response to a tropic TSH
stimulus. TSH activates adenyl cyclase in the basal surface of the thyroid epithelial
cell membrane. TSH also increases the prostaglandin concentration in the thyroid,
as do derivatives of 3',5'-cyclic adenosine monophosphate (cAMP) and theophylline,
an inhibitor of the enzyme that destroys cAMP. It is postulated that prostaglandins
modulate the effect of TSH on the thyroid.
    TSH causes the release of proteases at the apex of the thyroid cell. These proteases
cause droplets to form at the colloid-cell interface. The ensuing proteolysis results in
the liberation of iodinated tyrosines (MIT and DIT) and iodinated thyronines (T_3 and
T_4). The iodinated tyrosines are deiodinated within the epithelial cells by a specific

iodotyrosine deiodinase. The resulting iodide from this reaction is recycled within the cell. The iodinated thyronines are released by the epithelial cells at their basal ends directly into the vascular and lymphatic capillaries.

TSH exerts a biphasic effect on iodide accumulation by the thyroid. The initial action in response to an acute administration of TSH is to reduce this iodide accumulation by increasing the outward iodide flux through increasing the permeability of the cell membranes to iodide. About ten hours later, however, the iodide accumulation will be increased above that of untreated controls. TSH also affects subsequent phases of iodine metabolism. Iodide oxidation and organic iodinations are enhanced and occur during the period that iodide accumulation is depressed. Organic iodinations occur independently of new synthesis of thyroglobulin, since they take place while protein synthesis is inhibited by puromycin. Furthermore, TSH stimulates an increased activity of the coupling reaction.

BIOLOGIC ACTIONS OF THYROID HORMONES (T_3 AND T_4)

The administration of either T_3 or T_4 causes an increase in the oxygen consumption (calorigenesis) of the whole organism and of many, but not all, of its constituent tissues. T_3 initiates a quicker response and an earlier maximum response than does T_4.

The most careful studies on whole animals and their tissues were performed by Barker [1], who investigated the effects of administering T_4 in athyroid rats. The gastric mucosa showed an almost immediate increase in oxygen consumption, the heart responded in three hours, and the smooth muscle, skeletal muscle, liver, kidney, and diaphragm responded in six hours. The whole animal reached peak response in 12 to 24 hours. The spleen, brain, and testes were unaffected. Several factors cause decreased oxygen uptake; for example, the oxygen uptake by the whole animal decreases with aging. Adrenergic blockade is an effective inhibitor of increased oxygen uptake in response to T_4 administration, and thyroidectomy causes a reduction of the metabolic response to epinephrine.

METABOLISM OF NUTRIENTS

Protein metabolism is enhanced by T_3 and T_4 as is evidenced by an increase in the turnover of protein in tissues. Small to moderate doses of T_3 or T_4 increase protein synthesis, but large doses depress synthesis. Carbohydrate metabolism is probably affected secondarily by the induced synthesis of metabolic enzymes as well as by the permissive effect of catecholamines. The latter enhances the release of glucose by the liver and increases its utilization by the tissues. Lipid metabolism is uniquely affected as is manifested by an alteration of the plasma concentration of cholesterol. In hypothyroidism, plasma cholesterol levels are elevated, whereas in the hyperthyroid state, cholesterol levels tend to be depressed.

REFERENCE

1. Barker, S. B., Makoto, S., and Masao, M. Metabolic and cardiac responses to thyroxine analogs. *Endocrinology* 76:115, 1965.

SELECTED READING

Dumont, J. E. The action of thyrotropin on thyroid metabolism. *Vitam. Horm.* 28:287, 1971.

Gual, C., Kastin, A. J., and Schally, A. V. Clinical experience with hypothalamic releasing hormones. 1. Thyrotropin releasing hormone. *Recent Prog. Horm. Res.* 28:173, 1972.

Reichlin, S., et al. The hypothalamus in pituitary-thyroid regulation. *Recent Prog. Horm. Res.* 28:229, 1972.

Werner, S. C., and Ingbar, S. H. (Eds.). *The Thyroid: A Fundamental and Clinical Text* (3rd ed.). New York: Harper & Row, 1971.

Werner, S. C., and Nauman, J. A. The thyroid. *Annu. Rev. Physiol.* 30:213, 1968.

The Pancreatic Hormones and the Control of Carbohydrate and Lipid Metabolism 38

Harry Lipner

INTRODUCTION

Early studies showed that the pancreas in some manner regulated blood sugar concentrations and that when the concentrations of blood sugar were elevated, sugar also appeared in the urine.

The pancreas in the human is a pink organ lying transversely in the abdominal cavity. Its head is contained in the curve of the middle section of the duodenum on the right, and its body and tail are attached by the mesentery to the duodenum and stomach and extend to the spleen on the left.

The pancreas has two functions: it provides enzymes, produced in the acinar tissue, that are conveyed into the duodenum by one or more ducts; and it provides hormones that are produced and secreted by the islets of specialized cells, which were first described in 1869 by Langerhans. These hormones are directly involved in the regulation of carbohydrate metabolism and indirectly involved in lipid and protein metabolism. Malfunctioning of the hormonal secretions of the pancreas is the basic cause of the disease *diabetes*.

CELLULAR STRUCTURE OF THE PANCREAS

The islets of Langerhans are distributed through and imbedded in the acinar tissue. The cells composing the islets are thought to originate from cells in the acini by transverse, rather than longitudinal, division. Although four different cell types have been described, only three have been identified with specific secretions. The terminology applied to these cells is presently in a state of flux. The *alpha cells* are divided into two types, α_1 and α_2, based on tinctorial properties. The α_2 cell was formerly called the A cell, and the α_1 cell is the former delta or D cell. The α_2 cell is the source of glucagon, and the α_1 cell secretes gastrin. The *beta cell* or B cell is the source of insulin. The fourth cell type is an agranular cell that is seen in many histologic sections, but it has no known function. Except for their different types of secretory granules, the cells are almost indistinguishable. The α_1 cell reacts with immunofluorescent antibodies to gastrin and stains with silver stains. The α_2 cells are localized in specific areas of the islets of certain species (e.g., at the periphery in the rat and centrally in the horse), whereas in man and the guinea pig, they are randomly distributed. The beta cells are the most numerous of the islet cells and are contained in all the islets. The secretory granules are specific to each species and have distinctive crystalline-like patterns.

The mechanism of the secretion of insulin by beta cells has been intensively studied and is now described as a process of emiocytosis. Freeze-fracture techniques and electron microscopy have confirmed the details of hormone secretions. The hormone, in granular form, is contained in membrane-bound secretory granules. These granules move to the surface of the cells where they coalesce with the plasma membrane; they then open to release the hormonal granules into the surrounding extracellular fluid. The number of secretory granules in the membrane is an indication of the secretory activity of the cell.

THE CONTROL OF GLUCAGON AND INSULIN SECRETION

The blood sugar concentration ultimately reflects the combined secretory activity of the α_2 and beta cells, which thereby constitute a glucostat. Although a number of extrinsic factors are imposed on the activity of the pancreatic islet cells, it is their secretory activity that returns the blood sugar to a preexisting basal level. The extrinsic factors — such as the amino acid and free fatty acid concentration in the blood, physical activity, and autonomic nervous system inputs — modulate the intensity of the islet cell response to the prepotent stimulus, glucose.

GLUCAGON

The secretion of glucagon is closely tied to plasma glucose concentrations. Hypoglycemia induces an increase in plasma glucagon; conversely, hyperglycemia causes a prompt suppression of glucagon secretion (Fig. 38-1). Hyperaminoacidemia causes a pronounced increase in glucagon concentrations in the plasma (Fig. 38-2). Since

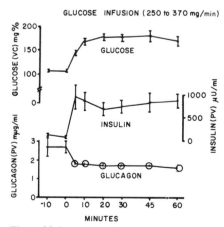

Figure 38-1
Mean level (± SE) of plasma glucose in the inferior vena cava and of insulin and glucagon in the pancreaticoduodenal vein of fifteen dogs during hyperglycemia. Each large circle on the mean curve of glucagon indicates that it differs significantly from the control value at zero times ($p <$ 0.01). (From A. Ohneda, E. Anguilar-Parada, A. M. Eisentraut, and R. H. Unger. *Diabetes* 18:1, 1969. Reproduced with permission.)

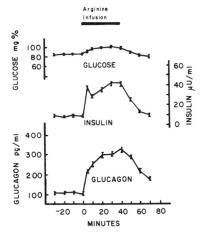

Figure 38-2
Response of insulin and glucagon concentrations to the infusion of arginine (mean ± SE is indicated). (From R. H. Unger. Reprinted by permission. *N. Engl. J. Med.* 285:443, 1971.)

hyperaminoacidemia also causes an increase in plasma insulin concentrations, a teleologic explanation may be that the increase in plasma glucagon is a protective device to prevent extreme hypoglycemia. Support for this concept is seen in the failure of hyperaminoacidemia to cause hyperglucagonemia if a hyperglycemia is induced by the infusion of glucose at the same time that the amino acids are being administered.

Increased physical activity, e.g., exercise, causes an increase in the secretion of glucagon. This effect is seen in the absence of any significant change in the plasma glucose levels. The consequence of increased glucagon secretion is an elevation of plasma free fatty acids (FFA) (Fig. 38-3). It is most probable that the apparent constancy of the glucose concentration is indicative of a steady-state situation and that utilization keeps up with secretion. Free fatty acids are utilized less quickly, and they consequently accumulate in the blood and show a rising concentration.

Epinephrine, a catecholamine, is also a potent stimulator of glucagon secretion. The administration of epinephrine results in both hyperglycemia and hyperglucagonemia. The latter probably reinforces the action of the catecholamine on the liver and supplements directly the effect of the glucagon (Fig. 38-4).

INSULIN

The secretion of insulin is regulated by a number of factors. Glucose has a primary role in this regulation; hyperglycemia is a potent drive that induces the rapid secretion of insulin (see Fig. 38-1). Insulin secretion, as a result of a pulse stimulus of glucose, appears as a responding surge. If the blood sugar is raised by means of a continuous infusion of glucose, the concentration of insulin in the plasma rises. If a pulse of glucose is then superimposed upon the continuous infusion, there is a responding surge of insulin (Fig. 38-5). This pattern has been interpreted as a

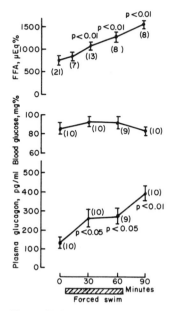

Figure 38-3
Influence of forced swim on blood glucose, plasma free fatty acid (FFA), and glucagon concentrations. The last group corresponds to animals that had been allowed to rest for 30 minutes after the forced swim period. Results are expressed as the mean ± SE. The numbers in parentheses correspond to the number of determinations. (From A. Luyckx and P. Lefebvre. *Postgrad. Med. J.* 49:620, 1973. Reproduced with permission.)

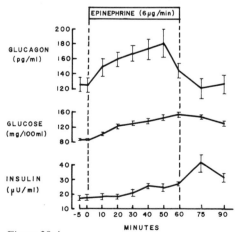

Figure 38-4
Effect of epinephrine on plasma glucagon, glucose, and insulin (mean ± SE). (From J. E. Gerich et al. *J. Clin. Endocrinol. Metab.* 37:479, 1973. Reproduced with permission.)

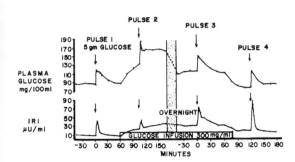

Figure 38-5
Evidence for two pools of insulin secretion in man: the effect of standard (5 gm) pulses of glucose upon insulin secretion before, during, and after a prolonged infusion of glucose in eight normal subjects (*IRI* = immunoreactive insulin). (From C. H. Goodner and D. Porte, Jr. *J. Clin. Invest.* 48:2309, 1969. Reproduced with permission.)

reflection of two pools of insulin. The acute response represents utilization from a pool of preformed insulin, and the second response to a chronic influx of glucose represents insulin secretion from a synthesis-related pool.

Epinephrine induces an inhibition of insulin secretion. This effect has been interpreted as being due to the stimulation of the alpha-adrenergic receptors of the islet beta cells, since it can be blocked by the prior administration of phentolamine, an alpha-adrenergic blocking agent. The prolonged administration of epinephrine causes an elevation of blood insulin above basal levels. This effect has been interpreted as being due to a beta-adrenergic action that can be blocked by the administration of propranolol, a beta-adrenergic blocking agent. During this phase of insulin secretion, a glucose surge causes no further insulin secretion. A model that fits this pattern of response is one that postulates that epinephrine stimulates an acute block of insulin secretion but still allows a small but continuous secretion of insulin from the synthesis-related pool (Fig. 38-6). Under the stress of a number of different conditions, all of which cause activation of the sympathetic nervous system, hyperglycemia unaccompanied by an elevation of insulin concentrations is found (Fig. 38-7).

Cholinergic agents stimulate the secretion of insulin by pancreatic islets in vitro, and stimulation of the vagus nerve causes the secretion of insulin in vivo. The effect of vagal stimulation is blocked by atropine. Stimulation of the pancreatic nerve causes insulin release, an effect that is also blocked by atropine. Glucose-stimulated insulin secretion is blocked by atropine administration and the stimulation of the pancreatic nerve to cause sympathetic activation in this mixed nerve. These observations establish a role of the autonomic nervous system in the regulation of insulin secretion: the parasympathetic branch induces the secretion of insulin, whereas the sympathetic branch induces inhibition. Stimulation of the pancreatic nerve in atropinized animals causes glucagon secretion, and its secretion is therefore probably due to sympathetic stimulation (Fig. 38-8).

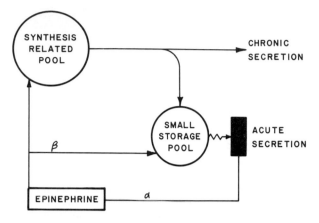

Figure 38-6

A proposed two-pool model for insulin secretion and its interaction with epinephrine. Note the dual α-adrenergic and β-adrenergic effects on the small storage pool. (From D. Porte, Jr., and R. P. Robertson. *Fed. Proc.* 32:1792, 1973. Reproduced with permission.)

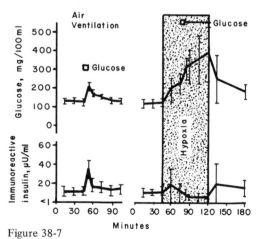

Figure 38-7

Effect of hypoxia on insulin and glucose levels and the insulin responses to a glucose pulse (*left*: control; *right*: hypoxia). (From D. Porte, Jr., and R. P. Robertson. *Fed. Proc.* 32:1792, 1973. Reproduced with permission.)

   It is now believed that almost any factor that can activate the adenyl-cyclase receptor in the membrane of the beta cell is capable of stimulating insulin secretion. This effect is enhanced in the presence of glucose.

   The glucagon derived from the α_2 cell as well as the glucagon-like immunoreactive factor from the intestinal mucosa (enteroglucagon) have been implicated in enhanced insulin secretion. Similarly, gastrin, pancreozymin, and secretin have also been suggested as possessing insulinotropic activity. The actions of these various factors in the control of insulin secretion are diagrammed in Figure 38-9.

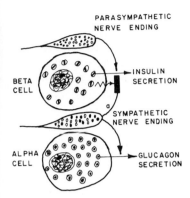

Figure 38-8
Neural input to the endocrine pancreas. (From D. Porte, Jr., and R. P. Robertson. *Fed. Proc.* 32:1792, 1973. Reproduced with permission.)

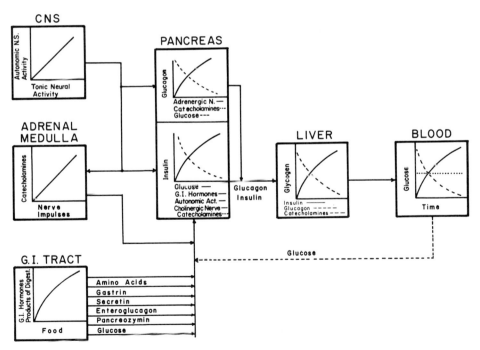

Figure 38-9
Control of pancreatic islet secretion.

DIABETES MELLITUS

The concepts of the pathophysiology of diabetes mellitus have changed considerably since von Mering and Minkowski in 1889 demonstrated the development of diabetic symptoms in dogs subsequent to pancreatectomy and Banting and Best found in 1921 that this condition could be alleviated with pancreatic extracts. It is now known that diabetes of the juvenile type need not be associated with inadequate insulin release; rather, it may be associated with inappropriate glucagon secretory patterns. In the diabetic, a meal may be accompanied by inadequate insulin secretion, but it may also be accompanied by an increase in glucagon secretion (Fig. 38-10). Furthermore, as Figure 38-10 also shows, the administration of insulin to such a subject may not suppress the glucagon secretion. In the normal individual, a meal is accompanied by an increased secretion of insulin and a suppressed secretion of glucagon.

New data are slowly accumulating that suggest that a defect in alpha-2 cell function may be associated with diabetes. In juvenile diabetics, insulin-induced hypoglycemia is unaccompanied by an elevation of glucagon secretion. The normal subject under similar conditions, however, shows a prompt secretion of glucagon. Intravenous arginine infusion in diabetics is accompanied by a rapid rise in the glucagon titers (Fig. 38-11), despite a significant existent hyperglycemia. The normal subject responds with an increased secretion of glucagon but with no change in blood sugar. Inappropriate secretion of glucagon could well be a partial explanation of the pathogenesis of diabetes. It should be emphasized, however, that there is probably no single answer regarding the nature of the disease entity called diabetes mellitus.

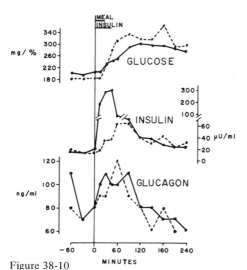

Figure 38-10

Failure of exogenous insulin administration to correct the unsuppressibility of glucagon secretion of a diabetic patient after the ingestion of a carbohydrate meal. The dashed line indicates values before insulin infusion, and the solid line, the values after insulin infusion. (From R. H. Unger. *N. Engl. J. Med.* 285:443, 1971. Reproduced with permission.)

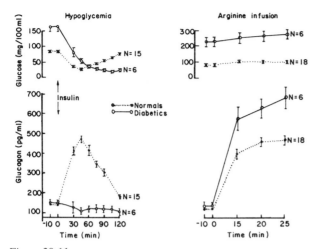

Figure 38-11
Plasma glucose and glucagon responses to insulin-induced hypoglycemia (*left*) and to intravenous arginine infusion (*right*) in juvenile-type diabetics (solid lines) and in normal controls (dotted lines). (Vertical bars indicate standard errors of the mean; *N* is the number of patients.) (From J. E. Gerich et al. *Science* 182:171, 1973. Reproduced with permission.)

SELECTED READING

Eisenstein, A. B., Strack, I., and Steiner, A. Glucagon stimulation of hepatic gluconeogenesis in rats fed a high-protein, carbohydrate-free diet. *Metabolism* 23:15, 1974.

Gerich, J. E., Karam, J. H., and Forsham, P. H. Stimulation of glucagon secretion by epinephrine in man. *J. Clin. Endocrinol. Metab.* 37:479, 1973.

Inersen, J. Adrenergic receptors and the secretion of glucagon and insulin from the perfused canine pancreas. *J. Clin. Invest.* 52:2102, 1973.

Marks, V. Glucagon and lipid metabolism in man. *Postgrad. Med. J.* 49:615, 1973.

Porte, D., Jr., and Robertson, R. P. Control of insulin secretion by catecholamines, stress, and the sympathetic nervous system. *Fed. Proc.* 32:1792, 1973.

Steiner, D. F. Proinsulin and the biosynthesis of insulin. *N. Engl. J. Med.* 280:1106, 1969.

Tager, H. S., and Steiner, D. F. Isolation of a glucagon-containing peptide: Primary structure of a possible fragment of proglucagon. *Proc. Natl. Acad. Sci. U.S.A.* 70:2321, 1973.

The Handbook of Physiology. Section 7, Editors: Greep, R. O., and Astwood, E. B. *Endocrinology,* Vol. 1, The Endocrine Pancreas. Washington, D.C.: The American Physiological Society, 1972.

Unger, R. H. Glucagon physiology and pathophysiology. *N. Engl. J. Med.* 285:443, 1971.

Wise, J. D., Hendler, R., and Felig, P. Influence of glucocorticoids on glucagon secretion and plasma amino acid concentrations in man. *J. Clin. Invest.* 52:2774, 1973.

Harry Lipner

THE ROLES OF CALCIUM AND PHOSPHATE IN THE ORGANISM

The adult human contains about 1200 gm of calcium. All except one percent is contained in the hydroxyapatite of bone; the remaining one percent is distributed in the extracellular body fluids, the plasma, and cells. By far the largest amount of soluble calcium is found in the extracellular fluid. At the pH of plasma, it exists in a steady-state equilibrium between adsorbed and chelated molecules and those in the ionic form:

$$Ca^{+2} \underset{}{\overset{pH = 7.4}{\rightleftharpoons}} Ca \overset{\displaystyle\diagup \text{citrate}}{\underset{\displaystyle\diagdown PO_4^{-3}}{-\text{protein}}}$$

Ionic	Bound	Total
1.29 mM	1.6 mM	2.9 mM
5 mg/100 ml	6 mg/100 ml	11.6 mg/100 ml

An increase in plasma pH causes a decreased concentration of Ca^{+2} and induces increased irritability as well as spontaneous neural and muscular discharge. The ionic state of calcium is necessary for many other functions in addition to membrane irritability, such as cell-cell adhesion, cell permeability, excitation-contraction coupling, stimulus-secretion coupling, and blood coagulation.

Phosphate is present in amounts of between 500 and 600 gm, of which 85% is present in the skeleton. The remainder is found primarily in intracellular pools. The concentration in the plasma is variable. The phosphate in cells is, in large part, present as the organic phosphates implicated in high-energy reactions (adenine and guanine nucleotides), in nucleoproteins, in phospholipids, in coenzymes, as well as in the inorganic form required for the utilization of glucose.

Calcium is absorbed from the duodenum and is used to replenish the calcium that is lost in the urine and feces. Phosphate is mainly absorbed in the jejunum. There is no preferential system for phosphate absorption, but the absorption of calcium is dependent on the presence of vitamin D_3 (cholecalciferol).

ENDOCRINE CONTROL OF CALCIUM METABOLISM

SOURCE OF HORMONES

The control of calcium metabolism is under the influence of three hormones: one that is secreted by the parathyroid gland (parathyroid hormone); one that is secreted by the parafollicular cells of the thyroid gland, which are also called "light" cells or C cells (calcitonin or thyrocalcitonin); and the last is cholecalciferol, which is primarily considered to be a vitamin. The parathyroid glands in humans are found on the surface of or partially imbedded in the lobes of the thyroid gland in the throat region. There are generally four of these glands, which are attached in pairs to each thyroid lobe. Accessory glands are occasionally found in the mediastinal tissues in the chest. The parathyroid gland is composed of two cell types: the principal cell is a relatively small cell with a centrally placed nucleus, and the oxyphilic cell has a purple-staining granular cytoplasm. The parathyroid hormone (PTH) is produced in the principal cell. The oxyphilic cell appears sometime between the ages 4.5 and 7 years.

Calcitonin is the product of the parafollicular cell in the thyroid. In the embryo, the parafollicular cells arise from the terminal branchial or ultimobranchial pouch and merge with the thyroid follicular cells, though remaining separate from these latter cells. In lower vertebrates, the parafollicular cells give rise to the ultimobranchial gland. In the human, the parafollicular cells stain palely, but they contain mitochondria, secretory vesicles, and membrane-bound secretory granules that average 100 to 200 mm in diameter. They are few in number and hard to identify with light microscopy. The parafollicular cell is also the source of a rare thyroid tumor that produces calcitonin.

Vitamin D_3 or cholecalciferol is derived from two sources: the diet and ultraviolet irradiation of the skin. The vitamin is transported to the liver, where it may be either stored or oxidized to a more active molecule, 25-hydroxycholecalciferol, which enhances the absorption of calcium from the intestine. The vitamin or its 25-hydroxy form may be transported to the kidney, where it is converted to its most active derivative, 1,25-dihydroxycholecalciferol.

CHEMISTRY OF THE HORMONES

Parathyroid hormone was isolated as a crude, active fraction from bovine parathyroid glands in 1925 by Collip [2]. A more effective extraction procedure that was developed in 1959 led to the isolation of highly purified hormone preparations. Subsequently, the application of rapidly evolving biochemical techniques led to the isolation of a polypeptide. Highly purified bovine, porcine, and human preparations of parathyroid hormone were used to show that it contained 84 amino acids; however, it was found to differ in the number of specific amino acids [1]. All the biologic activity of the molecule is contained in the 29 N-terminal amino acid units.

In 1962 Copp [3] noted that in the dog, the rate of decline of plasma calcium levels after the administration of calcium was greatly reduced by thyroparathyroidectomy. The presence of a hypocalcemic factor in the parathyroid gland was postulated, and it was called *calcitonin*. The source of the hormone was localized subsequently in the parafollicular cell of the thyroid, and the name *thyrocalcitonin* was applied to distinguish this factor from the assumed parathyroid factor. It is now recognized that only the parafollicular cells make the hypocalcemic factor, and the name "calcitonin" is replacing "thyrocalcitonin" in the literature.

Large-scale preparative extractions of thyroid tissue that were performed by several pharmaceutical firms finally allowed the isolation of sufficient material for the analysis and identification of the amino acid sequence of calcitonin. These studies led to the identification of the bovine, ovine, and porcine hormones. Human calcitonin was prepared from medullary tumors of the thyroid; however, the hormone from normal human tissue has not been isolated, because of its low concentration in this tissue.

The ultimobranchial gland of the Pacific salmon is a rich commercial source of calcitonin. It is especially important because the salmon calcitonin is 20 to 50 times more active than mammalian calcitonin (Fig. 39-1). In 1967. the amino acid sequence of porcine calcitonin was established by Potts [5]. Subsequently, other laboratories confirmed this sequence. The amino acid sequences for calcitonin that has been derived from several species are now known [1]. The molecule consists of 32 amino acids, and the entire molecule is necessary for biologic activity.

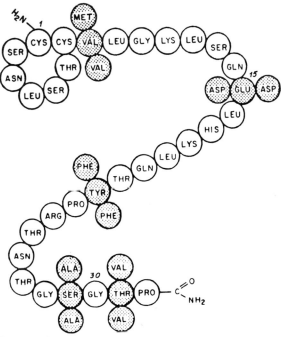

Figure 39-1
The amino acid sequence of salmon calcitonins I, II, and III. The main peptide backbone represents the sequence of the predominant form, salmon calcitonin I. Starting from the C-terminal group, the first four sets of dark circles adjacent to the main sequence indicate the four positions in which the salmon calcitonin II differs from calcitonin I. In addition to these differences, calcitonin III shows a fifth difference, the substitution of methionine for valine, as indicated by the dark circles at position 8 from the N-terminal group. (From J. T. Potts, Jr., et al. *Vitam. Horm.* 29:41, 1971. Reproduced with permission.)

Recent studies by DeLuca et al. [4] have stimulated renewed interest in the relationship between vitamin D_3 (cholecalciferol) and calcium metabolism. Vitamin D_3 is derived from the diet or formed in the skin by ultraviolet irradiation. A photochemical reaction induced by ultraviolet rays converts 7-dehydrocholesterol into cholecalciferol. The hypothesis developed by DeLuca is that cholecalciferol is a hormone. It is formed in the skin, stored in the liver, and converted to more active compounds in the liver and kidney (Fig. 39-2).

ACTIONS OF THE HORMONES

Parathyroid hormone (PTH) acts on two sites to contribute to calcium homeostasis: it acts on the kidney to enhance the resorption of calcium from the glomerular filtrate, and it acts on bone to cause the resorption of calcium in that location. In both cases, PTH exerts two effects: it increases the uptake of calcium into the cell, and it activates adenylcyclase. The consequence of adenylcyclase activation is an increase in cAMP, which causes an activation of protein kinase and an increase in the efflux of calcium from the mitochondria. The rise in the calcium levels in the cytosol causes

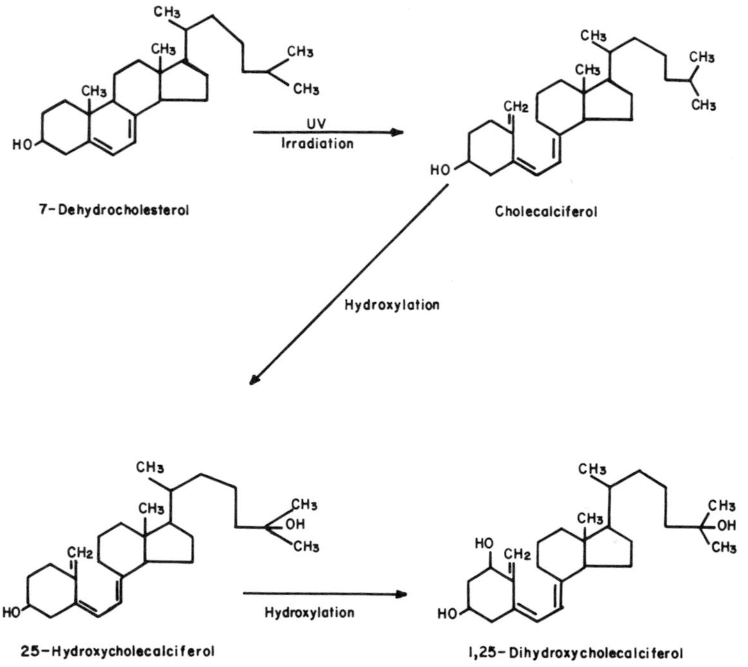

Figure 39-2
The structures of the protovitamin, 7-dehydrocholesterol, and the various naturally occurring, biologically active cholecalciferols that are now identified as hormones.

inhibition of adenylcyclase and either increased or decreased enzymatic activities. The result of these effects is the integration of the cellular response to PTH. Rasmussen [6] has suggested that the mechanism for the PTH effect on bone is due to the activation of osteogenic cells located on the surface of bone. These cells undergo several mitotic divisions and are converted to preosteoclasts, which fuse to give rise to multinucleated osteoclasts. The latter, in the process of remodeling the bone, cause the resorption of bone surface. The osteoclasts then undergo fission to become mononucleated cells again; they become preosteoblasts and then osteoblasts, the bone-forming cells (Fig. 39-3).

Calcitonin (CT) acts in an opposite manner to PTH. It inhibits the activation of osteo-progenitor cells and increases the rate of conversion of osteoclasts to osteoblasts. It also inhibits the renal synthesis of 1,25-dihydroxycholecalciferol, and it increases the urinary excretion of calcium. The net effect is to facilitate the removal of calcium from the blood and restrain the resorptive effect of PTH on bone.

The role of 1,25-dihydroxycholecalciferol in the absorption of calcium from the intestine is expressed by the induction of synthesis of a protein that facilitates the absorption of calcium. The action of this vitamin D_3 derivative on bone is apparently expressed through the action of PTH in an undefined manner.

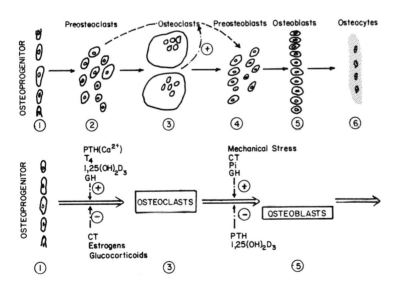

Figure 39-3
The sequence of cellular events within a bone remodeling unit upon the endosteal surface and some of the factors that regulate this sequence. Osteoprogenitor cells (*1*) are activated to undergo mitotic division to form daughter cells or preosteoclasts (*2*). These fuse to become osteoclasts (*3*), which, after carrying out the resorptive phase of skeletal remodeling, undergo fission to preosteo-blasts (*4*). These in turn become osteoblasts (*5*), which carry out the formative phase of the remodeling cycle and, in the process, become incorporated into bone as osteocytes (*6*). (From H. Rasmussen. *Mt. Sinai J. Med. N.Y.* 40:462, 1973. Reproduced with permission.)

CONTROL OF CALCIUM METABOLISM

The secretion of parathyroid hormone is closely related to the concentration of calcium in the blood. The hormonal control of calcium metabolism is illustrated in Figure 39-4. A decline in the blood calcium levels due to the administration of ethylenediaminetetraacetic acid (EDTA) causes an increased secretion of PTH. The latter acts on the kidneys to increase the reabsorption of calcium and on bone to increase its rate of resorption. The result is a shift of calcium into the blood and body fluids. Vitamin D_3 that is synthesized in the skin is stored in the liver, where it may be converted to 25-hydroxycholecalciferol. Under the influence of PTH, the 25-hydroxycholecalciferol is converted to the 1,25-dihydroxycholecalciferol. The latter acts on the intestine to increase the absorption of calcium more rapidly

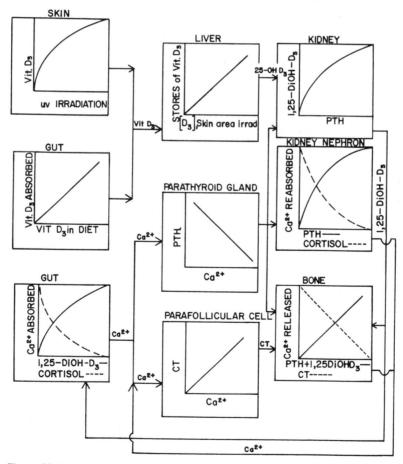

Figure 39-4
Hormonal control of calcium metabolism.

than does the 25-hydroxycholecalciferol. The PTH and the 1,25-dihydroxychole-calciferol have a synergistic effect in facilitating bone resorption of calcium. Secretion of calcitonin occurs when the blood calcium concentrations rise by 0.5 to 1 mg per 100 ml of plasma. The increased secretion of calcitonin antagonizes the actions of PTH. The integrated actions of the three substances bring about the homeostasis of blood calcium. Treatment with glucocorticoids (e.g., cortisol) or excessive secretion of glucocorticoids prevents calcium absorption in the intestine or its resorption in the kidney, and either may augment the effect of PTH on bone, thus favoring the resorption of bone calcium and leading to osteoporosis.

All the hormones that influence bone growth exert an effect on calcium metabolism. These effects may be sufficiently slow so that compensatory processes may prevent the disturbance of the homeostatic controls. The effect of hormones that influence bone formation is also exerted at the level of resorption of bone. Growth hormone, thyroxine, cortisol, and insulin must be present for the normal growth and maintenance of bone.

An insufficiency of growth hormone in development results in impaired linear growth at the ends of the bone, the epiphyseal cartilage, whereas an excess of growth hormone causes an increased linear growth. In the adult, growth hormone in excess causes the bone to increase in diameter — i.e., cortical bone formation continues — even though the bone cannot grow in length, because the epiphyses are closed. The bones of the face increase in size, causing enlargement of the nose, cheeks, skull, and jaw (prognathism). The bones of the fingers increase in diameter, giving them a spatulate appearance. Growth hormone causes the appearance in plasma of a sulfonation factor that stimulates the cellular activity of cartilage with increased synthesis of proteins, ribonucleic acids, and mucopolysaccharides.

Insulin plays no distinctive role in calcium metabolism, and it is probable that it influences bone growth and stability by contributing to the control of the synthetic activity of the cells in the bone. The thyroid hormones are essential for normal bone growth. Cretins manifest pronouncedly disturbed bone growth. The thyroid hormones modulate the secretion of growth hormone, and, in their absence, the pituitary content and the plasma concentration of growth hormone are depressed. Thyroidectomy depresses the response to parathyroid hormone and calcitonin. In the presence of excess thyroid hormone, excessive resorption of bone with increased blood calcium levels occur. This effect is probably mediated via other hormones. Thyroid hormones most likely regulate the metabolism of the osteoprogenitor cells.

Glucocorticoids in excess cause osteoporosis, but this effect is probably due to depressed growth hormone secretion and an inhibition of the absorption of calcium in the intestine. This effect may induce a secondary hyperparathyroidism and cause the secretion of parathyroid hormone, which further enhances resorption of bone.

The onset of puberty, with its increased rate of growth, is associated with an increased secretion of gonadal steroids. The closure of the epiphyses and the cessation of linear growth generally occur shortly after the adult level of gonadal steroid secretion is achieved. In the case of girls, growth ceases about two years after beginning of menses. Boys frequently continue to grow until about the age of 15 to 21 years.

The mechanism of growth control is indirect, and it probably depends on the regulation of the secretion of other hormones.

REFERENCES

1. Aurbach, G. D. Isolation of parathyroid hormone after extraction with phenol. *J. Biol. Chem.* 234:3179, 1959.
2. Collip, J. B. Extraction of a parathyroid hormone which will prevent or control parathyroid tetany and which regulates the level of blood calcium. *J. Biol. Chem.* 63:395, 1925.
3. Copp, D. H., et al. Evidence for calcitonin – a new hormone from the parathyroid that lowers blood calcium. *Endocrinology* 70:638, 1962.
4. DeLuca, H. F. The role of vitamin D and its relationship to parathyroid hormone and calcitonin. *Recent Prog. Horm. Res.* 27:479, 1971.
5. Potts, J. T., Keutman, H. T., Niall, H. D., and Tregeat, G. W. The chemistry of parathyroid hormones and the calcitonins. *Vitam. Horm.* 29:41, 1971.
6. Rasmussen, H. "Secondary hyperparathyroidism." *Mt. Sinai J. Med. N.Y.* 40:462, 1973.

SELECTED READING

Arnaud, C. D. Hyperparathyroidism and renal failure. *Kidney Int.* 4:89, 1973.

Aurbach, G. D., et al. Structure, synthesis and mechanism of actions of parathyroid hormone. *Recent Prog. Horm. Res.* 28:353, 1972.

Brewer, H. B., Jr., and Ronan, R. Bovine parathyroid hormone: Amino acid sequence. *Proc. N.Y. Acad. Sci.* 67:1862, 1970.

Deftos, L. J., Powell, D., Parthemore, J. C., and Potts, J. T., Jr. Secretion of calcitonin in hypocalcemic states in man. *J. Clin. Invest.* 52:3109, 1973.

DeLuca, H. F. The kidney as an endocrine organ involved in calcium homeostasis. *Kidney Int.* 4:80, 1973.

DeLuca, H. F. The kidney as an endocrine organ for the production of 1,25-dihydroxy vitamin D_3, calcium-mobilizing hormone. *N. Engl. J. Med.* 289:359, 1973.

Raisz, L. G., and Bingham, P. J. Effect of hormones on bone development. *Annu. Rev. Pharmacol.* 12:337, 1972.

Rasmussen, H., and Bordier, P. (Eds.). *The Physiological and Cellular Basis of Metabolic Bone Disease.* Baltimore: Williams & Wilkins, 1974.

Talmage, R. V., Cooper, C. W., and Park, H. Z. Regulation of calcium transport by parathyroid hormone. *Vitam. Horm.* 28:103, 1970.

Thorngren, K. G., and Hanssen, L. I. Effect of thyroxine and growth hormone on longitudinal bone growth in the hypophysectomized rat. *Acta Endocrinol.* (Kbh.) 74:24, 1973.

Thorngren, K. G., Hanssen, L. I., Menander-Sellman, K., and Stenstrom, A. Effect of dose and administration period of growth hormone on longitudinal bone growth in the hypophysectomized rat. *Acta Endocrinol.* (Kbh.) 74:1, 1973.

Welsch, U., Flitney, E., and Pearse, A. G. E. Comparative studies on the ultrastructure of the thyroid parafollicular C-cells. *J. Microsc.* (Oxf.) 89:83, 1969.

THE GASTROINTESTINAL SYSTEM

VII

Overview of the Gastrointestinal System   40

Daniel Weiner

INTRODUCTION

Most biologic systems are not in a true state of electrochemical equilibrium (see Chap. 1) but are open systems that tend to move in the direction of increased disorder or entropy. In a dynamic equilibrium, composition is time-invariant only at the expense of a continuous exchange of matter or energy with the environment. The function of the gastrointestinal (GI) system may be viewed as that of maintaining the delicate entropy balance that is demanded in biologic systems, and one of the primary roles of the nutrients (ignoring new protein synthesis) may be viewed as that of decreasing the entropy in these systems.

Figure 40-1 shows the major functional units of the gastrointestinal system. The lumen (cavity) of the GI tract may be considered to be topologically external to the body, since it is a tube open at both ends (mouth and rectum). There are several aspects to be considered in its function. Material as a supply of nutrients must be ingested and thus be made available in the lumen of the tract. Important in this regard is the regulation of food and water intake, deglutition, and esophageal function. By means of metabolic and chemical reactions, this material must be converted to a form that is suitable for absorption; this activity is facilitated by secretions. In this context, *absorption* will be considered as the transfer of material from the lumen, or external environment, to the blood, or internal environment. Other topics to be considered include motility and digestion, including secretion and its control. The mechanisms of absorption that control the movement of material from the external to the internal environment will be discussed, as will those that control the exclusion of unwanted contents of the lumen and excretory functions.

The two major factors that affect the variety and quantity of nutrients supplied by the GI tract are the regulation of oral intake and the regulation of absorption of nutrients at the level of the intestinal mucosa. The only substances whose absorption is known to be regulated by the intestinal mucosal cells are iron, calcium, and vitamin B_{12}. The absorption of these substances will be considered in detail later.

FUNCTIONAL ORGANIZATION

The input to the gastrointestinal system consists of the various nutrient materials ingested. The desired output consists of substances that may be utilized by the various in vivo chemical processes that are needed to supply the substrates for the metabolic and structural requirements of the internal environment of the body. Materials that are not needed are eventually excreted as waste products, thus providing

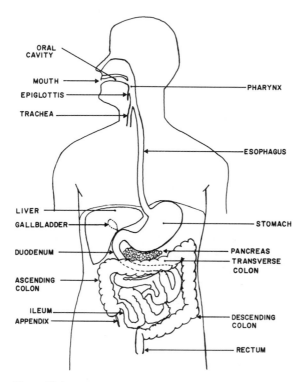

Figure 40-1
Major functional units of the gastrointestinal system.

an additional output. The secretions into the various organs of the GI system facilitate the chemical changes necessary for the generation of the desired outputs.

Figure 40-2 presents an expanded view of these basic concepts. Each of the organs of the GI tract is shown along with the internal sources of secretion. The output from one stage in the system forms the input to the next as material progresses through the tract. The principal output consists of the materials that are absorbed from the small intestine and colon into the interstitial spaces and the vascular and lymphatic systems. The excretory output is composed of the ingested materials that are not utilized by the system and the waste products that are secreted into the tract.

CHARACTERISTICS OF SMOOTH MUSCLE

The effector organ for motility in the gastrointestinal tract is, for the most part, the smooth muscle that is contained in the wall of the esophagus, stomach, small intestine, and large intestine. In general, this smooth muscle consists of regions of longitudinal muscle, circular muscle, and submucosal muscle. Except for the skeletal muscle found in the pharynx, proximal esophagus, and external anal sphincter, the GI tract is lined with an outer coat of smooth muscle.

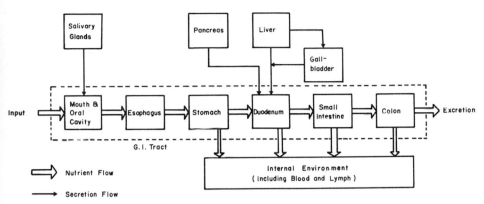

Figure 40-2
Functional operation of the gastrointestinal system.

There are several unique characteristics of smooth muscle that differentiate it sharply from skeletal and cardiac muscle (see Chap. 5). Table 40-1 lists these differences. The characteristics of smooth muscle uniquely adapt it to its role in the GI system. In contrast to the rapid, finely controlled movements of skeletal muscle, the movements of the viscous organs are slow and sustained. The one-to-one relationship between nerve fiber and muscle fiber that is present in skeletal muscle is not necessary for the slow propulsive and mixing movements of the stomach and intestine, and this arrangement is not present in smooth muscle. Rather, the function of nervous innervation in the GI tract seems to be that of modulating large groups of smooth muscle cells. The property of having an increase in fiber length without a concomitant increase in tension is called *receptive relaxation.* If the length-tension relationships of skeletal muscle and smooth muscle were similar, the expansion in volume of the stomach, for example, from between 50 and 100 ml to more than 2 liters would not be possible without rupture of its wall.

Because of the small size and slow contractility of the intestinal smooth muscle cells, it has been difficult to obtain precise electrophysiologic information from direct micropuncture experiments. Thus, many of the principles that will be discussed in this section have come from sucrose gap experiments and from measurements made in uterine and urinary tract smooth muscle, which is somewhat easier to study than is intestinal muscle. When available, however, data from intestinal smooth muscle have been used.

In smooth muscle, there is a transmembrane potential of approximately 40 to 60 mV, the inside of the cell being negative with respect to the outside. This transmembrane potential appears to be somewhat less in magnitude than the K^+ equilibrium potential predicted by the Nernst equation (see Chap. 5). Several explanations for this discrepancy have been offered, but at the present time the exact meaning of this difference has not been elucidated.

Spontaneous, slow, transient depolarizations in the transmembrane potential are observed in many types of smooth muscle. Some smooth muscle cells appear to have

Table 40-1. Comparison of Characteristics of Smooth and Skeletal Muscle

Characteristic	Smooth Muscle	Skeletal Muscle
Speed of contraction	Slow	Rapid
Capacity for sustained contraction	Good	Fair
Gradual increase in fiber length without significant increase in wall tension	Present	Absent
Oxygen consumption	Proportional to length	Proportional to tension
Innervation	Motor fibers end free among smooth muscle cells; one nerve fiber affects many cells	Each skeletal muscle cell is innervated by a somatic nerve fiber
Initiation of impulse	Pacemaker muscle cells present; impulse may be initiated by smooth muscle cell; cell-to-cell propagation of impulse	Only by nervous innervation; no cell-to-cell propagation of impulse

the slow depolarization characteristics that are associated with the pacemaker cells in atrial heart muscle. Mechanical deformation of smooth muscle cells has been shown to change the transmembrane potential, and it is affected by many neurotransmitters and hormones. Acetylcholine, serotonin, histamine, mechanical stretch, or electrical stimulation will decrease or depolarize the smooth muscle membrane, whereas epinephrine hyperpolarizes the membrane. Waves of hyperpolarization can also be seen in smooth muscle with certain types of nerve stimulation; the physiologic significance of these is unclear.

Spike potentials are associated with membrane depolarization, which may be accompanied by mechanical contraction. The strength and the duration of the mechanical contraction are apparently related to the frequency of the spike potentials (Fig. 40-3). There is a burst of spike-potential activity during the spontaneous depolarization of the membrane. The spike potentials do not ordinarily overshoot zero; this might not represent a Na^+ equilibrium potential, but Ca^{+2} may be involved. The mechanism of excitation-contraction coupling in smooth muscle is thought to involve calcium, but it is otherwise poorly defined [1]. The slow depolarization waves can be observed to spread from cell to cell, sometimes traveling over long distances. That this spread of depolarization is not due to simple mechanical deformation of the membrane can be shown by their presence even when mechanical contraction is blocked.

It was pointed out (Table 40-1) that a one-to-one relation between the nerve fiber and the muscle cell does not exist in smooth muscle; rather, the nerve fibers seem to terminate free among a group of smooth muscle cells. Conventional neural mediation of this spread of excitation, therefore, is not possible. The spread of excitation from cell to cell seems to occur at nexus, which are regions of fusion of the outer layers of the intercellular plasma membranes. There is little information about whether the

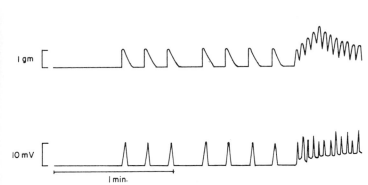

Figure 40-3
Tension and transmembrane potential (sucrose gap) in the isolated teniae coli of the guinea pig.

nexus is actually a region of low resistance between cells or whether some other special mechanism is responsible for cell-to-cell transmission; intercellular bridges have not been definitely identified. A hormonal basis for this phenomenon has been proposed, but this view does not have many adherents at present.

In summary, some smooth muscle cells act as pacemaker cells, and spontaneous, slow depolarizations of the cell membrane occur. Superimposed on this slow depolarization are spike potentials, which are associated with mechanical contraction of the muscle. This excitation and the subsequent mechanical activity can be induced by neurotransmitters, hormones, mechanical deformation of the cell membrane, or by actual cell-to-cell transmission of the wave of depolarization. Certain neurotransmitters, such as epinephrine, can hyperpolarize the cell membrane and abolish the activity that is induced by the above factors. Thus, the general level of motor activity in the smooth muscle can be regulated by the central nervous system by the controlled release of inhibitory or stimulatory neurotransmitters and hormones, while fine modulation is accomplished by local factors such as mechanical stimulation and the level of activity of the contiguous plexus.

NEURAL CONTROL MECHANISMS

The vagus, pelvic parasympathetic, and splanchnic nerves constitute the major afferent and efferent pathways innervating the gastrointestinal tract (Fig. 40-4). The skeletal muscle located in the pharynx and upper esophagus is innervated by the somatic nerve fibers.

The motor fibers in the vagus nerve include parasympathetic preganglionic fibers, which arise from the cells of the brain-stem vagal nuclei; sympathetic fibers, which arise from branches to the vagus from the cervical paravertebral ganglions; and special somatic motor nerves, whose source is not clear, that innervate the GI skeletal-type muscle without ganglionic synaptic interruption. The preganglionic parasympathetic fibers of the vagus nerve pass to the esophagus, stomach, small intestine, and the

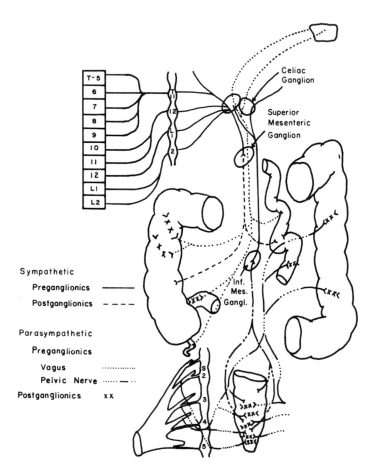

Figure 40-4
The extrinsic autonomic innervation of the intestinal tract.

proximal colon, making synaptic connections with secondary parasympathetic motoneurons of the enteric plexus. This distribution is characteristic of the parasympathetic
nervous system; i.e., the terminal synaptic connection is located in or adjacent to the
innervated organ. In contrast, the synaptic connections of the sympathetic nervous
system are located in the paravertebral and abdominal sympathetic ganglia. The motor
fibers in the splanchnic nerves are the preganglionic and postganglionic sympathetic
fibers that also originate in the paravertebral and abdominal sympathetic ganglia. Each
abdominal ganglionic plexus distributes its postganglionic branches to the GI tract along
with branches of the associated artery. The final termination of these fibers is not clear.
For the most part, these fibers terminate on the vascular musculature of the vessels of
the gut. A few sympathetic postganglionic fibers have been described that end among
the smooth muscle fibers and on the neurons of the enteric plexus.

These two sets of motor fibers have been classically viewed as reciprocal in function: the parasympathetic system being excitatory and the sympathetic system being inhibitory. This concept is probably not completely correct, since, for example, the two divisions have complementary effects on salivary secretions (see Chap. 41). Sensory nerves travel in both systems; those of the vagus nerve arise from the unipolar cells of the nodose ganglion.

As has been noted, the parasympathetic nerve fibers to the GI tract are preganglionic and terminate, in many cases, on postganglionic nerve cell bodies, which form plexus or networks containing solitary or clustered nerve cell bodies. This postganglionic neuronal network also appears to function as an "associative" or internuncial system in local reflex arcs. This neural sequence appears to be unique to the intestinal plexus. Some cells with sensory functions may also serve to modulate the neuronal activity in these plexus. The plexus are named according to the stratum of the wall that they occupy − subserosal, myenteric, submucosal, and mucosal − but they are interconnected, not structurally isolated from one another.

A recent discovery, which has not yet been fully interpreted, will be merely mentioned. It has been found that serotonin is synthesized in the myenteric plexus. The view has been advanced that serotonin is a neurohumoral transmitter in the plexus, but convincing evidence for this hypothesis is lacking. Some nerve cells in the plexus take up silver stain, and it is thought that these cells may be responsible for serotonin production.

REGULATORY MECHANISMS OF FOOD AND WATER INTAKE

In general, most adult, free-living, wild animals tend to maintain their weight at a constant level. In animals, caloric intake has been shown to be matched to expenditure over a wide range of daily exercise. Experiments in rats have shown that their body weight is controlled at a stable value over an approximately twofold range of caloric expenditure (Fig. 40-5). A similar phenomenon has also been observed in man [2].

It is clear that the regulation of food intake must function not only to control oral intake but also to insure that certain nutrients are available. The type of food-seeking behavior that is directed toward the ingestion of a particular nutrient has been called *specific appetite*. A basic distinction between hunger and appetite must be made: hunger is the basic drive that induces food-seeking behavior; appetite is a sensation, usually pleasant, that evokes a desire for food.

REGULATION OF FOOD INTAKE

In order to determine the input-output relationships that obtain in a well-ordered regulatory feedback system, it is necessary to identify the input and output state variables. Unfortunately, in the case of the mechanisms regulating food intake, the nature of the output is confused by the presence of appetite and conditioning, and

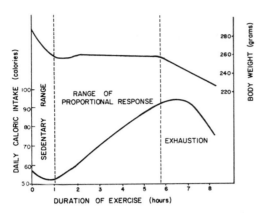

Figure 40-5
Voluntary caloric intake and body weight as functions of exercise in normal rats. (Modified from
J. Mayer. General Characteristics of the Regulation of Food Intake. In C. F. Code [Ed.], *Alimentary Canal*. Vol. 1, Food and Water Intake. *Handbook of Physiology*. Baltimore: Williams &
Wilkins, 1967. Reproduced with permission.)

it is not clear which of the many metabolic variations accompanying food deprivation
is the essential control signal that initiates food-seeking behavior.

The primary area involved in the central nervous system's regulation of food ingestion is in the hypothalamus. The ventromedial hypothalamic area seems to function
primarily as a satiety center. Stimulation of this area causes a cessation of eating or
aphagia, whereas ablation of this area increases food intake, resulting in hyperphagia
and obesity. Recent experiments have suggested, however, that the critical area may
be ventrolateral to the hypothalamic nucleus itself. The dorsolateral hypothalamus is
also involved in the activation of feeding behavior. Bilateral lesions in this area result
in hypophagia and loss of weight, and may lead to starvation.

Chemosensitive cells have been described in the ventromedial hypothalamus, or
satiety center, that are apparently sensitive to arterial-venous glucose differences or
to their own rate of glucose utilization. Electrical activity recorded from this area is
significantly and selectively affected by the blood glucose level; hyperglycemia increases
the frequency of activity and hypoglycemia slows the activity in the satiety center.
Recent studies have shown that when the degeneration of cells in the ventral medial
hypothalamus is artificially induced by gold thioglucose, the animals become significantly and uniformly obese.

Chemoreceptor mechanisms that are sensitive to glucose and amino acids have
been demonstrated in the intestine. Afferent discharges are evoked in the mesenteric
nerves by perfusion of the intestinal lumen with glucose and various amino acids under
controlled conditions of temperature, pressure, and osmolarity; however, the precise
nature of the function of these receptors remains obscure.

The early theories of the regulation of food intake emphasized the relationship of
strong contractions of the empty stomach to hunger (hunger pangs). Gastric distention
was noted to inhibit food intake or to induce a sensation of satiety. It has been shown,

however, that although gastric contractions may be associated with hunger, food-seeking behavior may be initiated in their absence. The role of gastric distention alone in the inhibition of intake is also open to question. It is generally accepted that gastric distention may play a role in the short-term regulation of food intake (i.e., meal-to-meal or day-to-day regulation), but long-term regulation must clearly operate through additional control mechanisms.

The presence of the glucose-sensitive receptors in the ventromedial hypothalamus has implicated these hypothalamic cells in the mechanism of control of food intake. Experiments have shown hunger can be induced by low blood glucose. Although the blood glucose level may be a factor in short-term regulation, the control of blood glucose within a narrow range by a variety of mechanisms, including glycogen break-down in the liver and gluconeogenesis, makes it unlikely that this mechanism could account for long-term regulation.

A thermoregulatory hypothesis has been proposed on the basis of experiments in animals that showed that direct, local cooling of the rostral area of the hypothalamus induces food ingestion and warming of the same area inhibits this behavior. The relationship between temperature changes and food intake seems to be a very complex one, and, for a variety of reasons too detailed to discuss here, the thermostatic mechanism cannot explain long-term regulation of food intake. Heat resulting from the specific dynamic action (SDA) of food has also been assigned a role in regulation, but this mechanism also cannot account for long-term control.

An overall surplus of energy intake over energy expenditure has been shown to cause the deposition of fat in lipid depots. It has been suggested that the long-term regulation of food intake is achieved by means of the sensitivity of the hypothalamic centers to varied concentrations of unspecified circulating metabolites that are related to the size of body reserves of fat. The hypothesis that the lipid stores are an important factor in long-term regulation is supported by the observations made in parabiotic rats with lesions in the satiety centers of one partner. The animal with the lesion developed hyperphagia and subsequent obesity, whereas the parabiotic partner with the intact satiety center displayed hypophagia and wasting that became most evident when the animal whose satiety center had been damaged became very obese. This suggests that the increased size of the lipid depots in the lesioned animal is able to bring about satiety in the normal animal.

A role for nonesterified fatty acids has been proposed in this central regulation, but the exact substance and the manner through which this information is provided to the central mechanisms that are related to feeding are still unknown. The lipostatic hypothesis is, however, the most attractive in terms of explaining the long-term control of food intake.

Specific appetite is the preference that is instinctively shown by animals and, to some extent, by humans for food that contains a nutrient for which the animals are specifically deficient. Substances for which this type of nutrient-seeking behavior have been identified are sodium chloride, calcium, and the B vitamin, thiamine. In humans, the ingestion of clay, starch, and ice has been noted in iron-deficient individuals. This peculiar pattern of ingestion has been termed *pica*.

Thus, it appears that the regulation of food intake is composed of several simul
taneously acting components. Food intake is probably regulated on a short-term
basis by means of the glucose-sensitive mechanisms and the input from mechanisms
that sense gastric contractions and distention as well as the volume ingested. Long-
term balance is probably maintained by a mechanism that is related to the levels of
nonesterified fatty acids or the size of lipid depots. In addition, food-seeking behavior
may be directed toward the ingestion of specific essential nutrients. The mechanism
underlying this last aspect of regulation still requires considerable investigation.

REGULATION OF WATER INTAKE

Precise control of the concentration and volume of water in the internal environment is required due to the continuous exchange of water with the external environment. The control of oral intake is necessary to balance the water loss that is incurred
by urine excretion and by evaporation for temperature regulation.

Figure 40-6 shows the major signals and modulating factors that affect water intake.
A decrease in plasma volume initiates an increase in both water intake and antidiuretic
hormone (ADH) production. As the plasma volume increases as a result of these measures, the neutral resting state is altered. Water absorption and retention stimuli are
decreased, and the plasma volume returns to a normal value. Figure 40-6 also illustrates
several other regulatory pathways, such as osmotic and temperature influences.

In contrast to the relatively well-defined hypothalamic areas associated with hyperphagia and hypophagia, the functional areas for the control of water intake are more
diffusely distributed throughout the hypothalamus and brain. Apparently, both the
osmolarity and the volume of plasma are sensed. Units of cells in the hypothalamus
that respond with increased electrical activity to the injection of hypertonic solutions
into the carotid artery have been identified. Changes in the extracellular fluid volume
and its plasma component may be sensed by stretch receptors in the left atrium.

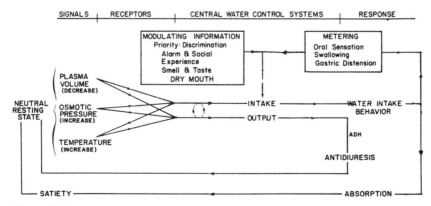

Figure 40-6
Feedback systems for the control of water intake and output; the major signals and some modulating factors are shown. (From J. A. Stevenson. Control Mechanisms Controlling Water Intake. In
C. F. Code [Ed.], *Handbook of Physiology*. Baltimore: Williams & Wilkins, 1967. Reproduced
with permission.)

REFERENCES

1. Christensen, J. Controls of gastrointestinal movement: Some old and new views. *N. Engl. J. Med.* 285:85, 1971.
2. Mayer, J. General Characteristics of the Regulation of Food Intake. In C. F. Code (Ed.), *Handbook of Physiology. Alimentary Canal.* Vol. 1. Food and Water Intake. Baltimore: Williams & Wilkins, 1967.

SELECTED READING

Anderson, R. G. G. Cyclic AMP and calcium ions in mechanical and metabolic responses of smooth muscle; influence of some hormones and drugs. *Acta Physiol. Scand.* [Suppl.] 382:1, 1972.
Bulbring, E., et al. (Eds.). *Smooth Muscle.* Baltimore: Williams & Wilkins, 1970.
Grossman, S. P. Hypothalamic and limbic influences on food intake. *Fed. Proc.* 27:1349, 1968.
Hoebel, B. G. Feeding: Neural control of intake. *Annu. Rev. Physiol.* 33:533, 1971.
Lepkovsky, S. Hypothalamic-adipose tissue interrelationships. *Fed. Proc.* 32:1705, 1973.

The Esophagus and Stomach

41

Daniel Weiner

INTRODUCTION

The esophagus transports masticated food mixed with saliva from the oral cavity to the stomach. This transport is initiated by the act of swallowing or *deglutition*. In this process, the food bolus is projected into the pharynx, where a complex reflex arc is initiated. Contraction of the constrictor muscle group of the pharynx is preceded by a brief drop in pressure at the pharyngoesophageal region. Simultaneously, the larynx moves upward and forward, the epiglottis descends, the vocal cords constrict, and there is inhibition of the respiratory cycle. These events insure that the food bolus does not enter the trachea. Similarly, the return of the food to the mouth and nose is prevented by the contraction of the lateral palatine pillars and the elevation of the soft palate and uvula. The increase in the pressure in the pharynx and the transient decrease in pressure at the pharyngoesophageal region result in the propulsion of the food bolus into the esophagus, through which it is transported to the stomach. More complete discussions of the reflex pathways associated with deglutition may be found in the literature [7].

The stomach is an organ that uniquely performs a variety of functions, including the storage of food, the initiation of a number of digestive processes, and the osmotic equilibration of its contents. In addition, the stomach secretes a variety of endocrine substances and an intrinsic factor that facilitates the absorption of vitamin B_{12} by the small intestine. The stomach can accommodate a 20-fold increase in volume with only a small increase in intraluminal pressure. This feature enables the stomach to function as storage organ, allowing time for the digestion of a rapidly ingested meal and the osmotic equilibration of the usually hypertonic gastric contents. Digestion in the stomach is initiated by secretions of digestive enzymes and hydrochloric acid.

STRUCTURE AND PERISTALTIC ACTION OF THE ESOPHAGUS

The upper esophageal sphincter and the proximal one-third of the esophagus consist of striated skeletal muscle and are innervated by special somatic fibers that probably arise in the central vagal nuclei. The motor innervation of the smooth muscle segment follows the general pattern for smooth muscle that was discussed in the previous chapter.

In the body of the esophagus, a propagated ring-like contraction, or peristaltic wave, sweeps the length of the esophagus following the opening of the upper esophageal sphincter. Although the motor innervation of the skeletal-muscle segment is different from that of the smooth muscle, the control of the skeletal muscle is such that it functions like smooth muscle. Transmission of the peristaltic wave from the skeletal to the smooth muscle portion occurs smoothly. The aboral propagation of the peristalt wave (from mouth to colon) in the smooth muscle segment is controlled by the vagal (parasympathetic) esophageal plexus [12]. The arrangement of neurons in the local plexus that is responsible for direction-oriented peristalsis is unknown. Occasionally, peristaltic waves may be seen to move in an oral direction, but the oral movement usually dies out in a few millimeters, whereas propagation in the aboral direction may travel many centimeters. Some investigators have suggested that a wave of relaxation precedes peristaltic waves; this has not been a constant finding and the mechanism is unclear.

Primary peristalsis originates in the skeletal-muscle portion of the esophagus and is initiated by the opening of the upper esophageal sphincter. Secondary peristalsis is initiated by mechanical distention of the body of the esophagus, such as that caused by a food bolus, and sweeps down over the remainder of the distal aspect of the esophagus. Tertiary contractions have also been described; these are spontaneous, weak, or nonpropulsive contractions, and they are seen most commonly in the aged [6].

The lower esophageal sphincter is a zone at the gastroesophageal junction with a resting wall tension that is high enough to close the lumen by creating an intraluminal pressure that is 10 to 40 mm Hg higher than atmospheric. The wall tension in this segment falls a few seconds after a peristaltic wave sweeping down the esophagus reaches this zone, but it is restored after a short period [5]. Controversy exists as to whether this zone is a true anatomic sphincter, and whether any special details of innervation account for the coordinated activity [8].

Figure 41-1 shows the sequence of pressure gradients that are developed following the initiation of propulsion of a food bolus from the oral cavity to the stomach. Pressures were measured by passing a pressure transducer down the oral cavity, pharynx, and esophagus. The transient relaxation at the pharyngoesophageal region can be seen to occur a few seconds after the initiation of deglutition.

The role of the adrenergic nerves in the function of the esophagus is still not known. The old idea of the reciprocal function of adrenergic and cholinergic nerves in smooth muscle led to the view that the adrenergic nerves are inhibitory. Against this view is the recent observation that norepinephrine increases the motor activity of esophageal smooth muscle. Several other examples of apparently similar effects of parasympathetic and sympathetic stimulation have been described as occurring elsewhere in the gastrointestinal tract, and the traditional concept of antagonistic action obviously needs reclarification.

SECRETIONS OF THE ORAL CAVITY AND ESOPHAGUS

Two major types of secretions are produced by the alimentary tract. Exocrine secretions are secreted into the mouth and the lumen of the tract. These are important primarily in nutrient digestion and absorption. Exocrine-secreting organs include the salivary glands, stomach, liver, small and large intestine, and, to some degree, the

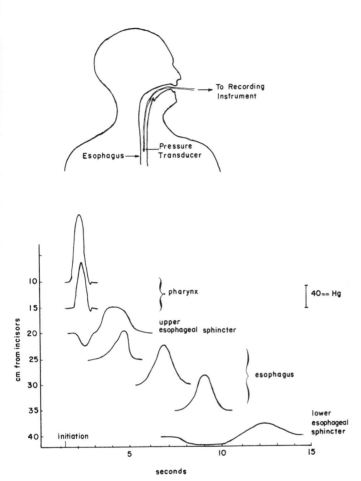

Figure 41-1
Propagation of an esophageal peristaltic wave.

pancreas. A second type of secretion includes those substances that are secreted into the blood. Those substances secreted by groups of cells into the blood that exert their characteristic action on remotely located cells are called hormones, or endocrine secretions. Usually excluded from this definition are products of metabolism (such as carbon dioxide and lactic acid) and neurotransmitters. No sharp distinction can be drawn, however, since some products of special metabolic pathways may meet all the above criteria. Endocrine secretions occur from cells in the gastric mucosa, pylorus, and duodenum (the area of heaviest secretion), certain cells located in the liver, and the alpha and beta cells of the pancreas. As analytic tools for identifying hormone-like substances become more sensitive, endocrine secretions from other organs and regions are continually being identified.

Saliva is produced by the parotid, submaxillary, and sublingual glands. In addition, several tiny buccal mucous glands produce small amounts of mucin. The primary function of saliva is to moisten food in order to stimulate taste receptors and to facilitate swallowing. The total salivary output per day in man is 1000 to 1500 ml. Saliva consists primarily of water, sodium and potassium chlorides, bicarbonates, and the digestive enzyme, amylase. Saliva also contains phosphates, thiocyanates, small amounts of trace metals, blood-group substances of the ABO and Lewis systems, and antibodies. The reaction of saliva is alkaline.

No major secretions are produced by the esophagus, but a small amount of mucin is produced by its mucosal epithelium. This mucin functions to facilitate the passage of the food bolus down the esophagus.

STRUCTURE AND MOTILITY OF THE STOMACH

The major subdivisions of the stomach are shown in Figure 41-2. The area immediately adjacent to the entry of the esophagus is known as the *cardia*. The *fundus* is the most superior aspect of the stomach, and the *pylorus* is located at the outlet of the stomach; because of its well-developed musculature, the terminal portion is often termed the *pyloric sphincter*, or, incorrectly, the pyloric valve. The stomach musculature consists of three external layers of smooth muscle: the outermost layer is longitudinal, the middle is circular, and the inner is oblique.

The volume of the fasting human stomach is relatively small (50 to 100 ml), and most of this volume is occupied by air. About 50% of the time, weak motor activity is seen in the fasting state. The resulting contractions occur at a rate of about three per minute and usually produce a luminal pressure of less than 5 cm H_2O. These

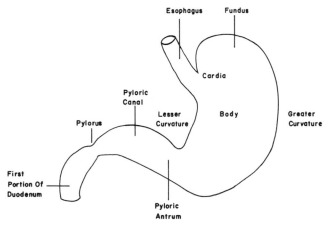

Figure 41-2
Subdivisions of the human stomach.

contractions have been labeled *type I contractions* [9] (Fig. 41-3). They are peristaltic in nature, usually starting in the cardia and sweeping down toward the antrum at a rate of 1 to 2 cm/sec. These contractions may last 10 to 20 seconds. Occasionally, a stronger contraction may be seen in the resting stomach that produces pressures of between 5 and 40 cm H_2O; these are *type II contractions.* They also occur at a rate of about three per minute and are differentiated from type I contractions by their higher pressure amplitudes. These contractions are also peristaltic in nature, moving from cardia to antrum. Four to five hours after a meal, a burst of strong contractions that lasts 10 to 15 minutes may be seen. In the past, an association between these contractions and the sensation of hunger was noted, and this activity has been termed "hunger contractions." More recent evidence suggests that there is no fixed relationship between hunger and these contractions, and it therefore seems reasonable to abandon this terminology.

In the antrum of the fasting stomach, the most common type of motor activity seen is the *type III*, or *tonic, contraction.* Tonic contractions occur as base-line pressure changes that are superimposed on type I or type II peristaltic contractions. Most type III contractions last about one minute, but some may be maintained for 4 to 6 minutes, producing pressure changes of 1 to 10 cm H_2O. The key aspect of type III contractions that differentiates them from the type I and II contractions is the tonic nature of the increase in base-line pressure that is found in the former case. Type I and II waves may be superimposed on a type III contraction, but, under normal conditions and in the absence of a type III contraction, the pressure in the stomach returns to zero after a type I or II contraction, whereas a type III contraction may produce a sustained increase in base-line pressure [3] (see Fig. 41-3).

Following the ingestion of a meal, motor activity in the stomach is quiescent for about one-half hour. In association with deglutition, a relaxation of the fundus occurs that probably serves to increase the capacity of the stomach. This prevents

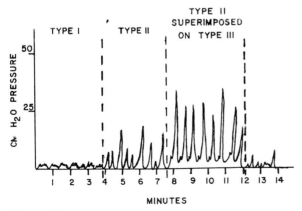

Figure 41-3
Types of stomach contractions.

an increase in intragastric pressure as the volume of the gastric content is increased by swallowed food. This phenomenon has been termed *receptive relaxation*; the mechanism appears to require an intact vagus nerve, but it has not been otherwise elucidated. At that time, weak type I contractions appear. These are subsequently followed by type II contractions, and both may be present together. The purpose of type I contractions, although they may sweep down over the stomach, appears to be primarily one of mixing rather than of propulsion; they serve to expose the food bolus to fresh gastric mucosal surfaces. Type II contractions are more clearly propulsive in nature. Type I and II contractions may arise at any point in the stomach, but they usually start in the cardia and move in an aboral direction. These contractions will continue if the vagus nerve is sectioned, but they require the intact intrinsic nervous system, i.e., the system of submucosal and myenteric plexus [3]. Although the three types of contractions have been considered as separate entities, it probably is the case that they are just different manifestations of the same basic motor patterns.

As a meal is ingested, the food that is layered up in the cardia of the stomach is gradually mixed with the stomach contents by the type I contractions. Subsequently, the contents are gradually propelled through the stomach into the duodenum by type II contractions, although there may be considerable to-and-fro motion produced by the type I contractions. The frequency of the type II contractions, and hence the rate of emptying of the stomach, is under both neural and hormonal control. The initial period of absence of motor activity following the ingestion of a meal is probably mediated by the vagus nerve. Much controversy exists regarding the state of the pyloric sphincter following ingestion of a meal, but it seems probable that the pyloric sphincter is closed. Type II waves gradually reappear, accompanied by type III tonic activity of the pyloric antrum. An occasional type II wave produces a propulsive peristaltic wave that results in the relaxation of the pyloric sphincter. The entry of food into the duodenum may initiate one of several hormonal or neural mechanisms that act to inhibit the type II contractions in the stomach and thus delay gastric emptying [9].

Figure 41-4 shows the various factors affecting motility. The presence of acid in the duodenum decreases type II motor activity in the stomach, as shown by the HCl-bulbogastrone feedback path of the figure. The inhibition of HCl release is controlled primarily by a neural mechanism, although recently a hormonal mechanism involving bulbogastrone has also been implicated. It has been found that bulbogastrone is secreted into the blood by the duodenal mucosa in response to acid pH, and that bulbogastrone is active in decreasing the secretion of HCl. It is interesting to note that bulbogastrone is also apparently active in inhibiting gastric motility. The dual inhibitory action on motility and acid secretion is also characteristic of the action of the hormone enterogastrone, which is secreted by the duodenum in response to the presence of absorbable fat and fatty acids in the lumen. Because enterogastrone is secreted by the jejunum and, perhaps to some extent, by the ileum, the presence of fat in the small intestine will inhibit gastric motor activity even after the passage of the partially digested food, or chyme, through the duodenum. Hyperosmolarity of the duodenal contents is another factor that inhibits gastric motor activity, and it exerts

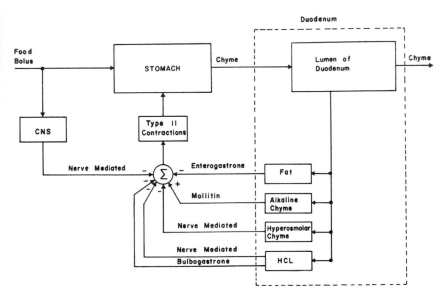

Figure 41-4
Neurohumoral regulation of gastric motility.

its effect by means of a neural reflex pathway. An additional duodenal hormone, mollitin, has been identified that increases the activity of type II contractions in the stomach. It has been suggested that mollitin may be secreted in response to an alkaline pH in the duodenum, but additional investigation is necessary to validate this claim.

The central nervous system integrative center that effects vomiting is located at the dorsolateral border of the lateral reticular formation of the medulla. There is, in addition, a chemoreceptive trigger zone located in the floor of the fourth ventricle where various toxic and pharmacologic (emetic) stimuli are received from the circulation. Afferent pathways also arise in the mucosa of the GI tract, especially in the region of the pylorus and duodenum, and in the pharyngeal mucosa. Other afferent stimuli for vomiting include the mechanical stimulation of the pharynx and palate, a rise in intracranial pressure, and rotation or acceleration of the head. Various sensory stimuli such as pain may also activate the vomiting center.

Following the stimulation of the vomiting center, a forced inspiration is taken that is followed by a general relaxation of the esophagus and stomach, including the sphincters. Subsequently, there is a vigorous constriction of the duodenum and pylorus. Vomiting itself, however, is brought about by the contraction of skeletal muscle, since smooth muscle cannot develop the pressure necessary to eject the gastric contents. During vomiting, the diaphragm descends, the abdominal muscles contract forcefully, respiration is inhibited, and the gastric contents are forced upward through the cardia into the esophagus and pharynx. When a conscious individual experiences vomiting, the glottis is contracted, the nasal passages are blocked by the elevation of the palate, and the gastric contents are ejected through the mouth. Active smooth muscle constriction is not essential for vomiting, and no evidence for antiperistalsis (i.e., peristalsis in the oral direction) has been noted.

MECHANISM AND REGULATION OF GASTRIC SECRETION

The epithelium of the stomach is covered with many small glandular openings, ranging from 1 to 4 mm in radius, that secrete gastric fluid. The structure of a typical secreting gland is illustrated in Figure 41-5, and the secretions of the various cell types are listed in Table 41-1. Mucous cells are located at the gastric pit and the neck portions of the gastric gland. The chief or peptic cells constitute the majority of cells lining the luminal wall. Cells of the remaining type, called the parietal or oxyntic cells, are scattered along the luminal wall and outer border of the gland. The glands of the cardiac and the pyloric areas of the stomach contain only mucin-secreting cells, whereas the glands of the main portion of the stomach, or fundus, contain all three types of secreting cells.

Pepsinogen is secreted as an inactive proenzyme and is converted to the proteolytic enzyme, pepsin, at a pH below 3.5 or in the presence of preformed pepsin by the splitting off of an amino acid residue. Pepsinogen has a molecular weight of 38,500, but after conversion to pepsin, the molecular weight is 34,000. During the process of pepsinogen activation, several basic peptides are split off, one of which seems to

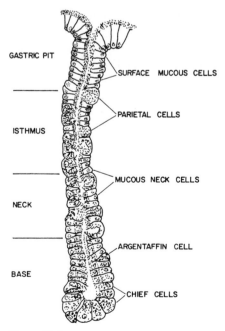

Figure 41-5
Gastric secreting gland.

Table 41-1. Secretions of the Cell Types of the Gastric Glands

Cell Type	Probable Secretion
Surface mucous cells	Mucin (may be accompanied by slightly alkaline fluid)
Parietal cells	HCl, electrolytes and H_2O, intrinsic factor,[a] blood-group substances
Mucous neck cells	Contain neutral mucopolysaccharides (function unknown)
Argentaffin cells	Function unknown (may contain serotonin)
Chief cells	Pepsinogen, other proteolytic proenzymes

[a]Intrinsic factor has been localized to the parietal cells in man, monkeys, cats, and rabbits. In the rat, it has been found in the chief cells.

function as a negative feedback inhibitor of pepsin. Pepsin is an endopeptidase, and it cleaves peptide linkages adjacent to free carboxylic groups. Small amounts of other enzymes are also secreted, but these are quantitatively unimportant.

The secretion of hydrochloric acid by the parietal cells is clearly an example of an active transport process that requires metabolic energy. Hydrogen ion is concentrated more than three million times in being transported from a concentration of 4×10^{-5} mEq/liter in the blood (pH = 7.4) to a concentration of 125 mEq/liter (pH = 0.9) in the gastric lumen [13]. The lumen of the gastric mucosa has a potential of 30 to 60 mV, and the lumen is negative with respect to the serosal or outer surface of the stomach [13]. Considerable evidence suggests that chloride ion is also actively transported from the blood to the lumen. During the stimulation of HCl secretion, the expected base excess (bicarbonate ion) in gastric venous blood has been observed. In preparations of isolated gastric mucosa, there is an absolute requirement for sodium ion, and probably for potassium ion as well, in the serosal bathing solution in order for the active transport of HCl to occur. Small amounts of Na^+ and K^+ are also present in secreted gastric juice. The mucosal permeability to these ions is low, and the significance of their presence in gastric juice is uncertain. The in vitro gastric mucosal resistance is in the range of 200 to 300 ohms·cm^2.

Several explanations for the mechanism of secretion of gastric acid have been proposed. These theories have to account for the observed electrical potential difference, the active transport of H^+ and Cl^-, and the requirement for the presence of Na^+ and K^+. A membrane-bound adenosinetriphosphatase (ATPase) that is stimulated by the presence of HCO_3^- has been obtained from gastric mucosa. This enzyme might be active at the site of Cl^- to HCO_3^- exchange in vivo. Gastric mucosal cells also contain a large amount of carbonic anhydrase; this enzyme facilitates the following reaction:

$$CO_2 + H_2O \rightleftharpoons H_2CO_3 \rightleftharpoons H^+ + HCO_3^- \tag{41-1}$$

This reaction probably supplies the excess HCO_3^- in gastric venous blood that is noted during HCl secretion. The present state of knowledge in this area does not allow a reasonable choice among the many mechanisms proposed. The reader is referred to one of a number of recent reviews listed at the end of this chapter for further study of this topic.

In the pyloric area of the stomach are located a histologically distinct group of glands, the pyloric glands, that may constitute one-fifth of the total area of the gastric mucosa. Only a few parietal (HCl-secreting) cells have been found in the pyloric glands; the major products of these glands are the exocrine secretion, mucin, and the endocrine secretion of an important hormone, gastrin. The G cell in the antrum and proximal duodenum secrete gastrin. The mucin-secreting cells of the gastric glands probably have an important role in the protection of the gastric mucosa from autodigestion.

Reference was previously made to the two types of secretions: exocrine and endocrine. The gastric exocrine secretions include pepsinogen, HCl, and intrinsic factor. The endocrine secretions of the GI tract serve the important function of regulating exocrine secretion and motility in the tract. Gastrin is one of the major hormones elaborated by the stomach. Histamine and serotonin are also present in the stomach and are largely confined to the mast cells.

The rate of secretion of gastric acid at any moment depends upon the interplay of many neural and hormonal factors. Since the neural and hormonal factors may be either stimulatory or inhibitory, it is difficult to account for a given secretory rate in terms of the relative contribution of each factor, but the major hormonal influences on gastric acid secretion can be identified (Table 41-2). Although histamine is a very potent stimulus to gastric acid secretion, its exact role in the normal regulation of gastric acid secretion is unknown. Histamine has been assigned a role as "the final common mediator of acid secretion"; recent evidence of a H_2-histamine receptor in the stomach supports this view [4]. Cyclic AMP has also been implicated in the control of HCl secretion [2].

Table 41-2. Gastrointestinal Hormones Affecting Gastric Acid Secretion

Hormone	Site of Production	Physiologic Stimulus	Effect on Acid Secretion
Gastrin	Pyloric antrum	1. Cephalic stimulation via vagal efferent pathway 2. Chemical and mechanical stimulation of antral mucosa (proteoses, amino acids, and distension)	Stimulatory
Histamine	Duodenal mucosa Gastric mast cells Oxyntic glands	Unknown	Stimulatory
Secretin	Small intestinal mucosa	1. HCl in duodenum (most potent stimulus) 2. Peptides, amino acids, fatty acids (questionable)	Inhibitory
Enterogastrone[a]	Small intestinal mucosa	Absorbable fats and fatty acids	Inhibitory
Bulbogastrone	Duodenal mucosa	Acid pH in duodenum	Inhibitory

[a]The existence of a new hormone, gastric inhibitory polypeptide (GIP), has been suggested. Whether this hormone is identical with enterogastrone is not yet known.

The control of gastric acid secretion is ordinarily divided into *cephalic, gastric,* and *intestinal* phases. In the cephalic phase, the stimulus for gastric acid secretion is the sight or smell of food or the stimulation of oral pharyngeal receptors by the food bolus (Fig. 41-6). The efferent arm of the reflex pathway shown in Figure 41-6 involves the vagus nerve. Postganglionic vagal fibers impinge on both the antral gastrin-secreting cells and the parietal cells in the area of the fundic glands. Vagal stimulation causes the cholinergically mediated release of gastrin, and it also has a direct stimulatory effect on the acid-secreting parietal cells. Because gastrin itself is a potent stimulator of parietal cell acid secretion, it serves as a hormonal link in a neural chain (Fig. 41-7). The direct vagal stimulation of the parietal cells will cause acid secretion in the absence of gastrin; evidence has also suggested that some tonic vagal stimulation of the parietal cells is essential for gastrin-induced acid secretion. The cephalic phase is primarily stimulatory to gastric acid secretion.

Figure 41-8 shows the gastric-phase system for acid secretion in the stomach. The gastric phase has the same two components that have been discussed for the cephalic

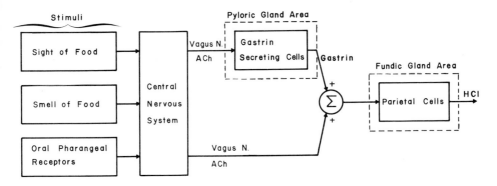

Figure 41-6
Stimuli of gastric acid secretions — cephalic phase.

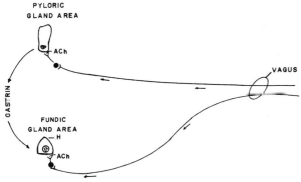

Figure 41-7
Neural and hormonal control of gastric secretion.

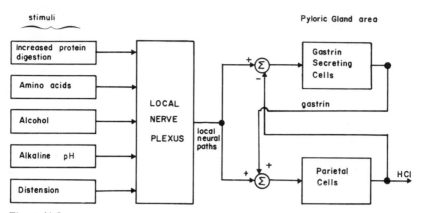

Figure 41-8
Stimuli of gastric acid secretions – gastric phase.

phase: namely, the direct cholinergic stimulation of the parietal cells and the choliner-gically mediated release of gastrin from the pyloric antrum. In addition, the inhibition of gastric acid secretion is achieved in this phase by factors that inhibit the release of gastrin. Incomplete products of protein digestion, amino acids, alcohol, alkaline pH, and gastric distention stimulate the secretion of gastrin. Recent evidence, however, has questioned the adequacy of alcohol as a stimulus. These stimuli evidently act through the local neural plexus, because the application of acetylcholine-blocking agents to the gastric mucosa stops gastrin release. A centrally mediated vasovagal reflex arc, as well as a local reflex arc, is apparently involved, since gastric distention has been shown to produce afferent impulses in the vagus nerve with a resultant, vagally induced secretion of acid. The chemical and mechanical stimulation of gastrin release, however, remains active after section of the vagus nerve, although the response is diminished. This sug-gests that the local plexus can maintain this reflex without central control. The presence of acid in the pyloric antrum causes a negative feedback inhibition of gastrin secretion; this inhibition of gastrin release is not interfered with by cholinergic-blocking agents and it also blocks gastrin release by acetylcholine. Therefore, acid inhibition of gastrin release must occur by means of some nonneural mechanism that prevents acetylcholine from causing the release of gastrin.

In summary, the major factors affecting the release of gastrin by the pyloric glands are:

1. Vagal stimulation (mediated by acetylcholine)
2. Distention and chemical stimuli acting through both local and central nervous system pathways on the antral pyloric glands
3. Inhibition of release of gastrin by the acid that bathes the pyloric antrum of the stomach

The intestinal phase of the control of gastric acid secretion has both stimulatory and inhibitory actions [10]. Figure 41-9 outlines the major factors that influence

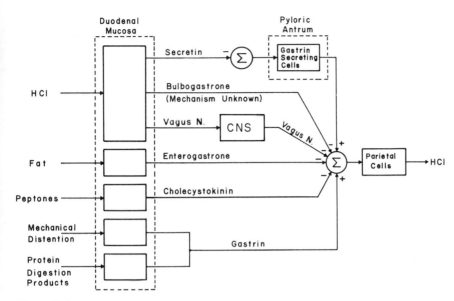

Figure 41-9
Intestinal phase of the control of gastric acid secretion.

acid secretion during this phase. The blank boxes of the figure represent where the
input parameters (HCl, fat, and so on) are converted to the corresponding output
parameters (secretin, bulbogastrone, and so on). These locations do not correspond
to any exact, known, anatomic areas. An intestinal source of gastrin is apparently
provided by the duodenal mucosa in response to protein digestion products and
mechanical distention. This mechanism, however, does not appear to be of major
importance in the normal regulation of gastric acid secretion. More important are
the inhibitory aspects of intestinal control. The presence of acid in the duodenum
provokes both hormonal and neural inhibition of gastric acid secretion [9]. Secretin
and bulbogastrone are humoral agents that are elaborated by the duodenal mucosa
in response to an acid stimulus. Secretin appears to suppress the release of gastrin
by the antral mucosa; the site of action of bulbogastrone is unknown. Bulbogastrone
appears to be the more potent of the two hormones. Acid in the duodenum also
inhibits gastric acid secretion by means of a neural reflex arc that apparently requires
the intact vagal system, since vagotomy abolishes this inhibition.

The presence of absorbable fat or fatty acids in the small intestine has been shown
to inhibit gastric acid secretion by a humoral mechanism. The hormone involved has
been called *enterogastrone,* although attempts at the chemical purification of this
substance have been unsuccessful. Recently, a polypeptide, named *gastric inhibitory
polypeptide* (GIP), of molecular weight 5105 has been isolated. GIP has a spectrum
of action when injected that is identical to the instillation of fats in the duodenum.
However, the identification of GIP as enterogastrone awaits further investigation.
Cholecystokinin (CCK) or pancreozymin, which is secreted by the duodenal mucosa

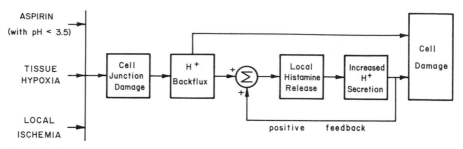

Figure 41-10
Mechanism of gastric mucosal cell damage by aspirin, tissue hypoxia, and local ischemia.

primarily in response to peptones, has a weakly inhibitory action on gastrin-stimulated acid production. This action appears to have a molecular structural basis, since the carboxyl terminal pentapeptide of gastrin is identical to that of CCK. Thus, CCK probably competes with gastrin for the receptor site on the parietal cell [1].

Pepsinogen secretion is stimulated primarily by vagal activity. Gastrin seems to produce a secretion that has low pepsinogenic activity. Histamine stimulates pepsinogen secretion, but the physiologic significance of this mechanism is unclear. No physiologic mechanism for the inhibition of pepsinogen secretion has been described.

Although many details of the control of gastric acid secretion are now reasonably well known, the methods for using this information in the understanding of the precise mechanisms of disordered function, such as gastric and duodenal ulcer, are still lacking. A few possible mechanisms responsible for ulcer production have been considered [11, 14] (Fig. 41-10); obviously, much detail remains to be filled in.

REFERENCES

1. Andersson, S. Secretion of gastrointestinal hormones. *Annu. Rev. Physiol.* 35:431, 1973.
2. Bieck, P. R., et al. Cylic AMP in the regulation of gastric secretion in dogs and animals. *Am. J. Physiol.* 224:158, 1973.
3. Bortoff, A. Digestion: Motility. *Annu. Rev. Physiol.* 34:261, 1972.
4. Code, C. C. New antagonists excite an old histamine prospector. *N. Engl. J. Med.* 290:738, 1974.
5. Cohen, S., and Harris, L. D. The lower esophageal sphincter. *Gastroenterology* 63:1066, 1972.
6. DiMarino, A. J., Jr., and Cohen, S. Characteristics of lower esophageal sphincter function in symptomatic diffuse esophageal spasm. *Gastroenterology* 66:1, 1974.
7. Glass, J. *Introduction to Gastrointestinal Physiology.* Englewood Cliffs, N.J.: Prentice-Hall, 1966.
8. Gorpel, R. K., Bauer, J. L., and Spiro, N. M. The nature and location of lower esophageal ring. *N. Engl. J. Med.* 284:1175, 1971.
9. Johnson, L. R., and Grossmann, M. I. Intestinal hormones as inhibitors of gastric secretion. *Gastroenterology* 60:120, 1971.

10. Katz, J. Gastrointestinal hormones. *Med. Clin. North Am.* 57:893, 1973.
11. MacDonald, W. C. Correlation of mucosal histology and aspirin intake in chronic gastric ulcer. *Gastroenterology* 65:381, 1973.
12. Pope, C. E. Esophagus: Physiology. In M. H. Sleisenger and J. S. Fordtran (Eds), *Gastrointestinal Disease.* Philadelphia: Saunders, 1973.
13. Rehm, W. S. Some aspects of the problem of gastric hydrochloric acid secretion. *Arch. Intern. Med.* 129:270, 1972.
14. Tran, T. A., and Gregg, R. V. Transmittal of restraint-induced gastric ulcers by parabiosis in rats. *Gastroenterology* 66:63, 1974.

Gastrointestinal System
Secretory Organs

42

Daniel Weiner

INTRODUCTION

The physical relationships of the liver, gallbladder, pancreas, and small intestine are shown in Figure 42-1. The secretion of the liver, which is called bile, passes down the bile duct to the gallbladder. If the sphincter at the end of the common duct is open, bile may pass directly into the duodenum. If the sphincter is closed, bile enters the gallbladder where reabsorption of water and some ionic exchange may occur. The pancreatic secretions enter the duodenum via the pancreatic duct.

The liver and pancreas produce exocrine secretions that are important in the digestion and absorption of proteins and fats. These secretions are also important in buffering the acidic chyme as it enters the duodenum from the stomach. The many metabolic functions of the liver are discussed at length in biochemistry textbooks and will not be discussed here.

BILIARY SECRETION

The hepatocytes form at least 500 to 1000 ml of bile daily in man. From the hepatocytes, bile passes through the intercellular canaliculi, into ductules, and then into the bile ducts. The liver produces bile at a relatively low pressure of 4 to 5 cm H_2O. A back pressure of 30 cm H_2O will stop its secretion, and, since the sphincter of Oddi can withstand pressures of 50 to 100 cm H_2O, it is probable that the formation of bile may be stopped by this mechanism under certain circumstances.

Bile is isosmotic with plasma (300 mOsm/liter), and its osmolarity is in part due to its containing approximately 170 mEq/liter of bicarbonate. As a result of this excess bicarbonate, hepatic bile is alkaline in reaction; its pH ranges from 7.5 to 9.5. In addition, bile contains an abundance of organic anions such as bilirubin and bile salts [1].

The cycling of bile (Fig. 42-2) is called *enterohepatic circulation.* This apparently functions to conserve the pool of bile acids and to provide a feedback mechanism for the control of secretion of liver bile [4]. The gallbladder fills from the cystic duct and is usually under a pressure of 20 to 25 cm H_2O. The major functions of the gallbladder are to concentrate and store bile and to deliver concentrated bile to the duodenum during meals. The hepatic bile is concentrated five to ten times in the gallbladder, with up to 90% of the water being removed as an isotonic solution. Since the gallbladder's capacity may be as much as 100 ml, the gallbladder may contain, after concentration, almost the entire 24-hour production of bile.

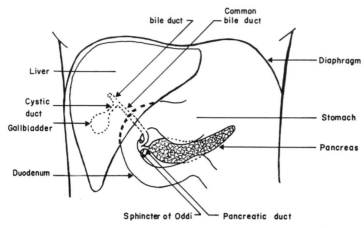

Figure 42-1
Physical relationship of the liver, pancreas, and gallbladder.

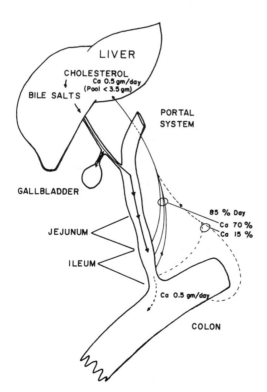

Figure 42-2
Enterohepatic circulation of bile salts.

Under usual circumstances, the contraction of the gallbladder and the relaxation of the sphincter of Oddi occur simultaneously. The presence of absorbable fat in the duodenum stimulates the release of the hormone cholecystokinin (CCK) or pancreozymin, the opening of the sphincter of Oddi, and the contraction of the gallbladder. The emptying of the gallbladder is then accomplished in a series of contractions. There also appears to be a parallel neural reflex mechanism that involves the vagus nerve and the intrinsic nervous system.

The secretion of bile by the liver is stimulated by a neural pathway with the efferent arm in the vagus nerve, as well by a hormonal mechanism that involves CCK, secretin, and gastrin. The most potent stimulus to the secretion of liver bile, however, is the presence of acids in the portal venous blood following the reabsorption of the previously secreted bile acids in the ileum. The reabsorption of bile acids is very efficient, with 90% to 95% of the acid being reabsorbed in a single passage through the intestine. Once bile salts are absorbed from the intestine, they are transported to the liver in the portal blood, where they are almost completely extracted in a single passage through the liver [3].

In addition to the bicarbonate ion previously mentioned, bile contains salts, mucin, albumin and globulin, cholesterol, lecithin, and some neutral fats and fatty acids. The salts play an important role in the absorption of fats; this will be discussed in detail in Chapter 43. Bile also has a role in the neutralization of the acidic gastric contents as they enter the duodenum.

PANCREATIC SECRETION

Pancreatic juice is an alkaline exocrine secretion, ordinarily of pH 8, that is composed of water, electrolytes, and enzymes. The rate of secretion may be very high; in man, up to 4.7 ml/min may be secreted. The volume of pancreatic juice produced per day is approximately 1 to 2 liters.

The pancreas is made up of large lobules; each lobule is made up of smaller lobules, which in turn are composed of small acini. The acini consist of groups of secretory cells that are probably responsible for the enzymatic secretions of the pancreas. Each acinus is drained by a small ductule, which drains successively into larger ducts and finally into the main pancreatic duct. Histochemical and electron microscopic studies suggest that the epithelial cells of the intercalated ducts and the centroacinar cells are the sites of origin of electrolytic and fluid secretions.

The pancreatic enzymes are secreted for the most part as proenzymes, which are subsequently converted into the active enzyme. Table 42-1 lists the major components and functions of the enzymes secreted by the pancreas. It is interesting to note that many of these enzymes have an endocrine representation; lipase, amylase, and trypsinogen, for example, are ordinarily found in the blood. Although the endocrine function of these enzymes is unknown, the blood levels of these and other pancreatic enzymes may rise, during abnormalities of pancreatic function, to the point that extrapancreatic tissues may be subject to enzymatic attack. The variety of peptide-bond cleaving enzymes is due to the relative specificity of these enzymes. Trypsin, for example,

Table 42-1. Pancreatic Digestive Enzymes

Enzyme	Proenzyme	Activator	Types of Bonds Attacked
Trypsin	Trypsinogen	Intestinal kinase; alkaline pH; Ca^{+2}; trypsin (autocatalytic)	Peptide (endopeptidase)
Parachymotrypsin Deltachymotrypsin	Chymotrypsinogen	Trypsin	Peptide (splits the carboxyl linkages of tyrosine, tryptophan, phenylalanine, and methionine; also attacks peptide bond of tyrosyl and glycine chains)
Carboxypolypeptidase	Procarboxypolypeptidase	Trypsin	Peptide (exopeptidase)
Leucine aminopolypeptidase	—	—	Peptide (splits the bond between leucine and glutamic acid)
Lipase	—	—	Ester linkages of triglycerides
Amylase	—	—	Splits 1,4-glycosidic linkages in starch, glycogen, and dextrins; also splits 1,6-glycosidic linkages
Ribonuclease	—	—	Depolymerization of ribonucleic acids
Deoxyribonuclease	—	—	Depolymerization of deoxyribonucleic acids
Elastase	—	—	Splits elastic fibers
Alkaline phosphatase	—	—	Phosphoric ester linkages
Lecithin esterase	—	—	Splits phosphoric acid from lecithin
Cholesterol esterase	—	—	Splits cholesterol esters into free cholesterol and acidic anions

cleaves proteins into smaller peptides, and other enzymes are involved in the ultimate degradation to free amino acids.

Pancreatic juice is isosmotic with plasma, and its Na^+ and K^+ concentrations are approximately equal to those of plasma. The anionic composition of pancreatic juice varies with the rate of secretion; the principal ions are HCO_3^- and Cl^-. The bicarbonate concentration, which varies directly with flow, ranges from 25 to 170 mEq/liter, whereas the Cl^- concentration varies indirectly with flow. The sum of the HCO_3^- and Cl^- concentrations tends to be constant. This observation suggests that pancreatic secretions, as formed, have a high concentration of HCO_3^-, and that HCO_3^- is exchanged for Cl^- as the secretion is exposed to the epithelial cells of the pancreatic ducts. Thus, at low secretory rates, a relatively long period of time will be available for exchange, and the fluid will have high Cl^- and low HCO_3^- concentrations. At high secretory rates, little time is available for HCO_3^- and Cl^- exchange, and the secreted fluid will have a high HCO_3^- concentration. This hypothesis also explains the constancy of the sum of the HCO_3^- and Cl^- concentrations, which is independent of the secretory rate, since only an exchange process is involved [5].

REFERENCES

1. Admirand, W., and Way, L. W. Bile Formation and Biliary Tract Function. In M. H. Sleisenger and J. S. Fordtran (Eds.), *Gastrointestinal Disease.* Philadelphia: Saunders, 1973.
2. Dowling, R. H. The enterohepatic circulation. *Gastroenterology* 62:122, 1972.
3. Erlinger, S., and D'Humeaux, I. Mechanism and control of secretion of bile water and electrolytes. *Gastroenterology* 66:28, 1974.
4. Hardison, W. G. M. Bile salts and the liver. *Prog. Liver Dis.* 4:83, 1972.
5. Howat, H. T. (Ed.). *The Exocrine Pancreas. Clinics in Gastroenterology,* Vol. 1. Philadelphia: Saunders, 1972.

The Small Intestine and Colon   43

Daniel Weiner

INTRODUCTION

The small intestine, which consists of the duodenum, jejunum, and ileum, is that segment of the alimentary canal where approximately 90% of the absorption of nutrient materials takes place. Some digestive processes are initiated in the stomach, but the bulk of digestion takes place in the small intestine. This primary digestive role of the small intestine is demonstrated by the fact that the complete removal of the stomach (gastrectomy) is consistent with the satisfactory maintenance of life.

The duodenum is not only an endocrine secretory gland, but it is also, to a lesser degree, an exocrine organ. It is instrumental in mixing the gastric chyme with pancreatic and biliary secretions. The jejunum is the main site of homogenization and mixing of the intestinal chyme. Most of the nutrient absorption — including the absorption of sugars, amino acids, and fats — occurs in the jejunum. The ileum is responsible for the absorption of bile salts, vitamin B_{12}, a small amount of water, and the excess nutrients that have not been absorbed in the jejunum.

The primary function of the large intestine or colon is to serve as a storage organ for the nonabsorbed food and the biliary excretions that enter from the small intestine. Upon leaving the small intestine, the waste products pass through the ileocecal valve and enter the large intestine. Here, the absorption of water and water-soluble substances, combined with the action of intestinal bacteria, results in the formation of the fecal mass.

MOTILITY OF THE SMALL INTESTINE

The motor activity of the small intestine serves to mix the chyme for facilitating the action of digestive enzymes, to expose the chyme to the mucosal surface, and to transfer the chyme from one area to another. Type I contractions, which are similar to those observed in the stomach, are segmental, ring-type constrictions that contract at fixed locations and do not propagate up or down the intestine. The pressure developed by these contractions is less than 10 cm H_2O, and they serve a mixing function. A second type of contraction, type II, that probably serves a similar function is the pendular contraction or rhythmic segmentation. Rhythmic segmentation occurring in adjacent segments of small intestine creates the impression of the segments having a to-and-fro motion. There is a gradient of frequencies of the type I and pendular contractions down the intestine; the highest frequency is seen in the duodenum, and the lowest in the ileum.

Peristaltic contractions are responsible for the propulsion of the chyme through the small intestine. As noted in the case of the esophagus and stomach, peristaltic contractions almost always travel in an aboral direction, although at the point of initiation of a peristaltic contraction, some chyme undoubtedly moves in the oral direction as well. There appears to be a pacemaker area in the duodenum that initiates some peristaltic activity. Most peristaltic contractions move at a rate of 1 to 2 cm/sec, and they die out in a few to 10 or 20 cm. Some peristaltic rushes are seen, however, that travel in a smooth, coordinated fashion from the duodenum to the ileum. Type III and IV contractions in the small intestine have been described by Code [3]. The type III waves are probably identical to peristaltic contractions. Since a more detailed discussion of the characteristics of the type III and IV contractions does not aid in understanding motility in the small intestine, they will not be further considered here.

Contraction of the lamina muscularis mucosae occurs at a rate of three to four times per minute upon mechanical stimulation by chyme in the lumen. Such a contraction may last 30 to 60 seconds, and it causes the formation of folds and ridges in the intestinal mucosa. Movements of the intestinal villi have also been described, and these might be initiated by an intestinal hormone, villikinin. These movements also aid in mixing the chyme and in exposing the intestinal mucosal surface.

Motility is primarily under the control of the extrinsic and intrinsic nervous systems. Mechanical stimulation of the small intestine causes a contraction of the segment 2 to 3 cm above the point of stimulation and a variable relaxation a few centimeters below it. This myenteric reflex operates through the smooth muscle and the plexus of the intrinsic nervous system. Distention of an intestinal segment initiates the peristaltic reflex, which also operates through the intrinsic nervous system and produces contractions that progress down the small intestine. Several other major reflexes are listed in Table 43-1.

It should be noted that these reflexes regulate motility in order to smoothly coordinate motor activity in the entire gastrointestinal system. Increases in intraluminal pressure, for example, inhibit gastric motility, which in turn serves to slow the entry of chyme into the duodenum and thus acts to decrease the distention of the small intestine. Similarly, the entry of food into the stomach initiates peristalsis in the ileum

Table 43-1. Major Gastrointestinal Reflexes

Reflex	Stimulus	Effect on Motor Activity
Gastroileal	Entry of food into stomach	Stimulates peristalsis in ileum; causes relaxation of ileocecal valve
Enterogastric inhibitory	Increase in intraluminal pressure in duodenum or jejunum Hypertonic solutions in small intestine	Reduction in gastric motility
Ileogastric	Distention of ileum	Reduction in gastric motility
Intestinointestinal inhibitory	Distention of intestinal segment	Inhibition of intestinal motility

and relaxes the ileocecal valve; these effects allow the propulsion of the contents of the small intestine into the colon in preparation for the movement of the gastric contents into the small intestine.

SECRETIONS OF THE SMALL INTESTINE

The small intestine secretes a volume of approximately 2 liters per day of fluid, termed *succus entericus,* that comes from two primary sources. The first source is the epithelial cells of the intestine. These cells undergo mitosis at the base of the intestinal villus and then migrate up to the tip of the villus, where they are eventually extruded into the lumen. This process takes about two days in man. Since this secretion actually consists of cellular debris, all the intracellular enzymes will be present.

A second secretion of the small intestine is that of Brunner's glands. These glands, which are located primarily in the duodenum, are composed of branched tubules into which acini open. It is difficult to collect the secretion of these glands because it is usually contaminated with mucosal cell debris. The composition of the secretions of Brunner's glands has not been well characterized, but it is known that the pH is in the range of 8.2 to 9.3 and the major constituent is a mucoprotein. Pepsinogen and a mucinase are probably the only enzymes that originate in Brunner's glands. It appears likely that most measurements of the volume of secretion of these glands are overestimates to the extent that net water flux into the small-intestinal lumen is induced by the hypertonic contents in the lumen. This factor may introduce an error into composition determinations as well.

DIGESTION AND ABSORPTION IN THE SMALL INTESTINE

Many of the special transport mechanisms that are utilized by the absorptive processes of the GI tract are found in other tissues. However, several important features, such as surface area and anatomic polarity, are involved that uniquely adapt the small intestine, the major site for most absorption, to its role.

SURFACE AREA

The mucosal surface of the small intestine is thrown up into folds, known as the *folds of Kerckring* or *valvulae conniventes.* Figure 43-1 shows the increase in surface area over a simple cylinder that results from these folds. In addition, the surface of the folds is thrown up into villi, which further increases the surface area. The most important factor in this increase in surface area is the presence of microvilli on the surface of the epithelial cells. The microvilli are about 0.1 mm wide and 1.0 mm long, and there are about 600 to 1000 microvilli per cell.

The surface area as a function of unit length is largest in the duodenum; it decreases gradually down the length of the jejunem and reaches a minimum in the ileum. This

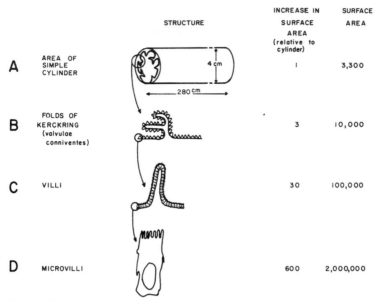

		STRUCTURE	INCREASE IN SURFACE AREA (relative to cylinder)	SURFACE AREA
A	AREA OF SIMPLE CYLINDER	4 cm 280 cm	1	3,300
B	FOLDS OF KERCKRING (valvulae conniventes)		3	10,000
C	VILLI		30	100,000
D	MICROVILLI		600	2,000,000

Figure 43-1
Three mechanisms for increasing the surface area of the small intestine. (From J. H. Wilson. *Intestinal Absorption.* Philadelphia: Saunders, 1962. Reproduced with permission.)

decrease in area seems to parallel the volume of material absorbed; the duodenum absorbs the largest volume per unit length and the ileum the least.

ANATOMIC POLARITY

The mucosal columnar epithelium of the small intestine has several features that confer the property of polarity of anatomic orientation on these cells. First, the microvilli in the mucosal surface increase the surface area 600-fold over the equivalent basal area of the cell, making a much larger surface area available for nutrient absorption. In addition, many special enzyme systems have been localized solely to the microvillus border of the mucosal cells; some of these have been implicated in special absorptive functions [7]. Finally, it has recently been postulated that the details of the attachment between mucosal cells may play a vital role in absorption. The role of the ultrastructure of the intercellular attachment in absorption will be subsequently considered.

MEMBRANE-TRANSPORT PROCESSES

In the lumen of the alimentary tract, the transfer of material occurs across a membrane layer of mucosal epithelial cells. These cells possess important specialized permeability and transport processes that result in the transcellular transport of many substances. Most of the subdivisions of the alimentary tract exhibit an electrical potential difference between the lumen and the serosal-facing (or inward-facing) surface of the mucosal cell layer that influences ionic transport. This potential appears to be a result of the active transport of sodium ion [12].

In the analysis of intestinal transport processes, it is important to determine whether a given transport process or a steady-state concentration distribution can be explained in terms of observable physical forces or whether some more complicated transport process, such as active transport, must be invoked. The principal transport mechanisms associated with the normal function of the GI system are (1) diffusion, (2) carrier-mediated transport, and (3) pinocytosis.

The Davson-Danielli model (Fig. 43-2) has been used to describe the mucosal epithelial membrane. This model is composed of phospholipids in which the hydrophilic or charged groups are exposed to the aqueous environment of the cell and the hydrophobic or water-insoluble groups are oriented toward the interior of the bilayer. Within the protein-lipid sandwich, the hydrocarbon tails of the lipids are in proximity to each other and are probably held in position by Van der Waal's forces. The protein coat of this membrane may act to stabilize the structure and act as the framework for carrier molecules that can traverse the membrane structure. The cell membrane is thought to contain water-filled pores, which are estimated to be 0.3 to 0.8 nm in radius [6]. The presumptive evidence for the existence of these aqueous pores in the mucosal epithelial cells of the small intestine is very compelling, and their existence is almost universally accepted at the present time.

Diffusion Across Membranes

The process of diffusion has been treated on a general basis in the Basic Sciences Unit and for the exchange of water and solute in Chapter 24. An analysis that emphasizes cross-coupling effects, solvent drag, and diffusion trapping (which are special diffusion

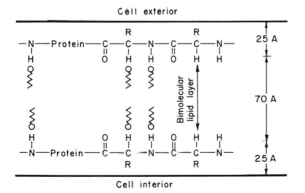

Figure 43-2
Davson-Danielli model of the cell membrane showing the lipid bilayer structure. (From F. M. Snell et al. *Biophysical Principles of Structure and Function.* Palo Alto, Calif.: Addison-Wesley, 1965. P. 191. Reproduced with permission.)

phenomena that apply to the transport mechanisms in the intestinal tract) is presented here.

In a system that is not too far from electrochemical equilibrium, it can be shown that there is usually a linear relation between fluxes and the corresponding forces. This relationship can be expressed as follows:

$$J_i = \sum_{k=1}^{m} (L_{i,k} X_k)$$

(43-1)

where $i = 1, 2, \ldots, n$
$k = 1, 2, \ldots, m$

The cross-coupling coefficient, $L_{i,k}$, expresses the effect of the "force" X_k on the flux J_i. This "driving force," for example, may be a pressure or a concentration gradient. The flux J_i of component i is defined as the number of moles of component i crossing a unit area of membrane per unit time. The set of n equations defined by Equation 41-1 expresses all the possible interactions between flows and forces in a system. If $L_{i,k} = 0$ for all k except $k = i$, the linear relation between the flow J_i and its conjugate force X_i is given by:

$$J_i = L_{i,i} X_i$$

(43-2)

For the case of the flow of a noncharged solute across a uniform membrane that separates two solutions of different compositions, the flux J_i will be given by a modification of Fick's first law (Eq. 2-2):

$$J_i = - \left(\frac{D_i}{\Delta h} \right) \Delta c_i$$

(43-3)

where $L_{i,i} = -D_i/\Delta h$
$X_i \ \ = -\Delta c_i$
D_i = diffusion coefficient
Δh = membrane thickness
Δc_i = concentration difference of solute across the membrane

In biologic systems, the profile of the concentration gradient, dc/dh, as a function of the distance through the membrane cannot be determined. For this reason, this gradient is usually assumed to be linear, and it is taken between points at the two surfaces of the membrane. Note that $-D_i/\Delta h$ is analogous to $L_{i,i}$, or membrane conductance, and Δc_i is analogous to X_i, the "force" in Equation 43-2. The flux, J_i, may then be considered to be the product of the membrane conductance and the force. In the case of water-soluble substances diffusing in a membrane with aqueous pores, D_i is the diffusion coefficient within the channel, and the effective area of the membrane is the pore area of the water-filled channels. For lipid-soluble substances, the entire membrane area is available for diffusion, and D_i is the diffusion coefficient of the diffusing substance in the membrane.

A linear relation between flows and their corresponding forces was assumed to exist in order to develop Equation 43-1. The special case for the cross-coupling coefficient $L_{i,k} = 0$ (except when $i = k$) has been considered. It has become clear from much experimental data, however, that this simplification may well be incorrect when applied to the situation that is illustrated in Figure 43-3.

If the solute flux in Figure 43-3 is from side 1 to side 2, $(J_x)_{1,2}$, and the solvent flux is from side 2 to side 1, $(J_y)_{2,1}$, and if the relative molecular dimensions of the system are approximately proportional to the sizes depicted in the diagram, then interaction or collision of components x and y will occur, and this will certainly produce a perturbation of the flux. In the case depicted in Figure 43-3, the presence of the flux $(J_x)_{1,2}$ would modify the flow of $(J_y)_{2,1}$ from what would be expected if the $(J_x)_{1,2}$ flux were not present. When the permeant molecule is about the same size as the channel in the

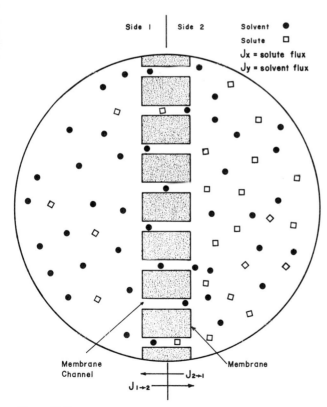

Figure 43-3
Membrane model illustrating the interaction of solute and solvent in membrane channels.

membrane, so-called *single-file diffusion* occurs. That these examples are not unreason-able for biologic membranes is supported by the data of Table 43-2, where the estimated pore or channel radius for some membranes and the calculated radius for a number of molecules are shown. The treatment of systems in which the molecular sizes of the

Table 43-2. Equivalent Radiuses for Aqueous Membrane Channels and Some Common Solutes

Cell or Tissue	Equivalent Pore Radius[a] (nanometers)
Human erythrocyte	0.35–0.41
Squid axon (axolemma)	0.85
Toad bladder	0.85
Human small intestine	
Jejunum	0.67–0.88
Ileum	0.30–0.38

Substance	Estimated Molecular Radius (nanometers)
H_2O[b]	0.18
Urea	0.27
Glucose	0.44
Sucrose	0.53
Inulin	1.2

[a] Calculated assuming channels are right circular cylinders.
[b] Assumed to be in dimer form.

solute and solvent molecules are of the same order of magnitude as the radius of the channels in the membrane is given in detail in Katchalsky and Curran [8]; only the general result will be considered here. Equation 43-1 can be written in the form:

$$J_i = L_{i,i} X_i + \sum_{j=i}^{m} (L_{i,j} X_j)$$ (43-4)

where $j = 1, 2, \ldots, m$, except $j = i$

This emphasizes the point that the flow of substance i, as well as depending on the conjugate force X_i, will also depend on any other forces in the system, X_j, if the cross-coupling coefficient $L_{i,j} \neq 0$.

Several surprising results follow from Equation 43-4. For the case of an uncharged substance diffusing through a membrane across which there is a potential difference, the flow of the uncharged molecule will be influenced by the presence of the poten-tial field. Since a potential difference exists across the membranes of many cells, this result is of great importance for biologic applications, and experimental verification that $L_{i,j} \neq 0$ has been established. A second result that has been verified is that the flow of one substance can be made to move in a direction opposite to its concentration gradient by the flow of a second substance in the direction of its concentration gradient; this phenomenon is termed *counterflow*.

A complete discussion of the conditions for flux coupling [1a, 8, 9a] is beyond the scope of this presentation. However, the case for a membrane separating a solution that contains a single, uncharged solute at two different concentrations will be considered (Fig. 43-4). The equations that describe the solute and solvent fluxes and the forces in this example (see Eq. 43-4) are:

$$J_V = L_P \Delta P + L_{PD} \left[RT(c_i'' - c_i') \right] \tag{43-5}$$

$$J_D = L_D \left[RT(c_i'' - c_i') \right] + L_{DP} \Delta P \tag{43-6}$$

where J_V = total volume flux
          J_D = difference between the flux of the solute and that of the solvent
          L_D = diffusion conductivity
          L_P = hydraulic conductivity of the membrane
          ΔP = difference in hydrostatic pressure across the membrane
$RT(c_i'' - c_i')$ = calculated osmotic pressure from van't Hoff's law (Eq. 2-21)

The cross-coupling coefficients are L_{DP} and L_{PD}, and, from the Onsager reciprocal relationship,

$$L_{DP} = L_{PD} \tag{43-7}$$

Note that in this system, there are two "forces" acting – diffusion and osmosis – and that the terms in Equations 43-5 and 43-6 are the membrane conductivity times its

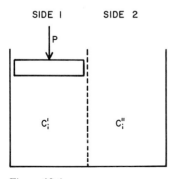

SIDE 1   SIDE 2

P

c_i' c_i''

Figure 43-4
Model for the development of the concept of the reflection coefficient. The solute is present at concentration c_i' on side 1 and c_i'' on side 2 of the membrane. Force is applied to the piston on side 1 to develop a pressure that will prevent solvent flow from side 2 to side 1 when c_i' is greater than c_i''.

conjugate force and the cross-coupling coefficient times each force. A very useful term has been introduced by Staverman, which he defined as the *reflection coefficient*:

$$\sigma = -\frac{L_{PD}}{L_P} \tag{43-8}$$

This term reflects the ratio of the cross-coupling coefficient to the hydraulic conductivity of the membrane, and it is a function of the solvent, solute, and the characteristic of the membrane. Its values may range from 1.0 for a membrane that is totally impermeable to the solute to zero for a membrane that has the same permeability to the solute as to the solvent; it may also assume negative values under certain special circumstances. By substituting σ, Equation 43-5 may be written as:

$$J_V = L_P (\Delta P - \sigma R T \Delta c_i) \tag{43-9}$$

If volume flow is prevented, as in the example of Figure 43-4, by the application of pressure on side 1 of the membrane, then Equation 43-9 reduces to*:

$$J_V = 0 = \Delta P - \sigma R T \Delta c_i \tag{43-10}$$

and

$$\sigma = \frac{\Delta P}{R T \Delta c_i} \tag{43-11}$$

The reflection coefficient can thus be seen to be approximately the ratio of the pressure necessary to oppose osmotic flow to the calculated osmotic pressure for an ideal semipermeable membrane (i.e., one in which there is no solute flow) for the type of system shown in Figure 43-4. This system also provides a convenient method for the determination of the value of σ. Note that as the solute penetrates the membrane, the applied pressure necessary to oppose the osmotic flow will gradually decrease. This effect can be corrected by making serial measurements of pressure with time and extrapolating to the pressure at zero time.

   Solvent drag is another transport process of biologic importance in the intestinal tract. In the presence of volume flows across a membrane, the interaction between the flow of the solvent and the flow of the solute may result in a transport process. This process is of special importance in intestinal absorption, where large, unidirectional fluxes of H_2O are regularly found. For a system containing a single, permeable solute at concentration c_i on both sides of a membrane, the phenomenon of solvent drag can be described quantitatively by the following equation:

$$J_i = c_i (1 - \sigma) J_V \tag{43-12}$$

---

*This simplification ignores the effect of solute flow on the volume of side 1. For most cases of biologic interest, this does not introduce a significant error.

where J_i = net solute flow
$\quad J_V$ = net volume flow
$\quad \sigma$ = the reflection coefficient as previously defined

Another case of a special diffusion process that is important in the small intestine involves the formation of a neutral molecule by the combination of ion pairs for the transfer of ions across a membrane with a low dielectric constant. An example of this phenomenon, known as *nonionic diffusion,* involves weak acids or weak bases and occurs in the GI tract. Consider the case of a membrane impermeable to ions that separates two solutions having different concentrations of a weak acid, HA, which dissociates as follows:

$$H^+ + A^- \rightleftharpoons HA \tag{43-13}$$

Since the membrane is permeable only to HA but not to H^+ or A^-, the flux J_A- will be equal to the flux of nonionized HA. Thus, J_A- will be dependent upon the pH as well as upon the concentration gradient of A^-. If the pH conditions favor the formation of the nonionized complex on one side and dissociation on the other side of the membrane, a net flow from the side favoring the nonionized complex to the side favoring the dissociated complex can be induced in the absence of any concentration gradient for A^-. Transport of this type has also been termed *diffusion trapping.*

Carrier-Mediated Transport

Studies of the permeability characteristics of mammalian cell membranes indicate that most membranes are relatively impermeable to water-soluble molecules having more than four carbon atoms or an effective radius larger than 0.4 to 0.5 nm. The presence of a charged moiety on a molecule may further decrease its permeability, depending on the fixed-charge structure of the membrane. Many water-soluble nonelectrolytes, such as sugars and amino acids, penetrate the mucosal epithelium of the small intestine so slowly by diffusion alone that other transport processes are needed to meet metabolic requirements. These processes are unique to biologic systems, and they are, in general, highly specific with respect to the substance they will transport. They also exhibit saturation and competitive-inhibition kinetics that closely resemble the kinetics of enzyme-substrate reactions. Further, these transport processes may be specifically inhibited by agents that do not cause a general inhibition of metabolic processes. These processes operate on an existing electrochemical gradient of the permeative substance and lead to the disappearance of this gradient.

The carrier system depicted in Figure 43-5 is an example of a facilitated transport system. *Facilitated transport* is defined as a process that results in penetration of a substance through a membrane at a more rapid rate than that expected as the result of free diffusion, but which is not directly coupled to metabolic reactions [8a]. In such a process, the substance S combines at one surface with a membrane component X to form a complex, XS. Both the membrane complex XS and the component X are

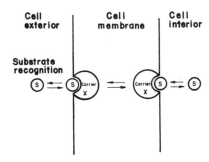

Cell exterior | **Cell membrane** | **Cell interior**

Substrate recognition

Figure 43-5
Model of the proposed mobile membrane-bound carrier system that demonstrates substrate recognition at the cell surface.

assumed to diffuse freely within the membrane. The complex then diffuses to the opposite side of the membrane, where it dissociates to S and X, with S being released to the solution. The free carrier diffuses back to the original side and eventually combines with another S molecule, so that the process is cyclic. The probability of an XS complex being formed on either side is related to the concentration of S on each side of the membrane.

A transport process of the type shown in Figure 43-5 will exhibit the phenomenon of *counterflow* or *counter transport*. This is the transport of a substance against its prevailing concentration gradient. To demonstrate counterflow experimentally, the concentration of the driven substrate, S_1, is set so that there is no net flux (i.e., its concentration gradient is zero). The driving substrate, S_2, is then added to side 1 of the membrane at a high concentration relative to its transport constant, K_{trans}. The resulting competition between the two substrates will prevent the movement of the driven substrate S_1 from side 1, but for as long as the driving substrate remains at low concentration on side 2 of the membrane, movement of the driven substrate from side 2 to side 1 will be largely unaffected. Thus, there will be a net movement of the driven substrate S_1 in the direction from side 2 to side 1, i.e., against the prevailing chemical gradient of this substrate. This phenomenon is most obvious when the concentration of the driven substrate is low relative to its K_{trans}. The presence of counterflow has been experimentally verified using monosaccharides as substrates across the red blood cell membrane and the small-intestinal mucosal surface.

A case of some importance that has been studied is one in which the free carrier does not diffuse in the membrane; for carrier movement to occur from one side of the membrane to the other, a carrier-substrate complex is required. This type of carrier transport has been designated *exchange diffusion*, and an example is the sodium-potassium exchange across the red blood cell membrane.

A membrane-bound carrier system has been proposed that is similar to the facilitated transport system previously described but with substrate affinities that are different at the two surfaces of the membrane. The affinity for the substrate, which is usually a nonelectrolyte, is thought to be changed by the presence of sodium or potassium ions

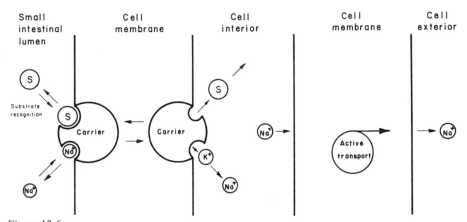

Figure 43-6
Model of the carrier system for the transport of organic molecules against a concentration gradient that employs differential substrate affinities on the two sides of the cell membrane.

(Fig. 43-6). On the outside of the cell, there is a high concentration of sodium ions, and the formation of the substrate-carrier complex is favored. On the inside of the cell, the sodium ion is displaced from the carrier molecule by a potassium ion, which decreases the affinity of the carrier for the substrate molecule and causes its release into the interior of the cell. In the case of the intestinal mucosal surface, the substrate diffuses out of the cell down its concentration gradient into the extracellular space. The sodium ion is extruded from the cell by an active transport mechanism. This system is capable of transporting the nonelectrolytic substrate against its prevailing concentration gradient by using the ionic conditions in the small intestine to change the affinity of the carrier for the substrate molecule on the two surfaces of the membrane. There are many data to suggest that a system similar to that depicted in Figure 43-6 is responsible for the absorption of glucose and other monosaccharides by the small-intestinal mucosal cells [10].

It seems likely that active transport systems involve a carrier mechanism similar to that shown in Figure 43-6, with the exception that the change in affinity for the substrate is more directly linked to metabolic processes than to ionic conditions. Active transport systems possess all the characteristics of a facilitated carrier system, i.e., the transport process demonstrates competitive inhibition, saturation kinetics, and substrate specificity. In addition, active cell metabolism is required for transport. The existence of many transport processes that fulfill the above criteria has been demonstrated; however, no theoretical model is sufficiently well established to be considered here.

Pinocytosis

One interesting example of a transport process is the phenomenon called *pinocytosis,* or literally, "cell drinking." This is a very old transport process phylogenetically, being present in paramecia. It has been identified in many mammalian systems, including the kidney and small intestine. In neonates, the transport of labeled proteins across the mucosal epithelial cell in pinocytotic vacuoles has been demonstrated. Pinocytosis must be considered as an example of an active transport process, since metabolic energy is involved in the formation of the new membrane that is pinched off to form the vesicle.

In the adult animal, pinocytosis does not appear to be an important transport mechanism from a quantitative point of view. Food allergies, however, may provide an example of the transport of intact proteins across the mucosal epithelial cell,* since it is clear that in general, no route other than pinocytosis is available for the penetration of intact protein molecules.

TRANSPORT OF WATER, ELECTROLYTES, AND NUTRIENTS

In other sections of this book, the various mechanisms that are available for the transfer of water, electrolytes, and nutrients across cell membranes in general have been discussed. The specific mechanisms of the GI tract that are now understood in some detail will be considered here for each class of substance.

WATER

In the stomach and small intestine, water is absorbed in response to osmotic gradients generated by solute transport or by the digestive splitting of complex nutrient substances. If the contents of the stomach or small intestine are hypotonic with respect to plasma, the net flow of water is from the lumen to the blood. Conversely, if the contents are hypertonic, the flow of water is directed from the blood to the lumen. The water fluxes in the small intestine can be of quite significant magnitude, and it appears that one of the functions of the stomach is to release the usually hypertonic chyme into the duodenum at a controlled rate. In patients who have undergone removal of a major portion of the stomach, the ingestion of a large hypertonic load may cause the loss of so much volume from the vascular space that transient hypotension may develop. This phenomenon is called the *dumping syndrome,* and it can sometimes be quite troublesome to the patient. Some water absorption does occur in the stomach if its contents are hypotonic, but a significant loss of blood volume in the presence of hypertonic contents is prevented by the relatively smaller mucosal surface area of the stomach.

The transport of water appears to be induced only by osmotic gradients. There is some question about the existence of an active transport mechanism for water in some lower forms of animals, but in mammalian systems, the active transport of water has not been demonstrated. A mechanism that has been proposed to exist in the small

---

*Protein molecules in the internal environment of the body are necessary to initiate the hypersensitivity-type reactions of the general class of food allergies.

intestine and gallbladder that links water transport to solute transport will be discussed in the next section.

SODIUM AND CHLORIDE ABSORPTION

It has been shown that sodium ion can passively diffuse across a mucosal epithelial border in response to a concentration gradient across the mucosal surface. In addition, active transport of sodium from the luminal-facing surface to the serosal-facing surface has been demonstrated. This active transport apparently involves a membrane-bound ATPase system that is located on the serosal-facing surface of the cell. There is much evidence suggesting that Cl^- diffuses passively in response to the electrical gradient created by the active transport of Na^+.

Curran [4] has proposed a model for the coupling of water transport to solute transport. Figure 43-7 shows the theoretical model and the anatomic correlate of this model. The mechanism proposed by Curran requires two membranes in series as illustrated. The solute is assumed to be actively transported across membrane α from compartment 1 to compartment 2. The reflection coefficients for the solute are assumed to be different at the two membranes, σ_α being greater than σ_β. The accumulation of solute in compartment 2 results in an osmotic pressure difference across membrane α equal to:

$$(\Delta \pi_{eff})_\alpha = \sigma_\alpha RT \Delta c_\alpha \tag{43-14}$$

The flux from compartment 1 to compartment 2 is given by:

$$(J_V)_\alpha = (L_P)_\alpha \sigma_\alpha RT \Delta c_\alpha \tag{43-15}$$

where $(L_P)_\alpha$ is the hydraulic conductivity of membrane α.

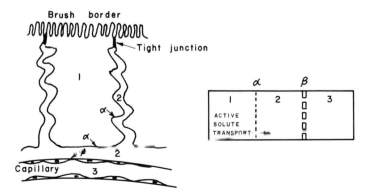

Figure 43-7
Model of the intestinal mucosal cell and the double-membrane model for solute-induced water transport.

Since σ_β is smaller than σ_α, the effective osmotic pressure difference across membrane β and the osmotically induced volume flow from compartment 3 to compartment 2 will be relatively small. If the volume of compartment 2 is constrained, the entrance of fluid from compartment 1 will lead to an increased hydrostatic pressure in compartment 2 that will cause a flow of fluid from compartment 2 to compartment 3 across the more permeable membrane β. The net result is a flow of water from compartment 1 to compartment 3 in the absence of a difference in water activities between these two compartments. It has been shown that the net volume flux in this system is given by:

$$J_V = L(\sigma_\alpha - \sigma_\beta)RT\Delta c_\alpha \qquad\qquad (43\text{-}16)$$

where L is the combined hydraulic conductivity of membranes α and β.

Many experimental observations dealing with intestinal water absorption can be explained by means of this double-membrane model. The dependence of water transport on the active transport of solute is readily understandable in terms of this model. In addition, the orientation of net transport from the lumen to the serosal cell surface can be explained. The net water flux that is induced by solute transport also serves to transport other small molecules by means of the mechanism of solvent drag.

CARBOHYDRATE ABSORPTION

Complex carbohydrates are split into disaccharides in the intestinal lumen by a number of the pancreatic enzymes discussed previously (see Chap. 40). Disaccharidases localized on the brush border of the mucosal epithelial cells split the disaccharides to monosaccharides at the surface of the cells. It is not clear whether this hydrolysis is also involved in a transport process.

The characterization of the transport mechanism for monosaccharides, such as glucose and galactose, is very complex. Glucose transport requires the presence of Na^+ in the luminal fluid, and it is inhibited by certain substances, such as phlorhizin, that do not generally inhibit metabolic processes. Specific differences between the transport rates of L-glucose and D-glucose are evident, with D-glucose having the higher transport rate. The best present evidence is that glucose absorption is accomplished by way of cotransport, which is a facilitated carrier system [1, 8a]. This carrier system utilizes the gradient for Na^+ from the lumen to the cell interior that is maintained by the active transport of Na^+ across the serosal-facing surface of the cell. Thus, glucose transport is linked indirectly to metabolic processes. It is difficult to determine whether or not to classify this process as active transport. Current usage seems to favor the restriction of this term to the transport of substances such as Na^+, which are more directly linked to metabolism. The importance of these details is inherent in the insight they provide toward an understanding of membrane physiology.

FAT ABSORPTION

In humans, approximately 50 to 100 gm of lipid is ingested in the diet each day. Most dietary lipid is in the form of triglycerides, which are long-chain fatty acids esterified to glycerol. The fatty acids are poorly soluble in water solution. The usual chain length of these fatty acids is 16 to 18 carbon atoms, although this may range from 8 to 22 carbon atoms. The triglycerides are hydrolyzed to β-monoglycerides and free fatty acids in the small intestine by the action of pancreatic lipase.

Lipid-soluble substances have the entire surface area of the mucosal epithelial cell available for their absorption. The effect of surface area on absorption can be seen by reference to Fick's equation for diffusion transport of a substance i across a membrane of surface area A as is shown by Equation 2-1, page 17, which indicates the total amount of substance i transported by the diffusion flux to the area available for diffusion. The problem, however, is that many fats and lipid substances are poorly soluble in the aqueous solution that is present in the intestinal lumen, and hence the lipids cannot easily gain access to the surface of the intestinal cell. In addition, according to the Davson-Danielli model of the cell membrane, the polar groups of the membrane lipids are oriented toward the surface of the cell, which tends to attract a layer of water along this surface.

Emulsification or solubilization of the monoglycerides and fatty acids must be accomplished by the detergent qualities of the bile salts (Fig. 43-8). Soluble aggregates, called *micelles,* are formed that have a diameter on the order of 4 to 5 nm. The solubilization of the fatty acids is important for absorption apparently because the fat is made more accessible to the mucosal epithelial cells. The fatty acids and monoglycerides diffuse passively into the epithelial cells,* where they are reesterified into triglycerides and combined with phospholipids in the endoplasmic reticulum and Golgi apparatus to form particles called *chylomicrons.*

The composition of chylomicrons is approximately 86% triglycerides, 8.5% phospholipids, 3% cholesterol and cholesterol esters, and 2% protein. The mechanism by which the chylomicrons leave the mucosal cell is still not clear. It has been proposed that the chylomicrons are discharged into the side of the epithelial cell and into the extracellular space by means of the phenomenon of reverse pinocytosis. The chylomicrons range in size from a fraction of a nanometer to about 3 or 4 nm. They enter the systemic circulation by way of the intestinal lymphatic circulation (the lacteals). The presence of these chylomicrons in the blood produces the opalescence of serum that may be observed after the subject has eaten.

The transport of lipids in the blood is an important area of study at the present time because disorders in lipid transport appear to be associated with the formation of arteriosclerotic plaques in the major arteries, with subsequent interference with the local nutritive blood supply. In the heart, this may cause myocardial infarction;

---

*The bile salts do not appear to be absorbed to any significant extent as the fatty acids and monoglycerides diffuse across the mucosal surface, although some investigators believe that minute quantities of bile salts are absorbed during fat absorption. The majority of bile-salt absorption occurs in the ileum, whereas fat absorption is usually completed in the duodenum and jejunum.

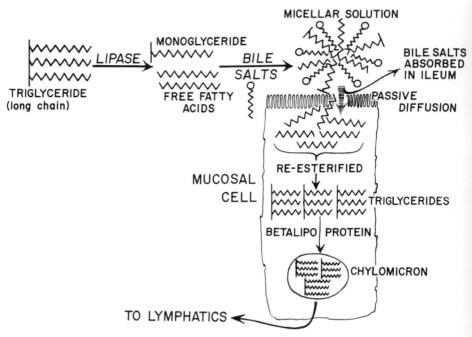

Figure 43-8
Sequence of steps in the normal absorption of ingested lipids.

in the brain, a "stroke" or the loss of an area of cerebral function may occur; and the loss of kidney function may result in renal failure or hypertension through the renin-angiotensin mechanism.

In a fortunately rare disease in which lymphatic blockage occurs (Whipple's disease), fat absorption is impaired because the chylomicron transport is via the intestinal lymphatic channels. It has been found that the administration of short-chain fatty acids (chain length of eight carbon atoms) in these patients bypasses the lymphatic block. These short-chain fatty acids are split from triglycerides by intracellular lipase and, being relatively water-soluble, gain access to the systemic circulation through the portal circulation rather than through the lymphatics.

PROTEIN ABSORPTION

The absorption of γ-globulins (immunologic proteins) has been demonstrated in the neonate, but, with this exception, proteins must be hydrolyzed to single amino acid units before absorption can take place. This hydrolysis is accomplished by a variety of gastric, pancreatic, and intestinal enzymes that have been previously discussed. It seems clear that amino acids are transported by a luminal, sodium-dependent, facilitated cotransport system that is very similar to the system described for facilitated mono-saccharide transport [1]. There are at least three separate carrier transport systems for amino acids that can be identified:

1. A system for the neutral amino acids, including histidine. This is the most efficient carrier; i.e., it has the highest transport rate and the greatest molecular specificity.
2. A second mechanism that is shared by the basic amino acids, ornithine, arginine, and lysine. Cystine may use this route as well as the mechanism for neutral amino acids.
3. A third transport mechanism involving the transport of proline, hydroxyproline, sarcosine, N-dimethylglycine, and betaine.

SUBSTANCES WITH SPECIAL TRANSPORT SYSTEMS

There are various other substances for which the absorption kinetics make it clear that a special transport system must be involved. In general, these substances are water-soluble rather than lipid-soluble, and they sometimes have a molecular size greater than 0.7 nm, which denies them access to the water-filled pore system for passive diffusion or transport by means of solvent drag. Examples of such substances are iron, calcium ion, vitamin B_{12} (extrinsic factor), folic acid, and thiamine.

IRON ABSORPTION

The control of the metabolic balance of iron is different from that of other metals. To a great extent, the total quantity of iron in the body is maintained within close limits by the regulation of the absorption of iron. Excretory mechanisms play a small role in iron loss, since the low capacity for iron excretion prevents the wastage of total body iron except in certain pathologic conditions. When the need for iron increases — e.g., during growth or after a period of bleeding — a signal is somehow transmitted to the intestine, and increased amounts of dietary iron are absorbed. After the increased requirement has been satisfied, intestinal iron absorption decreases to maintain a steady-state level of total body iron. Because the excretory mechanism is very limited, absorption in excess of that required leads to iron storage disease. This may result from faulty signal transmission or the inability of the small intestine to respond to the signal for decreased iron absorption.

The mechanism that regulates the rate of iron absorption has been the subject of much controversy. It is known that the block to iron absorption is present at an intracellular location [5]. In both normal and iron-deficient subjects, dietary iron rapidly enters the mucosal epithelial cell, but it remains largely within the cell until it is sloughed into the lumen during the normal mucosal regenerative cycle [13]. It has also been noted that there is a 48-hour lag between bleeding in an animal and the onset of increased iron absorption. No completely satisfactory hypothesis that is in agreement with all the experimental observations has yet been proposed; similarly, the nature of the signal that changes the rate of iron absorption by the intestinal mucosal cell is unknown. Figure 43-9 shows the known steps in iron absorption.

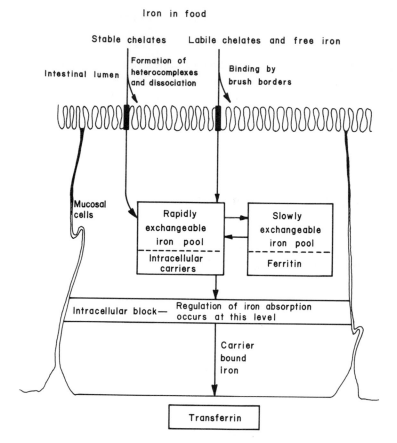

Figure 43-9
Processes involved in iron absorption by the intestinal mucosal cell.

The absorption of iron is also affected by precipitation with acid or complexing with chelating agents inside the lumen of the small intestine. These factors, however, do not function in a regulatory fashion.

VITAMIN B_{12} ABSORPTION

It was shown over forty years ago that a factor from the stomach, which was subsequently named *intrinsic factor,* was necessary for the absorption of vitamin B_{12}. This vitamin is absorbed in the ileum, and, in the absence of intrinsic factor, the daily requirement of a few milligrams cannot be absorbed even if several grams of vitamin B_{12} are present in the lumen of the small intestine [11]. Figure 43-10 illustrates the currently accepted hypothesis regarding the absorption of vitamin B_{12}. There is evidence that metabolic energy is required for the absorption of the complex of

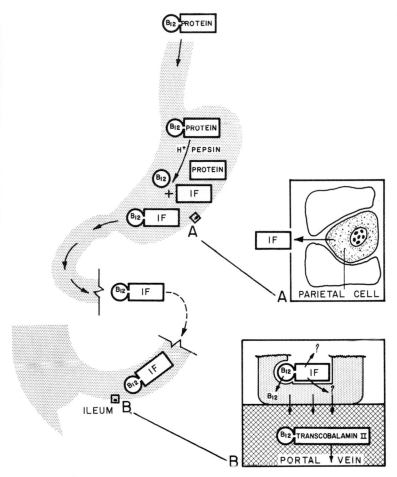

Figure 43-10
Major steps in vitamin B$_{12}$ absorption (*IF* = intrinsic factor).

vitamin B$_{12}$ and intrinsic factor; hence, the absorptive process can be tentatively considered to be an example of active transport.

FOLIC ACID AND THIAMINE

The kinetics of absorption of these two B vitamins suggests strongly that a carrier-transport system must be involved [2]. There is also some evidence that supports the concept that these vitamins are absorbed by an active transport process. It seems likely that specific transport mechanisms will be found for all the water-soluble vitamins, since their molecular size exceeds 0.7 nm. Data suggesting the existence of a special transport system for ascorbic acid have already been published.

CALCIUM AND PHOSPHORUS

Calcium absorption is known to be regulated by parathyroid hormone, and in vitro experiments support the existence of an active transport system for this divalent ion. Phosphorus absorption is regulated by vitamin D, although the fulfillment of the criteria for active transport has not been demonstrated.

HEAVY METALS

The heavy metals — such as copper, zinc, lead, and tin — may be absorbed via the same mechanism that is utilized by iron [9]. However, because of the technical difficulties involved in the analysis of small quantities of the heavy metals, precise information regarding their transport kinetics is generally not available.

STRUCTURE AND MOTILITY OF THE LARGE INTESTINE

The colon consists of the following anatomic parts: (1) the cecum with the attached appendix, (2) the ascending colon, and (3) the sigmoid (pelvic) colon. The ileum enters the colon by way of the ileocecal valve. This junction may have some sphincter-like properties, but it appears to function primarily as a valve that impedes the reflux of colonic contents into the ileum. However, if the colonic-ileal pressure difference is too great, reflux may occur.

The colonic motility pattern differs from that of the small intestine. Segmental contractions cause the colon to be divided into sausage-like segments; these segments are called *colonic haustra*. The haustral contractions appear about once in 30 minutes, and they apparently serve a mixing function. The contractions are strongest in the ascending and transverse colon, and they are weakest in the descending and sigmoid colon.

Propulsive peristaltic movements are ordinarily seen in the colon only three to four times a day. These peristaltic contractions are much stronger than those in the small intestine; intraluminal pressures may rise to 100 cm H_2O. These contractions pass over much of the colon and transport large portions of the fecal mass one-half to three-quarters the length of the large intestine within a few seconds. This mass movement of the colon is often evoked by the gastrocolic reflex, which is initiated by the distention of the stomach with food. This mass peristalsis may also be evoked by the duodenocolic reflex, which is often triggered early in the morning by the intake of breakfast or coffee. The pathways of these reflexes are unknown; they are not abolished by bilateral vagotomy, sympathectomy, or destruction of the communication between the intrinsic nerve plexus of the small and large intestine.

Mass peristalsis eventually moves the fecal mass into the pelvic colon, where it is stored until a defecation reflex results in its expulsion.

SECRETION AND ABSORPTION IN THE LARGE INTESTINE

The large intestine serves primarily as a storage organ for waste products, although a small amount of absorption occurs. Approximately 500 ml of water is absorbed per day, along with some active transport of Na^+. Although the quantity of water absorbed is small compared to the total quantity ingested and secreted into the GI tract, it serves to change the consistency of the stool to semisolid. A small amount of alkaline mucin is secreted into the large intestine.

Numerous vitamins are produced by the bacterial flora of the large intestine, and there is evidence that they are absorbed in the colon. These vitamins are of nutritional importance in some animals, such as rats, in which coprophagia is common, but in that case, the absorption of the vitamins then occurs in the small intestine. A number of lipid-soluble substances will passively diffuse through the mucosal epithelium of the large intestine, although it must be remembered that the villi, microvilli, and the brush border that are present in the small bowel, which increase its surface area, are lacking in the large bowel. Drugs with significant lipid solubility can be administered by rectal suppository or enema, however, with reasonably effective absorption. Examples of such drugs are phenobarbital, xanthines, and phenothiazine derivatives.

REFERENCES

1. Armstrong, W. M. (Ed.). *Intestinal Transport of Electrolytes, Amino Acids and Sugars.* Springfield, Ill.: Thomas, 1971.
1a. Bittar, E. E. (Ed.). *Membranes and Ion Transport,* Vol. I. New York: Wiley-Interscience, 1970.
2. Blair, J. A., Johnson, I. T., and Matty, A. J. Absorption of folic acid by everted segments of rat jejunum. *J. Physiol.* (Lond.) 236:653, 1974.
3. Code, C. F., et al. Motility of the alimentary canal in man. *Am. J. Med.* 13:328, 1952.
4. Curran, P. F., and Solomon, A. K. Ion and water fluxes in the ileum of rats. *J. Gen. Physiol.* 41:143, 1957.
5. Forth, W., and Rummell, W. Iron absorption, *Physiol. Rev.* 53:724, 1973.
6. Hendler, R. W. Biologic membranes. *Physiol. Rev.* 51:66, 1971.
7. Hopfer, V., et al. Glucose transport in isolated brush border membrane from rat small intestine. *J. Biol. Chem.* 248:25, 1973.
8. Katchalsky, A., and Curran, P. F. *Nonequilibrium Thermodynamics in Biophysics.* Cambridge, Mass.: Harvard University Press, 1965.
8a. Kotyk, A., and Janacek, K. *Cell Membrane Transport Principles and Techniques.* New York; London: Plenum Press, 1970.
9. Kowarski, S., Blair-Stanek, C. S., and Schacter, D. Active transport of zinc and identification of zinc-binding protein in rat jejunal mucosal. *Am. J. Physiol.* 226:401, 1974.
9a. Lightfoot, E. N. *Transport Phenomena and Living Systems.* New York: Wiley, 1974.
10. Schultz, S. G., and Curran, P. F. Coupled transport of sodium and organic solutes. *Physiol. Rev.* 50:637, 1970.

11. Toskes, P. P., and Deren, J. J. Vitamin B_{12} absorption and malabsorption. *Gastroenterology* 65:662, 1973.
12. Whittam, R., and Wheeler, K. P. Transport across cell membranes. *Annu. Rev. Physiol.* 32:21, 1970.
13. Yoshino, Y., and Manis, J. Iron-binding substance isolated from particulate fraction of rat intestine. *Am. J. Physiol.* 225:1276, 1973.

THE REPRODUCTIVE SYSTEM   VIII

Overview and Function of the Reproductive Units

Wayne S. Rogers

INTRODUCTION

The systems approach leads to a good understanding of the elements, the organization, and the processes of reproduction. The clinical specialist in reproduction (i.e., in obstetrics and gynecology) relies more and more on instrumentation for the diagnosis, therapy, and care of the patient. In this study of the reproductive system, we are concerned with contraception, conception, the development of the fetus, the physiologic changes that occur in the fetus and mother during pregnancy, the process of labor and delivery, and prenatal and postnatal care for the fetus, infant, and mother. This and the following chapters will be concerned with the physiologic aspects and some of the clinical aspects of the reproductive process, as well as with the instrumentation employed in the diagnosis of problems in normal and abnormal pregnancies.

In almost 20% of all pregnancies, the fetus is considered to be "at risk," depending on the state of the maternal unit and other criteria. In these individuals, the maternal-placental-fetal biologic unit (Fig. 44-1) may not function properly, and variations in the normal fetal environment may result. Saying that the fetus is "at risk" means that it is vulnerable to perinatal death or morbidity. The infants in the latter category may include those who develop physical, intellectual, or personality handicaps. The problems of intrauterine survival and the adaptation of the fetus to extrauterine life involve the services of many disciplines. Thus, the concept and the subspecialty of *perinatology* which is the study of the fetus from conception until it develops into the stabilized neonate, have evolved.

In the future, a pregnant patient may well be evaluated by an automated health-testing laboratory. A computerized obstetric profile may be obtained at the time of her initial prenatal visit. The criteria for the high-risk pregnancy, as programmed in the computer, will allow the segregation of those individuals that need specialized care. The data base for all the presenting pregnant women may be stored at the hospital. This information may then be recalled when needed or augmented when necessary. When the patient enters the hospital in labor, the entire profile may be easily obtained for review.

Although the reproductive system is described as a separate unit, it must be emphasized that it is also subject to the alterations that occur in the other biologic systems. As in all systems, it is necessary to be knowledgeable about the embryology, anatomy, and physiology of the system in order to appreciate the variations that may occur. Thus, this introductory chapter will present a brief treatment of these areas.

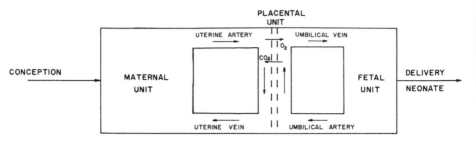

Figure 44-1
The biologic reproductive unit.

DEVELOPMENT AND ANATOMY OF THE REPRODUCTIVE SYSTEM

In the embryo, the reproductive system develops primarily from the genital thickenings noted on the posterior surface of the body cavity [1]. The germinal ridges that are seen as early as five to six weeks of gestation are in an undifferentiated sexual state until subsequent development is determined by the genetic information contained within the sex chromosomes (XX for female and XY for male). At this early state of development, three important structures become obvious: the *mesonephric ducts,* the *müllerian ducts,* and the *gonads* (Fig. 44-2). As determined by the sex chromosomes, only one system of ducts will develop, while the other regresses.

When the embryo is destined to become a female, the mesonephric tubules begin to degenerate and the müllerian ducts gradually begin to fuse to form the uterus, cervix, and upper vagina. The distal, unfused portions become the fallopian tubes. Thus, the genital and urinary systems are closely related anatomically and in function, so that disorders in development will frequently affect both systems. In the male, it is the müllerian ducts that regress, while the mesonephric tubules become the excretory and reproductive tracts. The steroid hormones of the developing testis (e.g., testosterone) probably initiate these changes in the male fetus.

The gonads begin to differentiate at approximately seven weeks following conception. Germ cells gradually migrate into the gonads from the yoke sac via the mesenchyme of the primitive mesentery. In the female, they enter the cortex of the ovary beneath a thickened portion of peritoneum called the *germinal epithelium.* At about four months of gestation, mesenchymal cells at the hilus of the ovary differentiate into specialized cells that eventually surround the oocyte and become the *primordial follicle.* When these cells mature into the granulosa cells, the formation of the primary follicle is complete. Thus, there is a month-long sequence in which the primordial follicle (the ovum and the surrounding cells) becomes a mature *graafian follicle.* The surrounding cells of the follicle increase in both number and size, differentiate into the theca interna and the theca externa, and form a fluid-filled sac with the ovum in the wall of the sac (Fig. 44-3).

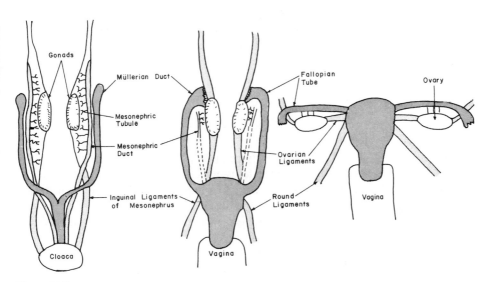

Figure 44-2
The embryologic development of the female reproductive tract.

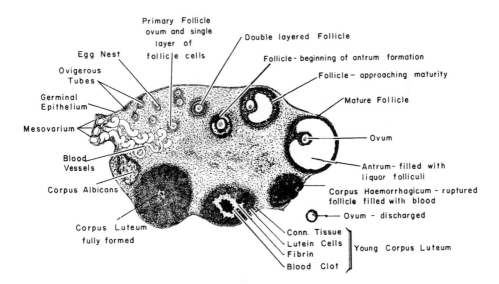

Figure 44-3
The ovary, showing the maturation of the follicles, ovulation, and the development of the corpus luteum. (From B. M. Patten. *Human Embryology*. New York: McGraw-Hill, 1968. Reproduced with permission of McGraw-Hill Book Company.)

With ovulation, the sac swells until the ovum is expelled. The follicular cells then become luteinized and form a mature *corpus luteum* that acts as an endocrine gland in producing steroid hormones. If fertilization does not occur, the corpus luteum degenerates to become the corpus albicans.

SPATIAL DISTRIBUTION

The appearance of the internal female pelvic organs viewed through an abdominal incision into the pelvic cavity is shown in Figure 44-4. The most inferior portion is occupied by the urinary bladder. Directly behind the bladder, separated by a space known as the anterior cul-de-sac, is the uterus, which is supported by the cardinal and uterosacral ligaments. Posterior to the uterus, there is another space, called the posterior cul-de-sac, which is lined by the peritoneum. The boundaries of this space are the uterus anteriorly, the uterosacral ligaments laterally, and the rectum posteriorly. The floor of this space is the posterior fornix of the vagina. This site can easily be entered surgically in order to observe the pelvic organs with a culdoscope. The fallopian tubes arise from the upper lateral portions of the uterine fundus and extend posteriorly. The distal portions of the fallopian tubes are adjacent to the ovaries, which are suspended by the ovarian ligaments.

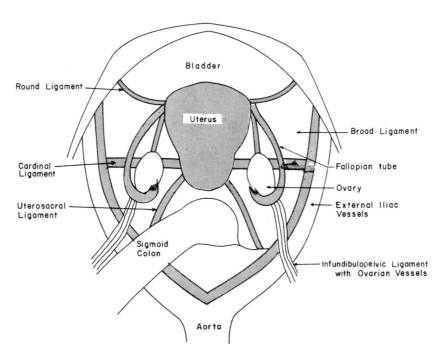

Figure 44-4
Female pelvic contents as viewed from above.

THE UTERUS

The uterus is shaped like an inverted pear. The upper portion is the corpus of the uterus, and the lower portion, the *cervix,* extends into the vagina. The cervic usually measures about 3 cm in length. Immediately above the cervix, there exists a small region known as the *isthmus* of the uterus. The uterine cavity is triangular in shape on cross section, and it measures approximately 3.5 cm in length (Fig. 44-5). At each cornu of the uterus, the cavity is continuous with the interstitial portion of the fallopian tube. The uterine wall is called the *myometrium* and consists essentially of muscle tissue. The uterine cavity is lined by a specialized tissue known as the *endometrium,* which consists of stroma and glands. The endometrium responds to the hormones produced by the ovary so that its structure becomes favorable for the implantation and development of the fertilized ovum.

　　The blood supply to the uterus is rather extensive. It is essential for the development of the placental and fetal units, and it is also concerned with fetal well-being during labor and delivery. The abdominal aorta bifurcates to form the two common iliac arteries. Each of the latter in turn divides and sends one branch known as the hypogastric artery or internal iliac artery into the pelvis. The latter sends large branches to the uterus and the cardinal ligament; these branches are the uterine arteries. The ovarian arteries arise directly from the abdominal aorta at a site where the embryonic gonads were originally located. They course down toward the pelvis in the fold of the peritoneum known as the infundibulopelvic ligament. After giving off branches to the ovaries and the fallopian tubes, the ovarian arteries then anastomose with the vessels

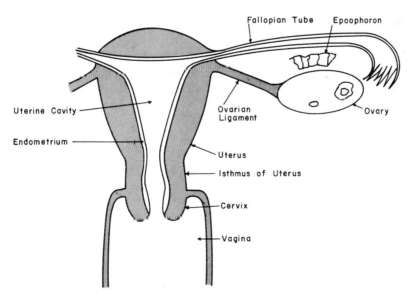

Figure 44-5
The uterus and its appendages.

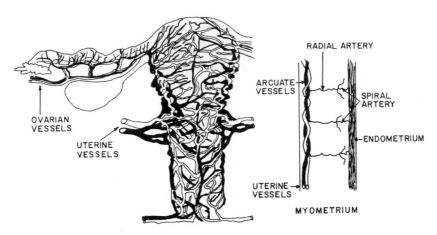

Figure 44-6
Blood supply of the uterus.

of the uterine arteries (Fig. 44-6). As noted in the cross section of the myometrium, these vessels traverse beneath the serosa or outer surface of the uterus and are termed the *arcuate vessels*. The radial arteries intertwine between the spaces formed by the spiral myometrial fibers. This arrangement is unique, inasmuch as contractions of the muscle fibers of the myometrium can occlude the blood flow and thus act as ligatures. As the radial arteries approach the endometrium, they again branch to form the spiral arteries and assure a rich supply of blood to this area.

THE FALLOPIAN TUBES

The fallopian tubes consist chiefly of inner, circular, and outer longitudinal muscular layers. The lumen of the tube is lined by a layer of columnar cells, many of which are ciliated. The most distal portion, the infundibulum, is shaped like a funnel; its fimbria have the ability to sweep down the ovum as it leaves the ovary. The fallopian tubes move the fertilized ovum into the uterine cavity by means of the peristaltic motion of the muscular walls, possibly assisted by the cilia in this function. Ordinarily, the egg is fertilized between the distal and midportion of the fallopian tubes.

THE OVARIES

The ovaries measure approximately 3 cm by 2 cm by 1 cm, but they may vary a great deal in both their size and shape. In general, they lie posterior to the uterus along the lateral pelvic wall and are attached to the broad ligament of the uterus by the mesovarium. Except for a small portion of the ovary located at the hilum, there is no peritoneum covering its surface.

The ovary consists of an outer layer, the cortex, and the medulla. The cortex, whose surface is covered by a single layer of cuboidal cells called germinal epithelium, contains the ova in their various stages of development, each surrounded by a specialized type of connective tissue. Although several hundred thousand oocytes are present at birth, the majority degenerate in later life, and only a few hundred ova are used for ovulation and reproduction. The remainder of the ovary, the medulla, consists chiefly of loose connective tissue, which contains the blood vessels that originate in the infundibulopelvic ligament.

The ovary and its contents are in a state of constant change throughout the life span of the female.

HORMONAL CONTROL OF THE REPRODUCTIVE SYSTEM

The central nervous system plays a primary and significant role in the hormonal control of the reproductive system by exerting an inhibitory effect upon the hypothalamus. The latter secretes various releasing factors that in turn act upon the anterior pituitary gland and cause it to release the gonadotropic hormones. Follicle-stimulating hormone releasing hormone (FSH-RH) and the luteinizing-hormone releasing hormone (LH-RH) stimulate the anterior pituitary gland to secrete the gonadotropins FSH and LH. These gonadotropic hormones stimulate the ovary and are responsible for the mechanism that eventually causes ovulation and the production of the steroid hormones.

FSH is probably responsible for the primordial follicle maturing into the graafian follicle. LH stimulates the interstitial cells of the ovary and is also referred to as interstitial cell–stimulating hormone. The latter cells, under the influence of both FSH and LH, are responsible for the secretion of the estrogens. When ovulation occurs and the corpus luteum is formed, both estrogens and progesterone are secreted. The endo metrium, which is the target organ of these sex hormones, responds in order to prepare itself for the possible implantation of a fertilized ovum. When radioimmunoassay techniques are utilized to determine these hormone levels, it has been noted that the LH level in the plasma rises abruptly prior to ovulation (see Fig. 35-8). The concentration of FSH in the plasma also rises, but to a much lesser degree. This sudden release of LH, known as the *LH surge*, is probably triggered by a rise in plasma estradiol. When this active estrogen is transported to the hypothalamus, it initiates the release of LH-RH. The sudden high concentration of plasma LH is necessary for the initiation of ovulation and the subsequent luteinization of the ruptured follicle. When ovulation occurs, it is followed by an increase in the plasma progesterone concentration, as well as by a second rise in the estradiol concentration, which reflects secretion from the mature corpus luteum.

Various feedback systems shown in Figure 44-7 control the entire mechanism. The concentration of the steroid hormones (estradiol or progesterone) may either stimulate the release of the gonadotropic hormones (positive feedback) or prevent the release of gonadotropins (negative feedback), depending upon the concentration of the steroids

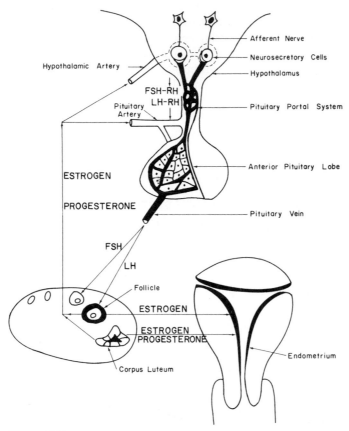

Figure 44-7
Interrelationship of the hypothalamus, the pituitary gland, and the ovary in the production of the hormones necessary for menstruation FSH-RH = follicle-stimulating hormone releasing hormone; LH-RH = luteinizing-hormone releasing hormone).

in the plasma. High concentrations of these steroids can inhibit the release of gonadotropins, which occurs when oral contraceptive tablets are used.

HORMONAL RESPONSES

The endometrium displays a specific response to the concentrations of the circulating steroid hormones. The endometrium is composed of a specialized connective tissue or stroma that contains the endometrial glands and small arterioles, and it can be differentiated into three strata. The *stratum basale,* the portion adjacent to the myometrium, is compact and responds very little to the steroid hormones. Immediately above this layer, a *stratum spongiosum* exists in which the stroma is less dense and in which the blood vessels and glands exhibit a greater degree of activity. The *stratum compactum* is immediately adjacent to the endometrial cavity itself. The stratum spongiosum and the stratum compactum are sensitive to the steroid hormones, and these layers are eventually shed during the process of menstruation.

The first day of the menstrual flow is considered to be the first day of the *menstrual cycle* (Fig. 44-8). When menstruation has ceased, there begins a period of reconstruction known as *proliferation,* which is due to an increase in the concentration of plasma estradiol. The endometrium at the beginning of this stage measures approximately 2 mm in thickness, but its thickness gradually increases to almost 6 mm toward the end of the cycle. This is due to anatomic changes in the stroma, the endometrial glands, and the vascularity of this tissue. Throughout the proliferative phase (or follicular phase), there is a rapid growth of the endometrial glands and the spiral arteries. The arterioles increase in size much more rapidly than does the endometrium, so that they quickly approach the surface of the endometrial cavity.

Following ovulation, estrogen and progesterone initiate the *secretory phase.* The endometrial glands become more tortuous, and secretions can be seen in the lumens of the glands. The stroma becomes less compact because of edema and the increase in vascularity. The coiled arterioles become increasingly irregular as they increase in length. Many small capillaries are formed in the stratum compactum as a result of this. Considerable glycogen is deposited in the cells of the endometrial glands. The entire endometrial cavity has thus prepared itself for the possible implantation of the fertilized ovum.

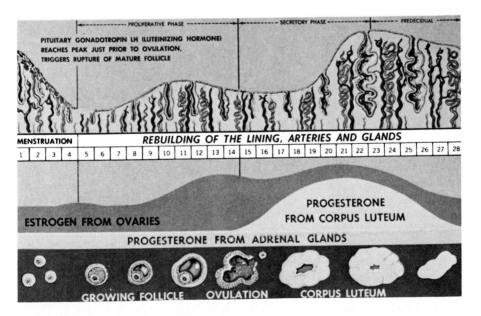

Figure 44-8
The menstrual cycle and the correlative changes in the ovary, the plasma steroid levels, and the endometrium. (Courtesy of Ross Laboratories).

If conception does not occur, then the process of menstruation is initiated. There is first a slowing of the circulation in the coiled spiral arterioles, which is probably a result of the increased resistance of the vessel walls. These blood vessels may then dilate, causing stasis of the contained blood. Prior to the actual process of menstruation, vasoconstriction of the coiled arteries occurs intermittently, with the result that the upper two strata of the endometrium become periodically ischemic. Subsequently, the process causes vessel wall damage, and bleeding occurs. Small hematomas develop within the endometrium that distend the superficial portions, causing a disruption in the lining and subsequent bleeding into the endometrial cavity. The bleeding ceases when the arterioles are repaired by further vasoconstriction. It is believed that most of the stratum compactum and a large portion of the stratum spongiosum are shed during menstruation. These changes do not occur throughout the entire endometrial cavity, but they do so in a sporadic fashion at different sites. The remaining strata reorganize to begin another cycle. The coiled and spiral arterioles are sensitive to the circulating steroid hormones, which cause their increase in growth.

Prior to menstruation, the decline in the plasma steroid level is responsible for the intermittent dilatation and constriction of these vessels. The menstrual discharge consists of both blood and fragments of endometrium. The usual menstrual blood does not clot, because of apparent fibrinolytic activity that liquefies any coagulated blood.

Of clinical interest is the basal body temperature of a female during the menstrual cycle. The temperature is taken daily on arising and prior to leaving the bed. If ovulation occurs, there is a significant rise in temperature that is caused by the progesterone release in the latter part of the cycle.

DETERMINATION OF HORMONES

It is possible to measure certain steroid hormones in small aliquots obtained from either human serum or tissue extracts. Two of the techniques that may be employed are:

1. *Radioimmunoassay.* This method has been used extensively for measuring the gonadotropic hormones that are released from the anterior pituitary gland [2, 3]. These hormones are bound to specific globulins for the purpose of transportation. A known amount of the hormone is labeled with radioactive iodine. When it is allowed to react with a particular amount of antibody, the radioactively labeled hormone competes with the unlabeled hormone for the specific antibody. The quantitative difference in the amount of the antibody-bound radioactivity is measured, and this is indicative of the quantity of the unlabeled hormone.

The use of radioimmunoassay (RIA) in the discipline of reproductive endocrinology has proved to be a significant diagnostic achievement. It is now possible to measure the concentration of many substances with relative ease and precise accuracy. This method will eventually replace the biologic assays and provide more rapid and economical measurements of specific substances which were not previously possible.

In order to evaluate the many disorders of reproductive physiology and endocrinology, extreme sensitivity and accuracy are frequently necessary. Radioimmunoassays are based on the immunological principle of antibody/antigen reaction. The latter provides for great specificity and requires a small aliquot of material to be tested. In most instances less than a milliliter of blood is needed so that this technique can also be readily utilized in small premature infants. The sensitivity for specific substances and its accuracy of measuring concentrations as minute as nanograms (10^9) and picograms (10^{12}) are the main advantages in clinical practice.

The hormone assays of primary interest in the reproductive system are LH, FSH, HCG and prolactin. Assays of these materials are necessary in the study of a wide variety of endocrine disturbances and for monitoring certain disease processes (trophoblastic disease) and their response to chemotherapy. Bioasay methods are cumbersome and without the specificity or accuracy needed to evaluate the above.

2. *Gas-Phase Chromatography.* Gas chromatography is used to perform the separation and subsequent measurement of various compounds. The basic principles of operation that are utilized by this technique are as follows. First, a sample of the material to be analyzed is injected into the top of the glass column, where it becomes instantly vaporized, and it is then carried down the column along with an inert carrier gas such as neon. The components of the sample traverse the column at different rates; these rates are determined by the differences among the partition coefficients of the components of the gaseous phase with respect to the stationary phase of the material, usually a relatively nonvolatile liquid, that is placed within the column. The presence of the sample components in the eluted gas from the column is detected by means of a flame ionization detector. The ionization detector operates on the principle that an organic compound is ionized by the heat of a gas flame, and the ions thus formed will produce an electrical current when passed between high-voltage electrodes. The magnitude of the current is proportional to the number of ions of the particular compound present, and this current is subsequently amplified and recorded on a strip chart.

REFERENCES

1. Arey, W. *Developmental Anatomy*. Philadelphia: Saunders, 1965. Chap. 18.
2. Jaffee, R. B., and Midgley, A. R., Jr. The Radioimmunoassay of Gonadotropins. In Sturgis, S. H. and Taymor, M. L. (Eds.), *Progress in Gynecology*, Vol. 5. New York: Grune & Stratton, 1970.
3. Kase, N. Advances in Steroid Assays. In Sturgis, S. H and Taymor, M. L. (Eds.), *Progress in Gynecology*, Vol. 5. New York: Grune & Stratton, 1970.

SELECTED READING

Patten, B. M. *Human Embryology*. New York: McGraw-Hill, 1968.
Segal, B. L., and Kilpatrick, D. G. *Engineering in the Practice of Medicine*. Baltimore: Williams & Wilkins, 1967.

The Placenta   **45**

Lawrence D. Longo, Wayne S. Rogers, Esther P. Hill
and Gordon G. Power

INTRODUCTION

The placenta is the coupling element between the mother and the fetus, and its primary functions are to provide for the transfer of gases and nutrients and the establishment of homeostasis in the fetus. The blood flow dynamics and the gas transfer functions will be treated on a quantitative basis in Chapter 46. The present chapter considers some aspects of the morphologic development of the placenta as well as the operation of some of its mechanisms, particularly that of the placental transfer of oxygen.

The placenta begins its development with the implantation of the ovum in the uterine wall. A detailed account of placental development may be found in the literature [9a]. The ovum begins the process of division into a mass of cells known as the *morula* while on its route down the fallopian tube to the uterine cavity. Seven days from the time it leaves the ovary, the ovum has entered the uterine cavity and has implanted itself, usually into the posterior wall of the uterus. The morula has become a spherical vesicle, called the *blastocyst,* that contains two components: the inner cell mass, which will eventually become the embryo, and a layer of ectodermal cells, which becomes the *trophoblast.* The trophoblast attaches the embryo to the uterine wall and, by invasion of the latter, supplies its nutrition.

In response to persistent hormonal stimulation from the corpus luteum, the cells of the stroma of the endometrium become enlarged and pale in appearance. These are now known as *deciduul cells.* It is thought that these cells have a dual function: they supply nutritive material as the process of growth and invasion takes place, and it is also believed that they limit the extent of the invasion of the trophoblastic cells. The latter begin to penetrate deep within the decidua, and they soon can be differentiated into two distinct cell layers, the *cytotrophoblast* and the more mature analog, the *syncytiotrophoblast.* As the trophoblastic cells invade the endometrium, they open up small vessels and form lacunae, or small lakes, that are filled with maternal blood. Eventually, a central core appears within these masses of cells that is known as the *mesoblast.* The primitive villi consist of this mesodermic center, which is covered by the trophoblastic cells. Blood vessels begin to form in situ within the mesoderm and eventually lead to the now-developing umbilical blood vessels that appear from the body stalk of the fetus. A primitive fetal circulation is thus established. These branching villi can project deeply within the decidua and almost reach the myometrium. This portion is termed the *chorion frondosum,* and it eventually becomes the mature placenta. The layer of the trophoblastic tissue adjacent to the endometrial cavity is called the *chorion laeve.*

The amnion enlarges rapidly and gradually surrounds the developing embryo. At approximately 12 weeks of gestation, the expanding amniotic sac and the overlying chorion laeve reach the opposite wall of the endometrial cavity. The chorion laeve gradually disappears, so that only a chorionic membrane and an amniotic membrane persist, which make up the wall of the amniotic sac.

The placenta is fairly well established at about the third month of pregnancy [9a]. As the columns of trophoblast reach the myometrium, they form anchoring villi that contain a central core of mesodermal tissue and the developing fetal blood vessels. In the meantime, other trophoblastic cells are tapping the endometrial spiral arterioles. When this occurs, the maternal blood spurts out at a pressure of approximately 70 mm Hg due to the pulsation of the arterioles. As the blood accumulates, a cleavage results, causing cavitation within the decidua. The fetal portion at the top of Figure 45-1 is known as the *chorionic plate*. The maternal surface of the placenta that is adjacent to the uterine wall and anchored to the latter by means of the anchoring villi is known as the *basal plate*. From the latter, septa made up of decidual tissue extend toward the chorionic plate at various locations. These septa do not reach the chorionic plate, and thus the maternal circulation is not confined to individual compartments. The latter form the major *cotyledons* of the placenta. There are about twelve such large cotyledons formed, which are bathed by the maternal blood. There also are about ten times as many rudimentary cotyledons. The blood vessels that are formed in the mesoderm of the villi gradually join those vessels from the chorioallantoic mesenchyme and the newly formed umbilical vessels, thereby completing the fetoplacental circulation. The placenta continues to mature by proliferation until approximately the seventh month of gestation, at which time cellular hypertrophy continues until term. The mature placenta thus contains two distinct circulations: the uteroplacental and the fetoplacental circulations (see Fig. 45-1).

The fetus requires oxygen on a moment-to-moment basis. Assuming that the fetal oxygen consumption is 6 ml/min per kilogram of fetal weight, a term fetus weighing 3 kg requires about 18 ml of O_2 per minute. In view of this vital requirement, the fetal oxygen stores are surprisingly small. Assuming that the blood hemoglobin concentration is 16.5 gm/100 ml of fetal blood (which is equivalent to an O_2 capacity of 22 ml/100 ml of blood), a 3-kg fetus with a blood volume of 300 ml has a total O_2 capacity of 66 ml. Since the average saturation of fetal hemoglobin is about 60%, the fetal O_2 stores are only about 40 ml, which provides barely a two-minute supply for the fetal tissues. Of course, this calculation is based on the assumption that the blood is the primary reservoir of O_2 in the fetus. Although myoglobin acts as an oxygen store in the muscles of man and most mammals, recent unpublished evidence indicates that there is little or no myoglobin in fetal skeletal muscle. Furthermore, all the O_2 could not be extracted from the blood, since this would require that the oxygen tension (Po_2) fall to zero, which is clearly impossible if the fetus is to remain alive. Thus, since the fetal O_2 requirement is rather large in view of its small stores, oxygen must be supplied without interruption.

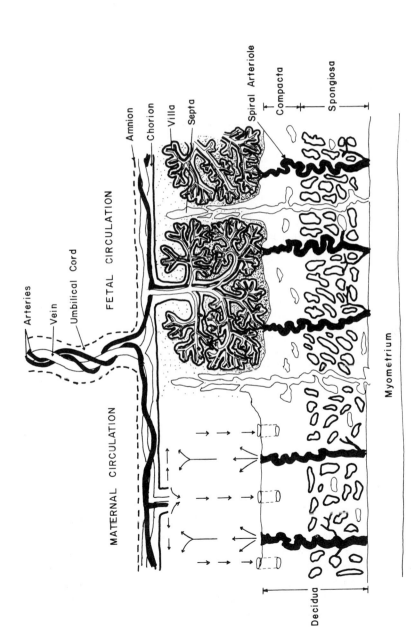

Figure 45-1
The maternal and fetal blood circulation through the mature placenta.

FACTORS AFFECTING THE PLACENTAL TRANSFER OF OXYGEN

A number of steps are involved in the transfer of oxygen from the ambient air to the cristae of the mitochrondia in fetal tissues where the O_2 is ultimately reduced; the transfer of O_2 across the placenta is but one link in this chain of sequential steps.

In the placenta, a sufficient volume of O_2 to supply the fetal requirements must diffuse across the membranes from the maternal to the fetal blood during the short time these two circulations are in close contact. A number of factors affect this transfer of O_2 (Table 45-1).

Table 45-1. Principal Factors Affecting Placental Oxygen Transfer

Factor	Causes
Placental diffusing capacity	Membrane diffusing capacity (area, thickness, solubility, diffusivity of tissues); capillary blood volume; diffusing capacity of blood (O_2 capacity, hemoglobin reaction rates, concentration of reduced hemoglobin)
Maternal arterial PO_2	Inspired PO_2; alveolar ventilation; mixed venous PO_2; pulmonary blood flow; pulmonary diffusing capacity
Fetal arterial PO_2	Umbilical venous PO_2; fetal O_2 consumption; peripheral blood flow; maternal arterial PO_2; maternal placental hemoglobin flow; placental diffusing capacity
Maternal hemoglobin O_2 affinity	pH; temperature; PCO_2; 2,3-diphosphoglycerate concentration; carbon monoxide concentration
Fetal hemoglobin O_2 affinity	pH; temperature; PCO_2; 2,3-diphosphoglycerate concentration; carbon monoxide concentration
Maternal placental hemoglobin flow rate	Arterial pressure; placental resistance to blood flow; venous pressure; O_2 capacity of blood
Fetal placental hemoglobin flow rate	Umbilical artery blood pressure; umbilical venous blood pressure (or maternal vascular pressure under conditions of sluice flow); placental resistance to blood flow; O_2 capacity of blood
Spatial relation of maternal to fetal flow	—
Amount of CO_2 exchange	—

Ideally, the exchange process would be studied by sampling the inflowing and end-capillary blood within a single exchange unit. However, it is not yet experimentally possible to sample the end-capillary blood directly or to study how these various factors affect the process of O_2 transfer within individual capillaries. Several problems account for this. The placenta may be compared to a "black box" with measurable maternal and fetal arterial inputs and mixed venous outputs. The venous outflow consists of blood from numerous exchange units in which the O_2 partial pressures have reached different degrees of equilibration because of the nonuniform distribution of maternal and fetal flows, the nonuniform distribution of the diffusing capacity to flow, and the placental O_2 consumption. Also, there are probably vascular shunts in both the maternal and the fetoplacental circulations that contribute to the fact that the uterine and umbilical veins contain a mixture of blood of various O_2 tensions. Another problem confronting the investigator is that when he changes a factor experimentally, various compensations mask its effect.

Normal values for the various determinants of O_2 transfer are necessary for the quantitative analysis of the exchange process. Some values — such as those for the maternal and fetal arterial O_2 tensions and O_2 affinities — are fairly well defined. Others — such as the diffusing capacity and the maternal and fetoplacental blood flows — are less well determined.

PLACENTAL DIFFUSING CAPACITY

The diffusion characteristics of the placental membrane may be described by Fick's first law [1]:

$$\frac{dQ}{dt} = \frac{AD\Delta c}{\Delta x} \tag{45-1}$$

where dQ/dt = volume of a given substance (e.g., O_2) crossing the placental
membrane per unit time
A = area of exchange
D = diffusion constant (cm^2/sec)
Δc = concentration difference (by volume) across the membrane
Δx = distance through which the diffusion takes place

The placental membrane is a complex structure, and its thickness and permeability vary with location. At a given location, however, the permeability, diffusivity, and thickness may be treated as constants and combined together into a single term that expresses the diffusion characteristics of the membrane:

$$\frac{dQ}{dt} \cdot \frac{1}{P_x} = D_P \tag{45-2}$$

where D_P is the placental diffusing capacity in milliliters per minute per difference in partial pressure (mm Hg) of gas x. For respiratory gases, the Bunsen solubility coefficient, α, and the partial-pressure difference (ΔP) are used rather than the concentration difference (Δc) in derivations from Fick's laws.

Carbon monoxide, a diffusion-limited gas, was used to study the diffusion characteristics of the placental membrane in sheep and dogs [7]. Catheters were introduced into the maternal femoral artery, a branch of the uterine vein, and the umbilical artery and vein. The animals were placed on a closed rebreathing circuit that had a carbon dioxide absorber and oxygen supply, and they were ventilated with a respiratory pump. Following a control period of 20 minutes to one hour, a given amount of carbon monoxide (CO) was introduced into the circuit. Serial samples were taken to follow the time-course of the change in maternal and fetal carboxyhemoglobin (COHb) concentrations. The partial pressure of carbon monoxide (P_{CO}) in mm Hg was calculated from the Haldane relation [2]:

$$P_{CO} = \frac{[COHb] \, P_{O_2}}{[O_2 Hb] \, M}$$

(45-3)

where M is the relative affinity of hemoglobin for CO as compared with O_2. The fetal uptake of CO that was calculated from the change in fetal carboxyhemoglobin concentration the mean maternal-to-fetal P_{CO} gradient, and the physicochemical relations of O_2 to CO were used to calculate D_p. This value was 0.54 (ml/min)/mm Hg per kilogram of fetal weight, or about 1.62 (ml/mm)/mm Hg for a 3-kg fetus [7]. Thus, the diffusing capacity f O_2 was found to be appreciably greater than previously suspected. From these calculations, the mean maternal-to-fetal P_{O_2} difference was calculated to be only 5 to 6 mm H which was some three times less than the previous estimate. Although the diffusing capacity calculated from the CO data may not seem significantly greater than the 0.5 (ml/min)/mm Hg that had been calculated from O_2 data, there is an important difference. The higher value strongly suggests that there is no significant barrier to the diffusion of O_2 through the placental membrane, contrary to what was suggested by the previous value. Furthermore, it suggests that the O_2 partial pressures equilibrate during the course of a single transit in the maternal and fetal placental exchange vessels.

RELATIVE RESISTANCE TO DIFFUSION OF MATERNAL AND FETAL BLOOD AND THE PLACENTAL MEMBRANES

A further consideration regarding placental O_2 transfer is the relative resistance to diffusion offered by maternal and fetal blood and the placental membranes. The total partial-pressure difference (Fig. 45-2) between maternal and fetal blood, $P_M - P_F$, is the sum of the pressure difference from the interior of the maternal erythrocyte to the maternal plasma, $P_M - (P_{pl})_M$, plus the pressure difference from the maternal plasma across the placental membranes to the fetal plasma, $(P_{pl})_M - (P_{pl})_F$, and the partial-pressure difference from the fetal plasma to the interior of the fetal erythrocyte, $(P_{pl})_F - P_F$. The reciprocal of the diffusing capacity — i.e., the total resistance to diffusion $(1/D_P)$ — is the sum of the resistance of the maternal blood $(1/\Theta_M V_M)$, the resistance of the placental membrane $(1/D_m)$, and the resistance of the fetal blood $(1/\Theta_F V_F)$ [6]. These resistances are depicted in Figure 45-2 and may be expressed as:

$$\frac{1}{D_P} = \frac{1}{\Theta_M V_M} + \frac{1}{D_m} + \frac{1}{\Theta_F V_F}$$

(45-4)

where Θ_M = absorbance capacity (rate of gas uptake) of maternal blood
[(ml O_2/min)/mm Hg/ml blood]
Θ_F = absorbance capacity (rate of gas uptake) of fetal blood
[(ml O_2/min) mm Hg/ml blood]

V_M = maternal blood volume (ml)
V_F = fetal blood volume (ml)
D_m = diffusing capacity of the placental membrane [(ml/min)/mm Hg]

Although D_m remains constant, the value of D_P varies along the capillary because Θ_M and Θ_F vary. The absorbance capacity of the red blood cells (Θ) is a function of both the rate of combination of O_2 with hemoglobin, which increases at increased O_2 tensions, and the concentration of reduced hemoglobin, which decreases at increased O_2 tensions [11]. The value of Θ is relatively constant at low O_2 tensions, but it decreases rapidly at high O_2 tensions (see below). By measuring the carbon monoxide diffusing capacity at various O_2 tensions, one may solve for the values of the maternal and fetal capillary blood volumes (V_M and V_F) and the membrane diffusing capacity (D_m).

The carbon monoxide diffusing capacity for sheep was measured at various O_2 tensions in a hyperbaric chamber using a method similar to that outlined above [6]. The values of D_P varied as a function of the oxygen tension. In Figure 45-3, the reciprocal of D_P is plotted as a function of the sum of the reciprocals of the absorbance capacities of the maternal and fetal red blood cells. From the plot, the slope is the reciprocal of the value of Vc, the maternal and fetal capillary blood volume, and the intercept is the reciprocal of D_m. From these studies, it was calculated that the resistance of maternal and fetal red blood cells was approximately *one-third* of the total resistance, whereas the resistance of the placental membrane per se was about *two-thirds* of the total resistance to O_2 diffusion [6]. This is in contrast to previous studies, which assumed that the placental membrane per se constituted the total resistance to diffusion.

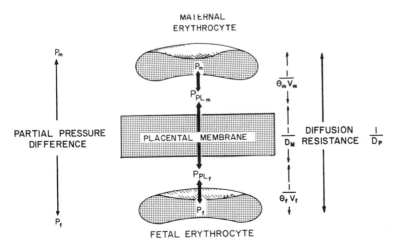

Figure 45-2
The maternal-to-fetal partial-pressure differences and resistances to diffusion.

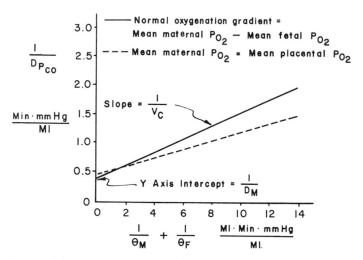

Figure 45-3

Plot of $1/(Dp)_{CO}$ (normalized to a fetal weight of 3 kg) as a function of $(1/\Theta_M) + (1/\Theta_F)$, assuming that $V_M = V_F$ and that the mean maternal placental PO_2 minus the mean fetal PO_2 equals that gradient during normal oxygenation. V_C is the maternal and fetal capillary blood volume.

MATHEMATICAL MODEL OF PLACENTAL OXYGEN TRANSFER

Based on a knowledge of the various parameters and mechanisms affecting O_2 transfer, the exchange process that occurs in the maternal and fetal exchange vessels may be mathematically simulated [4, 5, 9]. This mathematical model may be used to study how variations in the individual parameters affect the O_2 exchange process and thereby to calculate the relative importance of the various factors affecting oxygen transfer [5]. Moreover, the model may be a stimulus to design experiments that will either test the predictions of the model or provide more exact values for the parameters affecting the exchange process.

The model uses the values of the various parameters affecting exchange — such as inflowing oxygen tensions, the flow rates of maternal and fetal blood, and the placental diffusing capacity — in a forward integration procedure to calculate the time-course of O_2 transfer [5]. The model may be used to study how uncompensated changes of the individual factors (keeping all other parameters constant) affect the transient rate of O_2 transfer and the placental end-capillary O_2 tensions. In this case, the placenta is treated as an isolated preparation that is perfused under controlled conditions. The importance of different factors may then be compared with one another, and their individual effects on maternal and fetal end-capillary O_2 tensions and the O_2 transfer rate may be noted.

It must be recognized that this simplified model does not accurately represent the complexity of O_2 exchange in vivo. Furthermore, the model is no substitute for experimental measurements, but, until these can be obtained from single exchange vessels in the placenta, it would seem that this model provides the best approach to the study of the transfer process.

The model describes the diffusion of O_2 across a section of uniform tissue separating two parallel capillaries. The capillaries are assumed to have uniform diameters and equal lengths, and concurrent maternal and fetal flows are assumed to occur (Fig. 45-4). The blood flow is assumed to be nonpulsatile, with perfect radial mixing. The pH, the PCO_2, and the temperature are assumed to be constant along the capillary. The effect of all the capillaries in the placental exchange bed is represented by a single unit with capillary volumes V_M and V_F, and the blood flow rates $\dot{Q}_M$ and $\dot{Q}_F$ are assumed to be equivalent to those of the entire placenta. All references to diffusion denote that of O_2. As the blood moves along the capillaries, O_2 diffuses from the maternal capillary across the membrane to the fetal capillary. Thus, the variation of O_2 partial pressures with distance along the capillaries is equivalent to the change in partial pressure in an element of blood with time — a simple one-dimensional diffusion problem. The small amount of diffusion occurring along the length of the capillaries is ignored. The capillary transit time is equal to the capillary volume divided by the flow rate and is usually assumed to be the same in both vessels. Fick's first law of diffusion can be applied because the PO_2 remains constant (i.e., at a steady state) at each point along the capillary.

The differential equations describing the change in oxygen content on the two sides of the membrane are:

$$d[O_2]_M/dt = -D_P(P_M - P_F)/V_M \tag{45-5}$$

and

$$d[O_2]_F/dt = -(V_M/V_F)(d[O_2]_M/dt) \tag{45-6}$$

where $[O_2]_M = O_2$ concentration of maternal blood (ml O_2/ml blood)
$[O_2]_F = O_2$ concentration of fetal blood (ml O_2/ml blood)
P_M = maternal PO_2 at time t (mm Hg)
P_F = fetal PO_2 at time t (mm Hg)

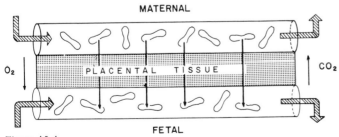

Figure 45-4
Maternal and fetal placental exchange vessels with uniform dimensions and concurrent flows. As the blood flows along the capillaries, O_2 diffuses in a one-dimensional plane from the maternal to the fetal blood.

Although these equations ignore placental tissue oxygen consumption, this component may be accounted for by assuming that the rate of loss of O_2 to the placental tissue is proportional to the partial-pressure difference between the capillary blood and the surrounding tissue; however, the effect of this component is relatively small.

The O_2 concentrations and the partial pressures are related by the hemoglobin dissociation curves for maternal and fetal blood. The modified Hill equation is used [3, 8]:

$$\log PO_2 = k_1 + k_2 \, (\text{pH} - 7.4) + k_3 \log_{10} [S_{\text{Hb}}/(100 - S_{\text{Hb}})] \tag{45-7}$$

where S_{Hb} is the percentage saturation of hemoglobin:

$$S_{\text{Hb}} = 100 \, [O_2]/(O_2 \text{ capacity}) \tag{45-8}$$

The O_2 capacity equals 1.34 times the hemoglobin concentration, which is a variable in the model. The constants used in Equation 45-7 that fit experimental dissociation curves for sheep maternal blood at 38°C [8] are $k_1 = 1.544$ when $S_{\text{Hb}} < 81$, $k_1 = 1.585$ when $S_{\text{Hb}} \geqslant 81$, $k_2 = 0.456$, $k_3 = 0.346$ when $S_{\text{Hb}} < 81$, and $k_3 = 0.283$ when $S_{\text{Hb}} \geqslant 81$. For fetal blood under the same conditions, $k_1 = 1.20$, $k_2 = 0.464$, and $k_3 = 0.348$. To obtain the value of Θ for Equation 45-4, we use an empirical equation that has been fitted to the data of Staub et al. [12] and scaled to correct for the difference between maternal and fetal reaction rates in both human and sheep blood:

$$\Theta = [k_1(1 - e^{-k_2 (100 - S_{\text{Hb}})}) + k_3 S_{\text{Hb}} - k_4] \times \tag{45-9}$$

$$(O_2 \text{ capacity}/10.2) \times [k_c(\text{sheep})/k_c(\text{human})]$$

where $k_1 = 3.287$
  $k_2 = 0.1117$
  $k_3 = 7.05 \times 10^{-3}$
  $k_4 = 0.8142$
  $k_c(\text{sheep}) = 130 \text{ mM}^{-1}\text{sec}^{-1}$ (maternal blood)
    $= 151 \text{ mM}^{-1}\text{sec}^{-1}$ (fetal blood)
  $k_c(\text{human}) = 164 \text{ mM}^{-1}\text{sec}^{-1}$

Figure 45-5 shows the variations in Θ_M and Θ_F and the resulting variations in D_P calculated with the model for a typical capillary transit. The initial low value of Θ_M is caused by the high oxyhemoglobin saturation in the maternal arterial blood.

Values of oxygen partial pressure in the plasmas, $(P_{pl})_M$ and $(P_{pl})_F$, can be obtained from the fraction of the total resistance due to the hemoglobin reaction rates:

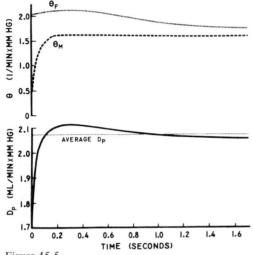

Figure 45-5
The calculated changes in the absorbance capacity of fetal and maternal blood (Θ_F and Θ_M) and the placental diffusing capacity (D_P) during a single capillary transit.

$$(P_{pl})_M = P_M - \frac{1/(\Theta_M V_M)}{1/D_P} (P_M - P_{I\cdot}) \qquad (45\text{-}10)$$

and

$$(P_{pl})_F = P_F + \frac{1/(\Theta_F V_F)}{1/D_P} (P_M - P_F) \qquad (45\text{-}11)$$

In Equations 45-4 through 45-11, t is the independent variable and $P_M, P_F, (P_{pl})_M,$ and $(P_{pl})_F$ are the dependent variables. The intermediate variables calculated during the solution of the equations are $[O_2]_M, [O_2]_F, D_P, \Theta_M,$ and Θ_F. The hemoglobin concentration, the initial values of P_M and P_F, and the values of $D_m, V_M, V_F, \dot{Q}_M,$ and $\dot{Q}_F$ are the parameters that are varied in a systems-analysis approach.

The above equations were solved with a fourth-order Runge-Kutta forward numerical integration technique [10] on an EMR-6130 digital computer. The integration begins with initial values of P_M and P_F that represent the arterial end of the capillary. The changes of these partial pressures are calculated for a small element of time. The changes are then used to compute new values of P_M and P_F. After many such successive steps, the cumulative time equals the capillary transit time, and the integration is terminated. The final P_M and P_F values represent those at the end of the capillary. The average D_P value is obtained by summing the D_P values at each step and dividing by the number of steps. The same procedure can be used to calculate the average values of P_M and P_F.

CONCLUSIONS

The model that has been presented uses equations that describe one-dimensional diffusion between two parallel capillaries to calculate the PO_2 changes that occur during the course of a single transit of blood in the maternal and fetal placental vessels. The relative resistances to diffusion offered by the intracellular hemoglobin reaction rates and the placental membrane produce slight variation in D_P during the capillary transit and permit the calculation of PO_2 values for the plasma and red blood cells. For representative values of D_P, uterine and umbilical PO_2, and capillary flow rates, the model predicts that the PO_2 reaches 99.4% equilibration by the end of a 1.7-second capillary transit. Moderate variation in almost any of the variables, however, can increase the end-capillary pressure difference considerably. Thus, although previous studies have indicated rapid and almost complete equilibration of inert gases, this model predicts that O_2 equilibration may not be so complete.

Studies with this model have shown that the effects of placental O_2 tissue consumption are a lowering of end-capillary PO_2 values and a decrease in the rate of O_2 exchange, whereas the effects of shunts are a lowering of the umbilical venous PO_2 and an elevation of uterine venous PO_2 due to mixing of the capillary blood with arterial blood that has bypassed the exchange area. In the presence of a placental tissue O_2 consumption of 8 ml/min per kilogram and 26% shunts, the venous PO_2 values agree well with experimental values, but the calculated O_2 exchange rate is still much larger than that obtained from experimental values. Thus, the concurrent flow model predicts a more efficient system than actually exists. Assuming either a countercurrent or crosscurrent flow pattern for the model would produce even higher O_2 exchange rates. Probably some degree of uneven distribution of capillary flow rates and diffusing capacities exists in addition to a complex combination of flow patterns.

It has been found with the use of this model that the main factors determining placental O_2 exchange are (1) the uterine and umbilical arterial O_2 tensions, (2) the maternal and fetal placental hemoglobin flows, (3) the placental diffusing capacity, (4) the characteristics of the oxyhemoglobin dissociation curve, (5) the amount of CO_2 exchanged, and (6) the spatial relation of maternal and fetal exchange vessels. The mathematical model was useful to study the effects of isolated changes in the first three of these factors individually. End-capillary O_2 tensions and the O_2 exchange rate are most sensitive to changes in the umbilical arterial O_2 tension, somewhat less sensitive to changes in the maternal and fetal placental blood flows or hemoglobin concentrations, and relatively insensitive to changes in the maternal arterial O_2 tension or the placental diffusing capacity.

The predictions of the model may be tested experimentally by using an isolated placental cotyledon preparation. Both the O_2 transfer rate and the outflowing PO_2 were found to be most sensitive to changes in the umbilical arterial PO_2; changes in the maternal arterial PO_2 and the umbilical flow rate had less effect in the physiologic range. Changes in the maternal PO_2, however, were found to become progressively more critical to fetal oxygenation as the level fell. The pattern of results was similar to that predicted theoretically. Evidence has been obtained from these studies that suggests that the umbilical arterial PO_2, rather than the umbilical blood flow, is the direct link that maintains the rate of placental O_2 transfer equal to the rate of fetal O_2 consumption.

REFERENCES

1. Fick, A. Über Diffusion. *Poggendorff's Ann. Physik* (Ser. 2) 94:59, 1855.
2. Haldane, J., and Smith, J. L. The absorption of oxygen by the lungs. *J. Physiol.* (Lond.) 22:231, 1897.
3. Hellegers, A. E., and Schruefer, J. J. P. Nomograms and empirical equations relating oxygen tension, percentage saturation, and pH in maternal and fetal blood. *Am. J. Obstet. Gynecol.* 81:377, 1961.
4. Hill, E. P., Power, G. G., and Longo, L. D. A mathematical model of placental O_2 transfer, with consideration of hemoglobin reaction rates. *Am. J. Physiol.* 222:721, 1972.
5. Longo, L. D., Hill, E. P., and Power, G. G. Importance of various factors limiting placental O_2 transfer. *Am. J. Physiol.* 222:730, 1972.
6. Longo, L. D., Power, G. G., and Forster, R. E., II. Placental diffusing capacity for carbon monoxide at varying partial pressures of oxygen. *J. Appl. Physiol.* 26:360, 1969.
7. Longo, L. D., Power, G. G., and Forster, R. E., II. Respiratory function of the placenta as determined with carbon monoxide in sheep and dogs. *J. Clin. Invest.* 46:812, 1969.
8. Meschia, G., et al. A comparison of the oxygen dissociation curves of the bloods of maternal, fetal and newborn sheep at various pH's. *Q. J. Exp. Physiol.* 46:95, 1961.
9. Power, G. G., Hill, E. P., and Longo, L. D. Combined diffusion and blood flow limitation in placental O_2 exchange. *Am. J. Physiol.* 222:740, 1972.
9a. Reynolds, S. R. M. Formation of fetal cotyledons in the hemochorial placenta. *Am. J. Obstet. Gynecol.* 94:425, 1966.
10. Richardson, L. F., and Gaunt, J. A. The deferred approach to the limit. *Trans. R. Soc. Trop. Med. Hyg.* (Lond.) 226A:300, 1927.
11. Staub, N. C., Bishop, J. M., and Forster, R. E. Importance of diffusion and chemical reaction rates in O_2 uptake in the lung. *J. Appl. Physiol.* 17:21, 1962.

SELECTED READING

Guilbeau, E. J., Reneau, D. D., and Knisely, M. H. A Distributed Parameter Mathematical Analysis of Oxygen Exchange from Maternal to Fetal Blood in the Human Placenta. In D. D. Reneau (Ed.), *Chemical Engineering in Medicine.* Washington, D.C.: American Chemical Society, 1973.

Meschia, G., Battaglia, F. C., and Bruns, P. D. Theoretical and experimental study of transplacental diffusion. *J. Appl. Physiol.* 22:1171, 1967.

Page, E. W., Glendening, M. D., Margolis, A., and Harper, H. A. Transfer of D- and L-histadine across the human placenta. *Am. J. Obstet. Gynecol.* 73:589, 1957.

Pryotowaki, II. Metabolism of gas exchange across the placental barrier. *Clin. Obstet. Gynecol.* 6:47, 1963.

Lawrence D. Longo, William W. Allen, and Gordon G. Power

INTRODUCTION

The fetal circulation is essential for the transport of nutrients, including oxygen, and the removal of various metabolites from fetal tissues. There are several anatomic differences from the adult circulatory system that serve to adapt the fetal circulation to intrauterine existence. Although numerous reports have described the fetal circulation in detail, there still are gaps in our understanding of many problems. These include the factors that control the cardiac output; its distribution to the various organs, particularly the brain, heart, and placenta; the dynamics of the parts of the circulation unique to the fetus; and the factors that control the changes in hemodynamics at the time of birth. This chapter will attempt to describe what is known, delineate important aspects about which little is known, and present a brief introduction to a systems-analysis approach for mathematically simulating this uniquely designed system. Several recent reviews have considered various aspects of the fetal circulation [3, 30, 51] and should be consulted for further details.

ANATOMIC CONSIDERATIONS

The major anatomic components of the circulation of the fetus are depicted in Figure 46-1. Although the various specialized vessels of the fetus were known to the ancients, it was William Harvey who first recognized their functional significance in 1628 [26]. The umbilical veins carry arterialized blood from the placenta to the fetal body. At the fetal liver, the vessel divides. A portion of this blood is carried through the *ductus venosus* (DV) to the *inferior vena cava* (IVC) and to the fetal heart. The remainder courses through the liver and thence through hepatic veins to the IVC. Upon reaching the heart, much of the blood flows through the *foramen ovale cordis* (FO) into the left atrium, left ventricle, and aorta to supply relatively well-oxygenated blood to the fetal brain. That portion of the blood returning through the *superior vena cava* (SVC) enters the right atrium and right ventricle, and it then flows into the pulmonary artery. A considerable fraction is shunted through the *ductus arteriosus* (DA) into the aorta and thence either into the placenta via the umbilical artery or to the abdominal organs and hindquarters. These various aspects will be considered in greater detail.

UMBILICAL VEIN

Arterialized blood from the fetal-placental capillaries flows through venules of increasing size to a single umbilical vein in humans and primates or to two veins in ruminants

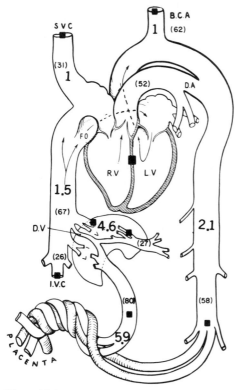

Figure 46-1
The fetal circulation in the lamb. The numbers without parentheses give the oxyhemoglobin saturation of the blood units and those in parentheses, the mean transit times in seconds between the various points indicated by the black squares (see Table 46-4). (Modified from G. V. R. Born et al. *Cold Spring Harbor Symp. Quant. Biol.* 19:102, 1954.)

and some other species. Although the major portion of the umbilical vein is extra-abdominal in the umbilical cord, a shorter segment is intraabdominal. The rate of umbilical venous flow has been reported to be from about 130 to 300 (ml/min)/mm Hg in the fetal lamb [1, 18, 33, 41, 44], with more recent and more reliable results tending to be higher.

There are numerous methodologic problems associated with these measurements of umbilical flow and other fetal physiologic studies. Most were done in animals in which the mother was acutely anesthetized, a procedure that may produce variable effects, depending upon the anesthetic agent used. In addition, the fetuses were exteriorized from their normal intrauterine station during most of these studies. The exteriorized preparation has limited value, since constriction of the cord vessels readily occurs. Those studies performed using "chronic" catheters in unstressed animals probably represent a more physiologic state, but even here, completely "normal" values for comparison are lacking. What is surprising is not the differences in values, but, considering that the studies were done over a period of years, by many different investigators, in different breeds of sheep, and using different methods of measurement and anesthesia, it is surprising that the results agree as well as they do. Dawes et al. [16] recently noted that the aortic blood flow changed as much as ±30% during fetal respiratory movements. Thus, the umbilical flow, which is a significant fraction of the aortic flow, probably varies within a given time period depending on the fetal activity.

Finally, it should be noted that almost all fetal cardiovascular studies have been, and probably will continue to be, performed in various animal species, since there is no way as yet that they can be performed safely in intact humans. Sheep and goats have served as the most useful animals for such studies because of the relatively large size of the fetuses, their ready availability, the ease of handling, and the large amount of physiologic data already available for these species.

The blood pressure in the umbilical vein is about 10 to 12 mm Hg [3]. The pressure drop across the placental circulation is from about 40 mm Hg in the cotyledonal arteries to about 10 mm Hg in the umbilical vein. The resistance in the umbilical vein from the placenta to its division at the liver is low, about 0.05 mm Hg·min·kg/ml.

Umbilical venous blood is relatively well oxygenated, with an oxyhemoglobin concentration of about 85% saturation. The oxygen tension (Po_2) is about 40 to 42 mm Hg, and the pH is about 7.35. Dawes [15] has noted that the blood gas values apparently have only a slight influence on the rate of umbilical blood flow.

DUCTUS VENOSUS

The single umbilical vein divides near the liver into (1) branches supplying the hepatic tissue and (2) the ductus venosus, which is a large, bare vessel, only 1.5 to 2 cm long, that lies buried in the liver parenchyma and connects directly with the IVC bypassing the liver [7]. Structurally, it resembles the abdominal portion of the umbilical vein, except that at its junction with the umbilical vein, its muscular coat is thickened into a sphincter-like structure with a small amount of smooth muscle [47]. There is evidence that this "sphincter" is innervated by postganglionic branches of the vagus nerve, but what its function may be remains unclear. Reynolds [43] has suggested that this sphincter in the ductus venosus may be important in the regulation of umbilical blood flow as well as of flow through the hepatic circulation. That the structure is not always necessary for fetal survival is indicated by its absence in the fetal horse and several other species.

Recent studies using dye-dilution techniques in the fetal lamb indicate that 53% (±0.05 SE) of the umbilical blood flows through this vessel (Fig. 46-2). This is to be compared with the value of 11% or greater reported by Barclay et al. [7] and that of 34% to 91% reported by Rudolph and Heymann [45]. The latter authors noted that the fraction of umbilical venous blood flowing through the ductus venosus increased with increasing umbilical flow. Occlusion of the ductus venosus in mature fetal lambs results in no significant change in arterial blood pressure or carotid arterial oxyhemoglobin saturation [2].

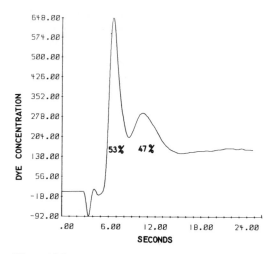

Figure 46-2
Indicator-dilution curve showing the fractions of dye passing through the ductus venosus (53%)
and the hepatic vessels (47%). The dye was injected into the umbilical vein and sampled in the
inferior vena cava immediately below the heart.

FORAMEN OVALE CORDIS

Much of the blood returning to the heart from the IVC flows on one side of the crista
dividens directly into the left atrium (LA) through the foramen ovale cordis (Fig. 46-3).
Blood returning from the SVC predominately enters the right atrium (RA), flows to
the right ventricle and pulmonary artery, and then flows to the lower half of the body,
the placenta, and the abdominal viscera. The exact magnitude of blood flow from the
IVC and SVC that flows through each of these pathways is not known, nor is it understood what factors regulate the flows through these separate routes.

DUCTUS ARTERIOSUS

Connecting the left pulmonary artery and the aorta is a vessel 6 to 8 mm in diameter
and about 2 cm in length. This is probably the most important of the vascular shunts
in the fetus, since it is intimately involved in the changes in systemic and pulmonary
hemodynamics before and immediately after birth. Assali et al. [4] estimated that
ductus arteriosus flow averages 101 (±30 SD) ml/min per kilogram, which is equivalent
to roughly 50% of the effective cardiac output. Morris et al. [40] reported that
asynchrony between the contractions of the right and left ventricles (the right occurring
first) is an important factor in maintaining the circulation of the ductus arteriosus,
particularly during early lung expansion. Innervation of the ductus arteriosus is from
both the right and left vagus nerves as well as from sympathetic nerves. The function
of these nerves in control of the circulation of the ductus arteriosus is not known.

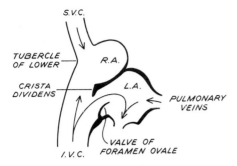

Figure 46-3
The entry of the inferior and superior vena cava into the fetal heart. The foramen ovale cordis lies between the inferior vena cava and left atrium, rather than between the right and left atria. (From G. S. Dawes. *Foetal and Neonatal Physiology. A Comparative Study of the Changes at Birth.* Chicago: Year Book, 1968.)

UMBILICAL ARTERIES

These large vessels (see Fig. 46-1) arise from the internal iliac arteries, but they may be considered terminal branches of the descending aorta that carry venous blood to the placenta. Their muscular walls are relatively thick, and various external stimuli result in severe segmental spasm. The roles of neurologic, chemical, and mechanical stimuli in triggering this vasoconstriction remain unclear.

PHYSIOLOGIC CONSIDERATIONS

FETAL BLOOD VOLUME

Not only does the fetal blood volume increase with gestational age, as does the extra-fetal volume of blood in the umbilical vessels and placenta, but so does the ratio between the fetal and the placental-umbilical volumes. Elliott et al. [19] concluded that the total volume of fetal and placental blood was about 9% of the fetal weight. Using a double-isotope technique, Creasy et al. [13] estimated that the total blood volume was 137 ± 3.1 ml per kilogram of body weight in fetal lambs during the last third of gestation. The red blood cell and plasma volumes were found to average 44.8 ± 1.8 ml and 89.9 ± 2.0 ml per kilogram, respectively, in their studies.

FETAL HEART RATE AND ARTERIAL PRESSURES

The fetal heart rate (FHR) has attracted widespread interest among clinicians as an indicator of fetal well-being, presumably because of its availability for monitoring and the sophistication in instrumentation to perform such monitoring. Although a significant decrease in the heart rate is often associated with compromised fetal oxygenation, it is not necessarily an index of either cardiac output (the product of heart rate and stroke volume) or the oxygenation of fetal tissues. The fetal heart rate is normally about 140 ± 20 beats/min in humans and 190 ± 25 beats/min in sheep [14].

The mean fetal arterial pressure rises during the course of gestation to about 55 mm Hg in humans and 65 mm Hg in lambs at term. The variation among species, however, is wide [14]. The human systolic and diastolic pressures are about 70 and 50 mm Hg, respectively. Table 46-1 lists some experimentally determined values for blood pressures and flows in the chambers and great vessels of the heart.

Table 46-1. Pressures and Blood Flows in the Chambers and Great Vessels of the Fetal Heart

	Blood Flow (ml/min per kilogram)	Pressures (mm Hg)		
		Mean	Systolic	Diastolic
Right atrium	—	3.5 ± 0.3[a]	—	—
Right ventricle	—	—	62 ± 11.2[a] 66[b]	8 ± 2.7[a] 10[b]
Left atrium	—	2.7 ± 0.5[a]	—	—
Left ventricle	—	—	58 ± 13.3[a] 64[b]	6 ± 1.7[a] 7[b]
Foramen ovale	60[a]	—	—	—
Pulmonary artery	138 ± 41[a] 148[b]	58 ± 8.4[a] 67[b]	—	—
Aorta (ascending)	97 ± 31[a] 110[b]	56 ± 8.6[a] 65[b]	—	—
Ductus arteriosus	101 ± 30[a] 108[b]	—	—	—

[a] Assali, Morris, and Beck [4].
[b] Assali, Bekey, and Morrison [3].

FETAL CARDIAC OUTPUT

In the adult, *cardiac output* refers to the output of the left ventricle, but, since the ventricles work in series, the outputs of both ventricles are the same. In the fetus, the heart works in parallel, and the effective cardiac output equals the output of the left ventricle plus that of the right ventricle that flows through the ductus arteriosus. To a certain extent, the variation in the results reported by various workers (see Table 46-2) is due to differences in what is measured as "cardiac output."

Barcroft et al. first measured fetal cardiac output in goats [9] and sheep [10] (Table 46-2) using an instrument that measured the changes in heart volume over a given time period. Barcroft [8] acknowledged that these results approached the "inferior limit" of normal. Studies using electromagnetic flowmeters in an exteriorized fetus with the chest opened and the vessels manipulated can hardly be expected to give results that are much more physiologic. More recently, studies using relatively noninvasive techniques with the fetus in utero have resulted in significantly higher values for cardiac output. These results suggest that fetal cardiac output is close to 600 ml/min per kilogram of body weight, a value six times that in the adult. This

Table 46-2. Combined Ventricular Outputs in the Fetus

Cardiac Output (ml/min per kilogram)	Author(s)	Method
200	Barcroft, Flexner, and McClurkin [9]	Cardiometer
245	Barcroft and Torrens [10]	Cardiometer
230	Dawes, Mott, and Widdicombe [18]	Calculation from blood O_2 contents
198	Assali, Morris, and Beck [4]	Electromagnetic flowmeters
361	Mahon, Goodwin, and Paul [34]	Indicator dilution
562	Rudolph and Heymann [46]	Microspheres
220	Assali, Bekey, and Morrison [3]	Electromagnetic flowmeters
515	Faber, Green, and Long [22]	Indicator dilution
658	Faber and Green [23]	Indicator dilution
640	Power and Longo [42]	Indicator dilution

value is three times greater when it is related to the output of a given ventricle. It seems that this relatively high output of blood is one of the most important adaptations of the fetus in utero to its environment of relatively low arterial O_2 tension.

The fetal systemic circulation, in comparison to that of the adult, is essentially a high-flow, low-resistance circuit. On the other hand, the fetal pulmonary circulation is a low-flow, high-resistance circuit. The fetal right ventricle is relatively prominent in relation to the left ventricle, in contrast to their relation in the adult. In the fetus, the right ventricular wall is as thick as the left, and the mean right ventricular pressures are slightly greater than those of the left ventricle. There is also some evidence that right ventricular ejection precedes that of the left ventricle.

DISTRIBUTION OF CARDIAC OUTPUT

Dawes et al. [18] first estimated the distribution of the fetal cardiac output on the basis of oxyhemoglobin saturations in various parts of the circulation (Table 46-3). Rudolph and Heymann [45] measured the distribution of flows by injecting radioactively labeled microspheres into the fetal circulation. This method is a reliable index of flow distribution, provided that the microspheres are trapped in the vascular beds of the various organs. Although this is true for most organs, recent evidence indicates that trapping may be incomplete in the brain. Thus, the true cerebral flow is probably greater than the value of 5% of cardiac output that was measured using microspheres [45].

FETAL CIRCULATION TIMES

The rapidity with which changes in O_2 tension in the placental exchange vessels is reflected in the vascular beds of the fetus is important. To understand this change,

Table 46-3. Distribution of Cardiac Output to Various Organs of Fetal Lambs

Organ	Dawes et al. [18][a]	Rudolph and Heymann [46][b]	Rudolph and Heymann [45][c]
Placenta	55 ml.	41.2%	$53.8\% - 0.08D$[d] $(\pm 5.2\%)$[e]
Brain	–	4.9	$2.14 + 0.008D$ (± 1.0)
Heart	–	4.6	3.2 (± 0.8)
Lungs	12	3.9	$-0.4 + 0.047D$ (± 1.7)
Gastrointestinal tract	–	5.4	$-0.6 + 0.04D$ (± 0.7)
Kidney	–	2.2	$3.2 - 0.012D$ (± 0.4)
Liver (via hepatic artery)	–	1.6	–
Spleen	–	1.6	1.05 (± 0.3)
Upper carcass	15	14.9	17.75 (± 2.5)
Lower carcass	18	19.9	18.26 (± 2.6)
Inferior vena cava	73	–	–
Superior vena cava	15	–	–
Foramen ovale	46	–	–
Right atrium	42 (15 from SVC) (27 from IVC)	–	–
Left ventricle aortic arch	58	–	–
Ductus arteriosus	30	–	–
Descending aorta	73	–	–

[a]Term.
[b]90–140 days gestation.
[c]60–150 days gestation.
[d]D = gestational age in days.
[e]Maximum standard error of mean in six age groups studied.

a knowledge of the circulation times in the various parts of the fetus is necessary. Table 46-4 presents data on these circulation times as determined in fetal lambs in utero using indocyanin green dye. Mean transit times are also shown in Figure 46-1. The time required for the circulation of blood from the umbilical vein, around the circuit, and back to the umbilical vein is about 13 seconds. This contrasts to a mean time of transit through the entire adult circulation of about 1 minute.

Fetal circulation times determine how fast the changes in O_2 tensions in the fetoplacental blood will affect the O_2 delivery to the fetal heart, brain, and other vital organs. For instance, during a severe uterine contraction, when the placental O_2 exchange may rapidly decrease 40% or more, the PO_2 in fetal arterial blood will be rapidly decreased, and the change will be reflected in the fetal tissues in about 5 seconds.

PULMONARY BLOOD FLOW

The fetal lungs receive about 4% of the total cardiac output (see Table 46-3). The pulmonary vascular resistance is about five times the systemic resistance, and it probably results from a combination of both mechanical tortuosity and chemical factors

Table 46-4. Circulation Times in Various Parts of the Fetal Vascular Bed

From	To	Time (seconds)
Umbilical vein	Carotid artery, 1st peak	1.9 (± 0.72, 0.17)
Umbilical vein	Carotid artery, 2nd peak	6.2 (± 1.41, 0.34)
Umbilical vein	Superior vena cava	21.5
Jugular vein	Atrium	1.0
Jugular vein	Carotid artery	2.2 (± 1.33, 0.27)
Jugular vein	Femoral artery	3.1 (± 0.83, 0.13)
Jugular vein	Umbilical artery	6.7 (± 0.90, 0.37)
Aortic arch	Umbilical artery	2.0 (± 0.41, 0.23)
Aortic arch	Femoral artery	2.5 (± 0.97, 0.40)
Femoral vein	Carotid artery	2.6 (± 0.81, 0.16)
Femoral vein	Femoral artery	3.1 (± 0.79, 0.14)
Umbilical artery	Umbilical vein	5.9 (± 2.2, 0.72)
Umbilical vein	Umbilical vein	13.0

such as the PO_2 and PCO_2 of the blood. Thus, the artificial ventilation of the fetal lung with a gas mixture containing 7% CO_2 and 3% O_2, which are values that were chosen to approximate the normal fetal blood gas tensions, resulted in the pulmonary vascular resistance falling about 33% [12]. Subsequent ventilation of the lungs with air resulted in the pulmonary vascular resistance decreasing to about one-third of its original value [12]. Gilbert et al. [25] recently showed that the fetal pulmonary vessels may be considered as a Starling resistor, with the arterial resistance being much greater than the venous resistance. These investigators further demonstrated that the flow through these vessels may vary as a function of the arterial blood pressure minus the surrounding alveolar pressures (so-called sluice flow), rather than as a function of arterial pressure minus the pulmonary venous pressure.

CEREBRAL BLOOD FLOW

To date, there are probably no accurate measurements of the blood flow to the fetal brain. Rudolph and Heymann [45] found that the brain receives about 5% of the cardiac output or about 150 ml/min per 100 gm of fetal weight, but these results may be invalid because of incomplete trapping of the microspheres in the brain. Both Makowski et al. [35], who used the microsphere method in sheep, and Barker [11], who used a direct outflow method in rats, reported that the fetal cerebral flow was about twice that of the adult on a weight basis. Attempts to measure the fetal cerebral blood flow using indicators and the Fick principle have not yet been reported.

UMBILICAL BLOOD FLOW

Umbilical blood flow represents the largest fraction of the cardiac output: about 40% to 55%, depending upon the gestational age (see Table 46-3). Near term, the values are probably 200 to 300 ml/min per kilogram. It is not clear whether the flow is relatively higher early in gestation and decreases to term; the evidence for this is not convincing.

The measurements of umbilical flow in the human fetus agree fairly well with the results that have been obtained from lambs. Assali et al. [5], using an electromagnetic flowmeter, reported values of 110 ml/min per kilogram in the human fetus. Stembera et al. [49], who employed a thermodilution method in newborn infants within 2 minutes of delivery, reported values of 75 ± 7 ml/min per kilogram. These values are undoubtedly low, because of the acute nature of the experiments and the manipulations involved.

MATERNAL VASCULAR PRESSURE EFFECTS ON FETOPLACENTAL BLOOD FLOW

In most organs, the rate of blood flow is a function of the hydrostatic driving pressure, i.e., the inflowing pressure minus the surrounding pressure. In the lung, this type of flow has been termed pulmonary "sluice" [6] or "waterfall" flow [49]. Thus, when the surrounding alveolar pressure is greater than the outflow pressure, the flow has been shown to vary as a function of the inflowing pressure minus the surrounding alveolar pressure, and the outflow pressure has no influence on the flow. These interrelations of alveolar pressure and blood flow in the normal lung have been analyzed theoretically and verified experimentally.

Sluice flow may also occur in the placenta so that the fetoplacental blood flow is a function of inflowing fetal arterial pressure minus the surrounding maternal vascular pressure. This relation has a number of implications for maternal-fetal respiratory gas transfer under certain conditions. There are at least two types of sluice flow that might exist in the placental vascular bed. In the first, the maternal vascular pressures in the capillaries or the intervillous space of the placenta might squeeze on the fetoplacental capillaries. In the second, during a uterine contraction, the amniotic fluid pressure might rise high enough so that the uterine vessels would collapse and result in a change in the maternal-placental blood flow.

Figure 46-4A illustrates conventional or nonsluice flow. Maternal blood enters through the uterine artery, flows close to the fetal vessels (remaining within vessels in sheep or entering the intervillous space in man), and exits through the uterine veins. Fetal blood enters the placenta from the umbilical artery, flows through the placental tissue in thin-walled capillaries (which are assumed to be freely collapsible), and drains into the umbilical vein. Figure 46-4A shows the fetal capillaries distended along their length, since the pressure within them is assumed to exceed the surrounding pressure ($P_{surrounding}$) of maternal blood. Fetal blood flow then depends upon the difference in pressure between the umbilical artery ($P_{umb.\ art.}$) and vein ($P_{umb.\ vein}$), since:

$$P_{umb.\ art.} > P_{umb.\ vein} > P_{surrounding} \qquad (46\text{-}1)$$

Nonsluice flow may be summarized very simply: whenever the umbilical venous pressure is greater than the surrounding pressure, the blood flow in the capillary is proportiona

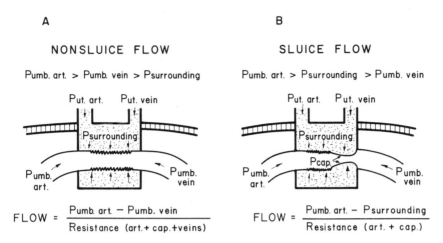

A

NONSLUICE FLOW

Pumb. art. > Pumb. vein > Psurrounding

$$FLOW = \frac{Pumb.\ art. - Pumb.\ vein}{Resistance\ (art. + cap. + veins)}$$

B

SLUICE FLOW

Pumb. art. > Psurrounding > Pumb. vein

$$FLOW = \frac{Pumb.\ art. - Psurrounding}{Resistance\ (art. + cap.)}$$

Figure 46-4
Nonsluice and sluice flow in the placenta. *A*. Fetal vascular pressure exceeds the maternal surrounding blood pressure, and conventional pressure-flow relationships prevail. *B*. Pressure of the surrounding maternal blood exceeds the umbilical venous pressure, and the fetal capillaries tend to collapse at their venous end; changes in the maternal vascular pressures affect the fetal blood flow.

to the difference between the umbilical arterial pressure and the umbilical venous pressure, and changes in the surrounding pressure have no influence on the flow.

Figure 46-4B shows the change in vascular relations under the conditions of sluice flow. In this case, it is assumed that the surrounding maternal blood pressure is higher than the umbilical venous pressure, or:

$$P_{umb.\ art.} > P_{surrounding} > P_{umb.\ vein} \qquad (46\text{-}2)$$

As fetal blood moves along the capillary, the intravascular pressure drops progressively. Toward the venous end of the capillary, it tends to become less than the surrounding pressure. The capillary begins to collapse (see Fig. 46-4B), and the blood flow is reduced. The capillary pressure then begins to rise, increasing toward the value of the inflowing (umbilical arterial) pressure. The vessel distends as a result and offers less resistance to flow. As flow is restored, there is once again a greater pressure drop along the length of the capillary, and the surrounding maternal blood pressure again exceeds the pressure within the venous end of the fetal capillary. The capillary then becomes further constricted at its downstream end, and the resistance to flow adjusts accordingly. Thus, there is an interplay between the pressures and flows within the two circulations. This does not mean that blood flow must necessarily be intermittent. It does require, however, that constriction at the downstream end must adjust automatically to maintain a pressure at that end that is equal to the surrounding pressure. Sluice flow may thus be summarized: whenever the surrounding maternal blood pressure is greater than the umbilical venous pressure, the capillary blood flow is proportional to the difference between the umbilical arterial pressure and the surrounding pressure, and changes in the umbilical venous pressure have no influence on the flow.

Finally, there is the simple case in which the surrounding maternal blood pressure exceeds the umbilical arterial pressure; in this instance, the capillary flow is zero.

EXPERIMENTAL EFFECTS OF INCREASED UTERINE VENOUS PRESSURE

The effects of increased maternal placental vascular pressure on the inflowing pressure to the perfused cotyledon were studied in 60 experiments carried out in 14 different sheep and in five experiments in one monkey. The principle of the method was to perfuse the umbilical circulation of an isolated segment of the placenta in situ at a constant rate while measuring the inflowing and outflowing pressures. The effects of changes in the maternal vascular pressures on the umbilical resistance to blood flow could then be measured, free from secondary compensations that the fetal circulation might otherwise make.

Figure 46-5 shows the pressures in various vessels following the complete occlusion of the maternal inferior vena cava (IVC) by a clamp in a typical experiment. Changes were detected first in the maternal circulation; the uterine venous pressure rose, over a period of several seconds, to about 40 mm Hg. About 8 seconds later, the inflowing pressure to the fetal circulation of the cotyledon rose 20 mm Hg from 58 to 78 mm Hg. Since this rise occurred at a constant flow rate, with a constant outflow pressure, and with constant amniotic and intraperitoneal pressures, it clearly indicates that a greater restriction to the umbilical blood flow occurred within the placenta. The fetal heart rate decreased 30 beats per minute (from 126 to 96 beats per minute) within 35 seconds after the IVC was clamped. When the clamp was released, the uterine venous pressure fell, and the fetal inflowing pressure returned to control levels. This illustrates the effect of the surrounding maternal vascular pressures on the fetal resistance to blood flow. It may be noted that the mean maternal arterial pressure fell 10 mm Hg (from 110 to 100 mm Hg). This was probably due to decreased venous return to the heart, and it tended to offset the increase in the maternal surrounding pressure caused by clamping the IVC.

In the series of experiments, the average pressure rise in the inflow cotyledonal vessel was 8 (± 4 SD) mm Hg in response to an average increase in the uterine venous pressure of 30 mm Hg. These responses were observed consistently, but there was a tendency for less pronounced responses to occur as the preparation aged. In only three instances did the inflow pressure fail to rise measurably when the uterine venous pressure rose. The time from the onset of IVC clamping to the onset of the rise in uterine venous pressure averaged 7 seconds, with a range of 1 to 20 seconds. The response-time of the pressure rise in the inflow cotyledonal vessel was more rapid in those instances when the uterine venous pressure rose rapidly, whereas when the latter rose slowly, the rise of the inflowing pressure was delayed. Systemic changes in the fetus included an average drop of 32 beats per minute in the heart rate and an average transient increase of 6 mm Hg in its blood pressure. The fetal bradycardia

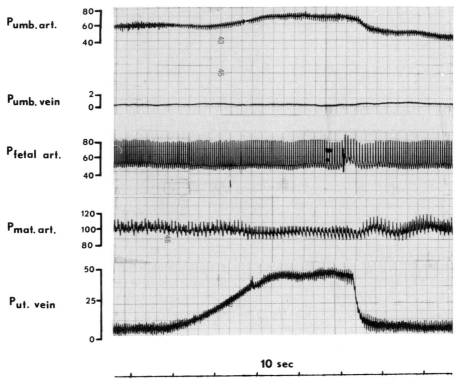

Figure 46-5
Pressure changes in various vessels following the clamping of the maternal inferior vena cava, which is signaled by the sudden rise in uterine venous pressure. The uppermost record shows the pressure rise in the inflowing cotyledonal artery. The rise indicates that the maternal "squeeze" tended to restrict flow through the fetoplacental vessels. A rise in fetal arterial blood pressure and a slowing of the fetal heart rate typically followed in about 5 to 20 seconds. All pressures are given in mm Hg with a common zero at the level of the perfused cotyledon.

became progressively more severe as the duration of clamping was prolonged. The average duration in which the IVC was clamped was 66 seconds, and the range was from 20 to 200 seconds. The rate of fetal perfusion had no significant influence on the results, and in several experiments in which the cotyledon was left exposed, the above effects were also demonstrated.

EXPERIMENTAL EFFECTS OF INCREASED OUTFLOW PRESSURE
ON INFLOW PRESSURE

Although changes in the surrounding pressure must lead to changes in the inflow pressure at a constant flow rate, such changes may occur under nonsluice conditions as well as under sluice conditions; therefore, they cannot prove per se that sluice flow is present. The criterion of sluice flow is the independence of the inflow pressure-flow relations under conditions of change in the outflow pressure. Several experiments were carried out to provide evidence that this criterion is fulfilled for the case of the umbilical circulation.

First, the inflow pressure was measured while the outflow pressure was raised in stepwise manner by elevating the height of the outflow reservoir. Figure 46-6 illustrates the average results of four sequential curves that were determined while the maternal blood pressures remained essentially constant. The fetal inflow pressure was 30.5 ± 1.3 mm Hg when the outflow pressure was zero, and, most importantly, it did not rise when the outflow pressure was increased to 7 and then to 15 mm Hg. Similar results were obtained in six curves for each of two other sheep. Once the outflow pressure exceeded 15 to 20 mm Hg, however, further raises in the outflow pressure resulted in increases in the inflow pressure. Above this level, the pressure-flow characteristics shifted from a sluice to a nonsluice pattern. Since the umbilical venous pressure is about 10 mm Hg in utero, these results strongly suggest that sluice flow characteristics normally prevail in the umbilical circulation.

In two experiments in each of two sheep, the change in the inflow pressure as a function of the outflow pressure was measured when the uterine venous pressure was 26 to 30 mm Hg. Under these conditions, there was a shift in the rise point to the right; i.e., the outflow pressure had no effect on the inflow pressure until the outflow pressure exceeded 20 to 22 mm Hg. This is additional evidence for the presence of sluice flow.

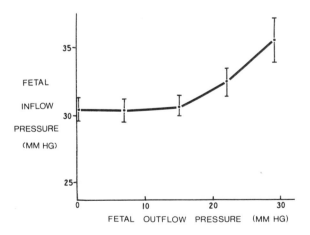

Figure 46-6
Relation of outflow pressure to inflow pressure in the fetal circulation of a cotyledon perfused at a constant flow rate. Results are averages (± SE) of four determinations in a sheep in which the maternal arterial and the uterine venous pressures remained constant. Note that the outflow pressure had no influence on the inflow pressure until its level exceeded about 15 mm Hg, suggesting that sluice flow characteristics occur in the umbilical circulation at normal pressure levels.

Second, if the response to IVC clamping were due to a sluice mechanism, it should be possible to block or attenuate the response by raising the outflow pressure from the cotyledon. This should distend the cotyledonal vessels and minimize their tendency to collapse from the surrounding maternal blood pressure. Therefore, measurements of the pressure responses in the inflowing cotyledonal artery were repeated after the height of the outflow reservoir was raised to increase the outflow pressure by 12 to 40 mm Hg. Whereas previously the fetal inflow pressure had increased, on the average, 2.7 mm Hg for a 10-mm Hg rise in the uterine venous pressure, the response now decreased to a 1.8-mm Hg rise, a decrease of 33%. The difference in the responses was statistically significant by the t test ($P < 0.05$). The block was essentially complete — i.e., the response was a pressure change of 1 mm Hg or less — in 12 of the 35 instances examined. In general, the block was more successful for mild increases in uterine venous pressures, as is clearly shown by the results of graded occlusions of the IVC in sheep (Fig. 46-7).

In Figure 46-7, the upper curve shows the mean increase in the inflowing cotyledon pressure in response to graded increases in the uterine venous pressure when the cotyledonal outflow pressure was zero. The lower curve shows the smaller responses that were observed when the cotyledonal outflow pressure averaged 25 mm Hg. It is evident that in this latter instance, the response of the inflowing cotyledonal pressure is minimal until the pressure surrounding the umbilical capillaries exceeds the cotyledonal outflow pressure. The rise point is only an indirect index of the pressure surrounding the fetoplacental vessels, since it is influenced not only by the surrounding pressure, but also by the pressure drop between the sluice flow and the reservoir.

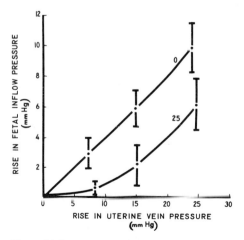

Figure 46-7
Results from studies of sheep showing the rise in the pressure of the blood perfusing a single cotyledon following stepwise increments in the uterine venous pressure. The upper curve shows the response when the fetal outflow pressure was zero, and the lower curve shows the attenuated response when the outflow pressure was increased to an average of 25 mm Hg. The experiment was repeated three or four times, and the average results (±1 SD) are shown.

PHYSIOLOGIC AND CLINICAL IMPLICATIONS

By affecting the balance of hydrostatic pressures between maternal and fetal blood, sluice flow mechanisms should thereby affect water movement across the placenta. Under sluice conditions, the fetal capillary pressure exceeds that of the maternal blood everywhere except at the downstream end of the vessel. Even here at the sluice point, the maternal and fetal blood pressures are equal, and thus, considering the entire length of the vessels, the mean fetoplacental capillary pressure would exceed the mean maternal capillary pressure.

One might expect that the fetus would therefore soon become dehydrated. This need not necessarily occur, however, since the osmotic pressures might not be equal on both sides of the placental membrane. The results of measuring osmotic pressure differences in maternal and fetal blood vary. Meschia and co-workers [37] reported values from 0 to 4 mOsm/kg water less in fetal sheep plasma than in maternal plasma, whereas others have found higher values (see [37]). The point remains, however, that an osmotic pressure of only 0.25 mOsm/kg water in fetal plasma would be sufficient to permit fetal hydrostatic pressures to exceed maternal pressures by 5 mm Hg. It is also true that the mean fetoplacental capillary pressures must exceed the maternal blood pressures during nonsluice flow if there is to be fetal blood flow at all, and thus it seems inevitable that a small osmotic pressure difference must exist.

There are a number of clinical conditions in which sluice flow may play an important role. For example, during uterine contractions, maternal placental blood flow might decrease, with a resultant decrease in the ratio of maternal-to-fetal blood flow (Fig. 46-8). This would result in a decrease in O_2 delivery to the fetus, and it is a likely explanation for the "cord compression" pattern of fetal bradycardia that is recorded during uterine contractions [29]. Thus, although this fetal heart rate pattern may be produced experimentally by cord constriction, it may, in vivo, actually result from the compression of the umbilical circulation within the placenta itself, rather than from a compression of the cord per se.

Another clinical condition that occurs in some pregnant women is the supine hypotensive syndrome [36]. The gravid uterus presses on the inferior vena cava when the woman lies supine, thus blocking the venous return to the maternal heart. Arterial hypotension then develops. During occlusion of the inferior vena cava, the outflow from the placenta is obstructed, and the pressure in the intervillous space surrounding the fetal capillaries presumably rises. This interrupts the umbilical part of the fetal circulation (see Fig. 46-8), which causes a rise in the fetal blood pressure and a consequent slowing of the fetal heart rate in response to baroreceptor signals. In fact, the slowing of the fetal heart rate has been observed clinically and documented electrocardiographically in cases of supine hypotensive syndrome, but its mechanism has not been well understood. Sluice flow phenomena have provided the first adequate explanation of this slowing of the fetal heart rate. Indeed, in the experiments described previously, clamping the inferior vena cava yielded results that duplicate the clinical findings in the supine hypotensive syndrome, including the fetal bradycardia that usually developed within 10 to 20 seconds. Thus, the initial slowing of the fetal heart

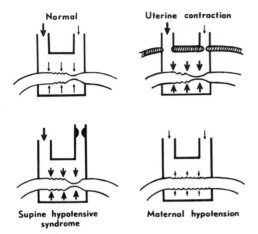

Figure 46-8
Clinical conditions that may affect the umbilical circulation by means of the sluice mechanism.
The size of the arrows illustrates the changes in vascular pressures relative to normal (*left upper
panel*). The diagram for uterine contraction (*right upper panel*) is meant to suggest that uterine
venous outflow is restricted by either (1) the direct action of myometrial muscle fibers on the
uterine veins or (2) the squeeze due to increased intrauterine pressure on the veins as they exit
from the uterus. The rise in fetal intrauterine pressures is also shown. The diagram for supine
hypotensive syndrome (*left lower panel*) is meant to suggest the collapse of the inferior vena cava.

rate may be attributed to fetal baroreceptor activity, while subsequently, as fetal
hypoxia develops, the fetal heart rate may be slowed to an additional extent.

Finally, the maternal arterial pressure may be reduced by anesthesia, blood flow,
and other causes (see Fig. 46-8). The maternal surrounding pressure would fall, and
even normal umbilical pressure would fully distend the intraplacental fetal vessels.
The umbilical flow would increase, and this adaptive mechanism would enhance the
delivery of O_2 and metabolites. This mechanism was demonstrated in experiments
in which the aorta was clamped. Its quantitative importance, however, must be
worked out in experiments that more closely simulate the blood pressure and flow
conditions that prevail during life.

INTERRELATIONS OF FETAL CARDIAC OUTPUT, UMBILICAL BLOOD FLOW, AND THE PLACENTAL EXCHANGE OF OXYGEN, CARBON DIOXIDE, GLUCOSE, AND WATER

UMBILICAL BLOOD FLOW AND PLACENTAL WATER TRANSFER

The fetus requires more water than oxygen, on a molar basis, for its growth [8].
During the last trimester of pregnancy, the fetal weight (W) of rhesus monkeys [14]
increases as:

$$W^{1/3} = 0.0594\,t - 1.49 \pm 0.00032 \text{ SE} \tag{46-3}$$

where t is the gestational age in days and SE is the standard error of the mean. The relation for the human fetus is similar. Thus, near term, the fetal weight gain is about 12 gm/day per kilogram, and the net water gain is about 9 gm/day per kilogram [50], or about 0.5 mole/day per kilogram. The rate of O_2 transfer is about 0.39 mole/day per kilogram (assuming an O_2 consumption of 6 ml/min per kilogram), or 8.64 liters/day per kilogram. Since water is produced on a mole-for-mole basis from O_2 when glucose is oxidized, i.e.,

$$C_6H_{12}O_6 + 6O_2 \rightarrow 6H_2O + 6CO_2$$

the rate of placental water transfer need only be about 0.11 mole/day per kilogram, assuming that all the water is produced by fetal metabolism is retained. Actually, the rate of placental water transfer is several hundred times this rate [27].

The rate of placental water transfer determines the fetal blood volume [24]. This in turn determines the central venous pressure and the fetal cardiac output, which determine the umbilical blood flow. Finally, the umbilical flow determines placental O_2 and CO_2 transfer, and, since CO_2 exchange can be linked to water exchange (see below), there is a circular interaction of all these various factors.

Faber [21] has developed a theoretical basis for linking umbilical blood flow and water exchange. Briefly, the rate at which the fetus gains water across the placenta, $\dot{V}_{H_2O}$, is:

$$\dot{V}_{H_2O} = (Pc_M - \pi_M + \pi_F - Pc_F)/R_P \tag{46-4}$$

where Pc_M and Pc_F are the maternal and fetal mean capillary hydrostatic pressures, respectively; π_M and π_F are the maternal and fetal osmotic pressures, respectively; and R_P is a proportionality constant representing the hydraulic resistance of the placental barrier. All pressures are referred to the same reference, the intrauterine arterial pressure, Po_M.

The fetoplacental blood flow, $\dot{Q}_F$, is determined by the difference in the umbilical arterial and venous pressures and the resistance to flow under nonsluice conditions. This is true for both the precapillary resistance, R_1, and the postcapillary resistance, R_2, when the appropriate pressure differences are used. It is useful to determine the fetoplacental blood flow in terms of the postcapillary resistance to flow between the placental capillaries and the fetal central venous pressure, Pv_F:

$$\dot{Q}_F = (Pc_F - Pv_F)/R_2 \tag{46-5}$$

Combining Equations 46-4 and 46-5 with the elimination of Pc_F gives:

$$\dot{Q}_F = [Pc_M - \pi_M + \pi_F - Pv_F - (\dot{V}_{H_2O} R_P)]/R_2 \tag{46-6}$$

Under equilibrium conditions with no net fetal water gain or loss, $\dot{V}_{H_2O}$ equals zero, and Equation 46-6 reduces to:

$$\dot{Q}_F = (Pc_M - \pi_M + \pi_F - Pv_F)/R_2 \tag{46-7}$$

Thus, when under equilibrium conditions, the fetoplacental flow has a unique value for a given combination of values of the other variables. It is of interest that under these conditions, the fetoplacental flow does not appear to be under the control of the cardiovascular regulatory mechanisms that are operative in the adult.

If equilibrium does not exist, a return to equilibrium conditions may occur in the following manner. If the fetoplacental flow is decreased, the product $\dot{V}_{H_2O} R_P$ becomes less negative, and the fetus will gain water from the maternal circulation, thereby increasing the fetal blood volume. This in turn will increase the fetal cardiac output by the Frank-Starling mechanism, which has been shown to be quite sensitive in the embryonic [20] and fetal [24] heart. This increase in cardiac output increases the mean fetal arterial blood pressure and the umbilical flow until Equation 46-7 is again balanced.

PLACENTAL EXCHANGE OF CARBON DIOXIDE AND WATER

Hill, Power, and Longo [28] recently studied the interactions of O_2 and CO_2 in maternal and fetal blood during a single capillary transit and the effects of various factors on the exchange process. It was calculated that although physically dissolved CO_2 equilibrates rapidly between the maternal and the fetal blood, the mean fetal bicarbonate ion concentration is about 1 mM greater than the mean maternal level. Other studies [31] suggest that the bicarbonate ion diffuses slowly across placental membranes if at all. Bicarbonate ion exerts an osmotic pressure that is roughly proportional to its concentration. Since the concentrations of most other ions are similar in maternal and fetal blood, the fetal blood should have an osmotic pressure that is 1 mOsm greater than that of maternal blood. This is roughly equivalent to a hydrostatic pressure of 20 mm Hg driving water across the placental membranes from mother to fetus.

Although the relation of the rate of water transfer across the placenta to the osmotic-pressure difference between maternal and fetal blood is probably linear, there is as yet no experimental data to confirm this. Under equilibrium conditions, the partial pressure of CO_2 is proportional to the osmotic pressure, since a 1-mM increase in bicarbonate ion results in an increase in osmolarity of 1.2 mOsm.

RELATION BETWEEN FETAL CARBON DIOXIDE PRODUCTION,
THE WATER TRANSFER RATE, AND FETOPLACENTAL BLOOD FLOW

Fetal production of CO_2 results from O_2 metabolism. The normal fetal respiratory quotient (ml CO_2/ml O_2) is about 0.9 ± 0.1 SD. Under steady-state conditions, the rate of placental CO_2 transfer matches the rate of CO_2 production.

Assuming that an increased rate of O_2 consumption and fetal CO_2 production were to occur, there would be an increase in the umbilical arterial PCO_2 and HCO_3^-. The resultant increase in the osmotic pressure of the fetoplacental blood would increase the rate of water flux from maternal to fetal blood, thus increasing the fetal blood volume. The resultant increase in central venous pressure would increase the fetal cardiac output and the fetoplacental blood flow. As the fetoplacental flow increased, the rate of CO_2 and O_2 exchange across the placenta would increase, which would decrease the fetoplacental end-capillary PCO_2 and increase the PO_2, since umbilical flow is one of the most important factors affecting the transfer of these gases. As fetal PCO_2, and thus the osmotic pressure, decreased, the rate of maternal-to-fetal water flux would decrease, and the fetal blood volume, central venous pressure, cardiac output, and umbilical flow would approach a new, quasi steady-state condition.

MATHEMATICAL SIMULATION OF FETAL OXYGENATION

FETAL ADAPTATIONS TO MAINTAIN TISSUE OXYGENATION

The primary functions of the fetal circulation are to provide O_2 and other nutrients and to remove CO_2 and various metabolites from fetal tissues. Under steady-state conditions, the fetal O_2 consumption is about 6 ml/min per kilogram, which is a value close to that found in the adult. There is no evidence of anaerobic metabolism in the fetus under normal conditions [17]. Misrahy et al. [38, 39] first compared the PO_2 of fetal and adult rabbit brain tissues using polarographic techniques. These workers found the values to be similar in both, namely 30 to 45 mm Hg. Longo and Power [32] estimate that the mean values of the intracellular PO_2 in both fetal and adult sheep cardiac muscle are also comparable, being about 3 to 4 mm Hg in each case.

Fetal arterial blood O_2 tension is about 28 to 30 mm Hg, a value that is similar to that found in an adult at about 22,000 feet elevation. The question arises as to what physiologic adaptations serve to maintain normal O_2 delivery to the fetal tissues under these conditions. A number of adaptations occur in man at high altitude, including an increased respiratory rate, increased circulatory erythrocyte number (thereby increasing the blood capacity for O_2), increased 2,3-diphosphoglycerate (which decreases the O_2 affinity of the blood), increased myoglobin concentrations in muscle, increased capillary density of various tissues, increased cardiac output, and perhaps other adjustments. Obviously, a change in the respiratory rate cannot pertain to the fetus. The fetal myoglobin concentrations are less than comparable adult values. The 2,3-diphosphoglycerate concentration in fetal blood is about the same as the adult level in humans, and it is much less than the adult level in sheep, but in both species, the fetal blood affinity for O_2 is higher than that of the adult under standard conditions.

In vivo, however, the oxyhemoglobin saturation curves are probably similar in both the fetus and the adult. Although much has been made of the fetal blood having a greater O_2 affinity than that of the adult, the physiologic significance of this in placental O_2 transfer has probably been grossly overrated. The fetal hemoglobin concentration is about 50% higher than the maternal level in humans, but this relation varies considerably among various species [14]. The few reported studies of the capillary density of fetal tissues are inconclusive, showing no clear increase in capillary density in embryonic or fetal tissues. Thus, probably one of the most important fetal adaptations to maintain O_2 delivery to its tissues is its much greater cardiac output, per unit weight, than that of the adult.

DESCRIPTION OF THE MODEL

The fetal circulation may be considered as a closed network of elements around which the blood advances (Fig. 46-9). Oxygenated fetoplacental blood mixes with

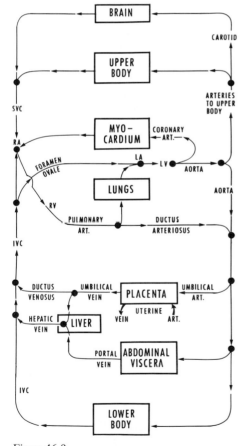

Figure 46-9
Schematic representation of the fetal circulation that was used in the development of the mathematical model (see text).

less oxygenated blood at several nodes, indicated by the closed circles in the figure. The changes in O_2 content are calculated by straightforward mixing equations. The distribution of this blood at the various shunts and other division points is again indicated by nodal points.

During steady-state conditions of constant fetal O_2 consumption and placental O_2 exchange, the calculation of the O_2 content and Po_2 in a given element of blood in the nonexchange areas (i.e., the vessels other than tissue capillaries) is relatively straightforward, using the equations noted.

On the other hand, non-steady-state conditions prevail during changes in the placental O_2 transfer rate in response to changes in the maternal or fetoplacental blood flows, the maternal arterial Po_2, the fetal O_2 consumption, or other conditions. Under these circumstances, the O_2 content in a given element of blood with differing transit-time profiles in the various parts of the circulatory system is a function of time and the flow rates through a given part of the vascular bed. An attempt to explain our solution of this problem is presented below.

The transit-time profiles, as represented by time-concentration profiles $c(t)$ (Fig. 46-10), express the outflow fractions of a nonpermeable substance, such as a dye or O_2, as a function of the time elapsed following a sudden injection (at time $t = 0$) of a unit quantity of the substance at the circuit element's inflow.

Each vascular element is provided with an inflow-content function, $I(t)$ (see Fig. 46-10), that is updated with each circuit of the network. This function extends back

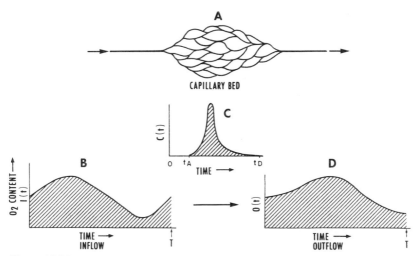

Figure 46-10

A typical circulatory system element. *A*. A capillary bed, with its inflowing and outflowing vessels and many different possible transit routes through the bed. *B*. The inflow concentration, $I(t)$, as it varies with time. *C*. The particular vascular bed requires time t_A for any substance, such as a dye, that is injected at the inflow to begin to appear at the outflow. After the elapse of time t_D, substantially all of the substance will have appeared in the outflow. The time-concentration function, $c(t)$, represents this distribution of transit times. *D*. The resulting outflow, $O(t)$, as it varies with time.

at least as many seconds prior to the current time as the time interval, t_D, that is required to include all but an insignificant fraction of the total area under the transit-time profile.

The outflow, $O(t)$ (see Fig. 46-10), from any vascular element is equal to the integral of the product of the inflow, $I(t)$, and the concentration, $c(t)$. The value of t_D is chosen such that the fraction of the substance taking longer than t_D seconds to transit the element is insignificant; therefore,

$$O(T) = \int_0^{t_D} I(T - t)c(t)dt \qquad (46\text{-}8)$$

A maximum circuit-time increment can be determined, and the transit-time profiles have a characteristic shape. From the injection time ($t = 0$) until a later "appearance time" ($t = t_A$; see Fig. 46-10), no dye from the sudden injection will appear in the outflow. Thus, any dye input at a time later than t_A will not affect its content in the outflow at time T, the time at which the content is measured, provided that $(T - t_A) \leqslant t_A$. Therefore, the maximum time increment for one complete circuit of the vascular system should be the least appearance time, t_A, for any $c(t)$ in the system. Since a diffusible substance, oxygen, is being represented, the amount of O_2 leaving the vascular system, as a function of content, must be subtracted from each element's outflow content. The O_2 content at any node upon which more than one element converges becomes the flow-weighted mean of the converging contents.

The model's flows and pressures may be expressed in terms of such parameters as resistances, inertances, and compliances. The resulting set of equations must be simultaneously satisfied. Flow, $\dot{Q}$, is represented as proportional to the difference in pressure across the element, $\Delta P = P_{in} - P_{out}$, and inversely proportional to its resistance, R:

$$\dot{Q} = \Delta P/R \qquad (46\text{-}9)$$

This relation holds until the flow velocity or flow acceleration becomes sufficiently high that the creation of kinetic energy (which is proportional to $\dot{Q}^2$) and the inertance, L, of the blood substantially affect the flow, as in the flow from the right and left ventricles of the heart, where the pressure, at any given instant, must be expressed as:

$$P = R\dot{Q} + k\dot{Q}^2 + L\ddot{Q} \qquad (46\text{-}10)$$

The rate of change of pressure with time, $\dot{P}$, in a compliant vessel (of compliance C) should be proportional to the difference between the flows into, $\dot{Q}_{in}$, and the flows out of, $\dot{Q}_{out}$, the vessel:

$$C(\Delta\dot{P}) = \Delta\dot{Q} = \dot{Q}_{in} - \dot{Q}_{out} \qquad (46\text{-}11)$$

The model of the fetal O_2 delivery system is represented in Figure 46-11. Obviously, during steady-state conditions, the rate of tissue O_2 consumption must be matched by

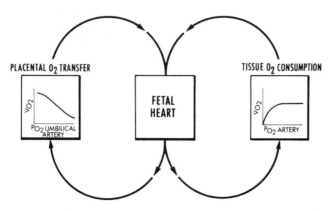

Figure 46-11
The fetal O_2 delivery system. During steady-state conditions, the placental O_2 transfer rate must equal fetal O_2 consumption.

the rate of placental O_2 transfer. Stainsby and Otis [48] showed that adult tissue O_2 consumption (see right side of Fig. 46-11) is constant over a wide range of arterial PO_2 values. Below a certain arterial O_2 tension (the critical PO_2), the O_2 consumption decreases. Placental O_2 transfer, although a function of a number of factors, is critically dependent upon the inflowing umbilical arterial PO_2 (see left side of Fig. 46-11). The umbilical arterial O_2 tension is not only an important factor in determining the rate of placental O_2 transfer, but is itself affected by the placental O_2 exchange rate.

An expanded representation of the fetal O_2 delivery system is shown in Figure 46-12. As in Figure 46-11, the area of placental transfer is shown on the left, the fetal heart is represented in the center, and the tissue exchange processes are illustrated on the right. The rate of placental O_2 exchange is a function of a number of factors, including the maternal and fetoplacental blood flow rates, which in turn are functions of the intrauterine pressures. We are currently developing a model of this process. The fetal cardiac output is critically dependent upon the mean systemic pressures and the blood volume. These, in turn, are dependent upon placental water exchange, the bicarbonate ion concentration, and PCO_2, as noted above. We have included these factors in our extended model, but these other interrelations are not shown in Figure 46-12 in an effort to avoid unnecessary confusion.

In conclusion, the fetal circulation is a complex, uniquely designed system for enabling the fetus to grow and develop despite a low arterial O_2 tension. Anatomically, several shunts and unique structures distinguish it from the circulatory system in the adult. Physiologically, the fetal systemic circuit is again unique in that it has a very high cardiac output and flow with low resistance. A major fraction of its cardiac output flows through the placenta for oxygenation. The fetoplacental flow is further distinguished by being influenced by maternal placental vascular pressures and changes in the intrauterine pressure. The fetal cardiac output and placental flow are probably linked to the fetal mean systemic pressure and blood volume, which in turn are linked

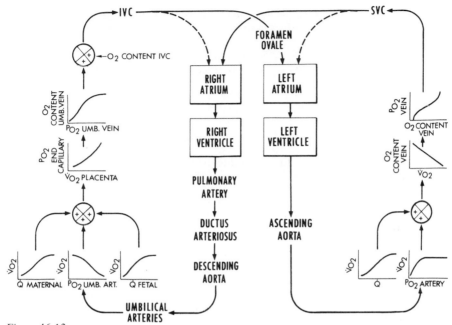

Figure 46-12
An expanded version of Figure 46-11. Represented are some of the basic components of the factors that affect O_2 transfer in the placenta and its delivery to the fetal tissues. The function curves indicate the controlling factor (independent variable) and the controlled factor (dependent variable), and they illustrate the concept that a controlled factor in one process becomes a controlling factor in a subsequent process. The inputs to the maternal and the fetoplacental flows are derived from a more complex version of this type of analysis.

to placental water exchange and the osmotic pressure of fetal blood. These are influenced by fetal tissue CO_2 production and O_2 consumption. Finally, mathematical models are proving to be useful in understanding these interrelations and the ultimate control of O_2 delivery to the fetal tissues.

REFERENCES

1. Acheson, G. H., Dawes, G. S., and Mott, J. C. Oxygen consumption and the arterial oxygen saturation in foetal and newborn lambs. *J. Physiol.* (Lond.) 135:623, 1957.
2. Amoroso, E. C., Dawes, G. S., Mott, J. C., and Rennick, B. R. Occlusion of the ductus venosus in the mature foetal lamb. *J. Physiol.* (Lond.) 129:64, 1955.
3. Assali, N. S., Bekey, G. A., and Morrison, L. W. Fetal and Neonatal Circulation. In N. S. Assali (Ed.), *Biology of Gestation*, Vol. 2. New York: Academic, 1968.
4. Assali, N. S., Morris, J. A., and Beck, R. Cardiovascular hemodynamics in the fetal lamb before and after lung expansion. *Am. J. Physiol.* 208:122, 1965.
5. Assali, N. W., Rauramo, L., and Peltonen, T. Measurement of uterine blood flow and uterine metabolism. VIII. Uterine and fetal blood flow and oxygen consumption in early human pregnancy. *Am. J. Obstet. Gynecol.* 79:86, 1960.

6. Banister, J., and Torrance, R. W. The effects of the tracheal pressure upon flow:pressure relations in the vascular bed of isolated lungs. *Q. J. Exp. Physiol.* 45:352, 1960.
7. Barclay, A. E., Franklin, K. J., and Prichard, M. M. L. *The Foetal Circulation and Cardiovascular System, and the Changes that They Undergo at Birth.* Oxford: Blackwell, 1944.
8. Barcroft, J. *Researches on Prenatal Life,* Vol. 1. Oxford: Blackwell, 1946.
9. Barcroft, J., Flexner, L. B., and McClurkin, T. The output of the foetal heart in the goat. *J. Physiol.* (Lond.) 82:498, 1934.
10. Barcroft, J., and Torrens, D. S. The output of the heart of the foetal sheep. *J. Physiol.* (Lond.) 105:22, 1946.
11. Barker, J. Fetal and neonatal cerebral blood flow. *Am. J. Physiol.* 210:897, 1966.
12. Cassin, S., et al. The vascular resistance of the foetal and newly ventilated lung of the lamb. *J. Physiol.* (Lond.) 171:61, 1964.
13. Creasy, R. K., et al. Determination of fetal, placental and neonatal blood volumes in sheep. *Circ. Res.* 27:487, 1970.
14. Dawes, G. S. *Foetal and Neonatal Physiology. A Comparative Study of the Changes at Birth.* Chicago: Year Book, 1968.
15. Dawes, G. S. Umbilical Blood Flow. In L. D. Longo and H. Bartels (Eds.), *Respiratory Gas Exchange and Blood Flow in the Placenta.* Washington, D.C.: Department of Health, Education and Welfare, National Institutes of Health, 1973.
16. Dawes, G. S., et al. Respiratory movements and rapid eye movement sleep in the foetal lamb. *J. Physiol.* (Lond.) 220:119, 1972.
17. Dawes, G. S., et al. Some observations on foetal and new-born rhesus monkeys. *J. Physiol.* (Lond.) 152:271, 1960.
18. Dawes, G. S., Mott, J. C., and Widdicombe, J. C. The fetal circulation in the lamb. *J. Physiol.* (Lond.) 126:563, 1954.
19. Elliott, R. H., Hall, F. G., and Huggett, A. St. G. The blood volume and oxygen capacity of the foetal blood in the goat. *J. Physiol.* (Lond.) 82:160, 1934.
20. Faber, J. J. Mechanical function of the septating embryonic heart. *Am. J. Physiol.* 214:475, 1968.
21. Faber, J. J. Regulation of Fetal Placental Blood Flow. In L. D. Longo and H. Bartels (Eds.), *Respiratory Gas Exchange and Blood Flow in the Placenta.* Washington, D.C.: Department of Health, Education and Welfare, National Institutes of Health, 1973.
22. Faber, J. J., Green, T. J., and Long, L. R. Permeability of rabbit placenta to large molecules. *Am. J. Physiol.* 220:688, 1971.
23. Faber, J. J., and Green, T. J. Foetal placental blood flow in the lamb. *J. Physiol.* (Lond.) 223:375, 1972.
24. Fouron, J. C., and Hebert, F. Cardiovascular adaptation of newborn lambs to hypervolemia with polycythemia. *Can. J. Physiol. Pharmacol.* 48:312, 1970.
25. Gilbert, R. D., et al. Site of pulmonary vascular resistance in fetal goats. *J. Appl. Physiol.* 32:47, 1972.
26. Harvey, W. *Exercitatio Anatomica de Motu Cardis et Sanguinis in Animalibus.* Frankfurt: G. Fitzeri, 1628.
27. Hellman, L. M., et al. The permeability of the human placenta to water and the supply of water to the human fetus as determined with deuterium oxide. *Am. J. Obstet. Gynecol.* 56:861, 1948.
28. Hill, E. P., Power, G. G., and Longo, L. D. A mathematical model of carbon dioxide transfer in the placenta and its interaction with oxygen. *Am. J. Physiol.* 224:283, 1973.
29. Hon, E. H., and Quilligan, E. J. The classification of fetal heart rate. II. A revised working classification. *Conn. Med.* 31:799, 1967.

30. Kaplan, S., and Assali, N. S. Disorders of Circulation. In N. S. Assali (Ed.), *Pathophysiology of Gestation*, Vol. 3. New York: Academic, 1972.
31. Longo, L. D., Delivoria-Paparopoulos, M., and Forster, R. E., II. Placental transfer of carbon dioxide following inhibition of fetal carbonic anhydrase or administration of THAM. *Am. J. Physiol.* 226:703, 1974.
32. Longo, L. D., and Power, G. G. The oxygen tensions in cardiac muscle of fetal and maternal sheep. *Am. J. Physiol.* (in preparation).
33. Lucas, W., Kirschbaum, T., and Assali, N. W. Cephalic circulation and oxygen consumption before and after birth. *Am. J. Physiol.* 210:287, 1966.
34. Mahon, W. A., Goodwin, J. W., and Paul, W. M. Measurement of individual ventricular outputs in the fetal lamb by a dye dilution technique. *Circ. Res.* 19:191, 1966.
35. Makowski, E. L., et al. Cerebral blood flow, oxygen consumption and glucose utilization of fetal lambs in utero. *Am. J. Obstet. Gynecol.* 114:292, 1972.
36. McRoberts, W. Postural shock in pregnancy. *Am. J. Obstet. Gynecol.* 62:627, 1951.
37. Meschia, G., Battaglia, F. C., and Barron, D. H. A comparison of the freezing points of fetal and maternal plasmas of sheep and goat. *Q. J. Exp. Physiol.* 42:163, 1957.
38. Misrahy, G. A., Beran, A. V., and Hardwick, D. F. Fetal and neonatal brain oxygen. *Am. J. Physiol.* 203:160, 1962.
39. Misrahy, G. A., et al. Fetal brain oxygen. *Am. J. Physiol.* 199:959, 1960.
40. Morris, J. A., et al. Dynamics of blood flow in the ductus arteriosus. *Am. J. Physiol.* 208:471, 1965.
41. Novy, M. J., and Metcalfe, J. Measurements of umbilical blood flow and vascular volume by dye dilution. *Am. J. Obstet. Gynecol.* 106:899, 1970.
42. Power, G. G., and Longo, L. D. Fetal circulation times and their implications for oxygenation. *Am. J. Physiol.* (In press).
43. Reynolds, S. R. M. Hemodynamic characteristics of the fetal circulation. *Am. J. Obstet. Gynecol.* 68:69, 1954.
44. Reynolds, S. R. M., et al. Qualitative nature of pulsatile flow in umbilical blood vessels, with observations on flow in the aorta. *Bull. Johns Hopkins Hosp.* 91:83, 1952.
45. Rudolph, A. M., and Heymann, M. A. The circulation of the fetus in utero. Methods for studying distribution of blood flow, cardiac output and organ blood flow. *Circ. Res.* 21:163, 1967.
46. Rudolph, A. M., and Heymann, M. A. Circulation changes during growth in the fetal lamb. *Circ. Res.* 26:289, 1970.
47. Salzer, P. Beitrag zur Kenntnis des Ductus venosus. *Z. Anat. Entwicklungsgesch.* 130:80, 1970.
48. Stainsby, W. N., and Otis, A. B. Blood flow, blood oxygen tension, oxygen uptake and oxygen transport in skeletal muscle. *Am. J. Physiol.* 206:858, 1964.
49. Stembera, Z. K., Hodr, J., and Janda, J. Umbilical blood flow in healthy newborn infants during the first minutes after birth. *Am. J. Obstet. Gynecol.* 91:568, 1965.
50. Widdowson, E. M. Growth and Composition of the Fetus and Newborn. In N. S. Assali (Ed.), *Biology of Gestation*, Vol. 2. New York: Academic, 1968.
51. Young, M. The Fetal and Neonatal Circulation. In W. F. Hamilton and P. Dow (Eds.), *Handbook of Physiology, Section 2, Circulation*, Vol. 2. Washington, D.C.: The American Physiological Society, 1963.

SELECTED READING

Dawes, G. S., Mott, J. C., and Vane, J. R. The density flowmeter, a direct method for the measurement of the rate of blood flow. *J. Physiol.* (Lond.) 121:72, 1953.

Fowler, K. T., West, J. B., and Pain, M. C. F. Pressure-flow characteristics of horizontal lung preparations of minimal height. *Respir. Physiol.* 1:88, 1966.

Lopez-Muniz, R., et al. Critical closure of pulmonary vessels analyzed in terms of Starling resistor model. *J. Appl. Physiol.* 24:625, 1968.

Maternal Adjustments During Pregnancy

47

Wayne S. Rogers

ANATOMIC CHANGES

Functional modifications occur in all the body systems of the maternal organism to accommodate for the growing conceptus and to prepare for the eventual delivery of the fetus. A successful pregnancy requires certain anatomic, endocrinologic, metabolic, and physiologic changes in the maternal organism. The most significant of these will be discussed in this chapter.

The most obvious maternal change is the increase in the size of the uterus in a very short period of time. The nonpregnant uterus is 6 to 7 cm in length and weighs approximately 60 to 75 gm. At term, it has become a 35- to 40-cm-long structure weighing over 1000 gm, and the myometrium has become a thin, muscular sac with a capacity of approximately 5 liters. These remarkable alterations are due to hypertrophy, elongation of the muscle cells, and an increase in the fibrous and elastic tissue elements. The vascular, lymphatic, and nervous tissues respond in a similar fashion. These anatomic changes do not occur in a symmetric fashion throughout the uterus, a factor that is of significance when labor starts. The size and vascularity of the cervix are also increased, which is associated with hyperplasia and hypertrophy of the endocervical glands. The vagina expands and shows vascular congestion. Glycogen is deposited in the cells of the vagina, and this is accompanied by a decrease in the pH of the vaginal secretions due to the presence of lactic acid. The abdominal cavity that contains this enlarging mass can accommodate it only by mechanically displacing the other bodily organs.

ENDOCRINE SYSTEM MODIFICATIONS

The endometrium prepares for the implantation of the fertilized ovum in response to steroid hormones. The increasing quantities of these and other hormones are necessary for the specific physiologic, metabolic, and anatomic changes that occur. During pregnancy, the maternal ovary and adrenal gland are not the only sites for steroid synthesis; the maternal organism, the placenta, and the fetus collaborate to produce the necessary steroid hormones and the other hormones found in pregnancy.

MATERNAL ENDOCRINE PRODUCTION

When ovulation occurs and the corpus luteum is formed, there results an increase in the plasma levels of both estrogen and progesterone. The estrogens are synthesized mainly in the adrenal glands and the ovaries. These steroid-producing structures are

869

able to convert acetate to cholesterol through a series of various complex biochemical reactions (Fig. 47-1). The first step is the breakdown of the long side-chain of cholesterol to form pregnenolone and then progesterone. From these two chemical compounds, all of the other steroid hormones are ultimately synthesized. As shown in the simplified illustration, various enzyme mechanisms convert these compounds to the C-19 steroids that are the three principal androgenic substances. Various oxidations and aromatization occur that produce the three principal estrogens: estrone (E_1), estradiol (E_2), and estriol (E_3). Estradiol (E_2) is the most functionally active of the estrogens. The plasma estradiol concentration varies between 100 and 300 μg per day, depending on the day of the menstrual cycle. Significant clinical disorders can occur in women who are not able to convert the androgenic C-19 steroids effectively to the estrogens.

The ovary becomes relatively inactive after 12 weeks of pregnancy, and thereafter the placenta becomes the principal site for estrogen production. The maternal adrenal gland continues to secrete a significant precursor compound, known as dehydroepiandrosterone (DHA) sulfate.

Progesterone is secreted by the corpus luteum as 17-hydroxyprogesterone. The precursor for this compound is the maternal cholesterol. At about 12 weeks of gestation, the plasma level of this compound falls, while the amount of progesterone produced by the placenta gradually increases.

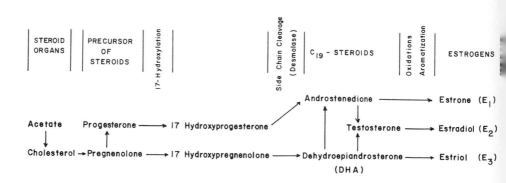

Figure 47-1
Steroidogenesis (simplified scheme).

PLACENTAL ENDOCRINE SYSTEM

The placenta is the chief source for several significant steroid and protein hormones. The major protein hormones are human chorionic gonadotropic hormone (HCG) and human placental lactogen (HPL).

Human Chorionic Gonadotropin (HCG)

This hormone is produced very early in pregnancy by the trophoblastic tissue, and it reaches its peak concentration at approximately 8 weeks of gestation (Fig. 47-2). The plasma level then falls rapidly and reaches a low plateau at about 13 to 14 weeks of gestation. It is thought that this hormone is necessary to maintain the viability of the corpus luteum during the first three months of pregnancy. The function of the corpus luteum begins to decrease after 8 weeks of gestation and becomes relatively inactive after the third month. HCG is rapidly excreted by the kidneys, so that the concentrations in the maternal blood and the urine are approximately the same at any given time. The assay of blood or urine HCG is the basis of the so-called pregnancy test. Thus, an apparently negative pregnancy test is not significant after 14 weeks of gestation when a normal pregnancy is present.

Human Placental Lactogen (HPL)

This protein hormone is synthesized by the trophoblastic tissue and was formerly known as human chorionic somatomammotropin (HCS). It is thought to have a

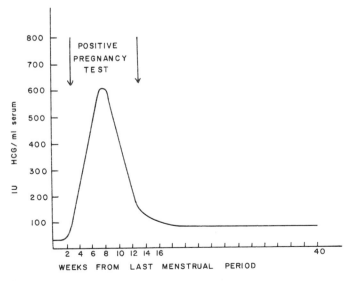

Figure 47-2
Human chorionic gonadotropic hormone (HCG) level during pregnancy.

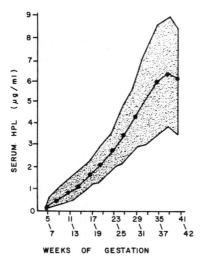

Figure 47-3
Characteristic pattern of placental lactogen (HPL) levels in 164 normal pregnant subjects during the course of pregnancy. The points connected by a solid line represent the mean values, and the shaded area represents ± 2 SD from the mean. (Modified from K. Varma et al. *Obstet. Gynecol.* 38:487, 1971. Reproduced with permission.)

growth-promoting potential similar to that of the pituitary growth hormone (GH) [4]. Lactogenic and luteotropic influences similar to those of prolactin have been described [10]. HPL can be detected by radioimmunoassay early in pregnancy, and its level rises progressively in the maternal serum as the pregnancy continues (Fig. 47-3). It has been estimated that at term, the placenta secretes from 1 to 2 gm of HPL per day. The half-life of plasma HPL is very short (20 to 30 minutes), and it quickly disappears from the maternal plasma following delivery [2]. This hormone is metabolically significant in that it has the ability to affect insulin activity and is thus concerned with carbohydrate and fat metabolism, as will be discussed later [12].

Thyrotropin

This hormone has been determined to be present in the placental tissue, but its significance has not yet been determined.

Estrogens

The major steroid hormones are the estrogens and progesterone. The placenta is the site for the synthesis of these hormones after the twelfth week of gestation. In the pregnant female, the estrogens cannot be produced from acetate and cholesterol, as they are in the adrenal and the ovary of the nonpregnant female. They depend upon certain precursors produced by the adrenal glands of the mother and fetus. The chief precursor substance, DHA sulfate, which is synthesized by the fetal adrenal gland, is

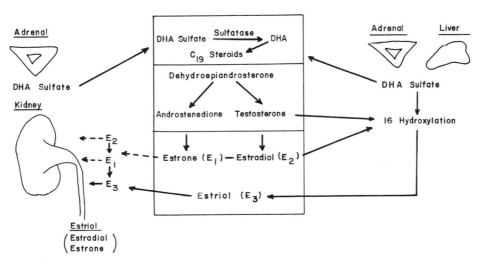

Figure 47-4
Steroid pathways in the fetoplacental unit.

responsible for approximately 90% of the estriol formed in a term pregnancy (Fig. 47-4).
The trophoblastic tissue of the placenta contains within it several of the potent enzyme
systems. DHA sulfate is acted upon by sulfatase to produce DHA. Various oxidation
and aromatization reactions take place in the placenta to synthesize the estrogens. An
unusually high production of estriol occurs during pregnancy; there is almost a 1000-fold
increase in its excretion, while the excretion of estrone and estradiol increases only
100-fold during pregnancy.

Progesterone

The syncytial cells are largely responsible for the synthesis of progesterone after the
twelfth week of pregnancy. Here again, the placenta must rely on the maternal plasma
cholesterol. During the latter months of pregnancy, about 300 mg of progesterone are
produced each day. This compound is then metabolized by the maternal and fetal liver,
and it is excreted in the maternal urine as pregnanediol.

FETOPLACENTAL ENDOCRINE SYSTEM

A properly functioning placenta as well as a viable fetal adrenal and liver are necessary
for the synthesis of the steroid hormones. As discussed previously, the fetal adrenal
gland is largely responsible for the synthesis of DHA sulfate, the precursor for the

placental conversion to estriol. The fetal zone of the adrenal cortex is also the site of formation of the 16-hydroxylated steroids that are necessary in the synthesis of estriol. The liver of the fetus may also contribute to this synthesis, since it has the ability to 16-hydroxylate DHA (see Fig. 47-4). Substances necessary for the conjugation of the steroids may also come from the fetal liver. The metabolism of placental progesterone to pregnanediol partially occurs in the fetal liver. Since the synthesis of the various steroid hormones is dependent upon the fetoplacental unit, assays of these compounds may be used as an index of the fetal well-being in utero.

METABOLIC CHANGES DURING PREGNANCY

The maternal organism is ultimately responsible for all the necessary materials that the growing fetus requires. Various metabolic adjustments take place so that proper biologic development can occur without jeopardy to the fetus.

Many of the metabolic changes in the maternal organism are due to the protein and steroid hormones that are produced during pregnancy. The various anatomic changes, such as those that occur in the pelvic organs and the breast, necessitate the ingestion and utilization of greater amounts of protein. Adequate quantities of glucose are required for the energy involved with protein synthesis.

These metabolic changes in the maternal organism are necessary so that the need for an increased supply of nutrients to the growing parasitic fetus can be satisfied. The mother has the ability to mobilize her stores of fat for energy utilization in order to conserve and continuously supply carbohydrates to the fetus, even through periods of prolonged maternal fasting. The glucose is stored in the liver and muscle as glycogen, and in the adipose tissue as triglycerides. The amino acids and dietary fatty acids are similarly stored in the cells of the muscle and adipose tissues. Increasing quantities of plasma insulin, which are due to the higher levels of these metabolites and the increasing plasma level of HPL, are largely responsible for controlling these mechanisms. The increase in plasma insulin facilitates the storage of glucose, and it enables the storage of free fatty acids because of the action of insulin on lipoprotein lipase. Protein synthesis is also maintained at adequate levels, since the increased insulin facilitates the transport of amino acids into the cells.

Plasma HPL also increases the plasma free fatty—acid concentration due to its lipolytic action.

PHYSIOLOGIC CHANGES IN PREGNANCY

Many of the physiologic changes in the maternal organism that occur in response to pregnancy are due to the increasing plasma levels of the various hormones previously described. The maternal organism attempts to compensate for these multiple and variable alterations.

CARDIOVASCULAR SYSTEM

The enlarging uterus is similar to a growing "tumor" that contains an arteriovenous shunt. This causes a decrease in peripheral resistance and subsequent alterations in the maternal cardiac output and blood pressure [8, 13].

The cardiac output promptly increases during pregnancy, rises rapidly in the first three months, and reaches a peak approximately in midpregnancy. When the mother is not in labor, the cardiac output can increase from 1000 to 3000 ml/min according to the Fick equation:

$$\text{Cardiac output} = \frac{\text{oxygen consumption}}{\text{arteriovenous oxygen difference}}$$

Thus, when the cardiac output is at its maximum, the arteriovenous oxygen difference is at its smallest. The oxygen consumption is not considerably increased at term (10% to 20%) [1]. Investigators have reported that this increase in cardiac output is primarily due to an increase in stroke volume and, to a lesser extent, to an increase in the cardiac rate [8, 13]. The usual peripheral resistance equals 1332 dyne·sec·cm^{-5} (the peripheral resistance is calculated from the mean brachial arterial pressure minus the mean right atrial pressure, in dyne/cm^2, divided by the cardiac output in ml/sec). The average value for the peripheral resistance at term, however, is 980 dyne·sec·cm^{-5}. It was thus concluded that the increase in cardiac output is associated with a decrease in the peripheral resistance [7].

There is an additional increase in cardiac output of about 15% to 20% that occurs with the onset of each uterine contraction during labor. When the contraction is initiated, approximately 500 ml of blood leaves the intervillous space and enters the maternal circulation. This increase in blood, which is primarily unsaturated uterine venous blood, results in an elevation of the right atrial pressure [5]. As the uterine musculature continues to contract, the arteriovenous shunt is eliminated, which results in a rise in the mean blood pressure and subsequently in a decrease in the cardiac output. As a result of these two opposing actions, the cardiac output at the peak of the contraction is approximately the same as that between the contractions.

The cardiac output is appreciably increased in the second stage of labor, when additional work is required of the mother to expel the fetus. A great deal of this increase in cardiac output is associated with apprehension and pain. The elimination of these with the administration of regional anesthetics, however, does not reflect this increase in cardiac output [2]. Thus, the cardiac output during labor is variable and depends on many factors, such as the intensity, frequency, and duration of the contraction as well as the amount of pain experienced and the work done by the mother. Following delivery of the infant, the uterus contracts and squeezes blood into the maternal circulation. This is compensated for by the normal amount of blood loss (300 to 500 ml) that occurs during the delivery of the infant and placenta.

There is a decrease in the maternal blood pressure that is first noted during midpregnancy. This is due to the decrease in the peripheral resistance caused by the

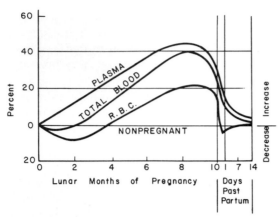

Figure 47-5
Changes in blood volume during pregnancy, labor, and the puerperium. (Modified from J. J. Bonica. *Principles and Practice of Obstetric Analgesia and Anesthesia* [1st ed.]. Philadelphia: Davis, 1967. Reproduced with permission.)

arteriovenous shunt. A further decrease to shock levels, known as *supine hypotensive syndrome,* can occur in some women in the supine position due to a decrease in the blood flow to the right atrium and an associated slowing of the maternal heart rate. This is a result of the compression of the inferior vena cava by the uterus and subsequent pooling of blood in the legs.

The total blood volume, the plasma volume, and the red blood cell volume all increase during pregnancy, depending upon the size of the mother and the fetus (Fig. 47-5). The plasma volume shows the greatest increase starting early in the pregnancy. This early onset is probably a result of the decreased vascular tone that occurs secondary to the increased plasma steroid levels. In addition, there is an increase in the vascular compartment that is largely due to the enlargement of the uterus and the intervillous spaces beneath the attached placenta. The increase in the red blood cell volume is less than that of the plasma volume. Laboratory studies reveal that the pregnant woman has a lower hemoglobin concentration and a lower hematocrit as a result of these changes; however, hemoglobin levels below 11.5 gm/100 ml are unusual in women on adequate diets with iron supplements.

The increase in blood volume is similar to the increase that occurs with the cardiac output. It is principally due to the increase in plasma volume, which occurs early and reaches a level that is approximately 50% greater than the nonpregnant level. This level is maintained until delivery has occurred. The hypervolemia is apparently necessary for the adequate exchange of blood gases and to protect the pregnant woman against the inherent blood loss that occurs during delivery. The increase in cardiac output is associated with an increase in the blood flow. This is most apparent in the uterus and the kidneys where flows of 400 to 500 ml/min are found, which are approximately 50% over normal. The extremities are another site where blood flow increases notably, which again is associated with the decrease in peripheral resistance that occurs in midpregnancy (Fig. 47-6).

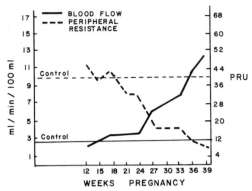

Figure 47-6
Blood flow and peripheral resistance in the forearm during pregnancy. (Modified and redrawn from S. Spaetz. *Acta Obstet. Gynecol. Scand.* 43:12, 1964.)

The total body fluids of a term pregnant patient are increased about 7 liters. Most of this increased fluid occupies the interstitial spaces. In spite of this increase, the osmolality of the fluids is maintained due to sodium chloride retention. There will be significant fluid retention when even small amounts of sodium are reabsorbed by the kidneys.

The blood coagulation system also shows significant changes. Several of the blood factors that are involved in the blood coagulation system are increased during pregnancy. Fibrinogen (Factor I) increases from 300 mg/100 ml to as high as 700 to 800 mg/100 ml at term. Due to an increase in these blood factors, a pregnant patient at term is in a state of potential hypercoagulopathy. The clotting mechanism may be activated by various disorders of pregnancy, so that disseminated intravascular coagulation (DIC) can occur. Excessive thrombosis is prevented, especially following delivery, by the presence of a potentially active fibrinolytic system. Most of the increased blood coagulation factors are consumed at the time of delivery in order to prevent excessive bleeding from the placental site.

The plasma steroid hormones are probably responsible for the increase in the coagulation factors in the maternal blood. Some of these blood factors are also increased in those nonpregnant women who are taking oral contraceptive tablets, which also suggests that these changes are due to the steroid hormones.

THE URINARY SYSTEM

Significant changes occur in the urinary system that are the result of the mechanical and endocrine alterations mentioned previously. Early in pregnancy, dilatation and elongation of the ureters occur as a result of the increased plasma progesterone level. The enlarging uterus may also compress the ureters, resulting in the mechanical dilatation of the ureter and kidney pelvis.

The renal plasma flow (RPF) and the glomerular filtration rate (GFR) are both increased during pregnancy [11]. The former increases approximately 25%, whereas the latter increases approximately 50%. These changes occur early and persist through-out the pregnancy, which again implicates the plasma hormones (possibly HPL and cortisol) as the responsible causative factors. The increase in the GFR results in a decrease in the blood levels of creatinine and urea. In a nonpregnant patient, most of the filtered glucose is reabsorbed by the renal tubules. In a pregnant patient, how-ever, clinical glucosuria develops in approximately 10% of the patients [3]. If the $(T_{gluc})_{max}$, the maximum ability of the kidney to reabsorb glucose, remains constant throughout pregnancy, glucosuria should develop in the majority of pregnant patients due to the increase in GFR. Apparently, this does not occur, since some of the nephrons have the ability to increase the $(T_{gluc})_{max}$ [6]. The lowering of the renal threshold for glucose is significant in the management of the pregnant diabetic patient. The mechanisms described above also hold true for certain amino acids.

All three constituents of the renin-angiotensin-aldosterone system are stimulated, resulting in an increase in their plasma levels. The blood pressure, however, falls in midpregnancy due to the decreased peripheral resistance, in spite of the elevated angiotensin. This lack of responsiveness to angiotensin is apparently a normal conse-quence of pregnancy. These factors might be of significance in the hypertensive disorders that occur during pregnancy. Sodium retention during pregnancy may be a response that results from the increase in aldosterone and the steroid hormones. A circadian rhythm of sodium excretion is also present during pregnancy [9].

THE PULMONARY SYSTEM

Modifications in pulmonary function occur in order to maintain an adequate exchange of both oxygen and carbon dioxide. A comparison of pulmonary function in non-pregnant women and women with a term pregnancy (Fig. 47-7) shows several distinct differences. In the latter, there is a significant increase in the tidal volume, whereas the expiratory reserve volume and the residual volume are both decreased. Since the expiratory reserve volume is decreased by 150 ml and the residual volume is decreased by 200 ml, then the functional residual capacity (FRC) is decreased by 350 ml. The progressive elevation of the diaphragm due to the enlarging uterus and the increase in the pulmonary blood volume are responsible for the decrease in the FRC. The respiratory minute-volume (the tidal volume times the number of respiratons per minute) increases progressively toward term, at which time it is approximately 50% above normal. The increase in the tidal volume is largely responsible for this, since the respiratory rate increases only slightly.

Alveolar ventilation is thus significantly increased, which facilitates the increase in O_2 consumption and CO_2 elimination that is necessary during pregnancy. Total O_2 consumption at term increases to about 50 ml/min (an increase of 20%). The P_{O_2} increases from a mean value of 95 mm Hg to a mean of 106 mm Hg. The hyperventila-tion results in a decrease in the P_{CO_2} from 40 mm Hg to 32 mm Hg. During pregnancy,

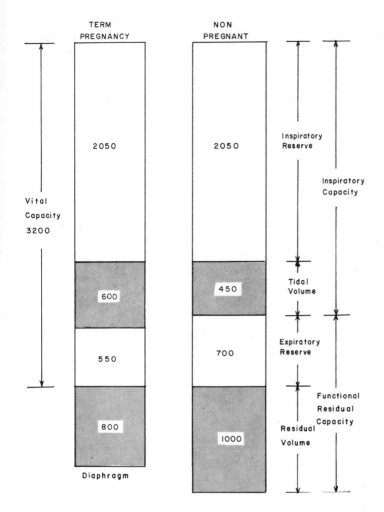

Figure 47-7
Pulmonary volumes and capacities at term and in the nonpregnant woman. (Modified from
J. J. Bonica. *Principles and Practice of Obstetric Analgesia and Anesthesia* [1st ed.]. Philadelphia:
Davis, 1967. Reproduced with permission.)

plasma progesterone readjusts the hypothalamus to this change; the respiratory centers
respond readily to small changes in P_{CO_2}.

In order to maintain a constant pH in the maternal blood, the plasma bicarbonate
and sodium concentrations are both reduced by means of excretion through the kidneys. The pregnant patient is thus in a state of chronic respiratory alkalosis that is
compensated for by a chronic metabolic acidosis. The osmolality of the plasma
decreases with the loss of sodium; this results in an increase in the excretion of urine,
which in turn causes thirst and the increased intake of fluids.

These physiologic adjustments result in an increase in the efficiency of gas transfer between the maternal blood and the alveolar contents. When ventilation is increased or decreased in a pregnant woman near term or when she is exposed to extremely high altitudes, changes in the maternal pH may occur. An exaggerated response may also occur during the administration of anesthetic gases. The action of progesterone on the respiratory center is largely responsible for the hyperventilation that occurs and the initiation of the events described above.

REFERENCES

1. Bader, R. A., Bader, M. E., Rose, D. J., and Braunwald, E. Hemodynamics at rest and during exercise in normal pregnancy as studied by cardiac catheterization. *J. Clin. Invest.* 34:1524, 1955.
2. Beck, P., and Daughaday, W. H. Human placental lactogen. Studies of its acute metabolic effects and disposition in normal man. *J. Clin. Invest.* 46:103, 1967.
3. Fine, J. Glycosuria in pregnancy. *Br. Med. J.* 1:205, 1967.
4. Grumbach, M. M., et al. Chorionic growth hormone-prolactin (CGP). Secretion, disposition, biologic activity in man and postulated function as the growth hormone of the second half of pregnancy. *Ann. N.Y. Acad. Sci.* 148:501, 1968.
5. Palmer, A. H., and Walker, A. H. C. Maternal circulation in normal pregnancy. *J. Obstet. Gynaecol. Br. Emp.* 56:537, 1949.
6. Renschler, H. E., and von Baeyer, H. Glucosuria of pregnancy, incidence and mechanism (Abstract) *International Congress of Nephrology* p. 260, 1966.
7. Rose, D. J., et al. Catheterization studies of cardiac hemodynamics in normal pregnant women with reference to left ventricular work. *Am. J. Obstet. Gynecol.* 72:233, 1956.
8. Rovinsky, J. J., and Jaffin, H. Cardiovascular hemodynamics in pregnancy. III. Cardiac rate, stroke volume in multiple pregnancies. *Am. J. Obstet. Gynecol.* 95:787, 1966.
9. Seldin, D. W. (Ed.). The physiology of diuretic agents. *Ann. N. Y. Acad. Sci.* 139:273, 1966.
10. Selenkow, H. A., et al. Measurement and Pathophysiologic Significance of HPL. In A. Pecile and C. Finzi (Eds.), *The Feto-placental Unit.* Amsterdam: Excerpta Medica Foundation, International Congress Series No. 183, 1969. Pp. 340–362.
11. Sims, E. A. H. Renal function in normal pregnancy. *Clin. Obstet. Gynecol.* 11:461, 1968.
12. Spellacy, W. N. HPL. Review of a protein hormone important to obstetrics and gynecology. *South. Med. J.* 62:1054, 1969.
13. Walters, W. A. W., MacGregor, W. G., and Hills, M. Cardiac output at rest during pregnancy and the puerperium. *Clin. Sci.* 30:1, 1966.

SELECTED READING

Hansen, J. M., and Uleland, K. Influence of caudal anesthesia on cardiovascular dynamics during normal labor and delivery. *Acta Anaesthesiol. Scand.* [Suppl.] 23:449, 1966.

Howard, B. K., Goodson, J. H., and Mengert, W. F. Supine hypotensive syndrome in late pregnancy. *Obstet. Gynecol.* 1:371, 1953.

Kerr, M. G., Scott, D. B., and Samuel, E. Studies of the inferior vena cava in late pregnancy. *Br. Med. J.* 1:532, 1964.

Sims, E. A. H. Kidney disease in pregnancy. *Annu. Rev. Med.* 16:221, 1965.

Spellacy, W. N., Carlson, K. L., and Birk, S. A. Dynamics of human placental lactogen. *Am. J. Obstet. Gynecol.* 96:1164, 1966.

Labor and Delivery

Wayne S. Rogers

48

INTRODUCTION

The mean duration of a pregnancy is 280 days from the onset of the last menstrual period or 266 days from conception. A full term pregnancy lasts approximately 40 weeks. *Labor* denotes the final sequence of events in pregnancy, which results in the ultimate delivery of the fetus. At present, the mechanism responsible for the onset of labor is not known, although many theories have been put forth to support the roles of various hormones as the initiating factor. It appears that the fetal adrenal cortex plays a role by way of steroidogenesis. This organ frequently does not develop in the anencephalic fetus, and in such a case, labor does not start until beyond the 42nd or 43rd week of gestation.

The prostaglandins, which are 20-carbon hydroxy fatty acids that appear in many forms, have also been implicated in initiating the process of labor [16]. Over a dozen of these substances have been isolated from human semen, and they are now known to be present in most tissues. Two of these, $PGF_2\alpha$ and PGE_2, have oxytocic properties (i.e., they have the ability to cause uterine contractions). Prostaglandin $F_2\alpha$ has been detected in the amniotic fluid of patients in labor, but it is not found in those patients prior to labor. Oxytocin secreted from the maternal pituitary gland can cause contractions in a term pregnancy. It can significantly affect the intensity and duration of a contraction, but it does not, however, seem to be be involved with the mechanism that initiates labor.

To effect a delivery, it is necessary for the uterus to contract in a coordinated and effectual manner. These contractions gradually thin out, dilate the cervix, and force the infant through the birth canal. The myometrium of the uterus contracts rhythmically during labor and following delivery of the fetus, causing the blood vessels that traverse the uterus to be partially or completely occluded.

The chemical mechanisms that provide the energy for myometrial contractions are similar to those utilized by skeletal muscles. Creatine phosphate, ATP, ADP, and actomyosin are all present in the myometrial cells. The synthesis of RNA increases progressively during the pregnancy, probably as a result of both uterine distension and plasma estrogen levels. The resting potential under these conditions approaches 50 mV.

MECHANICS AND PHYSIOLOGY OF THE UTERINE CONTRACTION

The myometrium contracts periodically throughout the entire pregnancy. These contractions are usually mild and of short duration. When the contractions increase in both intensity and duration so that progressive dilatation and effacement (thinning out) of the cervix occurs, the patient is clinically considered to be in labor. A convenient terminology has been adopted that divides labor into three stages. The *first* stage of labor begins with the onset of effectual contractions and ends with the complete dilatation of the cervix. The *second* stage of labor is that interval of time from the complete dilatation of the cervix to the actual expulsion of the infant. The *third* stage of labor is the interval between the delivery of the infant and the delivery of the placenta. The forces and work necessary to produce motion of the fetus against a plane of resistance are provided by the contractions of the upper, thick segment of the uterus and are supplemented by the increase in the intraabdominal pressure due to the bearing-down efforts of the mother in the second stage of labor. These forces must overcome the resistance of the cervix, the birth canal, and the muscles of the pelvic outlet.

The uterine contraction usually originates at a specific site and is initiated by uterine pacemakers [16]. These pacemaker myocytes do not differ anatomically from others present in the uterus. They are not limited to any particular area [8], but they are usually located near the insertion of the fallopian tubes. When the pacemaker action is initiated, successive myocytes are activated until the contraction spreads throughout the myometrium (Fig. 48-1). The pacemaker action must be synchronous; otherwise contractions may originate at various areas within the uterus. In order to maintain effectual labor, the gradient of contraction must be from the

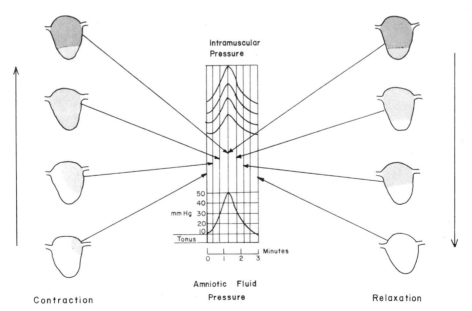

Figure 48-1
Normal contractile wave and its relation to pressure changes in the uterus. (Modified from R. Caldeyro-Barcia. *Clin. Obstet. Gynecol.* 3:394, 1960. Reproduced with permission.)

upper segment of the uterus toward the lower segment. An asynchronous contraction would increase the intraamniotic fluid pressure, but the intensity would be weak and irregular. The sum effect of such contractions would impede the progress of labor.

The electrical potential occurring within the uterus during labor can be measured by electrohysterograms. Two electrodes placed 10 cm apart at the upper portion of the uterine fundus or a unipolar recording with an exploring electrode over the fundus have been used [18, 19]. The signals are channeled through a direct-current amplifier system, which converts them to higher frequencies that are further amplified by high-gain amplifiers. When a multichannel tocodynamometer (TCD) is used over the uterus, a tocodynagraph may be obtained. This type of recording reveals that the force of contraction in the upper portion of the uterus is most prominent and that it decreases as it approaches the lower uterine segment. Labor would be unduly prolonged or no progress would occur if this were not true.

UTERINE WORK

The work done by uterine contractions is responsible for the motion that is necessary to expel the fetus from the uterus, and this work can be measured by a variety of methods. Graphic records of the contractions can be obtained (tocodynagraphy).

EXTERNAL METHODS

A TCD employs a transducer and three strain gauges placed in brass-ring mountings. The transducer is strapped across the abdomen of the mother over the fundus of the uterus (Fig. 48-2). As a contraction develops, the uterine fundus rises up and

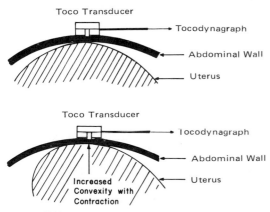

Figure 48-2
Action of the uterine contraction on the external transducer of a tocodynamometer.

the resulting convexity transmits the force against the transducer. The pressure recorded by the strain gauge is proportional to the degree of the convexity produced by the uterine contraction. This can be reproduced graphically through electrometric conversion. The tocodynagraph accurately records both the frequency and the intensity of the uterine contractions. There are many clinical applications for this technique. Multichannel tocodynagraphs can be obtained simultaneously at various sites within a labor and delivery unit.

INTERNAL METHODS

Recordings of the intensity and frequency of uterine contractions can be obtained directly from the uterine muscle at various sites within the uterus. A cannula can be inserted into the myometrium, and a small polyethylene catheter with a microballoon at its tip can be threaded through this cannula [5]. In a similar fashion, small metal capsules measuring 12 mm by 4 mm, which are constructed with a small aperture in the center that is covered by a membrane, can be inserted into the myometrium. The pressures that are exerted either upon the microballoon or upon the membrane can be transmitted through the polyethylene tube [17]. An electromanometer is then employed to obtain visual or graphic readings. These methods are not clinically feasible, but they have contributed a great deal to the knowledge and understanding of uterine contraction.

Tocodynagraphs can also be obtained by placing a small balloon through the cervix so that it rests between the myometrium and the amniotic sac. Amniotic fluid pressures can be measured directly by introducing a polyethylene tube into the amniotic sac. An open-tipped, fluid-filled polyethylene catheter can be inserted through the mother's abdomen and uterus directly into the sac; the most clinically useful method is to introduce the catheter through the partially dilated cervix and into the amniotic cavity. This intrauterine catheter must bypass the presenting part of the fetus after the rupture of the membranes. Such catheters have also been inserted into the amniotic fluid and, simultaneously, into the myometrium. With this method, it has been found that the pressure exerted by the myometrium approximates that in the amniotic fluid [2].

QUANTITATIVE MEASUREMENTS

Using the above described methods, uterine dynamics may be quantitated. The work done by the uterus is related to the product of the maximum intensity of the contraction times the number of contractions. The intrauterine pressure or the pressure within the amniotic fluid sac indicates the intensity of the uterine contraction. The resting pressure gradually increases during pregnancy, and, near term, approaches 10 mm Hg. Uterine activity, as defined in Montevideo units (mm Hg/10 min), is the product of the maximum intrauterine pressure and the number of contractions occuring over a 10-minute interval. The uterus contracts throughout the entire pregnancy, but these contractions are of low intensity and occur infrequently. When the contraction produce an intraamniotic pressure between 10 and 15 mm Hg, they are known as "Braxton Hicks" contractions. Although they occur sporadically throughout the pregnancy, they become clinically perceivable as term approaches. At this time, both the intensity and the frequency of contractions increase, causing most of the effacement of the cervix (Fig. 48-3).

Labor does not have a sudden onset but is a gradual progression of the uterine activity and work. The intensity of the uterine contractions increases from approximately 25 mm Hg at the onset of labor to over 50 mm Hg toward the end of labor, while the frequency increases to 3 to 5 contractions every 10 minutes. Between contractions, the intrauterine pressure returns to its resting level of 10 mm Hg. During the second stage of labor, the uterine-related work approaches 250 Montevideo units.

Many variations in these patterns may occur. If the uterine pacemakers are uncoordinated, the amniotic fluid pressures can become weak and irregular, thereby causing ineffectual labor. A complete inversion of the gradient of contraction can produce

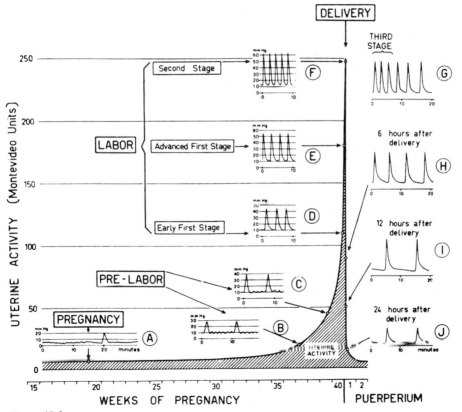

Figure 48-3
Uterine contractility throughout pregnancy, labor, and the early puerperium. (Tracing from R. Caldeyro-Barcia and J. J. Poseiro. *Ann. New York Acad. Sc.* 75:813, 1959. Reproduced with permission.)

high amniotic fluid pressures and yet achieve no progress during labor [7]. Hypertonus occurs when the resting amniotic fluid pressure is between 10 and 30 mm Hg. Hyperactive labor is present if a uterine contraction increases the amniotic fluid pressure above 50 mm Hg or if the period is less than 2 minutes between contractions. Hypoactive labor is defined as contractions of less than 30 mm Hg intensity or those in which the period is greater than 5 minutes between contractions. These variations may result in the complete cessation or the prolongation of labor, which is known as *dysfunctional labor* or *uterine inertia.*

For an average woman who has had previous children (multipara), approximately 80 to 160 contractions are necessary to efface and dilate the cervix [6]. If the average intensity of these contractions is 50 mm Hg, the uterine work necessary to dilate the cervix varies between 4000 and 8000 Montevideo units. During the first stage of labor, when uterine contractions occur frequently (up to 2 to 3 minutes), the cervix should be completely dilated in about 3 to 6 hours. In the second stage of labor, additional work is necessary to expel the fetus from the birth canal against the resistance of the bony and muscular components of the pelvis. This is accomplished by both voluntary and involuntary bearing-down efforts of the mother, which produce an additional 50 to 70 mm Hg.

Clinically, the magnitude and frequency of contractions are determined by palpation of the uterine fundus by the obstetrician and by the duration of the discomfort experienced by the patient. This determination may vary a great deal depending upon the experience of the obstetrician and the pain threshold of the patient. Intrauterine pressure recordings reveal that the duration and intensity of a uterine contraction are greater than those clinically observed (Fig. 48-4). A contraction can be perceived by palpation when the amniotic fluid pressure reaches approximately 20 mm Hg; thus, the clinician would describe the contractions as lasting 70 seconds. Above 40 mm Hg of intrauterine pressure, the uterus can no longer be depressed, so that the maximum elevation in pressure cannot be determined by palpation. The patient, on the other hand, does not perceive any discomfort until the amniotic fluid pressure rises to 15 mm Hg above the normal resting value of 10 mm Hg. The patient then experiences discomfort for 50 to 60 seconds, depending on the individual's pain threshold.

FETAL HEART RATE MONITORING

The fetal heart rate (FHR) can provide reliable information as to the status of the fetus during labor. It becomes increasingly important to monitor the FHR in those pregnancies where the fetus is already compromised, i.e., in high-risk pregnancies. When combined with chemical monitoring (to be discussed later), the evaluation of the fetus in utero can be determined with greater accuracy.

The time-honored method of listening to the fetal heart with a stethoscope (auscultation) and then averaging the number of beats over a specific interval (15 to 30 sec) is inadequate, since the FHR is only obtained for a specific interval of time. The limitations of intermittent auscultation are such that this method has been judged to be

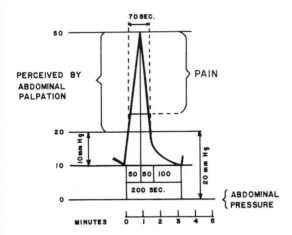

Figure 48-4
Relation between the pressure estimated by abdominal palpation and the actual recorded intra-uterine pressure. (R. Caldeyro-Barcia and J. J. Poseiro. *Clin. Obstet. Gynecol.* 3:392, 1960. Reproduced with permission.)

completely unreliable [3]. This method also has the disadvantage that the fetal heart-beat cannot usually be heard during a contraction, and the FHR is calculated only between contractions. The fetus is under its greatest stress, however, during the contraction, and the response of the FHR to the contraction cannot be accurately determined by this method. The fetal tolerance to the uterine contraction can only be judged by an instantaneous FHR recording.

Fetal distress has long been considered to be represented by any deviation in the FHR from its normal range of 120 to 160 beats/min. Fetal distress must be suspected when there is a continuous tachycardia (over 160 beats/min), bradycardia (less than 120 beats/min), or repeated irregularities in the FHR. Fetal anoxia usually causes an initial elevation in the FHR, and, with continuous deprivation of oxygen, bradycardia and arrhythmias can occur. In a term pregnancy, the basal heart rate is regulated by both vagal and sympathetic neural actions, and variations from this basic pattern are of significance in assessing the fetal condition during pregnancy and labor.

FHR RECORDING METHODS

Phonocardiography

To measure the heart sounds of the fetus, a phonotransducer is strapped externally over the mother's abdomen. The output of the transducer is amplified and transmitted to a speaker. This method can be used during pregnancy and probably in early labor. As the labor progresses, the recording of the beat-to-beat measurements usually becomes inaccurate.

Ultrasonography

The ultrasonic method employs the Doppler principle, and it incorporates a transducer with both a transmitter and receiver for a specific frequency. The signals obtained by this system are then amplified and recorded. This transducer has an advantage in that it can determine the fetal heart activity externally, through the mother's abdomen, without the rupture of the membranes. Ultrasonography is noninvasive and has a high degree of biologic safety. The ultrasound used is of low energy and nonaccumulative in its effect. The power level of ultrasound is usually less than $10 \ mW/cm^2$ [21]. The ultrasonic transducer is preferable to the phonotransducer in that with the former, there is little interference from other power sources or from the mother's own tissue potentials. The external transducer methods may be unsatisfactory if a great deal of maternal or fetal motion is present. Since ultrasonography is a valuable, safe, and relatively new method to be applied to study of biologic tissues, this subject will be discussed in a separate section (pp. 898–902).

Fetal Electrocardiography

Fetal electrocardiography (ECG) measures the electrical energy produced by the fetal heart rather than measuring the mechanical energy, as is done with phonocardiography or ultrasonography. The fetal ECG can be obtained either externally or internally by applying electrodes to the fetal scalp following the rupture of the membranes. The latter direct method offers accurate beat-to-beat measurements that can be processed quantitatively. An instantaneous FHR can be obtained with a recording system. This method can also be used to produce an accurate display of the shape of the ECG complex. Interference because of motion and noise due to the activity of the mother and fetus or the uterine contractions is less than that for signals obtained by external methods.

When the external method is used in which the recording is obtained through the mother's abdominal wall, it is necessary to interpose a high-gain, low-noise preamplifier prior to the attachment to the ECG recorder. The external method has several disadvantages even with the filtering of the fetal ECG. The signal-to-noise ratio, as well as the superimposed maternal tracings, may result in an unsatisfactory recording (Fig. 48-5). Electrical interference from the contracting abdominal wall of the mother, which is frequently associated with other maternal motion, can cause sufficient electrical noise to obscure the fetal ECG during labor. The external method has the advantage that it can be used before the rupture of the amniotic membranes and in early labor when the cervix is not sufficiently dilated.

The internal method of obtaining a fetal ECG via the vaginal route is the technique that is most commonly used. An amnioscope is passed through the cervix to visualize the presenting part. A clip electrode or a screw-type electrode can be attached to the presenting part of the fetus [15]. A silver-silver chloride unipolar electrode is attached to the fetus with either device, and it is connected to the negative input terminal of the amplifier. A similar unipolar electrode coil, which is wrapped around the insulated wire to the fetus, is connected to the positive input terminal of the amplifier. The latter

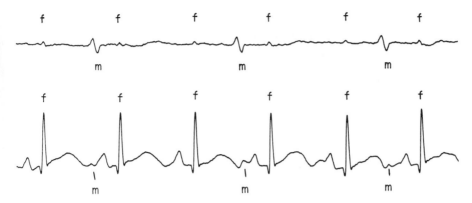

Figure 48-5
Comparison of fetal ECG obtained externally (*upper tracing*) and a simultaneous recording
obtained by direct methods (*lower tracing*). Fetal ECG is indicated by *f*; maternal ECG by *m*.
(From E. H. Hon. *An Introduction to Fetal Heart Rate Monitoring.* New Haven, Conn.: Yale
Co-operative Corporation, 1971. P. 71. Reproduced with permission.)

lies in the vaginal secretions and thus contact is made with the mother. The electrical
noise that is produced by the skeletal muscles is thereby largely eliminated (see
Fig. 48-5). The silver—silver chloride electrode has been found to be the most benefi-
cial in eliminating the level of noise interference [11]. An electrical hazard exists,
however, unless complete insulation is present around the electrode other than the
silver and silver chloride.

The use of a stainless steel spiral electrode has also been advocated when the FHR
is used only to provide triggering signals after proper amplification [12]. It is cheaper
and easier to manufacture than the silver—silver chloride electrode, and it can be used
again after proper sterilization.

THE FETAL MONITOR

The fetal monitor is an instrument that can simultaneously display the instantaneous
FHR and the uterine contraction pattern [13]. Since the fetus is under its greatest
stress during the contractions, an instantaneous FHR can indicate the fetal tolerance.
The frequency, the amplitude, and the duration of the uterine contractions as well as
the FHR are simultaneously recorded. The instantaneous FHR is computed by using
a digital, peak-detection, instantaneous cardiotachometer. The FHR is computed at
every individual time unit between beats. The electrical potential is inversely proportional to the period between successive beats. There is a one-beat delay in the computation, since it is necessary for the arrival of the next ECG complex to occur before
one can compute the interval.

A two-channel oscillograph and a strip-chart recorder allow the FHR and the
uterine contractions to be compared simultaneously. The fetal ECG signal must be
passed through a narrow band-pass filter (1.5 to 100 Hz) to provide the proper triggers,

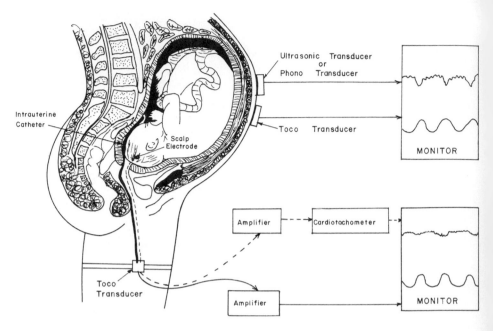

Figure 48-6
Methods employed in fetal monitoring.

and the signals from a phonotransducer or an ultrasonic transducer must also be filtered and amplified before the triggers enter the cardiotachometer. The fetal monitor is thus able to display a continuous record of both the FHR and the uterine contraction patterns obtained from the various transducers (Fig. 48-6).

INTERPRETATION OF FHR RECORDINGS

There are many changes in the FHR patterns that are associated with uterine contractions. The significance of all these changes is not completely clear, and standards are not uniform from institution to institution. Appropriate terminology has gradually been accepted [14]. *Bradycardia* or *tachycardia* refers to the slowing or speeding up of the FHR, with respect to the base-line FHR, over a period of more than 10 minutes. *Acceleration* and *deceleration* refer to changes in the FHR associated with uterine contractions.

The shape of the deceleration pattern when compared to the shape of the uterine contraction pattern is significant (Fig. 48-7). The FHR has a "uniform shape" when its pattern resembles that of the uterine contraction. The temporal relationship between the onset of the FHR deceleration and that of the uterine contraction is also significant. The FHR, however, fluctuates normally as much as 3 to 10 cycles per minute, and transient accelerations may occur in response to uterine contractions.

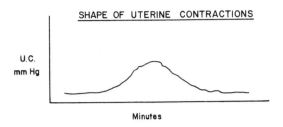

SHAPE OF UTERINE CONTRACTIONS

U.C.
mm Hg

Minutes

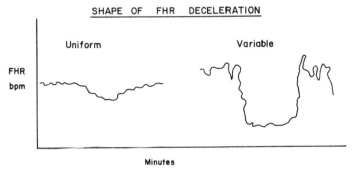

SHAPE OF FHR DECELERATION

Uniform   Variable

FHR
bpm

Minutes

Figure 48-7
Shapes of FHR pattern compared to the uterine contraction pattern. (From E. H. Hon. *An Intro-duction to Fetal Heart Rate Monitoring.* New Haven, Conn.: Yale Co-operative Corporation, 1971. P. 24. Reproduced with permission.)

Three patterns of deceleration have been described [14, 15], depending on their shape and their timing in relation to the uterine contraction.

Early Deceleration

In Figure 48-8, the FHR pattern is of uniform shape and correlates well with the uterine contraction pattern. Deceleration occurs early with the onset of the contraction and the increase in intrauterine pressure. The cause is thought to be due to the compression of the fetal head against the resisting forces of the pelvic canal. A similar pattern of deceleration is observed when one compresses the fetal head manually. These FHR changes apparently coincide with changes in cerebral blood flow, which stimulate the central vagus nerve to slow the conduction system of the heart.

Late Deceleration

This type of FHR deceleration (Fig. 48-9) is thought to be due to a uteroplacental insufficiency that results from a decrease in the blood flow in the intervillous space

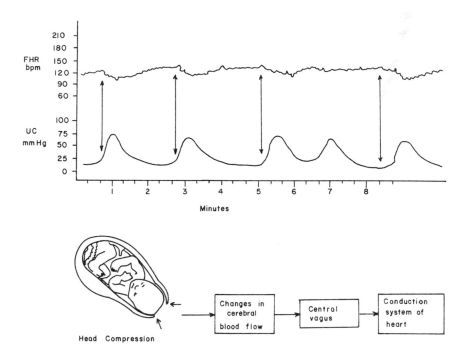

Figure 48-8
Early FHR deceleration patterns. (Modified from E. H. Hon. *An Introduction to Fetal Heart Rate Monitoring.* New Haven, Conn.: Yale Co-operative Corporation, 1971. P. 29. Reproduced with permission.)

during a uterine contraction. It is characterized by a uniform shape, which is similar to that of the uterine contraction pattern except that the onset of the deceleration takes place late in the contracting phase of the uterine contraction. Many factors can cause a decrease in the intervillous blood flow, such as maternal hypotension due to the supine hypotensive syndrome, blood loss, or other forms of obstetric shock. A vigorous, sustained, uterine contraction that compresses the vessels traversing the myometrium as well as frequent contractions also produce this type of deceleration. Maternal illness, such as diabetes or renal disease, limits the fetomaternal exchange of gases due to an improperly functioning placenta. These high-risk pregnancies should be monitored in order to determine if a uteroplacental insufficiency occurs with contractions. The fetus may be severely affected by the persistent hypoxia if the contractions are allowed to continue. Delivery by cesarean section is indicated in such cases.

Variable Deceleration

The FHR curve shown in Figure 48-10 is variable and does not reflect the shape of the associated intrauterine pressure curve. The onset of the deceleration also has no

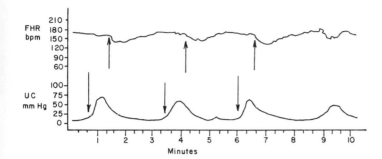

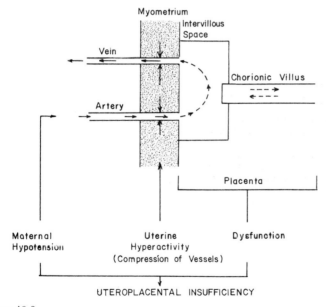

Figure 48-9
Late FHR deceleration patterns due to uteroplacental insufficiency. (Modified from E. H. Hon.
An Introduction to Fetal Heart Rate Monitoring. New Haven, Conn.: Yale Co-operative
Corporation, 1971. P. 29. Reproduced with permission.)

specific relationship to the onset of the contraction. This pattern may be produced by
compression of the umbilical cord, which would occur if it were lodged between fetal
parts and the uterine wall. Hemodynamic changes and hypoxia occur that stimulate
the central and peripheral vagus nerves, thus causing myocardial depression in the severe
or persistent cases. Many arrhythmias are also frequently associated with compression
of the umbilical cord.

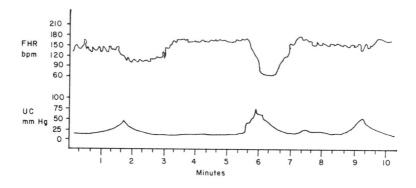

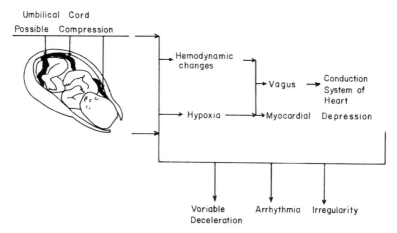

Figure 48-10
Variable FHR deceleration patterns due to cord compression. (Modified from E. H. Hon. *An Introduction to Fetal Heart Rate Monitoring.* New Haven, Conn.: Yale Co-operative Corporation, 1971. P. 29. Reproduced with permission.)

The interpretation and significance of other patterns have not been clearly defined and are beyond the scope of this text. It is generally agreed, however, that continuous FHR and simultaneous intrauterine contraction monitoring is a valuable tool in assessing the fetal condition. It should be used in the high-risk pregnancy and when auscultation reveals FHR irregularities.

BIOCHEMICAL MONITORING

Maternal or placental disorders may interfere with adequate oxygen and nutrient exchanges between the mother and fetus, and fetal acidosis may develop that interferes with cellular metabolism. The severity of damage to the infant depends upon the degree and the duration of these metabolic disturbances. When chronic placental insufficiency already exists, minor variations in the available oxygen may result in fetal damage. Although continuous FHR monitoring provides valuable information about the tolerance of these infants during labor, the exact assessment of the fetus can be determined by biochemical means. Some tests can be done during pregnancy in those cases in which maternal disease is present or dysmaturity is suspected.

Newer laboratory techniques employ microserologic examinations. These may be used during labor when there is evidence of fetal distress on the basis of either clinical or electronic monitoring methods. The basic technique was described in 1962 and has been elaborated upon since [20]. An endoscope is introduced through the cervix and placed against the presenting part of the fetus. The blood is collected by means of a glass capillary tube; approximately 0.1 to 0.2 ml is sufficient. The capillary tube is immediately sealed at both ends after mixing with heparin. The pH can be determined with a glass microelectrode immediately after collecting the specimen. The P_{CO_2} and the base excess or deficit can also be obtained from this sample.

These are the chief biochemical examinations of fetal blood. The determination of P_{O_2} is not reliable, because of technical and theoretical reasons. The measurement of blood lactate does not indicate acute hypoxia, since lactate is formed from anaerobic glycolysis in the fetus. The measurement of the base deficit is a more reliable indicator of metabolic acidosis, and the acid-base status of the fetus reflects the degree of fetal hypoxia. If, for some reason, the fetus is deprived of oxygen, the energy necessary to sustain the fetus is obtained from glucose. The lactic acid that is produced as a result of this form of metabolism decreases the pH of the intracellular and extracellular fluids. The acid-base balance can be readily understood when one reviews the Henderson-Hasselbalch equation (see Chap. 2; Eqs. 2-38 and 2-44). The pH of the blood is kept at a proper level by variations in pulmonary and renal function (see Chap. 17, pp. 283–284).

In the fetus, carbon dioxide readily crosses the placenta, so that the fetal P_{CO_2} is dependent upon the maternal P_{CO_2} [4]. The placenta transfers the bicarbonate ion very poorly, which may be due to the electrical charge of this ion. If the maternal P_{CO_2} level falls due to changes in the maternal blood, the fetus must regulate its own bicarbonate concentration in order to maintain a constant pH. Metabolic acidosis in a diabetic mother does not necessarily alter the fetal pH and bicarbonate levels, although fetal death can occur during such episodes. The pH of the fetal blood is usually 0.2 unit lower than the pH of the maternal blood. This is thought to be due to the slightly elevated P_{CO_2} that is produced in the fetal blood so that a pressure gradient will exist in order to eliminate this gas by diffusion. During labor, however, the pH of both the maternal and fetal blood falls (Fig. 48-11), but it returns to normal levels several hours after delivery [9].

In order for the fetus to maintain proper metabolic function, its blood pH must be maintained. It has been shown that the condition of the infant at birth is closely correlated with the pH: a fetal blood pH greater than 7.27 usually indicates a fetus in good condition, whereas a pH of less than 7.22 indicates a compromised fetus that should be delivered immediately. Several samples should be taken for confirmation.

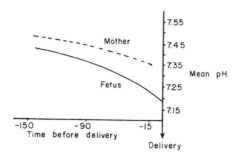

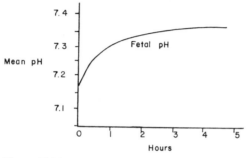

Figure 48-11
Mean pH of maternal and fetal blood during labor and after delivery. (Modified from F. W. Ekenhausen. *A Study of the Perinatal Acid-Base Equilibrium.* Thesis, Leiden University, 1969.)

In such cases, the maternal pH should also be checked for metabolic acidosis, since the hydrogen ion can be transferred across the placenta. Any maternal acidosis should be treated so that the hydrogen ion gradient is corrected. In cases where fetal acidosis was present, even with good electronic monitoring patterns, the infants frequently were born in poor condition. Conversely, in many cases when this pattern was poor and when acidosis was not present, these infants were born in good condition [10].

It has been suggested that microserologic examinations be done during labor in those pregnancies where the fetus is at high risk. Continuous FHR monitoring probably has a place in those pregnancies of lesser risk and in abnormal labors. Should abnormal FHR patterns develop in these cases, microserologic examinations are definitely indicated. Microserologic examinations can be easily done and frequently repeated.

ULTRASONIC MONITORING AND ULTRASONOGRAMS

Ultrasonography is becoming a valuable diagnostic tool in the study of biologic tissues. The energy form employed is high-frequency sound waves. They can be directed as a beam, in contrast to the diverse spreading that is inherent with lower frequency sound waves. These mechanical vibrations are nonionizing and noninvasive. The frequencies used in medical diagnosis range between 1 and 10 MHz. Since ultrasonography is free of any known adverse effects, it can be used continuously or repeatedly without the complications that exist with roentgenographic examination [1]. The source of ultrasonic energy depends on the piezoelectric effects of certain crystals, such as those of quartz or lead zirconate. These crystals are able to convert mechanical stress into an electrical signal or, conversely, an electrical stress into mechanical motion. These transducers resonate when struck by a high-voltage electrical discharge. The resulting ultrasonic beams are able to penetrate human tissue, and, as they pass through the various interfaces of the body, ultrasonic energy is reflected. A small amount of energy is dissipated as it passes through the interfaces. The echoes from each inter-face are reflected back to the transducer, which then converts them into electrical signals.

THE DOPPLER PRINCIPLE

The Doppler unit has been mentioned previously as an external transducer for detect-ing and calculating the FHR. In this case, the ultrasonic echo is used to detect move-ments of the fetal heart as the reflecting structure. A continuous beam of ultrasound originates in one transducer crystal, and the echoes are received by a second transducer that is adjacent to it (Fig. 48-12). Motion produced by the fetal heart beneath this beam alters the interval of time that the echo takes to return to the receiver. This difference in frequency can be processed electronically, amplified, and recorded. The cardiotachometer, using the proper filters to produce the triggers, can reproduce the FHR in analog form on a strip-chart recorder.

There are many transducers available for this purpose that operate at a frequency of 2 MHz. They must, however, be properly placed upon the maternal abdomen and remain in position for long periods of time. To overcome this, a multiple-element

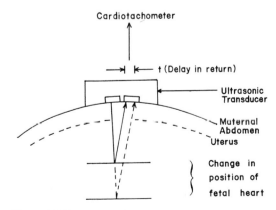

Figure 48-12
Doppler principle as applied to ultrasonography.

transducer is used to spread the beam from several transmitters and to survey a much larger area. Lenses have also been used for spreading the beam. At the present time, research is being done to develop lower input energy levels (2.5 mW/cm²) and to improve receiving circuits [21].

ULTRASONOGRAMS

As mentioned previously, piezoelectric crystals can produce electrical potentials as a result of the charge developed by mechanical deformation. These crystals are distorted when electrodes are placed on specific faces of the crystal and the force is applied in a specific direction. A voltage is produced between these electrodes that is linearly proportional to the mechanical displacement. When high voltages, on the other hand, are applied to these crystals, they respond by either bending, twisting, or shearing.

When a rapidly fluctuating electrical potential causes a small, brief distortion of the crystal, a high-frequency, directed, pulsating sonic beam results. When it is directed at biologic tissues such as the maternal abdomen, the beam can either be reflected or refracted from the tissue interfaces; some of these echoes return to the crystal, while others pass through the tissue interface. The intensity of the returning echo depends upon the density of the tissue and the spread of sound within it. This property of a tissue is known as its *acoustic impedance.* An ultrasonogram is actually a graphic display of what would result if a slice were taken through that particular portion of the abdomen. A pulsed-echo technique is used for this purpose. A single crystal rings for a fraction of a second and then rests for another brief period of time in order for the returning echoes to be converted into electrical signals. The actual ringing time is in the vicinity of 1.0 μsec, which is followed by a rest period of approximately 190 μsec. The same transducer is used for producing the ultrasonic beam and receiving the echo. As the beams penetrate into the body cavity, echoes are received sooner from those interfaces that are near the surface than from those deeper within the body. Echoes received from deeper interfaces are weaker than those received from near the surface, since some energy is lost within the various interfaces. The strength of these echoes is potentiated by sophisticated electronic devices.

A-Mode

In this mode (time-amplitude modulation), the signals from the various tissue interfaces are represented by vertical deflections from the base line of an oscilloscopic tracing; the height of these "blips" corresponds to the impedance of the particular tissue. Multiple echoes result, which represent the multiple interfaces. The distances between the interfaces can be obtained by measuring the distance between the blips on the oscilloscope. Thus, the A-scan technique measures only distances, such as diameters. When the interfaces of the fetal skull, for example, are at right angles to the beam, measurements within ±1 mm can be obtained. The anatomic points measured should be known, such as the parietal eminences of the fetal head, which should return blips of equal intensity (Fig. 48-13).

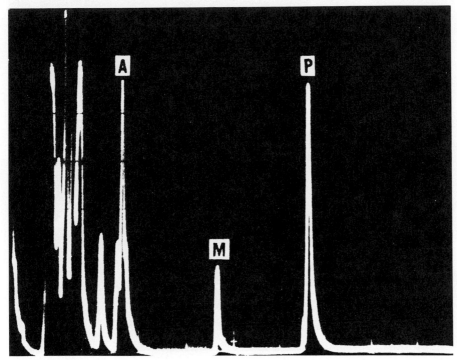

Figure 48-13
A-scan ultrasonogram. The probe is applied to the skin of the abdominal wall, which has previously
been covered with a film of oil to exclude any air. The beam is then directed so that it is at a right
angle to the interfaces of the fetal skull. The anterior (*A*) and posterior (*P*) skull echoes are recorded.
The biparietal diameter of the fetal head is the distance between these two peaks. A midline echo
(*M*) is noted between these two blips. (From L. M. Hellman and J. A. Pritchard. *Williams'*
Obstetrics [14th ed.]. New York: Appleton-Century-Crofts, 1971. P. 905.)

B-Mode

This technique, which is also known as *ultrasonic laminagraphy,* is used to obtain a
pictorial cross section of the abdomen. The transducer probe is passed across the surface of the abdomen with a rocking motion. Two-dimensional pictures are obtained
by sending the echo signal to the grid of the electron cone in the cathode-ray tube of
an oscilloscope. The electron beam intensity varies with the echo strength, and it
appears as a dot of light on the face of the tube. Its location corresponds with the
relative position of the interface within the body. This is accomplished by making a
time-base sweep that corresponds to the direction of the ultrasonic beam. This collection of dots of varying intensity produces a cross-sectional portrayal of the underlying
tissues that can be stored on the cathode-ray tube and photographed (Fig. 48-14).

The B-scan ultrasonogram can be used to diagnose multiple pregnancies, various
fetal anomalies, fetal size, the location of the placenta, extrauterine pregnancies, and

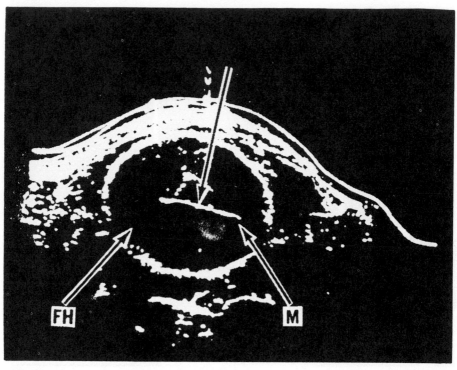

Figure 48-14
B-scan ultrasonic laminagram. The arrow is perpendicular to the linear midline structures of the fetal head. The biparietal diameter is at right angles to the midline echo. (From L. M. Hellman and J. A. Pritchard. *Williams' Obstetrics* [14th ed.]. New York: Appleton-Century-Crofts, 1971. P. 905. Reproduced with permission.)

so on. It can also be used in early pregnancies to differentiate between hydatidiform moles, uterine tumors, and intrauterine fetal deaths. It is rapidly becoming a significant tool for diagnostic studies of biologic tissues.

REFERENCES

1. Abdulla, U., et al. Effect of diagnostic ultra-sound on maternal and fetal chromosomes. *Lancet* 2:829, 1971.
2. Alvarez, H., and Caldeyro-Barcia, R. The normal and abnormal contractile waves of the uterus during labour. Amniotic fluid pressure same as in myometrial contractions. *Gynaecologia* 138:2, 1954.
3. Benson, R. C., et al. Fetal heart rate as a predictor of fetal distress. *Obstet. Gynecol.* 32:259, 1968.
4. Blechner, J. N. Fetal acid-base hemostasis. *Clin. Obstet. Gynecol.* 13:621, 1970.
5. Caldeyro-Barcia, R., Alvarez, H., and Reynolds, S. R. M. A better understanding of uterine contractility through simultaneous recording with an internal and a seven channel external method. *Surg. Gynecol. Obstet.* 91:641, 1950.

6. Caldeyro-Barcia, R., and Poseiro, J. J. Physiology of the uterine contraction. *Clin. Obstet. Gynecol.* 3:398, 1960.
7. Caldeyro-Barcia, R., Alvarez, H., and Poseiro, J. J. Normal and abnormal uterine contractility in labour. *Triangle* 2:41, 1955.
8. Carsten, M. E. Regulation of Myometrial Composition, Growth and Activity. In Assali, N. S. (Ed.). *Biology of Gestation, The Maternal Organism,* Vol. 1. New York: Academic, 1968.
9. Ekenhausen, F. W. *A Study of the Perinatal Acid-Base Equilibrium.* Thesis, Leiden University, 1969.
10. Gevers, R. H., Rhemrex, P. E. R., and Favier, J. *Foetal Heart Monitoring and Biochemical Examination of the Child During Labour.* Leiden, The Netherlands: Leiden University Press, 1971. P. 39.
11. Hon, E. H. Instrumentation of fetal heart rate and fetal electrocardiography. *Obstet. Gynecol.* 30:281, 1967.
12. Hon, E. H., Paul, R. H., and Hon, R. W. Electronic evaluation of the fetal heart rate. *Obstet. Gynecol.* 40:362, 1972.
13. Hon, E. H. The electronic evaluation of the fetal heart rate. Preliminary report. *Am. J. Obstet. Gynecol.* 75:1215, 1958.
14. Hon, E. H., and Guilligan, E. J. The classification of fetal heart rate. II. A revised working classification. *Conn. Med.* 31:779, 1967.
15. Hon, E. H. *An Introduction to Fetal Heart Rate Monitoring.* New Haven, Conn.: Yale Co-operative Corporation, 1971. p. 71.
16. Ivy, A. C. Functional anatomy of labor with special reference to the human being. *Am. J. Obstet. Gynecol.* 44:952, 1942.
17. Karlson, S. On the motility of the uterus during labour and the influence of the motility pattern on the duration of the labour. *Acta Obstet. Gynecol. Scand.* 28:209, 1949.
18. Larks, S. D., et al. Effects of oxytocin and analgesic drugs on the human electro-hysterograms. *Obstet. Gynecol.* 13:405, 1959.
19. Larks, S. D., et al. Electrical activity of the human uterus in labor. *J. Appl. Physiol.* 10:479, 1957.
20. Saling, E. *Prenatal and Neonatal Hypoxia in Relation to Clinical Obstetrical Practice.* Baltimore: Williams & Wilkins, 1969.
21. Wingate, M. D., and Ziedonis, J. G. Electronic monitoring of the fetal heart. Monitoring by ultrasound. *Int. J. Gynecol. Obstet.* 10:194, 1972.

Evaluation of the Perinate

49

Wayne S. Rogers

INTRODUCTION

The majority of women do well during the course of their pregnancy, labor, and delivery, and they do not require any unusual care or procedures. In approximately 20% of all cases, however, the fetus is considered to be at *high risk* and is vulnerable to perinatal death or morbidity. Statistically, the perinatal death and morbidity rates in such cases are almost twice those of pregnancies considered to be at low risk. The high-risk infant is more likely to develop a physical, intellectual, or personality handicap. Therefore, all pregnancies must be evaluated, and the high-risk pregnancy must be identified. This is accomplished by examining the patient's past medical and obstetric history, her physical condition, and laboratory test data. A description of the clinical aspects concerned with the identification of these high-risk pregnancies is beyond the scope of this text.

In some of these high-risk pregnancies, a continued intrauterine existence may be more hazardous to the fetus than exposure to the extrauterine environment. When such a situation is evident, an optimum delivery time must be chosen. For adequate care of high-risk pregnancies, sophisticated resuscitation techniques may be needed in addition to electrical, chemical, and biologic monitoring. In this chapter, the theory and techniques that are available to monitor the intrauterine environment of the developing fetus and the extent of fetal maturity will be treated.

THE INTRAUTERINE ENVIRONMENT

During development, the fetus is suspended weightlessly in an environment of amniotic fluid. The amniotic fluid and sac protect the fetus from injury, allow limited motion, and help to regulate fetal temperature. The volume of the amniotic fluid increases as pregnancy progresses, but this amount varies considerably with time [14]. The average uterus at term contains about 700 to 800 ml of amniotic fluid, but this value can normally vary between 500 and 1500 ml (Fig. 49-1). The greatest rise in volume appears to occur in the second half of the pregnancy, but the decline during the pregnancy varies widely with the individual patient. The determination of amniotic fluid volume can be made by injecting *p*-aminohippuric acid (PAH) and then measuring the spectrophotometric absorbtion peak of the fluid at 272 nm [16].

The origin, the composition, and the dynamics of the amniotic fluid at various stages of pregnancy have been extensively studied. It has been found that the amniotic fluid has a turnover rate of approximately 600 ml/hour in the human [11, 12]. Most of the fluid exchange (75%) occurs via the fetus, which plays a significant role in

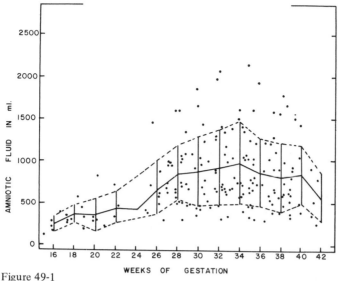

Figure 49-1

Intrauterine amniotic fluid volume during gestation. (From J. T. Queenan et al., Amniotic fluid volumes in normal pregnancies. *Am. J. Obstet. Gynecol.* 114:34, 1972. Reproduced with permission.)

maintaining the volume of the amniotic fluid by swallowing and absorbing approximately 450 ml of fluid during a 24-hour period [13]. The transfer rates of fluid from the fetus to the mother and from the fetus back to the amniotic fluid are also high. The exchange of fluid, however, from the maternal unit to the amniotic fluid is negligible.

The urine of the fetus has a lower osmolar concentration than the amniotic fluid, so that the osmolarity of the fluid near term decreases by approximately 20 to 30 mOsm/liter. Undoubtedly, the fetal urine is responsible for this decrease in osmolarity, since it is hypotonic as compared to the fetal plasma.

The method by which amniotic fluid is obtained for examination is known as *amniocentesis* [16]. This procedure is done quite easily and is rarely associated with any complications. The position of the fetus must be determined prior to the procedure in order to avoid possible trauma. Clinically, this is done by palpation of the mother's abdomen and by auscultation; roentgenography and ultrasonic laminagraphy are also frequently used for this purpose. A 20-gauge needle is inserted through the abdominal wall and the uterus after displacing the fetal head. Amniotic fluid may also be aspirated from the region of the fetal extremities, as long as the placental site is known to be high in the fundus.

FETAL GROWTH AND DEVELOPMENT

If the fetoplacental unit has not been functioning properly, the fetus may be predisposed to chronic malnutrition or chronic anoxia. Maternal diseases — such as diabetes,

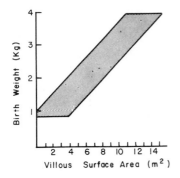

Figure 49-2
Relationship between the placental villous surface area and the fetal birth weight determined using planimetric techniques. (Modified from W. Aherne and M. S. Dunnill. *J. Pathol. Bacteriol.* 91:123, 1966.)

hypertension, renal disease, hormonal disturbances, placental abnormalities, and infectious diseases — are largely responsible for such dysfunction. Certain medications and drugs may also disturb the fetal development. When the genetic mechanism is defective, certain errors in development and metabolism can occur.

When the fetoplacental unit is functioning properly, the nutrition derived from the maternal organism is largely responsible for fetal growth and development. The rate of fetal growth can be expressed by the equation $W = (0.24 \times 10^{-6})(t - 36)^3$, where W = the weight of the fetus in kilograms and t = the period of gestation in days. In societies where malnutrition is prevalent, the constant 0.24×10^{-6} is decreased, and obstetric complications — such as toxemia, premature separation of the placenta, prematurity, and anemia — are common.

The available surface area of the placenta, which is necessary for the exchange of the various nutrients and gases, is a significant factor in determining fetal growth. A correlation exists between fetal size and placental size (Fig. 49-2). A decrease in the available surface area of the placenta may be due to a primary disturbance, but most frequently, it is secondary to maternal disease. Regardless of the causes, the impairment of blood flow in the uteroplacental unit results in chronic fetal malnutrition and subsequent fetal distress. When fetal dysmaturity is suspected because of the maternal history or clinical findings, the pregnancy may have to be interrupted to prevent fetal demise.

ASSESSMENT OF THE FETUS IN UTERO

There are various methods employed to assess intrauterine fetal development. The examination of one specific modality or the performance of one test is usually not adequate, and several different methods and repeated tests are usually employed. The following are most frequently used for this purpose.

MECONIUM-STAINED AMNIOTIC FLUID

Meconium is a greenish to black substance that is normally found in the intestines of the fetus. Its presence in the amniotic fluid may be an indication of fetal distress when the *vertex* is the *presenting* part (i.e., when the fetal head is in the lower uterine segment). The presence of meconium can be seen through the intact membranes with the aid of an amnioscope. An endoscope containing a light source is inserted through the cervix, the beam of light is directed toward a clear area of the presenting part, and if particles of meconium are present, they can be clearly seen floating in the reflected light.

ESTRIOL DETERMINATION

The maternal estrogens are synthesized chiefly by the ovaries and adrenal glands in the nonpregnant state. During pregnancy, however, the placenta becomes the chief site for estrogen production. As the placenta matures, the total mass of syncytium (trophoblast) increases, which results in almost a 1000-fold increase in the excretion of estriol. Since the precursors for the production of estriol are produced largely by the fetal adrenal, estriol determinations have been used to evaluate fetal well-being. The placenta is also the source of potent enzyme systems, such as the sulfatases and the aromatizing system that converts the C-19 steroids to estrogens. Estriol determinations, then, may also reflect placental function. Figure 49-3 shows the progressive increase in urinary and plasma estriol levels as pregnancy approaches term. Since there are obvious technical difficulties with 24-hour urine collection, serial estriol determinations are used more frequently in practice. An estriol excretion of 4 mg

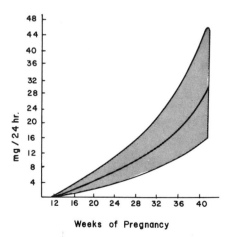

Figure 49-3
Estriol excretion in pregnancy. (Shaded area represents 95% of values; center line represents average excretion.) (Modified from V. G. Frandsen and G. Stakemann. *Acta Endocrinol.* (Kbh.) 44:183, 1963.)

per day in the last trimester usually indicates fetal death or a severely compromised fetus. When several determinations fall within the normal range, a subsequent drop in the level of estriol excretion is of greater significance than the absolute values. Constant low readings of less than 8 mg/24 hours during the last 8 weeks of pregnancy would indicate the need to terminate the pregnancy. Plasma estriol determinations are particularly valuable in those pregnancies that are affected by isoimmunization.

PREGNANEDIOL EXCRETION

As the mass of syncytial cells increases, so does the quantity of progesterone produced. The maternal and fetal livers reduce progesterone to 20-dihydroprogesterone and it is subsequently excreted as pregnanediol. It can be readily measured in the maternal urine and has been useful as an index of placental function (Fig. 49-4). The placental production of progesterone approaches 250 mg/24 hours at term, and a significant decrease from this rate may indicate placental dysfunction.

HUMAN PLACENTAL LACTOGEN (HPL)

This protein hormone is produced by the syncytiotrophoblast as early as the sixth week of pregnancy, and it is secreted in increasing amounts into the maternal circulation. Radioimmunoassay of maternal serum reveals secretion rates as high as 1 to 2 gm/24 hours. Very little of this hormone reaches the fetus. The maternal plasma HPL level probably reflects the function of the viable trophoblastic tissue and consequently the total functioning area available for exchange. This test appears to be of value in those mothers in whom toxemia develops or in whom chronic hypertension is present [15]. In mothers with toxemia, a greater number of serum HPL determinations fall within the fetal death zone shown in Figure 49-5. When the level falls below

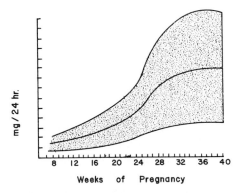

Figure 49-4
Pregnanediol excretion in pregnancy. (Shaded area represents 95% of values; center line represents average excretion.) (Modified from R. P. Shearman. *J. Obstet. Gynaecol. Br. Emp.* 66:1, 1959.)

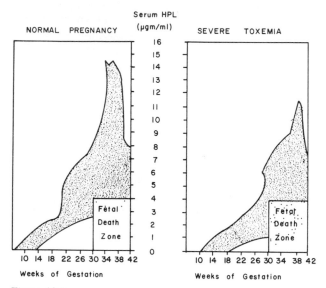

Figure 49-5
Serum placental lactogen (HPL) in normal pregnancy and severe toxemia. (Modified from statistics of W. N. Spellacy et al. *Am. J. Obstet. Gynecol.* 109:588, 1971.)

4 μg/ml after the 30th week of gestation, fetal death results in approximately one-half of the cases. The fetus does not appear to be jeopardized when the pregnancy is allowed to continue if the serum level is above 4 μg/ml. Low HPL levels are usually associated with meconium-stained amniotic fluid.

BILIRUBIN CONCENTRATION

The concentration of bilirubin in the amniotic fluid decreases progressively, so that after the 38th week of gestation, its presence cannot be detected. Prior to this time, it exists in the amniotic fluid in an unconjugated form. As the fetus matures, it develops the ability to conjugate bilirubin, so that the level in the amniotic fluid falls. This metabolite can be assayed by a modified diazo technique or by spectrophotometry. When the optical density of centrifuged amniotic fluid is measured at a wavelength of 450 nm, the typical bilirubin peak will decline with the duration of pregnancy (Fig. 49-6).

This determination is useful in prognosticating the status of the Rh-positive fetus in a Rh-negative sensitized pregnant woman. At the present time, all obstetric patients are screened for the Rh factor. The Rh-negative pregnant woman whose spouse is Rh positive is periodically examined for developing antibodies in her blood. The presence of such antibodies indicates that the fetus might be at risk. The degree of isoimmunization, however, is not necessarily correlated with the severity of risk to the fetus. Amniocentesis with spectrophotometric studies is far more reliable in determining the fetal condition. When hemolysis of the fetal red blood cells has occurred, a hump due to the

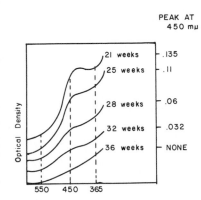

Figure 49-6
Decrease of the bilirubin peak at 450 nm in serial spectrophotometric analyses at progressively later periods of gestation. (From B. Mandelbaum, G. C. La Croix, and A. R. Robinson. *Obstet. Gynecol.* 29:471, 1967. Reproduced with permission.)

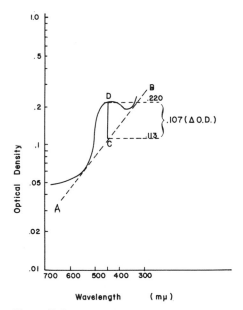

Figure 49-7
ΔOD 450. The line *AB* is expected in the absence of bilirubin, and *DC* is the amplitude of the increase in optical density at 450 mm caused by bilirubin in the amniotic fluid. The length of *DC* projected on the ordinate determines the ΔOD 450 value, which in this case is 0.107. (From D. E. Reid, A. J. Ryan, and K. Benirschke. *Principles and Management of Human Reproduction.* Philadelphia: Saunders, 1972. P. 819. Reproduced with permission.)

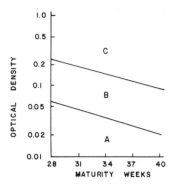

Figure 49-8
The ΔOD 450 at different weeks of maturity under different pathologic conditions (see text). (From W. Liley. In J. P. Greenhill [Ed.], *Year Book of Obstetrics and Gynecology.* Chicago: Year Book, 1965. Reproduced with permission.)

presence of bilirubin appears at the 450-nm wavelength, and the height of this peak from the expected base line is called the Δ*OD 450* (Fig. 49-7). Figure 49-8 illustrates how this determination can be used to prognosticate fetal blood hemolysis at different periods of gestation. If the ΔOD 450 falls into zone A, either there is no hemolytic disease or the fetus is mildly affected. Zones B and C represent more severe forms of hemolytic disease. Maturity studies would then have to be done to determine whether to deliver the infant or to resort to intrauterine transfusion.

OXYTOCIN CHALLENGE

The oxytocin challenge is performed to diagnose dysfunction of the fetoplacental unit. The test is performed when the fetus is thought to be at risk by administering a dilute solution of oxytocin intravenously to the patient after 28 weeks of gestation. This can be repeated as the pregnancy nears term. The oxytocin solution induces uterine contractions, which are carefully monitored for the presence of late deceleration patterns (see Chap. 48). Adverse patterns are associated with uteroplacental insufficiency and indicate that the infant may have to be delivered by cesarean section. Placental dysfunction that leads to fetal growth retardation may be regarded as representing the "nutritive" function of the placenta; the oxytocin challenge test indicates the "respiratory" function of the placenta, and it is related to the effect of the decreased uterine blood flow as it affects the fetus during labor.

ENZYME STUDIES

Still under a great deal of investigation are the determinations of the maternal blood levels and the amniotic fluid levels of the various enzymes. Oxytocinase may be decreased in toxemia of pregnancy, dysmaturity, and fetal death. Low serum levels are frequently associated with placental insufficiency, but as yet, the critical values and their significance have not been determined. The serum acid phosphatase level usually decreases, and any increase in this level may reflect cell damage. Maternal serum levels of diamine oxidase, monoamine oxidase, and lactic dehydrogenase increase during pregnancy. Many such enzymes are produced by the placenta, but their significance as an indicator of placental dysfunction is still not clear.

AMNIOGRAPHY

When some amniotic fluid is withdrawn and an aqueous contrast medium is instilled into the amniotic sac, various anomalies of fetal development can be seen on roentgenography. Since the fetus normally swallows amniotic fluid, the lack of the contrast medium within the intestinal tract of the fetus would indicate death in utero.

ASSESSMENT OF FETAL MATURITY

The concept of the identification of the high-risk pregnancy and the newer techniques for detecting the fetus at risk have made it necessary to deliver many of these infants prior to term. Fetal maturity must be accurately determined so that adaptation to extrauterine life can be predicted. The chance for fetal survival increases from 20% at the 28th week to 80% at the 34th week of gestation [8]. There is no single method that is adequate to determine the gestational age and thus fetal maturity. Several different procedures must be used and frequently repeated in order to make this determination.

The gestational age is usually determined from the patient's last menstrual period, expressed in weeks [14]. The number of weeks is counted from the first day of the last menstrual period, and 40 weeks is considered a term pregnancy.

Roentgenographic techniques can be employed to identify ossification centers and other bony marks. These ossification centers are recognizable when they are at least 3 mm in length. The proximal tibial epiphysis is usually visualized after 36 weeks and, when present, suggests fetal maturity. The distal femoral epiphysis can be seen as early as 32 weeks of gestation. The length of the lumbar spine of the fetus can also be measured, when allowing for the distance between the x-ray source and the receiving plate. Intrauterine fetal growth measurements are becoming important not only in assessing the period of gestation, but also as an index for fetal growth rate, especially when placental insufficiency exists. Since x-rays measure differences in radiopacity, it is necessary to wait for the fetal skeleton to develop (18 to 20 weeks of gestation) in order to employ this technique. The disadvantage of this method is that repeated roentgenographic monitoring during pregnancy may be hazardous to the developing fetus.

The use of ultrasonography in diagnosis differs from roentgenography in that it measures tissue density rather than radiopacity. When the A-scan method is used, the biparietal diameter can be accurately measured with an error of ±1 mm (see Chap. 48, p. 900). The diameter of the fetal skull is a good indicator of fetal weight and, in some respects, of fetal maturity (Fig. 49-9). When the biparietal diameter of the fetal skull is more than 9.8 cm, then the fetus is probably mature. Postmaturity should

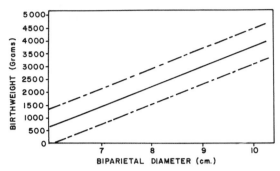

Figure 49-9
The relationship of the biparietal diameter of the skull to the birth weight in 489 newborn infants.
(From L. M. Hellman et al. *Am. J. Obstet. Gynecol.* 99:662, 1967. Reproduced with permission.)

be suspected when the biparietal diameter is over 10 cm. Ultrasonograms have also
been obtained using the B-scan method (see Chap. 48, pp. 901–902) for measuring
fetal head and fetal trunk diameters (Fig. 49-10).

Such fetal growth rate measurements can differentiate between dysmaturity and an
incorrect gestational age as determined from the last menstrual period. In Figure 49-10C,
a normal fetal growth pattern is seen, whereas in Figure 49-10D, the actual fetal age was
one month less than that determined from the patient's history [4].

Studies of the amniotic fluid can provide important information concerning fetal
maturity. This procedure is relatively innocuous, and it can be performed many times
with little risk to the mother or fetus. Certain precautions, such as determining the
placental site, however, are necessary.

When a drop of 0.1% Nile Blue sulfate is added to a drop of amniotic fluid, the
exfoliated cells of the fetal sebaceous glands will stain orange. These are then counted
with a microscope, and when there are between 20 and 30 of such cells present per
high-powered field, the fetus is considered to be mature. This simple method can be
done in almost any laboratory [2]. As the renal function of the fetus matures, the
concentration of creatinine in the amniotic fluid increases. A value greater than
2.0 mg/100 ml is used as an index of fetal maturity [10]. Spectrophotometric determinations can be made on amniotic fluid; as discussed previously, the bilirubin concentration peak at 450 nm disappears as the fetus matures (see Fig. 49-6).

The amniotic fluid is isotonic in early pregnancy, but with the continuation of
pregnancy, the protein concentration decreases. Various solutes are also reabsorbed,
and the amniotic fluid becomes hypotonic [9]. At term, an amniotic fluid osmotic
concentration of 250 mOsm/liter or less suggests fetal maturity. This figure is approximately 20 to 25 mOsm/liter lower than that of the maternal or fetal plasma. The
osmolarity continues to decrease after 40 weeks of gestation. The osmolality of the
amniotic fluid can easily be determined by an automatic freezing-point-depression
osmometer.

The maturity of the fetal lung can also be determined by examination of the
amniotic fluid. The fetal lung must be mature for extrauterine survival. The transition

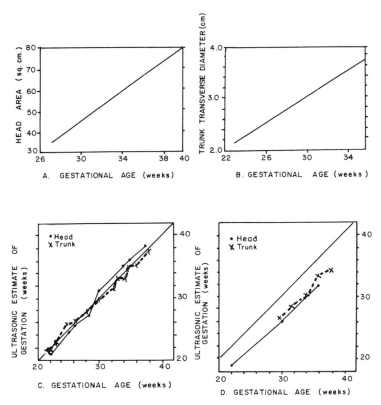

Figure 49-10
Fetal head and trunk dimensions as a function of gestational age. (From L. M. Hellman et al. *Am. J. Obstet. Gynecol.* 9:662, 1967. Reproduced with permission.)

from an aquatic environment to a gaseous environment on delivery is a radical and complex change. Once air has entered the alveolar sac, the air-fluid interspace is established. The pressure that is required to fill the alveolar space with air is greater than that needed with fluid as a vehicle. At the same time, the air-filled lung does not recoil with the same pressure, a phenomenon that has been described as hysteresis (see Chap. 19). This is due to the surface tension at the air-fluid interface. The surface-active agent, or surfactant, produced by specialized alveolar cells coats the alveolar lining. These molecules are able to lower the surface tension at the air-fluid interface. If this mechanism is not functioning properly, the smaller alveoli will empty into the larger ones and collapse.

The surfactant activity is largely due to compounds known as phospholipids [3, 7]. The most significant phospholipid in the lung effluent of the newborn infant is lecithin. The tracheal effluent of human infants is quantitatively similar to that found in a diluted form in the amniotic fluid. The lecithin concentration in the amniotic fluid comes principally from the fetal lung, and it can be used to determine alveolar stability

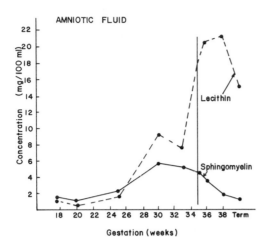

Figure 49-11
Mean concentrations in amniotic fluid of sphingomyelin and lecithin during pregnancy. (From L. Gluck et al. *Am. J. Obstet. Gynecol.* 109:440, 1971. Reproduced with permission.)

and lung maturity [6]. An increase in the concentration of the phospholipids in the amniotic fluid occurs as the pregnancy matures (Fig. 49-11). Although the phospholipids increase at approximately 35 weeks of gestation, this increase in their concentration is primarily due to the lecithin fraction; the sphingomyelin concentration begins to decrease after 35 weeks.

To determine fetal lung maturity, a small amount of amniotic fluid is centrifuged and the supernatant liquid is treated with 1:1 methanol:chloroform to extract the lipids. Both lecithin and sphingomyelin are then precipitated in cold acetone. These are then separated by thin-layer chromatography, and either visual estimates or densitometric measurements of the lecithin/sphingomyelin ratios (L&S) are calculated (Fig. 49-12). At maturity, the L/S ratio rises sharply, usually having a value of 2 or more.

The critical problem that confronts every newborn is its ability to stabilize the lungs so that normal respiration can ensue. The respiratory distress syndrome (RDS), or hyaline membrane disease, is the principal cause of death in premature infants. In RDS, the surfactant-producing mechanism is immature or has become defective during pregnancy or labor [1]. When the lungs are mature, the infant's chest on delivery is exposed to a high, negative thoracic pressure. The normal lung does not collapse during expiration, but is able to maintain residual air in the alveolus. With adequate surfactant, very little negative pressure is necessary for subsequent respiration. In contrast, the immature lung that lacks surfactant will eventually collapse during expiration. Monitoring of the amniotic fluid phospholipids when delivery is contemplated is therefore an important diagnostic method for predicting lung maturity and the avoidance of RDS.

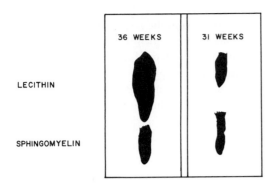

Figure 49-12
Lecithin-to-sphingomyelin ratios as determined by thin-layer chromatography (samples taken during the 36th and 31st weeks of pregnancy).

AMNIOCENTESIS FOR GENETIC STUDIES

Amniotic fluid examination may provide information concerning genetic or metabolic defects in the fetus (Fig. 49-13). The amniotic fluid normally contains various desquamated cells from the fetus. These can be used to predict the sex of the fetus as early as 14 to 16 weeks of gestation. Women who are heterozygous for X-linked recessive disorders, such as hemophilia and progressive muscular dystrophy, are thus able to know the sex of the unborn child. In these cases, amniocentesis can be performed early in pregnancy (14 weeks). If the fetus is determined to be a male, he is more likely to develop the above pathologic conditions. The parents are then encouraged to seek counselling concerning their desire for pregnancy termination.

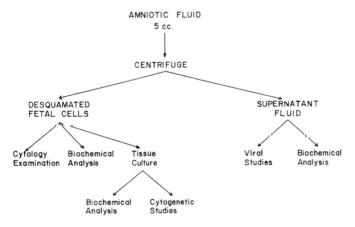

Figure 49-13
Amniotic fluid examination procedure.

Certain fetal cells, probably fibroblasts originating in the respiratory tract, can be grown in tissue culture. More complete, comprehensive cytogenetic and biochemical studies may be done with this material. Amniocentesis would then be done on the genetically high-risk pregnancy. As an example, elderly prospective parents are more likely to have an infant with Down's syndrome (mongolism). Biochemical studies of the various enzymes in the amniotic fluid of the susceptible fetus are equally valuable in predicting the fate of infants with metabolic disorders.

REFERENCES

1. Avery, M. E., and Mead, J. Surface properties in relation to atelectasis and hyaline membrane disease. *Am. J. Dis. Child.* 97:517, 1959.
2. Bishop, E. H., and Corson, S. Estimation of fetal maturity by cytologic examination of amniotic fluid. *Am. J. Obstet. Gynecol.* 102:654, 1968.
3. Clements, J. A. Surface phenomena in relation to pulmonary function. (Sixth Bowditch lecture.) *Physiologist* 5:11, 1962.
4. Garrett, W. J., and Robinson, D. E. Assessment of fetal size and growth rate by ultrasonic echoscopy. *Obstet. Gynecol.* 38:525, 1971.
5. Gluck, L. Biochemical development of the lung: Clinical aspects of surfactant development, RDS and intrauterine assessment of lung maturity. *Clin. Obstet. Gynecol.* 14:710, 1971.
6. Gluck, L., et al. Diagnosis of respiratory distress syndrome by amniocentesis. *Am. J. Obstet. Gynecol.* 109:440, 1971.
7. Gluck, L., et al. The biochemical development of surface activity in mammalian lung. I. The surface-active phospholipids; the separation and distribution of surface-active lecithin in the lung of the developing rabbit fetus and newborn. *Pediatr. Res.* 1:247, 1967.
8. Gruenwald, P. Perinatal death of full-sized and full-term infants. *Am. J. Obstet. Gynecol.* 107:1022, 1970.
9. Miles, P. A., and Pearson, J. W. Amniotic fluid osmolality in assessing gestational maturity. *Obstet. Gynecol.* 34:701, 1969.
10. Pitkin, R. M., and Awirek, S. J. Amniotic fluid creatinine. *Am. J. Obstet. Gynecol.* 98:1135, 1967.
11. Plentl, A. A. Dynamics of the amniotic fluid. *Ann. N.Y. Acad. Sci.* 75:746, 1959.
12. Plentl, A. A. Formation and circulation of amniotic fluid. *Clin. Obstet. Gynecol.* 3:386, 1960.
13. Pritchard, J. A. Deglutition by normal and anencephalic fetuses. *Obstet. Gynecol.* 25:289, 1965.
14. Queenan, J. T., et al. Amniotic fluid volumes in normal pregnancies. *Am. J. Obstet. Gynecol.* 114:34, 1972.
15. Spellacy, W. N., et al. Value of human chorionic somatomammotropin in managing high-risk pregnancies. *Am. J. Obstet. Gynecol.* 109:588, 1971.
16. Thompson, W., Lappin, T. R. J., and Elder, G. E. Liquor volume by direct spectrophotometric determination of injected PAH. *J. Obstet. Gynaecol. Br. Commonw.* 78:341, 1971.

SELECTED READING

Kubli, F. W., Kaeser, O., and Hinselmann, M. In A. Pecile and C. Finzi (Eds.), *The Fetoplacental Unit*. Amsterdam: Excerpta Medica Foundation, 1969. P. 323.

Pearlman, W. H. Progesterone metabolism in advanced pregnancy and in oophorectomized-hysterectomized women. *Biochem. J.* 67:1, 1957.

Ray, M., et al. Clinical experience with oxytocin challenge test. *Am. J. Obstet. Gynecol.* 114:1, 1972.

Spellacy, W. N., et al. Human placental lactogen levels and intrapartum fetal distress: Meconium-stained amniotic fluid, fetal heart rate patterns and Apgar scores. *Am. J. Obstet. Gynecol.* 114:803, 1972.

Taylor, E. S., et al. Estriol concentrations in blood during pregnancy. *Am. J. Obstet. Gynecol.* 108:868, 1970.

Index

INDEX

923